IFMBE Proceedings

Volume 87

The IFMBE Proceedings Book Series is an official publication of *the International Federation for Medical and Biological Engineering* (IFMBE). The series gathers the proceedings of various international conferences, which are either organized or endorsed by the Federation. Books published in this series report on cutting-edge findings and provide an informative survey on the most challenging topics and advances in the fields of medicine, biology, clinical engineering, and biophysics.

The series aims at disseminating high quality scientific information, encouraging both basic and applied research, and promoting world-wide collaboration between researchers and practitioners in the field of Medical and Biological Engineering.

Topics include, but are not limited to:

- Diagnostic Imaging, Image Processing, Biomedical Signal Processing
- Modeling and Simulation, Biomechanics
- Biomaterials, Cellular and Tissue Engineering
- Information and Communication in Medicine, Telemedicine and e-Health
- Instrumentation and Clinical Engineering
- Surgery, Minimal Invasive Interventions, Endoscopy and Image Guided Therapy
- Audiology, Ophthalmology, Emergency and Dental Medicine Applications
- Radiology, Radiation Oncology and Biological Effects of Radiation

IFMBE proceedings are indexed by SCOPUS, EI Compendex, Japanese Science and Technology Agency (JST), SCImago.

Proposals can be submitted by contacting the Springer responsible editor shown on the series webpage (see "Contacts"), or by getting in touch with the series editor Ratko Magjarevic.

More information about this series at https://link.springer.com/bookseries/7403

Ion Tiginyanu · Victor Sontea ·
Serghei Railean
Editors

5th International Conference on Nanotechnologies and Biomedical Engineering

Proceedings of ICNBME-2021, November 3–5, 2021, Chisinau, Moldova

Springer

Editors
Ion Tiginyanu
Academy of Sciences of Moldova
Chisinau, Moldova

Serghei Railean
Department of Microelectronics
and Biomedical Engineering
Technical University of Moldova
Chisinau, Moldova

Victor Sontea
Department of Microelectronics
and Biomedical Engineering
Technical University of Moldova
Chisinau, Moldova

ISSN 1680-0737 ISSN 1433-9277 (electronic)
IFMBE Proceedings
ISBN 978-3-030-92327-3 ISBN 978-3-030-92328-0 (eBook)
https://doi.org/10.1007/978-3-030-92328-0

This Springer imprint is published by the registered company Springer Nature Switzerland AG
The registered company address is: Gewerbestrasse 11, 6330 Cham, Switzerland

Preface

This volume presents the Proceedings of the 5th International Conference on Nanotechnologies and Biomedical Engineering (ICNBME), which was held on November 3–5, 2021, in Chisinau, Republic of Moldova. The event was held online due to the pandemic COVID-19. ICNBME-2021 continued the series of international conferences in the field of nanotechnologies and biomedical engineering with the main goal focused at bringing together scientists and engineers dealing with fundamental and applied research for reporting on the latest theoretical developments and applications in the fields involved.

The conference covered a wide range of subjects of primary importance for research and development such as nanotechnologies and nanomaterials; biomaterials for medical applications; biomedical instrumentation and signal processing; biomedical devices and sensors; health informatics, e-health and telemedicine; innovation, development and interdisciplinary research; molecular; molecular, cellular and tissue engineering; clinical engineering, health technology management and assessment; and excitations in condensed matter.

The papers included in the Proceedings reflect the results of multidisciplinary research undertaken by about one hundred of groups worldwide. Special attention is paid to the development of novel nanotechnologies and nanomaterials, in particular of bio-nanotechnologies and bio-nanomaterials, excitations in condensed matter. New biocompatible materials are proposed for use in regenerative medicine, cellular and tissue engineering. Interesting data on novel chemical and biosensors are reported which are based on nanostructured metal oxides and hybrid nanocomposite materials.

Considerable progress has been achieved at the intersection of nanotechnologies, information technologies and biomedicine as, for example, in health informatics, e-health, telemedicine, biomedical instrumentation and signal processing. New theoretical and experimental results are highlighted in such fields as metamaterials, aeromaterials, optoelectronic and photonic materials, photovoltaic structures, quantum dots, one- and two-dimensional nanomaterials, multifunctional hybrid materials like core–shell structures, etc. The Proceedings reflect the state of the art in controlling the properties of several classes of nanocomposite materials for important future applications in various fields.

We hope that the papers included in the ICNBME-2021 Proceedings will be of interest for established researchers working in multidisciplinary fields of science and technology, young scientists, students and broad community wishing to get up-to-date information on progress in the fast-developing areas of nanotechnology and biomedical engineering.

Ion Tiginyanu
Victor Sontea
Serghei Railean

Organization

5th International Conference on Nanotechnologies and Biomedical Engineering
ICNBME-2021, November 3–5, 2021, Chisinau, Republic of Moldova

Organizers

Moldavian Society of Biomedical Engineering
Technical University of Moldova
Nicolae Testemitanu State University of Medicine and Pharmacy

In Collaboration with

The International Federation for Medical
and Biological Engineering (IFMBE)

Academy of Sciences of Moldova

Sponsored by

– European Commission under the
Grant #810652 "NanoMedTwin"

Committees

Conference Chairs

Ion Tiginyanu	Academy of Sciences of Moldova, Republic of Moldova
Victor Sontea	President of the Moldavian Society of Biomedical Engineering, Republic of Moldova

International Advisory Committee

Adriana Velazquez Berumen	World Health Organization, Switzerland
Alexander Pogrebnjak	Sumy State University, Ukraine
Bogdan Simionescu	Romanian Academy, Romania
Boris Gorshunov	Moscow Institute of Physiscs and Technology, Russia
Emil Cebanu	Nicolae Testemitanu State Medical and Pharmaceutical University, Republic of Moldova
Franz Faupel	Institute for Materials Science, University of Kiel, Germany
Gert Baumann	Charité Hospital, University of Berlin, Germany
Hans Hartnagel	Technical University Darmstadt, Institute of Microwave Engineering and Photonics, Germany
Hidenori Mimura	Research Institute of Electronics, Shizuoka University, Japan
Jan Linnros	Royal Institute of Technology, Sweden
Lee Chow	University of Central Florida, Orlando, USA
Lorenz Kienle	Institute for Materials Science, University of Kiel, Germany
Nicolae Jula	Military Technical Academy, Romania
Nicolas Pallikarakis	University of Patras, Greece
Pascal Colpo	Joint Research Center, Italy
Peter Scharff	Technical University Ilmenau, Germany
Rainer Adelung	Institute for Materials Science, University of Kiel, Germany
Ratko Magjarević	University of Zagreb, Croatia
Şeref Komurcu	Anadolu Medical Center, Turkey
Sergey Gaponenko	National Academy of Sciences, Belarus
Serghei Cebotari	Hannover Medical School, Germany
Thierry Pauporte	Ecole nationale supérieure de chimie de Paris, France

Viorel Bostan	Technical University of Moldova, Republic of Moldova
Vladimir Fomin	Institute for Integrative Nanosciences, Germany
Yury Dekhtyar	Institute of Biomedical Engineering and Nanotechnologies, Riga Technical University, Latvia

International Program Committee

Adrian Dinescu	National Institute for Research and Development in Microtechnology – IMT Bucharest, Romania
Anatolie Sidorenko	Ghitu Institute of Electronic Engineering and Nanotechnologies, Republic of Moldova
Andrei Sarua	University of Bristol, UK
Artur Buzdugan	Technical University of Moldova, Republic of Moldova
Călin Corciova	Grigore T. Popa University of Medicine and Pharmacy, Romania
Dumitru Tsiulyanu	Technical University of Moldova, Republic of Moldova
Ghenadie Curocichin	Nicolae Testemitanu State Medical and Pharmaceutical University, Republic of Moldova
Grigor Tatisvili	R. Agladze Institute of Inorganic Chemistry and Electrochemistry of Ivane Javakhishvili Tbilisi State University, Georgia
Ian Toma	The George Washington University, USA
Leonid Kulyuk	Institute of Applied Physics, Republic of Moldova
Liliana Verestiuc	Grigore T. Popa University of Medicine and Pharmacy, Romania
Mircea Dragoman	National Institute for Research and Development in Microtechnology – IMT Bucharest, Romania
Nicolai Sobolev	University of Aveiro, Portugal
Oleg Lupan	Technical University of Moldova, Republic of Moldova
Radu Ciorap	Grigore T. Popa University of Medicine and Pharmacy, Romania
Roman Tomashevskyi	Kharkiv Polytechnical Institute, Ukraine
Stanislav Groppa	Nicolae Testemitanu State Medical and Pharmaceutical University, Republic of Moldova

Vasile Tronciu	Technical University of Moldova, Republic of Moldova
Veaceslav Ursaki	Academy of Sciences of Moldova, Republic of Moldova
Victor Vovc	Nicolae Testemitanu State Medical and Pharmaceutical University, Republic of Moldova
Viorel Nacu	Nicolae Testemitanu State Medical and Pharmaceutical University, Republic of Moldova

Organizing Committee

Sergey Railean (Head)	Technical University of Moldova, Republic of Moldova
Nicolai Ababii	Technical University of Moldova, Republic of Moldova
Alexandru Corlăteanu	State Medical and Pharmaceutical University "Nicolae Testemitanu", Republic of Moldova
Ulian Rotari	Technical University of Moldova, Republic of Moldova
Elena Darii	Technical University of Moldova, Republic of Moldova
Gheorghe Gorceag	Moldovan Society of Biomedical Engineering, Republic of Moldova
Nicolae Magariu	Technical University of Moldova, Republic of Moldova
Eduard Monaico	Technical University of Moldova, Republic of Moldova
Ion Pocaznoi	Technical University of Moldova, Republic of Moldova
Vasile Postica	Technical University of Moldova, Republic of Moldova
Alexandr Sereacov	Technical University of Moldova, Republic of Moldova

Reviewers List

Anatolie Sidorenko
Artur Buzdugan
Calin Corciova
Corneliu-Nicolae Druga
Daria Grabco
Eduard Monaico
Elena Achimova
Ghenadie Curocichin
Ian Toma
Ionel Sanduleac
Liliana Verestiuc
Mihai Macovei

Nicolae Jula
Nicolas Pallikarakis
Oleg Lupan
Postica Vasile
Leonid Kulyuk
Roman Tomashevskyi
Sontea Victor
Vasile Tronciu
Veaceslav Ursaki
Victor Vovc
Victor Zalamai
Viorel Nacu
Vladimir Fomin

Plenary Speakers, Abstracts

Applying the Negligible Mass of Graphene Aeromaterials: Repeatable Air Explosions and Instant Sterilization

Rainer Adelung

Institute for Material Science, Kiel University, Germany
ra@tf.uni-kiel.de

Aeromaterials in the form of aerographite have been discovered and created around ten years ago and were reported in advanced materials [1]. Beside the public excitement about the in those days labeled "lightest material," the material has unique properties that differ from other lightweight materials like aerogels, which is caused by their design. Aerographite is constructed in a template approach on a framework of tetrapodal ZnO that is shaped by sintering into a continuous macroscopic branched network. The ZnO network consists of ZnO rods with diameters of ~1–3 μm and length of typically 10–20 μm between the next intersections. Removal of the template follows in aerographite during the deposition process: while a carbon layer is wrapped around the ZnO network, the ZnO is etched away by hydrogen. Thus, a network of free-standing nanoscale thin films rolled up into microscale diameter tubes with tube length of 10–20 μm and macroscopic expansion on the cm scale is created. Later on, further CVD grown variants like the AeroBN [2] or aerogalnite [3] were created. By wet chemical deposition of 1D and 2D nanomaterials, carbon nanotube tube networks [4] or Aerographene samples [5] were manufactured as well.

Two structural features of the aeromaterial are most prominent: the low mass, mainly caused by the nanoscopic wall diameter and the large free volume which is interconnected and, in its way, special, as it is free from narrow restrictions. As a thought experiment, a sphere with an expansion of ~1 μm can be transported without collisions through the material from one side to the other on a relative straight pathway. The combination of low mass tubular micrometers and distances of tenth of micrometers has the immediate consequence that light is scattered in a very efficient manner. This explains the high light adsorption in aerographite [1] as well as the highly efficient laser light scattering in AeroBN [2].

The combination of the structural features of aeromaterials, the interconnected large free volume and the low weight are employed for a powerful pneumatic actuator. Low mass means low heat capacity, which results in reaching high temperatures with relative low power. Heating rates about 500.000 °C/s can be reached—consequently, even under ambient conditions, temperatures can be reached that result in a significant air expansion. These lead to explosion like air bursts from the aeromaterial. Other than combustion processes, no chemistry is involved, meaning the "explosion" can be repeated after fractions of seconds without any fuel, see figure below. The talk will present various applications reaching from aero-ear-headphones over actuators that carry the 10.000 times its own weight, miniaturized air/water pumps to disinfection of filters in fractions of seconds for complete sterilization.

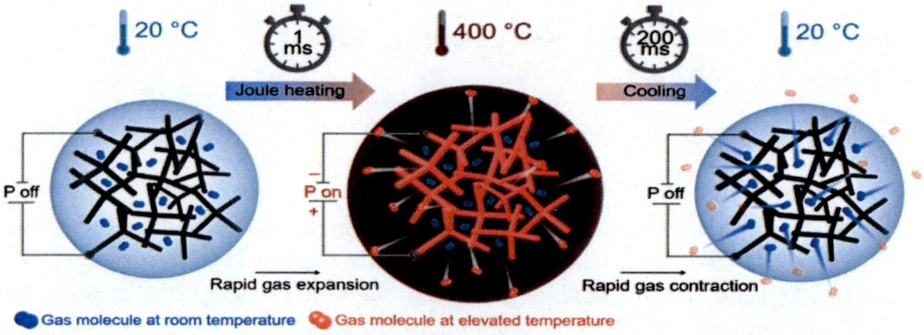

Illustration of an EPRAE (Electrically powered air explosion), see F. Schütt et al. Mat. Today, in press, doi.org/10.1016/j.mattod.2021.03.010.

References

1. Mecklenburg, M., et al.: Advanced Materials **24**, 3486 (2012)
2. Schütt, F., et al.: Nature Communications **11**, 1437 (2020)
3. Tiginyanu, I., et al.: Nano Energy **56**, 759 (2019)
4. Schütt, F., et al.: Nature Communications **8**, 1215 (2017)
5. Rasch, F., et al.: ACS Applied Materials and Interfaces **11**, 44652 (2019)

Terahertz Spectroscopy as an Effective Tool of Experimental Nanophysics

Boris Gorshunov

Laboratory of Terahertz Spectroscopy, Center for Photonics and 2D Materials, Moscow Institute of Physics and Technology (National Research University), Dolgoprudny, Russia
bpgorshunov@gmail.com

Since the development of coherent source (backward-wave oscillators-based) [1–3] and, later, pulsed time domain [4] terahertz spectroscopy techniques, the so-called terahertz gap or spectroscopic gap (0.1 – 10 THz) of the electromagnetic spectrum has been completely mastered. This paved the way for solving many fundamental and applied problems, the solution of which experienced enormous difficulties due to the lack of reliable experimental data on the electromagnetic properties of objects in the given frequency band. Nowadays, the powerful arsenal of terahertz spectroscopy available to researchers and developers is actively used in various fields of natural, material and applied sciences (see, e.g., [5] and references therein), and corresponding techniques and approaches are described in detail in many reviews and books (e.g., [6,7] and references therein). The laboratory of terahertz spectroscopy at the Moscow institute of physics and technology is equipped with a unique set of spectroscopic equipment that allows to conduct fundamental and applied research in various fields of condensed matter physics. The experiments can be carried out starting from sub-Hertz frequencies up to ultraviolet and at liquid helium up to room temperatures. Such a wide frequency interval allows for most in-depth studies of a wide variety of phenomena. At the same time, it is terahertz frequencies (quantum energies from fractions of meV to several meV) that are often key for understanding the microscopic nature of exotic properties that materials acquire when their size is reduced to nanoscale and which they do not possess in their “regular” macroscopic state. This happens when the “samples” dimensions become comparable to the spatial parameters that characterize their physical properties, such as correlation lengths of collective interactions and mean free paths of charge carriers. Understanding the nature of emerging phases is of great fundamental and technological interest but is presently still at its infancy. The talk provides a review of latest and recent results obtained by our group using terahertz spectroscopy on a number of nanostructured materials and systems that are most popular and most widely studied in recent years: carbon nanotubes, graphene, endofullerenes, nanoconfined water molecules and ultra-porous aerogalnite aero-GaN. Physical properties of the materials are discussed from both fundamental and technological viewpoints.

The research was partly supported by the Russian Science Foundation, grant RSF21-72-20050.

References

1. Volkov, A.A., Goncharov, Yu.G., Kozlov, G.V., Lebedev, S.P., Prokhorov, A.M.: Dielectric measurements in the submillimeter wavelength region. Infrared Phys. **25**, 369 (1985)
2. Volkov, A.A., Kozlov, G.V., Prokhorov, A.M.: Progress in submillimeter spectroscopy of solid state. Infrared Phys. **29**,747 (1989)
3. Kozlov, G., Volkov, A.: Coherent source submillimeter wave spectroscopy. Topics in Applied Physics vol.74. Millimeter and submillimeter spectroscopy of solids. Ed. G.Gruner. Springer, 1998.
4. Ralph, S.E., Grischkowsky, D.: THz spectroscopy and source characterization by optoelectronic interferometry, Appl. Phys. Lett. **60**, 1070 (1992)
5. Neu, J., Schmuttenmaer, Ch. A.: Tutorial: an introduction to terahertz time domain spectroscopy (THz-TDS). J. Appl. Phys. **124**, 231101 (2018)
6. Gorshunov, B., et al.: Terahertz BWO-spectroscopy. Int. J. Infrared Millimeter Waves, **26** 1217 (2005)
7. Gorshunov, B.P., Volkov, A.A., Prokhorov, A.S., Spektor, I.E.: Methods of Terahertz–Subterahertz BWO spectroscopy of conducting materials. Phys. Solid State, **50** 2001 (2008)
8. Siegel, P.H.: Terahertz technology. IEEE Trans. Microwave Theor. Tech. **50** 910 (2002)

Spin-Dependent Phenomena in Semiconductor Micro-and Nanoparticles for Biomedical Applications

Vladimir M. Fomin[1,2,3]

[1] Institute for Integrative Nanosciences, Leibniz IFW Dresden, Helmholtzstraße 20, D-01069 Dresden, Germany
[2] Laboratory of Physics and Engineering of Nanomaterials, Department of Theoretical Physics, Moldova State University, Strada A. Mateevici 60, MD-2009 Chisinau, Republic of Moldova
[3] Institute of Engineering Physics for Biomedicine, National Research Nuclear University "MEPhI," Kashirskoe shosse 31, 115409 Moscow, Russia
v.fomin@ifw-dresden.de

This talk represents an overview of spin-dependent phenomena in nonmagnetic semiconductor microparticles (MPs) and nanoparticles (NPs) with interacting nuclear and electron spins [1]. Its goal is to cover a gap between the basic properties of spin behavior in solid-state systems and a tremendous growth of the experimental results on biomedical applications of those particles. I will focus on modern achievements of spin-dependent phenomena in the bulk semiconductors from the theory of optical spin orientation under indirect optical injection of carriers and spins in the bulk crystalline silicon—via numerous insightful findings in the realm of characterization and control through the spin polarization—to the design and verification of nuclear spin hyperpolarization in semiconductor MPs and NPs for magnetic resonance imaging (MRI) diagnostics. Semiconductor MPs and nanoparticles NPs exhibit interesting electronic, optical and magnetic properties, which depend on a preferential orientation of electron and nuclear spins in those particles. These properties are essential for their biomedical applications. Spatial confinement of charge carriers (electron and holes) in a semiconductor nanostructure results in an increase of the spin–lattice relaxation time. Going from itinerant to immobile, fully localized electrons, while inducing the hyperfine dephasing, can be also beneficial in quenching the spin–lattice relaxation. The dynamic nuclear polarization in semiconductor nano- and microstructures opens fascinating prospects for creation of new efficient contrast agents in MRI, which is a powerful diagnostic tool in biomedicine. Perspective applications of silicon MPs and NPs in hyperpolarized ^{29}Si MRI are discussed. For instance, spin-dependent energy transfer from excitons confined in Si nanocrystals (nc-Si) to molecular oxygen in the ground triplet state is promising for application of nc-Si-based NPs and MPs in photodynamic therapy of cancer [2]. Both the bioimaging and therapeutic functionality of Si-NPs are being combined in theranostics of cancer [3].

I gratefully acknowledge fruitful and motivating collaborative work with V. Yu. Timoshenko.

References

1. Fomin, V.M., Timoshenko, V.Yu.: Appl. Sci. 10, 4992, 1–45 (2020)
2. Timoshenko, V.Y.: Porous Silicon in Photodynamic and Photothermal Therapy. In: Handbook of Porous Silicon, 2nd ed.; L. Canham, Ed.; Springer: Berlin−Heidelberg, pp. 1461–1469 (2018)
3. Kabashin, A.V., Timoshenko, V.Y.: Nanomedicine 11, 2247–2250 (2016)

3D-Printed Sensors of Nanostructured Semiconducting Oxides

Oleg Lupan[1,2,3]

[1] Center for Nanotechnology and Nanosensors/Department of Microelectronics and Biomedical Engineering, Technical University of Moldova, 168 Stefan cel Mare Av., MD-2004 Chisinau, Republic of Moldova
[2] Kiel University/Institute for Materials Science/Functional Nanomaterials, Kaiser Str. 2, D-24143, Kiel, Germany
[3] University of Central Florida/Department of Physics, FL 32816-2385, Orlando, USA
oleg.lupan@mib.utm.md, ollu@tf.uni-kiel.de

3D printed sensors will play an important role in the world of modern devices and are crucial nowadays due to complexity of mounting on various equipment for the development of biomedical or monitoring systems. Mixing semiconducting oxides directly during their printing/additive manufacturing makes them attractive for detecting applications by controlling their shapes, sensitivity and reliability. This overview is intended to summarize our recent results in this domain [1–4]. Developing devices for medical applications, one has to take into account numerous characteristics (toxicity, side effect or reactions, effect of humidity on response/performances, etc.). We will show how mixed or heterostructured oxides are built by a new 3D printing approach with acetone [2,3], VOC [3], lithium–ion batteries (LIBs) electrolytes containing, e.g., LiTFSI and LiNO3 [4], and other sensing performances of emerging oxides leading to their tuning for important applications in household sensors and alarms, automotive and biomedical engineering domains [4]. Developments in the last decade were done to enhance requirements of reliability and ultra-low-power consumption of the 3D sensors [3]. Thus, direct ink writing of microsensors can overcome the necessity for clean room technology. In this talk, we explain how mixed-metal oxide microsensors can be easily developed by printing of common metal nanomicroparticles. The advantages of such additive manufacturing are open porous semiconductor structure allows for sensitive VOC detection, and low base conductance leads to a low-power or energy-efficient microsensor. The combination of the possibility to detect electrolyte vapors used in LIBs and size control by the 3D-DIW printing method makes these heterostructures extremely attractive in controlling the safety of modern batteries. Also, these microdevices 3D printed directly on board or chip open new perspectives for nanoelectronics and biomedical applications. Dr. Lupan gratefully acknowledges the support of the Kiel University, Germany, and University of

Central Florida, USA, for an invited professor and visiting scientist positions. This work was partially supported by the Technical University of Moldova and through the ANCD-NARD Grant No. 20.80009.5007.09 at TUM.

References

1. Lupan, O., et al.: Tailoring the selectivity of ultralow-power heterojunction gas sensors by noble metal nanoparticle functionalization Nano Energy, **88**, 106241 (ISSN: 2211-2855) (2021)
2. Siebert, L., et al.: 3D-Printed Chemiresistive Sensor Array on Nanowire CuO/Cu2O/Cu Heterojunction Nets, ACS Appl. Mater. Interfaces, **11**, 25508–25515 (2019)
3. Siebert, L., et al.: Facile fabrication of semiconducting oxide nanostructures by direct ink writing of readily available metal microparticles and their application as low power acetone gas sensors. Nano Energy, **70**, 04420 (2020)
4. Lupan, O., et al.: Additive manufacturing as a means of gas sensor development for battery health monitoring, Chemosensors, **9**(9), 252 (2021). https://doi.org/10.3390/chemosensors9090252

Semiconductor and Plasmonic Nanoparticles for Biomedical Applications

Victor Yu. Timoshenko

Faculty of Physics of Lomonosov Moscow State University, Phys-Bio Institute of National, Research Nuclear University "MEPhI," Moscow, Russia
timoshen@physics.msu.ru

Semiconductor and plasmonic (NPs) exhibit unique optical properties for various biomedical applications. Silicon (Si) NPs (Si-NPs) are especially interesting because they are biocompatible, biodegradable and can be easily prepared by chemical and laser-assisted methods [1,2]. Nanocrystalline Si-NPs can exhibit the photoluminescence (PL) and Raman scattering, which are used for the optical bioimaging [2]. The conventional linear optical methods and nonlinear optical spectroscopy of the second harmonic generation and two-photon excited luminescence are demonstrated to be efficient to monitor Si-NPs in biosystems [3]. Si-NPs in aqueous media act as a light absorber in the visible and near-infrared spectral regions, and it is promising for photohyperthermia applications [4]. Si-NPs with impurities and intrinsic electronic states can act as labels in magnetic resonance imaging [5], which is widely used in the diagnosis of cancer. Halloysite nanotubes (HNTs) with immobilized plasmonic (gold and silver) NPs are explored as potential nanotemplates for surface-enhanced Raman scattering (SERS) for biosensorics and biophotonics [6]. HNTs with immobilized gold NPs are found to sensitize the photohyperthermia under continuous wave and nanosecond pulsed laser excitation with a photon energy close to the plasmonic resonance. These physical properties of semiconductor and plasmonic NPs are promising for the so-called theranostics (combination of therapy and diagnostics).

References

1. Timoshenko, V.Yu.: Handbook of Porous Silicon, Ed. L.Canham, Springer Publ., pp. 1461-1469 (2018)
2. Kabashin, A.V., Timoshenko, V.Yu.: Nanomedicine **11**, 2247 (2016)
3. Kharin, A.Yu., et al.: Adv. Opt. Mat. 1801728 (2019)
4. Oleshchenko, V.A., et al.: Appl. Surf. Sci. 145661 (2020)

5. Fomin, V.M., Timoshenko, V.Yu.: Appl. Sci. **10** (14), 4992 (2020)
6. Kornilova, A.V., et al.: Appl. Sci. **11** (10), 4565 (2020)
7. Kornilova, A.V., et al.: Appl. Surf. Sci. **566**, 150671 (2021)

Clinical Engineering: Invaluable Contribution in Modern Hospital Management

Nicolas Pallikarakis[1,2]

[1] Emeritus Professor of the University of Patras, Greece
[2] Institute of Biomedical Technology, Patras, Greece
nipa@inbit.gr

Today's hospitals are complex institutions requiring highly qualified managers to run critical operations not only within their facilities but also across the national or regional healthcare system they belong. Apart from the prerequisites for expertise on health care systems management and special skills to operate hospitals in an efficient, cost-effective and safe way, managers must deal with an exceptional high number of diverse professionals, that should apply teamwork, in a well-coordinated manner, in order to deliver the most precious service: health care.

Out of the almost one hundred specialties existing in a modern hospital environment, clinical engineering is a specialty which is often not receiving the recognition it deserves. In today's hospital environment, which is dominated by technology, clinical engineering professionals are among the most valuable contributors for hospital management. In fact, during the last fifty years, medical technology is reshaping the way health care is delivered in a continuous accelerating pace, and its reliable management is the cornerstone to safely pass the benefits deriving from its use, directly to the patients.

Clinical engineering departments (CEDs) are responsible for the overall hospital's health technology management (HTM), during all stages of the operational life cycle of medical equipment, starting from procurement and ending with decommissioning. Tasks that CEDs daily perform or are involved in, within the scope of their responsibilities, include: -Planning of new equipment acquisition, preparing technical specification for call of tenders, assessment of offers, acceptance testing, user training and put in service during the procurement process; keeping an updated inventory; follow service contracts, monitor or perform quality control and safety testing protocols; preventive and corrective maintenance; vigilance and decommissioning of obsolescence or overpassed technologies.

Today, using computerized medical equipment management systems, CEDs can provide direct information on medical equipment inventory, a series of costs associated to purchase and use of medical technology, data on quality and safety controls performed, maintenance, repair actions and many others, resulting to several (standard or customized) key performance indicators (KPIs), which are very critical for evidence-based management. Apart from HTM, clinical engineers are the most suitable employees to address medical devices' regulations issues, participate in health technology assessment (HTA) and the deployment of its findings.

From managers' perspective, which are responsible to take evidence-based and timely decisions on purchasing new equipment, decommissioning old, or overpassed and adopt innovative medical technologies, considering hospital's infrastructure, human resources, and budgets restrictions, CEDs are the most appropriate collaborators, to provide adequate input on such important issues.

Another very critical domain is patient safety. Many studies, published during the last 20 years, revealed a high risk for patients related with adverse events when being hospitalized. In most of the cases, it is due to "medical errors" which in a big percentage could be avoided. Although a large part is attributed to drug administration and complex operation under emergency circumstances, several of them are due to medical technology failures or misuse. Clinical engineering plays a key role in reducing the medical devices-related adverse events, through better safety and quality controls, preventive and corrective maintenance, user training and implementation of a reliable medical devices vigilance system.

However, despite their critical role in improving the efficiency, safety, cost-effectiveness and quality of care, it is difficult to even find clinical engineers in the list of professionals that work in a hospital! For example, in a list with more than 90 professionals working in NHS hospitals, including many that do not have direct contact with patients but considered vital for the efficient running of a hospital, clinical or biomedical engineers are not included!!

Successful management of hospital's processes must demonstrate tangible results in patient care improvement. For example: lower death rates among emergency patients, reduction of days of stay, increased access to high tech services, reduction of medical technology maintenance costs are some of them, most of which are medical technology depended.

This is why, clinical engineers that are responsible for the whole life cycle of medical devices in the hospital can provide an invaluable contribution in modern hospital management.

Improving Lifestyle of Elderly Through Wearable Devices and IoMT

Ratko Magjarević

University of Zagreb, Faculty of Electrical Engineering and Computing, Zagreb, Croatia
ratko.magjarevic@fer.hr

The projections of the share of the EU's population over 65 years old show an increase of 31,3% by 2100 while the working-age population (ages 15–64) is likely to decline. The healthcare costs are expected to grow at the rate of 5 to 6 percent annually, where the largest part of the growth is associated with aging and consequent higher incidence of chronic diseases and multiple impairments to physical functions. The consideration of biological age (compared to chronological age) of a person, based on several biomarkers with enable mathematical modeling, has been recently introduced for estimation of the aging process. Biological age is described by a number of physiological, psychological and behavioral variables that continuously change but may be acquired by continuous monitoring of vital signs and behavior patterns, detection of hazardous events, tracking of social inclusion of the person and even environmental data. The acquired data are then used for calculation of health risks and potential danger of unwanted events. Emerging technologies that are already on the market or being currently developing will enable substantial change in health care and social environment. They include wearable and IoT (or already well-defined Io Medical T) technology. Though many wearables are still considered consumer products, by adding smart algorithms for wearable data analysis, companies are developing them and associated apps in direction of increasing accuracy and reliability of the gathered data and transferring their functionality to the category of medical devices. Active implanted and stationary medical devices are being added the connectivity function so that they became networked as well. Artificial intelligence is added to process the "big data" and recognize relevant medical outcomes. From the point of view of the users, the concept needs to proof their minimal disruption and proven usefulness and benefits, whereas from the point of view of the healtcare system, the feasibility of integration and acceptable costs are the main concerns.

In the presentation, results of our research of monitoring activities of daily living, controlled physical activity during exercising, fall detection and short time glucose prediction will be presented.

Understanding the Structure of Gels Using X-ray Scattering Methods

Annela Seddon

HH Wills Physics Laboratory, Tyndall Avenue, University of Bristol, Bristol BS8 1TD, UK
annela.seddon@bristol.ac.uk

Many commonly used pharmaceutical and healthcare products are based on hydrogels, where an extremely small amount of a 3D fiber network supports a large volume of solvent leading to solid-like properties. Designing new hydrogel materials for specific applications such as tissue engineering is challenging; there is no definitive way of telling what systems will successfully form a gel, and many of the traditional methods for analyzing nanoscale structure such as TEM and AFM mean that the native gel structure cannot be properly elucidated in 3D. We have developed a number of tools using x-ray and neutron scattering to fully characterize novel hydrogel materials made from small molecules based on peptides, and in this, talk will demonstrate how scattering techniques can be used to better understand how gels are formed, the structural transitions which occur during gelation, and ultimately lead to the design of better biologically relevant gels for applications such as tissue engineering.

Surface Modifications of Biomedical Devices

Abdel Salam Makhlouf[1,2] and Abdalla Abdal-hay[3]

[1] Engineering, Metallurgy, Coatings & Corrosion Consultancy (EMC3), 78539 Texas, USA
[2] Central Metallurgical R&D Institute, Cairo, 11421 Egypt
[3] The University of Queensland, School of Dentistry, Herston, QLD 4006, Australia
asalam85@yahoo.com

Magnesium (Mg) is a biodegradable and bioresorbable metallic implant and has received massive interest in cardiovascular and bone applications as it has unique mechanical and biodegradable properties besides its excellent biocompatibility. However, because it possesses a rapid degradation rate, hydrogen gas evolution from Mg surface renders its clinical applications. The interactions between the surrounding tissues and biomaterials are directly associated with the surface characteristics of the biomedical devices. Surface modification is one of the most efficient ways to improve the surface properties of biomedical devices and endow them with new functions. Many attempts have been made for enhancing the biodegradable polymers functionalities, using surface topography and chemistry. In this study, electrospun biodegradable polymer nanofibers mats were fully masked with bioactive nanoceramics to stimulate bone formation capabilities. Because of the poor mechanical properties and wettability of poly (lactic acid) (PLA) electrospun fibers, a thin layer of polyvinyl alcohol (PVA) was deposited on each single PLA nanofibers to enhance the mechanical and wettability properties, thanks to its biocompatibility was improved. We have successfully depressed the rapid degradation of Mg by several coating strategies with further enhancement its biological functionality. Similarly, titanium surface was improved by enhancing osseointegration functionality with the surrounding natural bones.

Keywords: Biomaterials · Biodegradable polymers · Magnesium and its alloys · Titanium and its alloys · Bioactive ceramics

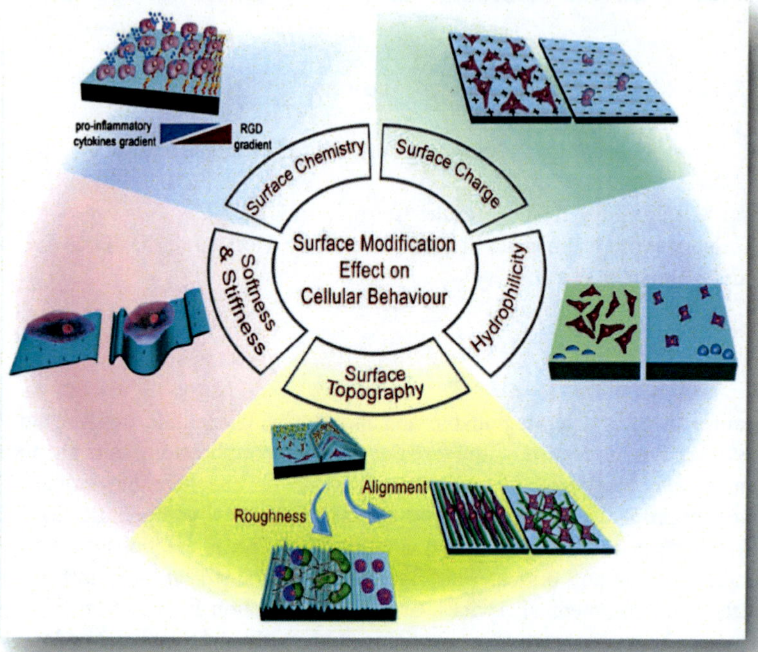

Graphical Abstract

Management of Medical Technologies for Ensuring the Safety, Efficiency, and Quality of Medical Services

Victor Sontea

National Center for Biomedical Engineering/Department of Microelectronics and Biomedical Engineering, Technical University of Moldova, 168, Stefan cel Mare av., MD-2004 Chisinau, Republic of Moldova
victor.sontea@mib.utm.md

Based on the evaluation of the global experience in the healthcare domain, it has been showed that high-performance medical devices represent an indispensable part of the medical act in the prevention, correct diagnosis, and treatment of diseases with high mortality and morbidity in the population. An efficient use of medical devices is expected to promote a significant increase of the number of cost-effective and qualitative investigations and treatment [1].

The maintenance, verification, and management of medical devices have therefore become a priority in the health policy of many states. There are many studies proving that suitable and coherent policies can improve the cost and effectiveness of the application of advanced medical technologies, and at the same time, they can increase patient safety and the overall quality of medical act [1].

The level of endowment of medical institutions with high-performance medical devices and an appropriate level of professionalism of medical resources are the key tools in ensuring the proper functioning of the health system and will have a direct impact on the functional effectiveness of the system, service quality, and degree to the satisfaction of the beneficiary [1].

Moreover, our international experience in the field has demonstrated that the allocation of the necessary resources and the creation, within medical institutions, of units responsible for the management of medical devices can significantly increase the performance of this equipment [2–4].

As a result, the quality of the health service is dependent both on the technical resources, including the endowment with advanced medical devices, and on the professional competence of the personnel involved.

In order to improve the current situation concerning safety, efficiency, and quality of medical services in the Republic of Moldova, five new Hospital Biomedical Engineering Departments (HBMED) were funded, and the curriculum on Management of Medical Technologies within our faculty was implemented [4].

The benefits of implementing this curriculum and the joint activities of the Department of Biomedical Engineering with HBMED will result in an optimal use of expensive medical devices, through qualified and trained personnel, and is expected to reduce user errors and maintenance costs [4].

The present research was financially supported by National Agency for Research and Development of Moldova (project 20.80009.8007.26).

References

1. Sontea, V., Morgoci, S., Turcanu, Gh., Pislaru C., Medical devices management strategy in Republic of Moldova, IFMBE Proceding, vol. 55, pp. 478–481. Springer, Singapore (2016)
2. Alotaibi, Y.K., Federico, F.: The impact of health information technology on patient safety. Saudi Med. J. 38 (12) 1173–1180 (December 2017). DOI: https://doi.org/10.15537/smj.2017.12.20631
3. Iadauza, E., et al.: Clinical Engineering Handbook, Academic Press (2019). ISBN: 978-012-813467-2
4. Sontea, V., Gatcan, S., Pislaru, C., Palii, V., Iavorschi, A.: Medical devices management effectiveness in a hospital (2015). http://repository.utm.md/handle/5014/8269

Contents

Nanotechnologies and Nanomaterials

Influence of Double Feedback on Stationary States of Quantum Dots Lasers

E. Grigoriev(✉), S. Rusu, and V. Tronciu

Department of Physics, Technical University of Moldova, Chisinau, Republic of Moldova
eugeniu.grigoriev@fiz.utm.md

Abstract. We report in this paper the results of theoretical investigations of the influence of double feedback on the stationary states of quantum dots lasers. The Bloch equations model was used to simulate and analyze these states. We have identified the distribution of external cavity modes varying the feedback strength.

Keywords: Double feedback · Quantum dots lasers · Stationary states · External cavity modes

1 Introduction

Semiconductor lasers have become indispensable in the modern society. Nowadays, every day activities are unimaginable without personal computers, internet access, laser printers, displays, etc., the operation of which is based on laser devices. Semiconductor lasers are widely used in a variety of fields such as medicine, optical communications, chemical industry, CD, DVD, BD, mechanics and measurements, spectroscopy, display systems and so on. It is worth mentioning that semiconductor lasers are of interest not only in terms of extremely useful applications, but also in terms of fundamental research, since they form nonlinear systems. Recent progress recorded in the development of new optoelectronic devices has given an impulse to the research of the growth technologies of quantum dot structures. In [1] the developments in quantum dot lasers over the past 20 years and the future prospects for these lasers for scientific and commercial applications are discussed. The review article [2] reports new findings on epitaxial quantum dot lasers on silicon and studies both theoretically and experimentally the connection between the material properties and the ultra-low reflection sensitivity that is achieved. The results show that such quantum dot lasers on silicon exhibit much lower linewidth enhancement factors than any quantum well lasers. Relative intensity noise behavior of a quantum dot laser at free running and under optical injection locking has been reported in [3]. Results demonstrate that an injection-locked quantum dot laser exhibit enhanced characteristic over free-running operation along with low-frequency noise behavior in a wide range of homogeneous broadening values.

The importance of these optoelectronic technologies is determined by the society's growing need of development and construction of devices with increased data transfer speed, with the lowest possible dimensions and costs, but also with high energy efficiency.

I. Tiginyanu et al. (Eds.): ICNBME 2021, IFMBE Proceedings 87, pp. 3–10, 2022.
https://doi.org/10.1007/978-3-030-92328-0_1

Thus, structures with quantum dots have become a fairly favorable active medium for lasers, replacing structures with quantum wells. The advantages of quantum dot lasers in an active medium were predicted about twenty years ago. These include: low threshold-current value; low temperature dependence, material amplification coefficients, larger differentials; frequencies higher than the modulation, weaker pulse-based noise, barely modulated signals and low sensitivity to the action of optical feedback [4–6].

In this paper we report the investigation of influence of double feedback on stationary states of quantum dots lasers based on Bloch equations. The paper is structured as follows. In Sect. 2 we present the setup and equations. Section 3 is devoted to results obtained for distribution of external cavity modes. The conclusions are given in Sect. 4.

2 Setup, Model and Equations

Figure 1 shows the schematic setup of the quantum dot laser under the influence of double feedback considered in this paper. The theoretical study is based on the structure of the quantum dot laser similar to that shown in [7]. This laser consists of the layer representing the active medium with quantum dots and a distant external resonator. It is well known that the model of rate equations is a standard method for describing and examining the dynamics of different semiconductor lasers. However, the model of Bloch equations is more complex, in which the rate of photon annihilation and polar degradation has a similar magnitude.

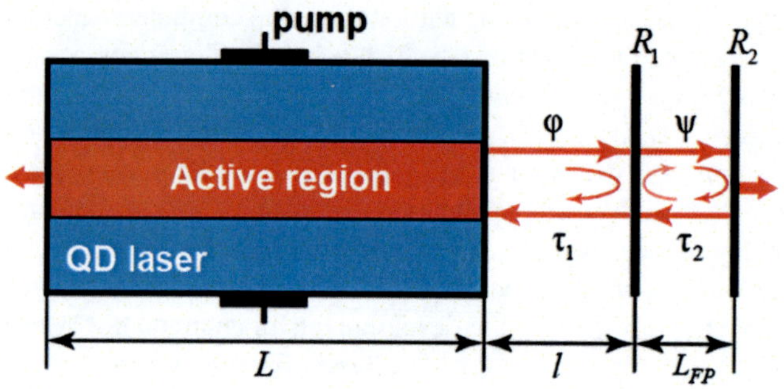

Fig. 1. Setup Diagram of the laser based on quantum dots: R_1 - the reflection coefficient of the mirror closer to the serum; R_2 - the reflection coefficient of the farthest mirror; L - laser length; l - distance between laser and resonator; L_{FP} - the length of the Fabry-Perot resonator. The current is injected only in the active region.

We consider the structure shown in Fig. 1 and the model of the quantum dot laser with an active medium using the following system of Bloch equations [8], in which E is the amplitude of the field, p is the polarization, N inversion, and the terms G_1 and G_2 describe the double feedback

$$\frac{dE}{dt} = -\kappa E + 2Z^{QD}\Gamma|g|p + \frac{Z^{QD}\Gamma\beta F_P}{\tau_{sp}E^*}\left(\frac{N+1}{2}\right)^2$$

$$+\Gamma_1 e^{i\varphi} E(t-\tau_1) + \Gamma_2 e^{i\psi} E(t-\tau_2), \tag{1}$$

$$\frac{dp}{dt} = -\gamma p + |g|EN, \tag{2}$$

$$\frac{dN}{dt} = -4|g|Ep + \frac{N_0(w_e) - N}{T_1(w_e)} - \frac{F_P}{\tau_{sp}}\left(\frac{N+1}{2}\right)^2. \tag{3}$$

The number of quantum dots in the active region of the laser is denoted by Z^{QD}. The feedback part is described by 2 last terms in (1), where $t_1 = 2l/c$ and $t_2 = 2L_{FP}/c_g$ are the delay times between the laser and the resonator and, respectively, in the resonator, c is the speed of light in a vacuum, and c_g in a resonator. The filling factor G represents the part of quantum dots in the volume of the modulus of the contributing laser emission. It is part of the active region which amplifies a certain mode, and is a constant parameter for a laser with a single mode. The lifetime of the $T_1(w_e)$ inversion depends on the density of charge carriers. κ is the annihilation rate of the photons, and g is the degradation parameter of the polarization. t_{eff}. is the effective rate of spontaneous emissions and is given by the Purcell F_p factor and spontaneous emission rate τ_{sp}, that is, $\tau_{eff} = F_p/\tau_{sp}$. g and β are the coupling factors and of spontaneous emission. The β factor describes the percentage of spontaneously emitted photons, which are emitted in the modulus of a resonant cavity wave [8].

For quantum dots in an active medium the parameters from the system of Eqs. (1)–(3) have the following values: $k = 10^{11}$ s^{-1}; $\tau_{sp}^{-1} = 10^{10}$ s^{-1}; $\gamma = 0{,}1$; $g = 3{,}5$ ns^{-1}; $T_1(w_e) = 8 \cdot 10^{10}$ s; $N_0(w_e) = 0.9$; $Z^{QD} = 10^{-6}$; $\tau_0 = 10^{-9}$ s.

For further numerical analysis of the system of Eqs. (1)–(3) it is more convenient to normalize the equations of this system to the lifetime of the carriers τ_0, including the following new parameters and dimensionless coefficients

$$\tilde{\kappa} = \kappa \cdot \tau_0; \quad |\tilde{g}| = |g| \cdot \tau_0; \quad \frac{1}{\tilde{\tau}_{sp}} = \frac{1}{\tau_{sp}} \cdot \tau_0; \quad \frac{1}{\tilde{T}_1} = \frac{1}{T_1} \cdot \tau_0;$$
$$A = 2Z^{QD}\Gamma|\tilde{g}|; \quad B = \frac{Z^{QD}\Gamma\beta F_P}{\tilde{\tau}_{sp}}; \quad C = \frac{F_P}{\tilde{\tau}_{sp}}. \tag{4}$$

The system of Eqs. (1)–(3) with (4) take the following shape

$$\frac{dE}{d\tau} = -\tilde{k}E + \frac{A\tilde{\gamma}}{\tilde{g}}p + \frac{B}{E*}\left(\frac{N+1}{2}\right)^2$$
$$+\Gamma_1 e^{-i\varphi} E(\tau-\tau_1) + \Gamma_2 e^{-i\psi} E_2(\tau-\tau_2), \tag{5}$$

$$\frac{dp}{d\tau} = -\tilde{\gamma}p + \tilde{g}NE, \tag{6}$$

$$\frac{dN}{d\tau} = -\frac{C\tilde{\gamma}}{\tilde{g}}E^*p + \frac{d_0 - N}{\tilde{T}_1} - \frac{1}{\tilde{\tau}_{eff}}\left(\frac{N+1}{2}\right)^2. \tag{7}$$

Dimensionalles Eqs. (5)–(7) represent a system of nonlinear differential equations describing the dynamics of the quantum dot laser under the influence of double feedback schematically represented in Fig. 1.

Next we will analyze the solutions of the uquation system (5)–(7) in the form of the so-called moduli of the outer cavity

$$E = E_s e^{i\omega_s t}; \quad p = p_s e^{i\omega_s t + i\alpha_s}; \quad N = N_s \tag{8}$$

For convenience we will use the following notations

$$\omega_s \tau_1 + \varphi \equiv \Phi; \quad \omega_s \tau_2 + \psi \equiv \Psi \tag{9}$$

In the stationare case ($dE_s/d\tau = dp_s/d\tau = dN_s/d\tau = 0$), from the system of Eqs. (5)–(7), using (8) and (9), we obtain the following set of equations

$$\begin{cases} i\omega_s E_s = -\tilde{k}E_s + \dfrac{A\tilde{\gamma}}{\tilde{g}} p_s(\cos\alpha_s + i\sin\alpha_s) + \dfrac{B}{E_s^*}\left(\dfrac{N_s+1}{2}\right)^2 \\ \qquad + \Gamma_1 E_s(\cos\Phi - i\sin\Phi) + \Gamma_2 E_s(\cos\Psi - i\sin\Psi), \\ i\omega_s p_s = -\tilde{\gamma} p_s + \tilde{g} N_s E_s(\cos\alpha_s - i\sin\alpha_s), \\ 0 = -\dfrac{C\tilde{\gamma}}{\tilde{g}} E_s^* p_s(\cos\alpha_s + i\sin\alpha_s) + \dfrac{d_0 - N_s}{\tilde{T}_1} - \dfrac{1}{\tilde{\tau}_{eff}}\left(\dfrac{N_s+1}{2}\right)^2. \end{cases} \tag{10}$$

From the system of Eqs. (10), using notations from (9) for the frequency modes of the external cavity we obtain the following transcendental equation for ωs:

$$\omega_s = -\Gamma_1 \sin(\omega_s \tau_1 + \varphi) - \Gamma_2 \sin(\omega_s \tau_2 + \psi). \tag{11}$$

Also we obtain the amplitude values of laser field strength and polarization vector

$$E_s = \sqrt{\frac{1}{CN_s}\left[\frac{d_0 - N_s}{\tilde{T}_1} - \frac{1}{\tilde{\tau}_{eff}}\left(\frac{N_s+1}{2}\right)^2\right]}. \tag{12}$$

and

$$p_s = \sqrt{\frac{\tilde{g}^2 N_s}{C\tilde{\gamma}^2}\left[\frac{d_0 - N_s}{\tilde{T}_1} - \frac{1}{\tilde{\tau}_{eff}}\left(\frac{N_s+1}{2}\right)^2\right]}. \tag{13}$$

in which the charge carrier density N_s is the solution of the cubic equation

$$\begin{aligned} &\tilde{T}_1 D N_s^3 + \left[\tilde{T}_1 \Gamma_f + 2\tilde{T}_1 D - 4\tau_{eff} A\right] N_s^2 \\ &+ \left[T_1 D + 2\Gamma_f (T_1 + 2\tau_{eff}) + 4 d_0 A \tau_{eff}\right] N_s \\ &+ \Gamma_f (T_1 - 4 d_0 \tau_{eff}) = 0, \end{aligned} \tag{14}$$

where

$$\Gamma_f = \tilde{k} - \Gamma_1 \cos\Phi - \Gamma_2 \cos\Psi; \quad D = \tilde{\tau}_{eff} BC - A.$$

Thus, we got a set of equations that describe the stationary states of the laser system shown in Fig. 1.

Finally, we consider that the results presented in this document provide a basis for future studies and especially for more detailed investigations on the operation of quantum dot lasers under the influence of external optical feedback.

3 Numerical Results and Discussions

We study the distributions of External Cavity Modes (ECM) in the plane of $(N_s - \omega_s)$. When the feedback is absent only one stationary state is present. The introduction of feedback leads to an appearance of additional other modes by pare. For this we solve the Eqs. (11) and (14). The Eq. (14) might have three solutions. Thus, we solve this

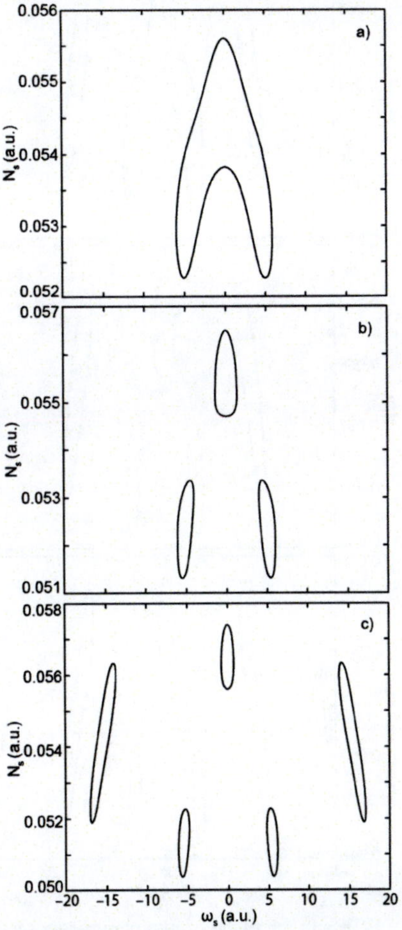

Fig. 2. The distribution of ECMs in the plane of two parameters $(N_s - \omega_S)$ for a) $\Gamma_1 = 5, \Gamma_2 = 5$, b) $\Gamma_1 = 5, \Gamma_2 = 10$, c) $\Gamma_1 = 5, \Gamma_2 = 15$.

equation, and for each real solution of N_s we have to obtain ω_s from (11). Figure 2a shows the distribution of modes for $\Gamma_1 = 5$, $\Gamma_2 = 5$. One ca see that the modes are not located on the ellipse but on the shape of "bell".

This is due to the presence of many solutions from Eq. (14). When we increase feedback in second branch the "bell" is deformed in 3 deformed ellipses (see Fig. 2b). A further increase of feedback strength $\Gamma_2 = 15$ new satellites appears in the distribution of ECMs (see Fig. 2c).

In Fig. 3 we show the results of an increase of feedback into first branch to $\Gamma_1 = 10$, and vary the feedback in second branch from 10 to 20. For $\Gamma_1 = 10$, $\Gamma_2 = 10$ the ECMs are located on so called "bell" and satellites ellipses (see Fig. 3a). When $\Gamma_2 = 15$ the

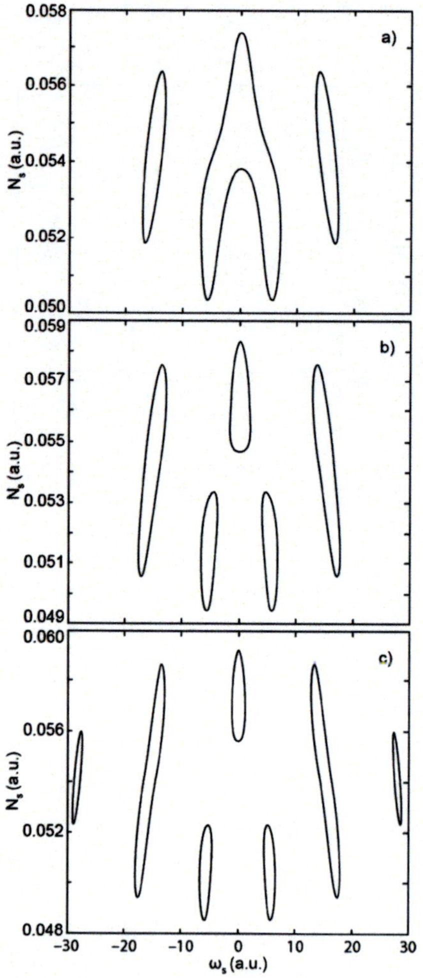

Fig. 3. The distribution of ECMs in the plane of two parameters ($N_s - \omega_s$) for a) $\Gamma_1 = 10$, $\Gamma_2 = 10$, b) $\Gamma_1 = 10$, $\Gamma_2 = 15$, c) $\Gamma_1 = 10$, $\Gamma_2 = 20$.

"bell" is deformed and ellipses are increased. For $\Gamma_2 = 20$ the new satellites deformed ellipses (see Fig. 3c). Thus, we observe that an increase of feedback strengths leads to a complicated distribution of external cavity modes.

Figure 4a shows the distribution of modes for higher feedback strength in first branch of external cavity $\Gamma_1 = 15$ and small feedback strength in second one $\Gamma_2 = 5$. One ca see a complicated distribution.

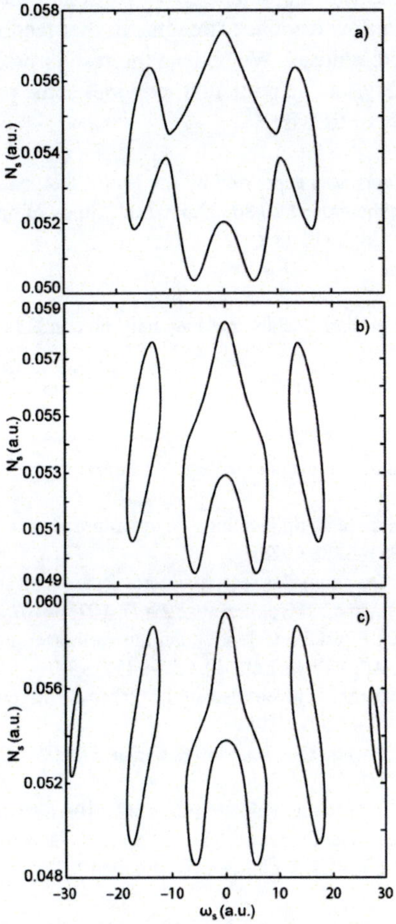

Fig. 4. The distribution of ECMs in the plane of two parameters ($N_s - \omega_s$) for a) $\Gamma_1 = 15$, $\Gamma_2 = 5$, b) $\Gamma_1 = 15$, $\Gamma_2 = 10$, c) $\Gamma_1 = 15$, $\Gamma_2 = 15$.

The shape of complicated distribution of modes is deformed when the feedback strength in second branch is increased and the "bell" shape figure and deformed satellite ellipses are present (Fig. 4c). Thus, we found that for large feedback strength the shape of location of external modes become more complicated.

4 Conclusions

In this document, we studies the stationary states of the laser with active medium quantum dots under the influence of optical feedback. For this we used the model of the Bloch equations, which helped us obtain the expressions for the amplitude values of the laser field strength and polarization vector, as well as a cubic algebraic equation for determining the density of charge carriers. We also obtained a transcendent equation for determining the moduli of the outer cavity. We found that the ECMs are located on the shape of "bell" for low feedback strength. Higher feedbacks implye the appearance of deformed satelite ellipses. We believe the results presented in this paper is a good basis for future study, and, in particular, provides some pointers for more detailed investigations of stability of ECMs.

Acknowledgment. This work was supported by the National Agency for Research and Development of Moldova within the project 20.80009.5007.08 "Study of optoelectronic structures and thermoelectric devices with high efficiency".

Conflict of Interest. The authors declare that they have no conflict of interest.

References

1. Justin, C., et al.: Quantum dot lasers—history and future prospects. J. Vacuum Sci. Technol. A **39**, 020802 (2021)
2. Grillot, F., et al.: Physics and applications of quantum dot lasers for silicon photonics. Nanophotonics **9**(6), 1271–1286 (2020)
3. Sheikhey, M.M., et al.: Quantum-dot semiconductor lasers with prominent relative intensity noise and spectral characteristics. Opt. Express **29**(7), 10236–10248 (2021)
4. Arakawa, Y., Sakaki, H.: Multidimensional quantum well laser and temperature dependence of its threshold current. Appl. Phys. Lett. **40**, 939–941 (1982)
5. Miyamoto, Y., et al.: Threshold current density of GaInAsP/InP quantum-box lasers. IEEE J. Quantum Electron. **25**, 2001–2006 (1989)
6. Bimberg, D., Ribbat, Ch.: Quantum dots: lasers and amplifiers. Micro-electronics J. **34**, 323–328 (2003)
7. Aust, R., et al.: Optical and quantum electronics. **48**(2), 109 (2016)
8. Luedge, K.: In Nonlinear Laser Dynamics - From Quantum Dots to Cryptography. Luedge, K. (ed.), chap. 1, p. 3. WILEY-VCH Weinheim, Weinheim (2012)

Quantum Photon Conversion via Coherently Driven Permanent Dipole Systems

Sergiu Carlig(✉), A. Mirzac, P. Bardetski, and M. A. Macovei

Institute of Applied Physics, Chişinău, Republic of Moldova
sergiu.carlig@ifa.md

Abstract. The photon quantum dynamics in a damped single-mode quantized cavity field coupled with a resonantly driven two-level system possessing nonzero permanent dipoles is investigated here. The interacting subsystems are very different frequencies (microwave and optical domains) and, in this highly dispersive regime, the emitter couples to the resonator mode via its diagonal dipole moments only. As a result, this interaction regime is responsible for the cavity multiphoton quantum dynamics and photon conversion from optical to terahertz ranges, for instance. Furthermore, enhanced terahertz photon correlations occur as well.

Keywords: Quantum dynamics · Permanent dipole · Photon conversion · Photon correlations

1 Introduction

The conversion of frequency is very interesting and actual problem in quantum optics. There are many schemas and possibilities to convert an input light beam into an output beam of a different frequency [1–6] with a lot of possible and promising quantum applications [7–10]. Since usually the frequency conversion investigations refer to resonant processes, we shall discuss a photon conversion scheme where non-resonant multiphoton effects are involved, respectively. More precisely, we investigate frequency downconversion processes via a resonantly laser- pumped two-level emitter possessing permanent diagonal dipoles, $d_{\alpha\alpha} \neq 0$ with $\alpha \in \{1, 2\}$, and embedded in a single-mode quantized resonator, see Fig. 1.

The two level emitter has an operating frequency in optical range, i.e. about 10^{15} Hz, as well as, the cavity frequency is at least at three order less. As a consequence, the two-level system interacts with the resonator mode via its permanent dipoles only. Also, the cavity's frequency or its multiples differs as well from the generalized Rabi frequency Therefore, this highly dispersive interaction regime leads to multiphoton absorption-emission processes in the resonator mode. Such kind of processes are mediated by the corresponding damping effects. We investigate and discuss here the corresponding cavity photon quantum dynamics in the steady state as well as its second-order photon-photon correlation function.

Particularly, the mean-photon number exhibits a plateau followed by an abrupt photon increase while its second-order photon-photon correlation function enhances

I. Tiginyanu et al. (Eds.): ICNBME 2021, IFMBE Proceedings 87, pp. 11–16, 2022.
https://doi.org/10.1007/978-3-030-92328-0_2

or decreases depending on the ratio $\eta = \frac{g}{2\Omega}$, see also Figs. (1, 2 and 3). Oscillatory behaviors of these quantities are observed too, as the qubit-cavity coupling strength g approaches the Rabi frequency Ω.

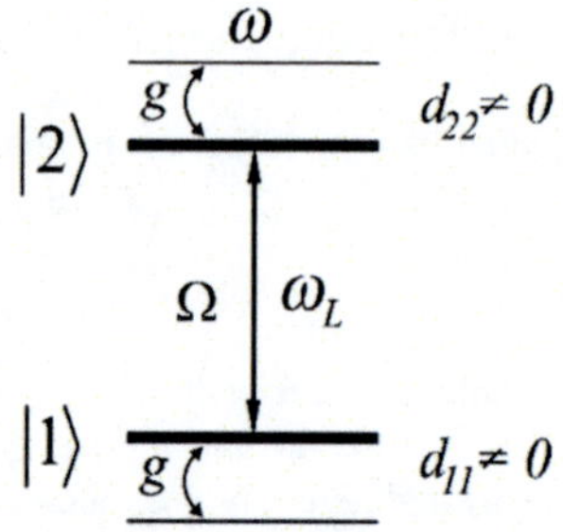

Fig. 1. The analyzed system: laser pumped two-level quantum emitter interacts with a resonator through its non-zero diagonal dipoles, $d_{\alpha\alpha} \neq 0$ with $\alpha \in \{1, 2\}$. Here, ω is the single mode resonator frequency, Ω is the corresponding Rabi frequency due to the off-diagonal dipole moment d_{21}. Laser frequency is ω_L, which respects the condition: $\omega_L \gg \omega$., and the emitter-cavity coupling strength is denoted by g.

Our setup can be implemented in real or artificially created systems having $d_{22} \neq d_{11}$, such as asymmetrical two-level quantum dots [11, 12] and molecules [13, 14], or, equivalently, spin or quantum circuits [15, 16], and taking advantage of the technological progress towards their coupling to various resonators.

The article is organized as follows. In Sect. 2 we apply the developed analytical approach to the system of interest and describe it, while in Sect. 3 we analyze the obtained results. The summary is given in Sect. 4.

2 Analytical Framework

The master equation describing, in the Born-Markov approximations, the interaction of a two-level emitter, having permanent dipoles, with a coherent electromagnetic wave of frequency ω_L, treated as classical field, as well as with a quantized single mode resonator of frequency ω, with $\omega \ll \omega_L$ (see Fig. 1), is given as follows:

$$\begin{aligned} &\tfrac{d}{dt}\rho(t) + \tfrac{i}{\hbar}[H, \rho(t)] = -\tfrac{\gamma}{2}\left[S^+, S^-\rho(t)\right] \\ &-\tfrac{\kappa}{2}\left[b^\dagger, b\rho(t)\right] - \tfrac{\kappa}{2}\bar{n}\left[b^\dagger, [b, \rho(t)]\right] + H.c. \end{aligned} \tag{1}$$

Here, the left right hand side is the coherent part of the master equation, and the right hand side describe the dissipation with, γ - the single-emitter spontaneous decay rate, and κ is the corresponding boson-mode's leaking rate. $S^+ = |2\rangle\langle 1|$ and $S^- = |1\rangle\langle 2|$ are the well-known levels transition operators, which respects the commutation relations for su (2) algebra, namely, $S_z = \frac{1}{2}\left[S^+, S^-\right]$ and $S^\pm = \pm\left[S_z, S^\pm\right]$, where $S_z = \frac{1}{2}\{\,|2\rangle\langle 2| - |1\rangle\langle 1|\}$ is the bare-state inversion operator. $|1\rangle$ and $|2\rangle$ are the ground and excited state of the emitter. $b^\dagger$ and b are the creation and the annihilation operator of the resonator electromagnetic mode. Commutation relations for bosonic operators are: $[b, b^\dagger] = 1$, and $[b, b] = [b^\dagger, b^\dagger] = 0$.

The Hamiltonian describing the laser driving two-level emitter possessing permanent dipoles, coupled to a single-mode resonator, in the dipole and rotating wave approximations, is [6]:

$$\begin{aligned} H = {} & \hbar\omega b^\dagger b + \hbar\omega_{21} S_z - \hbar\Omega\left(S^+ e^{-i\omega_L t} + S^- e^{i\omega_L t}\right) \\ & + \hbar g_0 (d_{22} S_{22} + d_{11} S_{11})\left(b^\dagger + b\right) \\ & + \hbar \bar{g}_0 \left(S^+ + S^-\right)\left(b^\dagger + b\right) \\ & - E_L (d_{22} S_{22} + d_{11} S_{11}) \cos(\omega_L t). \end{aligned} \tag{2}$$

The first and the second terms are free energies of the resonator and the two-level system the third and the sixth ones describe the interaction of the external laser field with the two-level emitter through its off-diagonal dipole moments d_{21}, $d_{21} = d_{12}$. The diagonal dipole moments are d_{22} and d_{11}, respectively. The fourth and the fifth terms denote the interactions of the cavity mode with the two-level emitter through diagonal and off-diagonal dipole moments, with the amplitude of the external driving field. $g_0 = \sqrt{\frac{2\pi\omega}{\hbar V}}$, where V is the quantization volume, and $\bar{g}_0 = g_0 d_{21}$, $S_{\alpha\alpha}$, $\{\alpha = 1, 2\}$ are the population operators, respectively.

Applying unitary transformation and neglecting rapid oscillating terms, Hamilton operator became

$$\begin{aligned} H = {} & \hbar\omega b^\dagger b + \hbar\Delta S_z - \hbar\Omega\left(S^+ + S^-\right) \\ & + \hbar g S_z \left(b^\dagger + b\right) + \frac{\hbar g_0}{2}(d_{11} + d_{22})\left(b^\dagger + b\right) \end{aligned} \tag{3}$$

where $g = g_0(d_{22} - d_{11})$ with $S_{22} = 1/2 + S_z$ and $S_{11} = 1/2 - S_z$. Further, performing an unitary transformation $V = \exp(\zeta b - \zeta^* b^\dagger)$ with $\zeta = \frac{g_0(d_{11}+d_{22})}{2(\omega + i\kappa/2)}$, in the whole master Eq. (1), containing the Hamiltonian (3), one arrives at the same form of the master equation with, however, the Hamiltonian (4), and where $\Delta \equiv \Delta - g_0^2(d_{22}^2 - d_{11}^2)/\omega$ when $\omega \gg \kappa$. The last term from the detuning's expression can be used to redefine the emitter's frequency $\omega_{21} \equiv \omega_{21} - g_0^2(d_{22}^2 - d_{11}^2)/\omega$.

Thus, the final Hamiltonian characterizing the respective coherent evolution of the considered compound system is:

$$H = \hbar\omega b^\dagger b + \hbar\Delta S_z - \hbar\Omega\left(S^+ + S^-\right) + \hbar g S_z\left(b^\dagger + b\right) \tag{4}$$

Here the first two terms describe the free energies of the cavity electromagnetic field and the artificial two-level system, with detuning $\Delta = \omega_{21} - \omega_L$ between two level transition frequency ω_{21} and the laser one. The third and the fourth component describe the laser interaction with the two-level system and the emitter-cavity interaction. Ω and g are the coupling strengths. We have to mention that the Rabi frequency Ω is proportional to the off-diagonal dipole moment d_{21} and the emitter-cavity coupling is proportional to the diagonal dipole moments, i.e. $g \propto (d_{22} - d_{11})$ [6].

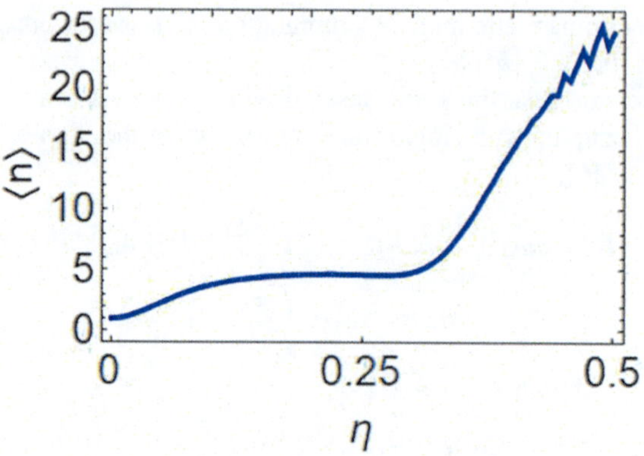

Fig. 2. The steady-state mean cavity photon number $\langle n\rangle \equiv \langle b^\dagger b\rangle$ as a function of $\eta = \frac{g}{2\Omega}$. Here $\bar{n} = 1, \gamma/\Omega = 10^{-2}, \kappa/\Omega = 10^{-5}, \omega/\Omega = 3$ and $\Delta/\Omega = 0.1$.

In what follows, we have numerically computed the Master Eq. (1) with the Hamiltonian (4) and the results are presented in the next Section.

3 Results and Discussion

Particularly, Fig. 2 shows the steady-state behavior of the mean-photon number $\langle n\rangle = \langle b^\dagger b\rangle$ as a function of $\eta = \frac{g}{2\Omega}$. One can observe that when $\eta = 0,\ \langle n\rangle = \bar{n}$, i.e., it is equal with the environmental mean thermal photon number. As η increases, the mean-photon number increases too, demonstrating a plateau followed by an abrupt positive change in the photon number. Since the resonance condition $2\Omega = k\omega, \{k \in 1, 2, 3, \cdots\}$, is not fulfilled here, it follows that *dispersive* multiphoton processes occur [6]. That is, the quantum photon dynamics is subject to absorption or emission of multiple photons into the cavity mode, although there are no any resonances in the system. The cavity mode frequency lies in the terahertz frequency domain which means that we have obtained an electromagnetic source of terahertz photons, above the thermal background given by $\bar{n}$, that is $\langle b^\dagger b\rangle \gg \bar{n}$. Furthermore, as the qubit-cavity coupling strength g approaches the Rabi frequency Ω, i.e.$g \sim \Omega$, photon oscillation behaviors are observed, see Fig. 2. On the other side, the photon statistics exhibits quasi- thermal features as η increases, see Fig. 3. Furthermore, the abrupt increases of the photon number is followed by increased photon-photon correlations, although correlation magnitude decays as the mean-photon number is maximal, compare Figs. 2 and 3, respectively.

So, we have shown the possibility to have a terahertz photon flux with increased photon-photon correlations. This is quite relevant since the photons usually do not interact mutually (for example, for a thermal light one has $g^{(2)}(0) = 2$, whereas for a coherent source - $g^{(2)}(0) = 1$).

Generalizing here, the presence of diagonal dipole moments, i.e. $d_{22} - d_{11} \neq 0$, in a resonance coherently pumped two-level system, makes possible the coupling to the resonator mode at a completely different frequency than the input one which drives the

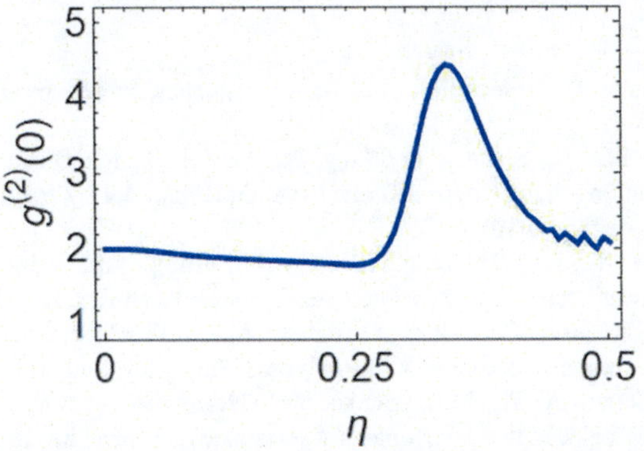

Fig. 3. The steady-state behavior of the second-order photon-photon correlation function $g^{(2)}(0)$ as a function of $\eta = \frac{g}{2\Omega}$. Here $\bar{n} = 1$, $\gamma/\Omega = 10^{-2}$, $\kappa/\Omega = 10^{-5}$, $\omega/\Omega = 3$ and $\Delta/\Omega = 0.1$.

two-state quantum emitter. Multiphoton emission into the cavity mode is feasible then. Photon correlations increases or decreases depending on the pumping parameter η.

4 Summary

Summarizing, we have investigated the feasibility to convert radiation from optical to terahertz domains through resonantly pumping two-level quantum system, interacting with single-mode cavities. Furthermore, the corresponding damping effects due to emitter's spontaneous emission and cavity's photon leakage were taking into account as well. Also, we have assumed that the transition frequency of the two- level emitter is much more than the cavity's frequency. In these far off resonant circumstances, cavity multiphoton absorption-emission processes occur. Particularly, we have computed the cavity mean-photon number as well as its second-order photon-photon correlation function. We have found that the photon statistics exhibits quasi- thermal photon statistics and photon correlations modifies as the pumping parameter η increases from zero. Moreover, oscillatory behaviors of these quantities are observed too, as the qubit-cavity coupling strength g approaches the Rabi frequency Ω.

Finally, as we already have mentioned, as a concrete system where the approach developed here can apply, can serve asymmetrical two-level quantum dots coupled to terahertz resonators as well as polar biomolecules, spin or quantum circuit systems, or pumped two-level quantum dots interacting with acoustical phonon resonators, respectively.

Acknowledgment. We acknowledge the financial support by the Moldavian National Agency for Research and Development, grant No. 20.80009.5007.07.

Conflict of Interest. The authors declare that they have no conflict of interest.

References

1. Huang, J., Kumar, P.: Observation of quantum frequency conversion. Phys. Rev. Lett. **68**, 2153 (1992)
2. Rakher, M.T., Ma, L., Slattery, O., Tang, X., Srinivasan, K.: Quantum transduction of telecommunications- band single photons from a quantum dot by frequency upconversion. Nat. Photonics **4**, 786 (2010)
3. Guo, X., Zou, C.-L., Jung, H., Tang, H.X.: On-chip strong coupling and efficient frequency conversion between telecom and visible optical modes. Phys. Rev. Lett. **117**, 123902 (2016)
4. Han, J., Vogt, Th., Gross, Ch., Jaksch, D., Kiffner, M., Li, W.: Coherent microwave-to-optical conversion via six-wave mixing in Rydberg atoms. Phys. Rev. Lett. **120**, 093201 (2018)
5. Kibis, O.V., Slepyan, G.Ya., Maksimenko, S.A., Hoffmann, A.: Matter coupling to strong electromagnetic fields in two-level quantum systems with broken inversion symmetry. Phys. Rev. Lett. **102**, 023601 (2009)
6. Mîrzac, A., Carlig, S., Macovei, M.A.: Microwave multiphoton conversion via coherently driven permanent dipole systems. Phys. Rev. A **103**, 043719 (2021)
7. Kimble, H.J.: The quantum internet. Nature (London) **453**, 1023 (2008)
8. Huang, Y.-P., Velev, V., Kumar, P.: Quantum frequency conversion in nonlinear microcavities. Opt. Lett. **38**, 2119 (2013)
9. Northup, T.E., Blatt, R.: Quantum information transfer using photons. Nat. Photonics **8**, 356 (2014)
10. Lake, D.P., Mitchell, M., Sanders, B.C., Barclay, P.E.: Two-colour interferometry and switching through optomechanical dark mode excitation. Nat. Commun. **11**(2208), 1 (2020)
11. Garziano, L., Macri, V., Stassi, R., Di Stefano, O., Nori, F., Savasta, S.: One photon can simultaneously excite two or more atoms. Phys. Rev. Lett. **117**, 043601 (2016)
12. Chestnov, I., Shakhnazaryan, V.A., Shelykh, I.A., Alodjants, A.P.: Ensemble of asymmetric quantum dots in a cavity as a terahertz laser source. JETP Lett. **104**, 169 (2016)
13. Macovei, M., Mishra, M., Keitel, C.H.: Population inversion in two-level systems possessing permanent dipoles. Phys. Rev. A **92**, 013846 (2015)
14. Anton, M.A., Maede-Razavi, S., Carreno, F., Thanopulos, I., Paspalakis, E.: Optical and microwave control of resonance fluorescence and squeezing spectra in a polar molecule. Phys. Rev. A **96**, 063812 (2017)
15. Greenberg, Ya.S.: Low-frequency Rabi spectroscopy of dissipative two-level systems: dressed-state approach. Phys. Rev. B **76**, 104520 (2007)
16. Yoshihara, F., Fuse, T., Ashhab, S., Kakuyanagi, K., Saito, S., Semba, K.: Superconducting qubit–oscillator circuit beyond the ultrastrong-coupling regime. Nat. Phys. **13**, 44 (2017)

Electrical Characterization of Individual Boron Nitride Nanowall Structures

Vasile Postica[1(✉)], F. Schütt[2], C. Lupan[1], H. Krüger[2], R. Adelung[2], and O. Lupan[1,2]

[1] Center for Nanotechnology and Nanosensors, Department of Microelectronics and Biomedical Engineering, Technical University of Moldova, Chisinau, Republic of Moldova
vasile.postica@mib.utm.md

[2] Institute for Materials Science, Chair for Functional Nanomaterials, Faculty of Engineering, Kiel, Germany

Abstract. In this work, the individual hexagonal boron nitride (*h*-BN) microtubular structures with different diameter (ranging from ≈0.2 to ≈2.5 μm) and a wall thickness below 25 nm were investigated for the first time by integration on SiO_2/Si substrate using a method based on focused ion beam deposition (FIB/SEM). The current-voltage (I-V) measurements were carried out in from a bias of −40 V to +40 V and in a temperature range from 25 to 100 °C. All fabricated devices showed excellent insulating properties and the resistance of ≈111 GΩ was calculated, which was attributed mainly to the top SiO_2 layer of the substrate measured without *h*-BN. The obtained results elucidate the excellent potential of the boron nitride microtubular structures with nanowalls to be used as high-quality shielding materials of other nano- and microstructures for application in nanoelectronics, nanophotonics and power electronics, where a relatively wide range of operating temperature is necessary.

Keywords: Hexagonal boron nitride · Microdevices · Microtubes · Nanomaterials · Electrical properties · Insulator

1 Introduction

Development of devices with superior performances based on nano- and microscale structures via "bottom-up" synthesis routes enable the controlled tailoring of electrical, optical and magnetic properties [1–4]. In this context, well-structured B – N – C systems are highly attractive as building blocks for nanoelectronics and microelectromechanical systems due to the possibility of controlling the electrical properties through the change in C content [6]. Also, the small lattice mismatch of BN with graphite (1.7%) [7] gives an excellent potential for development of high-quality graphene electronics [1, 5, 7]. For example, Levendorf *et al.* integrated layered *h*-BN using a scalable technique with patterned regrowth for application in thin circuitry [1]. *h*-BN is an attractive material due to the excellent mechanical properties and high thermal and chemical stabilities up to 800–1000 °C [2, 8, 9]. Furthermore, due to its bandgap of ~5.5 eV, BN nanotubes exhibit excellent insulating electrical properties with very low dependence on

I. Tiginyanu et al. (Eds.): ICNBME 2021, IFMBE Proceedings 87, pp. 17–23, 2022.
https://doi.org/10.1007/978-3-030-92328-0_3

tube diameter, helicity and number of tubular walls [2, 6], as well as high breakdown voltage ($V_{bd} \approx 0.7$ V nm^{-1}) [7]. Therefore, the tubular structures based on *h*-BN with higher diameter and different nanowall thicknesses can be used as protective shields for encapsulating of different metallic and non-metallic micro- and nanostructures [10, 11], as well as to fabricate typical Schottky diodes and transistors [12]. For example, Li *et al.* prepared BN-nanotube-encapsulated β-SiC nanowires using a simple chemical vapor deposition process [10]. Bando *et al.* synthesized insulating nanocables based on Fe-Ni alloy nanorods inside BN nanotubes using a two-stage process [11]. Such types of core-shell and core-multishell heterostructures demonstrated excellent properties for further integration in high-performance field-effect transistors [10, 13].

Nanomaterials from *h*-BN also demonstrated excellent potential for gas sensor devices, being a crystal with the chemical alternation of B and N atoms, while the ionic nature of the bonds is highly efficient for adsorption of gaseous species of different reducing and oxidizing gases [14]. For example, Sajjad and Feng reported on the synthesis of *h*-BN nanosheets and investigated the sensing properties to different reducing gases, showing possibility to detect CH_4 gas at 175 °C [14]. Lin *et al.* reported on gas sensing properties of *h*-BN nanosheets with several atomic layers, demonstrating a fast response to C_2H_5OH at 300 °C (optimal operating temperature) [15]. Recently, Schütt *et al.* synthesized randomly arranged and interconnected hexagonal boron nitride (*h*-BN) hollow microtubes with high porosity (>99.99%) and with a specific surface area of 900 m^2g^{-1} [3].

In this work, such type of individual microtubular structures with a wide range of diameters, ranging from ≈0.2 μm to ≈2.5 μm were integrated into devices in order to study their electrical and UV photodetection ability for possible future integration in nanoelectronics.

2 Experimental Conditions

Networks of interconnected *h*-BN microstructures were growth using a computer-controlled chemical vapor deposition (CVD) process and ZnO tetrapod networks with high porosity (up to 98%) as a template [16]. More details on synthesis procedure and growth conditions are presented in previous work [16]. The CVD process enables the growth of *h*-BN top layers below-on the ZnO template. The ZnO template is etched by hydrogen simultaneously with CVD process [16]. The *h*-BN CVD process gives a thicknesses of the microtubes diameter between 0.2 and 3 μm and nanowalls <25 nm [16]. The measurements of Raman spectra showed the complete removal of the ZnO from the synthesized material.

Several two-terminal devices based on individual BN microtubular structures with different diameters were made in FIB/SEM system using the method reported by Lupan *et al.* [3, 17, 18]. The devices were noted as BN1, BN2, BN3, BN4, BN5, and BN6 for microtubular structures with diameter (D) of ≈2.5, ≈1.2, ≈1.0, ≈0.65, ≈0.5, and ≈0.2 μm, respectively. The SEM micrographs of devices are shown in Fig. 1. Individual structures were connected to Au/Cr layers with complex based on platinum complex. The SiO_2 (350 nm)/Si were used as the substrates [3, 17, 18].

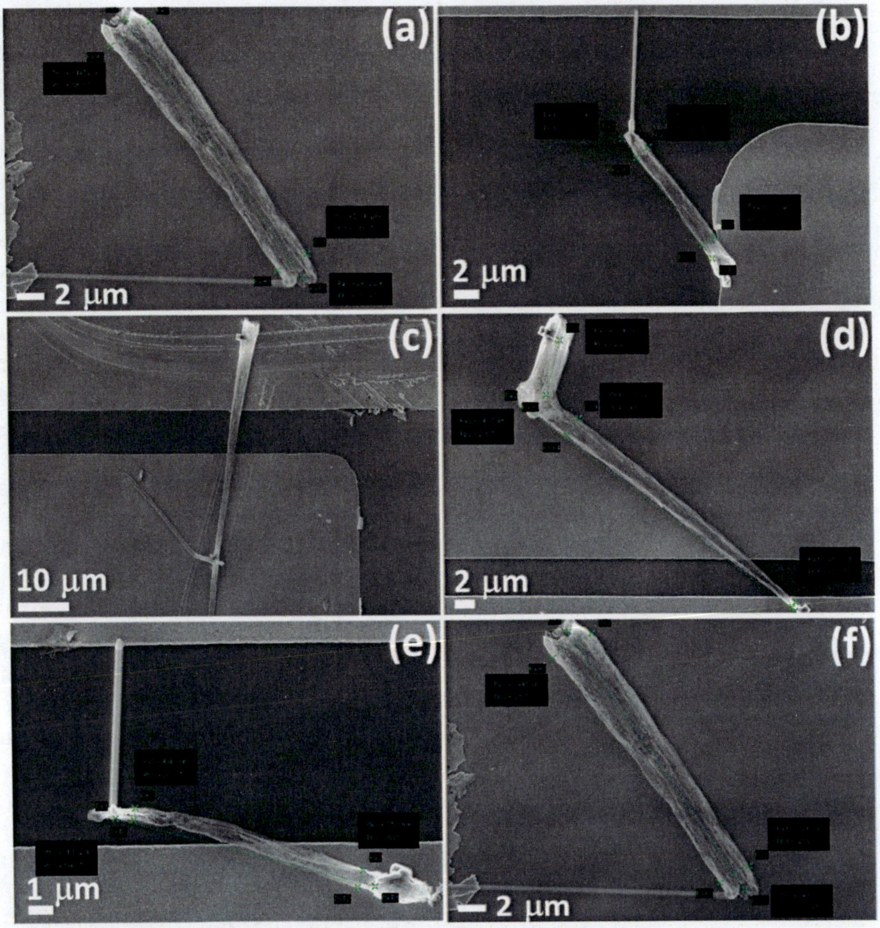

Fig. 1. SEM micrographs of devices based on individual microstructures of BN with different diameters: (a) BN1, D ≈ 2.5 μm; (b) BN2, D ≈ 1.2 μm; (c) BN3, D ≈ 1.0 μm; (d) BN4, D ≈ 0.65 μm; (e) BN5, D ≈ 0.5 μm; (f) BN6, D ≈ 0.2 μm.

3 Results and Discussions

BN is known to be an excellent insulating material with a direct band gap of 5–6 eV with high chemical stability up to 1000 °C [9]. In order to evaluate the high quality of synthesized microtubular structures of *h*-BN, the current-voltage (*I*-*V*) measurements of elaborated structures was performed in air and dark at normal atmosphere pressure and room temperature (≈25 °C). The electrical measurements were performed by two-point method. The results are presented in Fig. 2a. The measurements were performed in hysteresis mode, i.e. the applied bias was changed from 0 to +40 V (step 1), followed by a decrease to −40 V (step 2 and 3) and then was increased to 0 V (step 4). The *I*-*V* measurements of a chip without *h*-BN microtubular structure was also included, which represents the leakage current of the SiO_2 layer (see Fig. 2a). The observed peaks from Fig. 2a noted with arrows were observed in all measurements. The origin of these peaks is unknown, but most likely originate from the measurement unit (Keithley 2400 SourceMeter). As can be observed from Fig. 2a, the current of all devices, as well as for the SiO_2 layer (without *h*-BN) is practically similar, and is in a range of 360 pA at +40 V. The basically quasi-linear *I-V* curves indicate on insulating properties of devices. The observed hysteresis of *I-V* curves can be attributed to parasitic capacities of $Au/SiO_2/Au$ structures. In conclusion, the current of the devices does not exceed the leakage current of the SiO_2 layer indicating that no current passes through the BN microtubular structures. The resistance of devices was calculated to be ≈111 GΩ, which can be attributed mainly to SiO_2 layer (the resistivity of SiO_2 layer was calculated to be $\approx 2.2 \times 10^9$ Ω cm). This resistance is much higher than that of individual BN nanotube [6, 17]. Even by rise the temperature to 100 °C no considerable changes in current was detected (see Fig. 2b). Furthermore, the elaborated structures can support the bias voltage of 40 and 100 V for long time (see Fig. 2c), which is higher than for BN nanotubes [18]. These results indicate the excellent insulating properties and minor carbon contamination [19, 20], i.e. high quality of *h*-BN structures [9] and demonstrates promising properties in future for

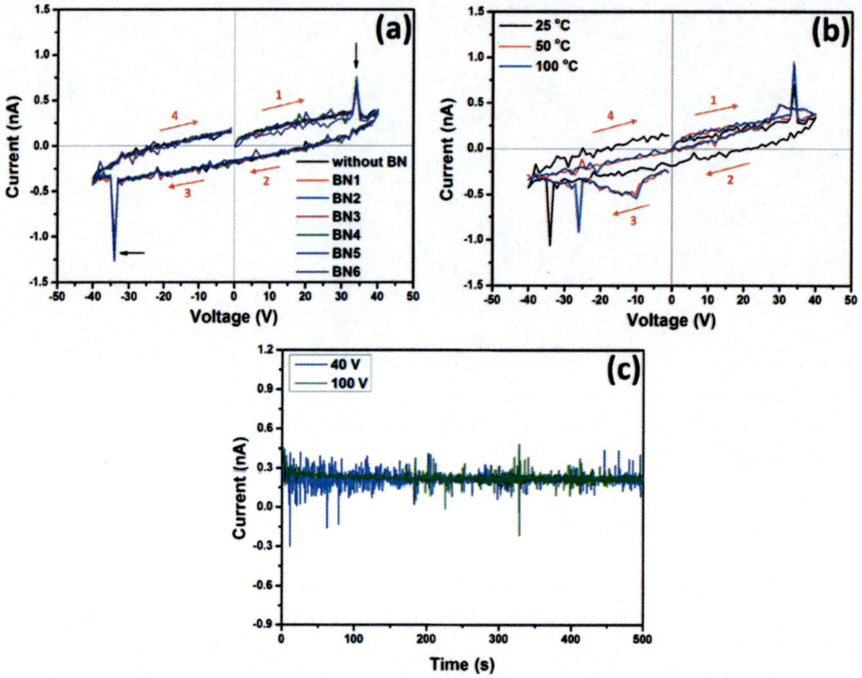

Fig. 2. (a) Current-Voltage (*I-V*) dependence of sensing structures based on BN microstructures. Also, the *I-V* characteristic of device without BN is presented for comparison. The step 1, 2, 3, and 4 indicate on the voltage sweep. (b) *I-V* characteristics of single device at different operating temperatures (up to 100 °C). (c) Measurement of current in time under 40 and 100 V applied bias voltage.

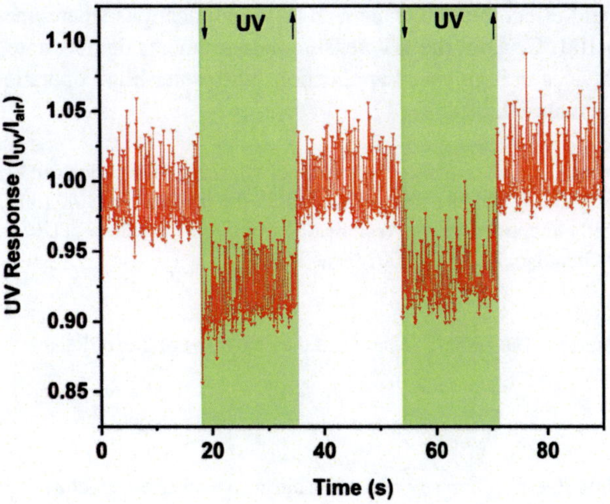

Fig. 3. The UV sensing properties of single *h*-BN microtube.

fabrication of power electronic devices, especially for improved graphene-based devices due to a small (1.7%) lattice mismatch with graphite [1, 7].

Figure 3 shows the UV dynamic response of individual N microtubular structures with diameter ranging from ≈0.2 μm. UV sensing measurements (λ = 365 nm) were carried out as described previously [18]. The UV response was determined as ratio of device current under UV illumination (I_{UV}) and in the dark (I_{air}). The light power density was set to 50 mW/cm^{-2}. The applied bias voltage was 40 V. The tested device showed and UV response of 0.9, i.e. the current was decreased under UV illumination. The signal-to-noise ratio was calculated using the standard deviation of the electrical signal before UV illumination and is ~3. It is important to highlight the fast response and recovery of the *h*-BN devices after switching on/off the UV light source. The calculated response and recovery times are ~100 ms and were limited only by the measurement unit (Keithley source meter 2400). This is very important for high-speed and solar-blind applications, such as space communications and harsh-environment detection systems. Also, due to the wide bandgap of ~5.5 eV, such devices can be an ideal candidate for fabrication of deep UV photodetectors (λ < 365 nm).

4 Conclusions

Herein, the successful integration of individual *h*-BN microtubular structures with diameters ranging from ≈0.2 to ≈2.5 μm into device with good stability and excellent insulating properties using FIB/SEM system was demonstrated. The bias voltage was changed from −40 V to +40 V, while operating temperature was increased up to 100 °C in order to investigate the electrical properties. The high electrical resistance of ≈111 GΩ was calculated, demonstrating minor carbon contamination. Such insulating microtubular structures with nanowalls can be easily integrated into graphene-based electronic devices and field-effect transistors, as well as for applications where operating temperature is up to 100 °C. Also, the UV sensing measurements demonstrated the potential to use such devices in high speed applications with solar-blind operation due to wide bandgap of the material.

Acknowledgment. Special thanks go to the Federal Ministry of Education and Research (BMBF) of Germany for partially funding this project within the project "PorSSi" (03XP0126 B). This work was partially supported by the Technical University of Moldova (TUM) and through the ANCD-NARD Grant No. 20.80009.5007.09 at TUM.

Conflict of Interest. The authors declare that they have no conflict of interest.

References

1. Levendorf, M.P., et al.: Graphene and boron nitride lateral heterostructures for atomically thin circuitry. Nature **488**, 627–632 (2012). https://doi.org/10.1038/nature11408
2. Wei, X., Wang, M.-S., Bando, Y., Golberg, D.: Post-synthesis carbon doping of individual multiwalled boron nitride nanotubes via electron-beam irradiation. J. Am. Chem. Soc. **132**, 13592–13593 (2010). https://doi.org/10.1021/ja106134s
3. Lupan, O., Chai, G., Chow, L.: Fabrication of ZnO nanorod-based hydrogen gas nanosensor. Microelectron. J. **38**, 1211–1216 (2007). https://doi.org/10.1016/j.mejo.2007.09.004
4. Lupan, O., et al.: Single and networked CuO nanowires for highly sensitive p-type semiconductor gas sensor applications. Physica Status Solidi (RRL) – Rapid Res. Lett. **10**, 260–266 (2016). https://doi.org/10.1002/pssr.201510414
5. Tang, C., et al.: Fluorination and electrical conductivity of BN nanotubes. J. Am. Chem. Soc. **127**, 6552–6553 (2005). https://doi.org/10.1021/ja042388u
6. Lupan, O., et al.: Individual hollow and mesoporous aero-graphitic microtube based devices for gas sensing applications. Appl. Phys. Lett. **110**, 263109 (2017). https://doi.org/10.1063/1.4989841
7. Dean, C.R., et al.: Boron nitride substrates for high-quality graphene electronics. Nat. Nanotechnol. **5**, 722–726 (2010). https://doi.org/10.1038/nnano.2010.172
8. Wang, J., et al.: Low temperature growth of boron nitride nanotubes on substrates. Nano Lett. **5**, 2528–2532 (2005). https://doi.org/10.1021/nl051859n
9. Kim, G., Jang, A.R., Jeong, H.Y., Lee, Z., Kang, D.J., Shin, H.S.: Growth of high-crystalline, single-layer hexagonal boron nitride on recyclable platinum foil. Nano Lett. **13**, 1834–1839 (2013). https://doi.org/10.1021/nl400559s
10. Li, Y., Dorozhkin, P.S., Bando, Y., Golberg, D.: Controllable modification of SiC nanowires encapsulated in BN nanotubes. Adv. Mater. **17**, 545–549 (2005). https://doi.org/10.1002/adma.200401266
11. Bando, Y., Ogawa, K., Golberg, D.: Insulating 'nanocables': invar Fe–Ni alloy nanorods inside BN nanotubes. Chem. Phys. Lett. **347**, 349–354 (2001). https://doi.org/10.1016/S0009-2614(01)01075-2
12. Sajjad, M., Morell, G., Feng, P.: Advance in novel boron nitride nanosheets to nanoelectronic device applications. ACS Appl. Mater. Interfaces **5**, 5051–5056 (2013). https://doi.org/10.1021/am400871s
13. Lauhon, L.J., Gudiksen, M.S., Wang, D., Lieber, C.M.: Epitaxial core–shell and core–multishell nanowire heterostructures. Nature **420**, 57–61 (2002). https://doi.org/10.1038/nature01141

14. Sajjad, M., Feng, P.: Study the gas sensing properties of boron nitride nanosheets. Mater. Res. Bull. **49**, 35–38 (2014). https://doi.org/10.1016/j.materresbull.2013.08.019
15. Lin, L., Liu, T., Zhang, Y., Sun, R., Zeng, W., Wang, Z.: Synthesis of boron nitride nanosheets with a few atomic layers and their gas-sensing performance. Ceram. Int. **42**, 971–975 (2016). https://doi.org/10.1016/j.ceramint.2015.08.109
16. Schütt, F., et al.: Conversionless efficient and broadband laser light diffusers for high brightness illumination applications. Nat. Commun. **11**, 1437 (2020). https://doi.org/10.1038/s41467-020-14875-z
17. Bai, X., et al.: Deformation-driven electrical transport of individual boron nitride nanotubes. Nano Lett. **7**, 632–637 (2007). https://doi.org/10.1021/nl062540l
18. Cumings, J., Zettl, A.: Field emission and current-voltage properties of boron nitride nanotubes. Solid State Commun. **129**, 661–664 (2004). https://doi.org/10.1016/j.ssc.2003.11.026
19. Ci, L., et al.: Atomic layers of hybridized boron nitride and graphene domains. Nat. Mater. **9**, 430–435 (2010). https://doi.org/10.1038/nmat2711
20. Arenal, R., Wang, M.-S., Xu, Z., Loiseau, A., Golberg, D.: Young modulus, mechanical and electrical properties of isolated individual and bundled single-walled boron nitride nanotubes. Nanotechnology **22**, 265704 (2011). https://doi.org/10.1088/0957-4484/22/26/265704

Tunable Ferromagnetic Nanomaterials for 6G Technology: Fundamentals and Prospects

Liudmila Alyabyeva[1(✉)] and Evgeny Gorbachev[2]

[1] Moscow Institute of Physics and Technology, Dolgoprudny, Russia
aliabeva.ln@mipt.ru
[2] Moscow State University, Moscow, Russia

Abstract. Modern subterahertz-terahertz (subTHz-THz) technologies, including 5G and 6G wireless telecommunication and THz optoelectronics, apply for working frequency range of tens of gigahertz to several terahertz (millimeter-submillimeter wavelengths). Demanded flexibility of various application calls for thorough investigation of materials with tunable functional characteristics. Understanding the physical mechanisms responsible for tunability of the specific parameters of materials paves the way for targeted controlling their properties and, in turn, allows to choose the material with desired parameters for the particular (and/or even multiple) applications when designing specific devices. Here, we discuss the influence of the variation of the form of material (nanoceramics, powders, single crystals) as well as of the grain sizes, and preparation conditions, on the dielectric and magnetic characteristics of the ferromagnetic hexaferrites and epsilon iron oxide that are among most perspective materials to be used for fabrication of elements and devices of future optoelectronics. The prospects of their potential applications are discussed in detail.

Keywords: Terahertz · Ferromagnets · Tunable materials · Nanoceramics

1 Introduction

Epsilon iron oxide, ε-Fe_2O_3, is desired, though still not broadly used due to certain technical handicaps [1], hard magnetic compound with superior values of magnetic coercivity and high frequencies of natural ferromagnetic resonance (NFMR) in subTHz range [2]. M-type hexaferrites, $A^{2+}Fe_{12}O_{19}$ (A = Ba, Pb, Sr) are highly demanded materials in magnetic device market with billions of dollars profit yearly. Albeit hexaferrites are thoroughly studied from viewpoint of their magnetic radio-frequency and DC characteristics [3], and despite their great potential for subTHz-THz technologies [4–6], the fundamentals of their THz properties remain a certain "*terra incognita*". Among vivid examples, Ca-Al substituted strontium hexaferrite is one of the two (the other one is ε-Fe_2O_3) rare-earth-free hard magnetic materials that exhibits gigantic values of coercive force (>20 kOe) [7]. $BaFe_{12}O_{19}$ reveals ferroelectric ordering that manifests itself in electrically polarized walls of magnetic domains, while admixture of only 10% of lead in chemical formula results in an absence of such polarization [8]. Admirable sensitivity

I. Tiginyanu et al. (Eds.): ICNBME 2021, IFMBE Proceedings 87, pp. 24–31, 2022.
https://doi.org/10.1007/978-3-030-92328-0_4

of hexaferrites' core characteristics (NFMR frequency, coercive force, dielectric constant, AC dielectric losses, etc.) to the internal and external influences (such as chemical composition, growth conditions, pressure or temperature) [9–11] make the materials attractive for the fundamental research and for the various application. On the other hand, same variability of properties can lead to difficulties in prognosis of the materials' electric and/or magnetic properties when one or several of such influences are applied. To overcome this predicament one needs to deeply understand the fundamental processes that govern the behavior of the materials in specific environment. Once the mechanisms responsible for the change in the properties are clear and corresponding models are proposed, the evolution of the characteristics, and consequently wanted parameters of the material can be prospected.

Another complication is a need for proper growing methods. ε-Fe_2O_3 is a metastable phase of the iron (III) oxide and is stable only in nanoscale form. The existing methods of its synthesis have shown to be energy consumptive and ineffective from the standpoint of output product. As for hexagonal ferrites, the bottleneck for successive investigation of their properties and further implementation into production is the lack of growth technology of large (exceeding $\approx 5 * 5 * 5$ mm^3) and high-quality single crystals.

We will discuss methods of synthesis of large amounts of ε-Fe_2O_3 and large-size hexaferrite single crystals and nanoceramics. We study the influence of the form of material, grain sizes, growth conditions (environment, temperature, etc.), and their chemical composition on the magnetic and dielectric properties of the compounds, discuss the nature of the phenomena observed in THz response and provide qualitative microscopic models for the origin of these phenomena. Detected tunable properties of studied compounds are also examined from the applicational point of view.

2 Materials and Methods

2.1 Synthesis and Characterization

ε-Fe_2O_3 nanoparticles were synthesized using method of thermal treatment of the TEOS xerogel performed by thermal treatment of the xerogel with the approach of sequent heating – exposure – air quenching. The obtained silica matrix with the ε-Fe_2O_3 nanoparticles was then mechanically crushed and treated with a 14 M aqueous solution of NaOH.

The Ca-Al substituted hexaferrite submicron particles ($Sr_{1-x/12}Ca_{x/12}Fe_{12-x}Al_xO_{19}$, $x = 4, 4.5, 5, 5.5, 6$) were obtained by citrate auto-combustion method. The morphology, phase purity, microstructure and magnetic hysteresis loops of the resulting ε-Fe_2O_3 and of Ca-Al substituted Sr hexaferrite single domain nanoceramics were studied using X-ray phase analysis (XRD), full-profile analysis by the Rietveld method, transmission electron microscopy and vibrational magnetometry.

Single-crystalline Ba/Pb hexaferrites were grown using the flux method. Their composition and unit cell parameters were determined using Energy-dispersive X-ray spectroscopy, powder X-ray diffraction and scanning electron microscopy; their orientation needed for optical measurements was checked using XRD mapping. A rich set of various experimental techniques were used for additional characterization and for studies of physical properties. These include dielectric and magnetic radiofrequency capacitance spectroscopy, microwave spectroscopy, THz pulsed time-domain and THz-subTHz monochromatic frequency-tunable backward-wave oscillator spectroscopy, infrared reflection spectroscopy, THz and IR spectroscopy in external magnetic fields, FMR spectroscopy. We performed measurements of temperature-dependent heat capacity. Magnetic and electric force microscopy (MFM, EFM) methods were used to study the magnetic/electric domain structure.

2.2 Terahertz-Infrared Spectroscopy

The measurements of the dielectric response were performed in the frequency range 2–8000 cm^{-1} and at temperatures 5 K–300 K utilizing a set of spectrometers. For the measurements of the infrared reflectivity at frequencies 50–8000 cm^{-1} with a 2 cm^{-1} resolution, a standard infrared Fourier-transform spectrometer Bruker Vertex 80v equipped with a microscope Hyperion 2000 was used equipped with cold finger cryostat. At THz frequencies (7–100 cm^{-1}, with a resolution of 1.2 cm^{-1}) spectra of real and imaginary parts of the complex dielectric permittivity were measured with the help of THz time-domain spectrometers TeraView TPS 3000 and Menlo Systems Tera K15 both equipped with exchange gas cryostats. The study of the lower-energy response, at frequencies 2–10 cm^{-1}, was carried out using a spectrometer based on continuously frequency tunable sources of radiation - backward-wave oscillators (BWOs); similar to the time-domain spectrometers, the BWO-spectrometer allows for direct measurements of real and imaginary parts of the dielectric permittivity [12]. Using Fresnel equations, the THz spectra of complex dielectric permittivity were recalculated into reflectivity coefficients that were merged with the measured IR reflectivity spectra. Finally, the broad-band spectra of complex dielectric permittivity were obtained by applying least-square fitting procedure to the spectra of reflectivity and real and imaginary permittivity.

3 Epsilon Iron Oxide

Exploited methods of synthesis of epsilon iron oxide, in general, could be divided into two groups. First group comprises silica-free methods, including hydrothermal [13], plasma dynamic [14] and microwave torch discharge [15] methods. Their main disadvantage is non-controllable admixture of impurities in outcome product, and as a result, low magnetic properties of the obtained compounds. Methods of the second group could be shortly described as stabilization of the iron oxide in mesoporous silica matrix. I.e., most of silica-involved techniques consist in an incorporation of Fe^{3+}-ions precursor into silica matrix with a subsequent crystallization. The incorporation can be performed via an impregnation of a mesoporous silica matrix by a solution of iron (III) salt or via a hydrolysis of tetraethoxysilane (TEOS) in a solution of iron precursor, such as pre-obtained crystalline [16] or amorphous [17] nanoparticles and iron (III) salts [1, 18].

Here, synthesis was performed via sol-gel method with rapid hydrolysis of TEOS. The main achieved advantages in comparison with other existing methods are short full-time cycle - about several hours (compare with 2–3 weeks for other existing techniques [1]), high yield of pure ε-Fe_2O_3 output without unintentional impurities, cost-effectiveness, ecological safety.

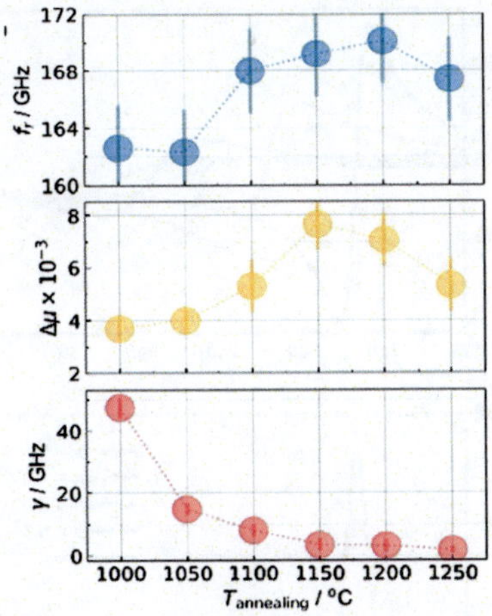

Fig. 1. Parameters of the NFMR of ε-Fe_2O_3 as a function of annealing temperature: resonance frequency f_r, contribution to magnetic susceptibility $\Delta\mu$ and damping factor γ.

The controllable variation of the pore sizes (achieved by variation of the annealing temperature in the growth process) resulted in variability of sizes of an output ε-Fe_2O_3 particles from 7 to 38 nm. The influence of the particle size on magnetocrystalline anisotropy paves the way to tune coercive force of the ε-Fe_2O_3 in range from 0.2 to 21 kOe, and to settle desired frequency of the natural ferromagnetic resonance in the range from 161 to 170 GHz (Fig. 1). Note, that the NFMR frequency increase is accompanied with the narrowing of its linewidth from 45 to 2 GHz.

4 M-Type Hexaferrites

4.1 $SrFe_{12}O_{19}$ Double Cation Substitution: (Ca and Al)

Strontium hexaferrite, $Sr_{1-x/12}Ca_{x/12}Fe_{12-x}Al_xO_{19}$, substituted with Ca and Al ions in proportional ratios, shows giant values of the coercive force when the grain size does not exceed the magnetic domain size. The variation of Al concentration induces the magnetic dilution of ferromagnetic Fe^{3+} ions in their cite-position. As a result, one realizes control of the coercive force and NFMR frequency (Fig. 2). Variation of the annealing temperature allows to control the grain size.

The change of the annealing temperature from 1300 C to 1400 C results in a jump of NFMR frequency by about 40 GHz; further increase of the temperature leads to monotonic increase of the frequency of NFMR. Combination of the two control mechanisms paves the way to vary the NFMR frequency of the $Sr_{1-x/12}Ca_{x/12}Fe_{12-x}Al_xO_{19}$ single-domain nanoceramics from 50 GHz to 300 GHz.

4.2 $SrFe_{12}O_{19}$ Single Cation Substitution: (Ga)

In contrast to Al ions, Ga ions have very similar ionic radius with Fe^{3+} ions, and the diamagnetic dilution of the $SrFe_{12-x}Ga_xO_{19}$ takes place for all five iron sublattices. Thus,

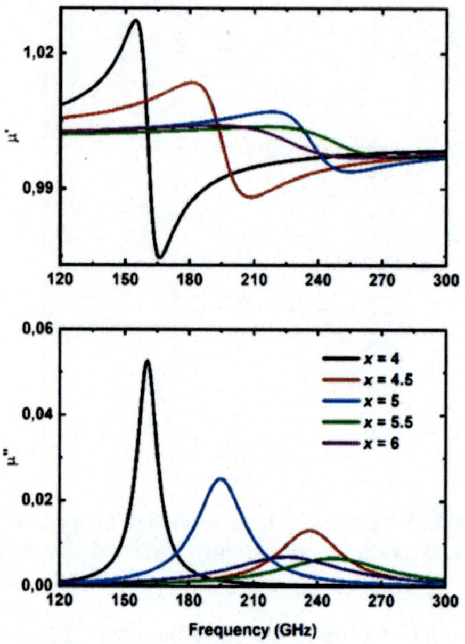

Fig. 2. THz spectra of complex magnetic permeability of $Sr_{1-x/12}Ca_{x/12}Fe_{12-x}Al_xO_{19}$ single domain nanoceramics for samples x = 4 … 6.

the influence of the Ga-substitution on magnetic crystalline anisotropy is not obvious. The variation of the gallium content x(Ga) from 1 to 6 leads to decrease in coercive force (Fig. 3). NFMR frequency changes from 49 GHz to 57 GHz with the maximum value observed for x (Ga) = 3.

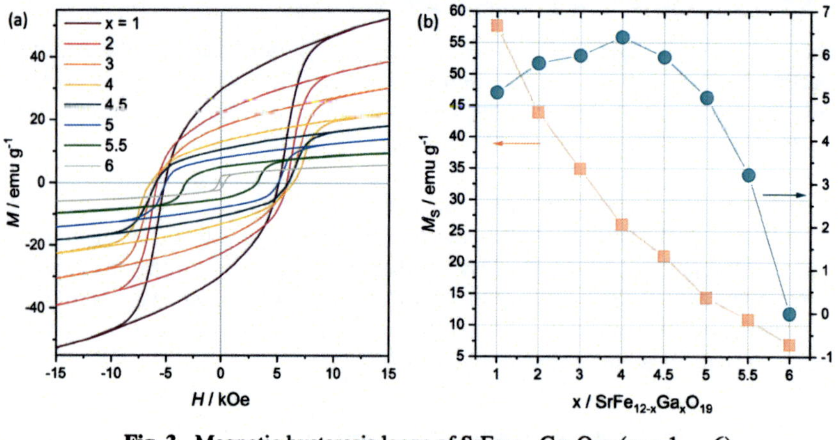

Fig. 3. Magnetic hysteresis loops of $SrFe_{12-x}Ga_xO_{19}$ (x = 1 … 6).

4.3 $(Ba/Pb)Fe_{12}O_{19}$ Single Crystals

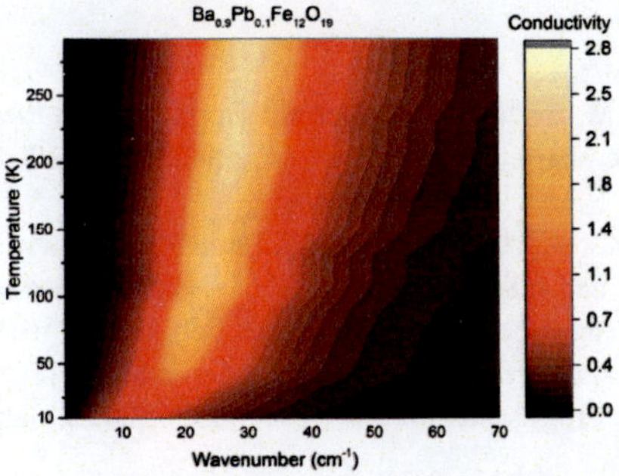

Fig. 4. THz soft mode in single crystalline Pb-substituted $BaFe_{12}O_{19}$.

In contrast to previously discussed compounds, Pb-substituted barium hexaferrite, $Ba_{1-x}Pb_xFe_{12}O_{19}$, is a soft magnetic material with NFMR frequency of about 50 GHz. The tunability of its THz-subTHz dielectric properties is expressed in a form of temperature unstable dielectric excitation (Fig. 4), that changes its peak frequency from 1 THz at 300 K down to 0.3 THz at 5 K. The corresponding spectral response is not affected by such severe environmental treatment as annealing in oxygen atmosphere. We will suggest mechanisms that can be at the origin of the observed soft excitation.

5 Conclusions

Thorough studies of substituted hexaferrites and ε-Fe_2O_3 reveal plenty of structure-, composition- and temperature-dependent phenomena of magnetic, electric and mixed magneto-electric nature. With the results presented here, we demonstrate that hexaferrites and ε-Fe_2O_3 are among the most promising candidates for materials that satisfy the ever-increasing requirements of the rapidly developing field of forthcoming THz electronics. Being planar, self-biased and low-loss materials, they meet the requirements for the development of microwave and THz circulators, phase-shifters, filters, isolators and millimeter wave resonators antennas. These compounds are considered as one of the most promising material classes for telecommunication systems. Possibility of the purposeful control of the resonance frequency from 49 GHz to 300 GHz and higher, that we discovered in the present study, makes these materials promising candidates for manufacturing self-biased devices with working frequencies meeting standards of the 5G and 6G generations in telecommunications.

Acknowledgment. The financial support of the Russian Science Foundation (grant 21-79-10184) is acknowledged. The work was performed in close collaboration with B. Gorshunov, L. Trusov, A. Sleptsova, E. Kozlyakova, A. Prokhorov, V. Lebedev, I. Roslyakov, A. Vasiliev, P. Kazin, M. Soshnikov, M. Wu, D. Myakishev, E. Anokhin, O. Brylev, M. Karpov, Ch. Xinming, R. Svetogorov, Ya. Mozharov, O. Magdysyuk, A. Sobolev, Ia. Glazkova, S. Beloshapkin, D. Vinnik, V. Anzin, A. Ahmed, P. Bednyakov, C. Kadlec, F. Kadlec, J. Prokleška, P. Proschek, S. Kamba, A. Pronin, M. Dressel, V. Abalmasov, V. Dremov, S. Schmid, M. Savinov, P. Lunkenheimer.

Conflict of Interest. The authors declare that they have no conflict of interest.

References

1. Popovici, M., Gich, M., Nižňanský, D., et al.: Optimized synthesis of the elusive ε-Fe2O3 phase via sol-gel chemistry. Chem. Mater. **16**, 5542–5548 (2004). https://doi.org/10.1021/cm048628m
2. Gorbachev, E., Soshnikov, M., Wu, M., et al.: Tuning the particle size, natural ferromagnetic resonance frequency and magnetic properties of ϵ-Fe2O3nanoparticles prepared by a rapid sol-gel method. J. Mater. Chem. C. (2021). https://doi.org/10.1039/d1tc01242h
3. Pullar, R.C.: Hexagonal ferrites: a review of the synthesis, properties and applications of hexaferrite ceramics. Prog. Mater. Sci. **57**, 1191–1334 (2012). https://doi.org/10.1016/j.pmatsci.2012.04.001
4. Ahmed, A., Alyabyeva, L., Torgashev, V., et al.: Effect of aluminium substitution on low energy electrodynamics of barium-lead M-type hexagonal ferrites. J. Phys. Conf. Ser. **1389**, 012044 (2019). https://doi.org/10.1088/1742-6596/1389/1/012044
5. Sleptsova, A.E., Alyabyeva, L.N., Gorbachev, E.A., et al.: Tuning the morphology and magnetic properties of single-domain SrFe8Al4O19 particles prepared by citrate auto-combustion route. Mend. Communs. **31**, 221–223 (2021). https://doi.org/10.1016/j.mencom.2021.03.025
6. Bowrothu, R., Kim, H.-I.I., Smith, C.S., Arnold, D.P., et al.: 35-GHz barium hexaferrite/PDMS composite-based millimeter-wave circulators for 5G applications. IEEE Trans. Microw. Theory Tech. **68**, 5065–5071 (2020). https://doi.org/10.1109/TMTT.2020.3022556
7. Gorbachev, E.A., Trusov, L.A., Sleptsova, A.E., et al.: Hexaferrite materials displaying ultrahigh coercivity and sub-terahertz ferromagnetic resonance frequencies. Mater. Today. **32**, 13–18 (2020). https://doi.org/10.1016/J.MATTOD.2019.05.020
8. Alyabyeva, L.N., Prokhorov, A.S., Vinnik, D.A., et al.: Lead-substituted barium hexaferrite for tunable terahertz optoelectronics. NPG Asia Mater. **13**, 63 (2021). https://doi.org/10.1038/s41427-021-00331-x
9. Karmakar, M., Mondal, B., Pal, M., Mukherjee, K.: Acetone and ethanol sensing of barium hexaferrite particles: a case study considering the possibilities of non-conventional hexaferrite sensor. Sens. Actuator B Chem. **190**, 627–633 (2014). https://doi.org/10.1016/j.snb.2013.09.035
10. Kumar, K., Pandey, D.: Quantum phase transitions in Ba(1−x)CaxFe12O19 ($0 \leq x \leq 0.10$). Phys. Rev. B. **96**, 024102 (2017). https://doi.org/10.1103/PhysRevB.96.024102
11. Almessiere, M.A., Slimani, Y., Güngüneş, H., et al.: Structure, Mössbauer and AC susceptibility of strontium nanohexaferrites: effect of vanadium ions doping. Ceram. Int. **45**, 11615–11624 (2019). https://doi.org/10.1016/j.ceramint.2019.03.033
12. Gorshunov, B., Volkov, A., Spektor, I., et al.: Terahertz BWO-spectrosopy. Int. J. Infrared Millimeter Waves. **26**, 1217–1240 (2005). https://doi.org/10.1007/s10762-005-7600-y
13. Ma, J., Chen, K.: Silica-free hydrothermal synthesis of ε-Fe2O3 nanoparticles and their oriented attachment to nanoflakes with unique magnetism evolution. Ceram. Int. **44**, 19338–19344 (2018). https://doi.org/10.1016/j.ceramint.2018.07.162
14. Sivkov, A., Naiden, E., Ivashutenko, A., Shanenkov, I.: Plasma dynamic synthesis and obtaining ultrafine powders of iron oxides with high content of ε-Fe2O3. J. Magn. Magn. Mater. **405**, 158–168 (2016). https://doi.org/10.1016/j.jmmm.2015.12.072
15. David, B., Pizúrová, N., Synek, P., et al.: ε-Fe2O3 nanoparticles synthesized in atmospheric-pressure microwave torch. Mater. Lett. **116**, 370–373 (2014). https://doi.org/10.1016/j.matlet.2013.11.057

16. Nakaya, M., Nishida, R., Hosoda, N., Muramatsu, A.: Preparation of monodisperse ε-Fe2O3 nanoparticles by crystal structural transformation. Cryst. Res. Technol. **52**, 1700110 (2017). https://doi.org/10.1002/crat.201700110
17. Namai, A., Yoshikiyo, M., Umeda, S., et al.: The synthesis of rhodium substituted ε-iron oxide exhibiting super high frequency natural resonance. J. Mater. Chem. C. **1**, 5200–5206 (2013). https://doi.org/10.1039/c3tc30805g
18. Jin, J., Ohkoshi, S.I., Hashimoto, K.: Giant coercive field of nanometer-sized iron oxide. Adv. Mater. **16**, 48–51 (2004). https://doi.org/10.1002/adma.200305297

Aerosol Spray Deposited Wurtzite ZnMgO Alloy Films with MgO Nanocrystalline Inclusions

Vadim Morari[1](✉), E. V. Rusu[1], V. V. Ursaki[2], K. Nielsch[3], and I. M. Tiginyanu[2]

[1] D. Ghitu Institute of Electronic Engineering and Nanotechnologies, Chisinau, Republic of Moldova

[2] National Center for Materials Study and Testing, Technical University of Moldova, Chisinau, Republic of Moldova

[3] Leibniz Institute for Solid State and Materials Research (IFW Dresden), Institute for Metallic Materials (IMW), Dresden, Germany

Abstract. In this paper $Zn_{1-x}Mg_xO$ thin films with composition range x = 0.00–0.80 have been obtained by aerosol spray deposition method on p-Si substrates by using zinc acetate and magnesium acetate as precursors. The produced thin films were characterized by scanning electron microscopy (SEM), energy dispersive X-ray (EDX) analysis, X-ray diffraction (XRD), and optical spectroscopy. SEM images revealed uniform nanocrystalline morphology of films, but the form of nanocrystals vary with variation of the Mg content. XRD analysis suggests that the produced films contain a wurtzite $Zn_{1-x}Mg_xO$ phase in the whole chemical composition range, with cubic phase MgO nanocrystalline inclusions with mean grain size around 20 nm. The optical bandgap was found to vary from 3.4 eV to 5.2 eV with increasing the Mg content from 0 to 60%.

Keywords: $Zn_{1-x}Mg_xO$ · Thin films · SEM · X-ray diffraction · Band gap

1 Introduction

The problem of developing optoelectronic devices for the ultraviolet (UV) region of the spectrum is currently very topical. The most common optoelectronic devices are radiation detectors and radiation emitters, such as light emitting diodes (LEDs) and lasers. A viable material for such devices is zinc oxide (ZnO), a material with a bandgap of 3.34 eV [1, 2]. In order to change the bandgap of the material and, respectively, to shift of the emission spectrum (in the case of light emitters) or the spectral range of sensitivity (in the case of radiation detectors) to shorter wavelengths, Mg doping can be applied for obtaining solid solution of ZnMgO [3–6]. The ZnO-$Zn_{1-x}Mg_xO$ system ensures the possibility of modeling the optical, luminescent and photoelectric properties in a fairly wide range, by adjusting the composition in the system (the value of the parameter x). The nanostructuring of these materials, in particular the production of nanostructured films, is an additional element for modeling specific properties. In this report we present results on the study of morphology, chemical composition and crystallographic structure of ZnMgO thin films [7–14] grown by aerosol spray deposition method [15, 16].

I. Tiginyanu et al. (Eds.): ICNBME 2021, IFMBE Proceedings 87, pp. 32–39, 2022.
https://doi.org/10.1007/978-3-030-92328-0_5

2 Thin Films Preparation of $Zn_{1-x}Mg_xO$

The scheme of the aerosol deposition method is illustrated in (Fig. 1). It requires a micro spray nozzle, a solution with the selected precursors, a substrate that is maintained at a constant temperature throughout the deposition process (for instance around of 500 °C), and a gas flow (in our case O_2). The distance (L) from the sprayer to the substrate was chosen experimentally, 19 cm being the optimal one for the most uniform coverage. The injection speed of 0.33 ml/min was used to obtain thin films with a thickness of 80–150 nm by varying the deposition time. The p-Si (100) substrate was chemically treated before starting the thin film deposition process. The solution for washing the silicon substrate surface consisted of distilled water, hydrogen peroxide and hydrochloric acid in volume fractions of $H_2O:HCl:H_2O_2 = 6:1:1$. The cleaning was performed keeping the substrate in solution at a temperature of 80 °C for 5 min, followed by rinsing in distilled water and drying over a heater.

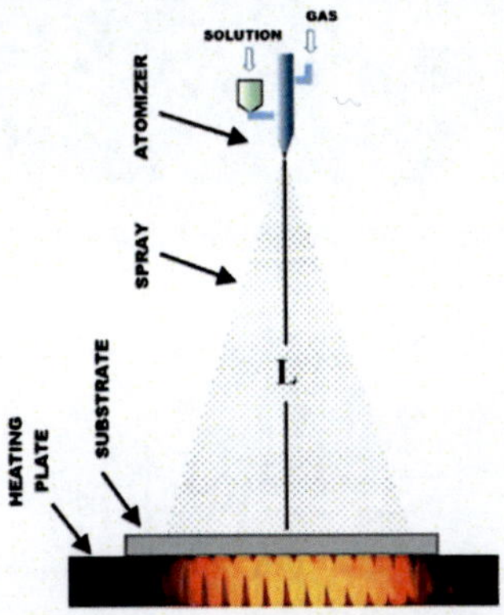

Fig. 1. The sketch of the aerosol deposition method

The solutions of zinc acetate and magnesium acetate (0.35M) were dissolved in ethyl alcohol (C_2H_5OH) and mixed in an ultrasonic bath for 30 min at a temperature of 50 °C for the aerosol deposition of $Zn_{1-x}Mg_xO$ thin films with different concentration of Mg.

3 Study of Morphologycal Properties

$Zn_{1-x}Mg_xO$ thin films with thicknesses between 80–150 nm and uniform morphology were prepared by aerosol deposition method at a temperature of 500 °C on p-Si (100) substrates. The characterization of morphology and chemical composition analysis of the thin films were performed with a LEO-ZEISS Gemini 1530 scanning electron microscope, equipped with an energy dispersive X-ray analysis (EDX) unit. X-ray diffraction (XRD) measurements were carried out on a Panalytical X'PertPro MRD diffractometer with CuKα radiation ($\lambda = 0.15406$ nm) in the 2Θ region of 20°–100°.

The morphology of $Zn_{1-x}Mg_xO$ films changes with increasing the Mg content from $x = 0.00$ to $x = 0.80$, as observed from SEM images presented in (Fig. 2). Hexagonal

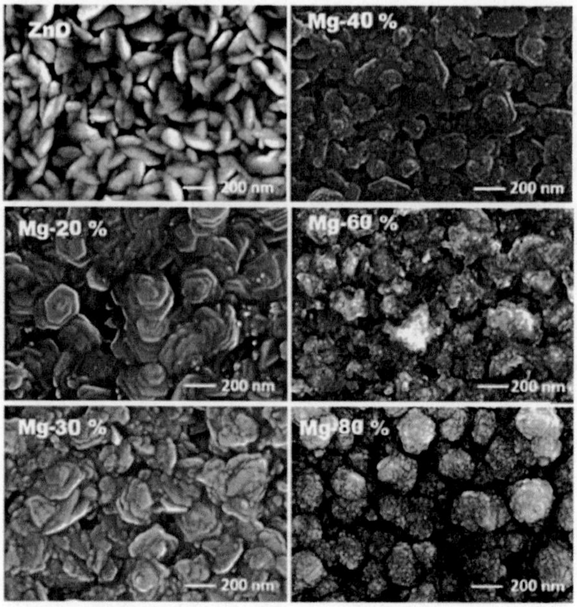

Fig. 2. Top view of SEM images of the $Zn_{1-x}Mg_xO$ thin films with the composition range x = 0.00–0.80 obtained by aerosol deposition

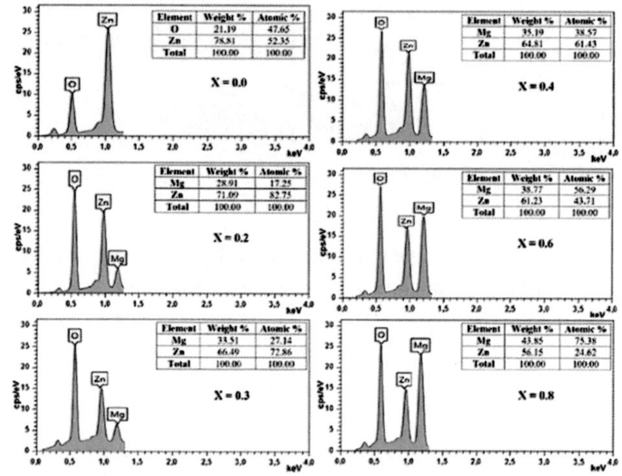

Fig. 3. Chemical composition determined from EDX analysis of $Zn_{1-x}Mg_xO$ thin films with the composition range x = 0.00–0.80 obtained by aerosol deposition method

structures with dimensions around 200–300 nm are formed with increasing the x value up to around 0.20, while the form of crystallites becomes more structureless with further increase of the Mg content. This behavior is explained by the increase of the concentration of MgO inclusions with cubic structure into the hexagonal wurtzite-type ZnMgO film.

The results of EDX measurements performed on top of thin films are shows in (Fig. 3). The Mg content increases from 0 to around 75 at.%, depending on the composition of the precursor solution.

The diagram of changing the chemical composition of ZnMgO film is illustrated in (Fig. 4). From this graph we can observe an increase in the concentration of Mg and a decrease in the concentration of Zn. So, we can see this linear dependence with increasing Mg concentration.

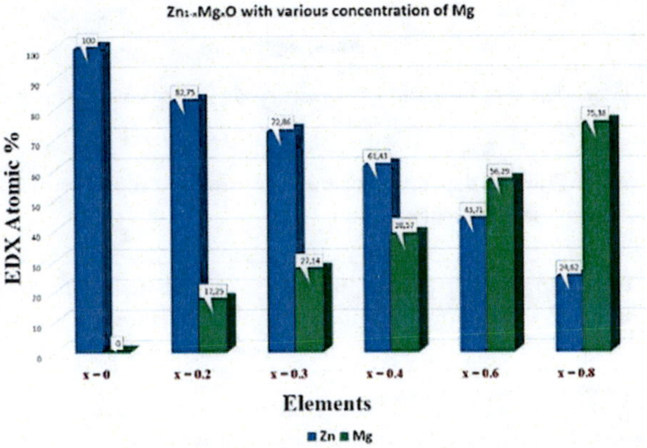

Fig. 4. ZnMgO diagram depending of the Mg concentration

4 Study of Bandgap and X-ray Diffraction Pattern

The optical properties of ZnMgO thin films were measured with a Jasco V-670 spectrometer at room temperature (300 K). The optical bandgap was determined from the Tauc plot of the absorption coefficient, and the results are shown in (Fig. 5).

One can see from (Fig. 5) that the sensitivity range of films can be tuned from UV-A to UV-C by changing the composition range x from 0.00 to 0.60. The increase of the optical bandgap with increasing the Mg content was previously reported in $Zn_{1-x}Mg_xO$ films deposited by spin-coating [17]. Apart from that, it was previously shown that a similar dependence of the optical bandgap upon the composition of $Zn_{1-x}Mg_xO$ films deposited by aerosol spray indicates on their wurtzite structure up to the Mg content of 60% [18].

The films composition determined from the EDX analysis is close the composition preset in the precursor solution (see Table 1), which suggests that Mg is well incorporated into the wurtzite ZnO lattice.

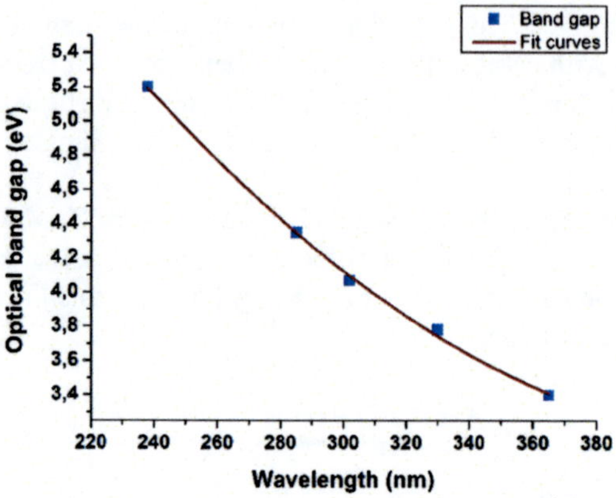

Fig. 5. The dependence of the band gap of $Zn_{1-x}Mg_xO$ with various Mg concentration

Table 1. Band gap of ZnMgO depending of the composition range x = 0.00–0.60 from EDX measurements and from the precursor solution

Thin films	Band gap position (nm)	Band gap energy (eV)	Mg content from the precursor solution (%)	Mg content from EDX analysis (%)
ZnO	*~365*	*3.4*	0.0	*0.0*
$Zn_{0.8}Mg_{0.2}O$	*~330*	*3.75*	0.2	*0.1725*
$Zn_{0.7}Mg_{0.3}O$	*~302*	*4.07*	0.3	*0.2714*
$Zn_{0.6}Mg_{0.4}O$	*~285*	*4.35*	0.4	*0.3857*
$Zn_{0.4}Mg_{0.6}O$	*~238*	*5.2*	0.6	*0.5629*

The XRD investigations (Fig. 6) suggest that the wurtzite structure is present in $Zn_{1-x}Mg_xO$ films even with Mg content of 80%.

Reflexes (100), (002), (101), (102), (103), and (104) from the wurtzite ZnO phase are observed at 30.79°, 34.29°, 35.35°, 46.74°, 64.20°, and 83.45°, respectively, according to the JCPDS Card No. 01-075-1533 and Card No. 36-1451.

The reflexes (100), (002), (101), and (102) are gradually shifted to higher 2Θ values with increasing the Mg content in films, they being placed at 31.74°– (100), 34.58°– (002), 36.62°– (101), and 47.72°– (102), in films with 60% Mg content. This shift indicates on the efficient incorporation of Mg into wurtzite ZnMgO structure (JCPDS Card No. 41-1415). However, not all the Mg is incorporated into the wurtzite ZnMgO, since reflexes (200) and (220) from a cubic MgO phase are clearly observed in films with Mg content higher than 40%. In films with 80% Mg content, the (111) reflex from this cubic phase is superimposed on the (101) reflex from the wurtzite ZnMgO phase. The position of the (200) and (220) peaks do not change with increasing the Mg content. This observation suggests that these reflexes come from the rock-salt MgO phase (JCPDS Card No. 75-1525), not from a possible rock-salt ZnMgO phase. The sizes of the MgO

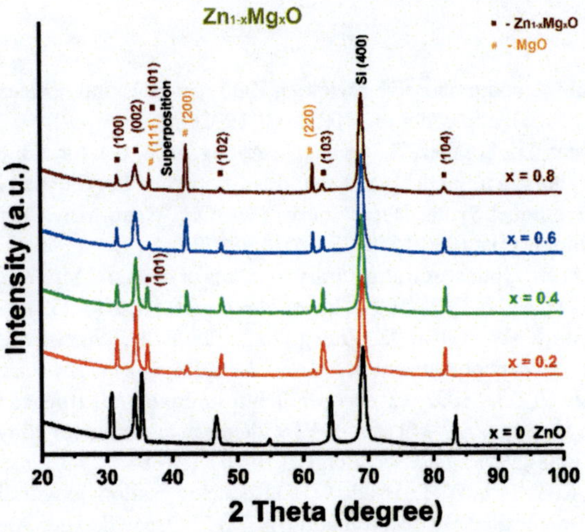

Fig. 6. XRD pattern of $Zn_{1-x}Mg_xO$ film deposited by aerosol deposition method

crystallites incorporated into the ZnMgO film are around 20 nm, as deduced from the width of the (200) reflex according to the Scherrer Eq. (1):

$$\tau = \frac{K\lambda}{\beta cos\theta} \tag{1}$$

Where, τ is the mean size of the ordered (crystalline) do-mains, which may be smaller or equal to the grain size; K is a dimensionless shape factor; λ is the X-ray wavelength; β is the line broadening at half the maximum intensity (FWHM); θ is the Bragg angle.

5 Conclusions

The results of this study demonstrate preparation of wurtzite phase $Zn_{1-x}Mg_xO$ films with x value from 0.0 to 0.8 on Si (100) substrates by the aerosol deposition method. The XRD analysis suggests that MgO nanocrystallites with sizes around 20 nm are incorporated into the wurtzite ZnMgO film at x values higher than 0.2. This incorporation results in changing the film morphology from hexagonal-type nanocrystallites to formless nanograins. The optical bandgap of the deposited material is tuned from 3.4 eV to 5.2 eV with increasing the Mg content from 0 to 60%.

Acknowledgments. This work was supported financially by the ANCD through grant no.20.80009.5007.02. V. Morari thanks to the DAAD scholarships for the research grant (Sandwich).

Conflict of Interest. The authors declare that they have no conflict of interest.

References

1. Özgür, Ü., et al.: A comprehensive review of ZnO materials and devices. Appl. Phys. Rev. **98**(4), 041301 (2005). https://doi.org/10.1063/1.1992666
2. Yilmaz, M., Tatar, D., Sonmez, E., Cirak, C., Aydogan, S., Gunturkun, R.: Investigation of structural, morphological, optical, and electrical properties of Al doped ZnO thin films via spin coating technique. Synth. React. Inorg. Met.-Org. Nano-Met. Chem. **46**(4), 489–494 (2015). https://doi.org/10.1080/15533174.2014.988795
3. Sharma, A.K., et al.: Optical and structural properties of epitaxial MgxZn1−xO alloys. Appl. Phys. Lett. **75**(21), 3327–3329 (1999). https://doi.org/10.1063/1.125340
4. Fan, M.-M., Liu, K.-W., Chen, X., Zhang, Z.-Z., Li, B.-H., Shen, D.-Z.: A self-powered solar-blind ultraviolet photodetector based on a Ag/ZnMgO/ZnO structure with fast response speed. RSC Adv. **7**(22), 13092–13096 (2017). https://doi.org/10.1039/c6ra28736k
5. Ohtomo, A., et al.: MgxZn1−xO as a II–VI widegap semiconductor alloy. Appl. Phys. Lett. **72**(19), 2466–2468 (1998). https://doi.org/10.1063/1.121384
6. Pan, C.-J., Lin, K.-F., Hsu, W.-T., Hsieh, W.-F.: Reducing exciton–longitudinal optical phonon coupling with increasing Mg incorporation in MgZnO powders. J. Appl. Phys. **102**(12), 123504 (2007). https://doi.org/10.1063/1.2820100
7. Yang, X.J., et al.: The origin of the triple-color photodetectors based on the ZnO/MgZnO films. J. Mater. Sci.: Mater. Electron. **30**(7), 6390–6394 (2019). https://doi.org/10.1007/s10854-019-00941-w
8. Fan, M.M., et al.: High-performance solar-blind ultraviolet photodetector based on mixed-phase ZnMgO thin film. Appl. Phys. Lett. **105**(1), 011117 (2014). https://doi.org/10.1063/1.4889914
9. Choopun, S., Vispute, R.D., Yang, W., Sharma, R.P., Venkatesan, T., Shen, H.: Realization of band gap above 5.0 eV in metastable cubic-phase MgxZn1−xO alloy films. Appl. Phys. Lett. **80**(9), 1529–1531 (2002). https://doi.org/10.1063/1.1456266
10. Chen, X., Kang, J.: The structural properties of wurtzite and rocksalt MgxZn1−xO. Semicond. Sci. Technol. **23**(2), 025008 (2008). https://doi.org/10.1088/02681242/23/2/025008
11. Hou, Y.N., Mei, Z.X., Liang, H.L., Ye, D.Q., Gu, C.Z., Du, X.L.: Dual-band MgZnO ultraviolet photodetector integrated with Si. Appl. Phys. Lett. **102**(15), 153510 (2013). https://doi.org/10.1063/1.4802486
12. Xie, X.H., Zhang, Z.Z., Shan, C.X., Chen, H.Y., Shen, D.Z.: Dual-color ultraviolet photodetector based on mixed-phase-MgZnO/i-MgO/p-Si double heterojunction. Appl. Phys. Lett. **101**(8), 081104 (2012). https://doi.org/10.1063/1.4746772
13. Wang, L.K., et al.: Single-crystalline cubic MgZnO films and their application in deep-ultraviolet optoelectronic devices. Appl. Phys. Lett. **95**(13), 131113 (2009). https://doi.org/10.1063/1.3238571
14. Ni, P.-N., Shan, C.-X., Li, B.-H., Shen, D.-Z.: High Mg-content wurtzite MgZnO alloys and their application in deep-ultraviolet light-emitters pumped by accelerated electrons. Appl. Phys. Lett. **104**(3), 032107 (2014). https://doi.org/10.1063/1.4862789
15. Morari, V., et al.: Photosensivity of heterostructures produced by aerosol deposition of ZnMgO thin films on Si substrates. In: Proceedings of SPIE - The International Society for Optical Engineering, vol. 11718, p. 1171818:1–8, (2020). https://doi.org/10.1117/12.2571189
16. Morari, V., et al.: Syntesis of Mg1Zn1-xO thin films by spin coating and aerosol deposition. The 9th ICMCS & The 6th CFM, Publications by Technical University of Moldova, Chisinau, 19–21 October 2017, p. 483 (2017). ISBN 978-9975-4264-8-0

17. Morari, V., et al.: Band tail state related photoluminescence and photoresponse of ZnMgO solid solutions nanostructured films. Beilstein J. Nanotechnol. **11**(2020), 899–910 (2020). https://doi.org/10.3762/bjnano.11.75
18. Morari, V., et al.: Injection Photodiode based on an Al–p-Si–n-$Zn_{85}Mg_{15}O$–n-$Zn_{65}Mg_{35}O$–Ag structure. Rom. J. Phys. **66**(609), 1–11 (2021)

Phase Transition in Laser Irradiated TiO_2 Thin Films

Ion Lungu[1(✉)], L. Ghimpu[2], T. Potlog[1], A. Medvids[3], and C. Moise[4]

[1] Research and Innovation Institute, Moldova State University, Chisinau, Moldova
[2] Institute of Electronic Engineering and Nanotechnologies, Academy of Sciences of Moldova, Chisinau, Moldova
[3] Department of Semiconductor Physics, Institute of Technical Physics, Riga Technical University, Riga, Latvia
[4] Center for Surface Science and Nanotechnology, University Politehnica of Bucharest, Bucharest, Romania

Abstract. In this study, the laser processing of thermally annealed TiO_2 thin films at 420 °C in hydrogen atmosphere, utilizing an pulsed fourth-harmonic generation Nd: YAG laser employing different laser intensities in the atmosphere at room temperature, has been reported. Further, the surface morphology and crystalline structure have been investigated by means of atomic force microscopy [AFM], X-ray diffraction [XRD], Raman analysis. The AFM images obtained show that the film's surface changes as the effect of the laser processes. Moreover, XRD and Raman analysis of the TiO_2 thin films indicate at the threshold laser intensity, I_{th} = 66 MW/cm^2 of the fourth-harmonic generation Nd: YAG laser phase transition from atanase-rutile to a crystalline 100% rutile.

Keywords: RF magnetron sputtering method · TiO_2 thin films · Laser processing · XRD · Raman analysis

1 Introduction

Titanium dioxide (TiO_2) can exist in three crystalline phases: anatase, rutile, and brookite as well as an amorphous phase. Recently, the optical and electrical properties of anatase and rutile phases indicated that the significant difference between these phases are the greater optical band gap and the smaller electron effective mass of anatase compared to the rutile phase [1]. Anatase TiO_2 have excellent photocatalytic [2] antibacterial properties when exposed to UV light (320–400 nm) [3] and photovoltaics properties [4]. The rutile phase of TiO2 films is suitable for antireflection (AR) coatings, multilayer optical coatings and optical wave-guides applications. Moreover, different deposition methods are utilised to prepare TiO_2 thin films: thermal evaporation, physical vapour deposition (PVD): DC or RF magnetron sputtering, E-beam evaporation, in addition to chemical methods such as sol-gel method, anodic oxidation, spray pyrolysis, chemical vapour deposition (CVD), atomic laser deposition (ALD). The best TiO_2 films for AR coatings are developed using the ALD method at low temperatures such as 80–150

I. Tiginyanu et al. (Eds.): ICNBME 2021, IFMBE Proceedings 87, pp. 40–47, 2022.
https://doi.org/10.1007/978-3-030-92328-0_6

°C and annealing at 200 °C [5]. Another key factor for modifying the microstructure of as-deposited thin films is laser irradiation. In comparison to thermal annealing, laser irradiation rapidly heats a thin film and brings about changes on the surface. Hence, in this paper, we perform an investigation regarding the evolution of the structural parameters of nanocrystalline TiO_2 thin films prepared by RF magnetron sputtering.in function of the laser intensity irradiation.

2 Film Fabrication

TiO_2 thin films were prepared by means of RF magnetron sputtering. For the preparation of TiO_2 thin films, a Ti target of 99.9% high purity was utilised in the sputtering process. The sputtering chamber was evacuated down to 5×10^{-6} torr using a turbo molecular pump. Subsequently, the distance between the target and the substrate was kept constant at 5 cm. The magnetron discharge was performed in a gas mixture comprising Ar and O_2, whereas the gas content was kept constant by introducing 20 sccm Ar (99.99%) and 8 sccm O_2 (99.99%) via independent mass-flow controllers. The working total pressure was maintained at 2×10^{-3} torr. During depositions, the RF power was kept constant at 100 W. Prior to deposition, the glass substrates were sequentially sonicated in acetone and ethanol, followed which they were rinsed with distilled water and dried. The substrate temperature was slightly higher than the room temperature (below 90 °C). After the deposition, the TiO_2 thin films were annealed at 420 °C in hydrogen atmosphere and subsequently laser-irradiated in air by pulsed fourth-harmonic generation Nd:YAG laser, with the intensities ranging from 1.0 to 290.0 MW/cm^2. Moreover, the samples were irradiated with three different laser intensities, namely 40 MW/cm^2, 66 MW/cm^2 and 100 MW/cm^2. The focused laser beam (spot diameter of 2.0 10^{-3} m) was scanned normally over the sample surface at a constant speed of 1.6 10^{-4} m/s, and the duration of the impulse was 3 ns. The scanned area was 0.8 10^{-4} m^2 for all the indicated intensities.

3 Results and Discussions

3.1 Surface Morphology

The AFM study of the samples allows us to determine the size of the grains and surface roughness more accurately. Moreover, topography image measurements were performed in an ambient atmosphere for the TiO_2 thin films – non-irradiated and laser-irradiated – with intensities such as 40 MW/cm^2, 66 MW/cm^2 and 100 MW/cm^2 in the tapping mode, using a silicon cantilever with a spring constant of 17 N/m and 230 kHz resonance frequency.

In addition, the sample was analysed by selecting three different scan areas: 10 μm × 10 μm, 5 μm × 5 μm and 1 μm × 1 μm. Local morphology features were measured (length and width of individual grains), and the roughness parameters of the surfaces, R_{MS} and R_a, were calculated using an image processing software (Nova Px) for topography. The AFM 2D and 3D scans for all the TiO_2 thin film presented in Fig. 1, with a scan area of 1 μm × 1 μm, show that the non-irradiated thin films possessed relatively highest roughness $R_{ms} = 6.300$ nm in comparison to the laser-irradiated ones. The 100 MW/cm^2

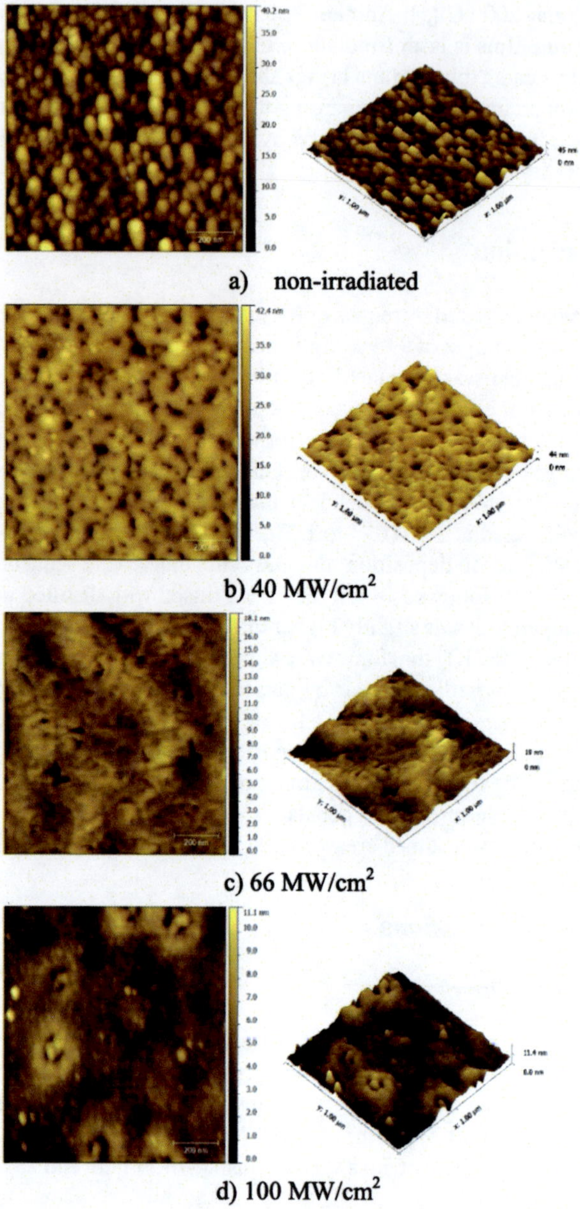

Fig. 1. The AFM 2D and 3D images (scanned area 1 μm × 1 μm) of non-irradiated and laser-irradiated TiO_2 thin films at different intensities (40 MW/cm^2, 66 MW/cm^2 and100 MW/cm^2)

laser-irradiated TiO_2 film presents the lowest $R_{ms} = 1.3392$ nm. Such exceedingly low values predict the uniform surface of the TiO_2 films following laser irradiation. With an increase in the laser intensity, the TiO_2 thin films demonstrate the compactness and well-adhesive nature of the glass substrates. Figure 1 (d) clearly shows that at 100 MW/cm^2, the laser intensity begins to form individual vertical TiO_2 nanocons.

Laser irradiation results in melting of the material, which is followed by solidification after the laser pulse. If the solid has a smaller density than the liquid, the melt is pushed upwards at the end phase of this process, leading to a characteristic conic shape of the surface. The roughness parameters of the thin films' surfaces have been presented in Table 1. These parameters depend on the intensity of the laser irradiation.

3.2 Structural Properties

Figure 2 presents the XRD patterns of the non-irradiated as well as the 40 MW/cm^2, 66 MW/cm^2, and 100 MW/cm^2 laser-irradiated TiO_2 thin films. The XRD pattern of the non-irradiated and the 40 MW/cm2 laser-irradiated TiO_2 thin films display diffraction peaks that correspond to the anatase and rutile phases. For the non-irradiated TiO_2 films, the more intensive peak situated at $2\theta = 25.4°$ is attributed to the (101) plane of the atanase phase. For the 40 kW/cm^2 laser-irradiated TiO2 thin film, the more intensive peak is situated at $2\Theta = 27.5°$ and is attributed to the (110) plane of the rutile phase. The other significantly less prominent peaks revealed in the XRD difractogramms present in the non-irradiated and 40 MW/cm^2 laser-irradiated samples are attributed to the mixture of the anatase and rutile phases. In case of the 40 MW/cm^2 and 66 MW/cm^2 laser-irradiated films, the diffraction peak is situated at $2\Theta = 27.5°$ with less intensity.

Our experimental XRD pattern corresponds with the JCPDS 302 71-1166 (anatase TiO_2) and JCPDS 72-1148 (rutile TiO_2). Hence, the XRD diffractograms presented in Fig. 2 showing the non-irradiated and 40 MW/cm^2 laser-irradiated TiO_2 thin films indicate the mixture of the anatase and rutile phases and broad diffraction peaks, which suggest a small-sized crystallite. On the other hand, the XRD pattern of the 66 MW/cm^2

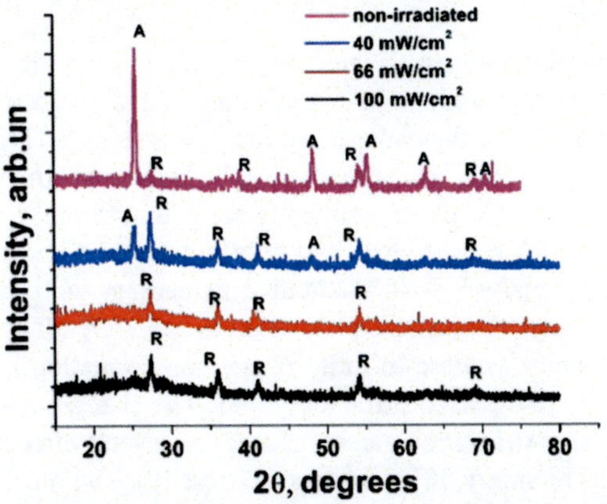

Fig. 2. XRD pattern of the non- irradiated and laser-irradiated TiO_2 thin films at different intensities

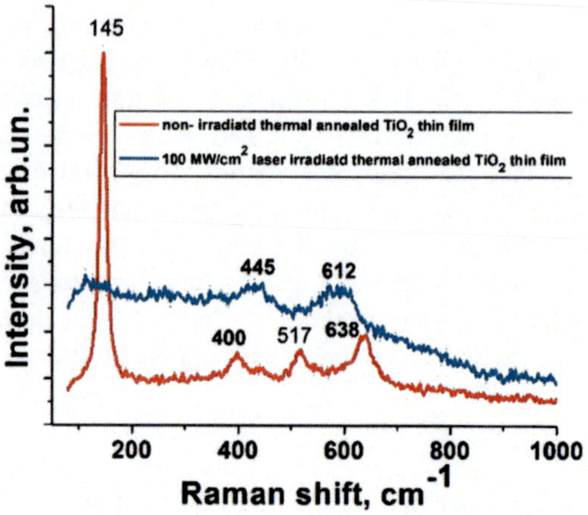

Fig. 3. Raman spectra of the non-irradiated and laser-irradiated TiO_2 thin films at different intensities

and 100 MW/cm^2 laser-irradiated TiO_2 films indicate a crystalline 100% rutile phase with the more (110) intensive plane. Based on the Scherrer equation obtained from the broadening of the anatase (101) and rutile (110) reflections, the crystalline sizes of the anatase phase decreases, whereas that of the rutile phase increases as laser irradiation increases, with the exception of the 100 MW/cm^2 intensity of irradiation. Figure 3 shows the Raman spectra of the 66 MW/cm^2 and 100 MW/cm^2 laser-irradiated TiO_2 thin films. The spectra of the non-irradiated 40 MW/cm^2 and 66 MW/cm^2 laser-irradiated TiO_2 films are similar and contain four vibrational modes that are approximately 145 cm^{-1}, 400 cm^{-1}, 517 cm^{-1} and 638 cm^{-1}. At the laser intensity of 100 MW/cm^2, the most intensive peak situated at 145 cm^{-1}diminishes and rutile peaks appear. Clearly distinguishable vibrations observed in these films at approximately 400 cm^{-1} are associated with the B_{1g} mode of the anatase phase, while the values 517 cm^{-1} and 638 cm^{-1} E_g (3) agree with those reported for the anatase phase in the region of the B_{1g} (2) or A_{1g} and E_g (3) modes [5].

The vibration around 145 cm^{-1} could be attributed to the anatase TiO_2. Moreover, the position and the full width at half-maximum of the A_{1g} and B_{1g} modes of the anatase phase demonstrated dependence on the laser intensity. Further increasing the laser intensity to 100 MW/cm^2 causes variation in the crystal structure, and the Raman spectrum exhibits two weak and broad bands at about 445 cm^{-1} (E_g) and 612 cm^{-1} (A_{1g}) (see Fig. 3), which can be attributed only to the rutile phase [6]. Hence, micro-Raman spectroscopy analysis is in qualitative agreement with the XRD results and could be used to monitor the phase composition of the TiO_2 films. At a laser intensity of 66 MW/cm^2, a quick anatase-to-rutile phase transformation is probably initiated, which is consistent with the XRD difractogramms. The changes observed in the Raman spectra following laser irradiation can be caused by several effects, such as grain size [7], oxygen nonstoichiometry, [8] and uniaxial stress [9]. The functional characteristics of non-irradiated and laser-irradiated TiO_2 thin films determined by Fourier transform infrared spectroscopy (FTIR) have been presented in the range 200–1600 cm^{-1}. The

Transmittance FTIR spectra of the non-irradiated and the laser-irradiated TiO_2 thin films observed at different intensities (Fig. 4) are similar. In addition, peaks around 759 cm^{-1} are attributed to the Ti-O bond of the rutile phase, as observed by other researchers [10]. The broad band of approximately 890 cm^{-1} observed for these films is attributed to the excitations of the symmetrical stretching mode ν_1. The vibrational frequency, which is lower in the excited state than in the ground state, is indicative of the fact that the bond length has changed.

Table 1. The diffracting angles of more intensive lines, phases and planes of TiO_2 thin films

Samples	2Θ, degree	Phase/plane	d, Å	D, nm	R_{ms}, nm	R_a, nm
Non-irradiated	25.45	A (101)	3.5215	6.17	6.300	4.939
	27.48	R (110)	3.2471	4.00		
40 MW/cm^2	25.29	A (101)	3.5145	4.30	4.495	3.500
	27.38	R (110)	3.2450	4.79		
66 MW/cm^2	27.41	R (110)	3.2362	6.72	1.800	1.435
	35.94	R (101)	2.4969	7.94		
100 MW/cm^2	27.41	R (110)	3.2362	5.11	1.339	1.034
	35.95	R (101)	2.4970	6.57		

d: interplanar distances, D: crystallite sizes, roughness parameters of the surfaces.

The slight shift in the peaks' position observed in the 66 MW/cm^2 and 100 MW/cm^2 laser-irradiated samples is probably caused by the transition from a mixture of atanase-rutile phases to a solely rutile phase. It is probable that compression of rutile and anatase leads to a reduction in the Ti-Ti and O-O bond distances. This reduction in bond distances can be compensated by rearranging the octahedra and transforming it to a distorted

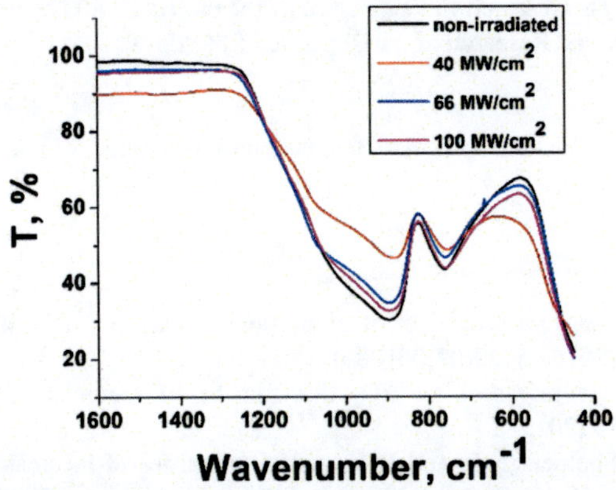

Fig. 4. The Transmittance FTIR spectra of the non-irradiated and laser-irradiated TiO_2 thin films at different intensities

square-pyramidal. In a recent study [10], it was found that one of the Ti-O distances exceeds the usual 2.20–2.25, signifying a range of single Ti-O bonds, and it was proposed that the environment of the corresponding Ti atom is distorted square-pyramidal rather than octahedral.

4 Conclusions

The results obtained on laser irradiation of the TiO_2 thin films, which were obtained by RF magnetron sputtering, have been presented in this paper. The changes pertaining to the morphology, structure and optical properties after laser irradiation have also been studied. The following detailed conclusions can be formulated in this regard:

1) This AFM study shows that the non-irradiated thin films relatively had the highest roughness $R_{ms} = 6.300$ nm in comparison to the 100 kW/cm^2 laser-irradiated films with $R_{ms} = 1.3392$ nm. By increasing the laser intensity, the compactness and well-adhesive nature of these films on glass substrates have been observed.
2) XRD and Raman analysis show that the laser irradiation conducted with three nanosecond pulses induces phase transformation from the mixture comprising atanase and rutile phases to 100% rutile phase at the laser intensity $I_{th} =$ 100 MW/cm^2, which was estimated as the threshold laser intensity of a fourth-harmonic generation Nd: YAG laser.
3) Furthermore, it was noted that the crystallite sizes of the grains of non-irradiated and laser-irradiated TiO_2 thin films do not change drastically. The crystallite size of the atanase TiO_2 phase decreases from 6.17 nm to 4.3 nm, whereas that of the rutile phase changed insignificantly.

Thus, by selecting a suitable laser intensity, the phase transition in TiO_2 thin films can be controlled, and a laser processing technique can be applied to fabricate AR coatings for solar cells.

Acknowledgment. This research was supported by the project 20.80009.5007.16 of the Ministry of Education, Culture and Research of the Republic of Moldova.

Conflict of Interest. The authors declare that they have no conflict of interest.

References

1. Diebold, U.: The surface science of titanium dioxide. Surf. Sci. Rep. **48**, 53 (2003). https://doi.org/10.1016/S0167-5729(02)00100-0
2. Patil, et al.: Recent advances on TiO_2 thin film based photocatalytic applications. Curr. Nanosci. **11**, 1–15 (2015)
3. Lin, H., et al.: Photocatalytic and antibacterial properties of medical-grade PVC material coated with TiO_2 film. J. Biomed. Mater. Res. B. Appl. Biomater. **87**, 425–431 (2008)

4. Mor, G., Varghese, O., Paulose, M., Shankar, K., Grimes, C.: A review on highly ordered, vertically oriented TiO_2 nanotube arrays: fabrication, material properties, and solar energy applications. Sol. Energy Mater. Sol. Cells **90**, 2011–2075 (2006). https://doi.org/10.1016/j.solmat.2006.04.007
5. Ponraj, J., Bosi, G.: Review on atomic layer deposition and applications of oxide thin films. Crit. Rev. Solid State Mater. Sci. **38**(3), 203–233 (2013). https://doi.org/10.1080/10408436.2012.736886
6. Adams, D., Murphy, R., Saiz, D., et al.: Nanosecond pulsed laser irradiation of titanium: oxide growth and effects on underlying metal. Surf. Coat. Technol. **248**, 38–45 (2014)
7. Strelchuk, V., et al.: Raman spectroscopy of the laser irradiated titanium dioxide. Semicond. Phys. Quantum Electron. Optoelectron. **13**(3), 309–313 (2010)
8. Sankapal, B., Lux-Steiner, M., Ennaoui, A.: Synthesis and characterization of anatase-TiO_2 thin films. Appl. Surf. Sci. **239**, 165–170 (2005). https://doi.org/10.1016/j.apsusc.2004.05.142
9. Bersani, D., Lottici, P., Xing, Z.: Phonon confinement effects in the Raman scattering by TiO_2 nanocrystals. Appl. Phys. Lett. **72**, 73 (1998). https://doi.org/10.1063/1.120648
10. Rambabu, Y., Jaiswal, M., Roy, S.: Effect of annealing temperature on the phase transition, structural stability and photo-electrochemical performance of TiO_2 multileg nanotubes. Catal. Today **278**, 255–261 (2016)

Check for updates

Comparative Analysis of Iron Oxide Nanoparticle's (Fe_3O_4) Cytotoxicity Synthesized by Chemical and Biogenic Methods

L. M. Farsiyan, Sh. A. Kazaryan, and Ashkhen A. Hovhannisyan(✉)

Institute of Biomedicine and Pharmacy, Department of Medical Biochemistry and Biotechnology, Russian-Armenian University, Yerevan, Armenia
ashkhen.hovhanisyan@rau.am

Abstract. For the use of iron oxide nanoparticles (NPs) for medical purposes, they must have the following properties: low cytotoxicity, bioavailability, the possibility of large-scale production, etc. Nowadays, there are many approaches for the iron oxide NPs synthesis, including chemical, biological, physical methods, etc. In this research, a comparative analysis of the cytotoxicity of iron oxide (Fe_3O_4) NPs synthesized with chemical and biological methods was carried out. For chemical Fe_3O_4 NPs oleic acid was used as a stabilizer, while for biogenic NPs were used various extracts of *O. basilicum*. As research results showed, the synthesized chemical and biogenic NPs do not have any pronounced cytotoxicity in relation to the studied bacterial strains and human erythrocytes, which allows them to use for further *in vivo* studies.

Keywords: Iron oxide (Fe_3O_4) nanoparticles · Chemical · Biogenic methods of syntheses · Cytotoxicity · Antibacterial activity

1 Introduction

The development of nanotechnology has opened up a wide range of prospective for the synthesis of structures with qualitatively new properties [1]. The use of these structures promisingly expands the capabilities of material production in different fields; however, their most promising applications are in the field of biomedicine: diagnostics, MRI imaging, therapy, drug delivery, etc. [2].

Fe_3O_4 NPs used in biomedicine must meet the criteria of biocompatibility and have minimal toxicity to the human body. It is proven, that the manifestation of toxic properties of any NPs, including Fe_3O_4 NPs, depends on their size, shape, stabilizing agents, dose, and duration of exposure [3]. Some of these toxicity issues can be resolved using biogenic synthesis methods, however, optimization of the biogenic method for obtaining NPs with the required parameters is still relevant. Thus, Fe_3O_4 NPs must meet a number of requirements, like other substances introduced into the body: biocompatibility, low values for therapeutic doses, high values of toxic doses high efficiency, and selectivity [4].

I. Tiginyanu et al. (Eds.): ICNBME 2021, IFMBE Proceedings 87, pp. 48–54, 2022.
https://doi.org/10.1007/978-3-030-92328-0_7

Hence, the use of Fe_3O_4 NPs and its complexes in theranostics is a matter of time, nowadays, the research is limited in finding Fe_3O_4 NPs with optimal properties (low reactivity, biocompatibility, minimal toxicity, high therapeutic index). Biotechnology is one of the advanced fields in the development of methods for the NPs synthesis, which methods allows synthesizing NPs with the least toxic properties, an environmentally safe, and economical manner on an industrial scale [5].

In this regard, the main goal of this research was the synthesis of Fe_3O_4 NPs by chemical and biological methods, as well as the comparative analysis of their biocompatibility and toxicity.

2 Materials and Methods

2.1 Plant Extracts Preparation

The Armenian plant *O. basilicum*, was collected during the flowering period of 2019 in Ararat province of Armenia. For plant extraction were used dried leaves of *O. basilicum*. Plants were extracted by using distilled water and 50% and 96% ethanol. Mechanically homogenized plant material was treated in ultrasonic homogenizer 75W (Ultrasonic Homogenizer, Sonic-150W, MRC, Israel) for 15 min. Afterward, the extracts were incubated for 24 h on a shaker (60–70 rpm) in dark conditions at room temperature. For purification of extracts from solid components, after incubation suspended matter was centrifuged 15 min at 3000 rpm, in a centrifuge GR412 3000 and filtered with filter paper [6].

2.2 Determination of Extracts Antiradical Activity

Antiradical activity (ARA) was determined by quenching the free stable radical of 2,2-diphenyl-1-picrylhydrazyl (DPPH).The IC_{50} value, was determined from the dose-dependent ARA curves [7].

2.3 Biogenic Synthesis of Iron Oxide NPs

For the iron (II; III) oxide NPs synthesis, a solution of Fe^{3+} and Fe^{2+} salts in a 2:1 M ratio was added to the *O. basilicum* extracts. Then, the 1.0 M of NaOH dropwise was added to the solution under continuous stirring at the ambient temperature. After completion of the reaction, the solution was then stirred for 1 h to homogenize the solution and also for the completion of the reaction. The samples were washed several times with distilled water. Then nanoparticles were dried in an oven for 24 h. The dried samples were stored in an air-tight container for further characterization.

2.4 Chemical Synthesis of Iron Oxide NPs

The Fe_3O_4 NPs coated with oleic acid were synthesized by a modified version of co-precipitation method. 1M $FeSO_4*7H_2O$ and 2M $FeCl_3*6H_2O$ solutions were added to 10 ml 4M NaOH and 1 ml oleic acid. Then, the mixture was thoroughly mixed and heated at 80 °C for 1 h and the color of the mixture turned from brown to black. Later the precipitate was separated, washed thrice with deionized water, and dried at room temperature. For further studies, NPs were dispersed in deionized water and treated for 40 min using an ultrasonic homogenizer (Sonic-150W, MRC, Israel).

Size determination of NPs was carried out based on our previous research [8].

2.5 Iron Oxide NPs Antibacterial Properties

The study of the antibacterial activity of the synthesized iron oxide NPs was carried out by the disk diffusion method against the wild-type *E. coli* strain DSM 1116 and the non-pathogenic *S. aureus* MDC5233. 0.1 ml of the bacterial suspension with OD 0.5 (McFarland criteria) was passaged on Petri dishes with agar nutrient media.

Antibacterial activities of NP's investigated solution (50 μM, 100 μM, 150 μM), the extract and a mixture of biogenic NPs (50 μM, 100 μM, 150 μM) stabilized with *O. basilicum* extract were investigated by the disc-diffusion method and bacterial growth rate.

As an indicator of antibacterial activity the inhibition zone formed after 24–48 h growth was taken. The square ($pixel^2$) was calculated by special program "Image Repair 3.19" [9].

The bacterial growth rate was determined by measuring the changes in optical density (OD) of bacterial suspension at a wavelength of 625 nm and monitored every hour till 8 h and at 24 h. The bacterial suspension was washed and concentrated by centrifugation at 3600 g for 15 min and transferred into appropriate medium.

2.6 Effect of Iron Oxide NPs on E. Coli Growth

Laboratory stock of *E. coli* DSM 1116 strain was grown in peptone medium (2% peptone, 0.5% NaCl) buffered with 0.1M K_2HPO_4 (pH 7.5) and 0.2% glucose was added. For studying the profile of the effect on *E. coli* growth, 0.1 ml of a bacterial suspension (optical density = 0.5) and 0.1 ml of iron oxide NPs were added to 100 ml liquid nutrient medium at concentrations from 50 μM to 150 μM.

The bacterial growth rate was determined by measuring the changes in OD of bacterial suspension at a wavelength of 625 nm and monitored every hour till 8 h and after 24 h [4].

2.7 Samples Cytotoxic Activity Determination Using RBC Test

The biocompatibility study of iron oxides NPs was carried out on the red blood cells of healthy male donors. Erythrocytes were preliminary twice washed in saline solution and then resuspended in phosphate buffer pH 7.4. Then, OD of erythrocytes suspension was brought to 0,7 which corresponds to approximately 6.2×10^6 cells/ml.

Resistance (stability) of erythrocytes was evaluated by changes of the optical absorption of the erythrocyte suspension at 680 nm for 10–15 min, detecting the results every 30 s on the UV/Vis spectrophotometer (JENWAY 6405).

The effect of extracts and nanoparticles on erythrocytes was assessed by their resistance (P), which was determined by the formula:

$$P = \frac{Ax}{A0 - Ah} * 100\%, \quad (1)$$

where Ax is the optical density (OD) of the sample after irradiation during x time; A_0 is the OD of an unhemolyzed suspension of erythrocytes, A_h is the OD of the sample after complete hemolysis, which does not reach zero due to the formation of hemi-chrome [10].

Each experiment was repeated 4 to 6 times, and for statistical analyses, 2–3 series of experiments for each case were carried out. The graphs and diagrams show the arithmetic mean and their standard errors ($p < 0.05$).

3 Results and Discussion

Initially, during the process of iron nanoparticles chemical synthesis, a black precipitate was formed with paramagnetic properties, which, according to the authors [11], indicates the correctness of the process and the formation of NPs. However, to prove the formation of NPs, determine the nature of the formed substance, and detect the size and shape of NPs, scanning emission microscopy (TEM) was carried out, according to the results of which it was proved that during the chemical synthesis Fe_3O_4 NPs were formed.

The average size of the synthesized Fe_3O_4 NPs was found to be 10.64 ± 4.73 nm, they have a rounded shape with a diameter from 4 to 24 nm and a single-crystalline structure. However, Fe_3O_4 NPs tended to permanently congregate, which indicates a possible replacement of oleic acid with a more aggressive stabilizer.

The electron diffraction patterns of the reference specimen Fe_3O_4 and Fe_3O_4 NPs coincided, indicating that the NPs have a single crystal structure. The images of Fe_3O_4 in the dark field also indicate that they have a single-crystalline structure [12].

A necessary criterion for biogenic NPs synthesis is the extracts high ARA, therefore, based on our previous research results, aqueous, 50%, 96% ethanol extracts of *O. basilicum* were selected for the synthesis of biogenic NPs [13]. The radical neutralizing activity of all *O. basilicum* extracts showed a dose-dependent character (Fig. 1). From the study of basil's dried leaves antiradical activity results it is clear that all extracts have a high ARA, which is most pronounced in 50% ethanolic extract of *O. basilicum*. Aqueous and 96% ethanolic extracts were less active. Synthesis and stabilization of biogenic NPs was carried out with those extracts.

During the biogenic synthesis the reaction mixture becomes from transparent to brown, which initially confirms that synthesis is done correctly. The SEM analysis confirm, that NPs belong to Fe_3O_4 type, have single-crystalline structure and average core size 10–15 nm.

All NPs, regardless from synthesis method (chemical or biological), showed paramagnetic properties.

The cytotoxicity of biogenic and chemical Fe_3O_4 NPs was tested on gram negative *E. coli* DSM 1116 strain and gram-positive *S. aureus* MDC5233 strain by disk-diffusion method. Extract stabilized biogenic and chemical Fe_3O_4 NPs in the concentrations of

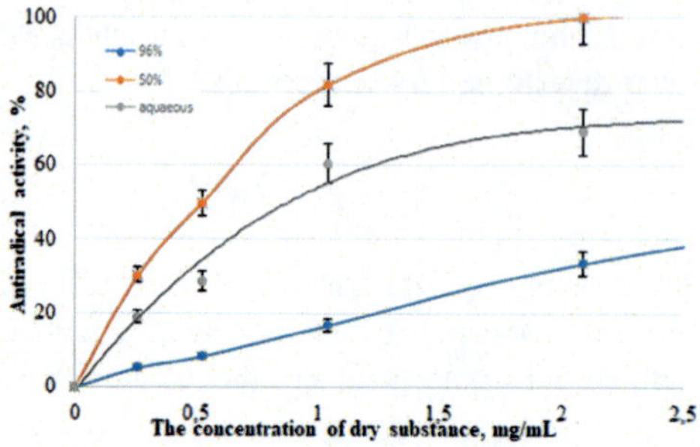

Fig. 1. Antiradical activity of *O. basilicum* dried leaves extracts

150 μM didn't show cytotoxic effect on *E. coli* DSM 1116 strain. It was also revealed that all Fe_3O_4 NPs, regardless of the synthesis method at a maximum concentration of 150 μM, did not show antibacterial effects against *S. aureus* MDC5233, while positive control (*O. basilicum* 50% ethanolic extract) showed cytotoxic properties, regardless the fact, that NPs stabilized with *H. perforatum* 50% ethanolic extract does not show cytotoxic properties.

As it is shown in Fig. 2, chemical Fe_3O_4 NPs in concentration 50–150μM had different regulating activities on the growth of *E. coli*. Fe_3O_4 NPs insignificantly increase both the growth rate and the number of bacteria at a concentration of 50μM, while in concentration of 100μM inhibits the growth after 5 h of cultivation. In concentrations of 150μM Fe_3O_4 NPs, the growth of bacteria coincides with their growth in the control medium [9].

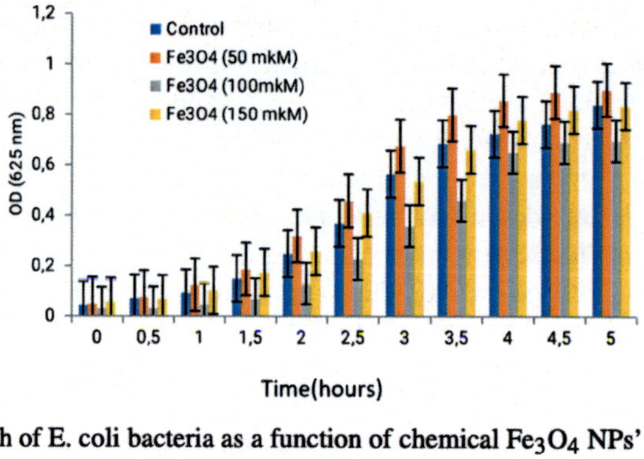

Fig. 2. Growth of E. coli bacteria as a function of chemical Fe_3O_4 NPs' concentration.

However, chemical Fe_3O_4 NPs at concentrations 50, 100, and 150 μM don't change the log phase duration, which determines the antibacterial activity of the active substance [14]. Thus, it was found that within the investigated concentrations, chemical Fe_3O_4 NPs do not exhibit growth inhibitory activity against *E. coli* growth.

As it is can be seen from the results in Fig. 3 biogenic NPs stabilized in *O. basilicum* aqueous and 96% ethanolic extracts of showed growth regulating activity, while it can't

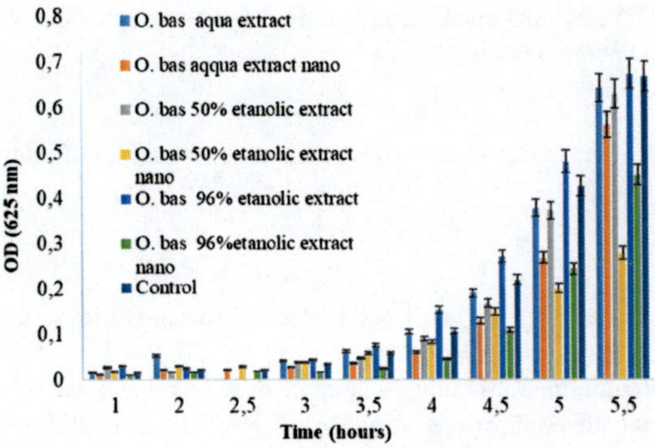

Fig. 3. Growth of E. coli bacteria as a function of chemical Fe_3O_4 NPs' concentration

be said for NPs stabilized in 50% extract of *O. basilicum*, which changes the duration of log phase of growth and shows insignificant antibacterial activity.

Thus, all NPs, regardless from synthesis method, don't show any antibacterial activity against strains of gram-positive and gram-negative bacteria in the scope of studied methods.

The toxicity of extracts and obtained biogenic and chemical NPs was also studied on erythrocytes, as a result of which it was determined that the NPs of iron oxides, as well as the studied extracts, did not have hemolytic properties. After 24 h in dark incubation conditions, the extracts of *O. basilicum* showed no hemolytic activity, and the erythrocyte resistance was 100%. In the conditions of dark incubation biogenic iron oxide NPs on the basis of *O.basilicum* aqueous extracts, after 24 h of incubation didn't lead to hemolysis and erythrocyte resistance was 100%.

From the given experimental results, it can be concluded that consequently, *O. basilicum* extracts as well as iron oxides NPs based on them and chemical NPs don't practically cause hemolysis of erythrocyte, and hence can be used for further *in vivo* studies.

4 Conclusions

Within the framework of this research, we obtained Fe_3O_4 NPs using both chemical and biotechnological methods. The average size of both types of nanoparticles was about 8–10 nm and they have single crystal structure.

The toxicity primary screening results showed, that both chemical and biogenic paramagnetic Fe_3O_4 NPs synthesized by us, do not exhibit pronounced toxicity at a maximum dose of 150 mM against human erythrocytes, studied gram-positive and gram-negative bacteria.

From the results it can be concluded that *O. basilicum* aqueous and 96% ethanolic extracts can be used for further studies.

These results, allow us to consider the chemical and biogenic paramagnetic Fe_3O_4 NPs synthesized by us, as potential theranostic agents for further studies and identification of the most suitable one.

Acknowledgment. The authors thank Rshtuni L (Russian-Armenian University) for the help with synthesis of chemical nanoparticles.

Conflict of Interest. The authors declare that they have no conflict of interest.

References

1. Abbaspour, N., Hurrell, R., Kelishadi, R.: Review on iron and its importance for human health. J. Res. Med. Sci. **19**(2), 164–174 (2014)
2. Dadfar, S.M., Roemhild, K., Drude, N.I., et al.: Iron oxide nanoparticles: diagnostic, therapeutic and theranostic applications. Adv. Drug Deliv. Rev. **138**, 302–325 (2019). https://doi.org/10.1016/j.addr.2019.01.005
3. Feng, Q., Liu, Y., Huang, J., Chen, K., Huang, J., Xiao, K.: Uptake, distribution, clearance, and toxicity of iron oxide nanoparticles with different sizes and coatings. Sci. Rep. **8**(1), 1–13 (2018)
4. Pan, B., Gao, F., Gu, H.: Dendrimer modified magnetite nanoparticles for protein immobilization. J. Colloid Interface Sci. **284**(1), 1–6 (2005). https://doi.org/10.1016/j.jcis.2004.09.073
5. Bhuiyan, M., Miah, M., Paul, S., Aka, T., Saha, O., et al.: Green synthesis of iron oxide nanoparticle using Carica papaya leaf extract: application for photocatalytic degradation of remazol yellow RR dye and antibacterial activity. Heliyon **6**(8), e04603 (2020). https://doi.org/10.1016/j.heliyon.2020.e04603
6. Farsiyan, L., Ghrejyan, E., Ohanyan, S., Tiratsuyan, S., Hovhannisyan, A.: Cytotoxicity assessment of different extracts and stabilized iron (III) oxide nanoparticles on E coli growth. ICBH **1**, 57–58 (2020)
7. Vardapetyan, H., Tiratsuyan, S., Hovhannisyan, A.: Antioxidant and antibacterial activities of selected Armenian medicinal plants. J. Exp. Biol. Agric. Sci. **2**(3), 300–307 (2014)
8. Gabrielyan, L., Hovhannisyan, A., Gevorgyan, V., Ananyan, M., Trchounian, A.: Antibacterial effects of iron oxide (Fe_3O_4) nanoparticles: distinguishing concentration-dependent effects with different bacterial cells growth and membrane-associated mechanisms. Appl. Microbiol. Biotechnol. **103**(6), 2773–2782 (2019)
9. Anush, K., Shushanik, K., Susanna, T., Ashkhen, H.: Antibacterial effect of silver and iron oxide nanoparticles in combination with antibiotics on E. coli K12. BioNanoScience **9**(3), 587–596 (2019). https://doi.org/10.1007/s12668-019-00640-0
10. Ohanyan, A, et al.: Antioxidant and hemolytic properties of different extracts from Prunella vulgaris L. leaves. Med. News North Caucasus **13**(3), 507–510 (2018). https://doi.org/10.14300/mnnc.2018.13091
11. Kekutia, S., et al.: A new method for the synthesis of nanoparticles for biomedical applications. Eur. Chem. Bull. **4**(1), 33–36 (2015)
12. Gabrielyan, L., Hakobyan, L., Hovhannisyan, A., Trchounian, A.: Effects of iron oxide (Fe3O4) nanoparticles on Escherichia coli antibiotic-resistant strains. J. Appl. Microbiol. **126**(4), 1108–1116 (2019). https://doi.org/10.1111/jam.14214
13. Kazaryan, S., Petrosyan, M., Rshtuni, L., Dabaghyan, V., Hovhannisyan, A.: Effects of green silver nanoparticles on CCl4 injured albino rats' liver. In: Tiginyanu, I., Sontea, V., Railean, S. (eds.) ICNBME 2019. IP, vol. 77, pp. 127–131. Springer, Cham (2020). https://doi.org/10.1007/978-3-030-31866-6_27
14. Hossain, M., et al.: Synergism of the combination of traditional antibiotics and novel phenolic compounds against Escherichia coli. Pathogens **9**(10), 811 (2020). https://doi.org/10.3390/pathogens9100811

Relaxation Parameters of Cu/substrate Type Coated Systems Under Nanoindentation

D. Grabco[1(✉)], C. Pyrtsac[1,2], and O. Shikimaka[1]

[1] Institute of Applied Physics, 2028 Chisinau, Republic of Moldova
[2] Technical University of Moldova, 2004 Chisinau, Republic of Moldova

Abstract. In this work, we studied the relaxation parameters, $h_{e\text{-}p}$ and h_{res}, of three composite structures Cu/LiF, Cu/MgO, and Cu/Si, which have different types of a chemical bond between the substrates (ionic (LiF), ionic-covalent (MgO), and covalent (Si)) and differ in hardness ($H_{Cu} = 0.6$ GPa, H_{LiF}, H_{MgO} and H_{Si} are 1.2, 7.5 and 8.2 GPa, respectively). For each type of substrates, coated systems (CSs) were fabricated with a following Cu film thickness: $t_1 = 85$; $t_2 = 470$ and $t_3 = 1000$ nm. The behavior of relaxation parameters was examined over a wide range of loads, $P = 2–900$ mN, during nanoindentation. The elastic-plastic parameters were shown to depend on the CS type, as well as on the film thickness and the magnitude of the applied load.

Keywords: Cu/substrate type coated systems · Depth-sensitive nanoindentation · Relaxation parameters

1 Introduction

In recent decades, the electronic industry continues to actively develop with a tendency towards a decrease in the geometric dimensions of the components of micro- and nano-electromechanical systems (MEMS and NEMS), sensors, microrobots, thin films and other devices in the nanoindustry. Most of the advanced devices that have contributed to the scientific and technological revolution contain thin films that are deposited on substrates of other materials with different properties. This has given rise to the need to study the electrical and magnetic characteristics of new structures and devices. At the same time, the resulting products must also have high mechanical parameters when used for a long term [1, 2]. Therefore, it is important to study the behavior of physical quantities characterizing the mechanical properties of new materials and compare these values with the bulk components already known.

The coated systems (CSs) of the "Cu/substrate" type have been extensively used in the electrical connections of new integrated circuits. Therefore, in the scientific literature there are many works in which the elastic and plastic properties of similar structures, such as hardness, Young's modulus, crack resistance coefficient, etc., are investigated [3]. At the same time, there are not enough works devoted to the relaxation parameters of Cu/substrate structures, although the material relaxation under deformation plays an essential role in the service life of electronic systems.

I. Tiginyanu et al. (Eds.): ICNBME 2021, IFMBE Proceedings 87, pp. 55–61, 2022.
https://doi.org/10.1007/978-3-030-92328-0_8

In this context, the task of the paper was to study the relaxation properties of several CS type Cu/substrate, namely Cu/LiF, Cu/MgO and Cu/Si.

2 Materials and Methods

The study of relaxation parameters was carried out using the indentation method. Depth-sensitive nanomicro-indentation is the main method for studying the mechanical properties of film/substrate materials. This method makes it possible to reveal the physical picture of the processes occurring in the tested sample during measurements, and residual stresses after the end of the experiment. In this work, the indentation procedure is used in the mode of a gradual increase in the load applied to the indenter from 0 to P_{max}, a short exposure under load, and then a gradual decrease to 0. Continuous recording of the penetration depth of the indenter into the material under study allows the loading/unloading process to be controlled. The values of P and h obtained during the indentation process are presented in the diagram $P = f(h)$ [3, 4].

From Fig. 1, the physical meaning of the parameters h_c, h_e, $h_{e\text{-}p}$ and h_{res} is seen, as the components of the maximum depth (h_{max}) of the indenter at the stages of loading-unloading: h_c is the depth of contact of the indenter with the sample at a maximum load; h_e is the depth of the section of elastic bending of the sample surface when the maximum load is applied; $h_{e\text{-}p}$ is a component of the indentation depth, which is restored elastically-plastically after the complete removal of the load; h_{res} - residual indentation depth after a complete removal of the load.

All CSs were obtained under the same conditions. Cu films were deposited on a freshly cleaved surface of single crystal substrates by magnetron sputtering using a Magnetron Sputtering RF device in the mode P = 200W, T = 500C. Microstructural studies were carried out applying the method of optical microscopy (OM) and the atomic force microscopy method (AFM). To reveal the contribution of the substrate to the deformation process, the Cu film was removed with a solution of 70% HNO3 + 30% CH3COOH; then, to reveal dislocation rosettes, the weakly concentrated aqueous solution of $FeCl_3$ was used for LiF crystal and the composition of 5p NH_4Cl + 1p H_2O + 1p H_2SO_4 for MgO one.

3 Results and Discussion

For the research, a logical selection of a series of materials was made, which can be classified according to the scheme below (Fig. 2). As shown in Fig. 2, the polycrystalline metal Cu served as the film for all CSs. Three representatives of the model monocrystalline materials were chosen as the substrate, whose mechanical properties have been studied in detail in the scientific literature [5]. It is seen that all crystals meant for substrates of “film/substrate” structures (LiF, MgO, Si) belong to the category of spatial structures, monolithic, three-dimensional single crystals, but possessing different types of crystal structure (LiF, MgO - type of NaCl; Si - type of diamond). Their chemical bond is respectively ionic, ionic- covalent, and covalent. Evaluation of the hardness of the selected samples showed that Cu and LiF belong to soft materials with $H = 0.6$ and 1.2 GPa, and MgO and Si are hard crystals with $H = 7.5$ and 8.2 GPa, respectively. As

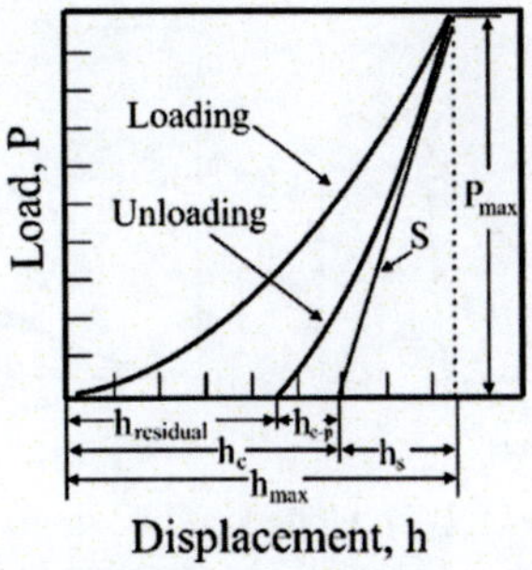

Fig. 1. Typical "load-displacement" diagram, *P*(*h*), which highlights the distribution of the maximum depth ($h_{\max}$) of the indenter in four component parts (h_c, h_e, $h_{e\text{-}p}$, h_{res}) during the loading-unloading process at depth-sensitive indentation.

follows from Fig. 2, similar CSs with a Cu film thickness: $t_1 = 85$; $t_2 = 470$ and $t_3 = 1000$ nm were fabricated for each type of substrate. Depending on the type of substrate, three CS groups were identified: 1 - "soft film/soft substrate, ionic", SSi, 2 – "soft film/hard substrate, ionic-covalent", SHic, and 3 - "soft film/hard substrate, covalent", SHc. As an example, Figs. 3, 4 show the curves *P*(*h*) for some of the studied structures, which reflect the characteristic form of the deformation curves inherent in one or another type of substrates and the CS obtained on their basis.

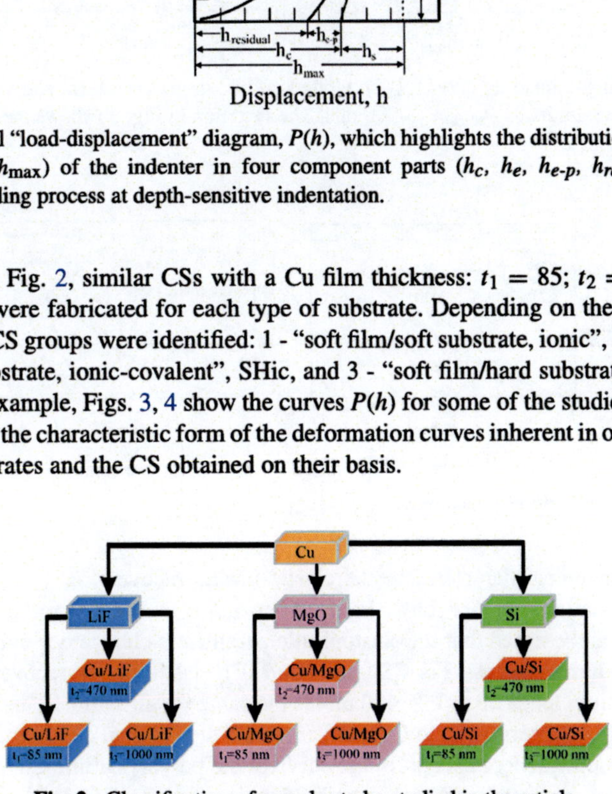

Fig. 2. Classification of samples to be studied in the article.

Based on the information in Fig. 1, the following pattern can be noted. The deformation curves obtained on a soft LiF substrate (Fig. 3a) and on a Cu/LiF CS (Fig. 3b) showed very weak relaxation properties. The sections of the curves at the unloading stage are located almost perpendicular to the abscissa axis. The curves for polycrystalline copper have the same shape (see insertion in Fig. 3b). In contrast, the deformation curves of hard crystals-substrates MgO (Fig. 4a) and Si, as well as composite structures Cu/MgO and Cu/Si (Fig. 4b), demonstrated noticeable relaxation properties.

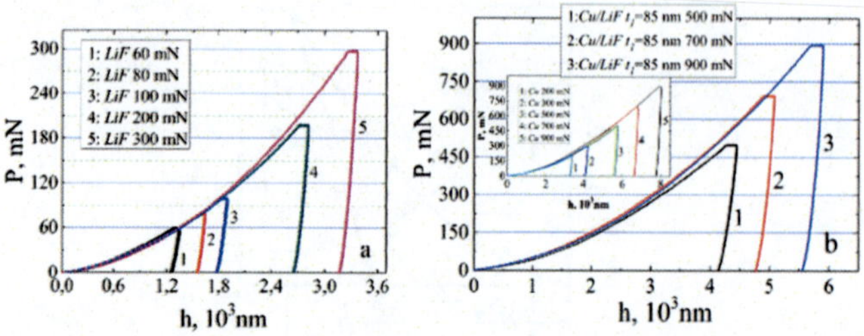

Fig. 3. Loading–unloading curves, *P*(*h*), plotted on LiF single crystal (a) and Cu/LiF CS (b) at different maximum loads P_{max} = 2–900 mN. The insertion in Fig. 3b shows the *P*(*h*) curves for polycrystalline Cu.

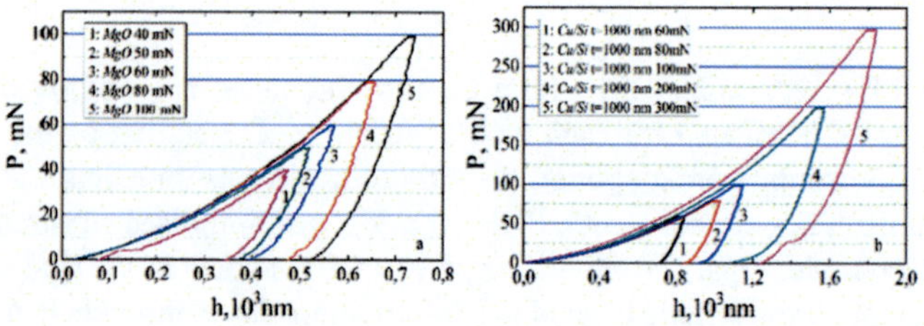

Fig. 4. Loading–unloading curves, *P*(*h*), plotted on MgO single crystal (a) and Cu/Si CS (b) at different maximum loads P_{max}, mN: a – 40–100; b – 60–300.

The parameters characterizing the elastic-plastic recovery, $h_{e\text{-}p}$, and the residual plastic deformation, h_{res}, are shown in Fig. 5 depending on the load for all of the studied samples. It can be noted that the elastoplastic parameters change depending on the CS type and the film thickness. For CS ionic (Cu/LiF), elastic-plastic recovery (Fig. 5a) at low and medium loads up to P $\approx$ 400 mN is the same for all samples, and at high loads it increases for CS compared to Cu and LiF, while the residual depth h_{res} is almost the same for all dependences except for the curve of the polycrystalline Cu (Fig. 5 d).

The difference manifested itself in both CS ionic-covalent (Cu/MgO) and CS covalent (Cu/Si). Here, for both parameters, the deviation of the Cu curve from other curves is observed even at the smallest loads: the greatest elastic-plastic recovery (Figs. 5b, c) and, accordingly, less residual depth (Figs. 5e, f). All the curves of the coated systems, $h_{e\text{-}p}$ and h_{res}, are located closer to the curves of the crystal-substrate (in many cases even overlapping with them), thereby showing a noticeable effect of the substrate on the mechanical behavior of the Cu/substrate systems.

The results obtained earlier in the works [6, 7] indicate that the depth-sensitive indentation process is multilevel. With an increase in the degree of indentation, depending on the value of P_{max}, various deformation mechanisms are included in the process, successively changing from elastic to elastic-plastic, plastically developed and sometimes brittle-plastic until the maximum load is reached. Throughout the penetration of the indenter, the deformation process is accompanied by another reverse synchronous process, which leads to relaxation of the defect structure formed in the material under the

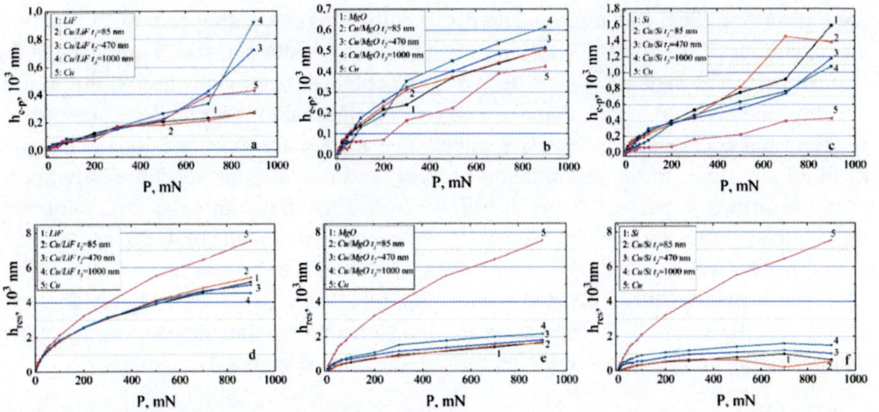

Fig. 5. Dependencies of the parameters that characterize the elasto-plastic restoration (a-c) and the residual depth (d-f) from the load P for the polycrystalline Cu, monocrystals (LiF, MgO, Si) used as a substrate and the composite structures (CSs) investigated after the complete extraction of the indenter from the sample.

load, which is recorded on the $P(h)$ curve at the unloading stage in the form of the above parameters h_c, h_e, $h_{e\text{-}p}$, h_{res}. After the complete removal of the indenter from the sample, the defect structure keeps on recovering, but much more slower over a long period of time.

The specificity of deformation of coated systems is that, unlike monolithic specimens, when deepening into the material, the indenter is exposed not only to the above factors, but also to the factors of changes in the structure and chemical composition of the deformed specimen. For example, when the indenter is introduced into CS of the "soft film/hard substrate" type, it more easily overcomes the film material, then it penetrates

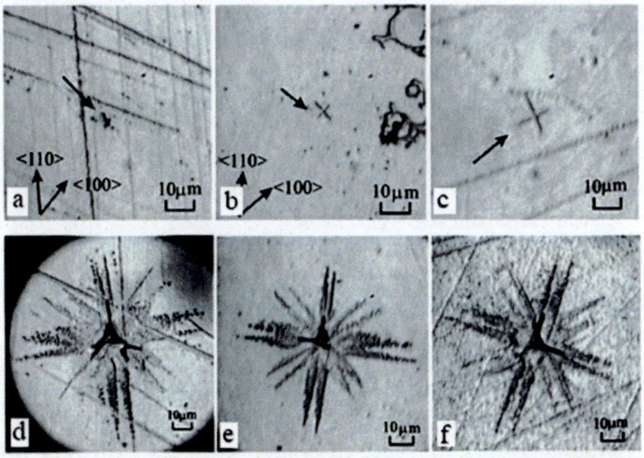

Fig. 6. Cu/MgO, 85 nm (a, d), Cu/MgO, 470 nm (b, e), Cu/MgO, 1000 nm (c, f). Dislocation rosettes revealed on the MgO substrate after removing the film Cu from CS. Loads, *P*, mN: a – 4; b – 20; c – 40. d, e, f – 500.

the substrate material with more and more difficulty. This process depends on two factors, the magnitude of the applied load, P_{max}, and the film thickness, t, which influence in the opposite way. The higher P_{max}, the deeper the indenter penetrates into the CS, and, hence, the substrate material begins to make a larger contribution to the total deformation.

Film thickness has the opposite effect. The thicker the film, the longer the film material takes part in the deformation process, and the later the substrate is included in the deformation process. Thus, it follows from Figs. 6 a–c that the first, minimum dislocation rosettes on the MgO substrate were recorded at loads of 4, 20, and 40 mN, respectively, for films of 85, 470, and 1000 nm. A similar trend was noted for two other CSs: 2, 3, 5 mN for Cu/LiF CS, and 80, 100, 200 mN for Cu/Si CS, respectively, for films of 85, 470, 1000 nm. With increasing load, dislocation rosettes develop significantly, demonstrating an increase in the effect of the substrate (Fig. 6 d–f).

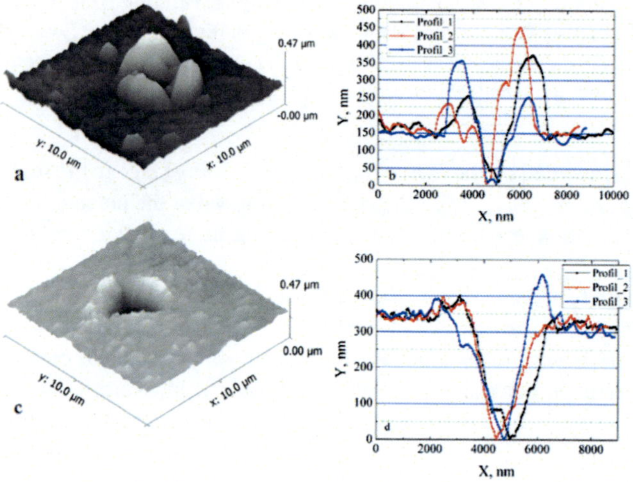

Fig. 7. AFM. Surfase relief near indentations on Cu/MgO; $t = 85$ nm (a,b) and $t = 1000$ nm (c,d). $P_{max} = 30$ mN

A similar result was confirmed for Cu/LiF and Cu/Si CSs. As a consequence, the elastoplastic parameters turn out to be closer in magnitude to the parameters of the film. At the same time, the deformation mechanism changes from the displacement of the film material onto the surface in the form of pile-ups on thin films (Fig. 7a,b) up to the mechanism of compaction of the material inside the film on thick films (Fig. 7c,d).

4 Conclusion

The relaxation parameters $h_{e\text{-}p}$ and h_{res} of coated systems Cu/LiF, Cu/MgO and Cu/Si having different film thickness ($t = 85$, 470 and 1000 nm) were examined in this paper. For all studied coated systems of the Cu/substrate type, the following regularity was noted: the greater the difference in hardness between the film and substrate material

($H_{Cu} = 0.6$ GPa, H_{LiF}, H_{MgO} and H_{Si} are 1.2, 7.5 and 8.2, respectively), the closer the elastoplastic parameters of the composite systems to the substrate parameters.

Acknowledgment. This work is supported financially by the National Agency for Research and Development of the Republic of Moldova under the grant no. **20.80009.5007.18**.

Conflict of Interest. The authors declare that they have no conflict of interest.

References

1. Volinsky, A.A., Gerberich, W.W.: Nanoindentation techniques for assessing mechanical realiability at the nanoscale. Microelectron. Eng. **69**, 519–527 (2003)
2. Yong, Z., et al.: Measurement of Young's modulus and residual stress of copper film electroplated on silicon wafer. Thin Solid Films **460**(1), 175–180 (2004)
3. Головин, Ю.И.: Введение в нанотехнику. Москва: Машино-строение (2007)
4. Pharr, G.M., Oliver, W.C., Clarke, D.R.: The mechanical behavior of silicon during small-scale. J. Electr. Mat. **19**(9), 881–887 (1990)
5. Grabco, D., Pyrtsac, C., Shikimaka, O.: Deformation under nano/microindentation of LiF, MgO, Si monocrystals stipulated as support materials for Cu/substrate structures. In: Proceeding of ICNBME-2013 Chisinau Moldova, pp. 102–106 (2013)
6. Grabco, D., Pushkash, B., Dyntu, M., Shikimaka, O.: Termal evolution of deformation zones around microindentations in different types of crystals. Phil. Mag. A **82**(10), 2207–2215 (2002)
7. Grabco, D., Shikimaka, O., Harea, E.: Translation-rotation plasticity as basic mechanism of plastic deformation in macro-, micro- and nanoindentation processes. J. Phys. D: Appl. Phys. **41**, 074016 (2008)

Controlling the Degree of Hydrophilicity/Hydrophobicity of Semiconductor Surfaces via Porosification and Metal Deposition

E. V. Monaico[1(✉)], S. Busuioc[1], and I. M. Tiginyanu[1,2]

[1] Technical University of Moldova, National Center for Materials Study and Testing, Stefan cel Mare avenue 168, MD-2004 Chisinau, Republic of Moldova
eduard.monaico@cnstm.utm.md

[2] Academy of Sciences of Moldova, Stefan cel Mare avenue 1, MD-2001 Chisinau, Republic of Moldova

Abstract. In this paper we present a systematic study of bulk GaAs wafers and gold-decorated GaAs surfaces exhibiting hydrophilic and hydrophobic behaviors. The wetting properties can be switched to superhydrophilicity and superhydrophilicity by simple electrochemical etching providing engineered porous morphologies. The results open interesting technological perspectives for the exploitation of GaAs surfaces.

Keywords: Wetting · Porous · Electrodeposition · Contact angle · Hydrophilic-hydrophobic

1 Introduction

The investigation of wetting properties (hydrophobicity and hydrophilicity) via measurements of contact angle (CA), give the information about the ability of liquids to flow on the surface or to form droplets. It is considered that a large value of contact angle (above 90 degrees) reflects a hydrophobic surface, while a low value of contact angle reflects a hydrophilic surface. As a rule, surfaces showing contact angle lower than 10 or higher than 150 are classified to superhydrophilic or superhydrophobic respectively. Numerous studies demonstrated that contact angle can be influenced by the micro- and nanoscale morphology of the surface, and by an engineered design of the surface roughness.

Among III-V semiconductor compounds, gallium arsenide (GaAs) represent a class of important materials for high frequency microelectronic and optoelectronics industries such as: telecommunication lasers [1, 2], imaging [3, 4], photodetectors [6], sensors [6] and solar cells [7, 8]. Nowadays, hydro insulation is very important technological step intended to protect electrical components embedded in consumer devices, such as computers, smartphones, smart watches, medical examination devices and more. To protect the microchips from contact with water, hydrophobic polymers such as polydimethylsiloxane (PDMS) are usually used, but they are characterized by low thermal conductivity

I. Tiginyanu et al. (Eds.): ICNBME 2021, IFMBE Proceedings 87, pp. 62–69, 2022.
https://doi.org/10.1007/978-3-030-92328-0_9

[9, 10] leading to poor heat dissipation [11]. The idea of III-V semiconductor impermeability has been previously reported, involving the transfer of epitaxially grown III-V structures to flexible and impermeable substrates such as PDMS [12]. However, these processes are complicated from the point of view of realization. An alternative approach could serve surface nanostructuring, which has been shown to be an effective way to control the hydrophobicity [13] and suppress ice formation [14].

Over the last two decades, it was demonstrated that electrochemistry represents not only a cost-effective approach for nanostructuring of semiconductor crystals at high etch rate, but also an environmentally friendly tool due to controlled electrochemical etching in salty water [15–18]. It is important to point out that GaAs is a material that has been shown to differ from other semiconductor compounds by the diversification of obtained morphologies. It should be noted that the intersection of pores was observed, for the first time in GaAs [19]. In addition, an absolutely new morphology namely, tetrahedron-like interconnected voids in GaAs were reported [20]. Moreover, GaAs with crystallographic orientation (111) exhibits polarity with two polar surfaces: (111)A (Ga-atoms side) and (111)B (As-atoms side). Recently, a comparative study of the electrochemical etching in neutral NaCl and acidic HNO_3 based aqueous electrolytes of (111)A and (111)B GaAs surfaces was performed, demonstrating the formation of tilted or perpendicular to the surface pores and even nanowires [21, 22].

It is worth to mention that functionalization of surfaces with metallic dots has attracted great research interest in the fields of medicine, chemistry and biology due to surface enhanced Raman scattering [23–25]. Deposition of nanodots can significantly modify the surface roughness leading to appearance of new properties. In case the deposition is performed on semiconductor substrates or porous matrices that possess good electrical conductivity, electroplating of metal dots proves to be the most efficient methods.

The goal of this paper is a systematic study of bulk GaAs wafers with two different crystallographic orientations and gold decorated GaAs surfaces, which shows hydrophilic or hydrophobic behaviors. The wetting properties can be switched by simple electrochemical etching leading to engineered porous morphologies.

2 Materials and Methods

Crystalline 500-μm thick Si-doped n-GaAs (001) and two side polished (111) wafers with the free electron concentration of 3×10^{18} cm^{-3} and 2×10^{18} cm^{-3} respectively, supplied by MaTeck GmbH, Germany were used as substrates. As a first step, the samples were sonicated in acetone for 5 min, cleaned in distilled water and dried. With the aim to remove the native oxide from the surface, the samples were dipped in a HCl/H_2O solution with the ratio (1:3) for 2 min.

Electrochemical deposition of gold was performed in a commercially available gold bath containing 5g/l Au (DODUCO, Germany). The electroplating of Au was carried out at T = 25 °C in a common two-electrode plating cell where the semiconductor sample served as working electrode, while a platinum wire was used as counter electrode. A pulsed rectangular shape negative voltage of – 15 V was provided by a home-made generator. The metal species were electrochemically reduced on the sample surface being in contact with the electrolyte, during the pulse time of 50 μs and 300 μs. A delay time as long as one second was kept after each pulse. The total time of the electroplating was 300 s.

Gold films were also deposited by sputtering for comparison purposes using a Cressington Sputter Coater 108 Auto instrument under current intensity of 40 mA for 15, 30, and 45 s at the distance of 45 mm between the sample and gold target, resulting in 7, 13, and 50 nm thickness of the deposited film respectively.

Electrochemical etching was carried out in potentiostatic regime at room temperature (T = 25 °C) at applied anodization voltage 3.5 V during 15 min, resulting in the formation of porous layer with the thickness around 50 μm. The electrical contacts to the sample were performed with silver paste, and then the sample was pressed against an O-ring in a Teflon cell with the 0.2 cm^2 area exposed to the 1M HNO_3 electrolyte. The experiments were performed in two-electrode configuration: a Pt mesh with the surface area of 6 cm^2 acting as counter electrode while the sample acted as working electrode.

The morphology of samples was investigated by using scanning electron microscope (SEM) TESCAN Vega TS 5130 MM equipped with an Oxford Instruments INCA Energy EDX system operated at 20 kV for chemical composition analysis.

Contact angle measurements were carried out with water using the KRUSS DSA25 (drop shape analyzer) equipped with a camera. The sample was illuminated from behind so that the shadow contour of the substrate-drop contact was visible and well highlighted to allow the software to determine the contact angle.

3 Results and Discussions

To establish how the deposition of metal influence the wetting properties of semiconductor crystals surface (GaAs in our case) we have investigated the contact angle of bulk GaAs substrates and bulk GaAs decorated with deposited gold nanodots.

The measured contact angle of bulk n-GaAs substrate is 80.4° as is shown in Fig. 1b. As can see from SEM image in Fig. 1c, the electrochemical deposition of gold on bulk n-GaAs substrate with pulse duration of 50 μs resulted in formation of thin layer consisting of closely packed Au nanodots. Thorough analysis of the SEM image presented in Fig. 1c shows that the transverse dimensions of the dots do not exceed 20 nm. The similar results were earlier reported for Au deposition on porous n-InP and n-GaP semiconductors compounds [26]. In this case, the measured contact angle decreased to 73.9° (see Fig. 1d) due to the small voids between the deposited Au dots.

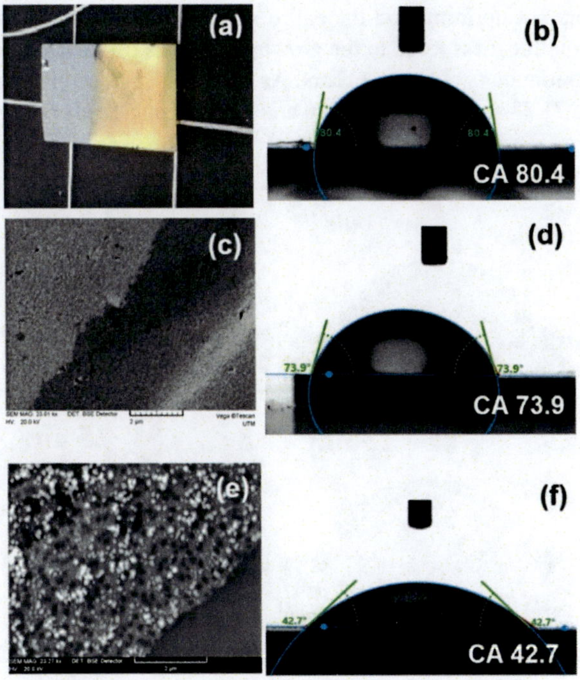

Fig. 1. Photo of electrochemically deposited Au film at pulse duration 300 μs (right side) on bulk GaAs (left side). SEM images of deposited gold film at pulse duration 50 μs (c) and 300 μs (e). Measured contact angle for bulk GaAs surface (b), deposited film with pulse duration 50 μs (d) and 300 μs (f)

The increase of the pulse duration up to 300 μs during electrochemical deposition leads to the formation of perforated Au membrane on the surface of GaAs semiconductor substrate with the thickness about 100 nm (Fig. 1e). The resulted contact angle about 42.7° was registered indicating hydrophilic properties of GaAs surface decorated by perforated gold membrane. The increase in hydrophilicity is mainly attributed to the difference in both the chemical properties of GaAs surface and its surface morphology.

The formation of pores in deposited Au film can be explained in the following way: applied voltage pulses with longer duration led to visible bubble formation at the GaAs surface during electroplating. It is considered that hydrogen or oxygen liberations are the major cause of these bubble formation. Formed bubbles on the surface of the sample reduce their effective surface area, resulting in fluctuations of current density and affecting the deposition process in these places. As a result, the gold is not deposited on places occupied by these bubbles on the GaAs surface.

To compare the impact of Au distribution during the deposition, an alternative method vas involved. Three samples with Au film thickness of 7 nm, 13 nm, and 25 nm were prepared using sputtering. No change in the morphology was evidenced after deposition of Au via SEM investigation. In spite of this, the decrease of contact angle from 80.4° for bulk GaAs to 70.9° for GaAs/Au with thicknesses of 7 nm was observed as is illustrated in Fig. 2b. It can be explained by the fact that deposition for such small thickness occurs not uniformly on the entire surface but via islands. Further increase of the deposition thickness leads to the growth of islands till their interconnection and formation of continuous gold layer occurs. As a result, the value of the contact angle is increased up to 71.3° and 72.8° for 13 nm and 25 nm respectively (see Fig. 2c,d).

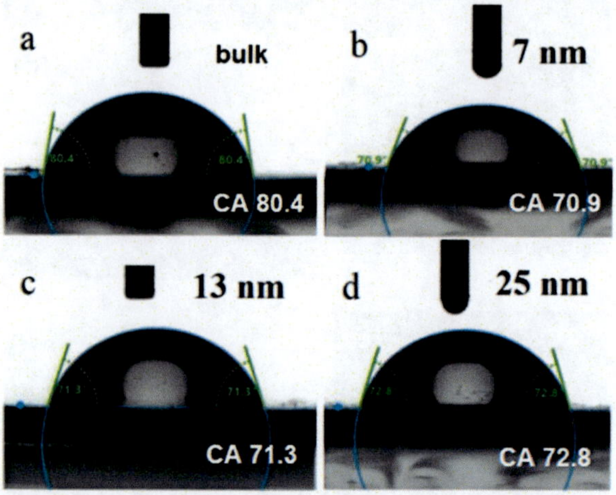

Fig. 2. Measured contact angle for bulk n-GaAs (a) and sputtered gold film with thickness of 7 nm (b), 13 nm (c), and 25 nm in (d) on GaAs surface

Higher uniformity of the Au film deposited via sputtering is explained taking into account the polarization curves measured at the beginning of anodization of bulk GaAs and bulk GaAs with deposited gold, reported in our recent work [27]. It was shown that anodization of bulk GaAs under 4 V leads to higher current density resulting in the formation of porous layer. However, in the case of bulk GaAs samples with electrochemically deposited Au nanoparticulate film with different durations of deposition leads to a significant decrease in the current intensity, especially for deposition at 50 μs width of the pulse [27]. It can be explained by formation of very small voids between the deposited nanodots, and thus a small amount of the electrolyte during electrochemical etching enters in contact with the surface of the bulk GaAs. This statement is confirmed by the investigation of deposited gold film via sputtering resulting in the formation of more continuous film leading to practically no reaction at the semiconductor-electrolyte interface [27].

Introducing porosity on the semiconductor surface is another way to control the wetting properties. We investigated porous layer obtained via anodization of both (111) GaAs surfaces. It was recently showed that anodization on different (111) GaAs surfaces resulting in different orientation of pores to the surface [21, 22]. On the (111)A GaAs surface the pores interconnect each other and are tilted (see Fig. 3a). Measured contact angle showed that the value of contact angle increased to 137.5° (see Fig. 3d), close to those of super-hydrophobic, comparatively to the bulk surface (95.4°). The situation is contrary in the case of nanostructured (111)B GaAs surface exhibiting pores and nanowires perpendicular to the surface (see Fig. 3b). This morphology gives a contact angle of 37.5° indicating strong hydrophilic properties (Fig. 3e).

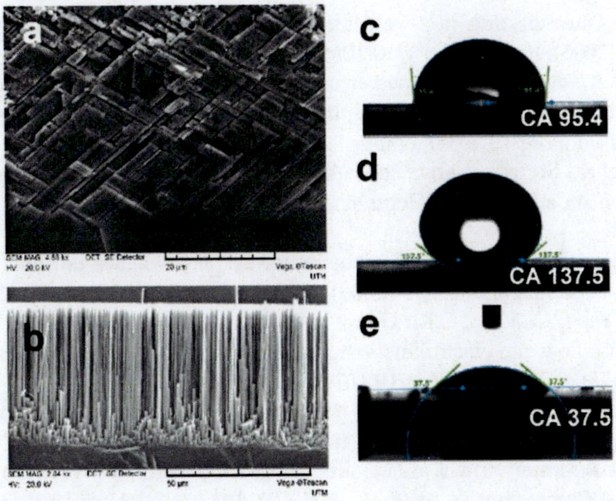

Fig. 3. SEM images of anodized n-GaAs (111)A and (111)B surfaces applied potential 3.5 V in 1M HNO_3 in (a) and (b) respectively. Measured contact angle for bulk (c), porous (111)A surface in (d) and porous (111)B surface in (e)

4 Conclusions

In this work we showed that the wetting properties of GaAs surface can be engineered by electrochemical deposition of gold nanodots. The increase in hydrophilicity was demonstrated for deposited gold with longer pulse duration leading to the formation of perforated Au membrane on the GaAs surface. The introduced porosity on the semiconductor surface gives the possibility to control the wetting properties of the GaAs surface. The morphology of pores, tilted or perpendicular to the surface, play an important role in switching from the hydrophobic to hydrophilic properties.

Acknowledgment. This work received partial funding from the European Commission under the H2020 grant #810652 'NanoMedTwin' and PostDoc Grant #21.00208.5007.15/PD.

Conflict of Interest. The authors declare that they have no conflict of interest.

References

1. Ustinov, V.M., Zhukov, A.E.: GaAs-based long-wavelength lasers. Semicond. Sci. Technol. **15**, R41–R54 (2000). https://doi.org/10.1049/ip-cds:20030959
2. Liu, H.Y., et al.: Long-wavelength InAs/GaAs quantum-dot laser diode monolithically grown on Ge substrate. Nat. Photon. **5**, 416–419 (2011). https://doi.org/10.1038/nphoton.2011.120
3. Ng, W.H., et al.: Broadband quantum cascade laser emitting from 72 to 84 μm operating up to 340 K. J. Appl. Phys. **101**, 046–103 (2007). https://doi.org/10.1063/1.2472196
4. Colley, C.S., et al.: Mid-infrared optical coherence tomography. Rev. Sci. Instrum. **78**, 123108 (2007). https://doi.org/10.1038/s41377-019-0122-5

5. Krishna, S.: Quantum dots-in-a-well infrared photodetectors. J. Phys. D: Appl. Phys. **38**, 2142–2150 (2005). https://doi.org/10.1088/0022-3727/38/13/010
6. Potyrailo, R.A., et al.: Towards outperforming conventional sensor arrays with fabricated individual photonic vapour sensors inspired by Morpho butterflies. Nat. Commun. **6**, 7959 (2015). https://doi.org/10.1038/ncomms8959
7. Holm, J.V., et al.: Surface-passivated GaAsP single-nanowire solar cells exceeding 10% efficiency grown on silicon. Nat. Commun. **4**, 1498 (2003). https://doi.org/10.1038/ncomms2510
8. Fan, Z., et al.: Three-dimensional nanopillar-array photovoltaics on low-cost and flexible substrates. Nat. Mater. **8**, 648–653 (2009). https://doi.org/10.1038/nmat2493
9. Tang, G.Y., Yang, C., Chai, C.K., Gong, H.Q.: Numerical analysis of the thermal effect on electroosmotic flow and electrokinetic mass transport in microchannels. Anal. Chim. Acta. **507**, 27–37 (2004). https://doi.org/10.1016/j.aca.2003.09.066
10. Erickson, D., Sinton, D., Li, D.: Joule heating and heat transfer in poly(dimethylsiloxane) microfluidic systems. Lab Chip **3**, 141–149 (2003). https://doi.org/10.1039/B306158B
11. Zhang, Y., et al.: Improvement of heat dissipation for polydimethylsiloxane microchip electrophoresis. J. Chromatogr. A **1057**, 247–251 (2004). https://doi.org/10.1016/j.chroma.2004.09.009
12. Kim, R.-H., et al.: Waterproof AlInGaP optoelectronics on stretchable substrates with applications in biomedicine and robotics. Nat. Mater. **9**, 929–937 (2010). https://doi.org/10.1038/nmat2879
13. Lu, Y., et al.: Robust self-cleaning surfaces that function when exposed to either air or oil. Science **347**, 1132–1135 (2015). https://doi.org/10.1126/science.aaa0946
14. Kim, P., et al.: Liquid-infused nanostructured surfaces with extreme anti-ice and anti-frost performance. ACS Nano **6**, 6569–6577 (2012). https://doi.org/10.1021/nn302310q
15. Monaico, E., Tiginyanu, I., Ursaki, V.: Porous semiconductor compounds. Semicond. Sci. Technol. **35**, 103001 (2020). https://doi.org/10.1088/1361-6641/ab9477
16. Tiginyanu, I.M., et al.: Pore etching in III-V and II-VI semiconductor compounds in neutral electrolyte. Electrochem. Solid-State Lett. **10**(11), D127–D129 (2007). https://doi.org/10.1149/1.2771076
17. Tiginyanu, I.M., Monaico, E., Albu, S., Ursaki, V.V.: Environmentally friendly approach for nonlithographyc nanostructuring of materials. Phys. Stat. Sol. (RRL) **1**(3), 98–100 (2007). https://doi.org/10.1002/pssr.200701007
18. Volciuc, O., et al.: Morphology, luminescence, and electrical resistance response to H2 and CO gas exposure of porous InP membranes prepared by electrochemistry in a neutral electrolyte. Appl. Surf. Sci. **257**(3), 827–831 (2011). https://doi.org/10.1016/j.apsusc.2010.07.074
19. Langa, S., et al.: Observation of crossing pores in anodically etched n-GaAs. Appl. Phys. Lett. **78**(8), 1074–1076 (2001). https://doi.org/10.1063/1.1350433
20. Langa, S., et al.: Formation of Tetrahedron-Like Pores during Anodic Etching of (100) Oriented n-GaAs. Electrochem. Solid-State. Lett. **5**(1), C14–C17 (2002). https://doi.org/10.1149/1.1423803
21. Monaico, E.I., et al.: Electrochemical nanostructuring of (111) oriented GaAs crystals: from porous structures to nanowires. Beilstein J. Nanotechnol. **11**, 966–975 (2020). https://doi.org/10.3762/BJNANO.11.81
22. Monaico, E.I., Monaico, E.V., Ursaki, V.V., Tiginyanu, I.M.: Evolution of pore growth in GaAs in transitory anodization regime from one applied voltage to another one. J. Surface Eng. Appl. Electrochem. **57**(2), 65–172 (2021). https://doi.org/10.3103/S106837552102006X
23. Yuanyuan, X., et al.: Two-step electrodeposited 3D Ni nanocone supported Au. J. Electrochem. Soc. **167**, 142502 (2020). https://doi.org/10.1149/1945-7111/abc0aa
24. Zhu, J., et al.: Green preparation of silver nanofilms as SERS-active substrates for Rhodamine 6G detection. Vacuum **187**, 110096 (2021). https://doi.org/10.1016/j.vacuum.2021.110096

25. Arfat, P., et al.: Review—nanopillar structure in the direction of optical biosensor on-chip integration. J. Electrochem. Soc. **168**, 057505 (2021). https://doi.org/10.1149/1945-7111/abfb3a
26. Tiginyanu, I., Monaico, E., Nielsch, K.: Self-assembled monolayers of Au nanodots deposited on porous semiconductor structures. ECS Electrochem. Lett. **4**, D8–D10 (2015). https://doi.org/10.1149/2.0041504eel
27. Monaico, E.V., Monaico, E.I., Ursaki, V.V., Tiginyanu, I.M.: Free-standing large-area nanoperforated gold membranes fabricated by hopping electrodeposition. ECS J. Solid State Sci. Technol. **9**, 064010 (2020). https://doi.org/10.1149/2162-8777/aba6a2

Variation of Acoustic Properties with Material Parameters in Layered Nanocomposites

S. Cojocaru(✉)

Horia Hulubei National Institute for Physics and Nuclear Engineering, Magurele, Romania
scojocaru@theory.nipne.ro

Abstract. We discuss some general features of the elastic vibration spectrum in layered nanocomposites based on the newly developed approach. The focus is on the compounds without any layering symmetry and on the dependence of their properties on various parameters. Some regular general features of the spectrum are revealed by a numerical and analytical study. In particular, existence of invariant characteristics which allow classification of the whole band structure in a way similar to the Sturm-Liouville theory. The spectral lines become multidimensional surfaces in the space which includes a variation of material parameters. It is demonstrated that confinement, layering and material selection produce qualitatively new properties which can be useful in applications.

Keywords: Acoustic wave spectra in nanocomposites · Ultrathin layered structures · Phonons in nanomaterials

1 Introduction

In addition to their inherent confinement at the nano-, or microscale, one of the key features of nanotechnologies is directly related to ever lower energies or temperatures at which micro-, and nano-devices are designed to operate. In these regimes the role of low energy excitations and their quantum nature come to the forefront, opening the way to applications which were previously inaccessible. For instance, it now becomes possible to detect single photons at record long wavelengths [1], or to process quantum information with single phonons [2], or to achieve record mK electronic refrigeration of a single microchip [3] etc.

In this context acoustic phonons are possibly the most ubiquitous type of such low energy excitations, which have a major contribution to equilibrium and transport properties of nanomaterials [4, 5]. Their ability to couple with most of the other excitation types creates rich opportunities for nanomaterial engineering and is an area of active research [6–8]. Quantization of the phonon field is based on the underlying classical model, which in our case is that of a thin layered elastic medium. It is the most commonly used setup in microelectronics and an extremely broad range of other areas, in particular, biology and medicine. Thus, in the last decades the medical use of ultrasound technology has shifted from bulk to shear polarized guided waves [9]. Respectively, it is required to determine the normal modes of confined vibrations, their amplitudes and

I. Tiginyanu et al. (Eds.): ICNBME 2021, IFMBE Proceedings 87, pp. 70–76, 2022.
https://doi.org/10.1007/978-3-030-92328-0_10

frequencies. These are known to form an intricate net of spectral lines even for "simpler" cases of regular or symmetric arrangement of layers. Such research is relying on advanced numerical analysis and a large number of elaborate theoretical methods, e.g., [10, 11]. However, properties of nanostructures unrestricted by such arrangement are significantly less explored and understood. In absence of symmetry it appears impossible even to qualitatively classify the vibration eigenmodes or anticipate the evolution of the spectral lines caused by variation in their structure or composition. This complexity is enhanced with the number of layers $\gamma = 1, 2, \ldots$ *etc.* by a rapid growth of the "degrees of freedom" spanned by material parameters: layer thicknesses, h_γ, mass densities, ρ_γ, elastic constants, μ_γ, λ_γ, or corresponding bulk shear, s_γ, and longitudinal, ℓ_γ, sound velocities. For such a multidimensional space even the advanced numerical simulation methods could not meet the challenge. The new method proposed in [12, 13] has allowed to obtain the complete analytic solution for the bi-layered structure in the long wavelength region of the spectrum, which allows to reveal some general properties of the system. It can serve as a reference for the numerical analysis in the whole range of wavenumbers for a specific material. In particular, analysis of the dispersion curves has indicated the existence of a one-to-one correspondence between the ordering of natural frequencies at long and short wavelengths, e.g., the fundamental flexural mode being connected to the lowest Rayleigh surface acoustic wave. In contrast to the case of symmetric stacking, no intersection points between the spectral lines have been found. These findings agree with the general arguments in [14] about absence of degeneracy in vibration spectra of non-symmetric multilayered structures and can lay the ground for a classification scheme analogous to the Sturm-Liouville theory. The natural ordering of the eigenfrequencies ω_n can then be generated by the frequency dependence of the secular determinant $\Delta(\omega)$, provided it is a differentiable function of ω. Unfortunately, the continuity requirement is not easy to be fulfilled by $\Delta(\omega)$. Thus, the main difficulty with finding the physical solutions from the dispersion equation in [12, 13] is to tackle the singularities of $\Delta(\omega)$ and to avoid the spurious solutions. This difficulty is shared with the other approaches. Moreover, the dynamic impedance matrix approach of [14] even relies on the presence of singular terms in the eigenfrequency equation to prove the fundamental result on the non-crossing of the acoustic spectral branches.

We also mention the previously unnoticed correlation between the signs of the vibration amplitudes on the external surfaces of the sample and ordering of the natural frequencies. This property could have the role that symmetric/antisymmetric classification of the Lamb modes plays for a single layered system. However, as will be shown below, the proof of such correlation also requires a deeper investigation of the band structure to overcome the limitation of the previous work.

2 The Secular Equation

The basis of the new approach is that solutions of the equations of elastodynamics are represented by a superposition of partial waves of a special form [12]: they have to automatically satisfy the boundary conditions on the external boundaries. To overcome the difficulties mentioned above we should refine the set of surface adapted partial waves. While propagation parallel to the (x, y) horizontal surface of boundaries is described by plane waves with a wavenumber k, the equations for the thickness variation in the z – direction of the displacement amplitude components are decoupled according to the polarization pattern into SH, u_y, shear horizontal,

$$\frac{\partial^2 u_y}{\partial z^2} - k^2 v_\gamma^2 u_y = 0,$$

and mixed P + SV, (u_x, u_z), or Rayleigh-Lamb waves:

$$s_\gamma^2 \frac{\partial^2 u_{x,\gamma}}{\partial z^2} - l_\gamma^2 k^2 w_\gamma^2 u_{x,\gamma} + \left(l_\gamma^2 - s_\gamma^2\right) k \frac{\partial i u_{z,\gamma}}{\partial z} = 0,$$

$$l_\gamma^2 \frac{\partial^2 i u_{z,\gamma}}{\partial z^2} - s_\gamma^2 k^2 v_\gamma^2 i u_{z,\gamma} - \left(l_\gamma^2 - s_\gamma^2\right) k \frac{\partial u_{x,\gamma}}{\partial z} = 0.$$

Here

$$w_\gamma = \sqrt{1 - (\omega/kl_\gamma)^2}, v_\gamma = \sqrt{1 - (\omega/ks_\gamma)^2} \tag{1}$$

Bottom and top surfaces of the $N-$ layered stack are at $z = d_\gamma = 0 = h_0 = 0$ and $z = d_\gamma = N$, respectively, and interfaces at $z = d_\gamma$ ($\gamma = 1, \ldots, N-1$), where

$$d_\gamma = \sum\nolimits_{\gamma'=0}^{\gamma} h_{\gamma'}. \tag{2}$$

The superposition of partial waves has the following form (the γ index is dropped)

$$\begin{aligned} u_y &= A \sinh((z-d)kv) + B \cosh((z-d)kv), \\ u_x &= Cw \sinh((z-d)kw) + Xv \sinh((z-d)kv) \\ &\quad + Y \cosh((z-d)kw) + F \cosh((z-d)kv), \\ iu_z &= Gw \sinh((z-d)kw) + Hv \sinh((z-d)kv) \\ &\quad + I \cosh((z-d)kw) + J \cosh((z-d)kv), \end{aligned} \tag{3}$$

and the relevant components of the stress tensor are obtained from

$$\begin{aligned} \sigma_{yz} &= \mu\left(\frac{\partial u_y}{\partial z}\right), \\ \sigma_{xz}(z) &= \mu\left(\frac{\partial u_x}{\partial z} + k(iu_z)\right) \\ i\sigma_{zz}(z) &= (\lambda + 2\mu)\frac{\partial (iu_z)}{\partial z} - \lambda k u_x. \end{aligned} \tag{4}$$

Note that in this new representation the partial waves on the right-hand-side of the equations are real single valued differentiable functions of all the variables, including frequency, wavenumber and material parameters. Indeed, one can easily verify that these are differentiable functions of v^2 and w^2 and we actually have only hyperbolic functions of polynomials. This is an essential aspect of the theory. In the above expressions z varies within the corresponding layer γ and continuity conditions on interfaces allow to cover the whole interval. We note that in this approach the problem is reduced only to these conditions. Indeed, when the superposition (3) is substituted into the wave equations of elastodynamics (1), the number of undetermined coefficients is drastically reduced. Thus, for the mixed polarization we get

$$I_\gamma = C_\gamma w_\gamma^2, G_\gamma = Y_\gamma, J_\gamma = D_\gamma, H_\gamma = F_\gamma / v_\gamma^2. \tag{5}$$

That reduces the number of coefficients by half and we obtain the following general equations of this approach valid for any number of layers (the index is again dropped and we show only the stress components, see (3–5)),

$$\begin{aligned} \sigma_{xz}(z)/k\mu &= 2Yw \sinh((z-d)kw) + F(1+v^2)\frac{\sinh((z-d)kv)}{v} \\ &+ 2Cw^2 \cosh((z-d)kw) + X\left(1+v^2\right)\cosh((z-d)kv), \\ i\sigma_{zz}(z)/k\mu &= \left(1+v^2\right)Cw \sinh((z-d)kw) + 2Xv \sinh((z-d)kv) \\ &+ (1+v^2)Y\cosh((z-d)kw) + 2F\cosh((z-d)kv). \end{aligned} \tag{6}$$

Finally, this number is again reduced by half for the coefficients with $\gamma = 0, N$ of the bottom and top surfaces. Thus, taking the clamped boundary conditions, we obtain from (3) that $B_{0,N} = 0$, $Y_{0,N} = -F_{0,N}$, $w^2 C_{0,N} = -J_{0,N}$. For the free standing compound we find from (3–5) that $A_{0,N} = 0$ and $C_{0,N} = -X_{0,N}(1 + v^2_{0,N})/2w^2_{0,N}$, $F_{0,N} = -Y_{0,N}(1 + v^2_{0,N})/2$. In particular, for the bi-layered structure we have 4 independent coefficients for the P + SV and 2 coefficients for the SH waves. Then the secular equation is determined solely by the continuity on the interface:

$$\begin{aligned} u_{x,y,z}\left(z = d_\gamma - 0\right) &= u_{x,y,z}\left(z = d_\gamma + 0\right), \\ \sigma_{(x,y,z)z}\left(z = d_\gamma - 0\right) &= \sigma_{(x,y,z)z}(z = +0). \end{aligned} \tag{7}$$

By substituting Eqs. (3–6) into (7), we can see that the resulting linear equations contain only variables h_γ from adjacent layers due to the definition in (2). It is then easy to prove by inspection that all the terms in this representation are continuously differentiable functions of their variables. Indeed, the "worst" terms, like *sinh(hkv)/v* or *sinh(hkw)/w*, are actually perfectly regular, real and single valued functions of wavenumber, frequency, thickness, sound velocities, etc.. Therefore, by increasing ω, with all the other quantities fixed, the determinant of the secular equation Δ(ω) will cross zero at the consecutive points $\omega = \omega_n$ in an ordered sequence, $n = 1, 2, 3, \ldots$ etc.. At the crossing points the slope $(\partial\Delta/\partial\omega)_n$ is finite and alternates in sign, i.e., $\Delta(\omega \approx \omega_n) \sim (-1)^n (\omega - \omega_n)$. Absence of singularities and of spurious solutions in this approach means that every crossing point ω_n corresponds to a physical eigenmode. It will be seen later that also the slope of Δ(ω) at this points has a physical meaning. The nodes of Δ(ω) depend continuously on parameters and can not disappear. Otherwise, it would imply that for some critical value of *k*, or any of the material parameters, the slope should vanish $(\partial\Delta/\partial\omega)_n = 0$. The latter condition can be visualized geometrically as some ω_n becoming a local extremum, i.e., $\Delta(\omega) \sim (\omega - \omega_n)^2$, and this, in turn implies that the respective root is degenerate, which would contradict the proof in [14]. We indeed observe this behavior in our numerical analysis of Δ (ω), including the quadratic behavior near the crossing points in symmetric structures.

3 Spectral Lines in the Parametric Space

We have thus established that the frequency-wavenumber or velocity-wavenumber spectrum for a given set of material parameters consists of indexed non-intersecting spectral lines, $\omega_n(k)$ or $c_n(k)$ respectively. Their single valuedness is an important property allowing establish an ordering of the separate branches, similar to the Stourm-Liouville theory. In particular, the band index, *n*, being an invariant characteristic of the spectral branch, is shared by the modes in the two regions, $kh < 1$ and $kh \gg 1$, which has been inferred earlier from numerical analysis [12]. By contrast, such property does not show up if, instead, one considers the solutions of $\Delta(k) = 0$ at a fixed frequency because the resulting inverse functions, $k(\omega)$ or $c(\omega)$, are generally multivalued. This difference is a consequence of the physical requirement that group velocity $c_g = \partial\omega_n(k)/\partial k$ in a normal acoustic medium can't diverge. However, it may and does become negative in many materials, so that for the same dispersion curve there exist two or more values of the wavenumber with the same frequency or phase velocity. We may now extend these considerations to the space of material parameters, which, as demonstrated above, preserves the same differentiability property as the k – space. The dispersion curves then evolve into separate non-intersecting multidimensional surfaces characterized by the same band index. In this extended space, the problem of applicative interest is that of engineering the properties of such a composite structure, which would support a vibration mode with a required frequency and wavelength. Following the preceding line of reasoning we may then consider the solutions of, e.g., $\Delta(\mathrm{h}_\gamma) = 0$ or $\Delta(\mu_\gamma) = 0$, etc., at a fixed value of ω, k and the rest of parameters.

It turns out that behavior of the dispersion surfaces in the extended space is very non-trivial. For a better illustration of the possibilities we show a representative example of a few lowest shear resonance eigenfrequencies and introduce the following scaled variables: $\delta = h_2/h_1$, $\rho = \rho_2/\rho_1$, $s = s_2/s_1$, $\Omega = \omega h/s_1$. The total thickness of the composite h and density ratio are kept constant, $\rho = 4$. In the Fig. 1 we see a gradual softening of the resonance frequency when part of the original material "1" is replaced by a softer layer "2", $s < 1$. This is what could be expected, along with our earlier result on the fundamental dilatational mode [12, 13].

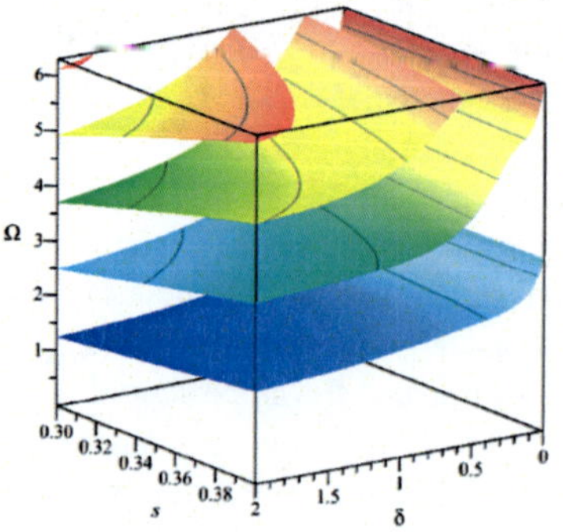

Fig. 1. Expected smooth softening of the shear resonance frequency Ω when the part δ of the original material is replaced by a softer one, $s < 1$. Mass density ratio is $\rho_2/\rho_1 = 4$.

However, when the replacement layer "2" is characterized by stiffer elastic constants (and even much stiffer $s \gg 1$), the resonance frequency, contrary to expectation, keeps softening until the space ratio occupied by the new layer becomes large enough to overturn the tendency towards the "normal", higher frequencies, Fig. 2.

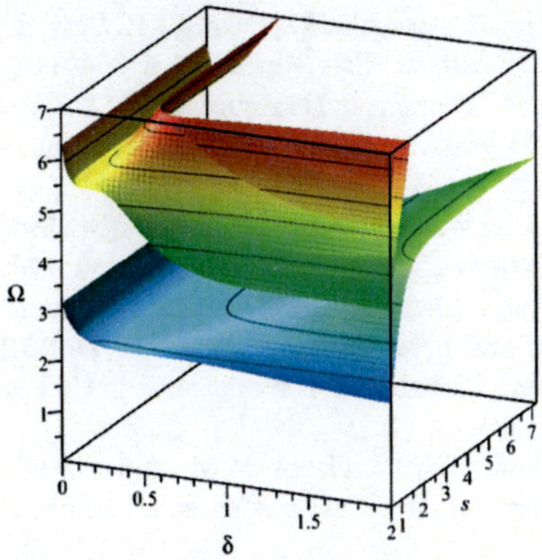

Fig. 2. Unexpected softening of the shear resonance frequency takes place when a small part $\delta \ll 1$ of the original material is replaced by a stiffer one, $s > 1$. The "normal" stiffening trend is recovered at larger δ. Mass density ratio is the same as in Fig. 1.

When the density ratio between the two materials is reversed, $\rho_2/\rho_1 < 1$, we observe similar non-trivial variation of the band structure. A more detailed analysis, which will be presented elsewhere, reveals more un-anticipated features of the parametric dependence.

We finally mention, that by applying our approach to the secular equation for the SH waves it can be rigorously proved that the alternating sign of the derivative $(\partial \Delta(\omega)/\partial \omega)$ at $\omega = \omega_n$ reproduces the sign of the surface amplitudes for the respective modes. This property can also be useful in applications for the identification of the specific modes by their wave-pattern on the external surfaces.

4 Conclusions

We have presented a new approach to the description of acoustic properties in confined layered systems in a more general setup, i.e., in absence of stacking symmetry. It demonstrates the existence of invariant characteristics of the band structure, ordering of the spectral lines and the slope of the secular determinant. Some new reported properties related to the dependence of the spectra on material parameters are qualitatively different from expectation and may be useful in engineering and design of nanocomposites.

Acknowledgment. This work was financially supported by ANCS Romania (project no. PN 19060101/2019-2022).

Conflict of Interest. The author declares having no conflict of interest.

References

1. Anghel, D.V., Kuzmin, L.S.: Cold-electron bolometer as a 1-cm-wavelength photon counter. Phys. Rev. Appl. **13**, 024028 (2020)
2. Bienfait, A., et al.: Quantum erasure using entangled surface acoustic phonons. Phys. Rev. X **10**, 021055 (2020)
3. Nguyen, H.Q., Peltonen, J.T., Meschke, M., Pekola, J.P.: Cascade electronic refrigerator using superconducting tunnel junctions. Phys. Rev. Appl. **6**, 054011 (2016)
4. Stroscio, M.A., Dutta, M.: Phonons in Nanostructures. CUP, UK (2004)
5. Cojocaru, S., Anghel, D.V.: Low-temperature electron-phonon heat transfer in metal films. Phys. Rev. B **93**, 115405 (2016)
6. Saira, O.P., Matheny, M.H., Wang, L., Pekola, J., Roukes, M.: Modification of electron-phonon coupling by micromachining and suspension. J. App. Phys. **127**, 024307 (2020)
7. Dong, X.L., Li, P.B., Liu, T., Nori, F.: Unconventional quantum sound-matter interactions in spin-optomechanical-crystal hybrid systems. Phys. Rev. Lett. **126**, 203601 (2021)
8. Ono, S.: Low-temperature electron-phonon relaxation in Cu and Ag thin films. Phys. Rev. B **101**, 201404(R) (2020)
9. Khodayi-mehr, R., Urban, M.W., Zavlanos, M.M., Aquino, W.: Plane wave elastography: a frequency-domain ultrasound shear wave elastography approach. Phys. Med. Biol. **66**, 125017 (2021)
10. Auld, B.A.: Acoustic Fields and Waves in Solids, 2nd edn. Krieger, FL (1990)
11. Graff, K.F.: Wave Motion in Elastic Solids. Dover Publications, NY (1991)
12. Cojocaru, S.: Surface adapted partial waves for the description of elastic vibrations in bilayered plates. Wave Motion **92**, 102430 (2020)
13. Cojocaru, S.: Unusual size dependence of acoustic properties in layered nanostructures. In: Tiginyanu, I., Sontea, V., Railean, S. (eds.) ICNBME 2019. IP, vol. 77, pp. 23–27. Springer, Cham (2020). https://doi.org/10.1007/978-3-030-31866-6_5
14. Kausel, E., Malischewsky, P., Barbosa, J.: Osculations of spectral lines in a layered medium. Wave Mot. **56**, 22 (2015)

Structural Characterization of Some As-S-Sb-Te Nanostructured Materials

Oxana V. Iaseniuc(✉) and M. S. Iovu

Institute of Applied Physics, Str. Academiei 5, 2028 Chisinau, Republic of Moldova

Abstract. Nanostructured As-S-Sb-Te semiconductors were synthesized and characterized by X-ray fluorescence analyzer (XRF), X-ray diffraction, and optical absorption methods. The X-ray diffraction patterns of ivestigated powders show the presence of amorphous and nanocrystalline phases with the structural units As_2S_3, Sb_2S_3 and Sb_2Te_3. The transmission spectra in the region of wavenumbers $\nu = 1000 \div 6000\ cm^{-1}$ show a high transparence just with a single weak absorption band at $\nu = 2340\ cm^{-1}$ caused of the presence H_2S impurity. For the alloys of $(As_2S_3)_x(Sb_2S_3)_{1\text{-}x}$ system, with increasing of the Sb_2S_3 trigonal structural units in the above mentioned system, the absorption edge is shifted toward lower photon energy, that corresponds to the optical band band gap about $E_g. = 2.34$ eV for As_2S_3, 2.1 eV for $(As_2S_3)_{0.65}(Sb_2S_3)_{0.35}$, 1.92 eV for $(As_2S_3)_{0.35}(Sb_2S_3)_{0.65}$ and 1.73 eV for Sb_2S_3.

Keywords: Nanostructured polycrystalline semiconductors · X-ray diffraction patterns · X-ray fluorescence analysis · Optical transmission and absorption

1 Introduction

Chalcogenide glasses have attracted much attention over the years due to their technological applications, infrared optical elements, as materials for image creation and storage of information, acousto-optic and memory switching elements [1, 2]. As_2S_3, As_2Se_3, As_2Te_3, and Sb_2S_3 are the most intensively studied chalcogenide glasses because of its ease of formation, its excellent infrared transmission and its resistance to atmospheric conditions and chemical stability. Even though As and Sb belong to the same group of the periodic table, As_2S_3, As_2Te_3, and Sb_2S_3 do not display the same glass forming tendency [3]. Addition of As_2S_3 to Sb_2S_3 enhances the glass forming ability and glasses in the mixed As-S-Sb system can be formed [4].

The three-dimensional network of As_2S_3 glassy is built of trigonal pyramidal units $AsS_{3/2}$, which are interconnected through As-S-As bridges. The basic structural units of Sb2S3 glassy are the trigonal pyramidal arrangement $SbS_{3/2}$ bonded to each other by S atoms [4]. The crystal structure of Sb_2Te_3 crystals exhibits the layered atomic arrangement in the rombohedral structure which consists of three quintuplet layers (QLs) and each quintuplet layer contain five atoms in the order of Te^1-Sb-Te^2-Sb-Te^1 [5].

I. Tiginyanu et al. (Eds.): ICNBME 2021, IFMBE Proceedings 87, pp. 77–83, 2022.
https://doi.org/10.1007/978-3-030-92328-0_11

This work report some experimental results concerning the structural characterization of vitreous and polycrystalline semiconductors in As_2S_3-Sb_2S_3-Sb_2Te_3 system prepared by rapid quenching technique.

The semiconductors containing elements As and S are high photosensitive amorphous materials, and widely are used for recording media of optical information. On the other hand, semiconductors containing elements Sb and Te are well known as "ovonic" materials for volatile memory. We shoes more complicated materials containing As, S, Sb, Te in order to expect and combine these two advantages using a single compositions. The investigations in this direction will be continued.

2 Experimental

The bulk chalcogenide glasses and polycrystalline semiconductors in As_2S_3-Sb_2S_3-Sb_2Te_3 system were prepared from the elements of *6N* purity (As, Sb, S, Te) by conventional melt quenching method. Initial chemical elements were placed and mixed in quartz ampoules which then were evacuated up to pressure of $P \sim 10^{-5}$ Torr and sealed. At last, the mixtures were melted at 850–900 °C in rocking furnace during 10 h for homogenization, and then quenched at the room temperature. The following nanostructured semiconducting samples were synthesized and studied: $(As_2S_3)_{0.65}(Sb_2S_3)_{0.35}$, $(As_2S_3)_{0.35}(Sb_2S_3)_{0.65}$, $As_{11.2}S_{48.0}Sb_{28.8}Te_{12.0}$, $As_{12.6}S_{54.0}Sb_{27.4}Te_{6.0}$, $As_{20.8}S_{48.0}Sb_{19.2}Te_{12.0}$, $As_{23.4}S_{54.0}Sb_{16.6}Te_{8.0}$.

The X-ray diffraction (XRD) measurements were performed by DRON-UM1 diffractometer with *Fe-Kα* radiation (λ = 1.936 Å), with *Mn* filter by $\theta/2\theta$ scanning method and was also characterized by X-ray fluorescence analyzer, which permits to estimate the presence of different chemical elements and phases in the investigated material.

For optical transmission spectra measurements, a UV/VIS (λ = 300 ÷ 800 nm) CARLZEISS Jena production and Spectrum 100 FTIR Spectrometer (PerkinElmer) (ν = 650 ÷ 7800 cm^{-1}, λ = 15 ÷ 2.5 μm) were used. The surface morphology of all samples was examined by the scanning electron microscopy (SEM).

3 Results and Discussions

3.1 XRF Results

The XRF represents a device which allows registering the presence of different chemical elements from the periodic table and estimating its quantity. The energetic spectrums of the emitted K_α or K_β, radiation, are unique for individual chemical element, and the intensity of the emitted radiation correlate with the concentration of the specific atoms in the analyzed materials.

As an example in the Table 1 and Figs. 2, 3 are shown the experimental results obtained for the investigated above mentioned semiconductors.

Table 1. The XRF results for the $As_{20.8}S_{48.0}Sb_{19.2}Te_{12.0}$ composition (in the right is the SEM image of the respective composition).

$As_{20.8}S_{48.0}Sb_{19.2}Te_{12.0}$

Element	Conc mg/cm^2	Mole (%)
As	7.96	7.35
Sb	52.10	29.61
S	25.61	55.27
Te	14.32	7.77
Total	100 %	100 %

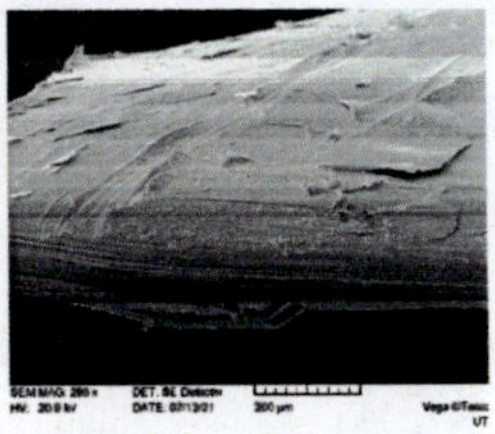

3.2 X-ray Diffraction Patterns and SEM Investgations

For the $As_{23.4}S_{54.0}Sb_{16.6}Te_{8.0}$ sample the peak with inter plane distances Sb_2S_3 with d = 3.22 Å and d = 2.86 Å have been detected. This peak has a maximum intensity. Another two maximum peaks with inter plane distances of d = 1.96 Å and d = 1.67 Å are characteristic for AsS_3 (Fig. 1).

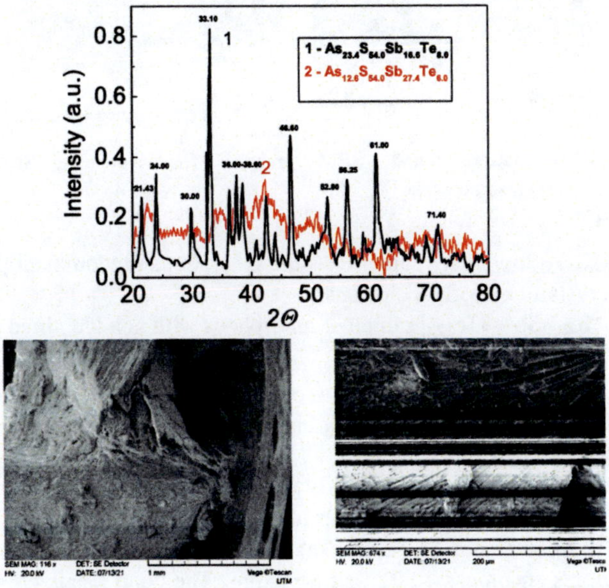

Fig. 1. The XRD diffraction pattern and SEM images for the $As_{23.4}S_{54.0}Sb_{16.6}Te_{8.0}$ and $As_{12.6}S_{54.0}Sb_{27.4}Te_{6.0}$ samples.

For the $As_{20.8}S_{48.0}Sb_{19.2}Te_{12.0}$ sample are presented the peaks with d = 3.21 Å ($2\theta \approx 35^0$) for Sb_2S_3 and d = 2.35–2.37 Å ($2\theta \approx 49° - 48°$, respectively) for Sb_2Te_3, and the maximum peak is for d = 3.157 Å ($2\theta \approx 34{,}4°$) (for Sb_2Te_3 units). The peaks with inter plane distances d = 3.23 Å (2θ≈34°), d = 3.13 Å (2θ≈35°), d = 2.1 Å (2θ≈55°) and d = 2.33 Å (2θ≈49°) are also refereed to the Sb_2S_3 units. A possible overlap of some peaks for Sb_2S_3 and Sb_2Te_3 is assumed.

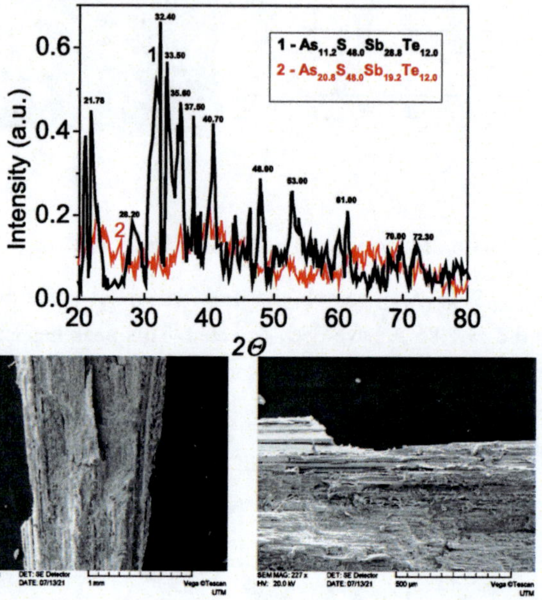

Fig. 2. The XRD patterns and SEM images for $As_{11.2}S_{48.0}Sb_{28.8}Te_{12.0}$ and $As_{20.8}S_{48.0}Sb_{19.2}Te_{12.0}$ samples.

The $As_{12.6}S_{54.0}Sb_{27.4}Te_{6.0}$ sample has the amorphous predominant phase with the peaks of the crystalline Sb_2S_3 inclusions.

The $As_{11.2}S_{48.0}Sb_{28.8}Te_{12.0}$ sample is amorphous with non identified peak with d = 5.72 Å (2θ≈20,2°).

The $(As_2S_3)_{0.35}(Sb_2S_3)_{0.65}$ sample has only peaks which belong to the Sb_2S_3 units.

Materials As_2S_3, As_2Se_3 and Sb_2Te_3 have been well studied and our studies are in good agreement with the literature data.

According to the [6] the monocrystals Sb_2Te_3, Sb_2Se_3 and Sb_2S_3 are well-known layered bulk structures with van der Waals interactions. It was shown, that, in the crystal structure of Sb_2X_3 (where is X = S, Se, Te) each Sb atom is surrounded by six X atoms and each X atoms encompassed by four Sb atoms. The bond lengths $d_{1,2}$ and $d_{3,4}$was determined to be 2.66/2.59 Å and 2.56/4.94 Å for Sb_2S_3 monolayer, and 2.95/3.13 Å and 2.99/3.02 Å for Sb_2Te_3 monolayer, respectively.

The SEM images show a layered structure of the investigated samples. Some layer has a dimensions of about of some micrometer (from 10 up to 50 μm thickness and about some cm in the lenght.

Some structural investigations of the chalcogenide glasses in the Sb_2S_3-As_2S_3-Sb_2Te_3 and $(As_2S_3)_{1-x}(Sb_2S_3)_x$ system was reported in [7–9].

3.3 Optical and Photoelectric Properties

Figure 3 illustrates the absorption spectra of some compounds of $(As_2S_3)_x(Sb_2S_3)_{1-x}$ system. With an increasing of the Sb_2S_3 structural units in the $(As_2S_3)_x(Sb_2S_3)_{1-x}$ system, the absorption edge is shifted to the lower photon energy, that correspond to the lower values of the optical band gap E_g.

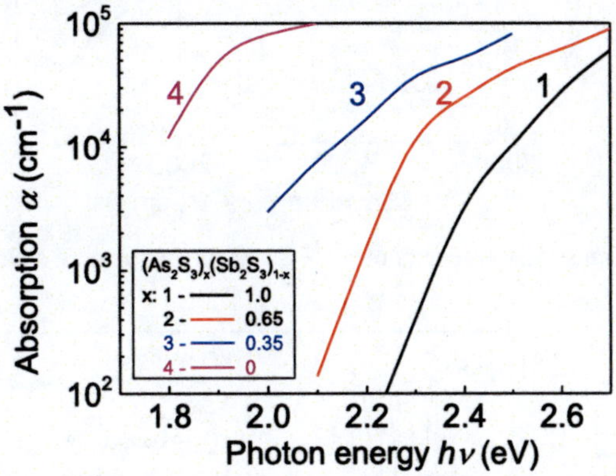

Fig. 3. Absorption spectra of some bulk compounds of the $(As_2S_3)_x(Sb_2S_3)_{1-x}$.system.

The optical gap E_g determined by extrapolation of the straight-line portions of the $(\alpha \cdot h\nu)^{1/2}$ vs. $(h\nu)$ graphs was found to be 2.34 eV for As_2S_3 [10], 2.1 eV for $(As_2S_3)_{0.65}(Sb_2S_3)_{0.35}$, 1.92 eV for $(As_2S_3)_{0.35}(Sb_2S_3)_{0.65}$, and 1.73 eV for Sb_2S_3. It was observed that in in chalcogenide glasses the absorption edge is broader than in crystalline analogues and this is caused by a broad energy distribution of electronic states in the band gap due to disorder and defects.

In the Urbach edge region ($\alpha \approx 1 \div 10^3$ cm^{-1}) the absorption coefficient spectra depend exponentially on the photon energy:

$$\alpha \propto \exp(\frac{h\nu}{\Delta_1}) \tag{1}$$

where Δ_1 is the parameter which characterizes the distribution of localized states in the band gap and it value for amorphous As-S material is about $\Delta_1 = 0.05$ eV [10]. No absorption bands was observed, besides the band located around $\nu = 2340$ cm^{-1}, generated by the presence of H_2S impurity.

Figure 4 shows the IR transmission spectra of investigated As-S-Sb-Te nanostructured semiconductors. The maximum transparency is observed for the chalcogenide glasses $(As_2S_3)_{0.35}(Sb_2S_3)_{0.65}$ and $(As_2S_3)_{0.65}(Sb_2S_3)_{0.35}$ (curves 6, 7). For the Sb_2Te_3 is observed reduced transparency (curve 1). It was established that in chalcogenide ternary system $(As_2S_3)_x(Sb_2S_3)_{1-x}$ introduction of crystalline Sb_2S_3 in vitreous As_2S_3, in the new formed alloys are present the trigonal structural units, and we observe a constant photosensibility in the region of photon energy $h\nu \geq E_g$ [11].

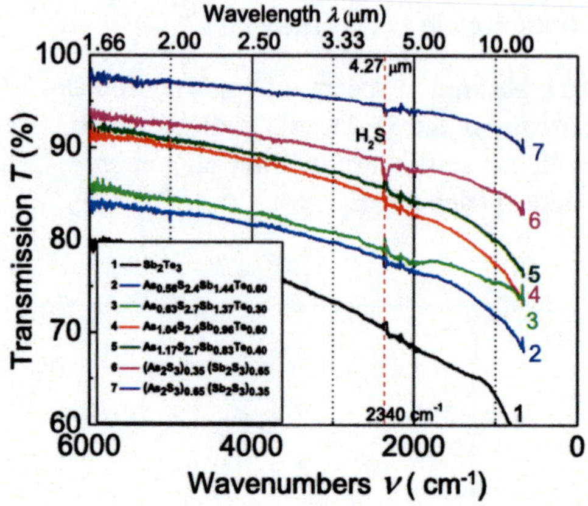

Fig. 4. Transmission spectra of As-S-Sb-Te nanostructured smiconductors.

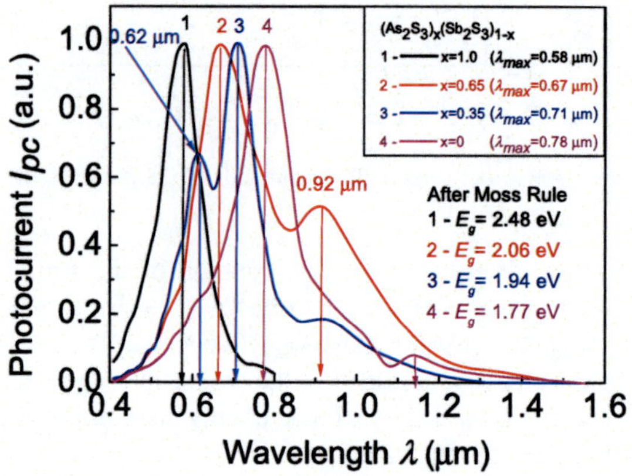

Fig. 5. Photocurrent spectra Ipc = f(λ) of some $(As_2S_3)_x(Sb_2S_3)_{1-x}$ chalcogenide glasses.

In the crystalline samples ($x \leq 0.15$) we observe an evident maximum in the region of absorption edge with a shoulder at the energies about $h\nu = 1.3$ eV ($\lambda = 0.92 \div 0.95$ μm), after that the photocurrent drop by the value 10^{-3} from the maxima at about 1.0 eV (Fig. 5). These particularities probably are provided by the new localized states induced by the introduction of atoms of Sb_2S_3 in the base matrix of the vitreous As_2S_3.

4 Conclusions

Some nanostructured layered semiconductors of As-S-Sb-Te system were investigated by XRF, XRD, SEM as well as by optical and photoelectric methods. The X-ray diffraction patterns and SEM images of studied materials show the presence of amorphous and nanocrystalline phases with the structural units AsS_3, Sb_2S_3, and Sb_2Te_3.

The transmission spectra in the region of wavenumbers $\nu = 1000 \div 6000$ cm^{-1} show a high transparence with only a single weak absorption band at $\nu = 2340$ cm^{-1}

caused of the presence H_2S impurity. The spectral characteristics of some alloys of $(As_2S_3)_x(Sb_2S_3)_{1-x}$ system indicate that with increasing of Sb_2S_3 in the system, the maximum of photosensibility is shifted in the region of long wavelength, and for some composition an additional maxima appear due to presence of crystalline Sb_2S_3 in matrix of vitreous As_2S_3.

Acknowledgment. This work was supported by the project ANCD 20.80009.5007.14.

Conflict of Interest. The authors declare that they have no conflict of interest.

References

1. Seddon, A.B.: Chalcogenide glasses: a review of their preparation, properties and applications. J. Non-Cryst. Solids **184**, 44–50 (1995)
2. Andriesh, A.M., Iovu, M.S.: Optical phenomena in chalcogenide glasses and their applications in optoelectronics. In: Lucovsky, G., Popescu, M. (eds.) Non-Crystalline Materials for Optoelectronics, INOE 2004. Optoelectronic Materials and Devices, vol. 1, pp. 155–210 (2004)
3. Cervinka, L., Hruby, A.: Structure of amorphous and glassy Sb_2S_3 and its connection with the structurevof As_2X_3 arsenic chalcogenide glasses. J. Non-Cryst. Solids **48**, 231–264 (1982)
4. Kamitsos, E.I., Kapoutsis, J.A., Culeac, I.P., Iovu, M.S.: Structure and bonding in As-S-Sb chalcogenide glasses by infrared reflectance spectroscopy. J. Phys. Chem. B **101**, 11061–11067 (1997)
5. Rajput, I., Rana, S., Jena, R.P., Lakhant, A.: Crystal growth and X-ray diffraction characterization of Sb2Te3 single crystal. In: AIP Conference on Proceedings, vol. 2100 (2019). https://doi.org/10.1063/1.5098624
6. Bafekry, A., Mortazavi, B., Faraji, M., et al.: Ab initio prediction of semiconductivity in a novel two-dimensional Sb_2X_3 (X=S, Se, Te) monolayers with orthorhombic structure (2021). www.nature.com/scientificreports. https://www.nature.com/articles/s41598-021-89944-4
7. N'dri, K., Coulibaly, V., Houphouet-Boigny, V., Jumas, C.: XRD and EDS characterization of some Sb_2S_3-As_2S_3-Sb_2Te_3 glasses prepared by rapid quenching method. J. Ovonic Res. **9**(4), 113–121 (2013)
8. Sava, F.: Structure and Properties of chalcogenide glasses in the system $(As_2S_3)_{1-x}(Sb_2S_3)_x$. J. Optoelectron. Adv. Mater. **3**(2), 425–432 (2001)
9. Stronski, A., et al.: The influence of composition on short-range order of amorphous As_2S_3-Sb_2S_3 chalcogenide alloys: a XRD and Raman study. Funct. Mater. **27**(2), 315–321 (2020)
10. Iovu, M.S., et al.: Spectroscopic studies of bulk As_2S_3 glasses and amorphous films doped with Dy, Sm, and Mn. J. Optoelectron. Adv. Mater. **3**(2), 443–454 (2001)
11. Popescu, M., Andriesh, A., Chiumach, V., Iovu, M., Shutov, S., Tsiuleanu, D.: The Physics of Chalcogenide Glasses. Ed. Stiintifica Bucharest - I.E.P.Stiinta, Chisinau (1996)

Photoluminescence Properties of $Eu(TTA)_3(Ph_3PO)_2$

O. Bordian[1](✉), V. Verlan[1], M. Iovu[1], I. Culeac[1], V. Zubareva[2], M. Enachescu[3], D. Bojin[3], and A. Siminel[1]

[1] Institute of Applied Physics, Chisinau, Republic of Moldova
olga.bordian@ifa.md
[2] Institute of Chemistry, Chisinau, Republic of Moldova
[3] CSSNT, University Politehnica of Bucharest, Bucharest, Romania

Abstract. Europium(III) coordination compound $Eu(TTA)_3(Ph_3PO)_2$ (1) {*TTA = thenoyltrifluoroacetonate*, Ph_3PO = *triphenylphosphine oxide*} has been prepared and characterized. Powder samples of the complex have been characterized by thermogarvimetric analysis (TGA), optical transmission and photoluminescence (PL) spectroscopy. PL have been registered for different temperatures in the range 11–300 K. The PL spectra was detected as specific narrow emission bands of internal transitions $4f \rightarrow 4f$ of the Eu^{3+} ion $^5D_0 \rightarrow {}^7F_j$ ($j = 0$–4). The major bands are centered at ca 580, 595, 615, 650 and 698 nm. PL data can be interpreted in the framework of the mechanism of energy transfer from the organic ligand matrix to Eu^{3+} ion.

Keywords: Eu^{3+} ion · Photoluminescence · Coordination compound · Life time

1 Introduction

Coordination compounds of rare earth metals are excellent materials for a wide range of applications because of their high efficiency, easy color tuning, temperature insensitivity, and high stability [1–3]. Among the most important rare earth compounds are the coordination complexes with Eu^{3+} ions [4, 5]. The narrow emission bands arising from the *4f* transitions, long lifetimes and relatively simple splitting pattern make Eu^{3+} ions attractive for PL spectroscopy and practical applications [5, 6]. Because of their excellent PL properties Eu^{3+} compounds are widely used in medicine, solar cells devices on flexible substrates, optical amplifiers, etc. [6–8].

Over the last years attention has been focused on studying the structure and luminescent properties of Eu^{3+} complexes with different donor ligands. Application of these donor ligands gives the possibility to modify the symmetry of the structure around Eu^{3+} ion and to allow transitions $4f \rightarrow 4f$ forbidden in symmetric spherical surround of ion Eu^{3+}, making luminescence uninterrupted without accumulation or loss of energy, and the application of the dissolution of coordination compounds in different solvents both polar and non-polar [9, 10].

I. Tiginyanu et al. (Eds.): ICNBME 2021, IFMBE Proceedings 87, pp. 84–91, 2022.
https://doi.org/10.1007/978-3-030-92328-0_12

In this paper we extend our previous study [9] of $Eu(TTA)_3(Ph_3PO)_2$ complex by expanded the temperature range of PL measurements 300–11 K, as well by carrying out TGA/DSC, TEM and SEM measurements.

2 Synthesis of Coordinating Compound $Eu(TTA)_3(Ph3PO)_2$

Synthesis of the coordinating compound $Eu(TTA)_3(Ph_3PO)_2$ **(1)** {*tris (thenoyltrifluoroacetonato) bis(triphenylphosphine oxide) europium(III)*}, was performed according to the scheme described in [11] with some modifications [9]. All chemicals of reagents were obtained from Sigma Aldrich and used without further purification. Reagents and solvents used in the synthesis of coordinating compounds with Eu^{3+} ions: *Europium trichloride* (*$EuCl_3$*, purity $\geq$ 99.9%), *triphenylphosphine oxide* (*Ph_3PO*, 98% purity), *Europium trichloride hexahydrate $EuCl_3 \cdot 6H_2O$*, 99.99% purity), 2-thenoyltrifluoroacetone/4,4,4-trifluoro-1-(2-thienyl)-1,3-butanedione (*TTA*, 99% purity), *Europium trinitrate hexahydrate $Eu(NO_3)_3 \cdot 6H_2O$*, 99.9% metal base).

Thermogravimetric (TGA) and derivative thermogravimetry (DSC) measurements, which was carried out in a temperature range 30–900 °C under N_2 gas atmosphere shows that the complex is relatively stable, which makes it a potential material for practical applications. The results of TEM/SEM microscopy indicate that the $Eu(TTA)_3(Ph_3PO)_2$ compound powder consists of nearly spherical grains of 5.7–20.3 nm in diameter. The non-uniform particle size distribution is possible cause by non-uniform distribution of temperature and mass during the synthesis. Absorption edge E_g for *TTA* (ionic ligand) and *TPPO* (non-ionic) ligand determined from transmission measurements are equals respectively to 3.06 eV and 3.72 eV [9].

The methods of measuring TGA, DSC, TEM / SEM and calculation of Fl parameters are described in detail in [10].

3 Photoluminescence Spectra and Discussion

Under near-UV excitation (405 nm) the $Eu(TTA)_3(Ph_3PO)_2$ complex exhibits a bright-red luminescence with dominant emission band at ca 615 nm. The PL spectra registered in the temperature range 11–300 K under excitation 405 nm are represented in Fig. 1. The PL emission spectra of the Eu^{3+} complex exhibit characteristic for these compounds narrow bands in the 575–725 nm region. These emission bands are related to the $^5D_0 \rightarrow {}^7F_j$ transitions of the Eu^{3+} ion. The five major peaks of the $^5D_0 \rightarrow {}^7F_j$ transitions are $^5D_0 \rightarrow {}^7F_{0\text{–}4}$. The bright red fluorescence under near-UV excitation is known to originate from $^5D_0 \rightarrow {}^7F_2$ transition of Eu^{3+} ion. While decreasing the temperature the maxima of the peaks slightly increase, as well as the resolution of the band splitting.

Figure 2 illustrates the high-resolution (0.2 cm^{-1}) emission spectrum for $^5D_1 \rightarrow {}^7F_{0\text{–}3}$ and $^5D_0 \rightarrow {}^7F_{0\text{–}1}$ transitions at 300 K. The bands, which are related to transitions from the higher excited state $^5D_1$ are commonly very weak in the Eu^{3+} compounds. The band at ca 580 nm, which refers to $^5D_0 \rightarrow {}^7F_0$ transition, belongs to the $4f \rightarrow 4f$ transitions with very small line width. In the case of the $Eu(TTA)_3(Ph_3PO)_2$ compound the line width of $^5D_0 \rightarrow {}^7F_0$ transition equals 0.48 nm.

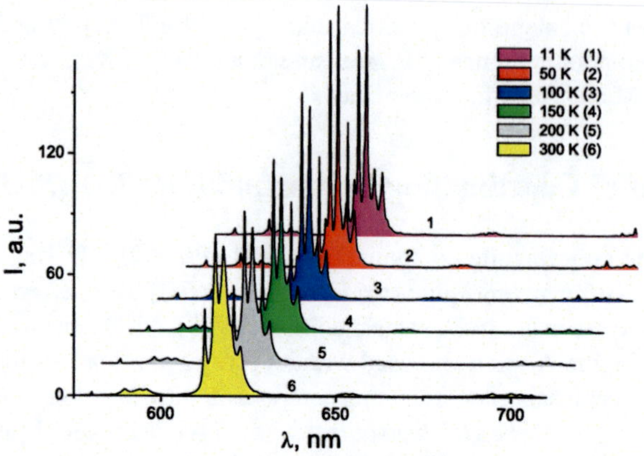

Fig. 1. PL spectra of thin sample layers of $Eu(TTA)_3(Ph_3PO)_2$ complex at different temperatures ($\lambda_{ex} = 405$ nm).

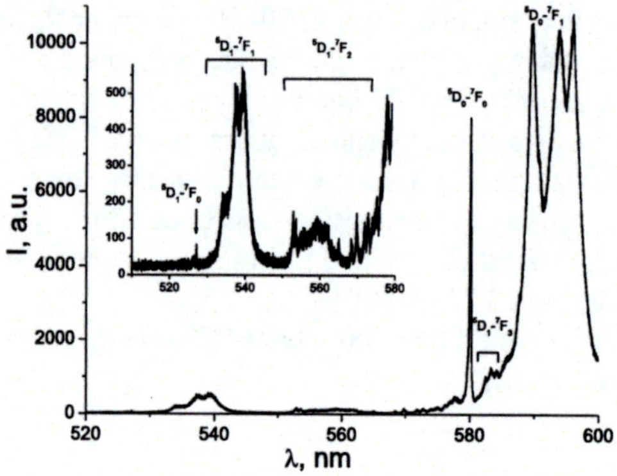

Fig. 2. High-resolution emission spectrum for $^5D_1 \rightarrow {}^7F_j$ and $^5D_0 \rightarrow {}^7F_{0\text{–}1}$ transitions at 300 K.

Figure 3 illustrates the PL emission band $^5D_0 \rightarrow {}^7F_0$ at different temperatures. The intensity of peak of $^5D_0 \rightarrow {}^7F_0$ transition slightly increases with temperature decrease, while the peak position to some degree shifts to lower energy. High-resolution emission spectrum of the $^5D_0 \rightarrow {}^7F_0$ transition (300 K) shows that the band actually consists of a major component A (ca 99% of integrated intensity) and a very weak component B. The component B (ca 1% of integrated intensity) is likely related to the defect states, and will be ignored in the discussion [12]. Observation of the characteristic $^5D_0 \rightarrow {}^7F_0$ transition suggests that the Eu^{3+} ion occupies a site with C_{nv}, C_n or C_s symmetry [13, 14]. Because both the excited state and ground state of the transition $^5D_0 \rightarrow {}^7F_0$ are nondegenerate, a single individual peak of the band $^5D_0 \rightarrow {}^7F_0$ is related to one distinct Eu^{3+} site [13].

The characteristic emission peaks of Eu^{3+} within the wavelength range from 585–600 nm correspond to the transition from the excited $^5D_0$ to $^7F_1$ levels, and are related

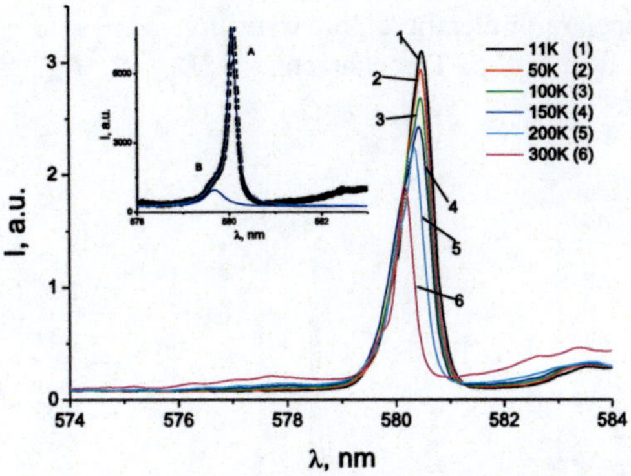

Fig. 3. High-resolution PL spectra of $^5D_0 \rightarrow {}^7F_0$ transition of $Eu(TTA)_3(Ph_3PO)_2$ complex at different temperatures ($\lambda_{ex} = 405$ nm). The insert shows the high-resolution spectrum of the transition for 300 K.

to magnetic dipole transition (Fig. 2). The magnetic dipole $^5D_0 \rightarrow {}^7F_1$ transition is considered insensitive to local site symmetry. The splitting pattern of the $^5D_0 \rightarrow {}^7F_j$ transitions in the luminescence spectrum of $Eu(TTA)_3(Ph_3PO)_2$ is compatible with low symmetry complex [13] (Fig. 4).

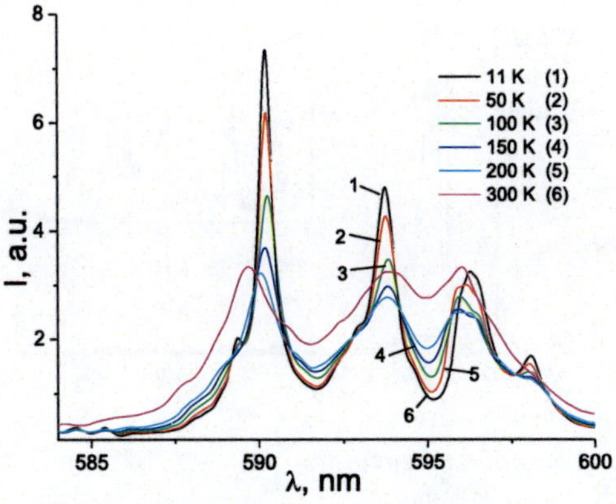

Fig. 4. PL emission spectra for magnetic dipole transition $^5D_0 \rightarrow {}^7F_1$ at different temperatures.

It is known that the $^5D_0 \rightarrow {}^7F_2$ transition is electric dipole allowed transition and is sensitive to a structural environment. The electric dipole transition $^5D_0 \rightarrow {}^7F_2$ is the dominant one in the PL emission spectrum of the complex $Eu(TTA)_3(Ph_3PO)_2$. The characteristic integrated intensity ratio (the symmetry parameter) $I(^5D_0 \rightarrow {}^7F_2)/I(^5D_0 \rightarrow {}^7F_1)$ ranges 17.4–17.8 in the temperature domain 11–300 K. Such high ration of the symmetry parameter commonly relates to complexes, where Eu^{3+} ion is not at an inversion centre. It also can be related to a highly polarizable environment around the Eu^{3+} ion [14, 15].

The emission spectra of electric dipole transition $^5D_0 \rightarrow {}^7F_2$ at different temperatures are illustrated in Fig. 5. The character of $^5D_0 \rightarrow {}^7F_2$ splitting points to low symmetry complex.

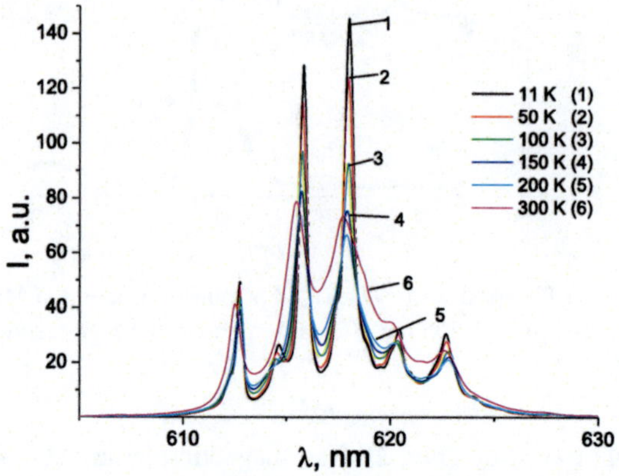

Fig. 5. PL emission spectra of electric dipole transition $^5D_0 \rightarrow {}^7F_2$ at different temperatures.

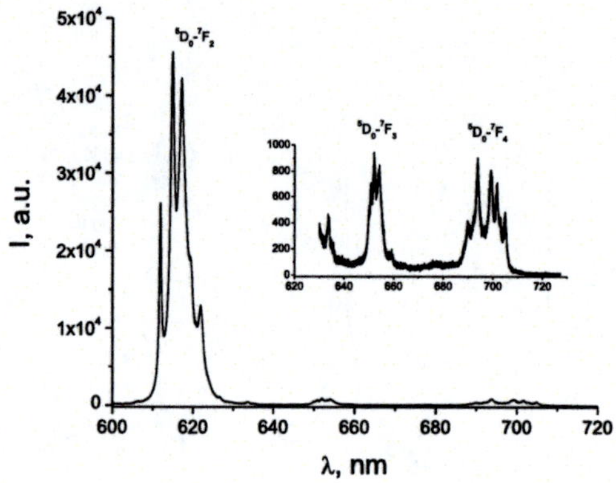

Fig. 6. High-resolution emission spectrum for $^5D_0 \rightarrow {}^7F_2$, $^5D_0 \rightarrow {}^7F_3$ and $^5D_0 \rightarrow {}^7F_4$ transitions. $T = 300$ K, $\lambda_{exc} = 405$ nm.

Two weak emission bands with barycenters at ca 650 and 698 nm are determined by the electric dipole transitions $^5D_0 \rightarrow {}^7F_3$ and $^5D_0 \rightarrow {}^7F_4$ (Fig. 6). Usually for Eu^{3+} compounds the $^5D_0 \rightarrow {}^7F_3$ transition is very weak, because it is forbidden according to the Judd–Ofelt theory [16, 17].

The other weak emission band with barycenter at ca 699 nm is related to $^5D_0 \rightarrow {}^7F_4$ transition. The number of the maximums at split of PL intensity of the $^5D_0 \rightarrow {}^7F_4$ transition is determined both by symmetry factors, as well as by the chemical composition of the host matrix [13].

Table 1 illustrates the integrated intensities of the $^5D_0 \rightarrow {}^7F_j$ ($j = 0$–4) transitions for different temperatures. It can be noted that in this temperature range the PL integrated intensity is almost constant.

Table 1. Integrated intensities of the $^5D_0 \rightarrow {}^7F_j$ ($j = 0$–4) transitions

T, K	Integrated intensity of Eu^{3+} transitions				
	$^5D_0 \rightarrow {}^7F_0$	$^5D_0 \rightarrow {}^7F_1$	$^5D_0 \rightarrow {}^7F_2$	$^5D_0 \rightarrow {}^7F_3$	$^5D_0 \rightarrow {}^7F_4$
10	2.1183	21.1348	367.9323	4.7150	12.5226
50	1.9608	20.7041	361.6521	5.3115	12.2244
100	1.8360	20.3745	356.3481	5.5462	12.2980
150	1.7759	20.0654	352.0475	5.3984	12.9192
200	1.3223	19.9642	361.8700	5.2203	13.1090
300	0.9097	20.0554	352.4475	5.2984	12.9192

Table 2 illustrates some optical and photoluminescent parameters of $Eu(TTA)_3(Ph_3PO)_2$ compound as determined from experimental data [10, 18]. These parameters are as follows: the radiative rate constant A_{rad}, the PL lifetime τ_{exp}, intrinsic quantum efficiency η, $\Omega_{0\rightarrow j}$ intensity parameters, and $\Delta\lambda$ – the FWHM of the transition $^5D_0 \rightarrow {}^7F_2$.

Table 2. Optical and PL parameters of $Eu(TTA)_3(Ph_3PO)_2$ (300 K)

E_g, eV	$\Delta\lambda$, *nm*	η, *%*	A_r, s−1	Ω_2, *cm*^{2}	Ω_4, *cm*^{2}
3.10	5.6030	53.46	1165.58	8.91×10^{-19}	9.63×10^{-20}

PL decay profile registered at 300 K under pulsed laser excitation 337 nm exhibits biexponential decay. Figure 7 illustrates PL kinetics profile registered for emission line at 615 nm. The characteristic life times are 0.18 ms and 0.36 ms. The most likely mechanism to explain this biexponential decay is a back transfer from the metal to the triplet state of the organic ligand [19]. A similar character of PL decay may be expected in the case of two different Eu^{3+} excited species [13, 20], which is not the case of $Eu(TTA)_3(Ph_3PO)_2$. For *RE* ions in the host matrix, there are four various excitation mechanisms: direct excitation of the RE ions through $4f \rightarrow 4$f transitions, and excitation through the matrix by charge transfer, or energy transfer [21].

High luminescence of $Eu(TTA)_3(Ph_3PO)_2$, narrow emission bands, along with long lifetimes make the compound attractive for potential various applications in optoelectronics and bio-medical needs.

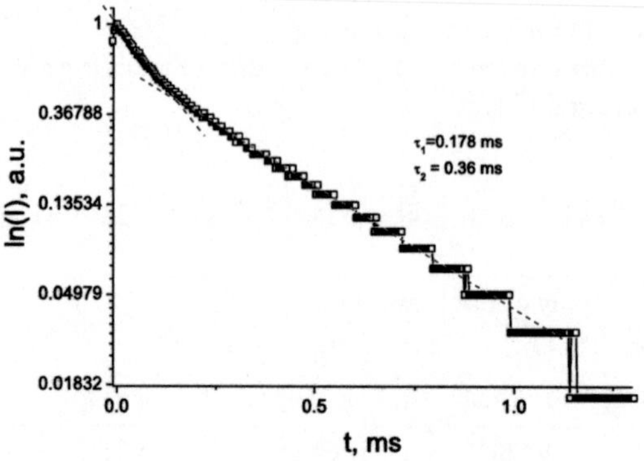

Fig. 7. PL decay profile in powder sample at 300 K measured at 611 nm under pulsed excitation 337 nm.

4 Conclusions

The coordination compound *Eu(TTA)*$_3$*(Ph*$_3$*PO)*$_2$ was synthesized and characterised. The compound shows high PL emission under excitation with near-UV radiation 405 nm. Analysis of the number of Stark components emission suggests a low symmetry coordination compound, however, it needs further confirmation by XRD investigations of the compound.

From the analysis of PL experimental data the luminescence parameters of the studied compound were determined: $\Delta\lambda$, emission probabilities of each transition, luminescence parameter, efficiency, etc.

The complex *Eu(TTA)*$_3$*(Ph*$_3$*PO)*$_2$ may find potential applications in optoelectronics, medicine and sensors developments.

Acknowledgements. This research is supported by the CSSDT/ANCD/MECC 15.817.02.03A, ANCD 20.80009.5007.1, ANCD 20.80009.5007.28 and Moldova - Belarus bilateral program NARD - GKNT.

Conflict of Interest. The authors declare that they have no conflict of interest.

References

1. Blasse, G., Grabmaier, B.C.: Luminescent Materials. Springer, Heidelberg (1994)
2. Bunzli, J.C.G., Choppin, G.R. (eds.): Lanthanide Probes in Life, Chemical, and Earth Sciences: Theory and Practice. Elsevier, Amsterdam (1989)
3. Spectroscopic Properties of Rare Earths in Optical Materials. Handbook on the Physics and Chemistry of Rare Earths, Amsterdam, The Netherlands, vol. 44, pp. 169–281 (2014)
4. Eliseeva, S.V., Bünzli, J.-C.G.: Lanthanide luminescence for functional materials and biosciences. Chem. Soc. Rev. **39**, 189–227 (2010)
5. Bünzli, J.-C.G.: On the design of highly luminescent lanthanide complexes. Coord. Chem. Rev. **293**, 19–47 (2015)

6. Jean-Claude, G.: Bünzli, lanthanide luminescence for biomedical analyses and imaging. Chem. Rev. **110**(5), 2729–2755 (2010). https://doi.org/10.1021/cr900362e
7. Binnemans, K.: Lanthanide-based luminescent hybrid materials. Chem. Rev. **109**, 4274–4283 (2009)
8. Armelao, L., et al.: Design of luminescent lanthanide complexes: from molecules to highly efficient photo-emitting materials. Coord. Chem. Rev. **254**, 487–505 (2010)
9. Bordian, O.T., et al.: Synthesis and optical characterization of *Eu(TTA)3(Ph3PO)2*. Moldavian J. Phys. Sci. **14**, N3-4 (2015)
10. Bordian, O., Verlan, V., Culeac, I., Bulhac, I., Zubareva, V.: Managing the luminescence efficiency of the organic compounds of *Europium(III)* through preparation technology. In: Proceedings of the SPIE 11718, Advanced Topics in Optoelectronics, Microelectronics and Nanotechnologies X, p. 1171817, 31 December 2020. https://doi.org/10.1117/12.2571186
11. Melby, L.R., Rose, N.J., Abramson, E., Caris, J.C.: Synthesis and fluorescence of some trivalent lanthanide complexes. J. Am. Chem. Soc. **86**, 5117 (1964)
12. Binnemans, K.: Interpretation of Europium(III) spectra. Coord. Chem. Rev. **295**, 1–45 (2015)
13. Coelho, A.C., et al.: Crystal structure and spectroscopic studies of a dimeric *Europium(III)* β-diketonate complex containing [3-(2-Pyridyl)-1-pyrazolyl]acetate. Eur. J. Inorg. Chem. 1284–1288 (2014)
14. Gorller-Walrand, C., Binnemans, K.: Rationalization of crystal-field parametrization. In: Gschneidner, K.A., Jr., Eyring, L. (eds.) Handbook on the Physics and Chemistry of Rare Earths, vol. 23, p. 121. North-Holland, Amsterdam (1996)
15. Zhang, X.-F., Xu, C.-J., Wan, J.: Mono- and dinuclear europium(III) complexes with thenoyltrifluoroacetone and 1,10-phenanthroline-5,6-dione. Monatshefte für Chemie – Chem. Monthly **145**(12), 1913–1917 (2014). https://doi.org/10.1007/s00706-014-1282-x
16. Judd, B.R.: Optical absorbtion Intensities of rare-earth ions. Phys. Rev. APS **127**(03), 750–761 (1962)
17. Ofelt, G.S.: Intensities of crystal spectra of rare earth ions. J. Chem. Phys. **37**(03), 511–520 (1962)
18. Verlan, V.I., Bordian, O.T., Iovu, M.S., Culeac, I.P., Zubareva, V.E.: Transfer of light energy from UV to visible domain in coordination compounds of *Europium(III)*. In: Luca, D., Sirghi, L., Costin, C. (eds.) INTER-ACADEMIA 2017. AISC, vol. 660, pp. 11–17. Springer, Cham (2018). https://doi.org/10.1007/978-3-319-67459-9_2
19. Reiller, P.E., Brevet, J.: Bi-exponential decay of Eu3+ complexed by Suwannee River humic substances: spectroscopic evidence of two different excited species. Spectrochim Acta A Mol. Biomol. Spectrosc. **75**(2), 629–636 (2010). https://doi.org/10.1016/j.saa.2009.11.02
20. Holz, R., Thompson, L.: Spectroscopically distinct geometrical isomers in a single crystal. Characterization of the eight-coordinate adducts of tris(dipivaloylmethanato)lanthanide(III) with 2,9-dimethyl-1,10-phenanthroline. Inorg. Chem. **32**(23), 5251–5256 (2002). https://doi.org/10.1021/ic00075a051
21. Banski, M., Podhorodecki, A., Misiewicz, J.: Influence of sol-gel matrices on the optical excitation of europium ions. Mater. Sci. Pol. **28**(1), 217–227 (2010)

Check for updates

Characteristics of Surface-Barrier Structures on Zinc Diarsenide with Hole Conductivity

I. G. Stamov(✉), D. V. Tkachenko, and Yu. Strel'chuk

Pridnestrovian State University, Tiraspol, Moldova

Abstract. The surface-barrier photosensitive structures based on zinc diarsenide crystals with metals have been produced. The electrical and photovoltaic properties of these structures were investigated. The photocurrent spectra and electrical characteristics are determined by the parameters of the semiconductor and the physicochemical properties of the contacting metals.

Keywords: Birefractive crystals · Schottky barriers · Photovoltaic properties · Electrical characteristics

1 Introduction

Photoactive structures based on birefractive crystals are of great interest for recording polarized radiation and as elements of a number of devices for polarization optoelectronics [1–3]. Currently, there are few works on heterojunctions and Schottky barriers based on semiconductors of groups 2-5 and 2-4-5, demonstrating the possibility of their application as detectors of linearly polarized radiation and current inverters controlled by light polarization [2, 4, 5]. Some of these crystals due to their individual characteristics are promising as solar energy converters and devices which use in the principles of their operation the birefractive properties of crystals in telecommunication systems and information processing on optical fibers (heterojunctions based on monoclinic diphosphide and zinc diarsenide with indium-tin oxides) [6, 7]. The high values of birefringence of monoclinic and tetragonal crystals, the specific rotation of the plane of polarization of the radiation of tetragonal crystals of group 2-5, the possibility of controlling their electrophysical properties in a wide range are undoubtedly promising objects for research from the point of view of designing on their basis compact devices for processing optical signals.

The results of studies of active structures on p-type zinc diarsenide are presented in this work. The hole type of conductivity and the concentration of charge carriers in undoped $ZnAs_2$ crystals produced by gas-phase methods and crystallization from the liquid phase are determined by their own defects [8, 9]. The energy of the acceptors in the band gap of the semiconductor is $E_a = 0.05$ eV.

There are not numerous data on the characteristics of rectifying and ohmic contacts based on materials A^2B^5 with hole conductivity [4, 5, 10]. The contacts of materials with large values of the work function of electrons $\chi > 4$ eV deposited on the surface of

I. Tiginyanu et al. (Eds.): ICNBME 2021, IFMBE Proceedings 87, pp. 92–99, 2022.
https://doi.org/10.1007/978-3-030-92328-0_13

zinc diarsenide do not have rectifying properties. The linearity of the I - V characteristic of such contacts is fulfilled for a wide range of applied voltages, up to the beginning of effective heating of crystals by Joule heat and in a wide temperature range (77–550) K. The fused contacts based on such materials have a lower noise level. The heights of barriers on $ZnAs_2$ with metals are low −0.2 eV for contacts with Ni and Cu and 0.35 eV with in [10].

The studies were carried out on crystals produced from the gas phase and by the Bridgman – Stockbarger method. The concentration of charge carriers was 10^{19}–10^{21} m^{-3}. The surface-barrier metal-semiconductor structures were produced by sputtering metals in high vacuum onto crystal surfaces treated in a polishing etchant [10]. The ohmic contacts to the structures were created by fusing tin at a temperature of 300 °C for 15 min. To produce photoactive structures, there were chosen metals Ca and Ce with low values of the work function of electrons. The metal films deposited in vacuum on the crystal surface had a thickness of 6–50 nm and were protected from the influence of atmospheric oxygen during measurements.

2 General Results

Figure 1 shows the temperature dependences of the resistivity of $ZnAs_2$ crystals obtained by different methods, from the gas phase and by the Bridgman - Stockbarger method. The depth of the acceptors from studies of the electrophysical properties of crystals is 0.05 eV.

The current-voltage (I-V curves) and capacitance-voltage (C-V curves) characteristics of the Ca - $ZnAs_2$ structure are shown in the Fig. 2. The features of the electrical characteristics are associated with the fact that the height of the potential barrier is much less than the band gap and the resistance of the base of the structure is large enough due to the low mobility of charge carriers. The current of the forward branch of the I - V characteristic is limited by the resistance of the base of the structure; the reverse current increases with increasing voltage, since it is determined by the charge transfer through the barrier, but not by generation - recombination processes.

Despite the small value of the barrier height and the significant part of the active component in the admittance, the capacitive component is easily determined and satisfies to the dependence $C \sim (\varphi_b - q\,U)^{1/2}$. The barrier height φ_b determined from this dependence is 0.53 eV.

The photoelectric effect in rectifying metal-semiconductor structures occurs as a result of the separation of nonequilibrium charge carriers, excited by light, by the field of a potential barrier. The photocurrent of the Schottky barrier consists of two components, electrons generated in the space charge region (SCR) and electrons that reach the SCR boundary by diffusion from the region adjacent to this boundary, with a width equal to the diffusion length of nonequilibrium electrons. The SCR width is determined by the formula [11]:

$$W = \left[2 \cdot \varepsilon \cdot \varepsilon_o \cdot (\varphi_b - q \cdot U)/q^2 \cdot N_a\right]^{1/2} \quad (1)$$

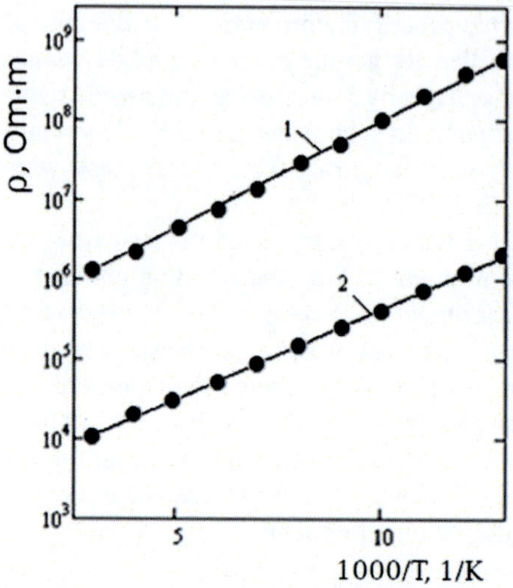

Fig. 1. The temperature dependences of the resistivity of $ZnAs_2$ crystals with hole conductivity produced: 1 - from the gas phase, 2 - by the Bridgman – Stockbarger method.

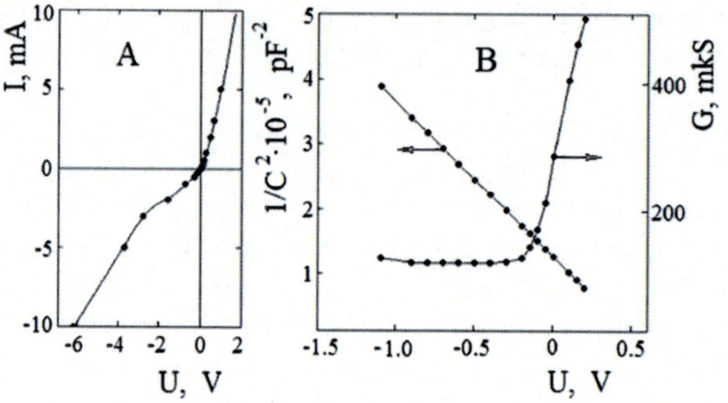

Fig. 2. A – I-V characteristic of Ca - $ZnAs_2$ barrier at room temperature, B - characteristics of admittance on the frequency - 465 Hz.

where: W is the width of the space charge region, ε is the dielectric constant of the semiconductor, ε_o is the dielectric constant of vacuum, φ_b is the height of the potential barrier, q is the electron charge, U is the voltage applied to the barrier, N_a is the concentration of ionized acceptors.

The photocurrent is determined as follows [12]:

$$I_{ph} = q \cdot g \cdot (1 - R) \cdot \eta \cdot \left[1 - e^{-\alpha} \cdot W/(1 + \alpha \cdot L)\right] \quad (2)$$

where: g is the rate of generation of photons, R is the reflection coefficient of light, η is the quantum yield of the photoelectric effect, α is the absorption coefficient of light, L is the diffusion length of minority charge carriers. If the condition $1 \gg \alpha \cdot L$, $I_{ph} \sim \alpha \cdot (L + W)$ is satisfied, i.e. the photocurrent is determined by the absorption coefficient and depends on the applied voltage if the SCR width is greater or comparable with the diffusion length of nonequilibrium charge carriers.

The photocurrent spectra of Ca-$ZnAs_2$ surface-barrier structures in polarized light are presented in the absence of a bias voltage applied to the barrier in Fig. 3. In the E||C polarization, the edge of the photocurrent is characteristic for the case of direct permitted transitions to the exciton band with a significant exciton binding energy (peak e_1 - 0.965 eV) [13–17]. In the depths of the fundamental absorption region, an increase of the photocurrent is observed due to the transitions e_2 - 1.399 eV and e_3 - 1.494 eV. For the polarization E⊥C, the energy peak e_5 - 0.965 eV has an insignificant intensity up to the edge of the photocurrent caused by this peak, a "step" e_4 (0.934 eV) appears. The beginning of this step corresponds to the absorption edge for the polarization E||C, the end of the step does to the beginning of the growth of the absorption for the polarization E⊥C. The edge of the photocurrent for the E⊥C polarization is shifted relative to the edge for the E||C polarization by 8.2 nm. The spectrum curve for the polarization E⊥C reflects the dependence of the absorption coefficient for direct forbidden transitions taking into account the features of the absorption edge for this polarization mentioned earlier. In the range 1.4..1.5 eV, two bends appear at energies a_2 - 1.399 eV and a_3 - 1.494 eV. These transitions correspond in energy to the transitions e_2 and e_3. The energy difference between these peaks corresponds to the gap (95 meV) between the *V*2.3 bands [17]. The transitions E_3, E_4, and E_5 were determined from the reflection spectra for polarized light [7] in the region under study, but our results coincide with a feature in the spectra e_3, which corresponds to electronic transitions in the band diagram 3 (*V*2.3) - *L*1.

The transitions e_1 and a_1 in the photocurrent spectra are associated with absorption bands in crystals for the polarization E⊥C (the a_1 band, as well as the features e_2, e_3, a_2, a_3 at high photon energies are not seen due to strong absorption in crystals for the polarization E||C in the transmission spectra). The "step" a_0 with an energy of 0.934 eV, a structured band in the region (0.896–0.903) eV, present in the absorption spectra is not detected in the photocurrent spectra.

As follows from the presented dependences, the region of linear dichroism in zinc diarsenide is 0.97–1.5 eV. The photopleochroism coefficient P [18] varies from maximum values in the region of the fundamental absorption edge to zero in depth. Other data are given for the spectral distribution of the photopleochroism coefficient in [4] from the spectral dependence of P and the region of linear dichroism. It is perhaps explained by measurements of the spectral characteristics of the photoconductivity rather than the photocurrent of the Schottky barrier.

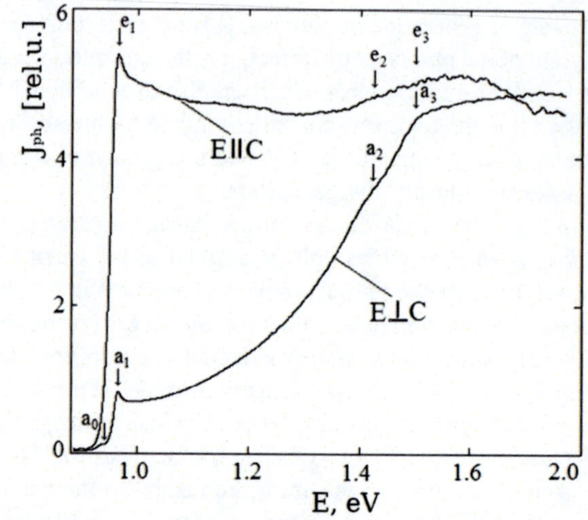

Fig. 3. The photocurrent spectra of the Ce - $ZnAs_2$ barrier in polarized light.

Figures 4 and 5 show the photocurrent spectra of Ca-$ZnAs_2$ rectifying structures for polarized light at several bias voltages.

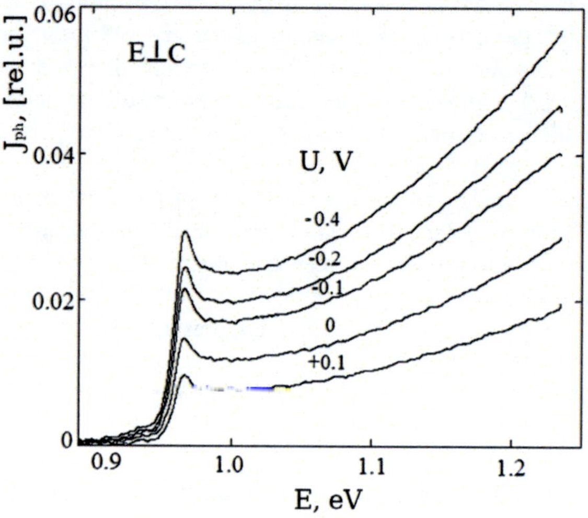

Fig. 4. Photocurrent spectra of Ce - $ZnAs_2$ in polarized light (E⊥C) and bias voltages in the range (+0.1 ÷ −0.4) V.

As the reverse bias on the barrier increases, the current increases due to the expansion of the space charge region in the semiconductor. The sign of the applied bias corresponds to the reverse bias for the barrier on the p-type semiconductor.

The dependences of the photocurrent on the applied voltage to the barrier are shown in Fig. 6. It follows from these dependences that the photocurrent is determined by the

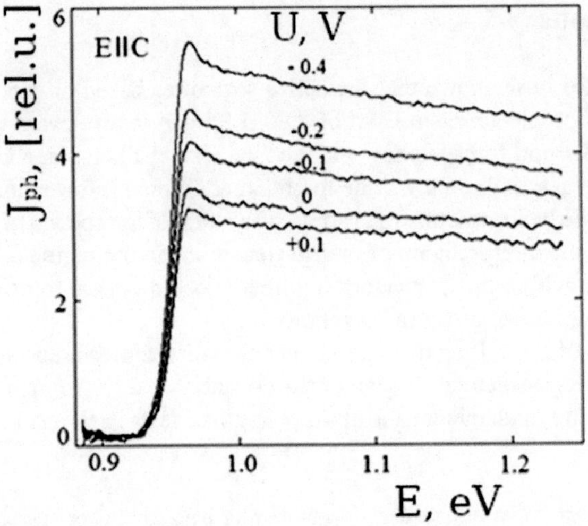

Fig. 5. The photocurrent spectra of Ce - $ZnAs_2$ in polarized light (E||C) and bias voltages in the range (+0.1 ÷ − 0.4) V.

width of the space charge region and its dependence on the voltage applied to the barrier. In addition, the diffusion length of minority charge carriers is less than the width of the space charge region.

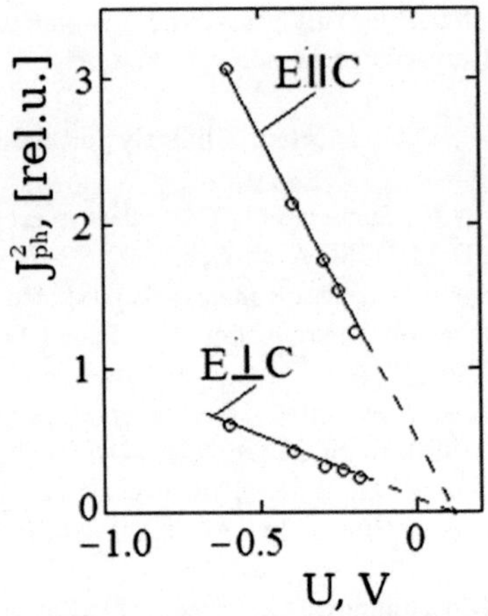

Fig. 6. The dependences of the photocurrent on the bias voltage of the C - $ZnAs_2$ barrier in polarized light at 1284 nm wavelength.

3 Conclusions

Thus, our studies have shown that the active structures based on zinc diarsenide have a fairly wide photopleochroism band of 0.86–1.35 μm in a relevant wavelength range - the first and second transparency windows of optical fibers. In addition, a study of the spectral characteristics allows one to obtain additional information on their optical properties and the band structure of crystals. The study of the spectra of the photoelectric effect confirms the interpretation of optical transitions in the region of the fundamental absorption edge within the framework of direct allowed optical transitions for E||c and direct forbidden transitions for polarizations E⊥c.

The spectra of crystals from a melt do not contain features associated with impurity absorption in the transparency region of the crystals, weakly pronounced fine structure in the region of the fundamental absorption edge manifests itself against the background of noise.

Conflict of Interest. The authors declare that they have no conflict of interest.

References

1. Lazarev, V.B., Shevchenko, V.Ya., Grinberg, Ya.Kh., Sobolev, V.V.: Semiconductor compounds A^2B^5. Moscow (1978)
2. Syrbu, N.N.: Optoelectronic properties of compounds of group A^2B^5. Chisinau (1983)
3. Marenkin, S.F., Trukhan, V.M.: Phosphides, zinc and cadmium arsenides. Minsk, publisher Varaksin A N, p. 224 (2010)
4. Nikolaev, Yu.A., Rud, V.Yu., Rud, Yu.V., Terukov, E.I.: Polarization photosensitivity of Schottky barriers on ZnAs2 crystals of monoclinic modification. Zhurnal tekhnicheskiy fiziki **79**(11), 36–39 (2009)
5. Syrbu, N.N., Stamov, I.G.: Photodetectors of linearly polarized radiation. FTP **25**(12), 2115–2125 (1991)
6. Radautsan, S.I., Syrbu, N.N., Stamov, I.G.: Photoelectric properties of ZnP_2 $(D^8{}_4)$ - ZnP2 $(C^5{}_{2h})$ heterojunctions. Dokl. USSR Acad. Sci. **23**(1), 72–74 (1977)
7. Mowles, T.: High efficiency solar photovoltaic cells produced with inexpensive materials by processes suitable for large volume production. U.S. Patent №: U.S. 6,541, 695 B1, pp 1–16, 1 April 2003
8. Ugai, Ya.A., Zyubina, E.A.: Anisotropy of thermal conductivity, thermoEMF and resistivity ZnAs2. Izv. Acad. Sci. USSR: Inorg. Mater. **2**(1), 9–16 (1966)
9. Nadtochiy, Yu.G., Pishchikov, D.I., Burtsev, Yu.N., et al.: Anisotropy of thermal conductivity, thermoEMF and resistivity of ZnAs2. Izv. Acad. Sci. USSR: Inorg. Mater. **98**(2), 293–297 (1991)
10. Yu, K.A., Kudintseva, G.A., Stamov, I.G., Syrbu, N.N.: Physical phenomena in metal-A^2B^5 Schottky diodes. Phys. Technol. Semicond. **19**(1), 28–31 (1985)
11. Roderick, E.H.: Metal-semiconductor contacts. Radio and Communication, Moscow (1982)
12. Out, I., Gentsov, D., Herman, K.: Photoelectric phenomena. Mir, Moscow (1980)
13. Khakimov, K., Vavilov, V.M., Marenkin, S.F., Chukichev, M.V.: Cathodoluminescence of $ZnAs_2$ crystals. Phys. Technol. Semicond. **21**(8), 1447–1451 (1987)
14. Marenkin, S.F., Pishchikov, D.I., Chumaevsky, N.A., et al.: Optical transmission of zinc diarsenide single crystals. Izv. Acad. Sci. USSR. Inorg. Mater. **25**(6), 1039–1040 (1989)

15. Morozova, V.A., Marenkin, S.F., Koshelev, O.G.: Determination of the parameters of the band structure of $ZnAs_2$. Moscow Univ. Bull. Series 3. Phys. Astron. **2**, 49–52 (2000)
16. Morozova, V.A., Semenenya, T.V., Marenkin, S.F., Koshelev, O.G., Chukichev, M.V.: Series of free exciton lines in the transmission spectra of zinc diarsenide. Moscow Univ. Bull. Series 3. Phys. Astron. **5**, 86–88 (1996)
17. Stamov, I.G., Syrbu, N.N., Zalamai, V.V., Dorogan, A.: Excitonic polaritons of zinc diarsenide single crystals. Physica B.: Condensed Matter **506**, 183–191 (2017)
18. Rud, Yu.V.: Photopleochroism and physical principles of creating semiconductor photodetectors. Izv. Universities USSR. Phys. **XXIX**(8), 68–83 (1986)

Nanomodification of the Activated Concrete Mixture in Magnetofluidized Layer

V. P. Gonciaruc[1](✉), O. A. Bolotin[2], M. K. Bologa[1], E. G. Vrabie[1], and A. A. Policarpov[1]

[1] Institute of Applied Physics, Chisinau, Moldova
[2] Institute of Geology and Seismology, Chisinau, Moldova

Abstract. The results of concrete hardening due to nanomodification using graphene of a mixture of sand and Portland cement, activated in a magnetofluidized layer, are presented. The magnetofluidized layer is a suspension of needle-shaped ferromagnetic elements in a rotating electromagnetic field. As a result of intensive movement and constrained impact between ferromagnetic elements and concrete particles, sand and cement are finely ground. The results of studies of structural changes in quartz sand and Portland cement during mechanical activation in a magnetically fluidized layer are presented and analyzed. Changes in the surface physical and mechanical properties of sand and cement in comparison with the initial samples were determined by X-ray diffractometry and IR spectroscopy. During the nanomodification of concrete, difficulties arise in the uniform distribution of graphene in the total volume due to the very small size of nanoparticles and its small amount. The magnetofluidized layer provides a high degree of mixing, which makes it possible to evenly distribute graphene in the volume of the concrete mixture components, while exerting an electromagnetic effect and crushing the raw components.

Keywords: Nanomodification · Graphene · Magnetofluidized layer · Concrete

1 Introduction

Graphene-based nanotechnology has been rapidly developing over the past decade. Graphene is a single-layer, one-atom-thick material consisting of carbon atoms packed into hexagons on a plane. In 2010, the Royal Swedish Academy of Sciences awarded the Nobel Prize in Physics to Andre Geim and Konstantin Novoselov (University of Manchester, UK) "for groundbreaking experiments regarding the two-dimensional material graphen" [1].

Graphene has become not only the subject of a huge number of scientific publications, but also many promising practical applications. Considerable attention is paid to the prospects for the use of nanotechnology in the production of building materials [2–8].

I. Tiginyanu et al. (Eds.): ICNBME 2021, IFMBE Proceedings 87, pp. 100–109, 2022.
https://doi.org/10.1007/978-3-030-92328-0_14

It is known [9] that the addition of carbon nanostructures ensures the formation of crystalline hydrates and the formation of a micron-level fibrillar microstructure, which leads to an improvement in the physical and mechanical characteristics of concrete.

It is important for the construction industry that graphene prossesses enormous mechanical strength. In England, for the construction of a new gym, concrete slabs were made from an experimental mixture of Concretene [10]. This is the first building experience in the world using concrete containing graphen. The new material is about 30% stronger than conventional concrete. According to Nationwide Engineering experts, replacing conventional concrete with Concretene can reduce CO_2 emissions from cement production by 25%.

A promising direction in the production of building materials is the use of two-stage nano-modification: at the first stage, the materials are mechanically activated (grinding), and at the second stage, carbon nanostructures are added to the mixture.

It is known [11] that grinding the components of a concrete mixture (cement and sand) leads to an increase in the strength and speed of concrete hardening, and an increase in the specific surface area of both inert and binding components contributes to an increase in their reactivity.

When using nanomaterials, their agglomeration occurs, which complicates their uniform distribution in the volume of the concrete mixture. As a result, the homogeneity of the mixture and the physical and mechanical properties of concrete are reduced. The problem arises of uniform distribution of a small amount of a finely dispersed substance in the volume of the modified material. Therefore, there is a need for efficient mixing of the composite.

The problem of mechanical activation of the components of the concrete mixture, as well as the uniform distribution of non-material in the volume, can be solved using a magnetofluidized layer.

If ferromagnetic needle-like elements are placed in a rotating electromagnetic field of sufficient strength, then they come into a complex, at first glance, chaotic motion, creating a kind of vortex layer. The authors named it magnetofluidized layer. In this layer, the medium being processed is intensively mixed and crushed with the simultaneous effect of an electromagnetic field, local high pressure of acoustic vibrations.

The works [12, 13] show the possibility of using a magnetofluidized layer in the processing of building materials, and the authors of [14] propose to use this layer to solve a number of problems associated with the difficulty. of uniform distribution of nanoparticles in the volume of concrete and mortar mixtures.

The aim of this work is to study a two-stage modification of a concrete mixture in a magnetofluidized layer.

2 Installation and Research Methods

The experimental installation (Fig. 1) includes a rotating electromagnetic field inductor 1 inside which a stainless steel pipe 2 is placed, which is closed with two covers 3.

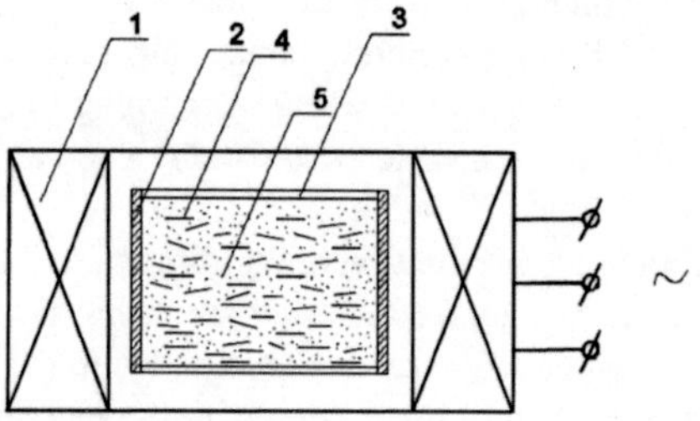

Fig. 1. Scheme of the experimental installation

The upper cover is equipped with a branch pipe for feeding bulk material, and the lower one - with a mesh and a branch pipe for removing the processed material (not shown in the figure). Inside the pipe 2, which is the working space, are placed the cylindrical ferromagnetic elements 4 and the particles of the processed material 5. As ferromagnetic elements 4, pieces of high-carbon steel wires with a hardness of 3,5- were used. 4.0 GPa, diameter 2.5 mm and ratio of length to diameter of the element $l/d = 10$. Portland cement PC 400-D20 manufactured by LafargeCimentMoldova S.A. was used as the processed material 5. and sand from local quarries, and after their mechanical activation, graphene was added.

The principle of operation of the installation is the following: at the moment of connecting the inductor 1 to the electric network in its coils a rotating electromagnetic field appears. Under the action of this external magnetic field in each ferromagnetic element 4 a magnetic moment of its own is induced (it is magnetized) and it begins to rotate. Due to the dipole-dipole interaction, as well as the collisions between them and with the pipe walls, the ferromagnetic elements are also involved in an intensive translational movement. This forms the magnetofluidized layer. At the same time, the layer as a whole performs a rotational movement in the direction of rotation of the external electromagnetic field. As a result of an intense chaotic movement, the ferromagnetic elements 4 fluidize and grind the processed material 5.

Ferromagnetic elements during their rotation in the magnetofluidized layer play the role of elementary stirrers. Therefore, in addition to mechanical activation, they also carry out intensive mixing of the material, and when adding graphene, they evenly distribute it in the volume of the components of the concrete mixture.

Structural changes in materials after their mechanical activation in a magnetofluidized layer were determined by X-ray phase analysis and IR spectroscopy on a DRON-3 diffractometer using FeKα radiation, the scanning interval was 3-74°2θ at a rate of 4°/min. and on the IR - Fourier spectrometer PE-100 Perkin - Elmer in the range 4000–650 cm^{-1}.

The efficiency of mechanical activation of the components of the concrete mixture, as well as their nanomodification by means of graphene, was evaluated by the change in the strength of concrete under compression ***fck, cube***, MPa. For this, specimens of a 20 × 20 × 80 mm beam were made from the components of the concrete mixture processed in the magnetofluidized layer and tested after 28 days of hardening according to GOST 10180-90.

3 Results and Discussion

Grinding of sand and cement in a magnetically fluidized layer is very intensive. The small contact surfaces during impact and the random position of the impact point relative to the symmetry axis of the colliding ferromagnetic elements contribute to the fact that even at low speeds, very large forces are developed. At the same time, the number and frequency of collisions between ferromagnetic elements are very high. This probably explains why the shredding is so intense. It is known that the grinding of a hard material is all the more intense, the higher the applied force, the shorter its duration of action and the higher the frequency of application of the force.

Initially, mechanoactivation was investigated when grinding sand in a magnetofluidized layer. It is shown that the average and maximum sizes of sand grains decrease rather quickly. Quartz sand is difficult to grind due to its high hardness (6–7 units on the Mohs scale). Because in the traditionally used equipment - ball mills, the abrasion grinding method is used, which is a less efficient grinding method for sand, grinding the sand in a magnetofluidized layer is quite appropriate. The peculiarities of the magnrtofluidized layer are the high frequency and force shock loads, as well as the friction, which lead not only to the crushing of the sand, but also to the considerable superficial activation of its particles due to the deformation of the crystalline lattice.

X-ray diffraction patterns (Fig. 2) of inactivated quartz sand samples demonstrated the presence of β quartz in general and a mixture of calcium carbonate and feldspar. After activation, the diffractogram (X-ray diffraction pattern) demonstrated a significant increase in the intensity of quartz sand reflections in the region of 0.333 nm and the widening of the quartz peaks to a change in activation time from 2 to 6 min. According to preliminary studies, it has been assumed that activation in the magmetofluidized layer leads to changes in the structure of quartz sand (partial amorphization, increase in specific surface area, manifestation of latent phases).

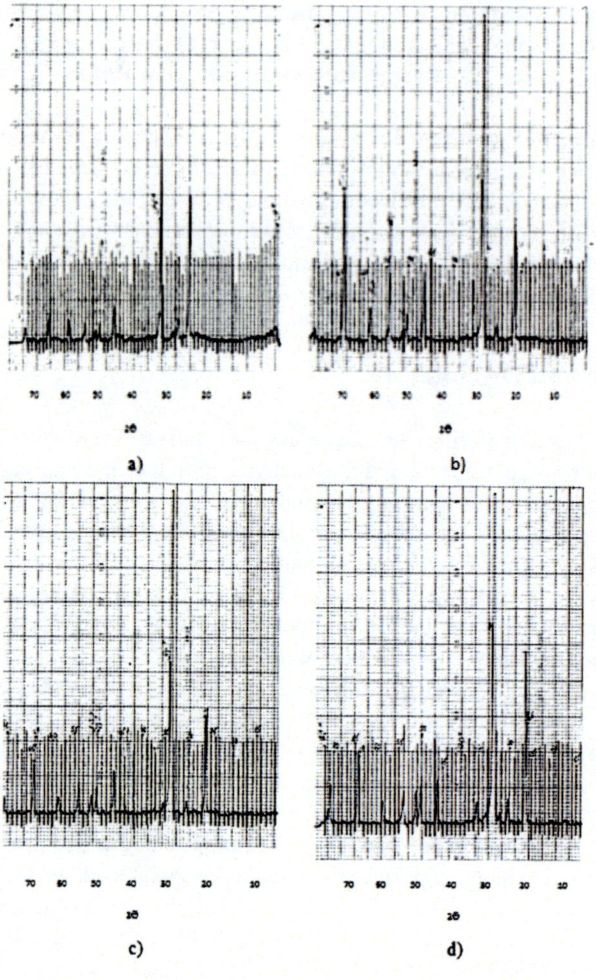

Fig. 2. Initial and activated quartz sand diffraction patterns in magnetofluidized layer: a) initially and activated: b) 2 min.; c) 4 min.; d) 6 min.

The diffraction patterns of the starting material show well-resolved systems of reflections, which indicate a high crystallinity of the phases in the state under study. Reflexes, which correspond to the initial crystalline state, demonstrate a significant decrease in their intensity in comparison with the samples subjected to mechanical activation in a magnetofluidized layer.

Another factor of the influence of mechanoactivation is, established by X-ray phase analysis, an insignificant expansion of the main peak of quartz, caused by deformation of the crystal lattice and the accumulation of internal stresses in it, which leads to amorphization of the quartz surface and contributes to an increase in its reactivity [15].

For a more complete understanding of the features of changes in quartz sand during mechanical activation, IR spectral studies of silica were of interest. For comparison, the IR spectra are shown in Fig. 3. The lines of the spectra of the samples are changed, which indicates the nature of the structural transformations, and the change in the properties of the obtained samples.

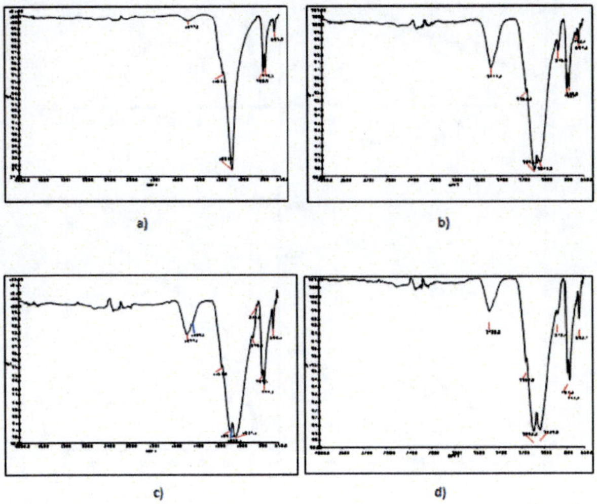

Fig. 3. IR spectra of initial quartz sand and activated in magnetofluidized layer: a) initially and activated: b) 2 min.; c) 4 min.; d) 6 min.

The researched samples show a spectral pattern of β-quartz: an intense band in the region 1167–1080 cm^{-1}, an average intensity of 803–780 cm^{-1} doublets and a weak band of 695 cm^{-1}.

Below are the isotherms of nitrogen adsorption on activated quartz sand in magnetofluidized layer (Fig. 4).

The isotherms obtained are typical for mesoporous sorbents and correspond to type IV according to the Brunauer, Deming et Teller classification. The shape of the hysteresis loop suggests the presence of slit-shaped pores, as indicated by the step on the desorption branch in the region of 0.5 P/P0 for the initial, inactivated sample. In the process of activating the quartz sand for 6 min., the edge of the top tip is smoothed. At P/P0 in region 1, the isotherms show a sudden increase in the sorption curve, which probably indicates the presence of large pores in the sand samples. This fact is also confirmed by the pore size distribution curves (Fig. 4 (1–4)). Their volume increases upon activation in the magnetofluidized layer.

The graphs show an increase in the number of particles with an effective pore radius of 20 Å at an activation time of up to 6 min. This, in turn, leads to an increase in the reactivity of quartz sand when it interacts with the cement, thus obtaining an active binder.

In the manufacture of ordinary Portland cement concrete, only a third of the volume of cement enters the hydration reaction, the rest of the mass remains a simple inert

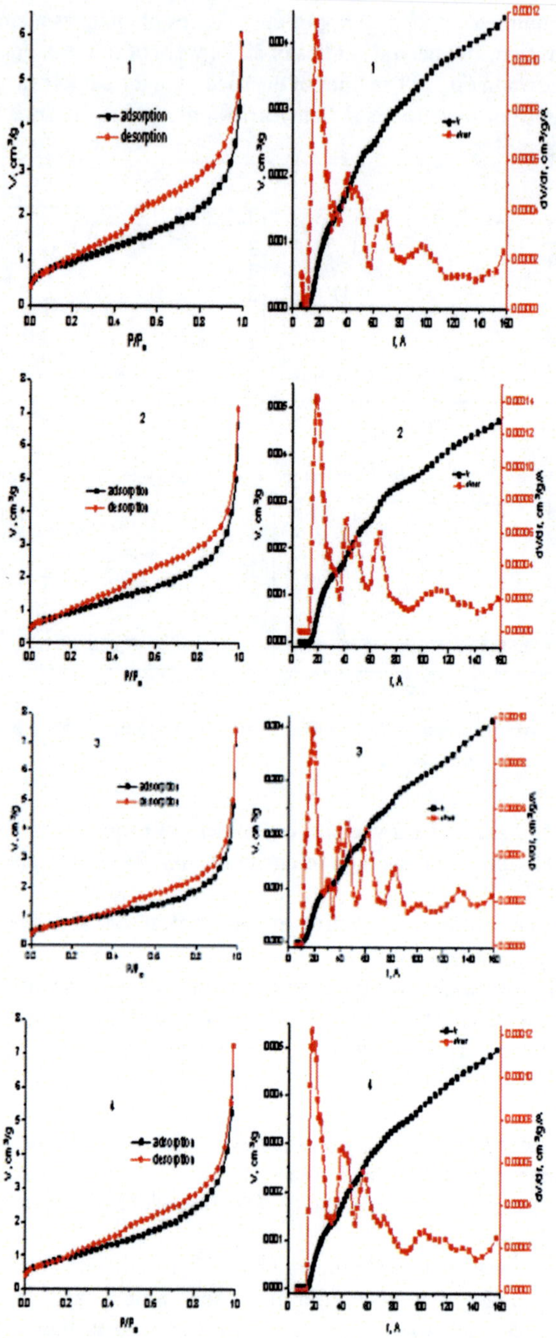

Fig. 4. Nitrogen adsorption isotherms on quartz sand initially activated in magnrtofluidized layer: 1- initially and activated: 2) 2 min.; 3) 4 min.; 4) 6 min

aggregate. Additional grinding of the cement allows the increase of the specific surface of the particles, which leads to an increase in the amount of cement that reacts with water. Consequently, less cement is needed to produce concrete with the desired characteristics.

Portland cement was treated in the magnetically fluidized layer with the same technological parameters as quartz sand. When grinding clinker, the crystal structure of the surface layers is distorted and the crystal lattice of the deep layers of minerals is deformed, the specific surface area of the binder grows, which leads to the activation of cement.

To evaluate the effectiveness of the activation of sand and cement in a magnetofluidized layer, compression tests were carried out on samples from ordinary concrete and concrete made from activated components.

For this, beams were made with dimensions of 20 × 20 × 60 mm, which were hardened for 28 days. The experimental results are shown in Fig. 5. It has been shown that the compressive strength of concrete made of activated components is almost 2 times higher than that of conventional concrete.

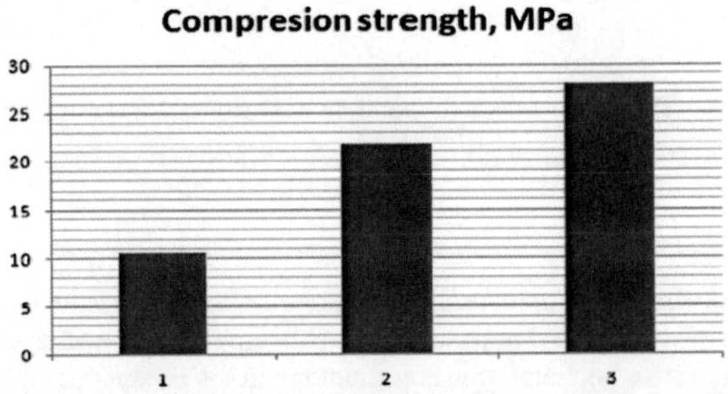

Fig. 5. Compresion strength of modified concrete. 1- control sample; 2- activated in a magnetofluidized layer; 3- activated in a magnetofluidized layer and nanomodified with graphene

The addition of nanomodifiers to Portland cement increases the strength of concrete products or saves cement while maintaining the strength of the concrete.

The surface energy of nanoparticles is very high, so that they form conglomerates. When the concrete mixture with the addition of nanomodifiers hardens, the nanoparticles negatively affect the formation of the cement matrix and the increase of the strength characteristics of the concrete. Therefore, nano-modification of building materials raises the issue of uniform distribution of nanoparticles in the volume of the dispersed mixture.

The grinding tasks of sand and cement, as well as the homogenization of the nanomodified concrete mixture can be solved with the help of the magnetofluidized layer. Due to the intensive movement and the impact between the ferromagnetic elements, the concrete components are finely crushed and the nanomaterials are evenly distributed in the volume of the concrete mixture.

It is shown that the addition of graphene even in small amounts (~0.01% of the volume of the mixture) provides an increase in the compressive strength of concrete by 23–28% compared to a concrete mixture activated only in a magnetofluidized layer. The results are shown in Fig. 5.

4 Conclusion

A two-step nanomodification of a concrete mixture was investigated. In the first stage, the sand and cement are ground in a magnetofluidized layer, and in the second stage a nanomaterial in the form of graphene is added.

When grinding sand and cement, structural changes occur in the particles of materials, which increases their reactivity during hydration. This affects the physical and mechanical characteristics of concrete. So, as a result of the activation of sand and cement in the magnetofluidized layer, the compressive strength of concrete increases almost 2 times.

The addition of graphene in the amount of 0.01% by weight relative to the cement leads to an increase in the compressive strength of the modified mixture by 23–28%. In this case, the magnetofluidized layer ensures a uniform distribution of the nanomaterial in the volume of the concrete mixture.

Acknowledgment. This research was carried out within the state program project ANCD 20.80009.5007.06 (2020–2023).

Conflict of Interest. The authors declare that they have no conflict of interest.

References

1. https://www.nobelprize.org/uploads/2018-06/popular-physicsprize2010.pdf
2. Rodionov, R.B.: Inovatsionnyi potentsial nanotekhnologii v proizvodstve stroitelinykh materialov [The innovative potential of nanotechnology in the production of building materials]. Stroitelinye materialy, oborudovanie, tehnologii XXI veka [Building materials, Equipment, Technologies of the XXI century], no. 8, pp. 72–75 (2006)
3. Rodionov, R.B.: Inovatsionnyi potentsial nanotekhnologii v proizvodstve stroitelinykh materialov [The innovative potential of nanotechnology in the production of building materials]. Stroitelinye materialy, oborudovanie, tehnologii XXI veka [Building materials, Equipment, Technologies of the XXI century], no. 10, pp. 57–59 (2006)
4. Rodionov, R.B.: Dostijeniya nanotehnologii v proizvodstve stroitelinykh materialov [Achievements of nanotechnology in the production of building materials]. Stroitelinye materialy, oborudovanie, tehnologii XXI veka [Building materials, Equipment, Technologies of the XXI century], no. 12, pp. 30–33 (2008)
5. Chistov, Yi.D., Tarasov, A.S.: Elementy nanotehnolodiy v proizvodstve betonov na osnove mineralinykh vyajushchikh veshchestv.Ch.1 [Elements of nanotechnology in the production of concrete based on binders. Part 1]. Stroitelinye materialy, oborudovanie, tehnologii XXI veka [Building materials, Equipment, Technologies of the XXI century], no. 3, pp. 69–72 (2007)
6. Chistov, Yi.D., Tarasov, A.S.: Elementy nanotehnolodiy v proizvodstve betonov na osnove mineralinykh vyajushchikh veshchestv.Ch.2 [Elements of nanotechnology in the production of concrete based on binders. Part 2]. Stroitelinye materialy, oborudovanie, tehnologii XXI veka [Building materials, Equipment, Technologies of the XXI century], no. 6, pp. 14–16 (2007)

7. Sakharov, G.P.: O kratkosrochnoy perspective nanotehnologiy v proizvodstve stroitelinykh materialov i izdeliy. Ch.1 [On the Short Term Perspective of Nanotechnology in the Production of Building Materials and Products. Part 1]. Tekhnologiya betonov [Technology of Concretes], no. 4(33), pp. 65–67 (2009)
8. Sakharov, G.P.: O kratkosrochnoy perspective nanotehnologiy v proizvodstve stroitelinykh materialov i izdeliy. Ch.2 [On the Short Term Perspective of Nanotechnology in the Production of Building Materials and Products. Part 2]. Tekhnologiya betonov [Technology of Concretes], no. 5(34), pp. 13–15 (2009)
9. Koroliov, E.V., Bajenov, Y., Beregovoy, V.A.: Modifitsirovanie stroitelinykh materialov nanouglerodnymi trubkami, fullerenami [Modification of Building Materials with Nanocarbon Tubes and Fullerenes]. Stroitelinye materialy [Constr. Mater.] **8**, 2–5 (2006)
10. https://m.hightech.plus/2021/05/26pervuyu-v-mire-betonnuya-plitu-s-grafenom-zalili-v-anglii
11. Lepilin, A.B., Korenyigina, H.V., Veksler, M.V.: Selektivnaya dezintegratornaya aktivatsiya portlandtsementa [Selective Disintegrator Activation of Portland Cement]. Stroitelinye materialy [Construction Materials], no. 7 (2007)
12. Logvinenko, D.D., Shelyakov, O.P.: Intensifikatsiya tekhnologicheskikh protsessov v apparatakh s vikhrevym sloem [Intensification of technological processes in the velocity-layer devices], 144 p. Tekhika, Kiev (1970). (in Russian)
13. Filonov, I.A., Yavruyan, Kh.S.: Mekhaniceskaya aktivatsiya portlandtsementa v apparate vikhrevogo sloya [Mecanical Actiation of Portland Cement in the Vortex Layer Apparatus]. Inzhenernyy vestnik Dona [Bulletin of Engineering News of the Don], no. 3 (2012). http://ivdon.ru/magazin/archive/n3y2012/969
14. Yavruyan, Kh.S., Filonov, I.A.: Gomogrnizatsiya nanomodifitsirovannykh tsementnykh system i podbor parametrov ikh obrabotki v ustanovkakh s vikhrevym sloem [Homogenization of Nano-modified Cement Systems and Selection of Parameters of Their Treatment in Vortex Layer Units]. Vestnik MGSU [Proceedings of Moscow State University of Civil Engineering], no. 2, pp. 130–136 (2013)
15. Demiyanova, L.P., et al.: Ftorirovanie nanorazmernogo amorfnogo kreneziyoma, poluchennogo iz silikatnogo syriya po ftoridnoi tehnologii[Fluorination of nanosized amorphous silica obtained from silicate raw materials by fluoride technology]. Fluorine Notes **82**, 3–4 (2012)

Highly Conductive ZnO Thin Films Deposited Using CVT Ceramics as Magnetron Targets

G. V. Colibaba[1,2(✉)], D. Rusnac[1,2], V. Fedorov[3], N. Costriucova[2], E. V. Monaico[4], and T. Potlog[1]

[1] Moldova State University, Chisinau, Republic of Moldova
gcolibaba@usm.md
[2] Institute of Applied Physics, Chisinau, Republic of Moldova
[3] Institute of Electronic Engineering and Nanotechnologies, Chisinau, Republic of Moldova
[4] Technical University of Moldova, Chisinau, Republic of Moldova

Abstract. Sintering of $ZnO + Me_2O_3$ (Me = Al, Ga, In) powder via chemical vapor transport based on HCl has been developed. The electrical properties of ZnO thin films obtained by DC magnetron sputtering of ZnO ceramic targets have been studied. Transparency, morphology, crystallinity and crystallite size of thin films have also been investigated. ZnO:Ga thin films with a resistivity of 2.5×10^{-4} Ω·cm have been successfully obtained. The films doped with Al have lower conductivity due to weak sputtering of insoluble Al_2O_3 dielectric inclusions in ceramics. In the case of sintering of ZnO together with In_2O_3, a significant loss of the doping material is observed.

Keywords: ZnO thin films · Magnetron sputtering · Electrical properties · New type of ceramic targets

1 Introduction

Wide band-gap zinc oxide is a semiconductor with multiple application opportunities. Deposited thin films, as well as nanostructures on their surface, can find eventual applications as light-emitting devices, photoconductive devices or gas sensors [1, 2]. DC magnetron sputtering is one of the simplest methods for growth of these films. However, the efficiency of this method is based on the presence of uniformly doped conductive ceramic targets [3]. The most extensively used sintering methods for ceramics have many disadvantages, including extremely high sintering temperatures of about 1300 – 1500 °C [4, 5]. Recently, chemical vapor transport (CVT) technique, which is successfully used for ZnO single crystal growth [6, 7, 8, 9], was proposed as an alternative method for sintering ceramics. A high pressure of doping gaseous species participated in CVT reactions, can contribute in obtaining ceramics uniformly doped in the gas phase at temperatures as low as 1070 °C [10, 11]. A lot of studies were focused on the obtaining of conductive ZnO thin films, with typical resistivity (ρ) of 4×10^{-4} Ω·cm. There are several methods to increase the conductivity of these films. The most conductive films deposited using buffer layers, strong magnetic field, or co-doping with F have a ρ value

I. Tiginyanu et al. (Eds.): ICNBME 2021, IFMBE Proceedings 87, pp. 110–116, 2022.
https://doi.org/10.1007/978-3-030-92328-0_15

in a range of (2.1 – 2.9) × 10^{-4} Ω·cm [12 – 14]. The present work is a comparative study of highly conductive ZnO thin films doped with Al, Ga and In and deposited using a new CVT ceramics containing Cl impurity.

2 Experiment

Ceramic samples ZnO:Me_2O_3 (Me = Al, Ga, In) obtained by CVT using HCl as a transport agent were used as targets. More detailed about sintering process was described in Ref. [10]. Thin films were grown on glass substrates by DC magnetron sputtering. The deposition temperature and film thickness were 200 °C and 800 nm, respectively. The magnetron power was varied within 1–12 W. Argon with a purity of 99.998% was used as the working gas. Post-growth annealing of films was not used. The resistivity, charge carrier concentration (n), and Hall mobility (μ) of ceramics and thin films were calculated from the Hall effect measurements using the van der Pauw method. To analyze the crystallinity of the samples, 2Θ X-ray diffraction (XRD) spectra, recorded using CuK_α radiation, were used. A Zeiss Sigma scanning electron microscope (SEM) and a TESCAN Vega TS 5130MM SEM equipped with an Oxford Instruments INCA energy dispersive X-ray analysis (EDX) system were used to study the morphology and chemical composition microanalysis.

3 Experimental Results

3.1 CVT Ceramics ZnO Doped with Ga, Al and In

The calculated total pressure of chlorides (P(MeCl) + P($MeCl_2$) + P($MeCl_3$)) of the most typical trivalent donors in ZnO (B, Al Ga, and In) is shown in Fig. 1. These calculations are presented for Me_2O_3–ZnO–HCl CVT systems at an initial HCl pressure of 2 atm. Details of calculations can be found in Ref. [15]. Boron and aluminum oxide should interact extremely weakly with the gaseous medium formed by the interaction of HCl + ZnO. Total pressure of Al chlorides in the Al_2O_3–ZnO–HCl CVT systems is as low as 10^{-8} atm at 1070 °C. As a consequence, a relatively slow dissolution of Al_2O_3 in the ZnO ceramics is expected. In the case of indium oxide, excessive interaction and partial loss of the doping material into the gas phase should be observed; total pressure of In chlorides is about 10^{-1} atm. The vapor pressure of Ga chlorides should have moderate values about 10^{-3} – 10^{-2} atm in the 700–1200 °C temperature range. Doping with gallium oxide seems to be the most suitable for sintering homogeneously doped ceramics without significant losses of the doping material.

Sintering of ZnO + Ga_2O_3 ceramics makes it possible to obtain a homogeneously doped material up to 3 mol% Ga_2O_3 concentration [11]. The obtained material is characterized by a low resistivity of 1.5 × 10^{-3} Ω·cm. The density and hardness of ZnO:Ga ceramics are close to those of single crystals: 94–95% relative density, hardness = 2 GPa. The resistivity of ceramics obtained using Al_2O_3 powder (1 mol%, 10–50 μm particle size) is much higher of about 10^{-2} Ω·cm. In this case, inclusions of undissolved dopant are observed in the target material. The use of a higher concentration of Al_2O_3 slows the sintering process; the resulting material does not have the required density and hardness. In the case of sintering of ZnO together with In_2O_3, a significant loss of the doping material (~90%) is observed.

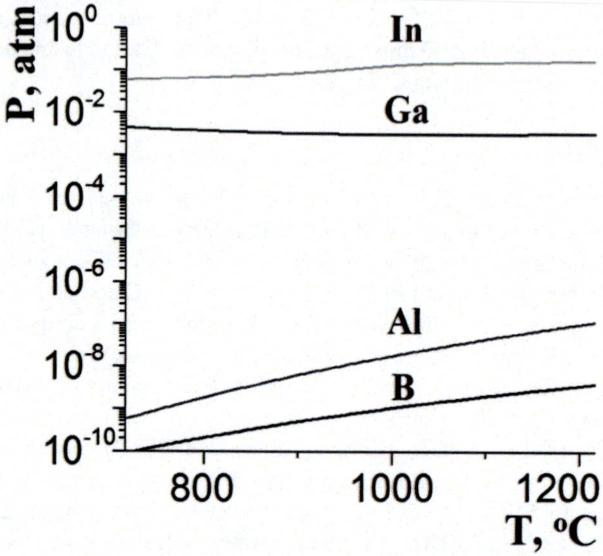

Fig. 1. Total pressure of III-valent metal chlorides vs temperature for $Me_2O_3 - ZnO - HCl$ CVT systems; $HCl^0 = 2$ atm.

3.2 Characterization of ZnO Thin Films Doped with Ga and Al

ZnO films deposited from obtained ceramics using magnetron sputtering are characterized by a high transparency in the visible spectral range (the average transparency is 87%, 79% with the substrate). The presence of Ga (Al) impurities is confirmed by EDX measurements. The Cl concentration is less than the detection limit of about 5×10^{19} cm^{-3}. The XRD spectra exhibit an intense peak of (002) reflection (Fig. 2); these films consist of crystallites oriented mostly along the *c* axis of a hexagonal structure [3]. For the most conductive films, no peaks associated with other reflections were observed (Fig. 2, inset) [16].

Adding 1 mol% of Ga_2O_3 to the ceramic target reduces the resistivity of the films by 5 times as is shown in Fig. 3.

A further increase in the Ga concentration has no significant effect. This behavior indicates on the solubility limit of the Ga impurity in ZnO. ZnO thin films doped with Al are characterized by lower conductivity, which is related to the fact that most amount of the impurity is not dissolved in the ceramics and remains in the form of Al_2O_3 dielectric inclusions, which are weakly sputtered. The increase in the Al content in the ceramic makes it not hard enough.

An increase in the growth rate of films leads to a significant decrease in the ρ value (to 2.5×10^{-4} Ω·cm in the case of Ga impurity), regardless of the impurity type (see Fig. 4). This is mainly due to an increase in the concentration of current carriers. High power of the magnetron contributes to the higher migration length of the impurity atoms on the surface of the growing film, and, consequently, to better incorporation of impurities into the crystal lattice as shallow donors [17]. An increase in the magnetron power also

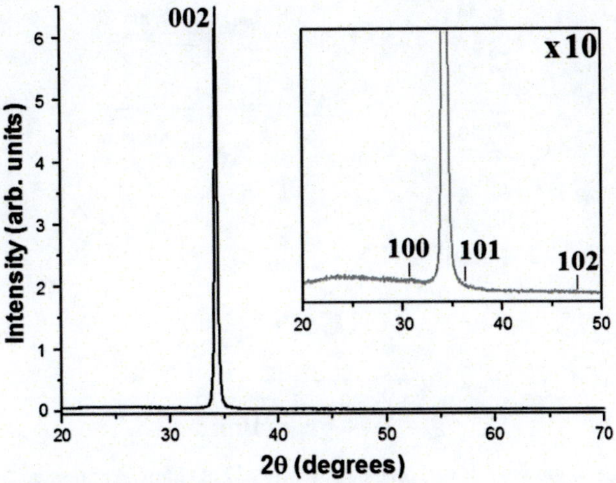

Fig. 2. XRD spectrum of ZnO:Ga thin film. Magnetron power 10 W, Ar pressure 2.6×10^{-6} atm. Inset illustrates this spectrum magnified by 10 times

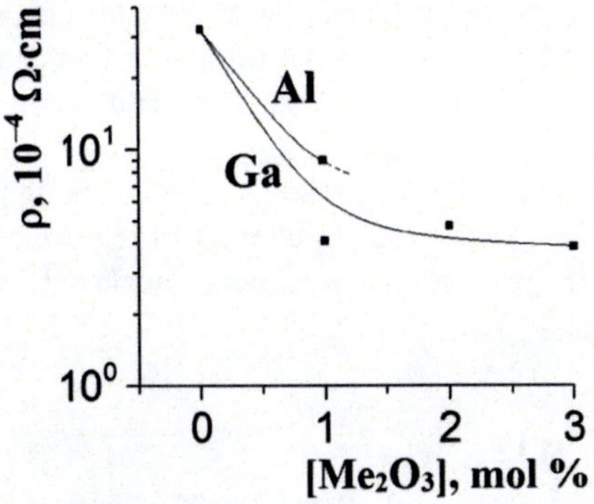

Fig. 3. Influence of dopant concentration in a target on the resistivity of ZnO films. Magnetron power 7 W, Ar pressure 2.6×10^{-6} atm.

contributes to a higher structural perfection of the films: an increase in the intensity of the (002) XRD peak and an increase in the crystallite size (see Fig. 4, inset).

An increase in the working gas pressure promotes the scattering of the sputtered particles and a decrease in the film growth rate [18]. Using this dependence (growth rate vs Ar pressure), the influence of working gas pressure on the value of resistivity can be predicted as shown in Fig. 5 with dotted lines. However, the real dependence of the resistivity ρ(Ar) is more complex. For the Ga impurity, there are obviously two competing factors. The first is an improvement of the structural perfection of the films. Films grown at low Ar pressure ($< 2.6 \times 10^{-6}$ atm) contain microcracks and cavities (Fig. 5, insets). With increasing pressure of the working gas, the deposited films are more compact. This promotes an increase in the mobility of free electrons up to 30 cm^2/Vs,

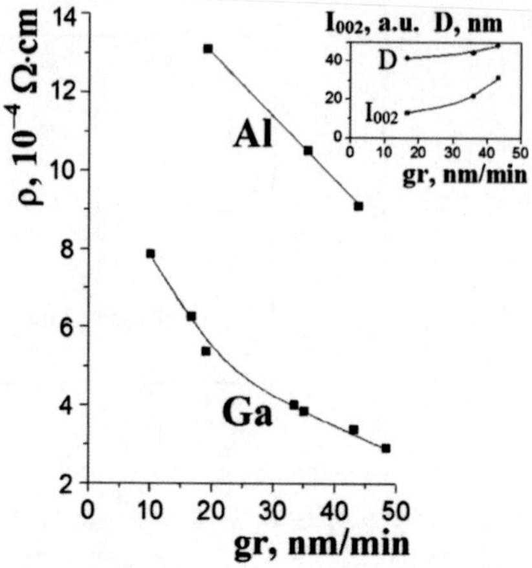

Fig. 4. Influence of growth rate (gr) on the resistivity of ZnO films. Ar pressure 2.6×10^{-6} atm, magnetron power varied in the range of 1 – 12 W. Inset illustrates intensity of (002) XRD peak and crystallite size (D)

which results in the presence of a minimum in the ρ(Ar) dependence. A further increase in the Ar pressure ($\geq 5 \times 10^{-6}$ atm) has no significant effect on the structural perfection of the films; and the conductivity depends on the second factor, namely by the decrease in the growth rate and migration length of impurity atoms. For films doped with Al, the third factor predominates: a high kinetic energy of the working gas ions is necessary for sputtering Al_2O_3 dielectric inclusions. Sputtering of these inclusions is possible only at high voltages of the magnetron system, which is achieved at the lowest working gas pressure.

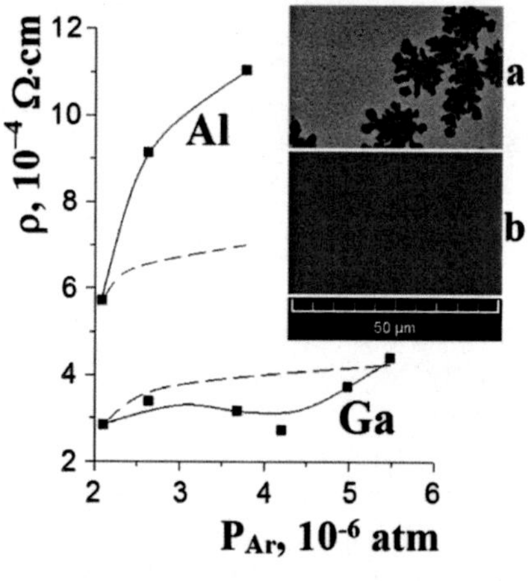

Fig. 5. Influence of the Ar pressure on the resistivity of ZnO films. Insets illustrate SEM images of ZnO films grown at Ar pressure of 2.1×10^{-6} (a) and 3.7×10^{-6} atm (b). Magnetron power 10 W

4 Conclusions

Sintering ZnO ceramics by means of CVT using HCl as a transport agent makes it possible to obtain material with high density, hardness and conductivity. The most suitable impurity is Ga; Ga_2O_3 almost completely dissolves in ceramics. The use of Al_2O_3 is limited by the low rate of solubility, and In_2O_3 by large losses of the doping material. The optimal growth conditions have been determined that make it possible to obtain ZnO:Ga thin films with a resistivity of 2.5×10^{-4} Ω·cm. The films doped with Al have lower conductivity due to weak sputtering of insoluble Al_2O_3 dielectric inclusions in ceramics.

Acknowledgment. This work was supported by the Ministry of Education, Culture and Research of Moldova under the project No. 20.80009.5007.16 (Photosensitizers for applications in pharmaceutical medicine and photovoltaics).

Conflict of Interest. The authors declare that they have no conflict of interest.

References

1. Özgür, Ü., et al.: A comprehensive review of ZnO materials and devices. J. Appl. Phys. **98**, 041301 (2005). https://doi.org/10.1063/1.1992666
2. Klingshirn, C.: ZnO: from basics towards applications. Phys. Status. Solidi. B. **244**, 3027–3073 (2007). https://doi.org/10.1002/pssb.200790012
3. Ellmer, K., Klein, A., Rech, B.: Transparent Conductive Zinc Oxide. Springer-Verlag, Berlin Heidelberg (2008)
4. Lee, J.-W., et al.: Microstructure and density of sintered ZnO ceramics prepared by magnetic pulsed compaction. Adv. Mater. Sci. Eng. **2018**, 2514567 (2018). https://doi.org/10.1155/2018/2514567
5. Boonyopakorn, N., Rangkupan, R., Osotchan, T.: Preparation of aluminum doped zinc oxide targets and RF magnetron sputter thin films with various aluminum doping concentrations. Songklanakarin J. Sci. Technol. **40**, 824–830 (2018)
6. Colibaba, G.V.: ZnO:HCl single crystals: thermodynamic analysis of CVT system, feature of growth and characterization. Solid State Sci. **56**, 1–9 (2016). https://doi.org/10.1016/j.solidstatesciences.2016.03.011
7. Colibaba, G.V.: Halide-hydrogen vapor transport for growth of ZnO single crystals with controllable electrical parameters. Mater. Sci. Semicond. Process. **43**, 75–81 (2016). https://doi.org/10.1016/j.mssp.2015.12.005
8. Colibaba, G.V.: Halide-oxide carbon vapor transport of ZnO: novel approach for unseeded growth of single crystals with controllable growth direction. J. Phys. Chem. Solids **116**, 58–65 (2018). https://doi.org/10.1016/j.jpcs.2018.01.009
9. Colibaba, G.V.: Halide-carbon vapor transport of ZnO and its application perspectives for doping with multivalent metals. J. Solid State Chem. **266**, 166–173 (2018). https://doi.org/10.1016/j.jssc.2018.07.019
10. Colibaba, G.V.: Sintering highly conductive ZnO:HCl ceramics by means of chemical vapor transport reactions. Ceram. Int. **45**, 15843–15848 (2019). https://doi.org/10.1016/j.ceramint.2019.05.087

11. Colibaba, G.V., et al.: Low-temperature sintering of highly conductive ZnO:Ga: Cl ceramics by means of chemical vapor transport. J. Eur. Ceram. Soc. **41**, 443–450 (2021). https://doi.org/10.1016/j.jeurceramsoc.2020.08.002
12. Miyazaki, M., et al.: Properties of Ga-doped ZnO films. J. Non-Cryst. Solids **218**, 323–328 (1997). https://doi.org/10.1016/S0022-3093(97)00241-X
13. Nomoto, J., et al.: Improvement of the properties of direct-current magnetron-sputtered Al-doped ZnO polycrystalline films containing retained Ar atoms using 10-nm-thick buffer layers. ACS Omega **4**, 14526–14536 (2019). https://doi.org/10.1021/acsomega.9b01761
14. Wang, F.-H., Chang, C.-L.: Effect of substrate temperature on transparent conducting Al and F co-doped ZnO thin films prepared by RF magnetron sputtering. Appl. Surf. Sci. **370**, 83–91 (2016). https://doi.org/10.1016/j.apsusc.2016.02.161
15. Colibaba, G.V.: ZnO doping efficiency by multivalent metals in complex CVT reactions. Solid State Sci. **97**, 105944 (2019). https://doi.org/10.1016/j.solidstatesciences.2019.105944
16. Liu, J., et al.: Comparative study of the sintering process and thin film sputtering of AZO, GZO and AGZO ceramics targets. Ceram. Int. **40**, 12905 (2014). https://doi.org/10.1016/j.solidstatesciences.2019.105944
17. Park, S.-U., Koh, J.-H.: Low temperature RF-sputtered In and Al co-doped ZnO thin films deposited on flexible PET substrate. Ceram. Int. **40**, 10021–10025 (2014). https://doi.org/10.1016/j.ceramint.2014.02.101
18. Colibaba, G.V., Rusnac, D., Fedorov, V., Monaico, E.I.: Effect of chlorine on the conductivity of ZnO:Ga thin films. J. Mater. Sci.: Mater. Electron. **32**(13), 18291–18303 (2021). https://doi.org/10.1007/s10854-021-06371-x

Direct Surface Patterning Using Carbazole-Based Azopolymer

O. Paiuk[1](✉), A. Meshalkin[2], A. Stronski[1], E. Achimova[2], K. Losmanschii[2], A. Korchovyi[1], Z. Denisova[1], V. Goroneskul[1], and P. Oleksenko[1]

[1] V.E. Lashkaryov Institute of Semiconductor Physics NASU, Kyiv, Ukraine
paiuk@ua.fm
[2] Institute of Applied Physics, Chisinau, Moldova

Abstract. This paper is devoted to the investigations of the recording of 1-D and 2-D holographic gratings using thin films of polyepoxypropylcarbazole (PEPC) obtained by deposition from solutions and their use as registering media for. For the direct recording azopolymer films based on polyepoxypropylcarbazole: methyl red with magnetic particles of Fe_2SO_4 were used. Diffraction efficiency in transmission of the recorded gratings consisted ~34%. Morphology of films surface and obtained surface relief's was investigated using AFM and good quality of films surfaces and obtained relief's was shown. For the first time simultaneously surface and magnetic relief were directly recorded using PEPC thin films as registering media.

Keywords: Polyepoxypropylcarbazole (PEPC) · Methyl red · Holographic gratings · Direct recording · Surface and magnetic relief

1 Introduction

Recently much attention has been paid to the study of organic polymer compounds (among them azopolymers), which could be used for surface relief patterning upon light exposure [1–7]. Under the light exposure of azopolymers the photoisomerization reaction of azobenzene molecules can induce substantial material motions. Due to the changes in transmission, reflection, and in thickness, high resolution, optical transparency, ability to form surface relief's under the influence of laser irradiation, carbazole-based azopolymer could be used for effective direct surface relief patterning (either with or no wet etching step using only light exposure) and production of optical elements such as diffraction gratings, microlens arrays, holograms, photonic crystals, nanostructured polarizers, plates, and broad-band antireflection coatings [6]. Sinusoidal modulated intensity pattern at the submicron or micron scale is most commonly used to form holographic diffraction gratings [8–10]. Usually surface relief is obtained using selective etching after exposure [5]. Direct fabrication of surface relief gratings by laser interference lithography was shown in [9, 11–16]. Physical phenomena used in fabrication of various optical elements are: photo-induced mass transport, which consists in lateral redistribution of polymer material under illumination [1–7]. Illumination by

I. Tiginyanu et al. (Eds.): ICNBME 2021, IFMBE Proceedings 87, pp. 117–123, 2022.
https://doi.org/10.1007/978-3-030-92328-0_16

linearly polarized light provides anisotropic changes, which are referred to as vectoral, which means that not only the photon energy value but also the direction of optical fields vector exerts some influence on the photostructural processes.

This report is concerned with obtaining and investigation of a carbazole-based azopolymer layers as registering media for direct recording of surface relief gratings.

2 Experimental Results and Discussion

Azopolymer based on polyepoxypropylcarbazole: methyl red with magnetic particles of Fe_2SO_4 was synthesized. The thin films based on obtained polymers were deposited by spin coating onto glass substrate. Room temperature transmission spectra of deposited polymer films were recorded using two-beam spectrophotometer SPECORD M40 in the range of wavelengths of 450–900 nm and showed good transparency in the region of wavelength used for measurement of diffraction efficiency (Fig. 1).

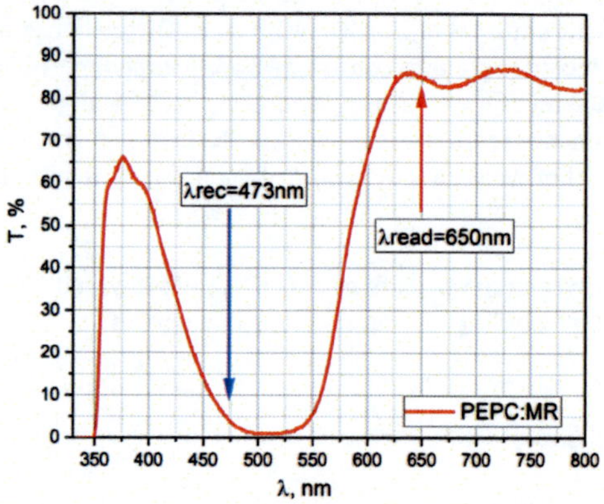

Fig. 1. Spectral dependence of transmission of polyepoxypropylcarbazole: methyl red fims.

The wavelengths used during grating recording and measurement of diffraction efficiencies are indicated by arrows.

Films thickness was measured using interference microscope MII-4, the obtained interference patterns were photographed and processed using Optic Meter program in order to estimate thicknesses of the films. Also films thicknesses were estimated using AFM microscopy. Thicknesses of films were within 1–3 мm. Morphology and surface relief of the obtained gratings were studied by atomic-force microscopy. One- and two-dimensional diffraction gratings were recorded by common two laser beams scheme using a linearly p-polarized light from a DPSS laser ($\lambda = 473$ nm and $I = 1700$ mW/cm^2) (Fig. 2).

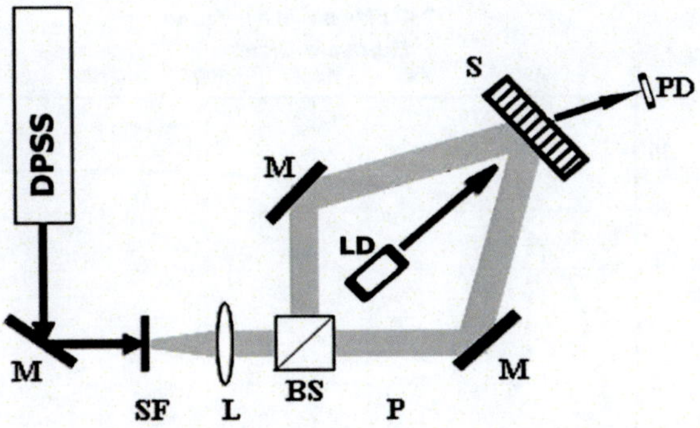

Fig. 2. Scheme used for the diffraction gratings recording: DPSS – laser, SF and L – collimator, BS – beam splitter, M – flat mirrors, S – registering media (sample). LD – light emitting diode, PD – registering unit.

Intensity ratio of beams consisted 1:1. The fringe spacing of the interference pattern, Λ, and thereby the periodicity of the surface relief gratings (SRGs), is determined by the equation $\Lambda = \lambda/(2\sin\theta)$, where λ is the wavelength of the writing beam and θ is the angle between the sample normal and the propagation axis of the writing beam. In this study, the periodicity was set to 5.9 μm, spatial frequency ν - 170 lines/mm. On the base of previous results which showed dependence of surface relief formation on polarization of recording light [2, 5] p-p- configuration of polarized light was used during recording process. The grating recording dynamics was followed by in-situ diffraction measurements (diffraction efficiency dependence on recording time). The time evolution of SRG formation was monitored by detecting the transmitted first-order diffracted beam from a normally incident 650 nm laser diode. Diffraction efficiency was calculated as ratio of the intensities of the first order diffracted beam and the transmitted beam out of the grating.

2.1 Gratings Recording

The first-order diffraction efficiencies for P:P polarizations during grating recording are presented in Fig. 3.

It can be seen from Fig. 3 that diffraction efficiency for P-P polarizations recording scheme saturates and reaches 33% within 30 min of recording.

In Fig. 4 AFM 3-D image of the holographic diffraction grating with period d ~ 5.9 μm recorded using polyepoxypropylcarbazole (PEPC) layers is shown. It can be seen from figure that the high quality of the surface relief is obtained, relief height of the gratings was ~0.4 μm.

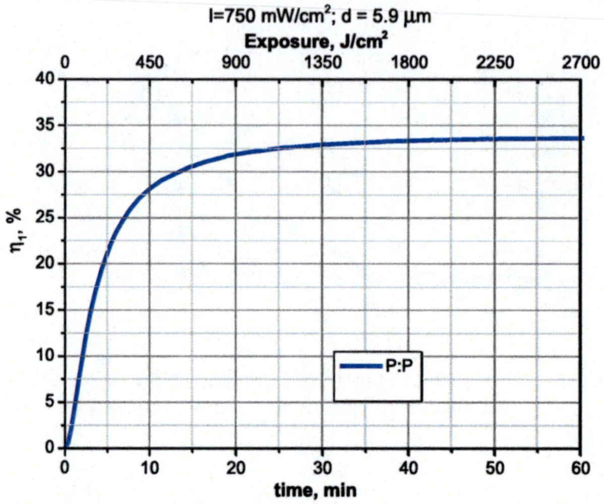

Fig. 3. Diffraction efficiency in transmission η dependence on recording time. P-P polarization recording scheme. I = 750 mW/cm^2, recording wavelength – 473 nm.

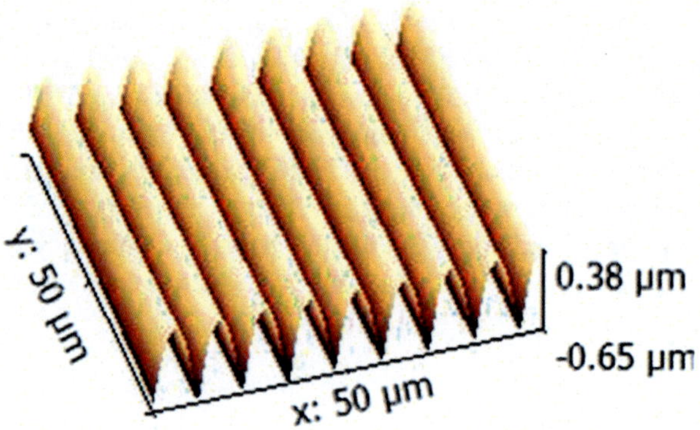

Fig. 4. AFM 3-D image of the holographic diffraction grating (d ~5.9 мm) recorded using polyepoxypropylcarbazole (PEPC) layers.

Figure 5a shows 2-D gratings recorded using polyepoxypropylcarbazole (PEPC) layers. High quality of the surface relief is obtained as can be seen from the figure.

Relief height of the gratings was ~0.36 μm. Relief profile is close to the sinusoidal one (Fig. 5b).

Diffraction efficiency (DE) of SRG can was calculated analytically and compared with experimental obtained one. A sinusoidal profile of (see Fig. 4 and 5) of grating leads to the well-known Raman-Nath expression $\eta_1 = J_1(\pi \Delta n h/\lambda)^2$, where J_1 are ordinary Bessel function of the first kind, Δn is the refractive index difference ($\Delta n = n_{polymer} - n_{air}$) and h is relief depth.

The good coincidence of experimental and theoretical values of diffraction efficiency in dependence on surface profile depth may be the evidence that the main contribution in the value of diffraction efficiency of recorded holographic grating brings the surface relief grating.

Theoretical estimated value of surface relief depth is about 400 nm for P:P polarizations by assuming the surface relief grating dominates in DE with slight influence of scalar refractive index gratings.

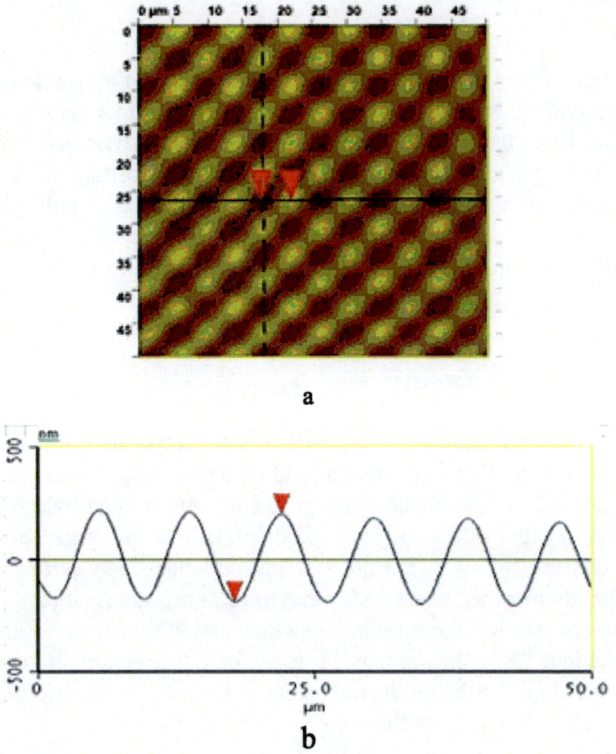

Fig. 5. a- AFM image of 2-D gratings recorded using polyepoxypropylcarbazole (PEPC) layers. b- relief profile is close to the sinusoidal one.

Recording of surface relief gratings using polyepoxypropylcarbazole (PEPC): methyl red with magnetic particles of Fe_2SO_4 layers was studied by MFM microscopy. Non-crystalline photoresponsive materials (for example, azobenzene-containing polymers and amorphous chalcogenides) can be easily doped [1, 13, 17]. Introduction of dopants can change optical, luminescent and magnetic properties of media [17]. This enables to record simultaneously magnetic and surface relief using such media [13]. Similar possibility as can be seen from Fig. 6 can be provided by using layers of polyepoxypropylcarbazole (PEPC): methyl red with Fe_2SO_4 magnetic particles.

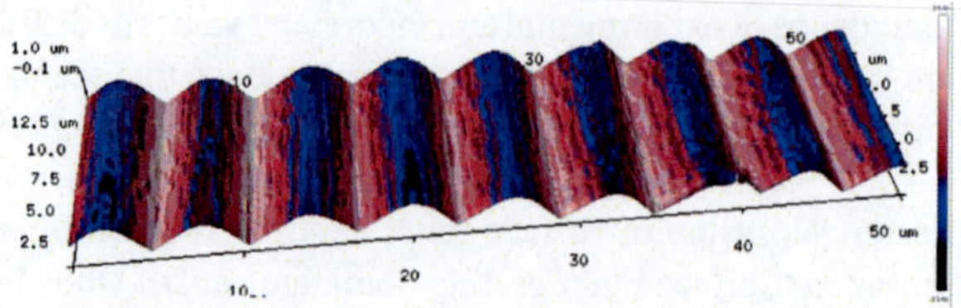

Fig. 6. AFM image of surface relief overlapped by MFM map (colour image, right scalebar) of grating directly recorded using polyepoxypropylcarbazole (PEPC) layers: methyl red with Fe_2SO_4 magnetic particles.

MFM images (Fig. 6) show that distribution and value of magnetic field correlates with period of grating surface relief. Mechanism of this effect may be similar to the observed in [18, 19] redistribution of nanoparticles in nanoparticle-dispersed photopolymers as a result of exposure by holographic light interference pattern. Possibility of direct one-step magnetic relief formation during grating recording using polyepoxypropylcarbazole (PEPC) layers: methyl red with Fe_2SO_4 magnetic particles. (or similar ones) can be used for creation of surface relief optical elements with unique properties for photonics and in magnetic memory applications.

3 Conclusions

The presented results demonstrated a direct, one-step process of holographic recording using p-p- configuration of polarized light by 473 nm light beam of surface relief structures (1-D and 2-D holographic gratings) using polyepoxypropylcarbazole (PEPC) layers: methyl red with Fe_2SO_4 magnetic particles. Diffraction efficiency of gratings in transmission consisted ~34.4%. For the first time simultaneously surface and magnetic relief were directly recorded using PEPC thin films as registering media.

Due to the changes in transmission, reflection, and in thickness under the influence of laser irradiation, PEPC layers may be used for effective amplitude-phase optical information recording, for the production of surface relief optical elements with unique properties for novel photonic applications.

Acknowledgment. The research was supported by the bilateral Moldova-Ukraine project (#17.80013.5007.03/Ua) and ANCD project (20.80009.5007.03).

Conflict of Interest. The authors declare that they have no conflict of interest.

References

1. Kryshenik, V.M., Azhniuk, Y.M., Kovtunenko, V.S.: All-optical patterning in azobenzene polymers and amorphous chalcogenides. J. Non-Cryst. Solids **512**, 112–131 (2019)

2. Podlipnov, V.V., Ivliev, N.A., Khonina, S.N., et al.: Formation of microstructures on the surface of a carbaseole-containing azopolymer by the action of laser beams. J. Phys. Conf. Ser. **1368**, 022069 (2019)
3. Podlipnov, V.V., Ivliev, N.A., Khonina, S.N., et al.: Investigation of photoinduced formation of microstructures on the surface of carbaseole-containing azopolymer depending on the power density of incident beams. Comput. Opt. **42**(5), 779–785 (2018)
4. Sorkhabi, S.G., Barille, R., Ahmadi, S., et al.: A new method for patterning azopolymer thin film surfaces. Opt. Mater. **66**, 573–579 (2017)
5. Achimova, E.A., Stronski, A.V., Paiuk, A.P., et al.: Recording of holographic diffraction gratings on carbazole containing polymer thin films. Optoelectron. Semiconductor Tech. **49**, 31–35 (2014)
6. Priimagi, A., Shevchenko, A.: Azopolymer-based micro- and nanopatterning for photonic applications. J. Polym. Sci. B Polym. Phys. **52**, 163–182 (2014)
7. Andries, A., Abaskin, V., Achimova, E., et al.: Application of carbazole-containing polymer materials as recording media. Phys. Status Solidi. (A) **208**(8), 1837–1840 (2011)
8. Stronski, A.V., Vlček, M.: Imaging properties of As40S40Se20 layers. Optoelectron. Rev. **8**(3), 263–267 (2000)
9. Stronski, A., Revutska, L., Meshalkin, A., et al.: Structural properties of Ag–As–S chalcogenide glasses in phase separation region and their application in holographic grating recording. Opt Mater. **94**, 393–397 (2019)
10. Andriesh, A., Sergheev, S., Triduh, G., et al.: Diffraction optical structures on the basis of chalcogenide glasses and polymers. J. Optoelectron. Adv. M. **9**(10), 3007–3012 (2007)
11. Vlček, M., Schroeter, S., Brueckner, S., et al.: Direct fabrication of surface relief images in chalcogenide glasses by eximer laser interference lithograpgy. J. Mater. Sci.: Mater. Electron. **20**, 290–293 (2009)
12. Stronski, A., Achimova, E., Paiuk, O., et al.: J. Nano Res. **39**, 96–104 (2016)
13. Stronski, A., Achimova, E., Paiuk, O., et al.: Direct magnetic relief recording using as40s60:mn–se nanocomposite multilayer structures. Nanoscale Res. Lett. **12**, 286 (2017)
14. Rahmouni, A., Bougdid, Y., Moujdi, S., et al.: Photoassisted holography in azo dye doped polymer films. J. Phys. Chem. B **120**(43), 11317–11322 (2016)
15. Meshalkin, A., Robu, S., Achimova, E., et al.: Direct photoinduced surface relief formation in carbazole-based azopolymer using polarization holographic recording. J. Optoelectron. Adv. Mater. **18**(9–10), 763–768 (2016)
16. Loshmanschii, C., Achimova, E., Abashkin, V., et al.: PC assisted waveplates application for one step holographic direct surface relief patterning on azopolymer. In: ICTEI Proceedings, pp. 128–129 (2018)
17. Stronski, A., Paiuk, O., Gudymenko, A., et al.: Effect of doping by transitional elements on properties of chalcogenide glasses. Ceram. Int. **41**, 7543–7548 (2015)
18. Tomita, Y., Suzuki, N.: Holographic manipulation of nanoparticle distribution morphology in nanoparticle-dispersed photopolymers. Opt. Lett. **30**(8), 839–841 (2005)
19. Tomita, Y., Suzuki, N.: Photopolymerizable nanocomposite photonic materials and their holographic applications in light and neutron optics. J. Mod. Opt. **63**(53), 511–541 (2016)

Biomedical Instrumentation and Signal Processing

Low Power Constant Current Driver for Implantable Electrostimulator of the Lower Esophageal Sphincter

Vladimir Vidiborschii[1(✉)], V. Sontea[1], S. Ungureanu[2], N. Sipitco[2], and D. Fosa[2]

[1] Technical University of Moldova, 168, Stefan cel Mare Bd., Chisinau, Republic of Moldova
vladimir.vidiborschii@scee.utm.md

[2] State University of Medicine and Pharmacy "Nicolae Testemitanu", Chisinau, Republic of Moldova

Abstract. Tone modulation of lower esophageal sphincters (LES) via electrical stimulation is a novel method of gastroesophageal reflux disease (GERD) treatment. Traditionally for output are commonly used constant current drivers, based on different schematic solutions. This allows delivering same stimulation energy even in case of impedance change during patient movements or electrode contact aging or encapsulation. The aim of our work was to design and implement simple constant current driver with digital current control and ultra-low power consumption. This driver was used in a prototype of implantable LES stimulator (WPLES), which effectiveness was confirmed during animal tests.

Keywords: Current driver · Implantable stimulator · WPLES

1 Introduction

Implantable electrostimulators were first presented in clinical practice in the mid 60th of the twentieth century, when a convenient pacemaker was created 0. From that point forward, the utilization of these devices in treatment and diagnosis of different diseases has been developing consistently.

One of the novel application of this method is gastric electrostimulation and direct modulation of lower esophageal sphincter (LES) tone as effective non-medicine treatment of gastroesophageal reflux disease (GERD) [1–4, 19]. Usually after surgical insertion of implantable stimulators in human body and mechanical tissue damaging, organism reacts to foreign body and releases a series of chemical and biological factors in the peri-implant space, resulting by encapsulation of the foreign matter with followed significant impedance increasing [5].

Implantable pulse generators usually uses either a constant current (CC) or a constant voltage (CV) power sources for theirs outputs to electrodes [1]. CC power sources adjust voltage in response to resistance (impedance) to ensure that consistent current is delivered to the patient. CV power sources do not adjust voltage in response to impedance; therefore, current delivered to the patient according to Ohm's law will vary in response to impedance changes.

I. Tiginyanu et al. (Eds.): ICNBME 2021, IFMBE Proceedings 87, pp. 127–135, 2022.
https://doi.org/10.1007/978-3-030-92328-0_17

As power source for electric stimulators most commonly are used primary chemical batteries or power transfer/wireless charging via magnetic field or electromagnetic waves. Taking in account smaller power sources of wireless powered devices, ultra-low power design considerations should be implemented for all components. In this article we will describe schematic solution of constant current output driver used in prototype of implantable LES stimulator (WPLES) of application project #15.817.04.19A "Electrostimulation of the lower esophageal sphincter with implantable microstimulator rechargeable by wireless energy transfer in patients with gastroesophageal reflux disease" with followed animal test evaluation in Centrul de chirurgie experimentală "Pius Brânzeu", P-ţa Eftimie Murgu Nr. 2, 300041 Timisoara, Romania [16–18, 22].

2 Design Review and Implementation

2.1 Literature search

During literature search authors found different schematic solutions for building of output constant current driver. Traditionally for this scope is used combination of Digital-to-Analog converter (DAC) for translation of digital code to voltage and operational amplifier (OA) for converting of voltage to current 21. DAC could be integrated in main controller or represents external device. For both type power consumption is usually relatively high, usually order of hundreds microAmpers.

For example, DAC model MAX57058 (used in research 6) has consumption current from 225 to 300 μA. Also power consumption of operational amplifiers needs to be considered. Gong et al. [6] in their work described interesting way of output driver design with using only of double OA for constant current delivery (Fig. 1):

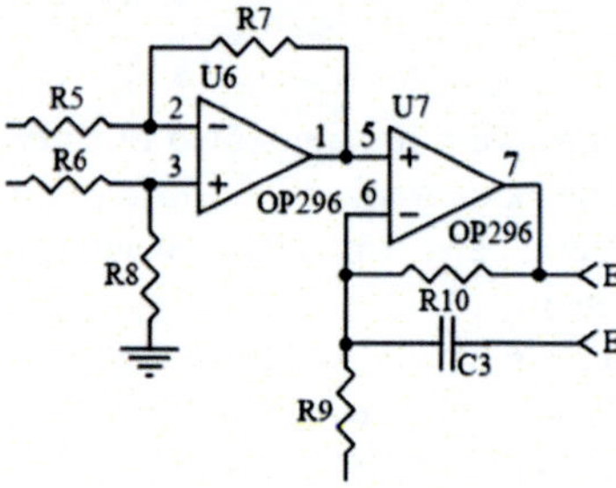

Fig. 1. Schematic of prosthesis circuitry [6]

In this type of driver 1st operational amplifier (U6) is responsible for generation of biphasic voltage pulse output, while 2nd operational amplifier (U7) converts voltage to current. Applying voltage to R5 or R6 inputs resulting generation of positive or negative output pulse.

Resistor R9 regulates the amplitude of output current pulse, for this circuit it is calculating according to Eq. 1:

$$I_{out} = \frac{250mV}{R9} \tag{1}$$

For compensation of net charge imbalance in circuit was also added 0.22 μF capacitor (C3), preventing net current delivery to the tissue and probable damage to the neural tissue. R10 (2 MΩ) is used to provide a leakage path for C3 and the capacitance of the electrode-tissue interface. E1 output is connected to the "ground" electrode, while E2 is connected to the stimulating electrode. This relatively simple schematics has advantage of ultra-low power consumption, especially for using of corresponding OA, like OPA2369 (dual op amp with only 700 nA of consumption current). From another hand, it has limited flexibility of establishing of output current, in this case only by changing of resistor's R9 value.

Our idea was to modify this type of circuit with exchanging of current-setting resistor to digital potentiometer.

2.2 Combined Driver with Op Amp and Digital Potentiometer

Firstly this idea was realized in Analog Devices™ Application note AN-120810 with next schematic solution (Fig. 2):

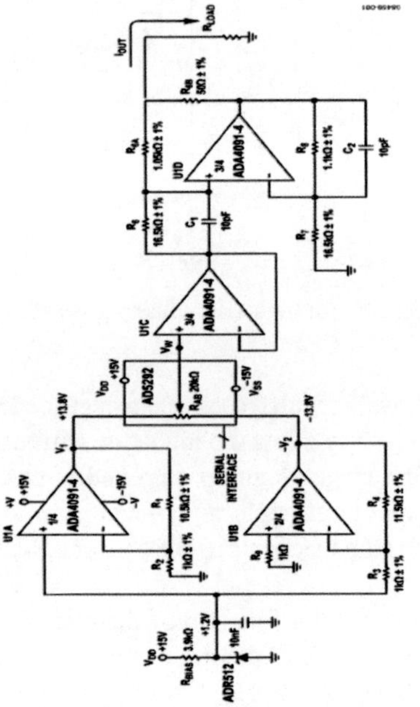

Fig. 2. Programmable bidirectional current source

It represents a programmable bidirectional Howland current source using the AD5292 digital potentiometer in conjunction with the quad ADA4091–4 op amp and the ADR512 voltage reference, offering a 10-bit resolution over an output current range of ±18.4 mA. The AD5292 is programmable over the SPI-compatible serial interface with 20-times programmable memory. Output current could be changed by digital wiper change of AD5292 via SPI-commands.

Having good advantage of digital current control, this type of driver has some weakness, like using of limited 20-times writable potentiometer's wiper settings, using of four operational OA and extra voltage reference, making circuit more sophisticated and increased minimal supply current up to about 1mA according to datasheets of ICs used [11–13].

2.3 Op Amp V-I Converter

Following literature search released promising source – "TI Precision Designs: Verified Design High-Side Voltage-to-Current (V-I) Converter SLAU502" from Texas Instruments™ 14.

Complete schematic of this driver is shown on Fig. 3:

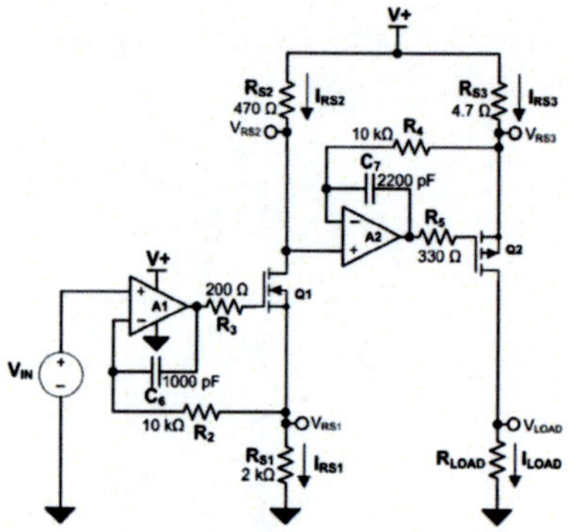

Fig. 3. Schematic for high-side voltage-to-current (V-I) converter

This circuit delivers a well-regulated current to a ground referenced load. The design utilizes a two-stage approach to allow the high-side current source to accept a ground-referenced input. The first stage us an op amp and N-channel MOSFET to translate the ground-referenced input to a supply-referenced sign. The supply-referenced signal drives the second stage op amp that controls the gate of a P-channel MOSFET to regulate the load current.

The voltage-to-current transfer function of the circuit is based on the relationship between the input voltage, V_{IN}, and the three current sensing resistors, R_{S1}, R_{S2}, and R_{S3}. The relationship between V_{IN} and R_{S1} will determine the current that flows through the first stage of the design. The current gain from the first stage to the second stage is based on the relationship between R_{S2} and R_{S3}.

So the output (load) current is calculated using Eq. 2:

$$I_{load} = \frac{V_{IN} \times R_{S2}}{R_{S1} \times R_{S3}} \tag{2}$$

Authors of SLAU502 also provided simulation sources to a SPICE-based electrical simulation software TINA-TI15.

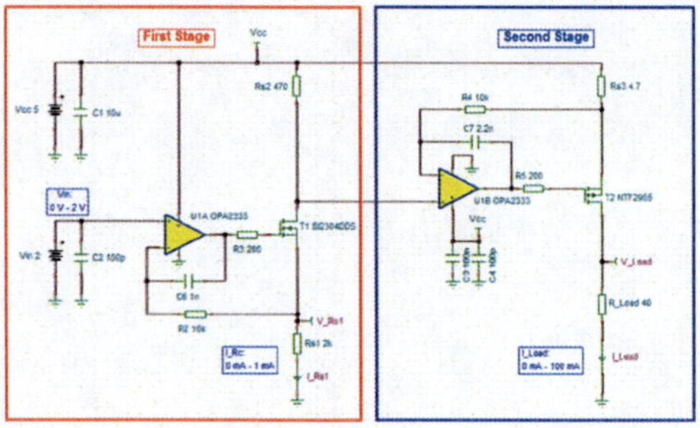

Fig. 4. Schematic TINA-TI™ model for 0–100 mA output

Simulated schematics in this SW packet is presented on Fig. 4:

Existing resistor value R_{S2} and R_{S3} are proposed for output in range 0 to 100 mA. Making values of R_{S2} and R_{S3} as constant and V_{in} equal to supply current (or to HIGH logic signal level in case of +3.3V or 5V power supply), we could change output current only by changing of value of $R_{S1.}$ Using the established Eq. (2) existing SPICE model in TINA-TI^{tm} simulation pack was made modification to own needs of output current in range 0–10 mA and driver control by 3.3V logic signal. So it was enough to modify R_{S3} value to 47 Ω and change R_{S1} to range 0 to 10 kOhm to get 0–10 mA output. Each modification test was evaluated by included SW tools, like "DC analysis" or "AC analysis" tools in Analysis tab.

For example, below is Voltages/Current tables for R_{S1} value of 9,92k and R_{S3} = 47 Ω, R_Load = 510 Ω (Fig. 5). Proposed model showed very good linearity response for load resistance changing in range from 1 Ω (I_out = 1.98 mA) to 1 kOhm (I_out = 1.98 mA) and 2 kOhm (I_out = 1.95 mA).

All simulations were done with Vcc = 4 V and Vin = 4 V as shown on Fig. 6:

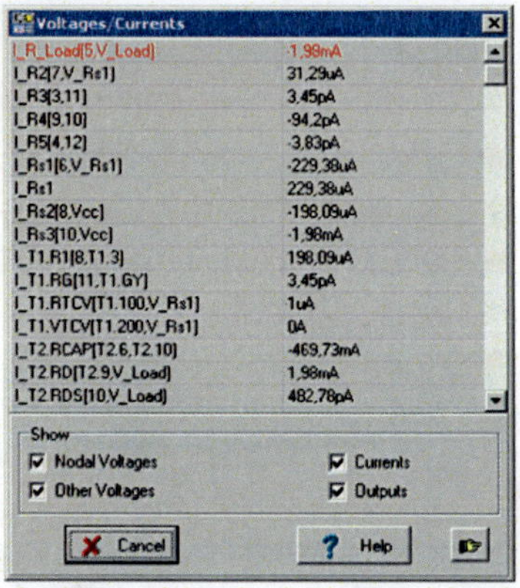

Fig. 5. AC analysis calculations results for test configuration

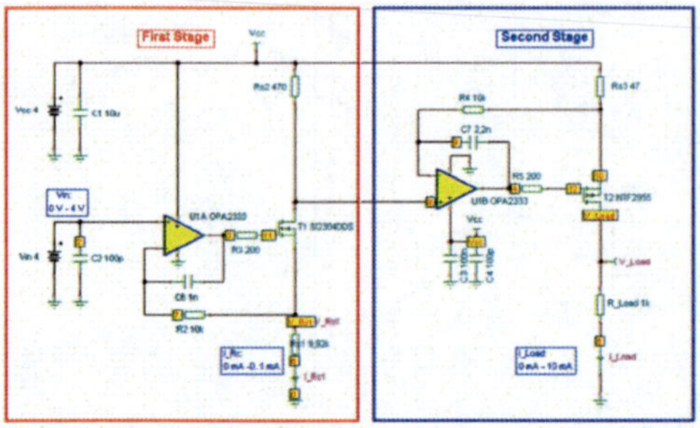

Fig. 6. Modified schematic model for 0–10 mA output

2.4 Proposed Output Driver Design

Basing on above mentioned designs authors developed own driver design with changing R_{S1} to a digital potentiometer, in our case 10 kOhm, 128 taps single-channel digital potentiometers with I2C Interface. Solution has advantages of dual op amp V-I converter, but has digital current setting available, see Fig. 7 for schematics:

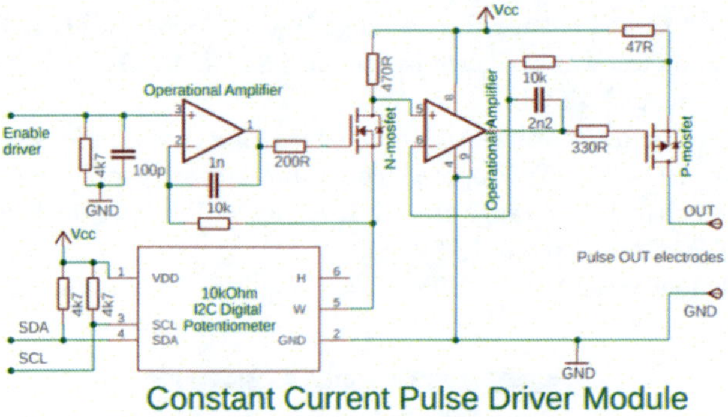

Fig. 7. Developed CC output driver schematics

For easy evaluation this schematics PCB was developed in form of evaluation module with test points, see PCB design and assembled module on Fig. 8:

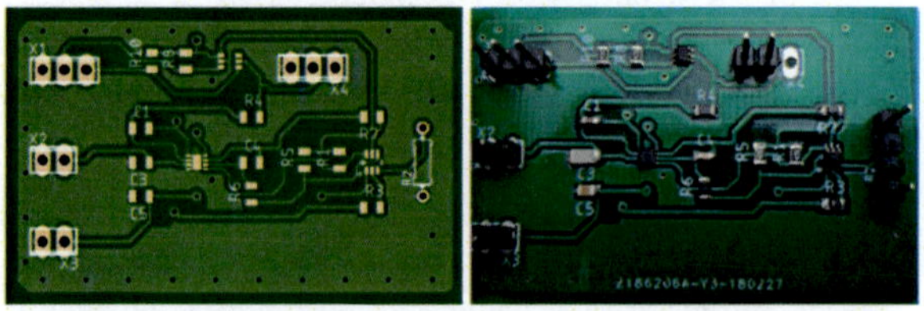

Fig. 8. Evaluation module for CC output driver

2.5 Output Driver Testing and Implementation

After assembling of evaluation module CC driver were performed different tests to confirm output characteristics.

As load capacity were used 1% 0.5W resistors, digital multimeter model VC61A and digital oscilloscope model SHS810 as measuring instruments. Driver showed real characteristics very close to simulated, only several % of difference, maybe due to instrumentation errors. Total current consumption registered were down to 18 μA (17 μA for used OPA2333 samples and about 1 μA for digital potentiometer), but this could be easily improved to about 1.5–2 μA in case of using nani-power op amp like OPA23699. Also this solution has advantage of using of very low PCB space, in case of using usual SON/SC-70 cases for ICs and 0201 passives it could be less than 5x5mm. Practically this driver was tested as part of WPLES prototype testing during animal studies at Centrul de chirurgie experimentală "Pius Brânzeu", P-ţa Eftimie Murgu Nr. 2, 300041 Timisoara, Romania in 20180.

After surgery procedure of WPLES implantation, with electrodes placement of external esophagus wall of a laboratory animal we've connected an oscilloscope to electrodes and registered signal during stimulation pulse (Fig. 9). This oscillogram was recorded for stimulation mode 5 (impulse 375 ms, 6 imp/min, 60 s), output voltage 3.3 V and set-up current 6 mA. A pulse amplitude of 1.6 V was recorded, which at 6 mA current is equivalent to 266.7 ohms of load impedance 0.

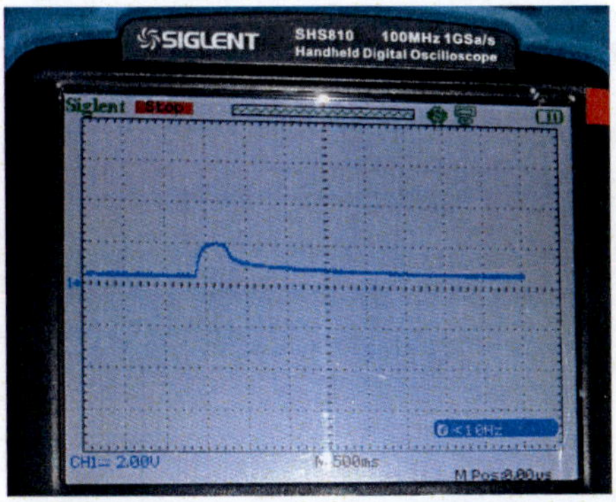

Fig. 9. Stimulation pulse oscillogram

3 Discussion

A simple, space-efficient constant current driver with digital control of output load was developed. Driver could be used in different biomedical applications, especially for ultra-low power applications like wireless powered electrostimulators. Additionally was developed and assembled an evaluation module for easy testing and debugging. Elaborated schematic solution was further used in development and assembling of prototype of implantable LES stimulator (WPLES), with confirmed effectiveness was during animal tests, project Nr. 20.80009.8007.26.

Conflict of Interest. The authors declare that they have no conflict of interest.

References

1. Bortolotti, M.: The "electrical way" to cure gastroparesis. Am. J. Gastroenterol. **97**(8), 1874–1883 (2002). PMID: 12190149
2. Hoppo, T., Rodríguez, L., Soffer, E., Crowell, M.D., Jobe, B.A.: Long–term results of electrical stimulation of the lower esophageal sphincter for treatment of proximal GERD. Surg. Endosc. **12**, 3293–3301 (2014)
3. Katz, P.O., Gerson, L.B., Vela, M.F.: Guidelines for the diagnosis and management of gastroesophageal reflux disease. Am. J. Gastroenterol. (2013)
4. Siersema, P.D., Bredenoord, A.J., Conchillo, J.M., et al.: Electrical Stimulation Therapy (EST) of the Lower Esophageal Sphincter (LES) for refractory gerd and two year results of an international multicenter trial. Gastroenterology **152**(5), S470 (2017)
5. Freire, M.A.M., et al.: Comprehensive analysis of tissue preservation and recording quality from chronic multielectrode implants. PLoS One **6**, 227554 (2011)
6. Gong, W., Merfeld, D.M.: Prototype neural semicircular canal prosthesis using patterned electrical stimulation. Ann. Biomed. Eng. **28**(5), 572–581 (2000)
7. Sun, X., et al.: System design and experimental research of lower esophageal sphincter stimulator for treatment of gastroesophageal reflux disease. In: 2017 39th Annual International Conference of the IEEE Engineering in Medicine and Biology Society (EMBC), pp. 2177–2180 (2017)
8. MAX5705 overview page. https://www.maximintegrated.com/en/products/analog/data-converters/digital-to-analog-converters/MAX5705.html. Accessed 18 Oct 2021
9. OPA2369 overview page. https://www.ti.com/product/OPA2369. Accessed 18 Oct 2021
10. AN-1208 application note. https://www.analog.com/media/en/technical-documentation/application-notes/AN-1208.pdf. Accessed 18 Oct 2021
11. AD5292 overview page. https://www.analog.com/en/products/ad5292. Accessed 18 Oct 2021
12. AD4091–4 overview page. https://www.analog.com/en/products/ada4091-4.html. Accessed 18 Oct 2021
13. ADR512 overview page. https://www.analog.com/en/products/adr512.html. Accessed 18 Oct 2021
14. Wells, C., Chan, D.F.: TI precision designs: verified design high-side voltage-to-current (V-I) Converter SLAU502–June 2013
15. TINA-TI SPICE-based analog simulation program page. https://www.ti.com/tool/TINA-TI. Accessed 18 Oct 2021
16. Ungureanu, S., Sipitco, N., Vidiborschii, V., Fosa, D.: Clinical study of the lower esophageal sphincter electrical stimulation. Glob. J. Res. Anal. **7**(1), 423–426 (2018)
17. Ungureanu, S., Sontea, V., Sipitco, N., Fosa, D., Vidiborschii, V.: Long distance wireless powered implantable electrostimulator. In: 1st International Scientific and Practical Conference Information Systems and Technologies in Medicine, ISM-2018, Kharkiv, Ukraine, 28–30 November 2018 (2018)
18. Ungureanu, S.N., Lepadatu, K.I., Sipitco, N.I., Vidiborschi, V.L., Gladun, N.V., Balica, I.M.: Influence of electrical stimulation on the function of lower esophageal sphincter in patients with gastroesophageal reflux disease. Exp. Clin. Gastroenterol. **128**(4), 51–55 (2016)

19. Vakil, N., et al.: The montreal definition and classification of gastroesophageal reflux disease: a global evidence-based consensus. Am. J. Gastroenterol. **101**(8), 1900–1920 (2006)
20. U.S. Food and Drug Administration (FDA): H990014 Summary of safety and probable benefit, Enterra Therapy System (2015). https://www.accessdata.fda.gov/cdrh_docs/pdf/h990014b.pdf. Accessed 18 Oct 2021
21. Medtronic Inc.: Enterra Therapy Fact Sheet (2003). http://www.medtronic.com/downloadablefiles/Gastro-EnterraFactSheet.pdf. Accessed 18 Oct 2021
22. Vidiborschii, V.: Efficiency of the LES implantable stimulator in animal tests. In: Technical Scientific Conference of Students, Master Students and PhD Students (with International Participation), 26–29 March 2019

Optoelectronic Devices for Blood Testing

Iryna Statyvka(✉) and Mykola Bohomolov

"Igor Sikorsky Kyiv Polytechnic Institute", National Technical University of Ukraine, Kyiv, Ukraine
statyvka.iryna@lll.kpi.ua

Abstract. The article is devoted to the improvement of technical solutions for laser diagnostics of formed elements of human blood using the method of spectrum interferometry, and the development of a method for analysis of the obtained speckle patterns. The developed laboratory stand was used to obtain images of speckles from human blood smears and described the main points of its further integration into the system of non-invasive blood analysis, as well as the development of the principle of image processing.

Keywords: Blood · Blood cells · Speckle interferometry · Speckle pattern · Non-invasiveness

1 Introduction

After the appearance on the world market of laser technologies and methods of their application, there was a need to improve the old and develop new means of optical diagnostics of biological micro-objects, which are widely used in practical and research fields of medicine. The study of blood cells is one of the main tests for the correct diagnosis of the patient and medical research. Development of technical means, methods, algorithms, software for analysis of the state of blood cells - erythrocytes based on modern optoelectronic and laser technologies provide high efficiency, accuracy, and non-invasiveness of research.

Important characteristics in the creation of modern hardware and software for them are the efficiency of the results, ensuring the non-invasiveness of the study and the compactness of the equipment itself.

2 Main Part

2.1 The Main Goals

The main objectives of this article are to solve the scientific problem - to create a scientific basis for the design and development of modern laser diagnostic systems (LDC), which includes the creation of a model of technical means to obtain images of blood cells, which will be able to further analyze, and the actual development of an algorithm that will process the resulting images [1, 2].

I. Tiginyanu et al. (Eds.): ICNBME 2021, IFMBE Proceedings 87, pp. 136–141, 2022.
https://doi.org/10.1007/978-3-030-92328-0_18

2.2 Theoretical Foundations

Methods of laser diagnostics are divided into methods of micro diagnostics (at the level of atoms and molecules) and macro diagnostics (at the level of cells and organs). Micro diagnostics uses all means of linear and nonlinear laser spectroscopy, and macro diagnostics uses methods of elastic and quasi-elastic scattering, interferometry, and holography [2].

For the analysis of homogeneous elements of blood, we are first of all interested in methods of macro diagnostics.

The most suitable in terms of speed of results and subsequent integration into non-invasive blood test systems is the method of speckle interferometry [2].

Speckle interferometry is one of the methods of spatial interferometry, based on the analysis of the granular structure of the image of the object. Proposed in 1970 by Antoine Labeiry. Speckle patterns are a carrier of information in speckle interferometry. They represent a complex granular structure. There are notions of objective and subjective speckles. An objective speckle interferogram is formed in the entire space in front of the illuminated surface. Such pictures are easy to observe with the use of laser radiation. However, if a snapshot of such an interferogram is taken, the resulting image is considered a subjective speckle interferogram. Subjective speckle interferogram occurs when the scattering surface is displayed on the screen using an optical system [3, 4].

The structure of the sinter is formed as a result of the intensity of the distribution of coherent light reflected from a rough surface, the roughness of which is proportional to the wavelength of light, or coherent light passing through the medium with random oscillations of refractive index.

The block diagram of obtaining speckle structures (Fig. 1) is as follows: Laser L forms a beam of coherent radiation, which, in turn, is directed to the focusing lens L1, the focused beam of light falls on the scattering surface SS, where the waves interfere with each other and fall on the lens L2, and from there on the registration screen S [2].

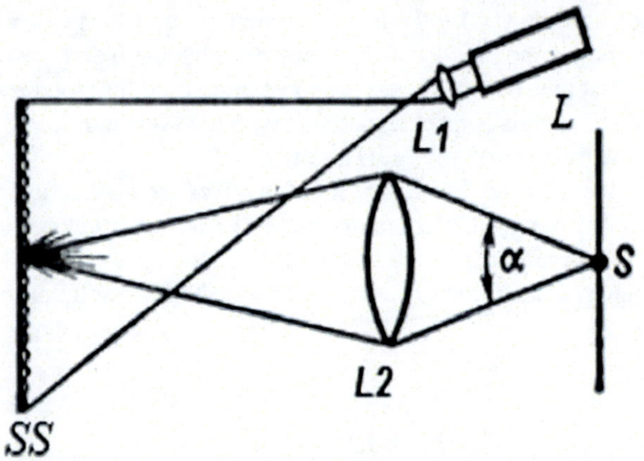

Fig. 1. The block diagram for obtaining speckle structures

In addition to obtaining the structure of blood cells, in combination with the statistical processing of the granular structure, information on the parameters of the object's motion can be obtained. This allows us to study the processes of blood coagulation and microcirculation by processing a series of images of the same sample recorded at different points in time [5, 6].

2.3 Obtaining Speckle Patterns

To be able to obtain speckle images of human erythrocytes in the laboratory, a stand designed based on a block diagram is required (Fig. 1) and an optical diagram (Fig. 2).

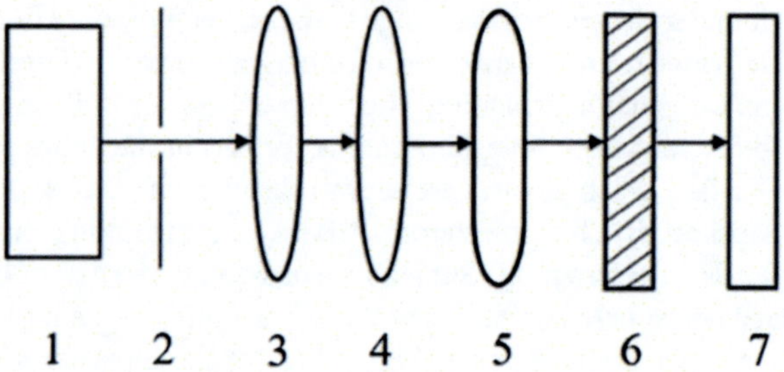

Fig. 2. Optical scheme used to study a blood sample: 1 - laser, 2 - diaphragm, 3, 4 - mirrors, 5 - lens, 6 - sample on a moving surface, 7 - screen

The principle of operation of the optical circuit: The generated flux of electromagnetic waves in the laser (1) is directed to the diaphragm (2), which is necessary for accurate light dosing and proper exposure. From the diaphragm (2) the beam of light falls on the mirrors (3, 4), and then on the condenser lens (5) for focusing rays and speckles - modulated electromagnetic waves, from there it is sent to the object (sample of human blood erythrocytes) and fall on the screen (7) on which we receive the speckle interferogram (Fig. 3 and Fig. 4,) of erythrocytes of human blood.

Due to the fact that He – Ne lasers are most often used for blood research, this type of lasers was used to obtain speckle patterns, with a wavelength $\lambda = 0.6328$ μm and a maximum power Pmax = 45 mW «ЛГ – 38».

Due to the fact that this laboratory stand involves the application of human blood samples in a thin layer on the laboratory glass, the device developed on its basis will allow the use of a minimum amount of sample [1].

Speckle patterns are recorded using a camera with a resolution of 3264 × 2448 pixels.

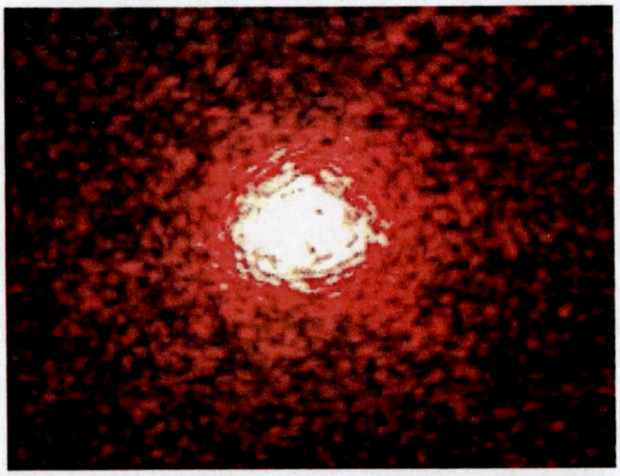

Fig. 3. Speckle–interferograms of the reference sample

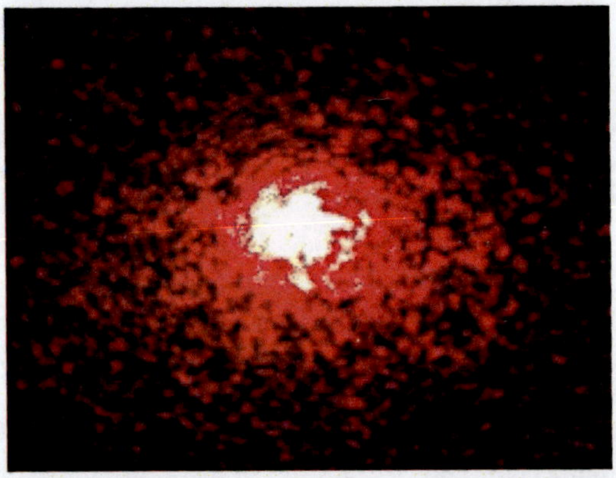

Fig. 4. Speckle–interferograms of the studied sample

2.4 Non-invasive Analysis

Speckle structures can also be obtained non-invasively. In this case, the laser radiation is directed to the area of human skin. In scientific works [5] and [6], the production of speckle patterns by laser irradiation of the human hand was observed. But this method of analysis still requires the presence of a qualified specialist to implement it.

When irradiating the hand, you need to consider many factors associated with the movement of the skin, behind which are the vessels. Therefore, the specialist must carefully fix the patient's arm in the laboratory stand, which complicates the procedure of analysis [5].

The design of the installation for analysis by speckle interferometry should exclude the possibility of obtaining inaccurate results due to errors in the analysis procedure.

Therefore, as an object of irradiation it is necessary to choose the area of skin that is the least mobile, but densely covered with blood vessels [5, 6]. An example of such an area is a human finger, from which blood is usually taken for invasive testing.

This solution will allow the analysis to be performed in outpatient clinics, public places, such as the airport or train station, and without the participation of highly qualified specialists.

The proposed scheme of analysis and obtaining its results is as follows:

1) the patient places a finger in a special hole on the device.
2) the finger is irradiated with a laser, or a series of lasers, to obtain a speckle of the picture, which is fixed by the camera;
3) after image processing, the device produces a result that can be displayed on the screen and sent to the doctor using a single medical network.

In order to avoid self-treatment, after image processing, the analyzer should not give reference values. The results should be given in quantitative terms, which are then sent to a specialist who uses them to diagnose or obtain data for an experiment or study.

2.5 Image Analysis

The development model of speckle pattern analysis involves the recognition of pathological erythrocytes and the creation of a counter that will give a quantitative result (meaning the number will count the number of pathological and normal erythrocytes).

To recognize pathological erythrocytes, the device must compare two images: the speckle pattern of the test and control samples.

To be able to compare two images, they should be modified first. Obtained on an experimental setup for invasive blood analysis using the method of speckle interferometry, has an uneven background, which causes additional errors. Therefore, to avoid this speckle patterns should be subjected to additional processing, scilicet to create a more uniform background.

To determine the number of objects (speckles) on the studied speckle interferograms, we need to perform the conversion of speckles of the image into a binary image.

The developed model, in the final version, will have the following structure:

1) the unit for reading data from the image (provides for the formation of an array);
2) the unit for converting the original data into other types (to ensure the possibility of further statistical processing);
3) a block in which based on the data transformed into the necessary format there is a construction of graphic representation of data;
4) a block of statistical transformed data with their subsequent interpretation as an array.

As an environment for practical testing in the development of the method of processing the results, it is advisable to use the MATLAB environment. This environment allows you to use built-in functions and modules to read graphic data and their subsequent writing to the data set. This is possible by recording the intensity of each pixel that makes up the analyzed image.

This approach allows us to create material in the form of numerical values for its further statistical processing, and the ability to build graphs of the distribution of intensities.

2.6 Checking the Correctness of the Results

In order to ensure the correctness and quality of development, it is necessary to combine few methods of analysis: analysis of speckle patterns obtained by irradiating blood smears; analysis of speckle patterns obtained by irradiating a human finger and analysis of blood cell structure using existing analyzers used in modern medical practice.

3 Conclusions

At present, laser diagnostic systems occupy a leading position among the devices and devices used in the medical field.

Speckle interferometry is a method that allows rapid analysis of blood cells by obtaining speckle patterns.

This method allows for a simple analysis procedure, fast results, and no need to use additional reagents for testing.

The laboratory stand developed on the basis of the block of the diagram of reception of speckle patterns allows to receive rather accurate and suitable for further processing of speckle patterns. This laboratory installation must be integrated into the non-invasive analysis device.

The object of study on such a device is a speckle pattern obtained from the skin of the finger.

To process the obtained speckle images, an algorithm is developed, which consists of several blocks and will be implemented using the MATLAB environment.

When checking the data obtained, it is necessary to compare in vitro and in vivo methods of obtaining images of speckles.

Conflict of Interest. The authors declare that they have no conflict of interest.

References

1. Priezzhev, A.V., Tuchin, V.V., Shubochkin, L.P.: Laser Diagnostics in Biology and Medicine. Nauka, Moscow (1989)
2. Ryabukho, V.P.: Spekl-Interferometry. Soros Educ. J. (2001)
3. Ulyanov, S.S.: What are speckles. Soros Educ. J. (5) (1999)
4. Franson, M.: Specklov Optics. Mir, Moscow (1980)
5. Shebalin, A.: Laser medical diagnostics of the body state by skin biospecles. Photonica (2008)
6. Deegan, A.J., Wang, R.: Microvascular imaging of the skin. Phys. Med. Biol. (2019). https://doi.org/10.1088/1361-6560/ab03f1

Smartphone-Based Pupillometer with Chromatic Stimuli to Screen Neuro-Ophthalmological Diseases

Ana Isabel Sousa[1(✉)], Carlos Marques Neves[2], and Pedro Vieira[1]

[1] Department of Physics, NOVA School of Science and Technology, NOVA University of Lisbon, 2829-516 Caparica, Portugal
ai.sousa@campus.fct.unl.pt

[2] Faculty of Medicine, University of Lisbon, 1649-028 Lisbon, Portugal

Abstract. Pupillometry technique has gained an increased interest over the years as it allows to objectively assess patient's consciousness and neuro-ophthalmological status by measuring pupillary light reflex (PLR). Chromatic pupillometry emerged with the discovery of intrinsically photosensitive retinal ganglion cells (ipRGCs), blue light sensitive and contributors to PLR. Automated pupillometers have been developed over the years to quantitatively measure PLR but have a reduced possibility to be a widespread screening tool as they are expensive and not portable. In this study, a smartphone-based pupillometer was developed, taking advantage of this technology accessibility, low-cost price, and portability. Chromatic stimuli were considered in this smartphone-based pupillometer to assess the contribution of ipRGCs to PLR and allow the screening of the neuro-ophthalmological diseases. In this preliminary study, six healthy individuals participated and pupillometric data was collected using the smartphone application developed, testing different protocols and background light conditions. The system presented good quality of eye images acquired and good behaviour in pupil data extraction. The acquisition protocols tested showed promising results to be used for chromatic pupillometry, although increasing the number of participants is mandatory and further research is needed. Nevertheless, this study shows the potential of the developed pupillometer to function as low cost, portable and accessible screening tool for neuro-ophthalmological diseases.

Keywords: Pupil · Pupillometry · Smartphone · Neuro-ophthalmological diseases

1 Introduction

Pupillometry is used to measure pupillary light reflex (PLR) as a response to a certain stimulus, usually light. This method has been used by clinicians to infer patient's consciousness and neurological status doing either the swinging flashlight test or using automated pupillometers.

I. Tiginyanu et al. (Eds.): ICNBME 2021, IFMBE Proceedings 87, pp. 142–148, 2022.
https://doi.org/10.1007/978-3-030-92328-0_19

Chromatic pupillometry, using coloured light stimuli, emerged over the last 20 years due to the discovery of melanopsin, a photopigment present in intrinsically photosensitive retinal ganglion cells (ipRGCs) [1, 2]. These cells are sensitive to the absorption of short-wavelength (blue) visible light. [3] These recently found cells brought renewed interest in pupillometry research and in its potential use to detect neuro-ophthalmological diseases [4–6].

Automated pupillometers, usually using near-infrared cameras, allow effective quantitative measurement of PLR. However, these types of systems are mainly expensive, not portable and with need of a trained operator. To have a screening tool, more accessible technologies need to be found. Smartphones could be a technological option to widespread pupillometry, reducing those mentioned limitations. The usage of smartphones for pupillometry started around 2013 [7] and there are some studies using this device as a pupillometer since then [7–9]. However, there is no evidence of a smartphone-based system built for chromatic pupillometry, apart from some preliminary results presented in a study from our group [10, 11].

In the present work, a smartphone-based pupillometer was tested and validated to be used for chromatic pupillometry, with the goal of understanding its viability and effectiveness to substitute traditional equipments used for this technique. Thus, this work is part of a validation stage of a medium range Android smartphone pupillometer app for chromatic pupillometry in healthy individuals. Acquisition protocol was also a part of the research in this study, to assess the protocols needed for chromatic pupillometry and neuro-ophthalmological diseases screening.

2 Methods

2.1 Study Participants

Six participants with no known visual abnormalities have been selected for this study. First, was determined the subject's dominant eye, information needed for the acquisition set up as explained in section B. Then pupillometry measurements were made to the opposite eye, for different conditions. All the recorded videos and participants data was anonymized and codified.

This study was approved by the Hospital Santa Maria (Lisbon – Portugal) ethics committee, and a written informed consent was obtained from all the participants.

2.2 Smartphone-Based Pupillometry System

The proposed system uses a smartphone for acquisition, through an Android application developed to support video acquisition and control of the flash light. This study used the rear facing cameras of a Nokia 7 Plus (Nokia Corporation, HMD Global, Finland), with Android version 9 Pie. The application uses the rear-facing cameras to acquire video and activates the rear-facing flash to work as a stimulus. Camera2 API from Android Developers was used for development to implement these requirements.

In terms of acquisition protocol, the app was prepared to enable the user to start recording, the flash activates automatically at a certain instant and with a certain duration and the recording continues for a defined period. When finished, the video file is saved in the smartphone. The app allows some parameters to be configured: time to stimulus (before stimulus), stimulus duration and total recording time.

While the traditional pupillometers use a near-infrared camera, in this study we are using only the smartphone cameras as they are. So, to have a sufficient image resolution, the subjects face was illuminated with an external light, working as a background light, which was placed in front of the face, in a lower position in comparison to the eye. In this way the eye and face get illuminated to obtain a good image quality and enough contrast between pupil and iris but without a direct light entering the retina.

As for the chromatic pupillometry, the rear-facing white flash light was filtered using a standard grade cellophane paper in front of it, in blue and red colors. In previous works these filters spectrum were referred and characterized [10], with the peak wavelengths of 451.9 nm for the blue light and 611.5 nm for the red one.

A chinrest was used in all the experiments to guarantee a steady support for the subject's face and a stabilization of the eye in front of the smartphone. It was asked for the individuals to focus with the dominant eye in a target image placed at the same high of the flash light at a distance of around 70 cm to the smartphone, near to the region where the individuals no longer saw the target due to smartphone occlusion. Accommodation of the non-dominant eye, which was the eye recorded, was then reduced and proper quantity of light entering the retina was expected when stimulus occurred.

2.3 Pupillary Data Processing

Data processing was performed in a computer after the acquisitions, using a Python algorithm and OpenCV library. First step was the extraction of all the frames from the video, followed by cropping each frame in the region of the eye and then a pupil detection algorithm is executed in each frame to get pupil characteristics.

After getting all the frames, the cropped eye image in each one of them is converted to gray, and the image is ready for pupil detection algorithms. Two algorithms were used for pupil detection, first *Pupil Reconstructor* (*PuRe*), developed by Santini et al. [12], was applied to the frame and in case it failed detecting the pupil, a threshold-based algorithm developed by us was run to detect the pupil.

PuRe is based on a novel edge segment selection and conditional segment combination schemes. It was designed for eye images acquired with near-infrared cameras and works purely based on edges through the selection of curved edges that could be significant points of pupil's contour. These selected segments are the conditionally combined to get pupil outlines candidates. An ellipse is fitted in each candidate, which are then evaluated through a confidence measure until the best pupil contour is found, considering its roundness and close to circular shape.

When *PuRe* fails, a second algorithm is used. This algorithm is threshold-based, meaning that the initial gray image, after a blur filter to remove noise, is binarized according to a certain threshold, manually chosen to properly segment the pupil. Then morphological operations are applied followed by the OpenCV *findContours* function. The detected contours are then filtered, by rejecting the contours with big extend that are not close to a pupil shape and the ones with a small area. Finally, the *minEnclosingCircle* function is used to find the best fitting circle and this is the detected pupil.

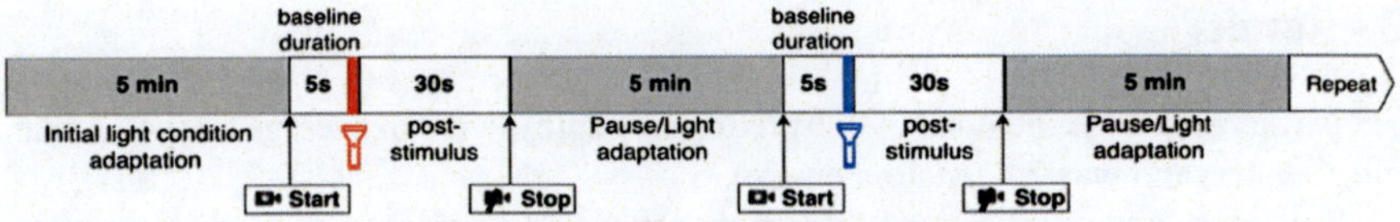

Fig. 1. Acquisition protocol schema [11].

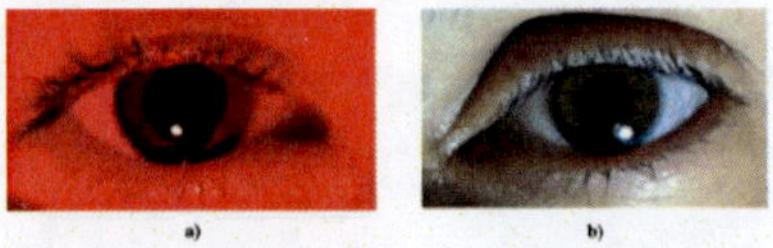

Fig. 2. Examples of eye frames in red (a) or white (b) background light.

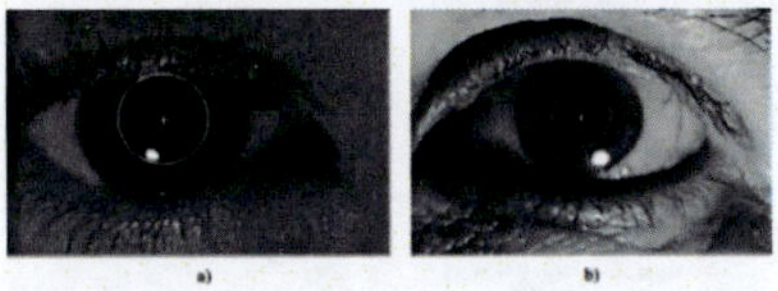

Fig. 3. Examples of eye frames in gray scale with pupil detected.

After having pupil data for each frame, pupil response signal is filtered to remove blinks and outliers using an exponential weighted moving average filter.

Pupil area was the parameter used in this study. Pupil size in pixels was then normalized by baseline, which is the mean pupil size before stimulus. Based in the equation mentioned in [13], pupil size was normalized according to the Eq. 1, where 100% means pupil's baseline size.

$$\text{pupil constriction} = 100 - \frac{\text{baseline} - \text{absolute size}}{\text{baseline size}} \times 100 \quad (1)$$

2.4 Chromatic Pupillometry Protocols

The acquisition protocols are influenced by the recording and stimuli durations. In this work a protocol was defined with initial light adaptation period, followed by 5 s of initial recording to get pupil's baseline, then the stimulus (with different durations) was shown and finally were 30 s of continuous recording to get pupil recovery. A pause was made in between measurements. Each scenario was repeated three times. A schema of this protocol is represented in Fig. 1.

Different acquisition scenarios were tested to understand what were the ones that would allow the activation of ipRGCs with the blue light stimuli, which in pupillometric data means a sustained and slower response using blue stimuli in comparison to the red ones [14]. Stimuli duration was varied: 1, 2, 3 or 10 s; background light was also varied: white and red lights. To each stimuli duration were made measurements with blue and red stimuli.

3 Results

Six participants were included in this preliminary study, with an average age of 33 years old. Gender ratio was 2:4 (Male: Female).

The smartphone application performed as expected allowing to do all the acquisition protocols desired and saving the video files in the smartphone. Video data processing occurred after the acquisitions in a computer. Figure 2 shows examples of frames with both white and red background lights to illustrate the quality of the acquired images. There is a clear contrast between iris and pupil. The pupil detection algorithms performed well in most of the cases; two examples of frames with pupil detected are shown in Fig. 3.

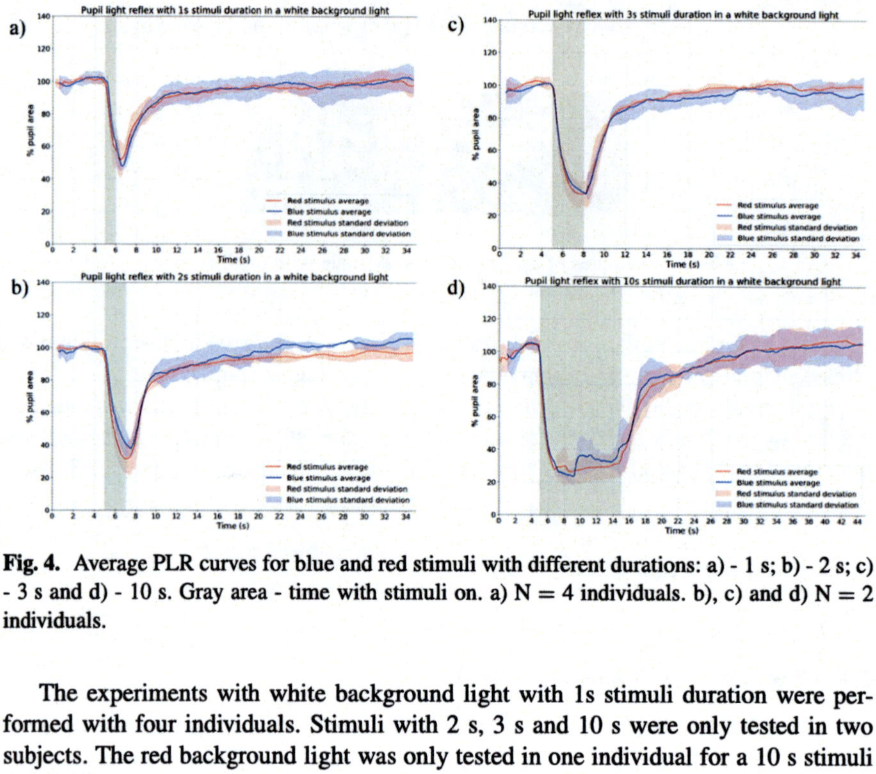

Fig. 4. Average PLR curves for blue and red stimuli with different durations: a) - 1 s; b) - 2 s; c) - 3 s and d) - 10 s. Gray area - time with stimuli on. a) N = 4 individuals. b), c) and d) N = 2 individuals.

The experiments with white background light with 1s stimuli duration were performed with four individuals. Stimuli with 2 s, 3 s and 10 s were only tested in two subjects. The red background light was only tested in one individual for a 10 s stimuli duration.

Graphic results for 1 s, 2 s, 3 s and 10 s stimuli duration with white background light are shown in Fig. 4. These graphs contain the average curves of the experiments for each scenario. Visually, one can observe similarities between red and blue stimuli curves for each different stimuli duration, particularly in the pupil recovery after stimulus.

As for the red background light, was tested only in one individual for now. Figure 5 shows its PLR curve for 10s stimuli duration for both white and red background light, respectively. In the first case both red and blue curves are similar; as for the red background light, there is a small difference between blue and red responses starting at around 18 s in the graph (3 s post-stimulus), with slower response for the blue stimulus.

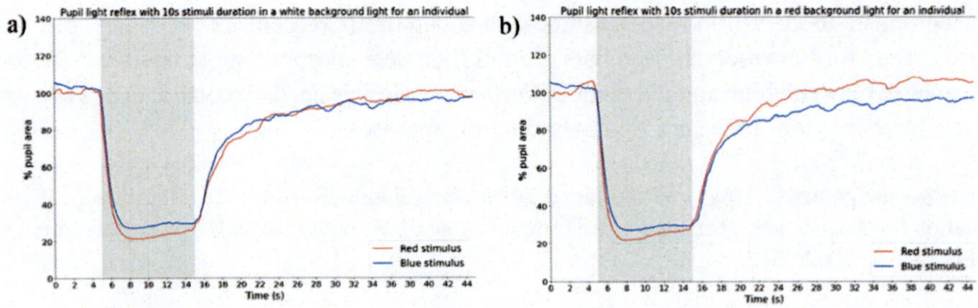

Fig. 5. Average PLR curves for blue and red stimuli for same individual for 10 s stimulus duration: a) – white background light, b) – red background light. Gray area represents the time with stimuli on.

4 Discussion

The pupillometric system developed in this work presented good results for eye image quality and pupil detection, indicating a sufficient quality of the acquired images with a smartphone for good chromatic pupillometric analysis using background light.

The acquisition protocol was under study in this work so that the activation of different photoreceptors was accomplished using chromatic stimuli and background light. With the goal to find a protocol that would activate ipRGCs, which means different pupil recovery curves for blue and red stimuli, stimuli duration was varied. The results obtained for the different stimuli durations show a similarity between blue and red PLR curves for all the durations using white background light. This similarity for both colours with background light adaptation may indicate the dominance of cones in the PLR, with almost no contribution of rods or melanopsin, according to Park et al. [14]. Although these protocols did not indicate to activate ipRGCs, it could be suggestive of a protocol to test cone function in PLR by using white background light, preferably with 1s duration for better comfort of the individual.

As the expected sustained response after blue stimulus offset was not achieved with the white light background, a red background colour was experimented to see if there was an indication of melanopsin (ipRGCs) activation. Red background test shows a small, sustained response 3s after blue stimulus offset (Fig. 5b). Although this result is from one individual and shows a small difference in pupil recovery curve, it can be an indicator that further research should be made and repeat this protocol with more individuals and clarify the viability of the red background light.

Future work should be increasing the sampling and the number of participants for all scenarios to have more significant results. After protocol validation in healthy individuals, tests should be made in patients with neuro-ophthalmological diseases to assess the viability of this system and protocols to be a screening tool.

5 Conclusions

In this study was developed and validated a smartphone-based pupillometer that allows chromatic stimuli with the main goal to be used as a screening tool for neuro-ophthalmological diseases. It was verified that a smartphone has technology sufficient to accomplish good pupillometric results. The acquisition protocol that activates ipRGCs still needs further research. Red background light seems promising to reach a different response between blue and red stimuli. Further research is needed to validate this system as a screening tool for neuro-ophthalmological diseases.

Acknowledgement. This work is funded by National Funds through FCT - Portuguese Foundation for Science and Technology and Future Compta S.A. under the PhD grant with reference PD/BDE/135002/2017.

Conflict of Interest. The authors declare that they have no conflict of interest.

References

1. Lucas, R.J., Douglas, R.H., Foster, R.G.: Characterization of an ocular photopigment capable of driving pupillary constriction in mice. Nat. Neurosci. **4**(6), 621–626 (2001)
2. Hattar, S., Liao, H.W., Takao, M., Berson, D.M., Yau, K.W.: Melanopsin-containing retinal ganglion cells: architecture, projections, and intrinsic photosensitivity. Science (80-). **295**(5557), 1065–1070 (2002)
3. Gamlin, P.D.R., McDougal, D.H., Pokorny, J., Smith, V.C., Yau, K.W., Dacey, D.M.: Human and macaque pupil responses driven by melanopsin-containing retinal ganglion cells. Vision Res. **47**(7), 946–954 (2007)
4. Hall, C.A., Chilcott, R.P.: Eyeing up the future of the pupillary light reflex in neurodiagnostics. Diagnostics **8**(1), 19 (2018)
5. Feigl, B., Mattes, D., Thomas, R., Zele, A.J.: Intrinsically photosensitive (melanopsin) retinal ganglion cell function in glaucoma, pp. 4362–4367 (2011)
6. Ba-Ali, S., Lund-Andersen, H.: Pupillometric evaluation of the melanopsin containing retinal ganglion cells in mitochondrial and non-mitochondrial optic neuropathies. Mitochondrion **36**(January), 124–129 (2017)
7. Kim, T., Youn, J.: Development of a smartphone-based pupillometer, **17**(3), 249–254 (20130
8. Shin, Y.D., Bae, J.H., Kwon, E.J., Kim, H.T., Lee, T.-S., Choi, Y.J.: Assessment of pupillary light reflex using a smartphone application. Exp. Ther. Med. **12**(2), 720–724 (2016)
9. McAnany, J.J., Smith, B.M., Garland, A., Kagen, S.L.: IPhone-based pupillometry: a novel approach for assessing the pupillary light reflex. Optom. Vis. Sci. **95**(10), 953–958 (2018)
10. Sousa, A.I., et al.: Development of a smartphone-based pupillometer for neuro-ophthalmological diseases screening. BIODEVICES 2020 - 13th International Conference on Biomedical Electronics Devices, Proceedings; Part 13th International Joint Conference on Biomedical Engineering System Technology, BIOSTEC 2020, no. Biostec, pp. 50–56 (2020)
11. Sousa, A.I., Neves, C.M., Pinto, L.A., Vieira, P.: Towards the development and validation of a smartphone-based pupillometer for neuro-ophthalmological diseases screening. In: Ye, X., et al. (eds.) BIOSTEC 2020. CCIS, vol. 1400, pp. 39–52. Springer, Cham (2021). https://doi.org/10.1007/978-3-030-72379-8_3
12. Santini, T., Fuhl, W., Kasneci, E.: PuRe: robust pupil detection for real-time pervasive eye tracking, pp. 1–11, December 2017
13. Kelbsch, C., et al.: Standards in pupillography. Front. Neurol. **10** (2019)
14. Park, J.C., Moura, A.L., Raza, A.S., Rhee, D.W., Kardon, R.H., Hood, D.C.: Toward a clinical protocol for assessing rod, cone, and melanopsin contributions to the human pupil response. Investig. Opthalmol. Vis. Sci. **52**(9), 6624 (2011)

The Anisotropy of Light Propagation in Biological Tissues

Elena Achimova(✉), V. Abaskin, V. Cazac, A. Prisacar, A. Mashalkin, and C. Loshmanschii

Institute of Applied Physics of Moldova, Chisinau, Moldova
elena.achimova@ifa.md

Abstract. In this paper we present a modified transmission digital holographic microscope that can be used to image the state of polarization of biological tissue. The resulting device, called polarization-sensitive phase-shifting digital holographic microscope (PS-DHM), records in on-axis geometry the interference between the reference and object beams with the same polarization, but were acquired at two orthogonally polarizations. The object wave transmitted by the biological tissue and magnified by a microscope objective. CCD camera records the two resulting holograms at vertical and horizontal polarizations. The PS-DHM system was upgraded with the liquid crystal variable retarder to perform phase shifts in the reference beam. The polarization-dependent phase-shifted holograms are recorded by rotating the half-wave plates. Using a single hologram, we reconstruct separately the phase map at each polarization, which are used then to represent the phase difference at two orthogonal states of polarization. This phase difference reflects the polarization-dependent refractive index associated to the anisotropy of biological tissue under study. The reconstruction and least-square unwrapping algorithms are used to extract phase information of biological tissues at different polarization states. The birefringence of tissue is obtained from the above-phase distributions. The proposed method is illustrated with applications to three samples: the cancer cells hematoblast, the pollen cells and the heart tissue. The results show that polarization sensitivity exists in the cancer cells hematoblast tissue, the *pollen cells* itself are not the anisotropic, but their walls are birefringent, and any significant anisotropy in the heart tissue was not watched. These results will provide reference for clinic diagnoses and pathological research.

Keywords: Digital holographic microscopy · Anisotropy · Refractive index · Polarization sensitive imaging

1 Introduction

Microscopic examination of cell morphology is one of the main research methods in many areas of biomedicine, such as cancer research, discovery of new drugs, cell behavior, phenotypic screening, the study of pathological processes, etc. [1, 2]. Digital holography (DH) has several features that make it an interesting alternative to conventional microscopy for observation of biological material: improved focal depth, possibility to

I. Tiginyanu et al. (Eds.): ICNBME 2021, IFMBE Proceedings 87, pp. 149–156, 2022.
https://doi.org/10.1007/978-3-030-92328-0_20

generate 3D images and phase contrast images. The light that crosses a biological material can contains amplitude (absorption), phase (refractive) and optical activity (state of polarization) information about the material itself. In common practice, the physical reference wave and object wave have the same polarization to provide the best fringes contrast [3].

Polarization imaging is useful in fields of life science for measuring cell's morphometry and structure, the intracellular content, orientation of the cell membrane, distributions of muscle fibers, and so on. While polarization is not sensitive to the chemical nature of the constituent molecules, it responds to the structural, anisotropic nature of macromolecular assemblies, such as the submicroscopic alignment of molecular bonds and filaments [4]. Thus, one of a further extension of digital holographic imaging is the possibility to quantitative measure the states of polarization modified by a sample and so visualize the structural details of target samples that have optical anisotropy.

In this work we present the polarization-sensitive digital holographic microscope based on the modified Mach-Zehnder interferometer with a liquid crystal variable retarder for producing all-optical phase-shits to obtain the birefringence distribution of biological tissues. By using the hybrid reconstruction algorithm and least square unwrapping, the phase distribution is extracted from the polarization-dependent phase-shifted holograms.

2 Experimental Procedure

The technique of DH implemented in a configuration of an optical microscope is formed Digital Holographic Microscopy (DHM). The objective lens produces a magnified image of the object and the interference between this image and the reference beam is achieved by the integration of the microscope to one of the arms of a Mach–Zehnder interferometer. The interference pattern is recorded by a digital camera. The phase shifting DHM (PS-DHM) was applied to obtain phase digital hologram of the object under study. The main advantage of PS-DHM is the high accuracy of the hologram acquisition. It is mainly due to the full-coverage of the camera sensor's active area. In addition, PS-DHM is performed without frequency filtering usually required for separating the conjugate and the zero-order images in the off-axis configuration. Our PS-DHM system was upgraded with the liquid crystal variable retarder (LCVR) to perform phase shifts in the reference beam. The merit of the developed PS-DHM is that LCVR executes optical phase shifting by driving voltages, with higher accuracy if compared to mechanical displacement, produced via a piezoelectric transducer. In particular, this is because LCVR is mounted in-line, vibration free, and functions more stable by producing high retardance uniformity [5]. Furthermore, PC automatic control of the LCVR adopted in LabVIEW software permits the holograms frame registration within 30 ms. Thus, fast hologram acquisition minimizes the periodic background noise and temperature variation typically influencing the image quality [6]. To acquire the polarization state of objects, the optical setup of DHM was modified by polarizing beam splitter and half-wave plates (Fig. 1).

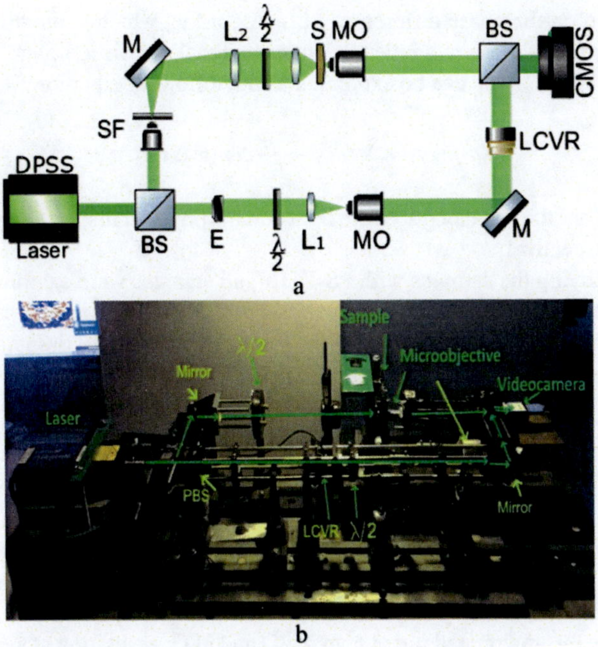

Fig. 1. Setup of DHM: laser single mode (λ = 532 nm, 100 mW), BS-polarized beamsplitters, λ/2 - half-wave plates, M - mirror, MO-microscope objective (magn.60x, NA = 0,65), S-sample, L1, L2 -lenses, CMOS camera, RVCL- LC variable retarder, SF-spatial filter, E- beam expander; b. Photo of the DHM.

The anisotropy in the birefringent material results in the polarization-dependent refractive index. In the uniaxial and linear birefringent materials, the refractive index is defined the extraordinary ray as n_e, the polarization state of which is parallel to the optical axis. Similarly, the refractive index of the ordinary ray is n_o, whose polarization state is perpendicular to the optical axis. Thus, birefringence is the difference between these two refractive indices, $\Delta n = n_e - n_o$. As with the refractive index, birefringence is dimensionless.

Phase differences between the extraordinary and ordinary rays are observed when the incident light passes through the sample, which is labeled $\Delta\varphi(m\Delta x, n\Delta y)$.

$$\begin{aligned} \Delta\varphi(m\Delta x, n\Delta y) &= \Delta\varphi\|(m\Delta x, n\Delta y) - \Delta\varphi\perp(m\Delta x, n\Delta y) \\ &= \frac{2\pi d}{\lambda} n_e - \frac{2\pi d}{\lambda} n_o = \frac{2\pi d}{\lambda}(n_e - n_o) = \frac{2\pi d}{\lambda}\Delta n \end{aligned} \tag{1}$$

where $\Delta\varphi\|(m\Delta x, n\Delta y)$ is the phase distribution of the extraordinary ray. $\Delta\varphi \perp (m\Delta x, n\Delta y)$ is the phase distribution of the ordinary ray. The birefringence is obtained by the phase difference $\Delta\varphi(m\Delta x, n\Delta y)$.

The biological tissue can be described as the spatial distribution of the refractive index in the microscopic level, which consists of fibers and cells. The tested tissue refracts the incident light into two orthogonal polarized lights. The birefringence $\Delta n(m\Delta x, n\Delta y)$ is obtained based on the phase difference $\Delta\varphi(m\Delta x,\ n\Delta y)$, which is the difference of the refractive indexes of these two independent orthogonal polarized lights. Therefore, the birefringence of the tissue can be expressed as the following equation:

$$\Delta n(m\Delta x, n\Delta y) = \frac{2\pi d}{\lambda}\Delta\varphi(m\Delta x, n\Delta y) \tag{2}$$

where λ and d are the wavelength of the light and the physical distance travelled through the sample, respectively.

Before studying the samples with PS-DHM, all samples were examined using light microscopy at different magnifications. It was required to develop a map of the studied area of the sample, on which we can see the features of biological tissue, which we will later identify with the phase map.

3 Results and Discussion

The results of the investigated *cancer cells of hematoblast* obtained from the light field microscope with 20^x and 60^x magnification of microscope objective and from the PS-DHM with 60^x Microscope objective are presented in Fig. 2. From light field examination one can see, that only at 60^x magnification some features of this tissue are visualized (Fig. 2a). Then the unwrapped phase map in 2D and 3D presentation at vertical (Fig. 2b) and horizontal (Fig. 2c) polarization in the reference and the object waves are shown. The unwrapped phase difference between the unwrapped phase map at two polarizations are presented in Fig. 2d. This phase difference reflects the birefringence of the tissue under study. For *cancer cells of hematoblast* samples we see that only large cells with several walls posses the birefringence.

The other group of investigated samples of *pollen cells* is presented in Fig. 3. Examination of these samples under white light showed that 8^x magnification was not enough, but 20^x was sufficient to visualize individual *pollen cells*. So the investigation of these samples was carried out by PS-DHM with microscope objective with 20^x.

From the unwrapped phase difference one can see, that *pollen cells* itself are not the anisotropic, but the walls are birefringent. However this peculiarity may be connected with physical sizes of walls which are strongly different in different directions.

Indeed, the anisotropy of the refractive index, i.e., birefringence, could be related to the presence of membranes, walls, filament arrays included in organelles and cells. To test this assumption, as the next sample for the study, we chose heart tissue samples, which are a range of filaments (Fig. 4a). The heart tissue images, acquired in the study in a light field at two magnifications of 8x and 90x, are shown in Fig. 4a. The filament structure of this tissue is visualized clear at both magnifications. But any significant anisotropy in this biological tissue was not watched independent on applied polarization of DHM imaging system.

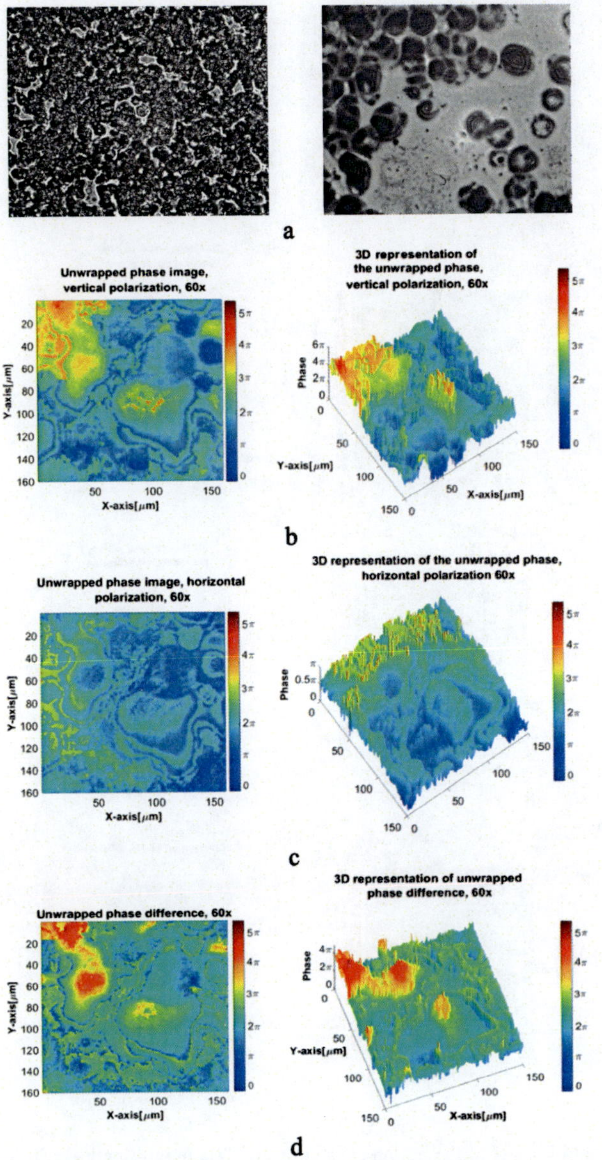

Fig. 2. Light field images of the cancer cells hematoblast at 20x and 60x magnifications; (b) Unwrapped phase image with vertical polarization of the DHM imaging system, 2D and 3D representation, respectively; (c) Unwrapped phase image with horizontal polarization of the DHM imaging system, 2D and 3D representation, respectively; (d) Unwrapped phase difference between vertical and horizontal polarization of the DHM imaging system, 2D and 3D representation, respectively.

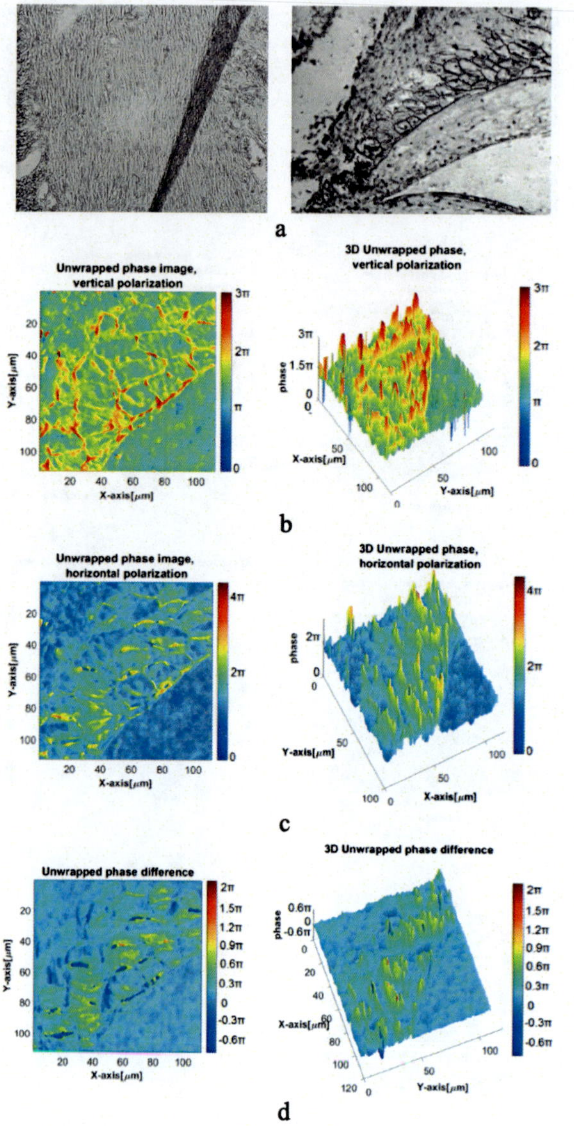

Fig. 3. Light field images of the pollen cells at 8x and 20x magnifications; (b) Unwrapped phase image with vertical polarization of the DHM imaging system, 2D and 3D representation, respectively; (c) Unwrapped phase image with horizontal polarization of the DHM imaging system, 2D and 3D representation, respectively; (d) Unwrapped phase difference between vertical and horizontal polarization of the DHM imaging system, 2D and 3D representation, respectively.

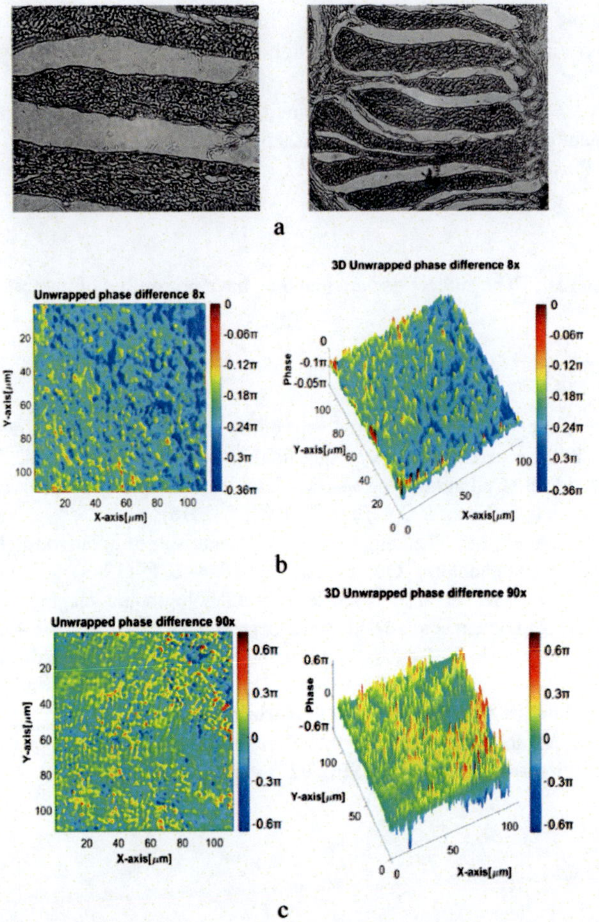

Fig. 4. Light field images of the heart tissue at 8x and 90x magnifications; (b) Unwrapped phase difference between vertical and horizontal polarization of the DHM imaging system at 8x magnification, 2D and 3D representation, respectively; (c) Unwrapped phase difference between vertical and horizontal polarization of the DHM imaging system at 90x magnification, 2D and 3D representation, respectively.

4 Conclusions

We present the polarization-sensitive phase-shifting digital holographic microscope (PS-DHM) set-up, which acquires phase-shifting holograms at two orthogonally polarizations. The PS-DHM system was upgraded with the LCVT to perform phase shifts in the reference beam. The polarization-dependent holograms are recorded by rotating the half-wave plates. This phase difference reflects the polarization-dependent refractive index associated to the anisotropy of biological tissue under study. The proposed method is illustrated with applications to the cancer cells hematoblast, the pollen cells and the heart tissue. The results show that polarization sensitivity exists in the cancer cells hematoblast tissue, the *pollen cells* itself are not the anisotropic, but their walls are birefringent, and any significant anisotropy in the heart tissue was not watched. These results will provide reference for clinic diagnoses and pathological research.

Acknowledgment. The work was partly supported by the project of National Agency for Research and Development of Moldova (ANCD 20.80009.5007.03) and International Project of Program ERA.Net Rus Plus (ANCD 21.80013.5007.1M).

Conflict of Interest. The authors declare that they have no conflict of interest.

References

1. Wallace, J.: A new laser-based bimodal waveguide interferometer sensor detects coronavirus via changes in the sensor's evanescent light field. Laser Focus World **56**(4), 25–27 (2020)
2. Moon, F., Javidi, B.: Cell morphology-based classification of red blood cells using holographic imaging informatics. Biomed. Opt. Exp. **7**(6), 2385 (2016)
3. Sankaran, V., Everett, M., Maitland, D., et al.: Comparison of polarized-light propagation in biological tissue and phantoms. Opt. Lett **24**(15), 1044–1046 (1999)
4. Coppola, G., Ferrara, M.: Polarizion-sensitive digital holographic imaging for characterization of microscopic samples: recent advances and perspectives. Appl. Sci. **10**, 4520–4542 (2020)
5. Bouchal, P., Celechovsk, R., Bouchal, Z.: Polarization sensitive phase-shifting Mirau interferometry using a liquid crystal variable retarder. Opt. Lett. **40**(19), 4567–4570 (2015)
6. Cazac, V., Achimova Abashkin, V., et al.: Polarization holgraphic recording of vortex diffractive optical elements on azopolymer thin films and 3D analysis via phase-shifting digital holographic microscopy. Opt. Exp. **29**(6), 9217–9230 (2021)

Cathodoluminescent UV Sources for Air Disinfection Applications

E. P. Sheshin[1], I. N.Kosarev[1], A. O.Getman[1], I. S. Savichev[1], A. Y. Taikin[1], M. I. Danilkin[2], and D. I. Ozol[1](✉)

[1] Moscow Institute of Physics and Technology, Dolgoprudny, Russia
[2] P.N. Lebedev Physical Institute of the Russian Academy of Sciences, Moscow, Russia

Abstract. Cathodoluminescent UV-light (peaked at 315 nm and 355 nm) sources have been created using commercially available phosphors, with photocatalytic activity observed at irradiation of titanium dioxide. The power density of UV radiation is higher than 10 mW/cm^2. The photocatalytic efficiency was estimated by measuring the oxidation rate of acetone vapors (reaction rate exceeds 2 ppm/min).

Keywords: Ultraviolet radiation · UV light source · Cathodoluminescence · Disinfection · Photocatalysis

1 Introduction

The SARS-CoV-2 pandemic situation makes air disinfection issues topical. Air disinfection with UV radiation can occur in two different ways [1]: either through direct destruction of dangerous entities with UV of germicidal range (250–300 nm) or through photocatalytic oxidation thereof at the surface of titanium dioxide (TiO_2) which acts as a catalyst at UV irradiation in the range of 310–370 nm [2]. Both ways are quite effective, although each of them has its pros and cons. The photocatalytic process is not selective, it is capable of oxidizing almost any organic matter, thus being useful against a wide range of microbes, i.e. bacteria, fungi, algae, protozoa, as well viruses, including the SARS-CoV-2 coronavirus [3]. Another advantage of photocatalytic air disinfection is the absence of undesired by-products, like ozone and nitrogen oxides.

Cathodoluminescent lamps with field emission cathodes [4, 5] provide a new option for environment-friendly mercury-free UV sources, especially bearing in mind the strong current trend of eliminating mercury-containing materials and devices from household and other usages, as ecologically dangerous mercury contaminators [6, 7]. Designing a cathodoluminescent source of the 'photocatalytic' range is less difficult than that of the germicidal range (in particular, uviol glass can be used instead of quarts, due to a good transmission of the former in the required range). The production technology of such UV sources is quite simple, it is well-known since the days of monochrome TV cathode-ray tubes, and such UV sources can be cheap in mass production. Effective UV LEDs in the 315–360 nm range, which are most suitable for photocatalysis, do not currently exist and are not expected to appear in the coming years [8].

I. Tiginyanu et al. (Eds.): ICNBME 2021, IFMBE Proceedings 87, pp. 157–162, 2022.
https://doi.org/10.1007/978-3-030-92328-0_21

We have created the samples of cathodoluminescent sources of UV light with emissions peaked at 315 nm and 355 nm. Commercially available phosphors were employed. We demonstrate the photocatalytic activity of the UV sources by measuring the photocatalytic oxidation rate of acetone vapors.

2 Results

The UV-lamp (Figs. 1,2) is a kind of a triode scheme that comprises a cathode, a modulator, and an anode. The cathode is manufactured of carbon fiber, treated in a particular manner [9, 10]. The carbon-fiber-based field-emission cathodes exhibit long service life (at least 10000 h of continuous operation without noticeable changes of their parameters). The anode is coated with a CRT-phosphor layer, and is a Lambertian radiation emitter (Fig. 3).

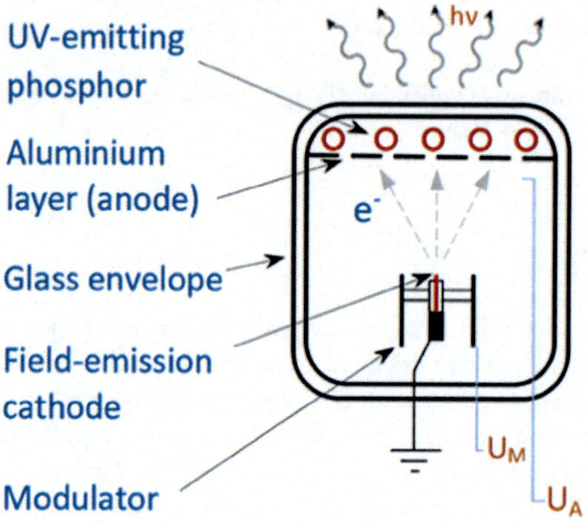

Fig. 1. Schematic view of a cathodoluminescent light source

The emission spectrum of a cathodoluminescent lamp depends only on the phosphor used. So, it is easy to obtain light sources of different spectral ranges with the same lamp design by simply replacing the phosphor (Figs. 4, 5).

There are many types of known UV- phosphors emitting in 300–370 nm range, with their cathodoluminescence (CL) efficiency being as high as 9% [11] and even 20%. Some of them are commercially available, but not optimized for electron beam excitation. The theoretical efficiency limit for CRT-phosphors is 30–40% for visible range [12, 13] and can reach 40–50% for UV-emitting phosphors [14, 15].

Cathodoluminescent lamps can provide the UV radiation power density of the order of 10 mW/cm^2 and higher at the surface of photocatalyst, with the radiation power being distributed uniformly.

Good uniformity of irradiation is the intrinsic feature of cathodoluminescent lamps, which are *large-area* light sources (in contrast to *point* light sources, i.e. LEDs). The

Fig. 2. Cathodoluminescent UV-lamp (switched on). Only a small part of the emitted light falls into the visible range

angular distribution of the cathodoluminescent lamp radiation is Lambertian [16] without the use of special diffusers (Fig. 3).

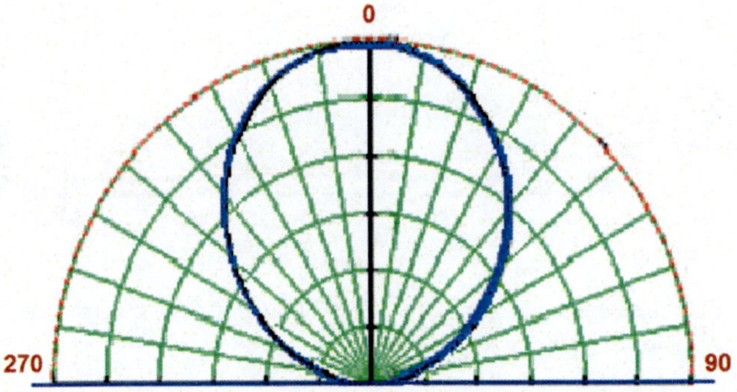

Fig. 3. The angular distribution of the cathodoluminescent lamp radiation

The light absorbed by a TiO_2 photocatalyst creates electron-hole pairs. Prior to transfer to recombination sites, charge carriers that are on the surface of the photocatalyst can be transferred to molecules contained in the air, thereby forming intermediate reaction products like hydroxyl radicals or oxygen ions. Thus, a chemical reaction is initiated, which is completed after a finite number of charge transfer acts. Due to photocatalysis, it is possible to carry out the oxidation of many organic compounds under ambient conditions. Hence, various microorganisms, as well as any other similar organic contaminators, can be destroyed by oxidizing their shells with the aid of photocatalysis.

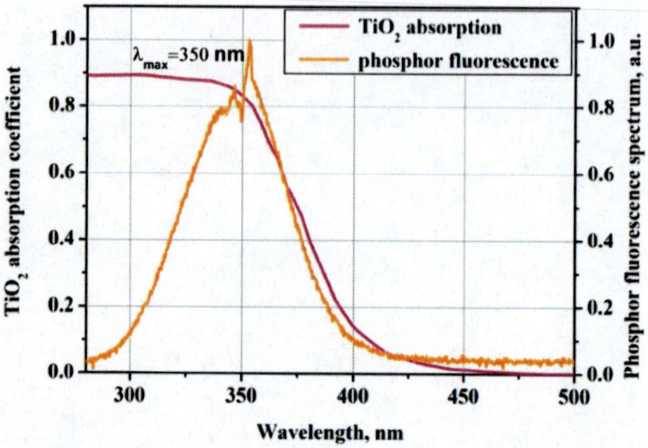

Fig. 4. Cathodoluminescence spectrum of the $BaSi_2O_5$:Pb phosphor versus the TiO_2 absorption curve: overlap at the fading part of the TiO_2 absorption.

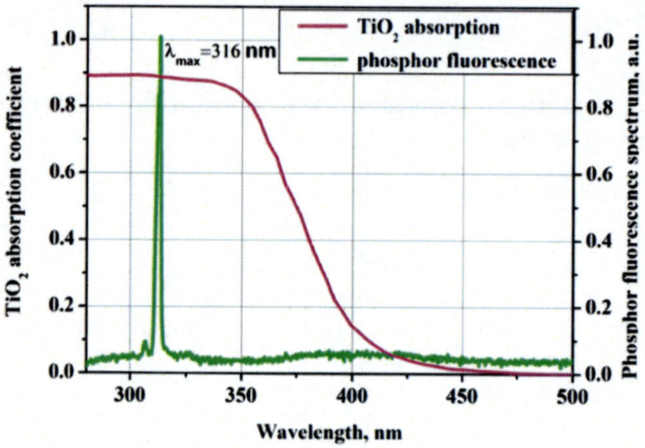

Fig. 5. Cathodoluminescence spectrum of the SrB_6O_{10}:Gd phosphor versus the TiO_2 absorption curve: efficient photocatalysis due to perfect overlap

According to [17], the activation threshold for photocatalysis at absorption of light in TiO_2 is 1 mW/cm^2.

The photocatalytic performance of our light sources was tested by measuring the oxidation rate of acetone vapors above the irradiated surface of the TiO_2 catalyst. The photocatalytic oxidation of acetone occurs as

$$C_3H_6O + 4O_2 \rightarrow 3CO_2 + 3H_2O \tag{1}$$

A sensor was used to measure the CO_2 concentration. The acetone oxidation rate, which was estimated by the CO_2 release rate, exceeds 2 ppm per minute. The emission intensity of the lamp with a silicate phosphor was about 2 times higher than that of the lamp with a borate phosphor, but the photocatalytic performance of the two lamps was comparable due to the smaller overlap integral between the emission spectrum of the former and the absorption spectrum of TiO_2 (compare Figs. 4, 5).

3 Conclusions

Prototypes of cathodoluminescent UV-radiation sources with field emission cathodes on the basis of carbon fiber have been manufactured. The use of field emission cathodes allows decreasing the power consumption of the device and ensures long service life thereof. Besides, the proposed lamps reveal high thermal stability. Notably, their production is similar to the production of CRT tubes, and it can be cheap and environmentally friendly.

Energy efficiency and power density of the tested cathodoluminescent UV sources are higher than those of the UV-LEDs of the same spectral range. The experimental UV light sources demonstrated high potential for photocatalytic purification and disinfection of air.

It is possible to further increase the power of a single lamp, for example, by using several cathodes [18]. Moreover, an additional increase (at least by a factor of 2–3) in the power and efficiency is possible by designing special UV-emitting CRT-phosphors optimized for electron beam excitation.

Conflict of Interest. The authors declare that they have no conflict of interest.

References

1. Kowalski, W.: Ultraviolet Germicidal Irradiation Handbook: UVGI for Air and Surface Disinfection. Springer, Heidelberg (2010). https://doi.org/10.1007/978-3-642-01999-9
2. Oppenländer, T.: Photochemical Purification of Water and Air: Advanced Oxidation Processes (AOPs)-Principles, Reaction Mechanisms, Reactor Concepts. Wiley, Hoboken (2007)
3. Foster, H.A., Ditta, I.B., Varghese, S., Steele, A.: Photocatalytic disinfection using titanium dioxide: spectrum and mechanism of antimicrobial activity. Appl. Microbiol. Biotechnol. **90**(6), 1847–1868 (2011). https://doi.org/10.1007/s00253-011-3213-7
4. Vereschagina N.Y., Danilkin M.I., Kazaryan, M.A., et al.: Cathodoluminescent UV-radiation sources. In: International Conference on Atomic and Molecular Pulsed Lasers XIII Proceedings, vol. 10614, p. 106141F. International Society for Optics and Photonics (SPIE) (2018)
5. Ozol, D. I., Sheshin, E. P., Danilkin, M. I., Vereschagina, N.: Cathodoluminescent UV sources for biomedical applications. In: Tiginyanu, I., Sontea, V., Railean, S. (eds.) ICNBME 2019. IP, vol. 77, pp. 313–317. Springer, Cham (2020). https://doi.org/10.1007/978-3-030-31866-6_60
6. Minamata Convention on Mercury at. http://www.mercuryconvention.org/
7. Coulter, M.A.: Minamata convention on mercury. Int. Leg. Mater. **55**(3), 582–616 (2016)
8. Amano, H., Collazo, R., De Santi, C., et al.: The 2020 UV emitter roadmap. J. Phys. D Appl. Phys. **53**(50), 503001 (2020)
9. Baturin, A.S., Yeskin, I.N., Trufanov, A.I., Chadaev, N.N., Sheshin, E.P., Tchesov, R.G.: J. Vac. Sci. Technol. B **21**(1), 354 (2003)

10. Egorov, N., Sheshin, E.: Field Emission Electronics. Springer, Cham (2017). https://doi.org/10.1007/978-3-319-56561-3
11. Broxtermann, M., Den Engelsen, D., Fern, G.R., et al.: Cathodoluminescence and photoluminescence of YPO_4:Pr^{3+}, Y_2SiO_5:Pr^{3+}, YBO_3:Pr^{3+}, and YPO_4:Bi^{3+}. ECS J. Solid State SC **6**(4), R47 (2017)
12. Levshin, V.L., Arapova, E.Ya., Popov, Yu.M., et al.: Trudy FIAN, vol. 23 (1963)
13. Ozol, D.I.: On the limits of the cathodoluminescence energy efficiency of phosphors. In: 2014 Tenth International Vacuum Electron Sources Conference (IVESC), pp. 1–2. IEEE (2014). https://doi.org/10.1109/IVESC.2014.6892049
14. Aluker, E.D., Lusis, D., Chernov, S.A.: Electronic Excitations and Radioluminescence of Alkali Halide Crystals. Zinatne, Riga (1979)
15. Ozol, D.I.: Cascade luminescence as a way to increase the energy efficiency of cathodoluminophores. In: 2016 29th International Vacuum Nanoelectronics Conference (IVNC), pp. 1–2. IEEE (2016). https://doi.org/10.1109/IVNC.2016.7551487
16. Sheshin, E.P., Kolodyazhnyj, A.Y., Chadaev, N.N., Getman, A.O., Danilkin, M.I., Ozol, D.I.: Prototype of cathodoluminescent lamp for general lighting using carbon fiber field emission cathode. J. Vacuum Sci. Technol. B **37**(3), 031213 (2019)
17. Fujishima, A., Zhang, X., Tryk, D.A.: TiO_2 photocatalysis and related surface phenomena. Surf. Sci. Rep. **63**(12), 515–582 (2008)
18. Sheshin, E.P., Melekescev, V.S., Taikin, A.Y., Ozol, D.I.: Multicathode field emission configurations and their optimization. In: 33rd International Vacuum Nanoelectronics Conference (IVNC), pp. 1–2. IEEE (2020)

A MEMS-INS/GPS Positioning Device for Urban Life Mobility Improvement

Teodor Lucian Grigorie(✉), N. Jula, I. R. Adochiei, C. M. Larco, R. V. Mihai, R. C. Pahonie, and S. Mustata

Military Technical Academy "Ferdinand I"/Aircraft Integrated Systems and Mechanics Department, Bucharest, Romania

Abstract. The paper presents a solution for a positioning device for urban life mobility improvement based on MEMS- INS and GPS data fusion. Two stages are used to process the MEMS-INS and GPS data. At a first stage, which runs off-line, the data fusion uses a method coming from the bio-signal processing domain, i.e. an extension of the Partial Directed Coherence (PDC) method. An optimized filtering procedure is applied here to the IMU components by using the wavelet transform, to process the inertial sensors signals. At the second stage, which runs online, a Kalman filter is used to fuse the INS and GPS data. Also, are shown the hardware structure of the system and the results of the system testing at the lab level.

Keywords: Urban life mobility improvement · Positioning device · MEMS-INS/GPS · Data fusion · Experimental model

1 Introduction

The studies in the last decades revealed that the world's population is ageing dramatically. Every country in the world is facing growth in the number, but also in the proportion of older persons in their population. At the global level, the population aged 65 and over is growing faster than all other age groups [1, 2]. This important segment of the population, together with the people with disabilities, face major problems related to mobility, which makes difficult and sometimes limits their participation to all aspects of urban life. In this context, one of the strategic objectives of the European Commission (EU), also taken over by most of the Member States, is to improve the access at urban life for people with disabilities and for elderly [3]. The EU works with cities and regions to develop a sustainable urban mobility policy. It supports initiatives to make cities accessible for all residents and commuters, including persons with disabilities and the elderly [4].

EU programs specific for this kind of actions offer various funding opportunities, but all of them target the improvement of the cities infrastructures. In in daily life, there are several PC, Android or iOS mobile applications helping people to live better lives, providing traveling guidance or information related to different accessibility issues. All these applications require a precise positioning of the user, which in all situations is provided by a GPS (global position system) receiver.

I. Tiginyanu et al. (Eds.): ICNBME 2021, IFMBE Proceedings 87, pp. 163–169, 2022.
https://doi.org/10.1007/978-3-030-92328-0_22

From another perspective, besides the aerospace applications, the last years shown an increasing trend in the use of positioning and navigation technologies in land-vehicle applications, including here automated car navigation, emergency assistance, fleet management, environment monitoring, asset tracking, and automotive assistance, positioning of elderly and disabled people [5]. The convergence of location, information management, and communication technologies has created a rapidly emerging market known as location-based service, which is a critical enabling technology using location as a filter to extract relevant information to provide value-added service such as location-aware billing, automated advertising services, and other location-based information sought by the user based on its own location. An important segment of this emerging market is reserved to location-based services for elderly and disabled people [6].

Our scientific goal is to propose a solution for a positioning device for urban life mobility improvement based on MEMS (Micro Electro Mechanical Systems) - INS (Inertial Navigation System) and GPS data fusion. Two stages are used to process the MEMS-INS and GPS data. At a first stage, which runs off-line, the data fusion uses a method coming from the bio-signal processing domain, i.e. an extension of the Partial Directed Coherence (PDC) method. An optimized filtering procedure is applied here to the Inertial Measurement Unit components by using the wavelet transform, to process the signals received from inertial sensors. At the second stage, which runs online, a Kalman filter is used to fuse the INS and GPS data.

The results here exposed are coming from an interdisciplinary project. The data acquired from a synergic MEMS-INS/GPS navigation system is processed using embedded systems based on nonlinear methods coming from the bio-signal processing domain. The correlation between the multivariable dynamic systems will be made in order to obtain a technological solution that, in real time, can provide accurate information on the position of the monitored person. The association INS/GPS refers to the use of GPS signal so as to correct or adjust the solutions provided by inertial navigation systems, depending on the movement parameters; in the unfortunate case in which the GPS system has no signal, the INS will continue to calculate the position and angle. For the indoor positioning GPS will be used only for pinpointing the initial position.

Current inertial navigation systems only use raw data obtained with high errors. Thus, advanced processing methods and improved technological solutions are required, both for the performance characteristics of the miniaturized inertial sensor sensors (miniaturization of the sensitive detection element), corroborated with the dynamic and environmental regimes in which the monitoring is performed, as well to improve the quality of the mathematical algorithms for the signal processing.

It must be mentioned that the indoor navigation is much more problematic than the outdoor navigation because indoor navigation is almost entirely dependent on the proposed synergic navigation system architecture that will only function according to the fusion of data from MEMS-INS/GPS subsystems.

2 Difficulty Elements of the Positioning Problem

Most of the positioning technologies used by the modern navigation systems have been available for over three decades. Widely used, the INS and GPS have become key technologies for many positioning and navigation applications. GPS however is still subject to severe performance degradation in the presence of signal blockage, diffraction and multipath and its application in signal-degraded environments such "urban canyons" remains a significant challenge. With characteristics complementary to GPS, the INS has been widely adopted to assist GPS-based navigation systems [7, 8].

GPS is the most widespread GNSS in the world and applies successfully in so many fields such as positioning, navigation, geodesy, mapping, timing and so on. Unfortunately, navigation accuracy and integrity of GPS are degraded in the presence of radio frequency interference, hostile jamming and high dynamical situations, when the satellite signals may get lost due to signal blockage [7, 9]. On the other hand, INSs can address this problem and overcome the non-availability of GPS signals for a short period of time due to the inherent sensors errors. INS is a self-contained positioning and attitude device, which since 1940s has become an important component in military and scientific applications. In fact, INSs are now standard equipment on most planes, ships, and submarines [7, 8]. The primary advantage of using INS is that velocity and position of the vehicle can be provided with abundant dynamic information and excellent short term performance. The main shortcoming is that the INS accuracy degrades greatly over time. In such case, INSs can benefit from aiding such as GPS. Therefore, there is a strong possibility that a GPS/INS integrated navigation system has superior performance in comparison with either a standalone GPS or INS because of their complementary operational characteristics [10, 11].

3 Proposed Solutions for Data Processing

3.1 First Stage in INS/GPS Data Fusion

One of the most important errors influencing the inertial sensors signals is the noise. Because the noise characterizing the inertial sensors is approximately constantly distributed over the low frequency spectrum its direct filtering by using classical methods is not recommended because that spectrum includes the dynamics of the monitored mobile systems. To reduce the noise level in accelerometers and gyro signals many denoising techniques have been used, between them being successfully used the wavelet based method [12, 13]. The great majority of these applications used the thresholding method for the wavelet denoising.

We propose a new time-frequency variable approach, an extension of Partial Directed Coherence (PDC) method, for assessing the multivariate dynamic systems coupling dynamics information [14, 15] by estimating the optimal level of decomposition for the wavelet filter [14, 16, 17].

The algorithm main idea is illustrated in Fig. 1, where the signals received from the inertial system (an accelerometer and a gyro) are processed and analyzed using the Wavelet transform until an optimal level of decomposition for the both signals is established and the useful signals are achieved. This structure is proposed in order to achieve a tuning method for inertial system denoising. In this structure, the reference signals are provided by GPS under the form of the positioning solution, and the disrupted input signals in PDC are the positioning outputs of the inertial navigation system (INS).

The optimal decomposition level of the wavelet filter optimal order is calculated using a correlation analysis function applied to the signals achieved from the accelerometers

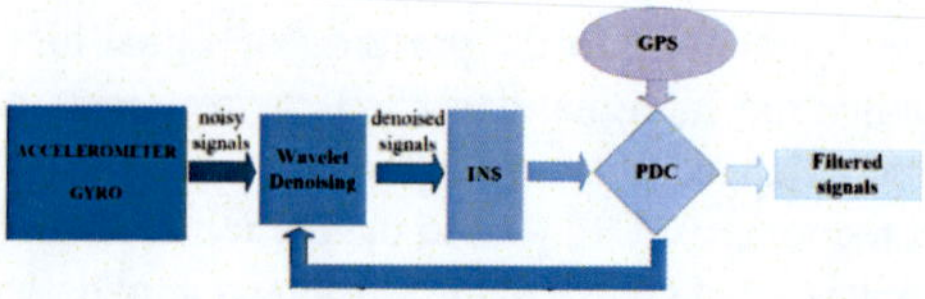

Fig. 1. The architecture of the general tuning method for the inertial system denoising with wavelet method

and gyro real signals, by introducing an optimized version of a time frequency approach, the PDC, to assess coupling dynamics information of the multivariate dynamic systems [18]. PDC approach can detect direct and indirect couplings between two time series. A correlation parameter ($\pi_{ij}(n)$), for a coupling estimation between two time series (X_i and X_j) based on the PDC approach was defined by Baccala et al. [19], with values between 0 and 1. These measures were considered to provide information on the presence and level of causal correlation between two time series (X_i and X_j) as follows: high values reflecting a directionally linear influence from X_j to X_i, meaning that, for values equal to 1, all the causal influences originating from X_j are directed towards X_i, and low values ($\approx$0) suggesting the absence of any causal correlation from X_j to X_i, meaning that X_j does not influence X_i. Another parameter calculation was proposed in order to estimate the coupling level (CL) between two time series belonging to the same system and to estimate the optimal level of the wavelet filter, by employing the following equations:

$$\begin{aligned} &a = mean\ PD(X_i \rightarrow X_j), b = mean\ PD(X_{i-1} \rightarrow X_{j-1}), \\ &CL = WoptLvl = WactualLvl + 1,\ \ if\ a - b > 0 \\ &CL = WoptLvl = WactualLvl,\ \ if\ a - b = 0 \\ &CL = WoptLvl = WactualLvl - 1,\ \ if\ a - b < 0 \end{aligned} \tag{1}$$

where, *WactualLvl* is the wavelet's actual level and *WoptLvl* is the wavelet's optimal level of decomposition.

These measures provide information on wavelet coefficient as follows:

a) if the previous value is lower than the current value then the order of the wavelet decomposition is equal to the previous value plus 1.
b) if the previous value is higher than or equal to the current value, then the optimal order of decomposition is equal to the previous value.

Figure 2 presents the results obtained through the tuning of the wavelet filters by using some experimental data: GPS solution and raw data from inertial sensors.

The differences between the filtered trajectory and reference trajectory are mainly due to the bias influencing the inertial sensors data. After the tuning, performed off-line (with recorded data), the wavelet filters are software implemented and continuously run in the same configuration during the online data fusion of INS and GPS.

3.2 Second Stage in INS/GPS Data Fusion

The second stage of the mathematical algorithm, which runs online, contains the development of the INS/GPS data-fusion techniques by incorporating a Kalman filter. The

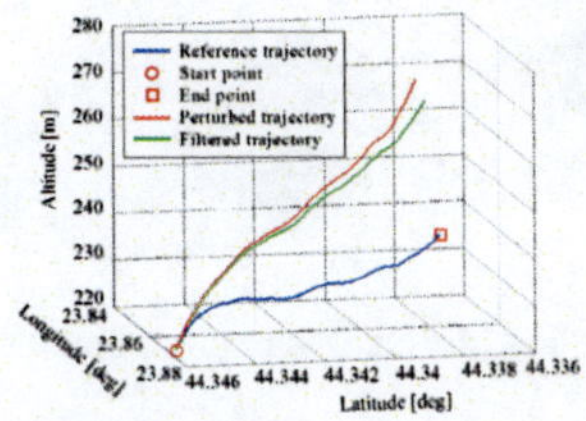

Fig. 2. Experimental filtering in three-dimensional positioning using PDC

chosen INS/GPS structure was based on a loosely coupled architecture, in which the data were fused with a Kalman filter according to the mechanism shown in Fig. 3. The navigation solution, provided by the integrated system, results from the solution correction provided by the INS with the estimates of the position errors, speed errors and attitude errors. The software implementation of the INS/GPS integrated systems considered the inertial navigators mechanization equations, their error dynamic equations and the Kalman filtering equations producing the data fusion of the two navigators, and the estimation of the positioning, speed and attitude errors.

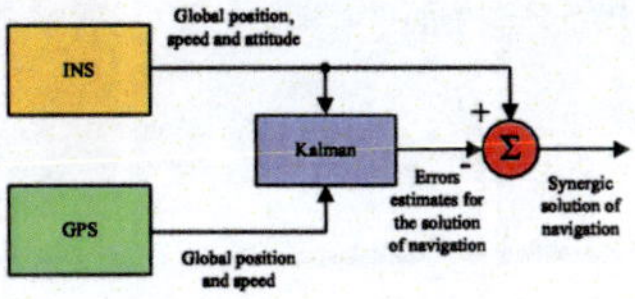

Fig. 3. Structural diagram of the second stage in INS/GPS data fusion

4 Hardware Structure

The technical solution is based on a MEMS-INS/GPS synergic system for high-precision global positioning. In the realization of the hardware structure three main components were used: 1. A module consisting of a GPS receiver and a magnetic field detection sensor; 2. An inertial measurement unit (IMU): a 3D gyro sensor and a 3D accelerometer (MEMS technology); 3. A data processing board.

The global positioning system and the compass (Fig. 4a) consist of a GPS receiver - NEO-M8N and a magnetometer HMC5883L, these being integrated in a module sold by U-blox. For the acquisition of data from the GPS receiver, UART serial communication was used, while for the magnetic field sensor the I2C interface.

The IMU-MPU6050 (Fig. 4b) includes an accelerometer and a gyroscope (3D MEMS sensors). The two sensors provide data on accelerations, angular velocities and system inclination, they communicate with the processing board via I2C communication. The data processing board includes a STM32F103 microcontroller (STMicroelectronics) (Fig. 4c), with a processor with ARM architecture. This allows reading the data provided by the sensors through various serial interfaces, such as UART and I2C communication, as well as the interaction with the system operator through UART serial communication, with a frequency of 72 MHz.

a) b) c)

Fig. 4. Hardware components

5 Results of the System Testing and Conclusions

Some images with the development and assembly of measurement hardware structures for INS/GPS navigator, but also with the commissioning and rough verification of the correctness of the signals provided by them in the on-board equipment laboratory of our institution, are presented in Fig. 5. It can be seen that the experimental setup contains the measuring hardware structures, an oscilloscope and a laptop. The structures were electrically powered using a battery that equips a drone developed in our laboratory. Through the developed software, the laptop allows both the visualization of the correct initialization of the system, and the real-time visualization of the sensors data.

Fig. 5. Development and testing of the system at the lab level

Figure 6 exposes the results obtained after the tuning of the Kalman filter by using experimental data recorded in the situation where the route followed involves a 180-degree turn of the carrier vehicle. As can be seen from the graphs shown, but also based on the results obtained from other tests, it can be concluded that this integration architecture works very well, the integrated solution following the solution of the GPS system, which enjoys a higher level of confidence in the Kalman filter.

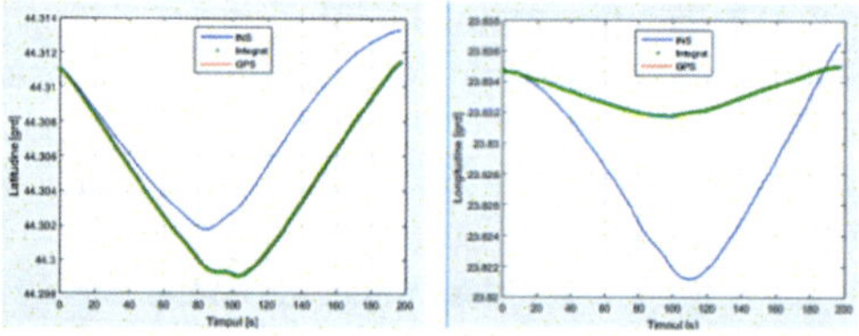

Fig. 6. Navigation solutions for latitude and longitude channels

Acknowledgment. This work was supported by a grant of the Romanian Ministry of Education and Research, CCCDI-UEFISCDI, project number PN-III-P2–2.1-PED-2019–5340, within PNCDI III.

Conflict of Interest. The authors declare that they have no conflict of interest.

References

1. https://www.un.org/en/global-issues/ageing
2. https://population.un.org/wpp/
3. A Renewed Commitment to a Barrier-Free Europe. https://eur-lex.europa.eu/legal-content/EN/TXT/?uri=celex:52010DC0636
4. EU Urban mobility and accessibility, Strategies and policies. https://ec.europa.eu/info/eu-regional-and-urban-development/topics/cities-and-urban-development/priority-themes-eu-cities/urban-mobility_en
5. Tewolde, G.S.: Sensor and network technology for intelligent transportation systems. In: IEEE International Conference on Electro/Information Technology, Indianapolis, USA, pp. 1–7 (2012)
6. Marco, A., Casas, R., Falco, J., et al.: Location-based services for elderly and disabled people. Comput. Commun. **31**(6), 1055–1066 (2008). https://doi.org/10.1016/j.comcom.2007.12.031
7. Mohinder, S., Lawrence, R., Angus, P.: Global Positioning Systems, Inertial Navigation, and Integration. JWS, Inc. (2001)
8. Titterton, D.H., Weston, J.L.: Strapdown Inertial Navigation Technology, 2nd edn. IET Digital Library (2004)
9. Wellenhof, B.H., Lichtenegger, H., Collins, J.: GPS Theory and Practice. Springer, Wien (2001). https://doi.org/10.1007/978-3-7091-6199-9
10. He, X., Hu, X., Wu, M.: Trends in GNSS/INS integrated navigation technology. Coordinates, III/ 3, March 2007
11. Schmidt, G.T.: INS/GPS technology trends, low-cost navigation sensors and integration technology. NATO RTO-EN-SET-116-2011, 28–29 March 2011, Bagneux, France (2011)
12. Kang, C.W., Kang, C.H., Park, C.G.: Wavelet denoising technique for improvement of the low cost MEMS-GPS integrated system. In: International Symposium on GPS/GNSS, Taipei, Taiwan, 26–28 October 2010 (2010)
13. Hasan, A.M., Samsudin, K., Ramli, A.R., et al.: Comparative study on wavelet filter and thresholding selection for GPS/INS data fusion. Int. J. Wavelets Multiresolut. Inf. Process **8**(3), 457–473 (2010). https://doi.org/10.1142/S0219691310003572
14. Adochiei, I.R., Grigorie, T.L., Adochiei, F.C.: Tuning of the wavelet filters for the IMU data based on the PDC method and the GPS solution in a bi-dimensional navigation application. INCAS Bull. **11**(3), 3–14 (2019)
15. Adochiei, F.C., Schulz, S., Edu, I.R., et al.: A new normalised short time PDC for dynamic coupling analyses in hypertensive pregnant women, BMT 2013, Graz, Austria, 19–21 September (2013). https://doi.org/10.1515/bmt-2013-4167
16. Edu, I.R., et al.: Tuning method of the wavelet function for gyro sensor signals denoising with wavelet transform. In: AEROSPATIAL 2014, Bucharest, Romania (2014)
17. Baccalá, L.A., Sameshima, K.: Partial directed coherence: a new concept in neural structure determination. Biol. Cybern. **84**(6), 463–474 (2001). https://doi.org/10.1007/PL00007990
18. Edu, I.R., Adochiei, F.C., Obreja, R., et al.: Inertial sensor signals denoising with wavelet transform. INCAS Bull. **7**(1), 57–64 (2015). https://doi.org/10.13111/2066-8201.2015.7.1.6
19. Baccala, L., Sameshima, K., Ballester, G., et al.: Studying the interaction between brain structures via directed coherence and granger causality. Appl. Signal Process. **5**(1), 40–48 (1998). https://doi.org/10.1007/s005290050005

In Vitro Analysis of Enamel Surfaces with Scanning Electron Microscope After Orthodontic Stripping Reduction Using Various Instruments

D. Rotarciuc[1], A. Țurcanu[2], E. Bud[1], and Eduard V. Monaico[2](✉)

[1] Faculty of Dentistry, George Emil Palade University of Medicine, Pharmacy, Science and Technology of Targu Mures, Targu Mureş, Romania
[2] National Center for Materials Study and Testing, Technical University of Moldova, Chişinău, Republic of Moldova
eduard.monaico@cnstm.utm.md

Abstract. Orthodontic stripping is used to reduce interproximal enamel tissue to solve aesthetic and occlusal problems. There is a wide variety of stripping instruments available on the market. In this study, a qualitative analysis was performed by means of scanning electron microscope with the aim of acquiring information about the efficiency of different manual, mechanical and optical enamel reduction techniques. The paper is especially dedicated to the *in vitro* use of the Er, Cr: YSGG laser for orthodontic purpose – an objective that has not been indicated and evaluated so far in clinical practice.

Keywords: Stripping · SEM · Electron microscope · Contemporary orthodontics

1 Introduction

Interproximal enamel reduction or stripping is a frequent procedure, which is widely used in orthodontic treatments according to the main therapeutic indications – mild or moderate dental crowding and morphology issues. Stripping is accomplished by the orthodontist who respects a rigorous protocol that includes special instruments able to reduce the right quantity of enamel of a desired interproximal place [1]. Plenty of stripping appliances evolved over time, from diamond – coated metal strips to oscillating diamond discs and Ortho Strip System [2].

Qualitative evaluation using scanning electron microscopy (SEM) showed a significant consequence of enamel morphology caused by stripping, leaving the surface full with grooves and ridges [3]. Thus, it is obvious the bacterial plaque accumulation, causing carious lesions and periodontal disease [2]. Er, Cr: YSGG laser was explored as an alternative method of selective elimination of mineralized oral tissue for coronal restauration [4]. Specialized literature does not offer specific information about using laser for interproximal stripping in orthodontics. However, the ablative principle of the

I. Tiginyanu et al. (Eds.): ICNBME 2021, IFMBE Proceedings 87, pp. 170–176, 2022.
https://doi.org/10.1007/978-3-030-92328-0_23

device deserves to be investigated in terms of the advantages provided – being minimally invasive and providing antibacterial effect. This research should offer justified recommendations to orthodontists using stripping methods.

The main goal of this study is *in vitro* evaluation of enamel surfaces reduced by different stripping techniques such as manual, mechanical and laser using systematic morphology analysis with SEM to emphasize the most efficient method from a clinical perspective.

2 Materials and Methods

2.1 Preparation of Dental Surfaces

The set of samples used in the study are composed from 20 teeth selected from dental institutions in Târgu Mureş in 2020. The teeth are selected according to the following inclusion criteria: intact, healthy, extracted for orthodontic or periodontal reasons teeth. Exclusion criteria are: presence of cavities, restorations, fractures, cracks, abrasions, fluoridations on the proximal surfaces.

The extracted teeth were cleaned of debris and soft tissue with diluted 2,5% sodium hypochlorite. After this, they were stored in 0.9% physiological serum. The selected samples were divided into 5 groups depending on the processing method: Group 1-surfaces were reduced with LM Cello abrasive strip; Group 2 - surfaces were reduced with Komet diamond disc; Group 3 - surfaces were reduced with Strauss diamond bur; Group 4 - surfaces were reduced with Er, Cr: YSGG laser Waterlase iPulse; Group 5 - control surfaces (intact enamel). The surface grinding was done according to the manufacturer's instructions using a permanent cooling water action and accomplished by a single operator.

2.2 Morphological Analysis with Scanning Electron Microscopy

After completing the stripping, the samples were individually placed in 0.9% physiological serum until exposed to optical analysis. The samples were prepared for analysis on the VEGA TS 5130 MM scanning electron microscope in the following way: sectioning the samples into approximate halves (see Fig. 1a), chemical fixation with 4% formalin solution at 48 °C for 12 h, washing with distilled water for 1 min and immersion in distilled water for 1 h with water changed every 20 min, serial dehydration with 25% (20 min), 50% (20 min), 75% (30 min), 96% (60 min) ethyl alcohol, drying, fixation with carbon two sided tape on the investigation supports. To ensure a uniform conductivity on the surface of the samples, a thin layer of Au was deposited from plasma at a current of 30 mA and duration of 15 s using the Cressington Carbon Coater/Sputter 108Auto, resulting in the formation of a continuous gold layer with a thickness of 20 nm. An Ag paste contact was deposited on top of the Au layer (see Fig. 1b).

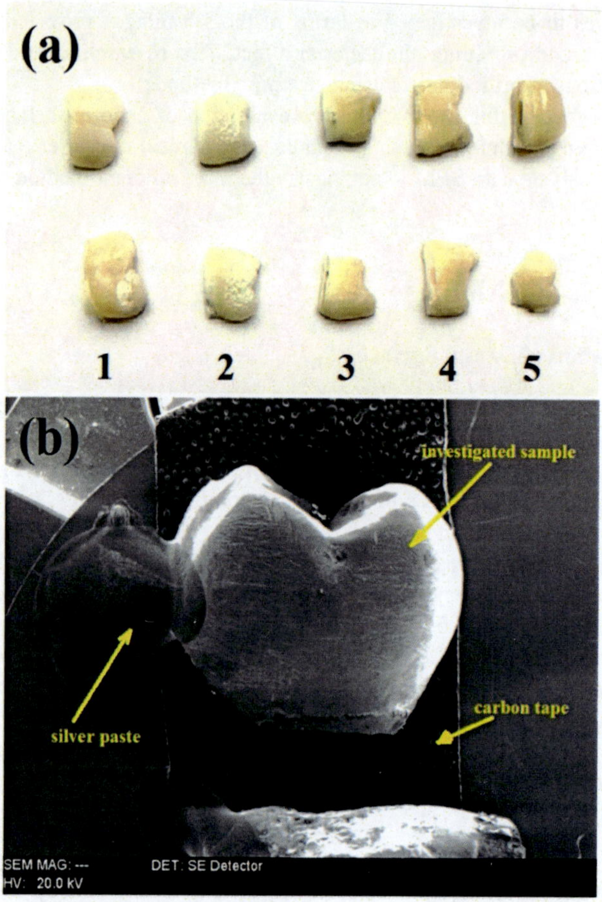

Fig. 1. (a) Photographic image of the sectioned samples. (b) SEM image of the sample fixed on carbon tape and deposited silver paste

3 Results and Disscusions

A range of magnifications of 50x, 150x, 350x, 1000x and 2000x were used to capture the morphology of the investigated surfaces. Standardized photographs of the morphological differences obtained from enamel reduction were taken. No statistical analysis was required as the observations were based on a qualitative analysis of the samples.

The SEM evaluation of the enamel surfaces is presented in Fig. 2. All surfaces subjected to enamel reduction show irregularities such as scratches, grooves and ridges compared to the area of the intact enamel. In particular, SEM analysis highlights the differences of three distinct enamel reduction systems: manual, mechanical and optical.

Stripping has been investigated over time by different authors, thus the results of the present study is in agreement with the observations of several studies, which confirm that interproximal enamel reduction has consequences on the structure of the hard

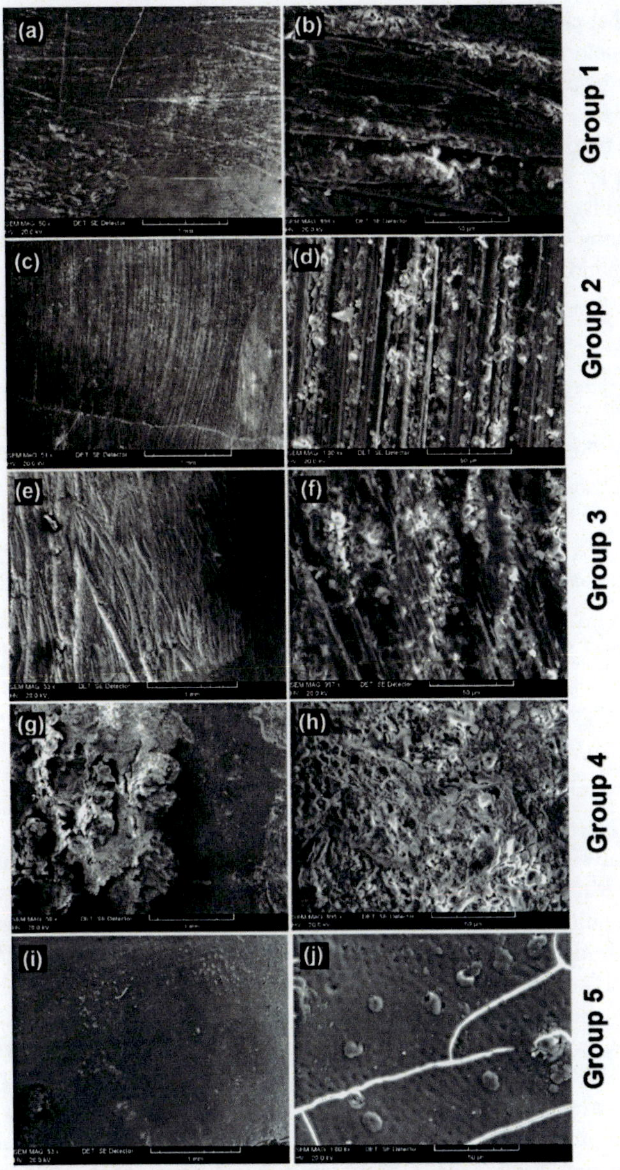

Fig. 2. SEM images at different surface magnifications processed with: (1) Manual abrasive strip (a, b), (2) Diamond disc (c, d), (3) Diamond bur (e,f), (4) Er, Cr: YSGG laser (g,h), and (5) Intact enamel (i,j)

dental tissue, as it produces significant morphological irregularities in both, manual and mechanical methods [5, 6]. The analysis conducted and systematized in Table 1 attests to the fact that classical mechanical stripping systems produce a regular but more pronounced structure, unlike manual methods, which achieve an asymmetric area. Based on the investigations of the present study, it was established that the smoothest area was achieved by the diamond disc (Group 2). This can be attributed to the shape and size of the abrasive particles of the instrument, but also to the adequate technique involving the use of water for cooling. This corroborate with other studies confirming the advantages of diamond discs, stating that this dental instrument produces a superior surface condition, meaning that a regular and slightly roughened area is obtained and the grooves created are minor and follow the same parallel direction [1].

Table 1. Systematized experimental results

Sample group	SEM analysis
Group 1 (Abrasive strip)	Surface characterized by non-parallel, extended grooves alternating with enamel ridges and irregular fragments
Group 2 (Diamond Disc)	More regular surface represented by a series of parallel lines with minor grooves and reduced debris of remaining enamel
Group 3 (Diamond Bur)	Irregular area with deeper and larger, relatively parallel grooves with persistence of enamel debris
Group 4 (Laser)	Irregular, honeycombed surface with small-sized crater formation
Group 5 (Control)	Unprocessed enamel occurs under normal conditions and is almost smooth, debris remains in some places and there are numerous fractures apparently caused by wearing or thermal, chemical and mechanical processing of the samples

SEM analysis explores the enamel morphological structure with certain advantages such as the resolution and capture of field's depth, which allows an excellent topographic evaluation. If the scanning electron microscope allows an accurate qualitative assessment, optical profilometry and atomic force microscopy also perform a quantitative analysis of the exposed samples. *Baumgartner et al.* confirms that the enamel has a rough texture respecting the working direction in comparison with the original intact enamel, which has a prismatic structure, using an analysis developed by optical profilometry based on interferometry techniques and 3D measurements [7]. The work of *Meredith et al.* focused on evaluation by anatomical force microscopy confirms the results of the present study by obtaining the roughest structure in the case of diamond burs [8]. It is not disputed that a quantitative data is advantageous and suitable for an accurate investigation, but the SEM analysis achieves the purpose of current research through definite accuracy and quality.

Although it is a practical and fast method, the Er, Cr:YSGG laser subjects the enamel to significant physical changes, including melting and recrystallisation, resulting in the formation of an area of pores and small bubble-like inclusions. These lasers produce specific, rough surfaces with the spread of numerous microcavities due to the vaporization of water trapped in the hydroxyapatite matrix of the enamel. This process causes roughness and irregularity similar to 37% orthophosphoric acid etching used in coronal restoration protocol [9]. The study by *Grippaudo et al.* investigates stripping using the chemical method in addition to the mechanical method, but finds that the results are unsatisfactory in forming a smooth enamel surface [6]. Recent literature mentions an important advantage of laser use, namely the prevention of demineralization following the application of the optical system on dental surfaces [10]. In order to use the laser correctly in the context of orthodontic stripping, an appropriate protocol for the ablative method is necessary, as currently the surface obtained does not meet the biological conditions for the patient's oral health. However, the procedure has the potential to be adapted to clinical conditions, due to the advantages of the Er, Cr:YSGG laser, defined by caries resistance by decreasing carbonate formation, pyrophosphate formation and calcium/phosphorus ratio exchange in the chemical structure of the tooth [4].

According to the existing literature and the obtained results in this study, it is noted the need for an appropriate enamel finishing method, in order to ensure long-term efficiency after the interproximal reduction of hard dental tissue [6]. *Kaaouara et al.* considers this step to be fundamental and recommends the use of abrasive discs (Sof-Lex, 3MESPE) with different granulations from rough to super fine, to reduce the traces left by stripping and to obtain a surface close to the intact enamel. Topical application of fluoride after interproximal enamel reduction has the advantage of obtaining a hard surface and less susceptible to carious disease [1]. Another recommendation is the use of sealants to obtain a smooth surface in the case of stripping with diamond burs, because even the most careful finishing will not flatten all the deep grooves created [6].

Limitations of this study are conditioned by certain false interferences, which would have affected the results. It is accepted that the amount of reduced enamel is influenced by the operator and the technical aspects used such as the pressure exerted, the hardness and size of the abrasive particles, the duration of the process, the access to the interproximal area and the variable morphological profile of the tooth [11].

Future studies should aim to evaluate *in vivo* the effectiveness of the methods investigated in this study, in order to verify the patient perception and cooperation. The study conducted in the paper allows the formation of valuable recommendations to orthodontists about the choice of stripping instrument according to the clinical indication, the biological risks associated with each instrument used in correlation with the uniformity of the obtained enamel and the possibility of new alternatives.

4 Conclusions

Different stripping methods have produced enamel surfaces with varying qualitative features according to SEM analysis. With respect to the classical methods, the areas processed with diamond burs were the roughest, followed by the abrasive diamond strips. Diamond discs produced the smoothest and least rough surface. Modern optical laser methods are not recommended at present stage due to accentuated porosity which is unacceptable from bioclinical perspective, but the potential of antimicrobial and demineralizing effects of the laser action on the dental tissue is a factor that encourages the progress of optical technology. It is worth to highlight the importance of properly finishing the enamel surfaces that undergo stripping.

The efficiency of stripping is marked by the development of old techniques and the invention of new ones, but maintaining the integrity and balance of the stomatognathic system is a fundamental condition of any therapeutic decision.

Acknowledgment. The authors acknowledge the support from the National Agency for Research and Development under the Grant #20.80009.5007.20.

Conflict of Interest. The authors declare that they have no conflict of interest.

References

1. Kaaouara, Y., Mohind, H.B., Azaroual, M.F., Zaoui, F., Bahije, L., Benyahia, H.: In vivo enamel stripping: a macroscopic and microscopic analytical study. Int. Orthod. **17**(2), 235–242 (2019). https://doi.org/10.1016/j.ortho.2019.03.005
2. Lapenaite, E., Lopatiene, K.: Interproximal enamel reduction as a part of orthodontic treatment. Stomatologija **16**(1), 19–24 (2014)
3. Lombardo, L., Guarneri, M.P., D'Amico, P., et al.: Orthofile®: a new approach for mechanical interproximal reduction: a scanning electron microscopic enamel evaluation. J. Orofac. Orthop./Fortschritte der Kieferorthopädie **75**(3), 203–212 (2014). https://doi.org/10.1007/s00056-014-0213-0
4. El Halim, S., Raafat, R., ElGanzory, A.: ESEM analysis of enamel surface morphology etched with Er,Cr:YSGG laser and phosphoric acid: in vitro study. Egypt. Dental J. **63**(1), 941–947 (2017). https://doi.org/10.21608/edj.2017.75250
5. Gazzani, F., Lione, R., Pavoni, C., Mampieri, G., Cozza, P.: Comparison of the abrasive properties of two different systems for interproximal enamel reduction: oscillating versus manual strips. BMC Oral Health **19**, 247 (2019). https://doi.org/10.1186/s12903-019-0934-y
6. Grippaudo, C., Cancellieri, D., Grecolini, M.E., Deli, R.: Comparison between different interdental stripping methods and evaluation of abrasive strips: SEM analysis. Prog. Orthod. **11**(2), 127–137 (2010). https://doi.org/10.1016/j.pio.2010.08.001
7. Baumgartner, S., Iliadi, A., Eliades, T., Eliades, G.: An in vitro study on the effect of an oscillating stripping method on enamel roughness. Prog. Orthod. **16**(1), 1–6 (2015). https://doi.org/10.1186/s40510-014-0071-8
8. Meredith, L., Farella, M., Lowrey, S., Cannon, R.D., Mei, L.: Atomic force microscopy analysis of enamel nanotopography after interproximal reduction. Am. J. Orthodontics Dentofacial Orthopedics **151**(4), 750–757 (2017). https://doi.org/10.1016/j.ajodo.2016.09.021
9. Vijayan, V., Rajasigamani, K., Karthik, K., Maroli, S., Chakkarayan, J., Haris, M.: Influence of erbium, chromium-doped: yttrium-scandium-gallium-garnet laser etching and traditional etching systems on depth of resin penetration in enamel: a confocal laser scanning electron microscope study. J. Pharm. Bioallied Sci. **7**, 616–622 (2015). https://doi.org/10.4103/0975-7406.163571
10. Lopes, D.S., Pereira, D.L., Mota, C.C., et al.: surface evaluation of enamel etched by Er, Cr:YSGG laser for orthodontic purpose. J. Contemp. Dent. Pract. **21**(3), 227–232 (2021). https://doi.org/10.5005/jp-journals-10024-2777
11. Livas, C., Baumann, T., Flury, S., Pandis, N.: Quantitative evaluation of the progressive wear of powered interproximal reduction systems after repeated use: an in vitro study Quantitative Untersuchung der fortschreitenden Abnutzung elektrisch angetriebener Interproximal-Reduktionssysteme bei wiederholter Anwendung: Eine In-vitro-Studie. J. Orofacial Orthopedics/Fortschritte der Kieferorthopädie **81**(1), 22–29 (2020). https://doi.org/10.1007/s00056-019-00200-x

Selective Ammonia Detection by Field Effect Gas Sensor as an Instrumentation Basis for HP-Infection Primary Diagnosis

Nikolay Samotaev[1](✉), M. Etrekova[1], A. Litvinov[1], and A. Mikhailov[2]

[1] National Research Nuclear University "MEPhI" (Moscow Engineering Physics Institute), Moscow, Russia
nnsamotaev@mephi.ru

[2] LLC "Research & Production Company INKRAM", Moscow, Russia

Abstract. The clinical tests of device based on Metal Insulator Semiconductor Field Effect (MIS-FE) gas sensor for Helicobacter pylori infection diagnostics (HP-infection) are done by breath test method. This method is based on detecting increase ammonia concentration in patient's exhaled air after reception of carbamide water solution. The device's stable operation is based on MIS-FE gas sensor's high sensitivity to ammonia, sensor's parameters stability as well as especially developed two channels gas sampling system.

Keywords: Helicobacter pylori infection · Ammonia · Breath (respiratory) test · Gas sensor · Field effect

1 Motivation

Recently, the so-called "breath test" technique has been intensively developed [1, 2]. The technique's essence is to human diseases identify by the one or another gas marker's presence in the exhaled air. This technique is attractive for its non-invasiveness and research procedure rapidity with the minimum possible patient's involvement in the sampling biological material process. Respiratory tests are used in gastroenterology, pulmonology, dentistry, but they have to use sophisticated and expensive equipment, since it's necessary to measure ultra-low concentration of the gas markers (biomarkers), against the significant interference background from a wide spectrum of the volatile organic compounds (VOC) - products of the human body metabolism. Now precision gas analysis instruments are used, operating on the chromatography, mass spectrometry, IR spectroscopy, ion mobility spectroscopy principles. Due to the high cost and complexity of maintenance (adjustment, calibration, etc.) such devices are available only for large medical centers with highly qualified medical specialists. One of the already established niches in the breath tests' area of gastroenterology is the detection of the HP-infection presence in the human stomach. The Helicobacter pylori effect on the gastric mucosa causes chronic inflammation and radical changes in the tissue micro-environment. This triggers so-called Correa's cascade of fatal diseases [3]:

I. Tiginyanu et al. (Eds.): ICNBME 2021, IFMBE Proceedings 87, pp. 177–184, 2022.
https://doi.org/10.1007/978-3-030-92328-0_24

- non-atrophic chronic gastritis,
- chronic atrophic gastritis,
- intestinal metaplasia,
- dysplasia,
- stomach cancer.

It is not the Helicobacter pylori bacterium itself that leads to stomach cancer, but the inflammation of the gastric mucosa caused by it. There is a "point of no return" for changes in the gastric mucosa under the influence of long-term inflammation associated with Helicobacter pylori, when the bacteria elimination itself may no longer prevent the development of cancer. As an alternative to the breath test, invasive methods are used, including the assessment of stomach material obtained by "invasion" into the body of the stomach in a direct way during endoscopic examination. In addition to ^{13}C, non-invasive methods include urease breath test and determination of HP-infection antigen in feces. Serological research methods are also applicable, which include the determination of IgG antibodies in the blood. The main methods for diagnosing HP-infection in the stomach and indications for their use are discussed in Table 1 [4–7].

Table 1. Methods for diagnosing HP-infection in the stomach and indications for their use

Diagnostic methods	Indications for use	Sensitivity, %	Specificity, %
Serological	Antibodies screening diagnostics of HP-infection in peripheral blood	90	90
Microbiological	Determination of the sensitivity of HP-infection to antibiotics (material - biopsy of the gastric mucosa)	80–90	95
Morphological	Primary diagnosis of HP-infection (biopsy examination)	90	90
Rapid urease test	Primary diagnosis of HP-infection (biopsy examination)	90	90
Breath (respiratory) test	Primary diagnosis and control of eradication	95	95
CITO TEST HP-infection Ag	Primary diagnosis and control of eradication	–	95

Currently, two different approaches are used for the breath test based on the biochemical method for determining the infection with the Helicobacter pylori bacterium by its urease activity, i.e. by the ability to quickly hydrolyze urea. The first approach is more high-tech [8–10], since it is associated with measuring the isotopic composition of the exhaled air after hydrolysis of carbamide - $(NH_2)_2CO$ (the presence of ^{13}C isotope in the sample is detected). The second approach is simpler and is associated with the detection of another hydrolysis product, ammonia (NH_3). In the second approach, the isotopic composition of urea, which the patient takes before the test, does not in any way affect the analysis result, but the detection of NH_3 at concentrations below 1 ppm (part per million) in the patient's exhaled air is a difficult physical task due to the unreliability of the existing cheap hardware design for this task. Currently, there are models of HP-infection research instruments on the market based on the "direct" detection of NH_3 through the use of electrochemical sensors. The duration of a breath test with such devices is no more than 10–20 min. The sensitivity of the electrochemical sensors used is metrologically poorly specified for micro concentrations (the sensor specification [11] indicates a resolution of 1 ppm with a zero drift of ± 3 ppm) and has cross-sensitivity to other gases that may be products of human metabolism organism, for example cross-sensitivity to hydrogen sulfide (H_2S), which can be a natural secretion of the oral (a disease of the digestive system accompanied by a pathological increase in the number of anaerobic microorganisms and an unpleasant odor in the oral cavity).

In essence, the above-described disadvantages of electrochemical sensors result from the concentration ranges of their application in the area of maximum permissible concentration of the working area for individual safety of a person - for NH_3 it is 28 ppm (20 mg/m^3) [12]. Since the main solvent demand in the field of electrochemical sensors is in the field of industrial safety, it is not economically profitable for sensor manufacturers to design and manufacture a small-scale electrochemical sensor, especially since electrochemical sensors based on NH_3 have a limited warranty period and storage (usually no more than 6–12 months). Our work is aimed at eliminating the described drawback in the instability of the sensors used for the breath test through the use of an original technique and a highly sensitive field-effect sensor.

2 Measurement

Figure 1 shows the appearance and MIS-FE sensor's scheme developed by our group based on the gas-sensitive capacitor-type MIS-structure Pd-SiO_2-Si [13–17]. The sensitivity of this gas sensor to ammonia in air is at a high level and allows us to measure NH_3 concentrations in the range from 0.1 to 5 ppm, which successfully solves the problem of monitoring the ammonia content in exhaled air (we have experimentally set the range from 0.4 to 2 ppm) during a urease breath test. The operation principle of the MIS-FE sensor based on the field effect is to change the electrophysical parameters of the semiconductor surface in the MIS structure as a result of the target gas molecules sorption at the metal-dielectric Pd-SiO_2 interface, which diffuse there from the environment through the porous film of the Pd gate. A fixed electric field voltage U_{sh} is applied to the MIS structure (Fig. 1, right, 5), which makes it possible to record the quantitative MIS-FE sensor's response (change in the differential capacitance ΔC) due to the shift of the capacitance-voltage characteristic (C-V) as a result of a flat zones voltage change in the MIS-structure under the gas concentration's C_{NH3} influence (Fig. 2, left). The operation principle and the model of the gas sensitivity mechanism are described in more detail in [13, 16].

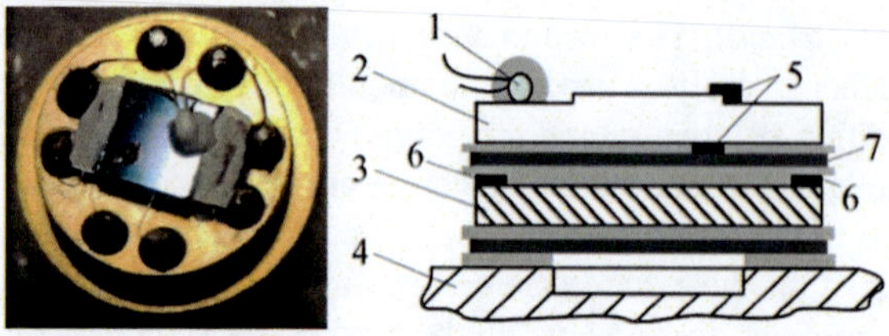

Fig. 1. (Left) MIS-FE sensor chip (Pd gate on SiO_2 insulator on Si substrate; Si-chip size is 5.0 × 5.0 × 0.4 mm) assembling on TO-8 holder and sensor in TO-8 package (diameter 11 mm). (Right) MIS-FE sensor's schema: 1 – thermometer; 2 – Si-chip; 3 – heater; 4 – TO-8 package; 5 – MIS-FE electrical contacts; 6 – electrical contacts for heater; 7 – binder composite electrical insulator

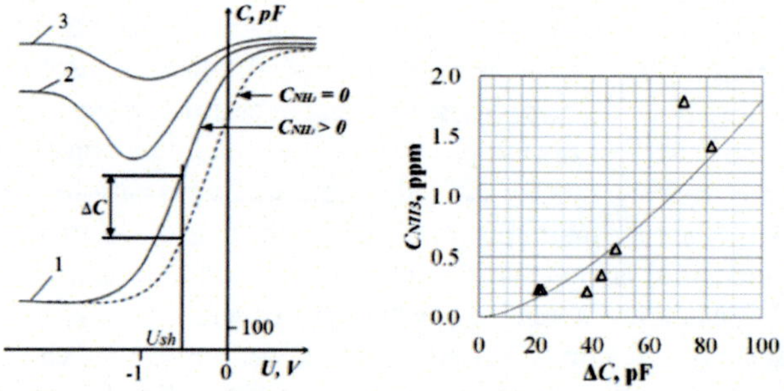

Fig. 2. (Left) The MIS structure's voltage-farad (C-V) characteristic at different heating temperatures: 1 - 100 °C; 2 - 150 °C; 3 - 190 °C; *Ush* - offset voltage; C_{NH3} - ammonia concentration; ΔC – sensor's response exposed to NH_3. (Right) The dependence of the sensor's response on the ammonia concentration (heating temperature is 100 °C)

It is known that a sensor's operating temperature increase reduces the time of its response to the target gas concentration by accelerating the gas molecule diffusion and sorption processes [17]. However, our experiments showed that the MIS-FE sensor's sensitivity decreases with increasing temperature, in particular, due to an undesirable change in the shape of (C-V) characteristics (Fig. 2) due to the semiconductor Si-substrate's properties. Acting within the framework of this contradiction, the optimal MIS-FE sensor's operating temperature was set at 100 °C. At the same time, the speed of the MIS-FE sensor (the response time to the beginning and end of the gas exposure) was no more than 10–20 min, which made it possible to maintain competitiveness. Figure 2 (right) shows the experimental dependence of the MIS-FE sensor's response ΔC with an operating temperature of 100 °C on the ammonia concentration C_{NH3} in the range from 0.25 to 1.7 ppm (calibration curve). Calibration was carried out on a specialized experimental setup using certified and verified equipment, including ammonia microconcentrations sources with a permissible relative error limit of ± 7%. The MIS sensor's signal was processed by an electronic circuit based on a capacitive-digital converter and transmitted to a computer for further processing. As can be seen from Fig. 2 (right), the relative error in ammonia concentration C_{NH3} measuring using the developed MIS-FE sensor is at the level of 20%.

As stated earlier, MIS-FE sensors are also sensitive to a number of other gases, such as hydrogen, hydrogen sulfide, and water vapor [17]. But at the same time, and very importantly, there is no sensitivity to organic substances, including VOCs, carbon dioxide and carbon monoxide. In [15], it is shown how the moisture influence can be successfully and easily compensated. Thus, of all substances contained in the air exhaled by a person [18], the MIS-FE sensor has the highest sensitivity only to ammonia and hydrogen. In this case, ammonia selectivity is achieved by a constructive method and the use of an absorber filter as the device's part, see description below.

The device for HP-infection detection consists of two parts: a gas analyzer based on the MIS-FE gas sensor and gas sampling system (Fig. 3, 4).

Measurements of NH_3 concentration during the analysis are spent against the accompanying gases containing in exhaled air of the patient, such as H_2, O_2, H_2O, CO_2, etc. which concentration can change uncontrollable in time. Therefore, such gas analyzer should possess selectivity only to NH_3. Selectivity is achieved by using differential system of gas concentration measuring with alternate passing gas test through two channels. One channel contains the filter almost completely absorbing NH_3, but passing other gases, and another without the filter, i.e. passes the gas sample without change of its structure. Thus, measuring a difference signal from channels by one MIS-FE sensor, it is possible to define concentration only of ammonia in the gas sample. The second part of the diagnostic device is the patient's gas probe system. It has been specially developed for carrying out the respiratory test on NH_3. The system consists of the tank made of material which is weak adsorbing NH_3. In this tank a gas sample of the air exhaled by the patient through a glass tube is selected. At use of such scheme, the sensor permanently stays in the gas sample atmosphere which has been selected from the patient. Thus, sampling selection is made only twice: before and after carbamide reception.

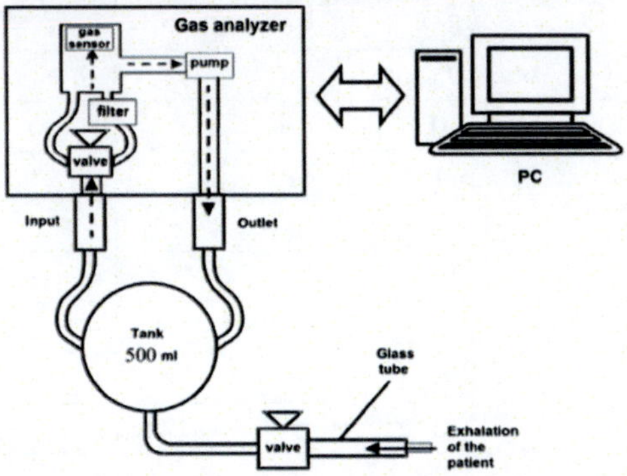

Fig. 3. Block diagram of the device for HP-infection diagnostics by respiratory test

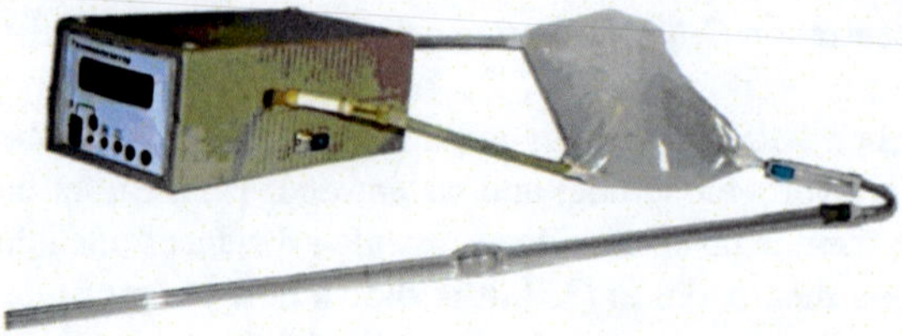

Fig. 4. Photo of the developed device for diagnostics HP-infection by respiratory test during experiments with patients

Twenty patients with different gastric diseases have been surveyed during developed device tests. For comparison, the efficiency of the device tests at the same patient's rapid urea test and cytological screening has been carried out. The inspection results are

Table 2. Results of the HP-infection diagnostics. Designation: «+» - patient is infected by bacterium HP, «–» - patient is not infected by bacteria HP, «№» - number of the patient, «test 1» - rapid urea test, «test 2» - cytologic screening, «test 3» - respiratory test

Gastric diseases	No	Test 1	Test 2	Test 3
Chronic gastritis	1	+	–	+
	2	+	+	+
	3	+	+	+
	4	+	–	–
	5	+	+	+
Gastroduodenitis	6	–	+	+
	7	–	–	–
	8	+	+	+
	9	+	–	–
Duodenal ulcer	10	–	+	+
	11	+	–	–
	12	–	–	–
	13	+	–	–
	14	–	–	–
	15	+	+	+
	16	–	–	–
Gastroesophageal reflux	17	+	+	+
	18	–	–	–
	19	–	–	–
Barrett's syndrome	20	+	–	–

presented in the Table 2. Tests show that long-term stability and high sensitivity to NH_3 especially developed MIS-FE sensor combination of the high selective filter of NH_3 created a reliable and chip tool for medical research of HP-infection presence.

The particular reliability of the developed method was that the sensor used is a solid-state semiconductor device and is not subject to degradation characteristic as of a liquid electrochemical sensor. Degradation of an electrochemical sensor can be caused not only by long-term use and storage, but also by extreme climatic influences. Typical, environmental parameter for electrochemical sensor electrolyte (like example, temperature range minus 10 °C to plus 50 °C for sensor [11]) is relatively narrow in comparison with Russian climate and specific of HP-infection devices using for primary diagnosis outside of clinics. The device can be used to diagnose diseases, residents of remote settlements in the Arctic zone, which can only be reached by helicopter. If for an electrochemical sensor several cycles of freezing an acidic electrolyte will be critical for the loss of the calibrating characteristics of a device for HP-infection diagnosing, then in our case the developed device retains all its metrological settings.

We can conclude that at the moment the developed breath test device for HP-infection (biomarker NH_3 gas), like the market segment located is between electrochemical sensors (the sensitivity of the MIS-FE sensor is better than that of electrochemistry) and chromatography (the filter in the separation of the components is more stable and cheaper than GS-column).

3 Conclusions

An experimental response dependence of the MIS-FE sensor based on a gas-sensitive capacitor type MIS structure Pd-SiO_2-Si with an operating temperature of 100 °C on ammonia in air in the range from 250 ppb to 2 ppm has been established. Due to the high MIS-FE sensor's ammonia sensitivity, a sensor device has been developed that is available for mass production and implementation using the breath test method for the primary non-invasive diagnosis of HP infection in humans. A constructive method for achieving selectivity and stability of analysis by optimizing the MIS-FE sensor's operating parameters and using a two-channel sampling scheme is proposed. Primary clinical tests have been carried out, confirming the competitiveness of the developed breath test method using a device based on a MIS-FE sensor, including along with such research as cytological screening.

Acknowledgment. This work was supported by a grant from the Russian Science Foundation No. 18–79-10230.

Conflict of Interest. The authors declare that they have no conflict of interest.

References

1. Plavnik, R.G., Nevmerzhitsky, V.I., Butorova, L.I., et al.: Use of breath testing in gastrointestinal disorders. Herald of North-Western State Med. Univ. Named After I I Mechnikov **12**(1), 53–62 (2020). https://doi.org/10.17816/mechnikov26270

2. Rezaie, A., Buresi, M., Lembo, A., et al.: Hydrogen and methanebased breath testing in gastrointestinal disorders: the North American consensus. Am. J. Gastroenterol. **112**(5), 775–784 (2017)
3. Robinson, K., Argent, R.H., et al.: The inflammatory and immune response to Helicobacter pylori infection. Best Pract. Res. Clin. Gastroenterol. **21**, 237–259 (2007)
4. Tryapitsyn, A.V., Mal'kov, V.A.: The role and diagnostic value of the most common methods for diagnosing Helicobacter pylori infection. Herald of North-Western State Medical University Named After I I Mechnikov **11**(4), 59–66 (2019)
5. Ivashkin, V.T., Mayev, I.V., Lapina, T.L., et al.: Treatment of helicobacter pylori infection: mainstream and innovations. Russ. J. Gastroenterol. Hepatol. Coloproctol. **27**(4), 4–21 (2017). (In Russ.)
6. Saez, J., Belda, S., Santibáñez, M., et al.: Real-time PCR for diagnosing helicobacter pylori infection in patients with upper gastrointestinal bleeding: comparison with other classical diagnostic methods. J. Clin. Microbiol. **50**, 3233–3237 (2012)
7. Gatta, L., Ricci, C., Tampieri, A., et al.: Accuracy of breath tests using low doses of 13C-urea to diagnose helicobacter pylori infection: a randomised controlled trial. Gut **55**, 457–462 (2006)
8. https://www.otsukael.com/product/detail/productid/10
9. http://hepyscreen.com/tech
10. http://www.amamed.ru/index.php?i=9#naz
11. http://www.membrapor.ch/sheet/Ammonia-Gas-Sensor-NH3-MR-100.pdf
12. Sanitary rules and regulations of the Russian Federation 1.2.3685-2, "Hygienic standards and requirements for ensuring the safety and (or) harmlessness to humans of environmental factors" at. https://docs.cntd.ru/document/573500115
13. Bolodurin, B.A., Korchak, V.Y., Litvinov, A.V., et al.: Comprehensive research on the response of MIS sensors of Pd-SiO_2-Si and Pd-Ta_2O_5-SiO_2-Si structures to various gases in air. Russ. J. Gener. Chem. **88**(12), 2732–2739 (2018)
14. Kalinina, L.N., Litvinov, A.V., Nikolaev, I.N., et al.: MIS-field effect sensors for low concentration of H_2S for environmental monitoring. Procedia Eng. **5**, 1216–1219 (2010)
15. Samotaev, N., Litvinov, A., Etrekova, M., et al.: Prototype of nitro compound vapor and trace detector based on a capacitive MIS sensor. Sensors (Switzerland) **20**(5), 1514 (2020)
16. Nikolaev, I., Litvinov, A., Yemelin, E.: A model of MIS sensors sensitivity mechanism to gas concentration. Sensors Syst. (Russ. Sci. J.) **7**, 66–73 (2006)
17. Etrekova, M., Litvinov, A., Samotaev, N., Filipchuk, D., Oblov, K., Mikhailov, A.: Investigation of selectivity and reproducibility characteristics of gas capacitive MIS sensors. In: Velichko, E., Vinnichenko, M., Kapralova, V., Koucheryavy, Y. (eds.) International Youth Conference on Electronics, Telecommunications and Information Technologies. SPP, vol. 255, pp. 87–95. Springer, Cham (2021). https://doi.org/10.1007/978-3-030-58868-7_10
18. Sukul, P., Schubert, J.K., Zanaty, K., et al.: Exhaled breath compositions under varying respiratory rhythms reflects ventilatory variations: translating breathomics towards respiratory medicine. Sci. Rep. **10**(1), 14109 (2020)

Identifying the Level of Ionizing Radiation Using a Device Implemented on the Arduino Development Board

Alexandru C. Tulică(✉) and I. Şerban

Product Design and Environment, Transilvania University, Braşov, Romania

Abstract. This article discusses how to build an inexpensive radiation identification device. The device can be used in the X-ray environment and can be a real-time indicator of large variations in ionizing radiation. The device is made of two compatible data acquisition boards, the main sensor involved is SBM-20-Geiger Muller. Current devices, such as photographic dosimeters or Geiger-Muller sensors, are not digital and do not alert quickly if there is significant radiation exposure. The device developed and presented in this article makes it possible to quickly identify and prevent ionising radiation. The device was calibrated using a Gamarad DL7 device, with a radiation source: Americium 241, by identifying the radiation level, the presented study was performed.

Keywords: Arduino · SBM-20 · Radiation · Protection · Geiger-Muller

1 Introduction

Since humans have lived on Earth, they have been subject to environmental factors, including ionizing radiation. Even though before technology radiation was natural, after human evolution, artificial radiation began to appear, because of their activities. In the 20th century, nuclear energy was discovered as one of the greatest realizations. Although artificial radiation has provided many benefits to humans, its harm endangers human life. As a result of the Chernobyl nuclear accident in Ukraine in April 1986, there has been an increase in human radiation activity. Researchers, as well as experts in the field, have been looking for ways to reduce or eliminate radiation [1].

Placing the body close to a radioactive source is exposure. To evaluate the hazard of such exposure, it is necessary to calculate the absorbed dose. This is defined as the energy required for a certain mass of tissue. In general, the dosage is not consistent throughout the body. A radioactive substance may be selectively absorbed by various organs or fabrics [2].

Radiological dosimetry refers to methods for the quantitative determination of the energy stored in a given environment by direct or indirect ionising radiation. Certain quantities and units have been defined to describe beam radiation, as well as the most commonly used dosimetric quantities and their units defined below. A simplified discussion of cavity theory, the theory that deals with calculating the response of a dosimeter in an environment [3].

I. Tiginyanu et al. (Eds.): ICNBME 2021, IFMBE Proceedings 87, pp. 185–192, 2022.
https://doi.org/10.1007/978-3-030-92328-0_25

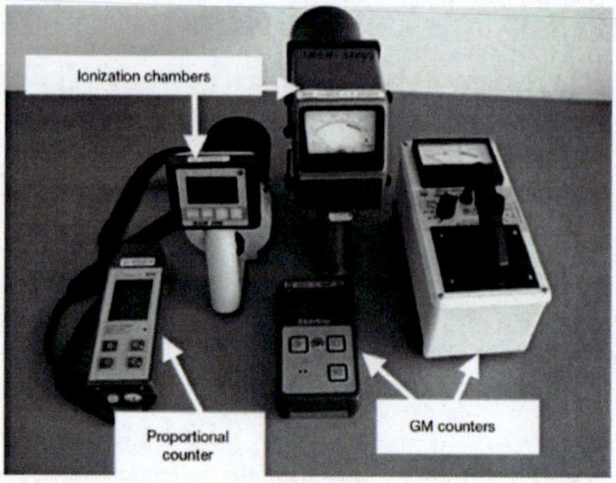

Fig. 1. Counter radiations [3]

The area meters commonly used for measuring radiation levels are of the following types: ionization chambers, proportional meter and GM meters, in Fig. 1.

2 Device Design and Realization

Arduino is known as a physical platform or embedded computing, which means that it is an interactive system, which through the use of hardware and software can interact with its environment. For example, simple use of the Arduino would be to turn on a light for a certain period, 30 s, after a button has been pressed. Arduino would wait patiently for the button to be pressed. When the button is pressed, the lamp will turn on and counting will begin. Once it has counted 30 s, it would deactivate it and then continue to wait to press another button. You can use this setting to control a lamp in a closet under stairs, for example [4].

Fig. 2. Arduino Uno [5]

The radiation detector compatible with Arduino Uno, in Fig. 2 has the following components: 1. Terminal Block DG301, has screws for fixing the battery wires, for connecting the wires the polarity of the wires must be checked, 2. Slide switch, 3. Toner Trimmer, marker 103, 4. Base for installing CD4011 IC (14-pin) with CD4011BE processor; 5. LED (Tube Event Indicator); 6. Buzzer; 7. Radial inductor; 8. SBM-20 sensor; 9. Pins for connection between Arduino Uno board and Radiation Detector compatible with Arduino Uno; 10. Electrolytic capacitor (10 μF) (Fig. 3).

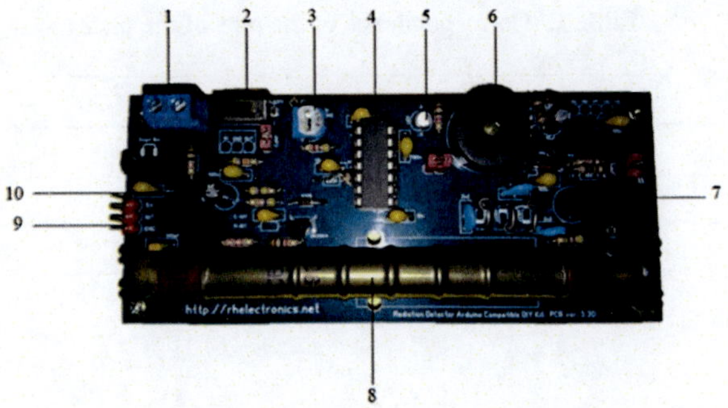

Fig. 3. Radiation detector compatible with Arduino Uno

To make a dosimeter implemented on an Arduino Uno R3 board, we needed: development board - radiation detector compatible with Arduino Uno, Arduino One board, SBM-20 sensor for radiation detection, connecting wires between the two components, the communication cable between the device and the laptop, the Arduino program, and the program will record and show all the measurements made by the device.

In the first part, to achieve the digitized dosimeter, the components of the radiation detector compatible with Arduino Uno (resistors, capacitors, diodes and others) were assembled. After identifying the components of the radiation detector compatible with Arduino Uno, they were glued using a soldering iron (soldering gun) and tin (Fig. 4).

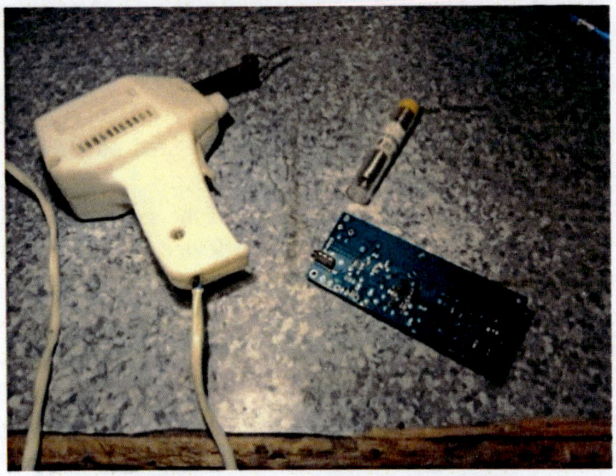

Fig. 4. The process of gluing the elements

After the Arduino compatible detector was made, the connection pins with the Arduino Uno board were identified.

Each wire was connected between the boards, following Table 1.

Table 1. Correspondence of the pins of the plates

Arduino Uno	Compatible Board- with SBM-20
5 V	5 V
2	INT
GND	GND

3 Measurements on the Source of Americium 241. Device Calibration

Americium-241 (Am^{241}) is an isotope of americium. Like all American isotopes, it is radioactive. Americium-241 is the most common isotope of americium.

It is the most widespread American isotope in nuclear waste. Americium-241 has a half-life of 432.2 years. It is commonly found in ionization smoke detectors. It is a potential fuel for long-lived radioisotope thermoelectric generators (RTG). Americium-241 is fusible, and the critical mass of an empty sphere is 57.6–75.6 kg and a sphere with a diameter of 19–21 cm.

Americium-241 has a specific activity of 3.43 Ci/g (Curie per gram or 126.9 gigabequerels (GBq) per gram). It is usually found in the form of Americium-241 (Am^{241}). This isotope also has a meta-state, with output energy of 2.2 meV and a half-life of 1.23 μs. Its presence in plutonium is determined by the initial concentration of plutonium-241 and the age of the sample. Due to the low penetration of alpha radiation, americium-241 poses a health risk only when ingested or inhaled.

Older plutonium-241-containing samples contain an accumulation of Am-241. Chemical removal of americium 241 from the redesigned plutonium (e.g., during the process of processing plutonium quarries) may be necessary in some cases [6].

The calibration, as well as the measurement of the radiation level with the SBM-20 sensor, was possible with the Gamarad DL 7 device, shown in Fig. 5, which has a radiation source with Am-241 (Americium^{241}, with 12 μCi (microCurie)).

The dosimeter was subjected to the Am^{241} source, to measure the radiation level given by the Gamarad DL7 device.

It is known that 1 Sv = 100 REM, and for the cosmic background, there are 2 * 10–5 REM, so we find that the cosmic background has 0.2 μSv/h, dose rate.

In the case of the Am^{241} source, with a test at 0.25 mREM, we have 1 Sv = 100 REM, and the source has 0.25 * 10–3 REM, so the source indicates a dose rate of 2.5 μSv/h.

Fig. 5. Gamarad DL7

Another free service from Nuke Professional

Rad Alert Network on web	Rad Inspector Tool	Annual Amounts of Radiation	ACTION
CPM Clicks Per Minute	Micro Sievert Per Hour	Milli Sievert Per Year	Action / Results Based on NO INTERNAL EXPOSURE
12	0.10	0.876	This is low, it doesn't get much lower
25	0.21	1.825	Pretty normal
50	0.42	3.65	Happens once in a while with no real reason, just keep an eye on it
100	0.83	7.3	ALERT - no need to panic, but try to figure out what is going on, stay out of the rain, avoid unnecessary trips
150	1.25	10.95	Real Risk of Cancer if exposed for a year
500	4.17	36.5	Real Risk of Cancer if exposed for 90 Days
		20	Is the Limit for Normal Nuclear Plant Workers
		100	Is the Limit for Fukushima Workers

Sievert Calcs based on Cesium 137 as the Radio-Isotope

IF the radiation gets inside your body it is 20 times more POWERFUL!

Uranium and Plutonium are only mildly radioactive, they don't create many clicks per minute, and yet they are INCREDIBLY dangerous because they are not just radioactive they are poisonous heavy metals that get into your body and then go into your bones and your most important organs and then they stay there. And there is no way to take them out effectively.
IF you suspect uranium or plutonium in the air, you should immediately leave or shelter in place, EVEN if your Geiger counter is not showing high clicks.

Fig. 6. Conversion CPM- μSv [7]

Table 2. Measurements on the Am^{241} source

	CPM	μSv/h
Average measurements	273.7241	2.27191
The MINIMUM value recorded by the achieved dosimeter	24	0.1992
The MAXIMUM value recorded by the achieved dosimeter	9312	77.2896
Conversion factor (CPM in μSv/h)	0.0083	

For the Am^{241} source, the average value recorded is 0.25 mREM, ie 2.5 μSv/h. It is found that the average of the measurements, according to Table 2 is 2.27191 μSv/h, so the meter registers within normal limits.

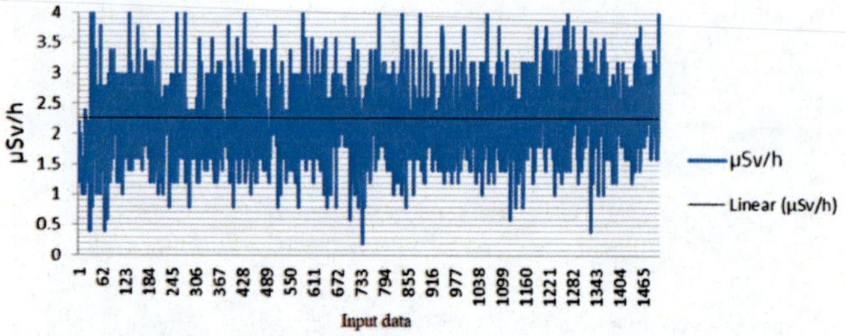

Fig. 7. Variation in radiation level (source of Americium-241)

For the interface to be friendly and to correctly and concretely identify the measured values, the device will be connected to the "Radiation Logger" program.

Keep in mind that, during the measurements, the device will be connected to the laptop, to a single port, therefore, it is impossible to run two programs, choosing only the "Radiation Logger" program to work.

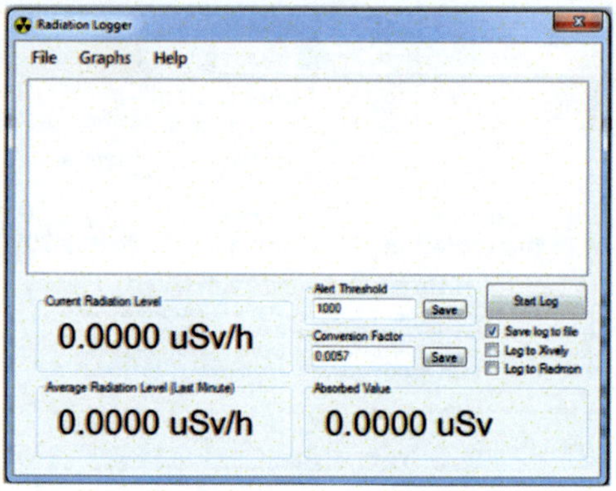

Fig. 8. "Radiation Logger" program interface

The "Radiation Logger" program shows the value area, the conversion factor of the CPM unit (counter per min or decays/minute) in µSv/h (values indicated when measuring the radiation level), the average of the values recorded in the last minute, the alert level (when the radiation dose is exceeded) and the area of the last value (Fig. 8).

On the OY axis, the measured radiation level [CPM] is observed, and the OX axis indicates the time when the measured radiation level was measured and recorded (Fig. 9).

That is why it was important to perform the conversion from the CPM unit of measurement to µSv/h, identified in Fig. 6.

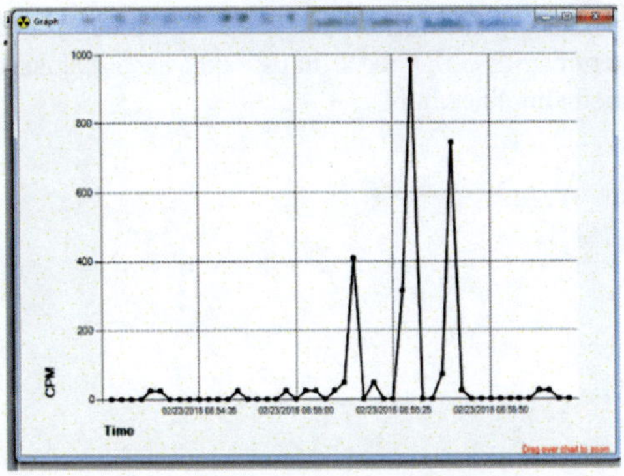

Fig. 9. Graph (example) made with "Radiation Logger"

4 Conclusions

Given that we are exposed to radiation of any kind (ionizing or non-ionizing), this must indicate the need to have a radiation monitoring device. The device used to identify and measure radiation levels is the dosimeter, in this article we have presented with can be made such a device, low-cost. The dosimeter is connected to a laptop, being digitized on friendly software (a pleasant interface) and understood by anyone. In addition to the values transmitted to the software installed on the laptop, it can emit an audible signal when the radiation is identified.

It is found that the Geiger-Muller tube registers values up to 0.2 μSv/h, this being the normal dose in the cosmic background. The average value of the Americium 241 source (2.5 μSv) is identified as approximately equal to what the device identifies, by averaging the values read by the Geiger-Muller sensor on the Arduino Uno compatible board (linear function in Fig. 7).

Conflict of Interest. The authors declare that they have no conflict of interest.

References

1. An Introduction to Radiation & Radiation Safety Science Buddies: An Introduction to Radiation & Radiation Safety (2017). https://www.sciencebuddies.org/science-fair-projects/references/radiation-introduction-radiation-safety
2. Measuring Radiation: Terminology and Units - Institute for Energy and Environmental Research Ieer.org. (2017). https://ieer.org/resource/classroom/measuring-radiation-terminology/
3. Podgorsak, E.B.: Radiation Oncology Physics: A Handbook for Teachers and Students, p. 45, 104. International Atomic Energy Agency (2005)
4. McRoberts, M.R.: Arduino starters kit manual-a complete beginners guide to the Arduino, Earthshine Design, pp. 9–10 (2009)
5. Arduino Uno at. https://ro.wikipedia.org/wiki/Arduino

6. **Americium at.** http://hps.org/documents/background_radiation_fact_sheet.pdf
7. **Geiger Counter interpretation simplified at.** https://2knowabout.blogspot.com/2014/02/geiger-counter-interpretation-simplified.html

Analysis of Mechatronic Devices or Systems that Identify the Biomechanical Parameters of the Lower Limb

Alexandru C. Tulică(✉), I. C. Roşca, and C. N. Drugă

Product Design and Environment, Transilvania University, Braşov, Romania

Abstract. This article will identify some of the most advanced systems for identifying biomechanical parameters in the lower limbs and the entire human body.

Technology has evolved, from here, we have devices such as Kistler plate, with which we can identify forces and accelerations in the limbs, FootScan plate, this device shows the plantar position, center of gravity-pressure of the human body, foot pressure and last but not least, the XSens device, with the help of which various biomechanical parameters can be identified. At the same time, the article offers a multicriteria analysis, through which we can realize the performance of the mentioned devices.

Keywords: Kistler · FootScan · XSens · Biomechanics · Parameters

1 Introduction

General terms of guidance are used to define spatial relationships and the parts of the human body. Thus, terms such as axes and reference planes are brought.

The starting point for their definition is the anatomical position of the human body, respectively the vertical or orthostatic position, particular to man, with the limbs hanging near the trunk, the head and eyes looking forward and the palmar face of the hand anteriorly, supine. By definition, the human body is three-dimensional, so it has three axes and three main spatial planes.

Flexion is the movement of bending (bending) in the sagittal plane of a segment of the human body from the initial orthostatic position, by moving away from this position, having the characteristic decrease of the angle in the joint to a part considered immobile body.

The extension is the movement opposite to flexion, by increasing the angle in the joint to the part considered immobile of the body, which takes place in the sense of returning to the orthostatic position, or in the sense of exceeding it (rotating the trunk backward exceeding the orthostatic position).

These movements also have specific names. Thus, in the shoulder, the flexion movement is called anteprojection or anteduction, and the extension movement is retroprojection or retroduction. Also, in the foot, the term dorsiflexion is preferred instead of the flexion of the foot and the term plantar flexion instead of the extension of the foot.

I. Tiginyanu et al. (Eds.): ICNBME 2021, IFMBE Proceedings 87, pp. 193–199, 2022.
https://doi.org/10.1007/978-3-030-92328-0_26

Abduction is the movement of a segment of the human body by moving away from the sagittal plane.

Adduction is the movement opposite to abduction, by bringing a segment of the human body closer to the sagittal plane.

Circumduction is the movement of a segment of the human body that describes a cone with a variable base, the tip of which is represented by the axis of the joint, such as rotating the arm in a circular motion around the shoulder, staying on the side surface of a cone.

Rotation is the movement of a segment of the human body around its longitudinal axis, so it is a "pivoting" movement. For the trunk, the rotation takes place around the main vertical axis (passing through the center of gravity of the human body) [1].

2 Devices for Biomechanical Measurement of Forces, Pressures

2.1 Kistler Plate

The multi-component force plate, identified in Fig. 1 provides a dynamic and quasi-static measurement of the 3 orthogonal components of a force (Fx, Fy, Fz) that act from any direction on the upper plate [2].

Fig. 1. Kistler plate [2]

Forces can be measured using ring discs made of piezoelectric material, sensitive to compressive forces (Fz) and shear forces (Fx, Fy), used to make the sensor (a); multicomponent piezoelectric sensor for measuring reaction forces on the Ox, Oy, Oz axes (b); the Kistler force plate which has embedded in its body the four piezoelectric sensors (c), positioned in the 4 corners of the Kistler plate, according to Figs. 2 and 3 [3].

Piezoelectric effects and quartz crystal axes: longitudinal effect (a); transverse effect (b); tangential effect (shear) (c) in Fig. 4, these effects on the force plate, underlie the operation of this device [4].

The measurements made with the help of this force plate and moments are noted in real-time, in a table of software specific to this device, so that it will be possible to observe: the forces on the three axes, their minimum and maximum; the average of these forces, the measurement times of the forces, according to Fig. 5.

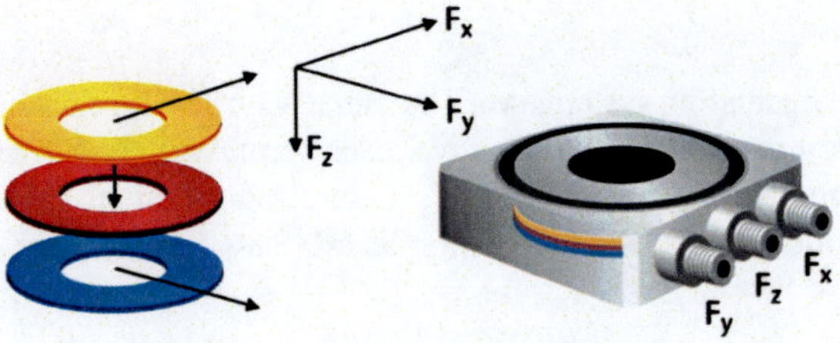

Fig. 2. 3-component force sensor [2]

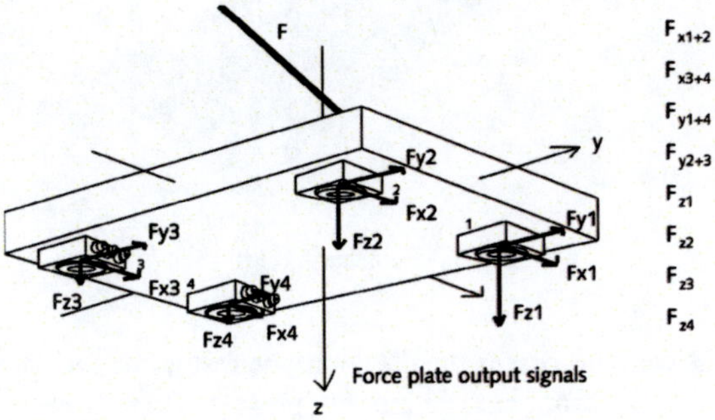

Fig. 3. Forced plate output signals [2]

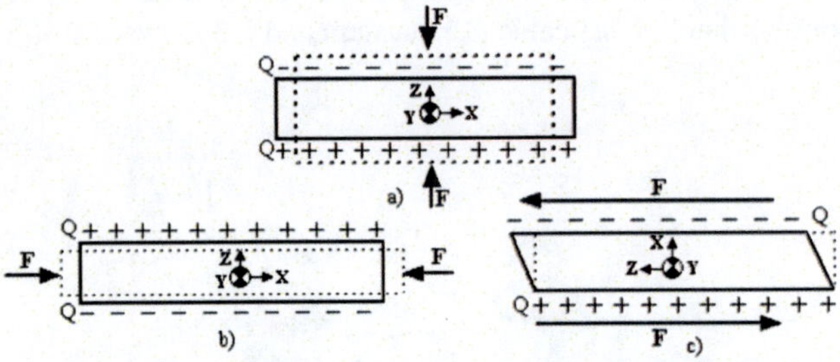

Fig. 4. Piezoelectric effects and quartz crystal axes [4]

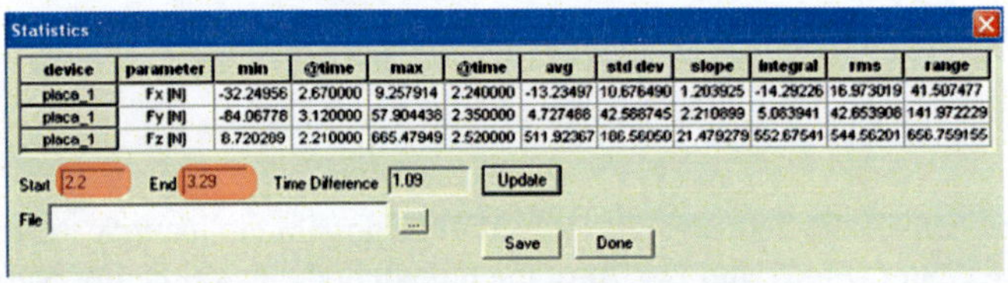

device	parameter	min	@time	max	@time	avg	std dev	slope	integral	rms	range
placa_1	Fx [N]	-32.24956	2.670000	9.257914	2.240000	-13.23497	10.676490	1.203925	-14.29226	16.973019	41.507477
placa_1	Fy [N]	-84.06778	3.120000	57.904436	2.350000	4.727466	42.588745	2.210899	5.083941	42.653906	141.972229
placa_1	Fz [N]	8.720289	2.210000	665.47949	2.520000	511.92367	186.56050	21.479279	552.67541	544.56201	656.759156

Fig. 5. Statistics- Kistler plate [3]

2.2 Footscan

In general, all foot scanning systems work the same way: The foot scanning plate (Fig. 6) measures plantar pressure using a sensitive resistive pressure X-Y matrix with sensors that scan sequentially. The system records pressure data when the subject is standing or showing locomotion on the data acquisition board.

Fig. 6. Footscan- 1.5 m [5]

The device software processes the data, and the result is a foot pressure image. The color of the image varies from blue (minimum pressure) to red (maximum pressure).

Furthermore, the system calculates some physical properties for further analysis of the measurement.

Footscan input level systems are available in three sizes: 0.5 m, 1 m and 1.5 m. The boards connect directly to a USB2 port on the computer via the fixed device (0.5m and 1m system) or detachable USB cable (1 m system and 1.5 m system) [5].

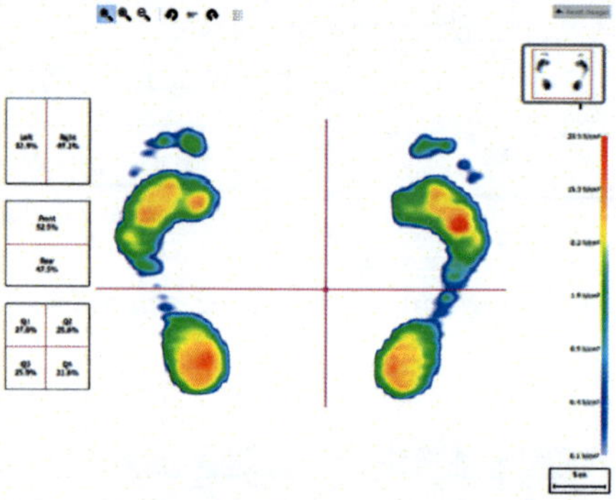

Fig. 7. Foot pressure image [5]

With the help of the Footscan device, it will be possible to identify the movement of the pressure center (Fig. 7 and Fig. 8) of the human body, in a time interval, also on the same graph, the initial position of the analyzed subject will be observed on the Footscan plate.

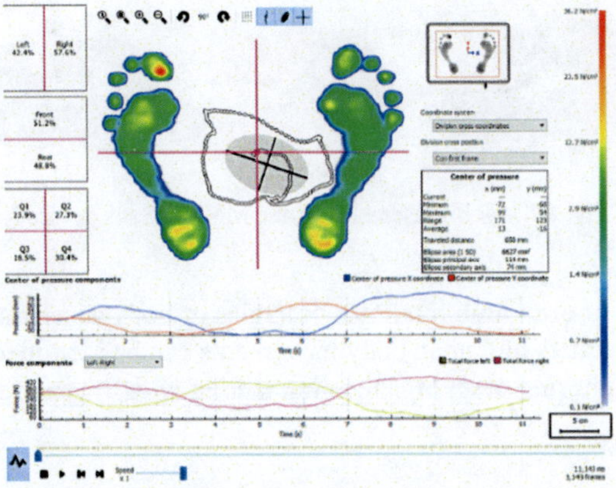

Fig. 8. Pressure center identified with Footscan

2.3 XSens- MVN System

The MVN-XSens device is a system of inertial kinematic measurement of the whole body, which incorporates synchronized video data. With the help of this device, we can identify angles of the joints of the human body, forces with which the subject interacts with the environment.

Depending on the system, either a Lycra suit or a set of mounting straps are provided.

There are two types of system, suit-Lycra type, and configured on belts that are mounted on the joint areas of the human body (Fig. 9).

Fig. 9. MVN-suit system (left) and joints mounted at the joints (right) [6]

The movement components are fixed to the extremities - head, hands and feet, using a band, gloves and foot pads, as can be seen in Fig. 10.

The locomotion of a human body can be easily identified with this device, the accuracy with which it takes information from the environment determines important data for analysis, whether we are talking about the analysis of locomotor movements such as

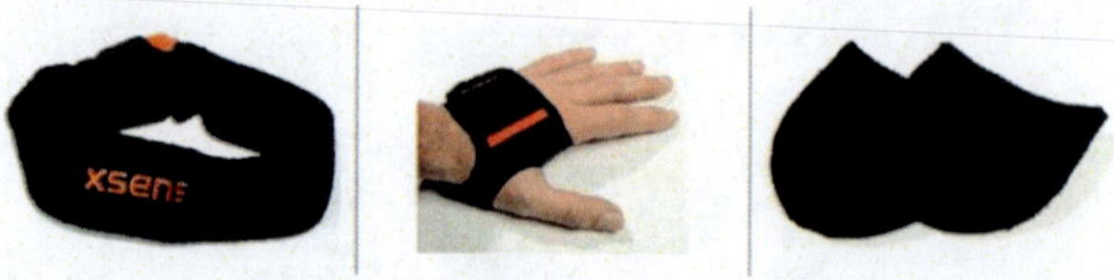

Fig. 10. Mobile elements of the MVN-XSens device [6]

flexion-extension or even limb rotations, or angles or body acceleration; variations can be observed, difficulties of human body movements can be identified.

Real-time monitoring of an organism interacting with the environment can be seen in Fig. 11.

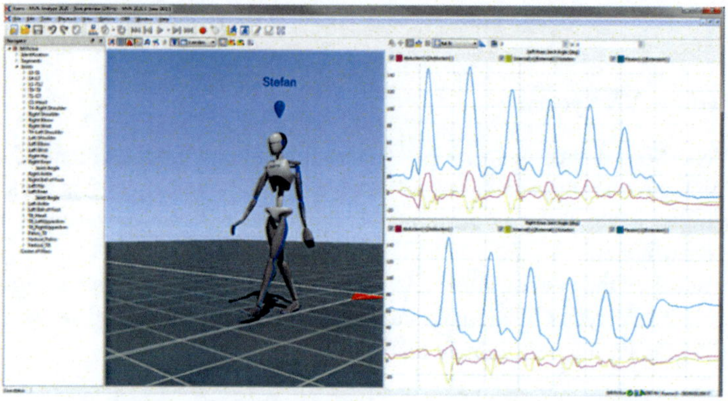

Fig. 11. Real-time locomotion analysis using the XSENS MVN system [7]

3 Multicriteria Analysis of Mechatronic Devices or Systems: Kistler, Footscan and Xsens MVN

For the analysis of these devices, criteria were chosen: reliability, accuracy, the interface of the results, the space of the measurements, the difficulty of use.

The table (Table 1) with the criteria is made, and the values offered for each device per criterion can be: 0- if the criterion is not valid for the respective device, 1/2 - if the criterion is valid, but not in full, and 1- is the fully valid criterion. If the score is lower, we can see that the device provides information on the biomechanical parameters of the lower limbs, but not as much as we would like.

The Kistler plate of forces and moments presents only a few biomechanical parameters that can be analyzed, the surface on which their measurement can be made is relatively small.

The Footscan device provides accurate information, but only at the level of the plantar portion of the foot and only a few parameters of locomotion.

Table 1. Device analysis- criteria

Criteria	Kistler	Footscan	XSenes MVN
Reliability	½	½	1
Accuracy	½	1	1
The interface of the results (soft)	1	1	1
The space of the measurements	0	½	1
The difficulty of use	½	½	½
Total score	2.5	3.5	4.5

But the Xsens MVN device is of the latest generation, with body analysis, interacting with the environment, the interface is friendly, the graphics are quick and easy to interpret.

4 Conclusions

Biomechanical analysis of the human body can be performed with various mechatronic devices, by various methods, for example, Kistler Board, RS Scan-Footscan Board, VICON, XSENS. Mechatronic devices in the biomechanical field use pressure sensors, forces, high-quality video cameras, all for the determination of biomechanical parameters at the highest precision.

Multicriteria analysis wants to show that each device is built on various topics, the information given by them is different, so to analyze the biomechanics of the lower limbs, we must use XSens MVN or the combination of Kistler Board and Footscan system.

Conflict of Interest. The authors declare that they have no conflict of interest.

References

1. Ileana- Constanța, R., Ionel, Ș.: Fundamentals of biomechanics, p. 9. Transilvania University of Brașov Publishing House (2015)
2. Kistler plate force at http://analisedemarcha.com/papers/manutencao/manuais/Kistler_Portable%20Multicomponent%20Force%20Plate%20for%20Gait%20and%20Balance%20Analysis%20in%20Biomechanics.pdf
3. Șerban, I.: Studies and research on the influence of the environment on human stability and locomotion (2011)
4. Medved, V.: Measurement of Human Locomotion, p. 130. CRC Presss, New York (2001)
5. Installation Guide and User Manual footscan System with footscan, vol. 9, p. 8
6. MVN User manual at https://www.xsens.com/hubfs/Downloads/usermanual/MVN_User_Manual.pdf
7. Xsens Technologies B.V.: MVN User Manual (2021)

Minimally Invasive, Fully Implantable Left Ventricular Assist Device: Concept, Design, and Early Prototyping

Florin Alexandru Pleşoianu[1], Carmen Elena Pleşoianu[2,3](✉), Andrei Ţăruş[4], and Grigore Tinică[2,4]

[1] Faculty of Biomedical Engineering, "Grigore T Popa" University of Medicine and Pharmacy, Iaşi, Romania

[2] Faculty of Medicine, "Grigore T Popa" University of Medicine and Pharmacy, Iaşi, Romania
carmen-elena.plesoianu@umfiasi.ro

[3] Department of Clinical Cardiology, Institute of Cardiovascular Diseases "Prof. Dr. George I.M. Georgescu", Iaşi, Romania

[4] Department of Cardiovascular Surgery, Institute of Cardiovascular Diseases "Prof. Dr. George I.M. Georgescu", Iaşi, Romania

Abstract. Advanced heart failure is an increasing prevalent pathology with important socio-economic impact, justifying the tremendous worldwide effort to develop and improve valuable alternatives to heart transplantation. Mechanically circulatory support is now into the spotlight with evidence showing improved survival and quality of life in an extremely fragile population. However, severe side effects such as thrombosis, bleeding or infection still limit its clinical use. Against this background, our multidisciplinary team (bioengineer, cardiologist, and cardiovascular surgeons) designed an innovative left ventricular assist device suitable for patients with advanced heart failure as bridge to cardiac transplant, temporary support when myocardial recovery is possible or as destination therapy for patient not eligible for heart transplantation. The device consists of an electronic circuit, internal battery, wireless energy receiving antenna and an axial flow pump that mobilizes blood from the heart to the arterial system, unloading the failing heart. The device is designed to provide multiple features: flow of blood with minimal turbulence and without areas of mechanical stress, energy autonomy and wireless power supply, possibility of fully implantation by minimally invasive techniques, and biocompatible amorphous tetrahedral carbon coating that may provide the premises to reduce the negative effects of known devices.

Keywords: Mechanical circulatory support · Left ventricular assist device · Minimally invasive · Advanced heart failure · Axial pump

1 Introduction

Advanced heart failure is becoming a major socio-economic issue. In the context of population aging and of the technological progress which enables better care in acute

I. Tiginyanu et al. (Eds.): ICNBME 2021, IFMBE Proceedings 87, pp. 200–207, 2022.
https://doi.org/10.1007/978-3-030-92328-0_27

cardiovascular settings with devices that increase survival (primary angioplasties, pacemakers, defibrillators, etc.), heart failure has become a global pandemic [1]. There is a percentage of up to 10% from the overall heart failure population [2] who remains severely symptomatic despite optimal guideline-directed management for which cardiac transplantation or mechanically circulatory support (MCS) are a last standing resource [2]. Because of the limited availability of heart transplants, the MCS has come into the spotlight with evidence associating it with increased survival and quality of life in a very fragile population [3]. In the evolution of MCS, continuous flow rotary pumps proved that may successfully replace the pulsatile ones which were initially used. Commercially available left ventricular assist devices (LVAD) with continuous axial flow pump are represented by Jarvik 2000 (Jarvik Heart, New York, NY, USA), first implanted in human in April 2000 [4] and HeartMate II (Abbott, Chicago, IL, USA) first implanted in human in July 2000 [5]. Important improvements [6] were developed creating these systems but unfortunately, severe complications [7] such as thrombosis, bleeding [8] or infection [9] still limit their clinical use justifying intensive research to create better ones.

2 Materials and Methods

Our multidisciplinary team consisting of bioengineer, cardiologist and cardiovascular surgeons designed an innovative minimally invasive fully implantable LVAD.

The concept of the device was created using SysML modeling language in Enterprise Architect (v12 Sparx Systems Ltd. Creswick, Victoria, Australia) taking into consideration the needs of all actors involved in the mechanical circulatory support: patients, health care providers and manufacturers.

Mathematical modeling was used to calculate the pump characteristics to deliver 5 l/min at a differential pressure of 100 mmHg under the size constraints imposed by minimally invasive implantation of the device in a subcutaneous pocket.

The conceptual design of the pump was created using turbomachinery design software CFturbo (2020 R2.4 CFturbo GmbH, Dresden, Germany).

The detailed design of LVAD was further refined, using Solidworks (Dassault Systèmes SolidWorks Corporation, Waltham, MA, USA).

3 Results

Our team designed a minimally invasive fully implantable circulatory assist device, wirelessly powered with internal rechargeable battery, and an axial flow helical pump that mobilizes blood from the left atrium to the subclavian artery to compensate for flow deficit of the failing heart with minimal turbulence, at physiological pressures, without areas of mechanical stress with low risk of cell damage and with autonomous operation at the interruption of the external source.

3.1 Concept

We have created a LVAD for partial support in patients with NYHA III-IV heart failure, that might pump up to 5 L/min of blood from the left side of the heart to the arterial system at a differential pressure of 100 mmHg. The characteristics of the device, the actors and the interfaces involved were defined.

The ventricular assist system contains an implant, a controller, rechargeable batteries, a charging station, and a console (Fig. 1).

The implant placed subcutaneously wirelessly receives the energy from the external wearable controller (Fig. 2).

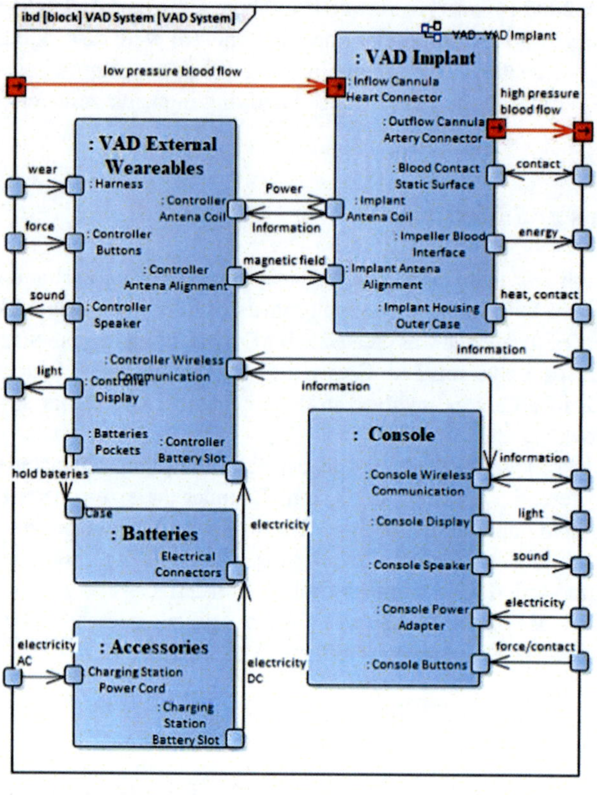

Fig. 1. First decomposition layer of the ventricular assist system.

The device is designed to be minimally invasive, fully implantable in a subcutaneous pocket in the pectoral area. The inflow cannula is connected to the left atrium (either endovascularly via the subclavian vein, superior vena cava, right atrium, fossa ovalis or by endoscopic surgery) while the outflow cannula is connected to the subclavian artery, proximal to the implantation incision.

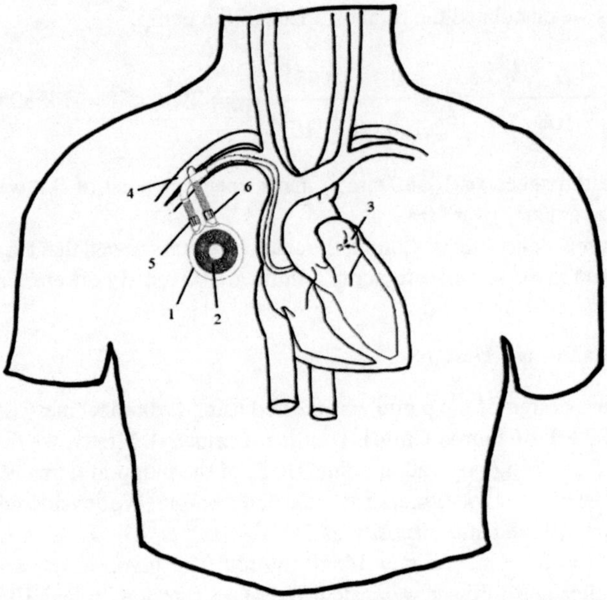

Fig. 2. The circulatory assist device – example of implantation in a subcutaneous pocket with inflow cannula connected endovascularly to the left atrium and the outflow cannula in the subclavian artery. Legend: 1 - device, 2 - antenna, 3 - inflow cannula, 4 - subclavian artery, 5 - inflow connection, 6 - outflow connection.

3.2 Mathematical Modeling of the Pump

Mathematical modeling was performed aiming to calculate the pump characteristics to comply with size constraints. Dimensional/functional calculus of the hollow axial flow pump were made.

From the requested pressure of 100 mmHg = $1.33 \cdot 10^4$ N/m^2 and blood density, we calculated the specific work "y".

$$y = \frac{\Delta p}{\rho} = \frac{1.33 \cdot 10^4}{1.06 \cdot 10^3} = 12.55\frac{m^2}{s^2}$$

For impeller diameter of 10 mm and flow of 5 l/min, we derived the specific diameter δ.

$$\delta = 1.054 \cdot D \cdot \frac{y^{\frac{1}{4}}}{V^{\frac{1}{2}}} = 1.054 \cdot 10^{-2} \cdot \frac{12.55^{\frac{1}{4}}}{(83 \cdot 10^{-6})^{\frac{1}{2}}} = 2.17$$

On Cordier diagram middle line of optimal efficiency interval we identified that for $\delta = 2.17$ we obtain the value of specific speed $\sigma = 0.7$ in the second interval of axial flow pumps.

From this we calculated the rotation speed of the pump:

$$n = \frac{\sigma \cdot y^{\frac{3}{4}}}{2.108 \cdot V^{\frac{1}{2}}} = \frac{0.7 \cdot 12.55^{\frac{3}{4}}}{2.108 \cdot (83 \cdot 10^{-6})^{\frac{1}{2}}} = 243\text{sec}^{-1} = 14580\text{min}^{-1}$$

With rotation speed of 14580 min^{-1} and a specific speed of 0.7 we are in a high efficiency area slightly over 90%.

For the pump's diameter of 12 mm, the same calculus showed that a n = 8340 min^{-1} must be applied to the impeller to pump 5 l/min at 100 mmHg differential pressure.

3.3 Turbomachinery Design

The conceptual design of the pump was created using turbomachinery design software CFturbo (2020 R2.4 CFturbo GmbH, Dresden, Germany). Firstly, we defined the fluid properties and the design operating point (DOP) of the pump in terms of total pressure difference, flow rate and rotational speed calculated before. We considered blood density at 1060 kg/m^3 y, kinematic viscosity as 3E*10^{-6} m^2 /s. The DOP was set to $\Delta P =$ 100 mmHg, Q = 5 l/min, and n = 14580 rot/min. The pump inner part consisting in inducer, impeller and diffuser was calculated at 13 mm length and 10 mm diameter. The inducer is an inlet stator with 3 blades parallel with pump's axis for directing blood flow (Fig. 3). The impeller is a 3-blade helical rotor with a wrapped angle of 235° and 225° at the hub and shroud sections of the pump, respectively that transforms rotational movement into hydrodynamic energy. The diffuser contains 3 blades that contribute to the axial flow of blood and improve efficiency.

Fig. 3. Conceptual design of the pump realized with Cfturbo

3.4 Design

The implant consists of an electric motor, an axial flow pump, an electronic circuit, a rechargeable battery, and a housing (Fig. 4).

The motor of the device is an electric brushless motor composed by rotor, stator, and magnetic bearings. Specific to this motor are the coreless windings of the stator, the tubular magnetic rotor, and the magnetic bearings.

The rotor is a two layers tube: the inner one with structural and shielding role and the outer one with rotational and bearing role. The tubular shape of the rotor allows the construction of the pump inside it.

The magnetic bearings consist of two components. The inner component attached on the exterior of the rotor and the outer component attached to the housing. The magnetic bearings keep the rotor suspended inside the stator allowing the rotational movement around the central axis. To stabilize the bearing, another force (electromagnetic or hydrodynamic) had to be involved.

The pump contains an impeller embedded inside the rotor, an inducer, and a diffuser. Inside the pump the blood does not encounter areas of significant occlusion, turbulence, or strong electromagnetic fields.

The surface of the pump and the motor may be covered with amorphous tetrahedral carbon.

The electronic system contains a power management block, a microcontroller, a motor driver, and a communication block. The energy wirelessly transmitted from an external source is received through the antenna and distributed by the power management block for both device operation and battery charging which is used as an internal power source when the external source is interrupted.

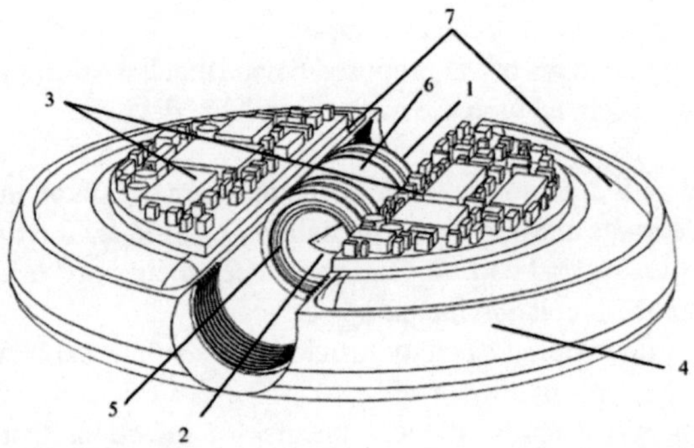

Fig. 4. Section through the housing of the early variant of the device. Legend: 1 – electric motor, 2 - pump, 3 – electronic system, 4 - housing, 5 - rotor, 6 - stator, 7 – rechargeable battery.

3.5 Early Prototyping

The design of the pump created in CFturbo was used as the input geometry for detailed design of the system in Solidworks. Several variants of the housing geometry were created and 3D printed with a fused filament fabrication technique (Fig. 5). The solid models were examined, and the chosen one was materialized with a clear resin stereolithography technique of 25 microns resolution.

Fig. 5. Design variant of the devices made in Soliworks (left) and the 3D printed impeller

4 Discussions

We presented a circulatory assist device characterized by minimally invasive implantation, pumping blood parallel to the heart through a pump built inside the rotor of a brushless electric motor with magnetic bearings powered by an electronic circuit which wirelessly receives energy from an extracorporeal source. Internal rechargeable battery ensures autonomy in operation when the external source is interrupted. The pump-rotor assembly levitates magnetically inside the stator. All surfaces in contact with the blood are covered with amorphous tetrahedral carbon.

There are several features of the proposed device that have the potential to minimize the complications associated with the market available devices:

- the main blood path does not cross the space between the stator and the rotor,
- the magnetic bearings allow increased reliability and reduce the risk of cell damage,
- the amorphous tetrahedral carbon coating reduces the risk of rejection and reduces friction between the blood and the pump surfaces,
- the autonomous functioning capacity provided by the internal battery contributes to the increase of patient's quality of life,
- the wireless energy reception reduces the risk of infections that accompanies the penetration of the patient's skin by electrodes or tubes common to many devices.

Further investigation must be performed to confirm the mathematical calculus of the pump by computed fluid dynamics and in vitro testing.

Furthermore, the concerns regarding stabilization of the bearing system and the risks associated to the transcutaneous energy transmission must be addressed.

5 Conclusions

In multidisciplinary team we have created an innovative, fully minimally invasively implantable left ventricular assist device capable to deliver 5 L/min of blood at 100 mmHg differential pressure using a 12 mm internal diameter axial flow pump turned at 8340 rot/min.

Conflict of Interest. The authors declare that they have no conflict of interest.

References

1. Ambrosy, A.P., Fonarow, G.C., Butler, J., et al.: The global health and economic burden of hospitalizations for heart failure: lessons learned from hospitalized heart failure registries. J. Am. Coll. Cardiol. **63**(12), 1123–1133 (2014)
2. Crespo-Leiro, M.G., Metra, M., Lund, L.H., et al.: Advanced heart failure: a position statement of the Heart Failure Association of the European Society of Cardiology. Eur. J. Heart Fail. **20**(11), 1505–1535 (2018)
3. Vieira, J.L., Ventura, H.O., Mehra, M.R.: Mechanical circulatory support devices in advanced heart failure: 2020 and beyond. Prog Cardiovasc Dis. **63**(5), 630–639 (2020)
4. Frazier, O.H., Myers, T.J., Gregoric, I.D., et al.: Initial clinical experience with the Jarvik 2000 implantable axial-flow left ventricular assist system. Circulation **105**(24), 2855–2860 (2002)
5. Siegel-Itzkovich, J.: Israeli surgeons implant first permanent artificial ventricle. BMJ **321**(7258), 399 (2000)
6. John, R., Kamdar, F., Liao, K., et al.: Improved survival and decreasing incidence of adverse events with the HeartMate II left ventricular assist device as bridge-to-transplant therapy. Ann. Thorac. Surg. **86**(4), 1227–1234 (2008)
7. John, R.: Current axial-flow devices–the HeartMate II and Jarvik 2000 left ventricular assist devices. Semin. Thorac. Cardiovasc. Surg. **20**(3), 264–272 (2008)
8. Eckman, P.M., John, R.: Bleeding and thrombosis in patients with continuous-flow ventricular assist devices. Circulation **125**(24), 3038–3047 (2012)
9. Angleitner, P., Matic, A., Kaider, A., et al.: Blood stream infection and outcomes in recipients of a left ventricular assist device. Eur. J. Cardiothorac. Surg. **58**(5), 907–914 (2020)

Near-Threshold Electron Emission Spectroscopy to Characterize Nanoobjects for Biomedical Applications

Yuri Dekhtyar(✉)

Riga Technical University, Riga, Latvia
jurijs.dehtjars@rtu.lv

Abstract. Rapid implementation of nanoobjects for biomedical applications needs simple techniques to characterize them. Great attention is paid to the surface of nanoobjects. Its electrical charge has a significant influence on the interaction of the nanoobjects with biostructures.

The article reviews the fundamentals and applications of near-threshold electron emission spectroscopy to characterize and control the electrical charge deposited on the surfaces of nanoobjects.

Keywords: Nanoobjects · Surface · Electrical charge · Near threshold weak electron emission

1 Introduction

The electrical charge located at the surface of materials has a significant influence on its interaction with bioobjects [1, 2]. Therefore, characterization of the electrical charge at the nanoobject surface aimed to interact with biostructures must be identified to control the reproducibility of the surface engineering. Rapid implementation of nanoobjects for biomedical applications needs simple techniques to characterize them. The growing interest in surface charge is demonstrated with the pace of publications. Following SCOPUS the annual number of publications dedicated to the charge of the biomaterial surface increased twice during the last decade [3]. Nevertheless, the knowledge formation dedicated to the charge deposition, its characterization, and interaction with bioobjects is still at the facts collection phase. Because of this, reviews on the achievements in the field are necessary.

To do not disturb gentle nanoobjects and prevent their influence on measurements the contact less characterization is preferable.

Therefore, the present article is devoted to the contactless near-threshold electron emission spectroscopy (NTES) for characterization of the surface electrical charge.

2 Fundamentals

To emit from a material an electron is supplied with energy (E). The electron, when it is already in a vacuum, has kinetic energy (Ek). Because of emitting the chemical couples

I. Tiginyanu et al. (Eds.): ICNBME 2021, IFMBE Proceedings 87, pp. 208–214, 2022.
https://doi.org/10.1007/978-3-030-92328-0_28

of the emitter can be changed or/and broken. For this, the corresponding energies, Er and Eb are needed. In this respect:

$$E = Ek + Er + Eb + Es,$$

Es- the energy barrier (threshold) that the electron needs to emit.

To prevent the surface of the emitter from impairing, Es can be delivered in a contactless manner (flow of photons, for instance). If $E \to Es$, and however, emission takes place, the couples are not damaged, and the material does not significantly change its properties. At the condition $E \to Es$, $Ek \to 0$, and near-threshold electron emission is achieved. In this mode E is equal to just of several eV [4], and a mean free pass (L) of the electron in the solid corresponds to the nanodimensional scale (10–1000 nm) [5].

Therefore, the NTES can be in use as the contactless and undamaged technology aimed to analyze nanoobjects.

NTES can be employed in several modes: photoelectron emission (PE), dual photoelectron emission (DPE) and exoelectron emission (EEE).

When PE is in use, a current (I) of emitting electrons is described with the generally known formula

$$I = K(h\nu - \varphi)^m, \quad (1)$$

where K - coefficient of proportionality, $h\nu$ - energy of exciting photon, φ - electron work function, m - power index; typically $m > 1$ [6].

For dielectrics and semiconductors having an energy gap (Eg)

$$\varphi = E_g + \chi,$$

where χ is an electron affinity directly connected with the surface charge. When χ changes, I responds very strongly as $m > 1$.

There are some limitations to use PE for characterization of nanoobject [7]. The size of the latter must not be larger than the mean free path of the emitting electron. The geometrical uncertainty to identify the nanoobject is not less than 0.2 nm [7].

To supply the single photon electron emission mode and to avoid multi photon effects as well as heating of a tested object, the flux of the photons should be rather weak. The details of the measurement technique are described in [7, 8].

DPE can be only applied for materials having Eg [7]. Electrons and holes are generated, when photons have energy $h\nu_{ad} \geq E_g$. The excited charge carries are mobile and therefore are able to compensate a surface charge, the later depending on the Fermi level pinned at the surface [9] and adsorbed ions/dipoles [10]. When PE is provided and $h\nu_{ad}$ is switched on/off, keeping the condition $h\nu_{ad} < h\nu$, an increment (ΔI) of I is induced because of the χ shift under the $h\nu_{ad}$ photon flux.

If heat is supplied to material, when PE is measured

$$I(T) = K(T)[h\nu - \varphi(T)]^{m(T)},$$

T - temperature of the material.

This phenomenon is named as EEE [8]. The EEE can be detected, if, for instance, heat induces relaxation/annealing of point-like imperfections located in the measured material [11].

EEE typically is delivered because of:

a. thermoionisation of local states trapping electrons and belonged to the imperfections [12]; such a mechanism is available for materials with wide E_g and small χ;
b. Auger transitions of electrons [8], from local electron traps of imperfections; this channel is available when $E_g > \chi$;
c. field emission from the electron traps by imperfections because of heat induced electrical polarisation/depolarisation of the emitting surface layer [13].

The EEE total emitted charge (Q) is directly proportional to a concentration of point defects in material [7, 8, 11].

3 Applications

PE was demonstrated (Fig. 1, 2) to identify dependence of the surface charge of hydroxyapatite (it is widely in use for bio-implants) on its nanoparticle dimension (X) [14] and doping [15].

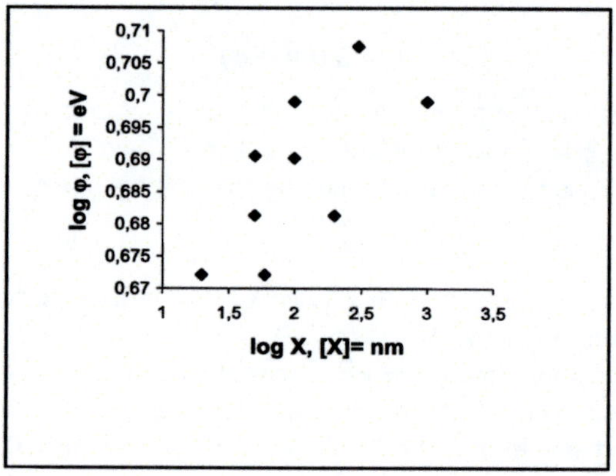

Fig. 1. The surface charge identified via φ of the hydroxyapatite nanoparticles in dependence on their dimension [14].

The results in Fig. 1 motivated ones to provide nanoscaled measurements (Kelvin probe atomic force spectroscopy) of the hydroxyapatite surface irregularities sharpness influence on its φ [16] (Fig. 3)

DPE assisted with luminescence spectroscopy was employed to identify the energy gap and localized states of the bone [17] (Fig. 4).

DPE was employed (Fig. 4) for measurements of the relaxation time of the charge excited on the bone local states to demonstrate a possibility of photoinducted therapy of the bone fracture [18].

Modern radiation therapy (RT) technologies and the needs of molecular and micro radiobiology necessities inspired detection of radiation absorbed by micro/nano volumes. However, there are still no detectors that measure radiation in nanovolumes. To fix the problem, a nanosized detector should be reached; nanoobjects are great candidates to become the dosimeter sensitive element.

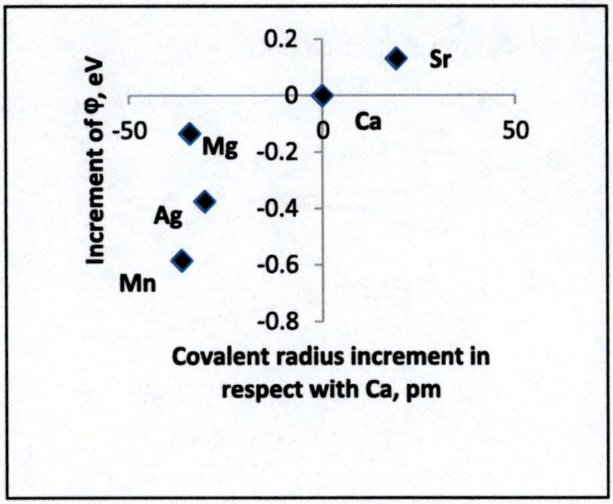

Fig. 2. The surface charge identified via φ of the hydroxyapatite in dependence on its dopant [15]

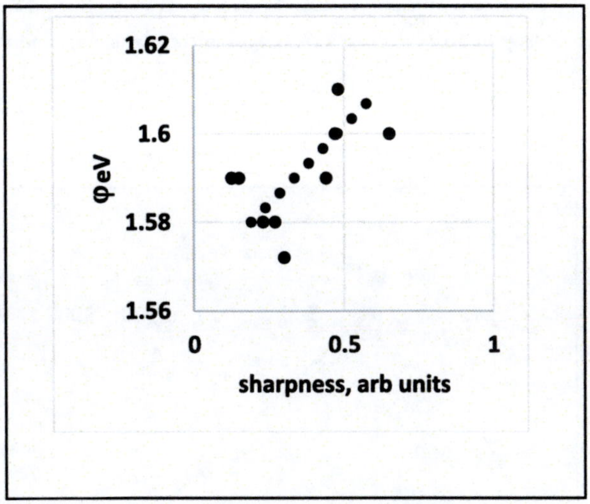

Fig. 3. The charge (identified via φ) of the hydroxyapatite surface in dependence on its irregularity sharpness [16].

EEE mode demonstrated a capacity to detect electrical repolarization of the poly(vinylidene fluoridetrifluoroethylene) (PVDF nanofilms [19] that were intended to reach a nanodosimeter [20].

Figure 5 demonstrates the possibility to apply EEE detected from the PVDF nanofilm for the nanodosimetry purposes [20]. The increment of the emitted electrons amount (ΔQ) depends on the delivered ultraviolet radiation exposure.

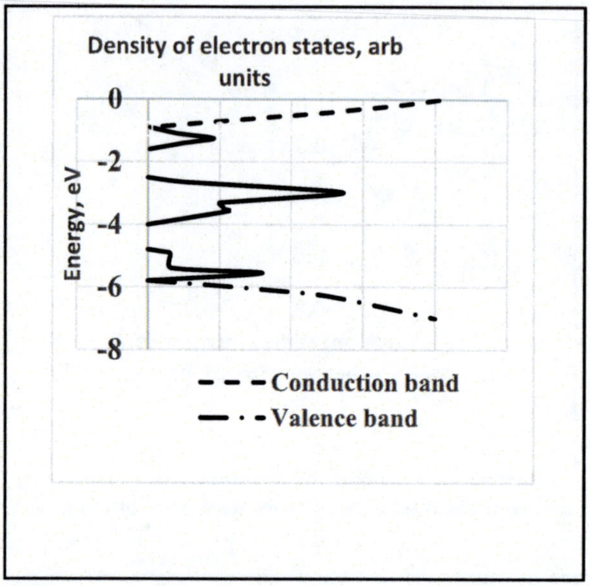

Fig. 4. Electron density of states in bone [17] (edges of the conduction and valence bands are indicated).

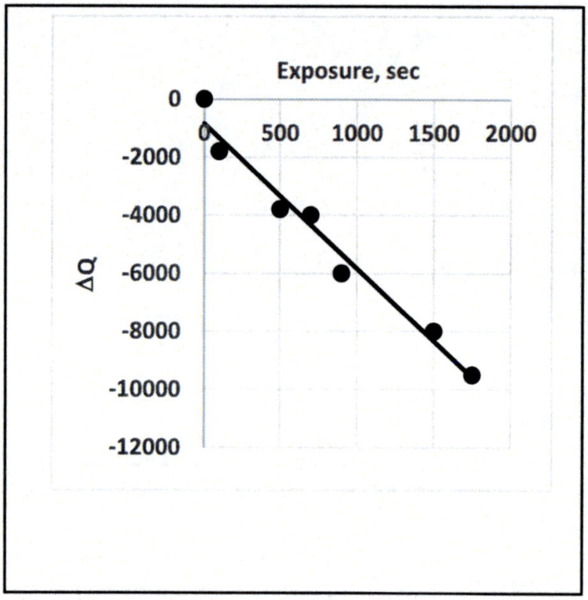

Fig. 5. Correlation of PVDF EEE ΔQ with exposure of delivered ultraviolet radiation [20]

4 Conclusion

Near -threshold PE, DPE and EEE analyses are suitable instruments for sensitive, contactless techniques to characterize nanoobjects aimed for biomedical applications. Influence of doping and size of the nanoobjects on their surface electrical charge can be identified, when PE is employed. DPE assists to recognize density of states within the energy gap. EEE is capable for nanodosimetry use.

Conflict of Interest. The author declares that no any of conflicts of interest he has.

References

1. Ratner, B.D., Hoffman, A.S., Schoen, F.J., Lemons, J.E.: Biomaterials Science: An Introduction to Materials in Medicine. Academic Press, Cambridge (1996)
2. Bystrov, V., Bystrova, A., Dekhtyar, Y.: HAP nanoparticle and substrate surface electrical potential towards bone cells adhesion. Adv. Coll. Interface. Sci. (2017). https://doi.org/10.1016/j.cis.2017.05.002
3. Baltacis, K., et al.: Physical fundamentals of biomaterials surface electrical functionalization. Materials **13**, 4575 (2020)
4. Fomenko, V.S.: Emission Properties of Materials. Naukova Dumka, Kiev (1981). (In Russian)
5. Ibach, H. (ed.): Electron Spectroscopy for Surface Analysis. Springer, New York (1977). https://doi.org/10.1007/978-3-642-81099-2
6. Dobrecov, L.N., Gomoyunova, M.V.: Emission Electronics. Nauka, Moscow (1996). (In Russian)
7. Dekhtyar, Yu.: Weak electron emission current for characterization of nanomaterials, gas and radiation sensing towards medical applications. Proc. Estonian Acad. Sci. **63**, 3 (2014). https://doi.org/10.3176/proc.2014.3. www.eap.ee/proceedings
8. Kortov, V.S., Shifrin, V.P., Gaprindashvili, A.I.: Exoelectron spectroscopy of semiconductors and insulators. Microelectronics **8**, 28–49 (1975). (in Russian)
9. Nesterenko, B.A., Cnitko, O.V.: Physical Properties of Atomic Clean surface of semiconductors. Naukova Dumka, Kiev (1983). (In Russian)
10. Volkenstein, F.F.: Electron Processes on Semiconductors Surface During Chemosorption. Nauka, Moscow (1987). (In Rusian)
11. Dekhtyar, Y., Vinjarskaja, J.A.: Exoelectron analysis of amorphous silicon. J. Appl. Phys. **75**(8), 4201–4207 (1994)
12. Nassenshtein: Electron emission from the solid state surface after mechanical treatment, Exoelectron emission, Inostrannaya literature, Moscow, pp. 72–95 (1962). (in Russian)
13. Rosenman, G.I., Rez, I.S., Chepelev, Y., Angert, N.B.: Exoemission of defected surface of lithium tantalite. J. Techn. Phys. **51**(2), 404–408 (1981). (in Russian)
14. Bystrov, V., et al.: Size depended electrical properties of hydroxyapatite nanoparticles. In: IFMBE Proceedings, vol. 14, pp. 3149–3150 (2006)
15. Bystrov, V.S., et al.: Computational study of hydroxyapatite structures, properties and defects. J. Phys. D: Appl. Phys. **48**, 195302 (20pp) (2015). https://doi.org/10.1088/0022-3727/48/19/1953
16. Sharkeev, Yu.P., Popova, K.S., Prosolov, K.A., Freimanis, E., Dekhtyar, Yu. Khlusov, I.A.: Electrical potential and topography of the surface of a calcium-phosphate coating deposited with RF-magnetron discharge. J. Surf. Invest.: X-ray Synchrot. Neutron Tech. **14**(1), 200–206 (2020)

17. Arvin, H., et al.: Electronic transitions and structural changes in bone. Latvian J. Phys. Technol. Sci. **6**(s), 50–55 (2000)
18. Derjugina, I., Katasheva, J., Katashev, A., Tatarinov, A.: Usage of ultrasound for local bone injuries monitoring. J. Biomech. **34**(1), 82–83 (2001)
19. Bystrov, V.S., Paramonova, E.V., Dekhtyar, Y., Pullar, R.C., Katashev, A., et al.: Polarization of poly(vinylidene fluoride) and poly(vinylidene fluoridetrifluoroethylene) thin films revealed by emission spectroscopy with computational simulation during phase transition. J. Appl. Phys. **111**, 104113 (2012)
20. Bystrov, V., Kliem, H., Bystrova, N., Paramonova, E., Dekhtyar, Yu., Polyaka, N.: Modelling of ferroelectricity in PVDF thin films and composite. In: Proceedings of International Conference Electroceramics XI, H-01-P (2008)

A Less Traditional Approach to Biomedical Signal Processing for Sepsis Prediction

Victor Iapăscurtă(✉)

Department of Anesthesia and Intensive Care, N. Testemiţanu State University of Medicine and Pharmacy, Chişinău, Moldova
victor.iapascurta@usmf.md

Abstract. Most of the data generated by monitors in a clinical setting represent time series data which can be visualized and subsequently used for decision making. This usually is the simplest part. A more challenging aspect is using this data for more complex task like machine learning with the same goal – computer assisted decisions. Within this challenge raw biomedical signal data need to be preprocessed before being passed to the machine learning algorithm. This can be done by a multitude of methods. A number of such methods comes from the field of Algorithmic Complexity and although of a promising nature, these particular methods are poorly explored yet. The current research presents an example of applying the Block Decomposition Method to data routinely generated by patients in a modern Intensive Care Unit. The final goal of a larger research, the actual research being part of, is building a system for early sepsis prediction.

Keywords: Signal processing · Block decomposition method · Sepsis

1 Introduction

Data generated by monitors in a clinical setting usually represent various biological signals like heart rate (HR), blood pressure (BP), oxygen saturation (SaO_2), respiratory rate (RR), etc. over time (i.e., time series) and can be of importance in predicting, diagnosing and treating a particular disease. A recent supplement to the management of this data are systems based on machine learning (ML). Before being passed to ML module the data usually need to be preprocessed. Among techniques for such preprocessing there is a less explored non-linear approach coming from the field of algorithmic complexity.

This paper describes an attempt at exploring the possibility of using the principles and metrics of algorithmic dynamics in the representation of medical data, first of all, in the form of time series, which describe the clinical condition of the patient.

After preprocessing using the Block Decomposition Method (BDM) as a mean for generating specific features from raw monitoring data, it can be used for ML.

The final goal of a research, the current research being part of, is creation of an ML-based system for early sepsis prediction with its subsequent testing in clinical conditions and the current paper describes the first stages of this journey.

I. Tiginyanu et al. (Eds.): ICNBME 2021, IFMBE Proceedings 87, pp. 215–222, 2022.
https://doi.org/10.1007/978-3-030-92328-0_29

2 Materials and Methods

2.1 Data

The data set used in the research represents public access data from the 'Early Prediction of Sepsis from Clinical Data - the PhysioNet Computing in Cardiology Challenge 2019'. According to the rules of the event, these data can be used in eventual publications after the date of the closing conference, September 8–11, 2019 (https://physionet.org/content/challenge-2019/1.0.0). The data come from two geographically distinct US hospital systems: Beth Israel Deaconess Medical Center (set A) and Emory University Hospital (set B). These data were collected in the last decade with the approval of the institutional review boards, were de-identified and labeled using Sepsis-3 clinical criteria [1]. The data consist of a combination of summaries of vital signs per hour, laboratory values and static patient descriptions. In particular, the data contain 40 clinical variables: 9 variables of vital signs and vital support (FiO2), 26 laboratory variables and 6 demographic and logistical variables. In total, these data include over 1.5 million time-windows and over 10.4 million data points (non-missing physiological and laboratory variables). The data extracted from the electronic systems of the respective hospitals underwent a series of preprocessing steps before the formal analysis. All patient characteristics were condensed into hourly baskets; for example, several measurements of heart rate in a time window were summarized as a measure of average heart rate.

Since there are missing values in the data for the research had been selected subsets with minimum missing values, out of which were extracted/generated 5416 septic samples and 14172 non-septic samples. Out of this, 608 samples served as test set.

2.2 Main Steps of Data Preprocessing

Original data is in '.psv' (pipe separated value) format. For further processing it is converted to '.csv' (comma separated value) format using the "rio" (R) package. Initial data sets are explored using standard statistical methods: mean, standard deviation, median, percentage ratio. After experimenting with different parameters/biomedical signals available in the dataset, six physiological parameters were selected for further exploratory data analysis and processing: heart rate (HR), oxygen saturation (SaO_2), systolic blood pressure (SBP), diastolic blood pressure (DBP), respiratory rate (RR) and temperature (Temp).

For further processing data was grouped in "Circulatory" (HR, SBP, DBP) and "Respiratory-Metabolic" (SaO_2, RR, Temp) and reshaped to 3×3 matrices. An example of such matrix is presented in the subsection 'D' below.

2.3 BDM as the "Core" Technique for Data Processing

Block Decomposition Method, proposed by Algorithmic Information Dynamic (AID), is a novel approach that can be used for data representation and processing.

AID [2, 3] is an emerging field of complexity science based on algorithmic information theory (AIT), which comprises the literature based on the concept of Kolmogorov–Chaitin complexity and related concepts such as algorithmic probability, compression, optimal inference, the universal distribution, Levin's semi-measure and others.

Central to AIT is the definition of algorithmic (Kolmogorov– Chaitin or program-size) complexity (Kolmogorov, 1965; Chaitin, 1969) [4]:

$$K_T(s) = min\{|p|, T(p) = s\},$$

that is, the length of the shortest program p that outputs the string s running on a universal Turing machine T.

AID strives to search for solutions to fundamental questions about causality: why a particular set of circumstances leads to a particular outcome. In this aspect it essentially differs from traditional statistics. As an applied science, AID is a new type of discrete calculus based on computer programming and aimed at studying causation by generating mechanistic models to help find first principles of physical phenomena, building up the next generation of machine learning [5].

In the AID toolkit, there is a special tool for providing reliable estimations to uncomputable functions, namely the online algorithmic complexity calculator (OACC) [4], which provides estimations of algorithmic complexity and algorithmic probability for short and long strings and for two-dimensional arrays better than any other traditional tool, none of which can capture any algorithmic content beyond simple statistical patterns. The OACC uses the BDM method [4], which is based upon algorithmic probability defined by the coding theorem method (CTM) [6, 7]:

$$BDM = \sum_{i=1}^{n} CTM\,(block_i) + log_2(|block_i|)$$

The OACC is available as an online version [2] as well as standalone packages in R and a number of other languages, including the Wolfram Language [8], and it is used for respective calculations for the scope of the current work.

2.4 Data Processing Flow

Because the intensity of calculations is quite high (due to the volume of data and the very essence of the calculations) the use of OACC in the described form is insufficiently productive. To accelerate these calculations, the core of the respective package [8], which can be applied on two-dimensional structures (matrices), was extracted and integrated in a program that performs the processing of the data flow in the current research. In order to make raw biological signal datasets appropriate for OACC and calculate BDM, a mapping procedure illustrated below was applied.

a. Original data (HR, SBP, DBP) as 3 × 3 matrix.

88	83	74.5
95.5	100.25	90.25
51	52	50.5

b. Binarized data (using the threshold value by row).

1	1	0
1	1	0
0	1	0

c. Calculated BDM value for the respective matrix.

17.4911 bits

A more detailed description of similar mapping is presented in [9, 10].

The final format of the data (Table 1) includes 14 columns: two columns with BDM value for the matrices and twelve columns with the difference of respective parameters value between 2 consecutive hours. The table below presents 3 data subsets (clinical cases).

Table 1. Format of data passed to ML.

BDM1	BDM2	V1	V2	V3	V4	...	V12
16.956	17.491	−4.5	1.5	0	−1.5	...	0.5
18.328	14.815	−8.5	0	0.5	−10	...	0
18.456	15.942	1	0	0.2	28	...	0

2.5 Machine Learning Stage

The final data organized in the described format were ultimately delivered to a model based on the GBM algorithm (Gradient Boosting Machine – an algorithm based on decision trees), which performed the prediction by classification (sepsis vs non-sepsis). Optimal hyperparameters (learning rate, number of trees, depth, number of branches) were initially identified by automated machine learning (AutoML) using the H2O platform. These hyperparameters were then justified in the context of the present data by a 10-fold cross-validation grid search.

3 Results

3.1 ML System Creation

As an orientation point for the creation of the sepsis prediction system in this research was the InSight system, reported in the literature as one of the best in the field [11]. Using fairly general information from [12] this system was replicated in the R language and tested for performance on the data set in the current study. Results were obtained that correspond to those reported by the authors. These results further served as milestones in the creation and evaluation of our own system. Table 2 presents the comparative performance metric of these two systems.

Table 2. Comparative metrics of two sepsis prediction systems (InSight vs the system created in the current research).

Metrics	InSight	Our system
AUC	0.95 (0.01)	0.95 (0.008)
Accuracy	0.91 (0.01)	0.91 (0.007)
Recall	0.82 (0.01)	0.84 (0.03)
Specificity	0.94 (0.02)	0.93 (0.01)

The metrics used are traditional for such systems. The mean values and the standard deviation are indicated: AUC - area under the curve (FPR on X axis vs recall on Y axis).

The performance of these two systems is quite close. A higher "Recall" in the case of the created system would indicate a more balanced system, which detects sepsis cases a little better, than the InSight system. The resulting system differs from InSight in the initial data used for machine learning, the data processing methods and the final data format, which are ultimately delivered for learning and validation. For example, the final data in the case of the created system represent vectors with length 14 (numeric strings made up of 14 numbers/values) compared to 30, in the case of InSight, i.e., a reduction in dimensionality more than 2 times. The appearance of the final data used by the created system is illustrated in Table 1.

3.2 ML System Performance

The system validation (i.e., 10 folds cross-validation) and testing on the test set were performed. The test set represents 608 samples which did not participate in the model training. The performance of the created system run on Wolfram Language 12.3 engine is illustrated in the following figures (Figs. 1 and 2):

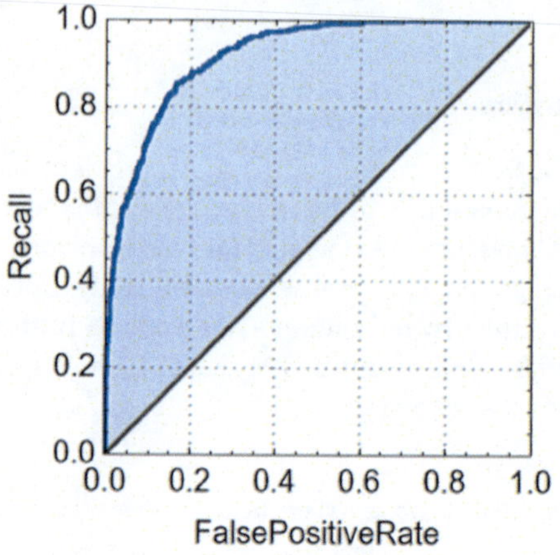

Fig. 1. ROC (receiver operating characteristics) curve

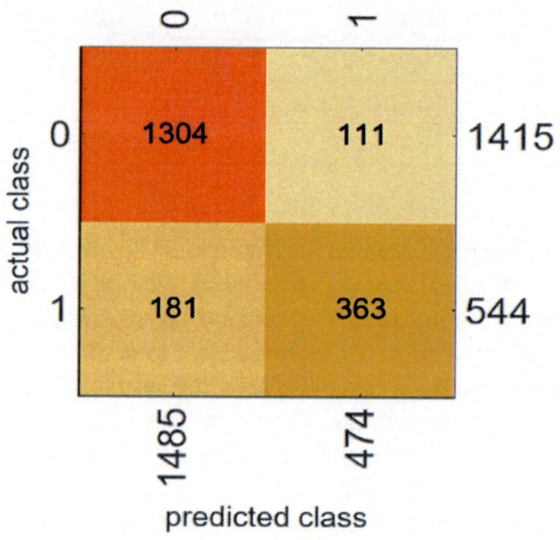

Fig. 2. Confusion matrix

3.3 Incorporation of the ML-System into an Application for Clinical Use

For the clinical use of the system, it has been embedded in a Shiny application (R programming language) which is supposed to be easy to use by ICU physicians. A demo-version of the application is available at: https://viapascurta.shinyapps.io/ISAAC_sepsis_demo/

The application can be used as follows: (1) entering the data describing the case/patient (the 6 parameters) - can be entered hourly; (2) the data can be viewed graphically in the form of time series. With the accumulation of 3-time windows, the risk of sepsis in the targeted patient is automatically determined ("High risk" vs "Low risk"); (3) as the time progresses, after each hour the patient's condition (in relation to the risk of sepsis) is re-evaluated with the display of the risk for each evaluation. The risk dynamics can be viewed in a separate window.

4 Conclusions

Current research presents a concept that is less traditional for biomedical signal processing and use for a medical purpose on the example of sepsis prediction. The algorithmic complexity can be seen as a metric of the processes that take place in a human subject and the representation of the data obtained with its help (e.g., applying BDM) has a potential still little explored. The preliminary results presented here seems to be promising. However, it makes sense to avoid too optimistic forecasts until the long-term application of the system in clinical conditions, taking into account possible technical difficulties, which could not be foreseen at this stage.

Acknowledgment. This research is part of the activity within Trello based "Algorithmic Information Dynamics" research group headed by Hector Zenil (Oxford Immune Algorithmics, Oxford University Innovation, Department of computer science, University of Oxford/Alan Turing Institute, London, UK/Algorithmic Dynamic Lab, Karolinska Institute, Stockholm, Sweden).

Conflict of Interest. The authors declare that they have no conflict of interest.

References

1. Shankar-Hari, M., Phillips, G., Levy, M., et al.: Developing a new definition and assessing new clinical criteria for septic shock: for the third International consensus definitions for sepsis and septic shock (Sepsis-3). JAMA **315**, 775–787 (2016)
2. Zenil, H., Kiani, N.: Algorithmic information dynamics: a computational approach to causality and living systems from networks to cells. In: MOOC by Complexity Explorer, Santa Fe Institute, Santa Fe, NM (12 June 2018 to 13 Oct 2018) (2018). www.complexityexplorer.org/courses/63-algorithmic-informationdynamics-a-computational-approach-to-causality-and-livingsystems-from-networks-to-cells-2018
3. Zenil, H., Kiani, N., Tegnér, J.: Methods of information theory and algorithmic complexity for network biology. Semin. Cell Dev. Biol. **51**, 32–43 (2016). https://doi.org/10.1016/j.semcdb.2016.01.011
4. Zenil, H., et al.: A decomposition method for global evaluation of shannon entropy and local estimations of algorithmic complexity. Entropy **20**(8), 605 (2018). https://doi.org/10.3390/e20080605
5. Zenil, H., et al.: An algorithmic information calculus for causal discovery and reprogramming systems. iScience (in press). https://doi.org/10.2139/ssrn.3193409
6. Soler-Toscano, F., Zenil, H., Delahaye, Jean-Paul., Gauvrit, N.: Calculating kolmogorov complexity from the output frequency distributions of small turing machines. PLoS ONE **9**(5), e96223 (2014). https://doi.org/10.1371/journal.pone.0096223
7. Delahaye, Jean-Paul., Zenil, H.: Numerical evaluation of algorithmic complexity for short strings: a glance into the innermost structure of randomness. Appl. Math. Comput. **219**(1), 63–77 (2012). https://doi.org/10.1016/j.amc.2011.10.006

8. Zenil, H., et al.: The online algorithmic complexity calculator (OACC) v3.0. Algorithmic Dynamics Lab, Science for Life Laboratory (SciLifeLab), Unit of Computational Medicine, Center for Molecular Medicine at the Karolinska Institute in Stockholm, Sweden (2018). www.algorithmicdynamics.net/software.html
9. Iapăscurtă, V.: Detection of movement toward randomness by applying the block decomposition method to a simple model of the circulatory system. Complex Systems Journal **28**(3), 59–77 (2019)
10. Iapascurta, V.: Block decomposition method and traditional machine learning for epileptic seizure prediction. In: Cellular Automata and Discrete Complex Systems, 26th IFIP WG 1.5 International Workshop, AUTOMATA 2020, Stockholm, Sweden, 10–12 August 2020, Proceedings, Special Session: Algorithmic Information Dynamics, p. xxii (2020). https://www.automata2020.com/videos--material.html
11. Burdick, H., et al.: Effect of a sepsis prediction algorithm on patient mortality, length of stay and readmission: a prospective multicentre clinical outcomes evaluation of real-world patient data from US hospitals. BMJ Health Care Inf. **27**, e100109 (2020). https://doi.org/10.1136/bmjhci-2019-100109
12. Mao, Q., et al.: Multicentre validation of a sepsis prediction algorithm using only vital sign data in the emergency department, general ward and ICU. BMJ Open **8**, e017833 (2018). https://doi.org/10.1136/bmjopen-2017-017833

Influence of Change in Cardiac State on Probable Properties of Rhythmograms

Y. I. Sokol, P. F. Shapov, Mykhailo A. Shyshkin(✉), and R. S. Tomashevskyi

Department of Industrial and Biomedical Electronics, National Technical University «Kharkiv Polytechnic Institute», Kharkiv, Ukraine

Abstract. The miniaturization of medical electronics in recent years, like electronics in general, provides new opportunities for creating miniature and subminiature monitoring systems that are capable of monitoring important human biophysical signals. In particular, to carry out long-term monitoring of parameters of cardiac activity.

In this case, a wide field opens up for the development of new effective wearable devices. However, the circuitry implementation of such devices is impossible without significantly new methods of processing and analyzing such complex signals as the signal of the electrical activity of the heart. The paper proposes a study of the probabilistic properties of rhythmograms depending on the change in the patient's cardiac state. The results of these studies make it possible to obtain information about changes in the internal structure of the conduction of electrocardiographic impulses and significantly increase the information content of rhythmograms.

Keywords: Cardiac condition · Atrial fibrillation · Rhythmogram · Cumulant analysis

1 Introduction

The main methods of heart rate analysis at the moment are statistical, geometric and nonlinear methods, autocorrelation, spectral and fractal analyzes [1, 2]. Statistical methods are used to quantitatively analyze a signal. With this method, the signal is processed as a set of consecutive time intervals. The disadvantage of the statistical methods proposed today is the impossibility of determining the internal structure of a number of mechanisms for the generation of a pathological state. Despite this, statistical indicators characterize the formation of a signal under the influence of random factors quite fully. Time analysis includes a group of methods based on the use of statistical calculations. Its difference from statistical analysis is the use of specific indicators that are used only for the analysis of biomedical signals, in particular, heart rate. Timing analysis has a high predictive value, high reproducibility, and the ability to improve the reliability of results with increasing signal sample length [3, 4]. Spectral analysis allows you to decompose a series into its constituent components and quantify the impact of each of them. Mathematically, this method is carried out using the discrete Fourier transform. Currently,

I. Tiginyanu et al. (Eds.): ICNBME 2021, IFMBE Proceedings 87, pp. 223–230, 2022.
https://doi.org/10.1007/978-3-030-92328-0_30

some systems for analyzing daily heart rate variability provide images of spectrograms at certain intervals, which makes it possible to assess the dynamics of spectrum components in time. [5, 6]. Autocorrelation analysis is used to evaluate the internal structure of a signal. Geometric methods are based on the analysis of graphs and shapes that allow you to visualize the heart rate, as well as evaluate some of the indicators of these shapes. The main advantage of geometric methods is insensitivity to the analytical quality of the signal time sampling, and the main disadvantage is the presence of a signal of an acceptable length (usually at least 20 min). When analyzing cardiac arrhythmias, it is advisable to use the methods of correlation rhythmogram [7, 8].

In nonlinear methods, to present the results, the following are used: the senior Lyapunov exponent, the entropy of a dynamical system, attractor graphs, and other methods [9].

As can be seen, despite the widespread use of various methods for studying the heart rhythm, they are rather applicable for post-factor analysis of rhythmograms in stationary conditions, for determining long-term predictions of changes in the state of patients. However, they are practically not used for direct assessment and detection of cardio conditions in the online mode.

The study proposed by the authors makes it possible to assess the information content of the probabilistic parameters of rhythmograms and to single out the most significant ones for detecting changes in the cardiac state. In this case, two significantly different cardiac states were used Normal state (NSR) Atrial fibrillation (AF) state.

2 Analysis of the Probable Properties of the Rhythmogram for Different Cardiac States

2.1 Cumulative Analysis of the Rhythmogram

Any rhythmogram can be interpreted in two ways:

a) as a sequence of random durations of heart contraction in time (random continuous process with discrete time). The duration of the RR interval can be considered a random continuous variable for such a process.
b) as a stream of random events inside the cardiac conduction system, the result of which is the contraction of the ventricles (R waves of the electrocardiogram). A random discrete value for such a stream of events can be conventionally considered the heart rate (averaged over a fixed time interval)

Both variants of this presentation of the rhythmogram allow considering its probabilistic properties using either distribution rules for continuous or discrete random variables.

The choice of the type of random variable in the study of the probabilistic properties of the rhythmogram does not matter, since a change in these properties for one variant of the description of the rhythmogram entails a change for another variant.

If we use option a), then it should be noted that full information about the probabilistic properties of a random process (regardless of whether it is continuous or discrete) is carried by the probability distribution $f(x)$ of the random variable x. Such a probability distribution in the general case can be represented in the form of the inverse Fourier transform of the characteristic function $\Theta(u)$ of the random variable x

$$f(x) = \frac{1}{2\pi} \int_{-\infty}^{\infty} \Theta(u) e^{jux} du, \quad (1)$$

where u – real type variable;

$j = \sqrt{-1}$;

$\Theta(u)$- characteristic function of a random variable x, depending, in general, on the cumulants of this quantity.

Expression (1) mathematically represents the density $f(x)$ as a linear transformation $\Theta(u)$, establishing the adequacy between probabilistic properties $f(x)$ and $\Theta(u)$. In turn, $\Theta(u)$ can be represented as an infinite series, whose elements linearly depend on the initial moments α_k. [10].

$$\Theta(u) = 1 + \sum_{k=1}^{\infty} \frac{\alpha_k}{k!} (ju)^k,$$

where α_k –k-th order initial moments.

On the other hand, the initial moments can be expressed in terms of centered standard numerical characteristics: central moments or cumulants (semi-invariants). Cumulative analysis of random variables is more adequate for the formal description of their probabilistic properties, since it is based on direct and inverse Fourier transforms using the mathematical apparatus of characteristic functions. Moreover, such an analysis makes it possible to visually assess the closeness of the distribution of random variables to the normal distribution. For this distribution, all cumulants of the 3rd and higher orders are equal to zero. Let us give a notation system for the numerical characteristics of a random variable:

m- mean of a random variable x (1st order cumulant);
σ^2 – variance (2nd order cumulant);
γ_a – asymmetry coefficient (3rd order cumulant coefficient);
γ_k – kurtosis coefficient (4th order cumulant coefficient).

The dependencies between the initial moments and the listed numerical characteristics can be expressed in the form of the following system of equations:

$$\left\{ \begin{aligned} \alpha_1 &= m \\ \alpha_2 &= \sigma^2 + m^2 \\ \alpha_3 &= \gamma_a \sigma^3 + 3m\sigma^2 + m^3 \\ \alpha_4 &= \gamma(_э + 3)\sigma^4 + 3\sigma^4 + 4m\gamma_a\sigma^3 + 6m^2\sigma^2 + m^4 \end{aligned} \right. \quad (2)$$

From expressions (2) that the initial moments are combinations of the above standard characteristics.

To study the probabilistic properties of the rhythmogram, we will consider it as a non-stationary random process, the numerical characteristics (cumulants) of which are formally functions of time. This formalization assumes that the dependence of the numerical characteristics of the rhythmogram on time can also be non-stationary processes, the non-stationarity model of which can change when the type of cardiac state changes. The detection of such a change can be carried out using standard statistical models of ANOVA (one- or multifactor).

In Fig. 1 shows the dependences on the observation time of the estimates of the above numerical characteristics of the rhythmogram in a sliding observation window for two fixed cardiac states (S0 – normal (NSR), S1 - atrial fibrillation (AF)).

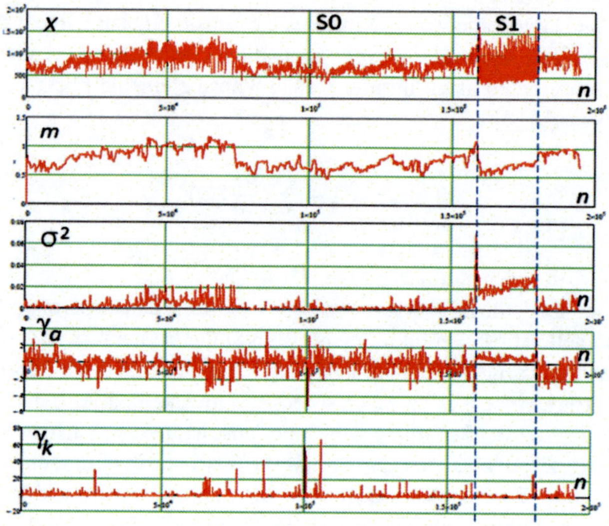

Fig. 1. Dependencies of four moments of the rhythmogram in a sliding observation window

When using a discrete model (option b)), it is convenient to study changes in such an informative parameter about changes in the properties of a stream of random events as the connectivity coefficient (the ratio of the square of the mean value to the variance) RR interval in a sliding observation window.

$$\gamma_{con} = \frac{m^2}{\sigma^2} \tag{3}$$

Figure 2 shows the dependence of the connectivity coefficient RR of the interval for two alternative cardio states (S0 / S1).

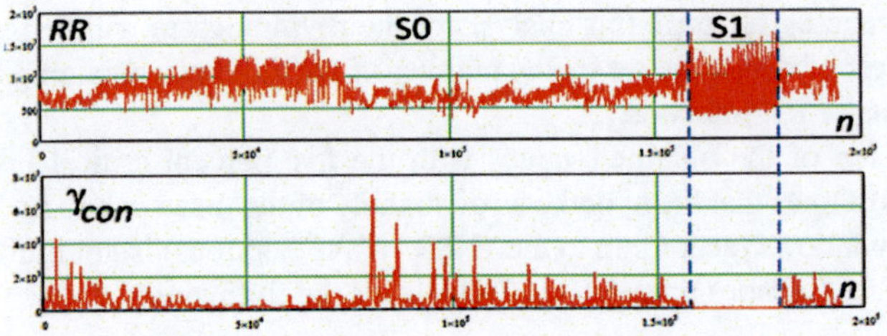

Fig. 2. Connectivity coefficient of RR interval

Figures 1 and 2 well illustrate the significant variability of the used numerical characteristics when changing the cardiac state. It should be noted that such variability is characterized by important diagnostic properties:

- a change in the cardiac state entails a change in all, without exception, numerical characteristics, which indicates a change in the probabilistic model when the cardiac state changes;
- since option b) uses a complex indicator (connectivity coefficient) that depends on the mean, variance and their interaction, it is more visual and informatively preferable for constructing a probabilistic model of the dynamics of the electrical activity of the heart.

2.2 ANOVA of the Numerical Characteristics of the Rhythmogram

Figure 3 shows a part of the rhythmogram, including the areas corresponding to the normal rhythm and atrial fibrillation. Each of areas has 1200 observations (N = 2400). The results for each of the indicated areas of the rhythmogram were divided into six groups (k = 6). Each group had 200 observations (n = 200).

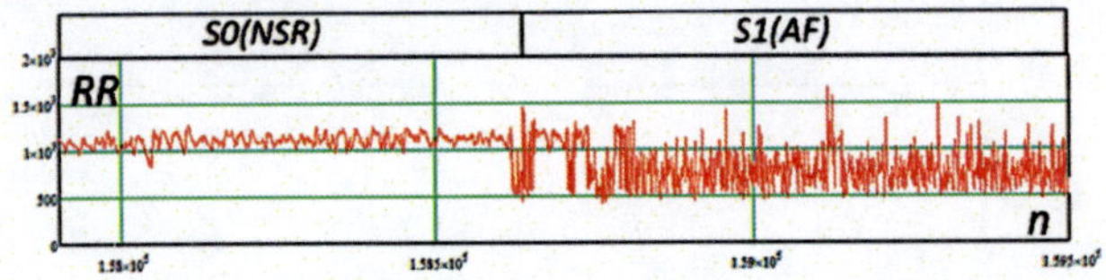

Fig. 3. Part of the rhythmogram used in ANOVA with NSR and AF areas

Use of one-factor ANOVA for the results of each of the rhythmogram areas at the level of significance $\alpha = 0.05$ made it possible to calculate Fischer's criterion statistics with degrees of freedom numbers 5 and 194 for each of the studied variants of cardiac state:

$$F_{5,194} = 224.94(\text{NSR});$$

$$F_{5,194} = 28.74(\text{AF}).$$

The influencing factor in this analysis of the rhythmogram will be the observation time, the levels of which are set by the number of the window containing 200 results of measurements of RR intervals.

Comparison of the obtained values with the five percent critical Fisher statistics ($F_{crit} = 2.29$) shows that the hypothesis of equality of the mean values in each of the six observation windows (according to the "NSR" - "AF" options) is not fulfilled, since the obtained Fisher statistics exceedF_{crit}. This means the following:

a) the rhythmogram, regardless of the cardiac state, is a non-stationary, in terms of the mean value, a random process (224.94 > 2.29, 28.74 > 2.29);
b) nonstationarity for the "NSR" state is much higher than for the "AF" state (224.94 > 28.74).

ANOVA was carried out using the Data Analysis tools of the Microsoft Excel.

The conclusion about the nonstationarity of the rhythmogram does not exclude studies to verify its ergodicity. The importance of such a check is due to the fact that any observation window has a limited width, the choice of which is determined by the autocorrelation function of the rhythmogram. If such a function tends to zero with increasing window width, then the initial process is ergodic (although not stationary). If the limit of the autocorrelation function tends to a constant value (greater than zero), then the rhythmogram is not ergodic. The last statement means that in order to obtain adequate statistical conclusions, it is necessary to work with an ensemble of rhythmograms, and not with one rhythmogram of a sufficiently long duration. But ensure the last condition is practically quite difficult.

In Fig. 4 shows the autocorrelation functions for the states of NSR (2.4a) and AF (2.4b).

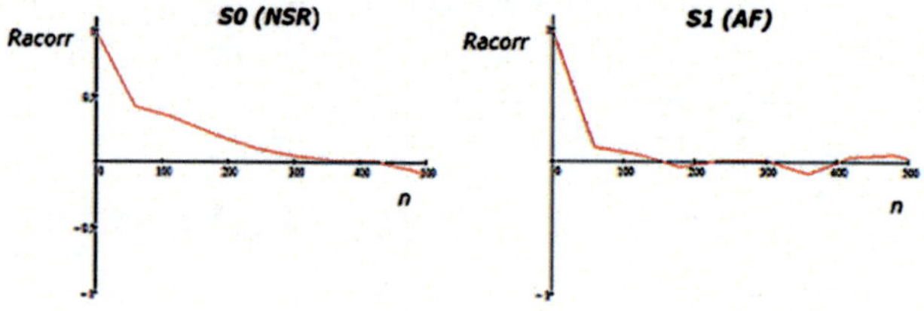

a) b)

Fig. 4. Autocorrelation functions of the rhythmogram (a - NSR, b - AF)

The autocorrelation functions of the rhythmogram presented in Figs. 4a and 4b allow us to draw the following conclusions:

- regardless of the type of cardiac state, the rhythmogram is an ergodic process;
- the autocorrelation function for the NSR is wider than for the AF. This means that the spectra of the rhythmogram for NSR and AF have different widths;
- the minimum allowable observation width should not be less than n = 200, which provides an autocorrelation level not exceeding 0.2.

To perform a two-factor ANOVA without repetitions, we will take "Cardiac state" as an influencing factor. It has two quality levels "NSR", "AF".

Initial data are presented for each of the studied cumulants in Table 1.

Table 1. Initial data for ANOVA

	#1	#2	#3	#4	#5	#6
Mean						
NSR	0,944	1,066	1,080	1,116	1,117	1,121
AF	0,914	0,748	0,799	0,762	0,758	0,682
Variance						
NSR	0,0048	0,0024	0,0025	0,0063	0,0036	0,0046
AF	0,0802	0,0306	0,0363	0,0283	0,0270	0,0313
Asymmetry coefficient						
NSR	−0,574	−0,072	−0,0224	−0,8872	0,0355	-2,5742
AF	−0,221	0,659	0,92312	1,08105	0,61778	0,73337
Kurtosis coefficient						
NSR	−0,639	−0,717	−0,5192	1,0849	-0,5909	19,121
AF	−1,541	0,5035	2,2983	2,0716	0,7867	0,9084
Connectivity coefficient						
NSR	183,39	459,89	464,19	194,97	342,29	270,76
ФП	10,431	18,250	17,560	20,521	21,280	14,848

Table 2 shows the results of calculating Fischer's statistics, which are responsible for the influence of the factor "Cardio-state". Number of degrees of freedom: 1 and 5.

Table 2. The results of calculating Fisher's statistics for the factor "Cardiac condition". Critical statistics $F_{1;5;0.05}= 6{,}6079$

Cumulant	Fischer's statistics $F_{1;5}$
Mean	26,5784
Variance	17,8326
Asymmetry coefficient	8,18693
Kurtosis coefficient	0,4235
Connectivity coefficient	36,1098

Analysis of Table 2 led to the following conclusions:

3 Conclusions

1) Most of all, the factor "Cardiac state" affects the coefficient of connectivity, since its F statistic is maximal ($F_{1;5}$= 36,1098).
2) *F* factor has a lesser effect on the rest of the numerical characteristics, and for the mean, variance and asymmetry coefficient, the factor influence is statistically significant (their F statistics are higher than the critical value $F_{1;5;0.05}$= 6,6079).
3) In fact, the space of informative parameters for the formation of the procedure for recognizing an AF attack includes four basic parameters listed in descending order of their diagnostic properties: connectivity coefficient, mean, variance, asymmetry coefficient.

Conflict of Interest. "The authors declare that they have no conflict of interest".

References

1. Task force of the European society of cardiology and the North American Society of Pacing and Electrophysiology. Heart rate variability: standards of measurement, physiological interpretation and clinical use. Circulation **93**, 1043–1065 (1996)
2. Cerutti, S., Baselli, G., Bianchi, A.M., Mainardi, L.T., Signorini, M.G., Malliani, A.: Cardiovascular variability signals: from signal processing to modeling complex physiological interactions. Automedica **16**, 45–69 (1994)
3. Box, G.E.P., Jenkins, G.M., Reinsel, G.C.: Time Series Analysis, Forecasting and Control, 3rd edn. Prentice-Hall, Englewood Cliffs (1994)
4. Bianchi, A.M., Mainardi, L., Petrucci, E., Signorini, M.G., Mainardi, M., Cerutti, S.: Time-variant power spectrum analysis for the detection of transient episodes in HRV signal. IEEE Trans. Biomed. Eng. **40**, 136–144 (1993)
5. DeBoer, R., Karemaker, J., Strackee, J.: Comparing spectra of a series of point events particularly for heart rate variability data. IEEE Trans. Biomed. Eng. **31**(4), 384–387 (1984)
6. Yamamoto, Y., Hughson, R.L.: Coarse graining spectral analysis: new method for studying heart rate variability. J. Appl. Physiol. **71**, 1143–1150 (1991)
7. Woo, M.A., Stevenson, W.G., Moser, D.K., Trelease, R.B., Harper, R.M.: Patterns of beat-to-beat heart rate variability in advanced heart failure. Am. Heart J. **123**(3), 704–710 (1992)
8. Sokol, Y., Shyshkin, M., Butova, O., Akhiezer, O., Dunaievska, O.: Improved graphical analysis of atrial fibrillation based on Holter measurement data. In: 2019 XXIX International Scientific Symposium "Metrology and Metrology Assurance" (MMA), Sozopol, Bulgaria, pp. 01–06 (2019). https://doi.org/10.1109/MMA.2019.8936003
9. Henriques, T., Ribeiro, M., Teixeira, A., Castro, L., Antunes, L., Costa-Santos, C.: Nonlinear methods most applied to heart-rate time series: a review. Entropy (Basel) **22**(3), 309 (2020). https://doi.org/10.3390/e22030309. PMID: 33286083
10. Dubkov, A.A., Malakhov, A.N.: Cumulant analysis of a functional nonlinear transformation of non-Gaussian random processes and fields. Dokl. Akad. Nauk SSSR **222**(4), 793–796 (1975)

A Brain-Computer Interface for Controlling a Mobile Assistive Device by Using the NeuroSky EEG Headset and Raspberry Pi

Oana-Andreea Rușanu(✉)

Product Design, Mechatronics and Environment Department, Transilvania University of Brasov, B-dul Eroilor No. 29, Brașov, Romania
oana.rusanu@unitbv.ro

Abstract. The brain-computer interface (BCI) constitutes an excellent solution for people with neuromotor disabilities that need an alternative communication and control channel with the outside environment. This paper proposes a Python-based BCI system for controlling a mobile assistive device using the NeuroSky EEG portable headset and the Raspberry Pi microcontroller board. Thus, disabled persons could enjoy the opportunity of controlling a mobile robot by using commands based on voluntary eye-blinking. The original implementation of the proposed BCI system consisted of Python programming for raw EEG signal analysis on the computer (Spyder IDE) for detection and counting of intentional eye-blinks and Python coding for executing the robot movement commands by the Raspberry Pi (Thonny IDE). The WebSockets protocol facilitates wireless communication between the computer (Windows) and Raspberry Pi. The presented BCI system is an experimental prototype for a better understanding of simulation and testing of the BCI technology by people with neuromotor disabilities.

Keywords: BCI · Eye-blink · Python · EEG headset

1 Introduction

The brain-computer interface (BCI) focuses mainly on providing support for people with neuromotor disabilities. The BCI proved to be a fantastic system by enabling an alternative communication and control channel. Employing neuronal biopotentials thanks to different techniques (electroencephalography – EEG, functional magnetic resonance imaging – fMRI, near-infrared spectroscopy – NIRS, and implantable sensors) and leveraging signal processing methods and artificial intelligence algorithms led to the inception of brain-computer interfaces capable of translating thoughts into actions. The brain-computer interface encounters several difficulties related to variable cognitive abilities, massive data dimensions, instantaneous time response, real-time processing, and complex detection of a specific neuronal pattern. The BCI comprises advanced systems [1] in specialized research laboratories and portable, simple, quick experimental devices designed for affordable home use [2].

I. Tiginyanu et al. (Eds.): ICNBME 2021, IFMBE Proceedings 87, pp. 231–238, 2022.
https://doi.org/10.1007/978-3-030-92328-0_31

Among the most spread fundamental BCI applications is the mobile robot controlling by using the voluntary eye-blinking bio-signal. According to previous studies [3], portable EEG headsets were preferable to robust EEG acquisition systems to increase the set-up and programming convenience. Then, to implement and control a mechatronic assistive device [4] in a straightforward manner, microcontroller-based multifunctional boards (Arduino, Raspberry Pi, NI myRIO) proved to be advantageous. Nevertheless, significant challenges arise from optimized integration between hardware components (EEG acquisition device and microcontroller-based board) and software modules (Python, LabVIEW, Matlab, and Arduino) to develop an innovative assistive application that is simple to use and reliable and accurate for people with neuromotor disabilities.

This paper proposes an original Python application for controlling a Raspberry Pi-based assistive robot by using the voluntary eye-blinking detected across the raw EEG signal from the NeuroSky biosensor. In contrast to most previous BCI projects involving EEG processing only by Raspberry to detect voluntary eye-blinking, the current work provides improved accuracy by analyzing the raw EEG signal on the computer and implementing a novel Python algorithm to capture the exact moment of performing the eye-blinking. Accordingly, it resulted in the achievement of the new Python solution for enabling WebSockets-based Wireless communication between the Windows computer and Raspberry Pi. The commands determined by voluntary eye-blinks for controlling the mobile robot are smooth, rapid, and reliable.

2 Hardware System – Neurosky and Raspberry Pi

The proposed BCI prototype is based on the following systems: the NeuroSky Mindwave Mobile second edition EEG portable headset, the Raspberry Pi Model B 4 GB board, and the necessary components to build a mobile robot. NeuroSky headset is an improved version of the portable ThinkGear device initially released in 2011. It offers several benefits for developers and users: low price, easy set-up, embedded filters and amplifiers, Bluetooth connectivity, code examples in different programming languages, comfortable wearing, raw EEG acquisition, and eSense meters (attention and meditation level, eye-blinking strength). The Raspberry Pi is a single-board computer embedding high-performance functionalities: Wireless LAN, Bluetooth 5.0, video/audio output, quad-core processor of 1.5 GHz, 40 general-purpose input/output pins (GPIO), micro-HDMI, microSD, and camera serial interface. The current work implies constructing a mobile robot (Fig. 1) composed of a compact chassis, two DC motors, two researchable batteries, an MG90S servo-motor for switching movement direction (turn to the left or turn to the right), and the L298N driver based on a dual H bridge.

3 Software System – Python Application

The software system of the proposed BCI prototype consists of two primary levels: the computer level – Python implementation of the raw EEG analysis for voluntary eye-blinks detection and the Raspberry Pi level – Python coding of the movement commands used to control the mobile robot.

Fig. 1. The hardware system of the proposed BCI – NeuroSky EEG headset and the Raspberry Pi-based mobile robot

A. The Python Application Running on the Computer Level

The computer level comprises a new approach to optimize voluntary eye-blinking detection. The NeuroSky chip transmits data to a computer by using Bluetooth protocol. Then, the acquired raw EEG signal is analyzed by the Python application running on the computer. After that, the corresponding command is sent to the Raspberry Pi using WebSockets protocol after counting the voluntary eye-blinks. The WebSockets protocol successfully enables wireless communication between the two primary levels. Thus, Raspberry Pi controls the mobile robot by setting the proper movement direction: forward, backward, left, or right.

Figure 2 shows the workflow of the Python program running on the computer in the Spyder development environment. The first stage supposed the import of the necessary Python libraries: Neuropy – EEG data acquisition from NeuroSky biosensor, Tkinter – creating graphical user interfaces, Matplotlib – graphical displaying the time – amplitude variance of the raw EEG signal, Numpy – working with multidimensional arrays, Scipy – calculating statistical metric (mean of values), and time – offering the sleep function for waiting a specific time interval. The second stage is essential because it contains the code instructions enabling the WebSockets-based wireless data exchange between the computer and Raspberry Pi, and both of them should be connected to the same Internet Network. Therefore, the IP address received by Raspberry should be used in the Python program running on the computer. The Raspberry Pi accomplishes the role of server, while the computer is the client.

The third stage involves adding the settings for proper communication between the computer and NeuroSky (COM port and samples per second) and enabling the optimized voluntary eye-blinks detection algorithm (previously-stored samples and current recorded samples).

The fourth stage is related to real-time EEG data acquisition and voluntary eye-blinks counting. Thus, the main algorithm is based on the following steps: reading one raw EEG value, storing each value to an array called 'a', waiting for a predefined Time Interval (2 or 10 ms), and incrementing the number of current recorded samples. Setting time Interval = 2 ms results in acquiring 512 raw EEG samples during one second. Setting time Interval = 10 ms results in acquiring 100 raw EEG samples during one second.

It should be emphasized that it is challenging to get waiting time = 2 ms in Windows-based computers. The time resolution response of the Windows operating systems is standardized to a value between 10–25 ms. Nevertheless, some computers can quickly achieve the waiting time interval = 2 ms. Otherwise, a lower time interval should be set: 10 ms. Also, using particular programs should be possible to adjust the Windows time resolution response.

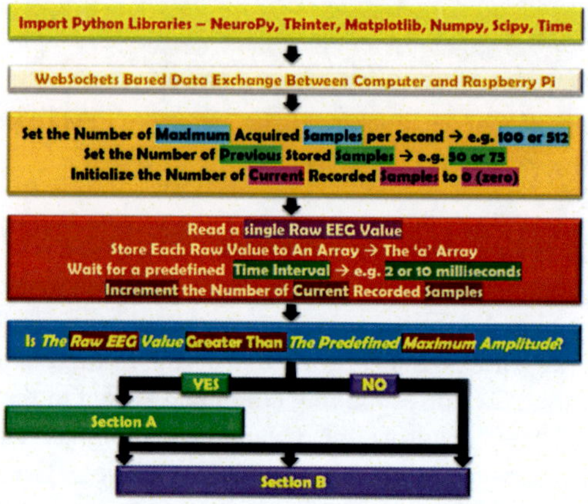

Fig. 2. The workflow of the Python application running on the computer

The fifth stage implies comparing each acquired raw EEG value to a predefined maximum amplitude that should be set to a user-customized value before running the Python program. Python executes section A (Fig. 3 and Fig. 4) of code instructions if the acquired raw EEG value exceeds the amplitude threshold. Otherwise, if the acquired raw EEG value is lower than the amplitude, Python executes section B (Fig. 5) of codes.

Figure 3 shows the workflow of the first sequence included by section A of Python code. The initial phase checks if the number of current recorded EEG samples is greater than or equal to the number of previously stored EEG samples. Accomplishing this condition involves removing as many samples as equal to the difference between current and previous acquired samples until getting a raw EEG value greater than the predefined maximum amplitude. Then, the main 'a' array is reorganized to store only the previous recorded samples, whose number was initially predefined. Avoiding the above-stated condition results in getting a lower number of currently recorded samples than the number of the previously stored samples at the moment when a specific amplitude value exceeded the established threshold. Therefore, it defines a new array by setting to zero all its samples, whose number is equal to the difference between previous and current acquired EEG samples. Then, the 'a' array is reorganized to store both these first elements set to zero and the previously recorded samples until exceeding the maximum amplitude. After handling the above-described condition, the current number of acquired samples is assigned the value of the previous number of stored samples.

Figure 4 shows the workflow of the second sequence included by section A of Python code. This sequence checks that the current recorded samples should be lower

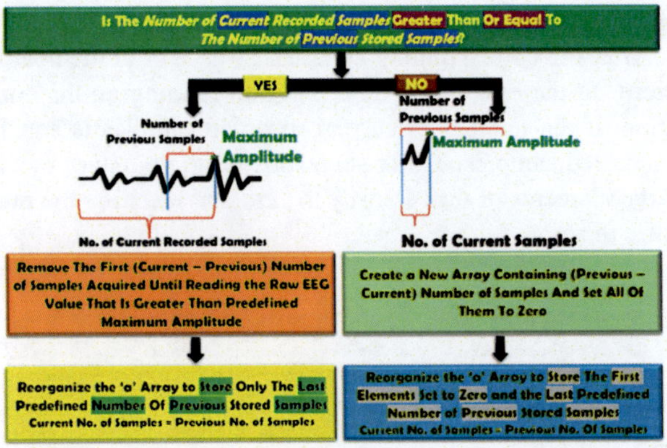

Fig. 3. The workflow of the first sequence included by section A of Python application running on the computer

than the initially set number of maximum acquired samples. Averaging the 512 or 100 values calculates the arithmetic mean based on the 'a' array. The arithmetic mean is a simple measurement able to identify if a voluntary eye-blink was performed or not. Therefore, checking the mean value between a minimum and a maximum threshold result in detecting an intentional eye-blink. Each positive response to this condition is equivalent to a voluntary eye-blink occurrence, and the corresponding counter gets incremented. Further, the same steps described above are executed.

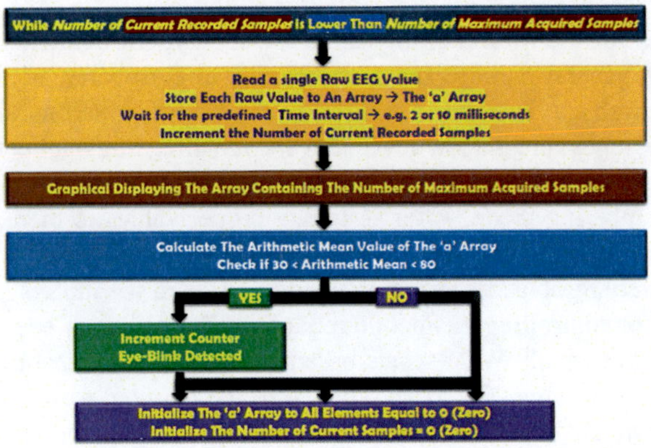

Fig. 4. The workflow of the second sequence included by section A of Python application running on the computer

Figure 5 shows the workflow of section B of Python code implemented on the computer level in Spyder IDE. The section B is mainly aimed at handling the situation given by achieving the two conditions: 1 – the number of the current recorded samples should be equal to the number of the initially set maximum acquired samples and 2 – the latest recorded raw EEG value should be lower than the predefined maximum amplitude.

If both conditions are accomplished, then it follows the displaying and checking of the counter value. After the corresponding command was sent to the board, the counter is initialized to zero. In the end, the Python program running on the computer assesses the last condition: if the number of current recorded samples is equal to the number of maximum acquired samples. This case results in no voluntary eye-blink detection, initializing all the elements of the 'a' array to zero and resetting the number of current recorded samples to zero.

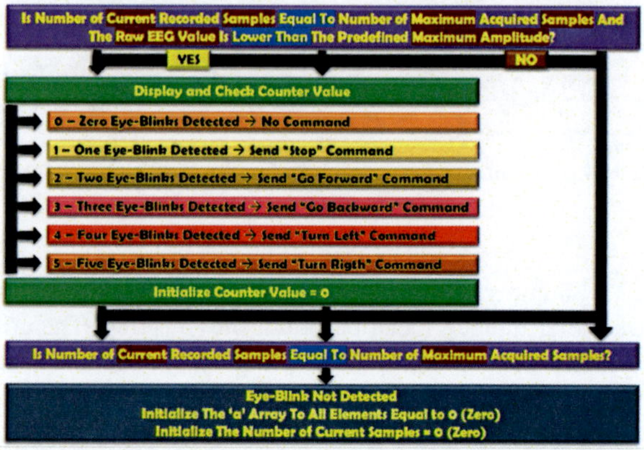

Fig. 5. The workflow of section B of Python application running on the computer

B. The Python Application Running on the Raspberry Pi
The first phase consists in importing the necessary Python libraries: RPi.GPIO – for controlling the general-purpose input-output pins of Raspberry, bottle – is a micro-framework for web applications, time – for adding delays or waiting intervals, socket – for sending messages across a low-level network and threading – for concurrent running some code sequences. The second phase is supposed to set up the BCM (Broadcom GPIO) pins numbering mode. After assigning proper values to them, all the pins for changing the movement direction and adjusting the speed of the left and right-side DC motors. For safety reasons, by changing the duty cycle of the motors, a low movement speed was set for controlling the mobile robot by using voluntary eye-blinks. The third phase is related to enable WebSockets-based wireless communication between Raspberry Pi and the computer. The same settings for the required parameters (header, port sever, address) should be set for Python applications running on the computer, and Raspberry Pi connected to the same network. Also, a function should be defined for enabling the server to listen for new threads or network clients. The last phase allows receiving and checking the network messages.

4 Results

This paper proposes an experimental prototype based on a brain-computer interface that is helpful for people with neuromotor disabilities to test and get familiar with this technology. The two video presentations of the initial experimentation based on a different Python-based approach by turning on and off different LEDs and the real-time running of the Python BCI proposed system are available at the following Unlisted YouTube links: https://youtu.be/KnVpvXKPenA and https://youtu.be/cImOpV_Dc1g.

Voluntary eye-blinking constitutes a precise and straightforward control signal necessary to command the presented mobile assistive robot. The time response is approximately equal to one second and can be impacted by graphical displaying of the time domain of the acquired raw EEG signal. Depending on the computer performance, an array containing 100–512 raw EEG samples is analyzed by a customized algorithm for detecting the voluntary eye-blinking pattern. Among the limitations of the current Python-based BCI implementation, before running the computer application, it is necessary to set user profile customized thresholds for maximum amplitude of the raw EEG signal and the arithmetical mean of the acquired array of raw EEG values. Nevertheless, the proposed BCI application for controlling the mobile robot provides the following benefits: code instructions entirely implemented in Python open-source programming language, smooth wireless communication based on WebSockets between computer and Raspberry Pi, the use of the most portable and inexpensive EEG headset, and an optimized algorithm to detect the exact real-time moment when a voluntary eye-blink is executed.

5 Conclusions

This paper proposes a brain-computer interface experimental system based on integrating the NeuroSky portable EEG headset and the Raspberry Pi board for controlling a mobile robot by implementing two new Python applications. Regarding the novelty of the proposed BCI solution, an improved algorithm for voluntary eye-blinking detection and WebSockets-based wireless communication between the computer and Raspberry Pi were implemented in the Python open-source programming language. As a future research direction, an improved compact hardware design is still necessary for building the mobile robot by using an expansion board to control the motors and power the Raspberry Pi board. In addition, an improved software framework should automatically adjust the motor's speed, avoid obstacles, and offer intuitive feedback for voluntary eye-blinks execution.

Conflict of Interest. The authors declare that they have no conflict of interest.

References

1. Sahal, M., Dryden, E., Halac, M., Feldman, S., Heiman-Patterson, T., Ayaz, H.: Augmented reality integrated brain computer interface for smart home control. In: Ayaz, H., Asgher, U., Paletta, L. (eds.) AHFE 2021. LNNS, vol. 259, pp. 89–97. Springer, Cham (2021). https://doi.org/10.1007/978-3-030-80285-1_11
2. Madona, P., Mujiono, R.R., Wijaya, Y.P.: Controlling the direction of wheelchair movement using Raspberry-Pi based brain signals. In: 2019 2nd International Conference on Applied Engineering (ICAE), pp. 1–4 (2019). https://doi.org/10.1109/ICAE47758.2019.9221680

3. Chaurasia, V., Mishra, V., Jain, L.: Brain-bot: an unmanned ground vehicle (UGV) using Raspberry Pi and brain computer interface (BCI) technology. In: Singh, M., Gupta, P.K., Tyagi, V., Sharma, A., Ören, T., Grosky, W. (eds.) ICACDS 2016. CCIS, vol. 721, pp. 252–261. Springer, Singapore (2017). https://doi.org/10.1007/978-981-10-5427-3_27
4. Katona, J., Ujbanyi, T., et al.: Speed control of Festo Robotino mobile robot using NeuroSky MindWave EEG headset based brain-computer interface. In: 2016 7th IEEE CogInfoCom, pp. 000251–000256 (2016). https://doi.org/10.1109/CogInfoCom.2016.7804557

4-Quadrant Interpretation of the Speed Spot Plot Asymmetry for Arrhythmia Detection

Y. I. Sokol[1], Mykhailo A. Shyshkin[1(✉)], O. A. Butova[1], O. B. Akhiiezer[2], and O. I. Dunaievska[2]

[1] Department of Industrial and Biomedical Electronics, National Technical University «Kharkiv Polytechnic Institute», 2, Kyrpychova str., Kharkiv 61002, Ukraine

[2] Department of Computer Mathematics and Data Analysis, National Technical University «Kharkiv Polytechnic Institute», Kharkiv, Ukraine

Abstract. Today, The World Health Organization estimates that by 2030, about 25 million people will die from cardiovascular disease every year, meaning that heart disease will remain the leading cause of death. One of the most common types of cardiovascular disease is arrhythmia caused by abnormal electrical activity in the heart. An effective method for studying the nonlinear characteristics of HRV on an ECG, in particular arrhythmias, is the Poincaré Plot - a diagram in which each R-R interval is displayed as a function of the previous R-R interval. This work is a continuation of studies of a modified graphical method for displaying HRV, called Speed Spot, in which both the current value of the rhythm and the rate of its change are graphically displayed. In particular, the asymmetry of the location of the points is interesting, which shows the degree of imbalance in the processes occurring in the heart. The paper considers asymmetry in each of the four quadrants of this graphical representation as a possible mechanism for detecting various arrhythmias.

Keywords: Speed Spot · Cardiac arrhythmias · Poincare Plots · Heart rate variability · Electrocardiography

1 Introduction

The study of HRV heart rate variability gives an idea of the nature of the dynamics of the human physiological system and can be used to identify the presence and progression of heart diseases, including arrhythmias.

One of the simplest and most commonly used methods for examining HRV and detecting arrhythmias is the electrocardiogram (ECG). The main problem of automatic ECG analysis is the nonlinearity of ECG signals, the presence of background noise and artifacts of electrode movement, a large amount of ECG data (24-h monitoring), and a large scatter of ECG morphology. Therefore, despite a wide range of studies in this area, new methods and algorithms for recognition (including automatic) and classification of various heart diseases are currently an urgent task.

One of the methods of such analysis is the graphic interpretation of the heart rhythm, in particular scatterography, or in other words, the analysis of Poincaré Plots.

I. Tiginyanu et al. (Eds.): ICNBME 2021, IFMBE Proceedings 87, pp. 239–245, 2022.
https://doi.org/10.1007/978-3-030-92328-0_32

Methods for quantifying Poincaré Plots are constantly evolving and improving to speed up and improve the accuracy of the detection and occurrence of arrhythmias.

The autonomic nervous system (sympathetic (SNS) and parasympathetic (PNS)) plays a particularly important role in the regulation of heart rate. Often, heart rate variability (HRV) is caused by an imbalance in the levels of activity of the sympathetic and parasympathetic nervous systems. The SNS is responsible for increasing the heart rate, while the parasympathetic tone decreases the heart rate. The frequency with which the pulse increases or decreases is variable, that is, the periods of increasing or decreasing intervals are also not equal. As a result, heart rate asymmetry (HRA) is a common occurrence in healthy hearts.

Asymmetry confirms the presence of complex nonlinear dynamics of physiological signals. This is due to the nonlinear, chaotic nature of the physiological processes occurring in the body, and especially with the pronounced chaotic dynamics of the heart rate.

The aim of this work is to study the asymmetry of Speed Spots in a four-quadrant representation for arrhythmias such as atrial fibrillation.

2 Methods

A. Poincaré Plot

A Poincaré Plot (Lorentz plot) is a plot in which each RR interval is plotted as a function of the previous RR interval. where the values of each pair of consecutive RR intervals are plotted at one point on the graph.

These points usually tend to build one or more clusters. The shape, size and position of these clusters are the main characteristics that are used to analyze HRV and heart rate [1, 2]. Using dynamic RR interval information as a visualization tool, specialists can extract unique information about heartbeat patterns and classify heart diseases with greater accuracy.

The main studies of Poincaré Plot are based on the analysis of the physiological basis of ECG time series or data analysis (parameters SD1, SD2, SD1/SD2 ratios), showing that the emergence of nonlinear behavior can be a distinguishing feature between health and disease. The standard descriptors SD1 and SD2 represent the minor and major semiaxes of this enclosing ellipse, based on the variance distributions [3].

In some cases, the diagnosis of heart disease according to the Poincaré Plot is based on the study of the relationship between various forms (diagrams) of the Poincaré Plot and heart failure.

The classification of Poincaré Plot patterns is given in [4], where it is shown that certain arrhythmias form certain shapes on two-dimensional Lorentz plots. This definition was based on a study of the geometrical and taxonomic parameters of the Poincaré Plot sections: Lmax is the maximum projection length of a cluster of data points on the bisector, Wmax is the maximum length of the projection of a cluster of data points on the perpendicular to identity line; CC - central cluster of data points; ECC is an eccentric cluster of data points.

A number of works have been devoted to the study of the asymmetry of Poincaré Plots, which show the relationship between the asymmetry of Poincaré Plots and nonlinear dynamics of the heart rate [5, 6]. Any imbalance between increasing RR intervals and decreasing RR intervals corresponds to points above and below the identity line, which indicates asymmetry in the time series of RR intervals. The asymmetry of Poincaré Plots is quantified using three different indices: the Guzik index (GI) [7], the Porta index (PI) [8] and the Ehlers index (EI) [9].

According to research [7], HRV is defined as an imbalance between the acceleration and deceleration of the heart rate. In [10], the asymmetry of HRV was analyzed for different signal lengths of 5 min and 30 min, and indices for calculating the asymmetry of GI and PI were proposed. When calculating these indicators, the location-position of points on a two-dimensional map is determined relative to the line of identity, and not information about the number of points in the distribution.

B. Speed Spot

The idea of the visualization method is that the coordinate of the state point SSi (x, y) in the orthogonal coordinate plane xy is formed as the current value of the RR interval (x coordinate) and the value of the rate of change in the duration of RR intervals near the current i-value RR (y coordinate) - dRR. The y coordinate was calculated by different expressions of numerical differentiation with different time divisions (sample length) [11]. The drawing of the Speed Spot showed that the method of differentiation affects the shape of the resulting Speed Spots. Differentiation by central differences was chosen as the most acceptable from the point of view of the invariance of the form with respect to the time partition.

Graphical interpretation of heart rate using Poincaré Plots and Speed Spots is shown in Fig. 1.

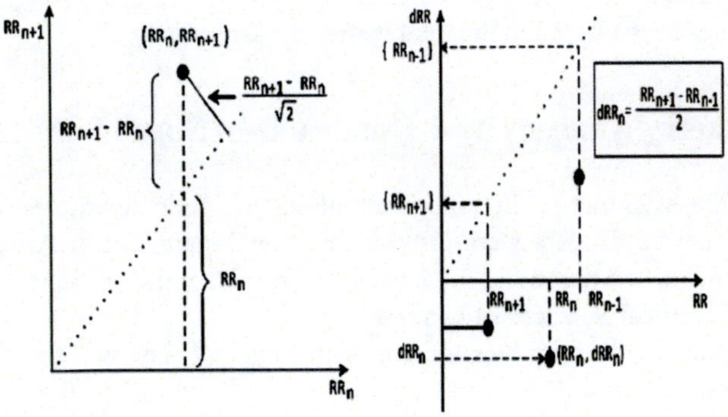

a) b)

Fig. 1. Graphic interpretation of Poincaré Plots (a) and Speed Spots (b)

Figure 2 below shows images of Poincaré Plots and Speed Spots for normal rhythm (NSR) and atrial fibrillation (AF).

The early studies of the asymmetry of the Speed Spots [12] showed that, as well as for the Poincaré Plots, they are characterized by an imbalance of the acceleration and deceleration processes.

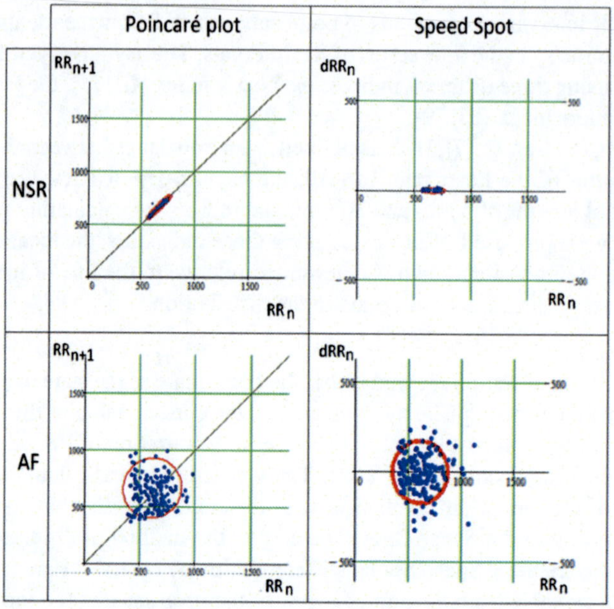

Fig. 2. Poincaré plots and Speed Spots for normal rhythm and atrial fibrillation

In this case, the coefficients of the degree of imbalance for the Speed Spots are the ratio of the variance of the array of points with positive values of the speed and with negative values of the speed to the total variance, respectively.

3 Asymmetry Analysis by 4-Quadrants of Speed Spot Graph

To analyze the asymmetry, rhythmograms of sixteen patients with paroxysmal atrial fibrillation were used. Data were provided by the Department of Arterial Hypertension Government Institution "L.T. Malaya Therapy National Institute of the National Academy of Medical Sciences of Ukraine".

Speed Spots partition enables formally appreciate the nature of the change of heart rate on the follow types:

a) the value of the rhythm is greater than the average and the speed is positive (acceleration of the rhythm) - the first quadrant (Q1);
b) the value of the rhythm is greater than the average and the speed is negative (acceleration of the rhythm) - the second quadrant (Q2);
c) the value of the rhythm is less than the average and the speed is negative (slowing down of the rhythm) - the third quadrant (Q3);
d) the value of the rhythm is less than the average and the speed is positive (deceleration of the rhythm) - the fourth quadrant (Q4).

To obtain the possibility of such analysis, the primary data for plotting the Speed Spot is "normalized" to the average value of the corresponding parameter (rhythm or the rate of its change).

In this case, the initial expressions for determining the main descriptors of the Speed Spot (major and minor axes of the enclosing ellipse) SD1 and SD2

$$SD1_{SS} = STDEV(dRR) = \sqrt{\frac{1}{N-1}\sum\nolimits_{n=0}^{N-1}\left(dRR_n - \overline{dRR}\right)^2}, \quad (1)$$

$$SD2_{SS} = STDEV(RR) = \sqrt{\frac{1}{N-1}\sum\nolimits_{n=0}^{N-1}\left(RR_n - \overline{RR}\right)^2} \quad (2)$$

are changed to the following:

$$SD1^*_{SS} = STDEV(dRR) - MEAN(dRR), \quad (3)$$

$$SD2^*_{SS} = STDEV(RR) - MEAN(RR). \quad (4)$$

For expressions (1) and (3) we have $dRR_n = \frac{RR_{n+1} - RR_{n-1}}{2}$ as derivative of RR for n interval.

In the case of such a normalization, the cloud of points of the Speed Spot will be located around the origin of coordinates and will take the form as in Fig. 3.

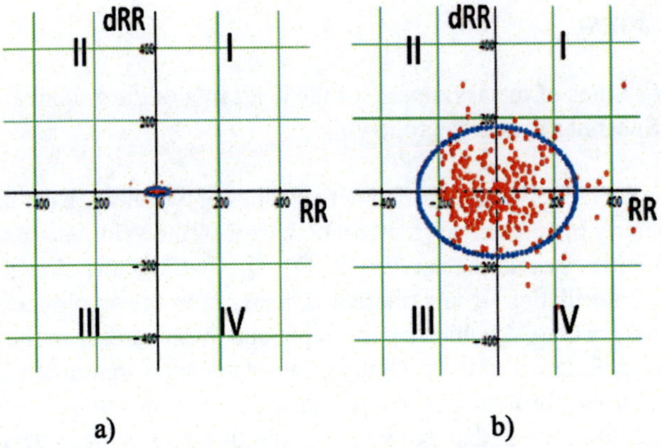

Fig. 3. Normalized Speed Spots with enclosing ellipses for the state of normal rhythm (a) and atrial fibrillation (b)

Asymmetry was assessed by the number of points in each quadrant for normal and atrial fibrillation states. The values of these points correspond to the coefficients Q1–Q4.

Graphs of changes in the number of points in each of the quadrants for a window of 300 points are shown in Fig. 4. (NSR - normal sinus rhythm, AF - atrial fibrillation episode).

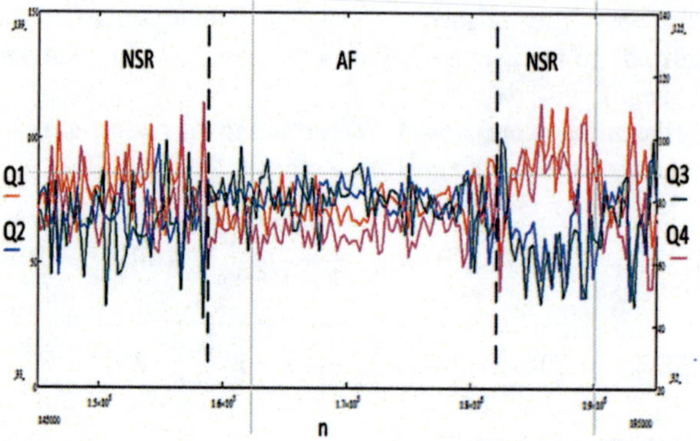

Fig. 4. Plot of Speed Spots points distribution in quadrants coefficients Q1-Q4 for normal rhythm (NSR) and atrial fibrillation (AF)

Figure 4 shows that the almost uniform alternation of acceleration and deceleration areas at a normal rhythm preceding atrial fibrillation and a decrease in the scatter of parameters during atrial fibrillation is replaced by a sharp imbalance in the rates of change immediately after the end of atrial fibrillation. This is indicated by a significant discrepancy between the curves Q1, Q3 and Q2, Q4, which means a sharp change in the sign of the velocity. This behavior of heart rate characteristics appears to be the factor leading to an increased likelihood of stroke.

4 Conclusions

The results of studies of the asymmetry of the heart rate in the graphical representation of the Speed Spots has shown the following:

A) at a normal rhythm, the distribution of the values of the durations of the RR intervals has a regularly manifested asymmetry both in absolute value and in the rate of their change. In this case, all the coefficients Q1–Q4 rhythmically change their sign;
B) when atrial fibrillation occurs, there is a decrease in the amplitudes of changes in the coefficients Q1-Q4, which can be explained by a decrease in the rate of cardiac output and a violation of the cyclicity of waves of acceleration and deceleration;
C) the end of atrial fibrillation is accompanied by a significant divergence of the Q1, Q3 and Q2, Q4 curves, that is, there is a significant asymmetry accompanied by a sign change in both the heart rate parameter and the parameter of its rate of change;
D) the 4-quadrant Speed Spot analysis can be used to analyze the degree of arrhythmia. Of interest for further research is the analysis of the asymmetry of Speed Spots for various types of cardiac abnormalities.

Conflict of Interest. "The authors declare that they have no conflict of interest".

References

1. Brennan, M., Palaniswami, M., Kamen, P.: Do existing measures of Poincaré plot geometry reflect nonlinear features of heart rate variability? IEEE Trans. Biomed. Eng. **48**, 1342–1347 (2001). https://doi.org/10.1109/10.959330
2. Kitlas, A., Oczeretko, E., Kowalewski, M., Urban, M.: Poincaré plots in analysis of heart rate variability. Physica Med. **XX**(Suppl. 1), 76–79 (2004)
3. Kitlas-Golińska, A.: Poincaré plots in analysis of selected biomedical signals. Stud. Log. Grammar Rhetoric **35**, 117–127 (2013). https://doi.org/10.2478/slgr-2013-0031
4. Esperer, H.D., Esperer, C., Cohen, R.J.: Cardiac arrhythmias imprint specific signatures on Lorenz plots. Ann. Noninvasive Electrocardiol. **13**, 44–60 (2008)
5. Porta, A., Casali, K.R., Casali, A.G., Gnecchi-Ruscone, T., Tovaldini, E., Montano, N., et al.: Temporal asymmetries of short-term heart period variability are linked to autonomic regulation. Am. J. Physiol. Regul. Integr. Comp. Physiol. **295**, R550–R557 (2008)
6. Karmakar, C.K., Khandoker, A.H., Palaniswami, M.: Heart rate asymmetry (HRA) in altered parasympathetic nervous system activity. In: Computing in Cardiology, vol. 37, pp. 601–604 (2010)
7. Guzik, P., Piskorski, J., Krauze, T., Wykretowicz, A., Wysocki, H.: Heart rate asymmetry by Poincaré plots of RR intervals. Biomed. Tech. **51**, 530–537 (2006)
8. Porta, A., Guzzetti, S., Montano, N., Gnecchi-Ruscone, T., Malliani, A.: Time reversibility in short-term heart period variability. In: Computers in Cardiology 2006, pp. 7–80 (2006)
9. Ehlers, C.L., Havstad, J., Prichard, D., Theiler, J.: Low doses of ethanol reduce evidence for nonlinear structure in brain activity. J. Neurosci. **18**, 7474–7486 (1998)
10. Piskorski, J., Guzik, P.: Geometry of the Poincaré plot of RR intervals and its asymmetry in healthy adults. Physiol. Meas. **28**, 287–300 (2007)
11. Sokol, Y., Shyshkin, M., Butova, O., Akhiiezer, O., Dunaievska, O.: Improved graphical analysis of atrial fibrillation based on Holter measurement data. In: 2019 29th International Scientific Symposium "Metrology and Metrology Assurance 2019" (MMA), pp. 178–183 (2019)
12. Sokol, Y., Shyshkin, M., Butova, O.: Asymmetry of «Speed Spot» clouds as a marker of hidden cardiac abnormalities. In: 2020 IEEE KhPI Week on Advanced Technology (KhPIWeek), Kharkiv, Ukraine, pp. 283–286 (2020). https://doi.org/10.1109/KhPIWeek51551.2020.9250106

Internet of Things (IoT) in Monitoring Physiological Parameters

Robert Fuior[1,2](✉), Andra Cristiana Băeşu[2], and Călin Corciovă[2]

[1] "GheorgheAsachi" Technical University of Iasi, Bulevardul Profesor Dimitrie Mangeron 67, Iasi, Romania

[2] University of Medicine and Pharmacy "Grigore T. Popa", Iasi, Romania

Abstract. With the rapid development of the world economy and the significant improvement in living standards, people's average life expectancy has increased, which has led to substantial changes in the types of diseases encountered in the daily population. These diseases are found in both older and younger people. At the same time with the development of informatics in the field of health, the clinical data of the patients can be registered in different servers that are found under the name of Cloud. They allow the visualization and interpretation of recordings from subjects using a device connected to these servers, an important role for both medical staff because it can easily keep track of certain patients with chronic diseases.

The system uses an Arduino development platform to purchase data from attached sensors. The ECG module is based on an instrumentation amplifier with several levels of filtering, so as to acquire a series of signals corresponding to a physiological chart in a single channel, which are sent to an Atmega microcontroller. The pulse detection part was performed using a pulse oximetry module and determines the levels of oxygen saturation in the blood.

The advantages of this system consist in: efficient, fast acquisition and storage, real-time remote monitoring, user-friendly interface, accessibility for patients and caregivers.

Keywords: ECG · Heart rate · Arduino · Microcontroller · Internet of Things · Cloud

1 Introduction

According to the latest research conducted in the field, they highlight for example the fact that in China are cardiovascular diseases that have become the main cause of death and the prevalence of these diseases in China continues to increase. According to the summary of a 2018 research report on cardiovascular disease in China, the number of patients with cardiovascular disease is about 290 million [1, 2].

Thus, with the development of health informatics, clinical data - including those obtained from electrocardiograms (ECG), electroencephalograms (EEG), magnetic resonance imaging, monitoring of vital parameters are digitized and monitored continuously. In order to be able to interpret these data collected from different devices for

I. Tiginyanu et al. (Eds.): ICNBME 2021, IFMBE Proceedings 87, pp. 246–253, 2022.
https://doi.org/10.1007/978-3-030-92328-0_33

monitoring vital parameters, a computer storage system has been implemented in which the history of each patient can be seen [3].

This device comes to the aid of people at home who are in certain difficulties. Being designed to prevent various cardiovascular diseases, because they are most often found among the elderly. It is designed to monitor vital physiological parameters [4].

2 Materials and Methods

The system is designed from two structures for acquiring patient data. One having the role of "master" who will receive the data from the transmission module located on the patient, called "slave" through the wireless connection. So the working architecture is described in the block diagram of the system shown in the Fig. 1.

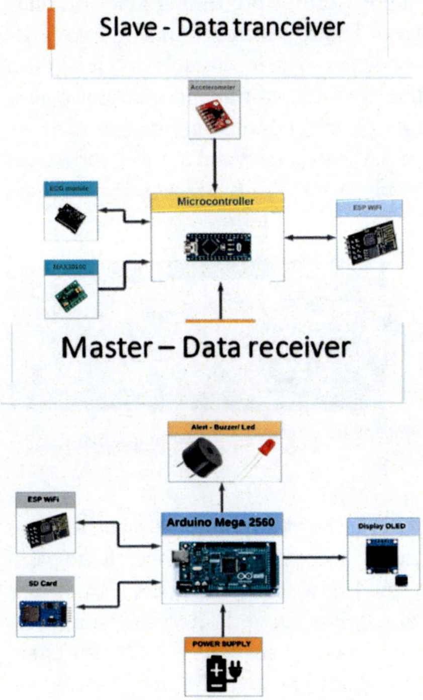

Fig. 1. Block diagram of the system

The Arduino Mega development board contains a microcontroller that operates at a supply voltage between 3.5–5.5 V, also through a voltage regulator of 5 V the platform can be powered up to a voltage of 12 V. It contains 54 external input and output ports, 15 PWM ports through which we can acquire/transfer data with a sampling rate set according to the program as well as a set of 16 analog ports shown in the Fig. 2 [5].

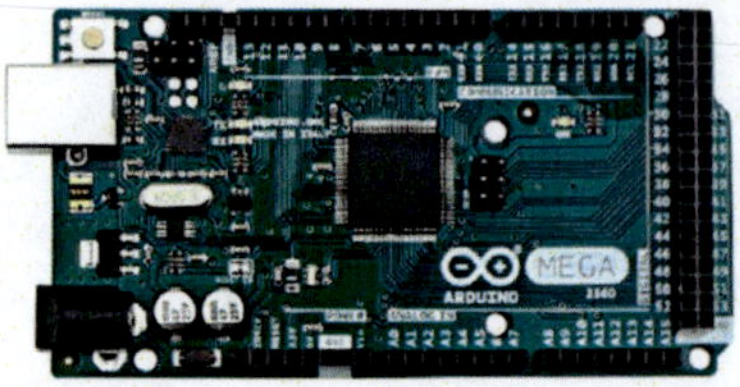

Fig. 2. Arduino Mega 2560 [5]

Four of those ports are dedicated to the UART protocol generally used for a set of sensors/modules with ports specific to this protocol, as well as to some OLED displays. The flash memory of the microcontroller is 256 KB, of which 8 KB are occupied by the bootloader with an operating frequency of 16 MHz. The Arduino Mega 2560 can be powered via the USB connection or with an external power supply.

Arduino Nano is an open-source processing platform that can be used for a variety of applications. In order to be programmed, the Arduino IDE application will be used, through which it is possible to program, compile and load the program in the microcontroller. The microcontroller is able to run and interpret code sequences specific to the C++ programming language. It has 14 digital inputs/outputs of which 6 are PWM type, analog inputs/outputs, a 5 V power port and a 3.3 V shown in the Fig. 3. Because there is no external power supply, a Mini USB connection can be used [6].

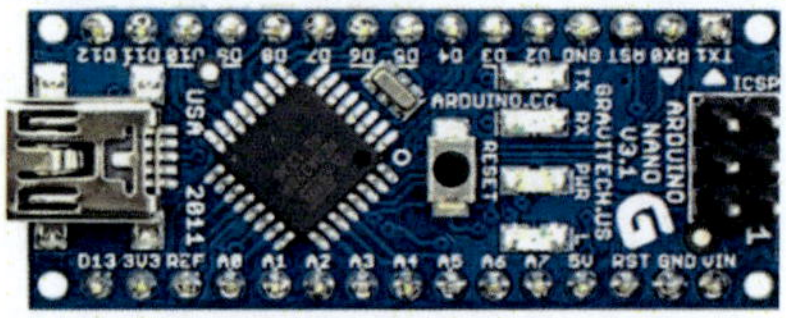

Fig. 3. Arduino Nano [6]

This module has the role of transmitting the data from the patient to the recording system in the Cloud. Having a fixed IP through which the pairing between the "Master" module and the "Slave" module will be performed, it can work independently as long as it can connect to a local wireless network. Reconnection between modules is very fast because it has a specific protocol. It requires a 3.3 V power supply and has a set of I2C protocol ports that will connect together with the Arduino development platforms [7] (Fig. 4).

In order to be able to acquire and process the biological signal (ECG) we used an instrumentation amplifier together with an analog-to-digital converter ADS1292R. The converter has two channels with a bandwidth of 24 Bits. The first channel is for displaying the electrocardiogram, while the second is for determining the respiratory function. It works as an instrumentation amplifier that measures the electrical activity of the heart, while also determining the value of respiratory impedance with 24-bit broadband output signals. It also contains filters for smoothing and attenuating the noise that may occur

Fig. 4. WiFi module ESP8266 [7]

during the measurement. It is designed to extract and amplify low value biopotentials [8].

Fig. 5. Instrumentation amplifier ADS1292R [9]

ADS1292 has a set of ports through which it is possible to make connections between different acquisition platforms. We used pins: VDD and GND for the power supply part (3.3 V–5 V), the START pin will be used to initialize the acquisition part, the CS port being able to select and optimize the desired signal input. The DRDY and MISO ports are the digital output ports of the mode, also the GPIO1, GPIO2 ports can be used, shown in the Fig. 5 [9].

The MAX30100 pulse sensor was used to determine the patient's pulse level and oxygen saturation. This sensor has several input and output ports through which connections can be made between different development platforms. Also the sensor contains an internal microcontroller that can help the values purchased from translators, it can also return both digital and analog signal depending on the desired application.

The receiving signal from the sensor will connect to port SCL and SDA of the Arduino development platform. Pulse monitoring and oximeter is done in real time, shown in the Fig. 6 [10].

The SD shield is powered by a 5 V power supply and also has a set of ports through which the connection between the module and the Arduino development platform is made. Data recorded from subjects can be easily processed by those conducting the related investigations. Due to its ability to store information for a long time, it is often used to collect patient data for analysis shown in the wiring diagram to Fig. 7 [11].

The recording data were used separately in monitoring the patient, both healthy and unhealthy. It is very useful to record medical data on the SD card as it is able to store a variety of data over a long period of time without the need for a full inspection by a user.

Fig. 6. Max30100 – pulse oximeter sensor [10]

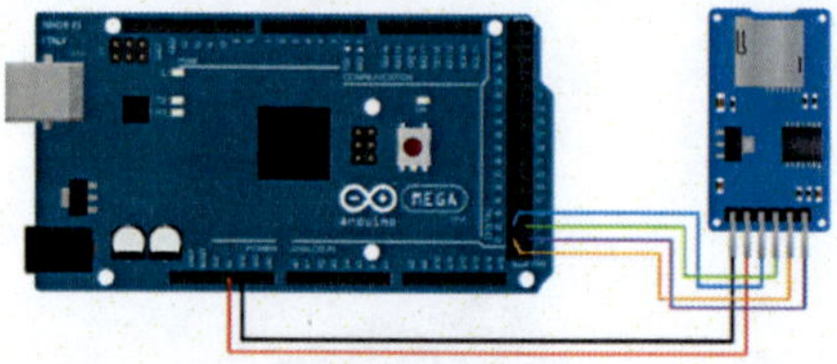

Fig. 7. Wiring diagram to SD module [11]

Accelerometer is used can detect the patient's movements, relative to the 3 axes. The mobility of the 3 axes is determined according to the position it will take at the time of calibration, allowing the measurement of inclination changes of less than 1°. It works at a supply voltage between 2.8 and 3.3 V, and the output signal is returned in analog form to the microcontroller via three analog ports shown in the Fig. 8.

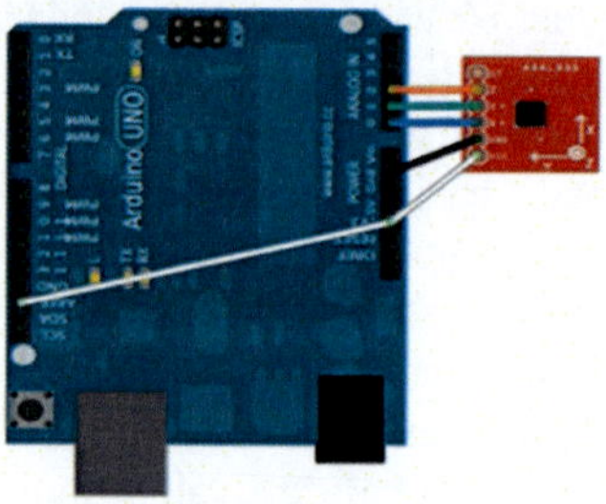

Fig. 8. Wiring diagram to ADXL335 [12]

This LCD Display OLED is easy to use together with an Arduino UNO R3 development board, the connection of the pins being clear and accurate being made by simply overlapping the LCD over the development board and the data library necessary to process the information using the code to be loaded on the board is created to use this model of Arduino board shown in the Fig. 9.

However, it is also possible to use this LCD on other models of development boards such as the Arduino MEGA 2560 R3 and a data library will have to be created specifically for this type of board that will allow other LCDs to use the ports available on that board. being placed differently from the Arduino UNO board.

Fig. 9. Display LCD OLED

3 Results

The source code of the system in addition to the functions related to the acquisition and processing of biomedical signals contains a series of lines of code dedicated to creating the user interface. It was desired to obtain a device that does not require complex technical knowledge in use. The graphical interface is made in a simplistic way so that it is practical in terms of functionality [12].

A first step in the structure of the code was to check each module if it is connected correctly, and then the calibration of the attached sensors will be implemented. After this verification stage, certain code sequences were structured, through which the wireless connection/pairing between the "master" and "slave" modules will be made. Having a specific protocol, the connection between the modules is fast. These code sequences are shown in the Fig. 10 [13, 14].

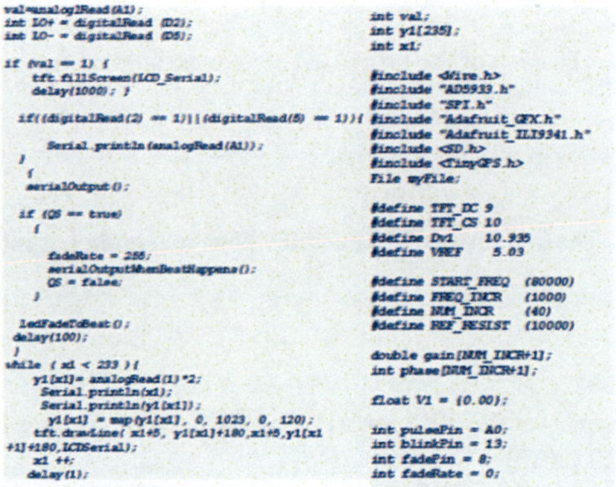

```
val=analog1Read(A1);
int LO+ = digitalRead (D2);
int LO- = digitalRead (D5);

if (val == 1) {
    tft.fillScreen(LCD_Serial);
    delay(1000); }

  if((digitalRead(2) == 1)||(digitalRead(5) == 1)){

      Serial.println(analogRead(A1));
  }
   {
   serialOutput();

  if (QS == true)
    {

      fadeRate = 255;
      serialOutputWhenBeatHappens();
      QS = false;
    }

  ledFadeToBeat();
 delay(100);
 }
while ( x1 < 233 ){
    y1[x1]= analogRead(1)*2;
     Serial.println(x1);
     Serial.println(y1[x1]);
      y1[x1] = map(y1[x1], 0, 1023, 0, 120);
    tft.drawLine( x1+5, y1[x1]+180,x1+5,y1[x1
+1]+180,LCDSerial);
    x1 ++;
   delay(1);
```

```
int val;
int y1[235];
int x1;

#include <Wire.h>
#include "AD5933.h"
#include "SPI.h"
#include "Adafruit_GFX.h"
#include "Adafruit_ILI9341.h"
#include <SD.h>
#include <TinyGPS.h>
File myFile;

#define TFT_DC 9
#define TFT_CS 10
#define Dv1     10.935
#define VREF     5.03

#define START_FREQ  (80000)
#define FREQ_INCR   (1000)
#define NUM_INCR    (40)
#define REF_RESIST  (10000)

double gain[NUM_INCR+1];
int phase[NUM_INCR+1];

float V1 = (0.00);

int pulsePin = A0;
int blinkPin = 13;
int fadePin = 8;
int fadeRate = 0;
```

Fig. 10. Initialization, declaration and electrocardiogram display function [13]

The electrocardiography part was analyzed using the ECG mode used, and in the first phase a series of signals were generated that correspond to the physiological chart. These being generated with a patient stimulator, later to display the signal on the screen were introduced a series of libraries and processing functions, processing the acquired signal in order to eliminate the artifacts that may occur at the time of recording [15].

4 Conclusions

The optimization of the device consists in obtaining data as accurate as possible from patients, but also to facilitate their transmission in the shortest possible time. The data transmission is influenced by the connection between the available WiFi networks in the surroundings.

The realization and development of such a system represents a wide field of research that involves both knowledge in the biomedical field and in the engineering field. Also, as future research directions, improvements can be made on both the hardware and the software side.

Optimization of the device is necessary to achieve accuracy of real-time data but also a patient's safety that can be medically supervised remotely, any alteration of its status being immediately signaled to allow a rapid medical response show in Fig. 11.

Fig. 11. Display of the processed electrocardiogram on the screen

Conflict of Interest. The authors declare that they have no conflict of interest.

References

1. Hu, S., et al.: Summary of the 2018 report on cardiovascular diseases in China. Chin. Circ. J. **2019**(34), 209–220 (2015)
2. Yanowitz, F.G.: Introduction to ECG Interpretation. LDS Hospital and Intermountain Medical Center, Salt Lake City (2012)
3. Hong, S., et al.: CardioLearn: a cloud deep learning service for cardiac disease detection from electrocardiogram. In: Companion Proceedings of the Web Conference 2020, Taipei, Taiwan, 20–24 April 2020 (2020)
4. Johnson, K.W., et al.: Artificial intelligence in cardiology. J. Am. Coll. Cardiol. **71**(23), 2668–2679 (2018)
5. Arduino Mega2560. https://www.arduino.cc/en/pmwiki.php?n=Main/arduinoBoardMega2560
6. Arduino Nano. https://www.arduino.cc/en/pmwiki.php?n=Main/ArduinoBoard
7. WiFi Module - ESP8266. https://www.sparkfun.com/products/17146
8. Module ECG. https://www.ti.com/product/ADS1292R#tech-docs

9. Module Pulse - SpO2 Max30100. https://www.maximintegrated.com/en/products/sensors/MAX30100.html
10. Gibson, T.C., Miller, S.W., Aretz, T., Hardin, N.J., Weyman, A.E.: Method for estimating right ventricular volume by planes applicable to cross-sectional echocardiography: correlation with angiographic formulas. Am. J. Cardiol. **55**(13), 1584–1588 (2012)
11. Corciova, C., Ciorap, R., Zaharia, D., Salceanu, A.: Influence of ambient temperature on central and peripheral impedance measurements of the human body. Environ. Eng. Manag. J. **10**(4), 511–517 (2011). https://doi.org/10.30638/eemj.2011.072
12. Gao, H.Q., Duan, X.H., Guo, X.Q., Huang, A.P., Jiao, B.L.: Design and tests of a smartphones-based multi-lead ECG monitoring system. In: 35th Annual International Conference of the IEEE Engineering in Medical and Biology Society (EMBC) (2013)
13. Gatouillat, A., Badr, Y., Massot, B., Sejdić, E.: Internet of medical things: a review of recent contributions dealing with cyber-physical systems in medicine. IEEE Int. Things J. **5**(5), 3810–3822 (2018)
14. Lau, D.T., Liu, J., Majumdar, S., Nandy, B., St-Hilaire, M., Yang, C.S.: A cloud-based approach for smart facilities management. In: Proceedings of the 2013 IEEE Conference on Prognostics and Health Management (PHM), Gaithersburg, USA (2013)
15. Ciorap, R., Hritcu-Luca, C., Corciova, C., Stan, A., Zaharia, D.: Home monitoring device for cardiovascular diseases. In: Vlad, S., Ciupa, R.V., Nicu, A.I. (eds.) International Conference on Advancements of Medicine and Health Care through Technology, vol. 26, pp. 49–52. Springer, Heidelberg (2009). https://doi.org/10.1007/978-3-642-04292-8_11

Developing of Algorithms for Improving Accuracy of Search for Biomarkers Within Results of the Computed Tomography

O. S. Medvedev[1], A. A. Birillo[1], A. N. Dudzich[2], V. L. Krasilnikova[2], and V. S. Asipovich[1](✉)

[1] Department of Human Engineering and Ergonomics, Belarusian State University of Informatics and Radioelectronics, Brovki 6, Minsk, Belarus
{o.med,v.osipovich}@bsuir.by

[2] Department of Ophthalmology, Belarusian Medical Academy of Postgraduate Education, Minsk, Belarus

Abstract. Provided results of research for improvement of recognition accuracy of eyeballs and bone structures of the eye sockets. For the goal achievement was used deep learning of the neural network with and without use of augmentation. Shown, how expanding the training set improves neural network accuracy.

Keywords: Eyeball · Bone structures of the eye sockets · Augmentation · Marking · Shift · Deformation · Rotations · Noises

1 Introduction

Analysis of images in format DICOM [1] (obtained through multi-slice computed tomograph MSCT [2, 3]) and 3-dimensional reconstruction of facial skull bones allows the surgeon to assess with greater reliability the anatomical aspects of an individual patient, localization, limits and spread of pathology process to plan the upcoming surgery [4].

The biomarkers were taken as main element of software processing for the neural network to work with computed tomograph obtained results.

Currently training set for further network learning is limited, which does not allow to improve the neural network accuracy. To expand the original data set we will use augmentation. That is, the construction of additional data from the original by introducing distortions, noises and the use of other techniques.

The goal is development of software, capable to unambiguously identify biomarkers limiting the bone orbit, as well as, analyzing results of augmentation during the neural network training.

2 Materials and Methodology

Source data. As source data for neural network training were used results of multi-slice computed tomography of patients with broken orbital socket of varying degrees of severity.

I. Tiginyanu et al. (Eds.): ICNBME 2021, IFMBE Proceedings 87, pp. 254–260, 2022.
https://doi.org/10.1007/978-3-030-92328-0_34

Preparing images for neural network learning was done through per-layer marking of those. Conversion of DICOM images to RGB have been done in advance. As tool for data marking was used application VGG Image Annotator (that is an application for manual annotation of images with ability to carry out multiple markup) [5]. As a result of marking we have got the files in csv and json format, that stores information about coordinates of the points, limiting orbital socket (Fig. 1).

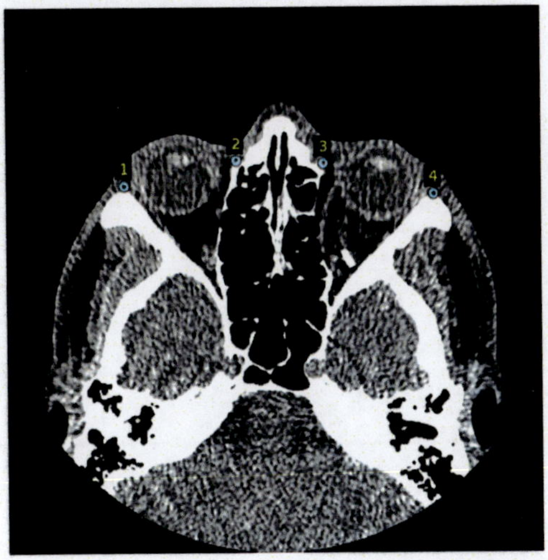

Fig. 1. The result of source data marking for one layer

The source data was split into training and test data in proportion of 90% and 10% respectively. Test source data was used to verify that an implementation of a neural network works correctly after learning. Also, test data was used for controlled experiment to compare the calculations of eye socket volume based on neural network marking, with results based on manual marking.

Neural network learning. The achievement of the target has been implemented through Python programming language with use of Anaconda platform. We used TensorFlow and Keras as main frameworks to work with neural networks. The number of images, used for neural network training was constantly increasing, starting from just a hundred and ending with 8000.

3 Improving Software Accuracy

Since initial number of images failed to meet minimum acceptable value, it was decided to use methods increasing the number of images for learning, such as augmentation.

Also was developed new architecture of the neural network (Fig. 2).

For example, it could be various image brightness, variety of noises, stretching or shrinking, shift from the center, image rotation or totality of above mentioned in varying proportions [6].

Therefore, it was decided to create following types of augmentation and mix of those:

Layer (type)	Output Shape	Param #
conv2d_1 (Conv2D)	(None, 497, 497, 16)	12304
max_pooling2d_1 (MaxPooling2	(None, 248, 248, 16)	0
conv2d_2 (Conv2D)	(None, 241, 241, 32)	32800
max_pooling2d_2 (MaxPooling2	(None, 120, 120, 32)	0
conv2d_3 (Conv2D)	(None, 116, 116, 32)	25632
max_pooling2d_3 (MaxPooling2	(None, 58, 58, 32)	0
conv2d_4 (Conv2D)	(None, 57, 57, 64)	8256
max_pooling2d_4 (MaxPooling2	(None, 28, 28, 64)	0
conv2d_5 (Conv2D)	(None, 26, 26, 128)	73856
max_pooling2d_5 (MaxPooling2	(None, 13, 13, 128)	0
conv2d_6 (Conv2D)	(None, 11, 11, 256)	295168
flatten_1 (Flatten)	(None, 30976)	0
dropout_1 (Dropout)	(None, 30976)	0
dense_1 (Dense)	(None, 512)	15860224
dense_2 (Dense)	(None, 256)	131328
dense_3 (Dense)	(None, 8)	2056

Total params: 16,441,624
Trainable params: 16,441,624
Non-trainable params: 0

Fig. 2. Neural network architecture

1. Image shift (bias augmentation).
2. Changes in image brightness (brightness augmentation).
3. Noises (noise augmentation) – creation of noises of various intensity on the whole image.
4. Image rotations (rotate augmentation).
5. Image deformation.

The algorithm of data set preparation is structured so, that augmentation takes place prior image feeding to the neural network. This ensures: each time the network will have on its input pseudo new, altered image, which would increase network accuracy.

Consider, for example for separate layer (Fig. 3) following options of transforming the image (Fig. 4).

Image shift could be done through finding of dimension of the image matrix (num_rows, num_cols), composition of the matrix for the affine transformation. The angle alpha is equal to 90, since the image shifting only along the axis x and y. Then, optimized function (cv2warpAffine [7]), that automatically create required image size, applied. Resulting image is shown on Fig. 4(A).

Image brightness modification is obtained through creation of the initial matrix of the image with same dimension. Then we are passing through each RGB pixel and multiplying the value for each channel by entered brightness coefficient, checking, at the same time, that resulting value belongs to (0,255) range. Thus, passing through the whole image we are getting “lightened” image (Fig. 4(B)).

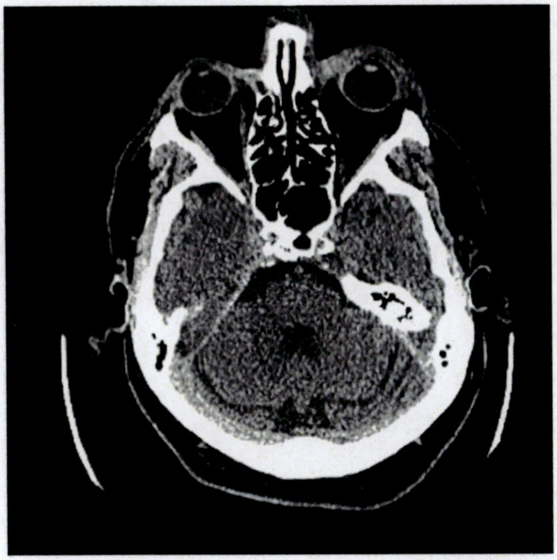

Fig. 3. Initial image without any augmentation.

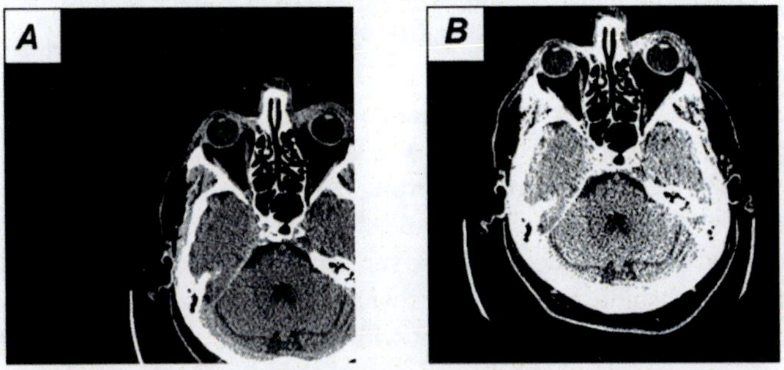

Fig. 4. Image shift (A), Image brightness modification (B).

Noises could be obtained through arbitrary traverse for each pixel, adding random number in (-noise_level, noise_level) range, if at least one of the channels isn't equal to zero. Then we are verifying RGB channels values to belong to (0,255) range, if above isn't true we are cutting values to acceptable.

Image rotation is based on affine transformation function. Sine modulus and angle cosine are determined for matrix. Then, value, which must be added to height an width resolution (cause image dimension isn't matching with source matrix), is found. New image is got applying affine transformation to the initial one.

Image deformation is also obtained through affine transformation. Nonzero elements of fairly simple matrix M, are stretch/shrink coefficients on the X and Y axis. Then, the transformation is used itself. Result is shown on Fig. 5(A).

Scaling is based on deformation, but with same coefficients on both axes. Result is shown on Fig. 5(B).

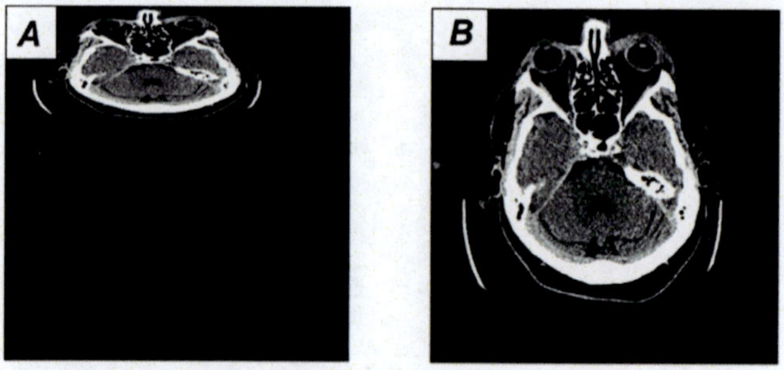

Fig. 5. Deformation (A), scaling (B).

4 Results and Discussion

As result of neural network learning, weight coefficient matrix for each layer was got. Obtained weight coefficients was used for finding eye sockets entrance coordinate on per-layer results of eye sockets bones scans from test set.

As accuracy indicator of the neural network work was used MAE (Mean Absolute Error) value. In fact, it represents an average deviation of biomarker search and marking results, made by neural network, with reference marking, made by human. Lower MAE value means higher neural network accuracy, since, network-made marking most closely matches with human-made [8].

Since neural network was trained iterative during 50 epochs, 50 values, representing network accuracy per epoch, exists. However, there is a high probability of "memorizing" training images by network, therefore getting correct results only on images belonging to the training set. To avoid this, validation set of images, never seen by network before, exists. At the end of each epoch network verified on such set. So, the value Val_MAE (Validation Mean Absolute Error), better characterizing neural network, comes into play.

As result of processing of the test set with use of trained neural network with (and without) augmentation, was got a dependency of neural network error from exact epoch. (Fig. 6, 7). As can been seen, network error (MAE) gradually decreases with epoch pass, reflecting that network optimization function works properly (Fig. 6).

As can been seen on figures, network accuracy after learning with use of augmentation has almost doubled, while, according to the results of testing, the network has proved to be more suited to the non-standard and atypical input data. So, the error of network, trained without augmentation, was in range of 5–6 conventional units (Val_MAE), while, the network trained with use of augmentation got accuracy about 2–3 (Val_MAE).

This represents a high eye socket marking accuracy achieved by network, trained with use of augmentation. Error margin, as 2–3, could be ignored, because it represents only a few pixels, that is not important for our task. The proposed software is feasible to use for eye socket volume calculation process automation during orbit bones replacement surgery preparatory stage, as well as, for evaluating results of such a surgery.

The orange curve behavior (Fig. 7), that is significant fluctuations of Val_MAE are caused by the fact that during validation the network face some images not belonging to the training set. Considering that, the weight coefficients optimization using gradient

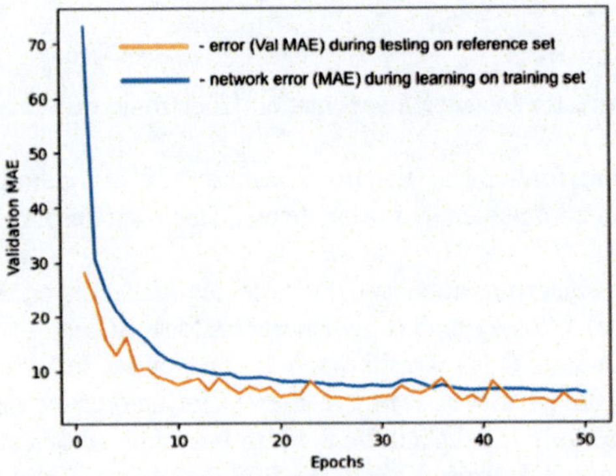

Fig. 6. Training result without use of augmentation: blue line – network error (MAE) during learning on training set; orange line – error (Val_MAE) during testing on reference set

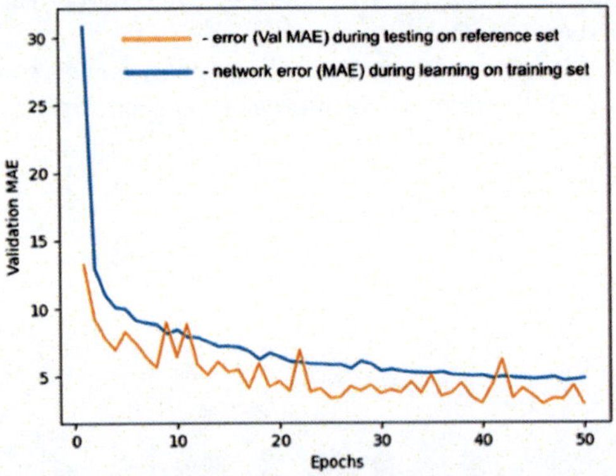

Fig. 7. Training result with use of augmentation: blue line – network error (MAE) during learning on training set; orange line – error (Val_MAE) during testing on reference set

descent doesn't always find the best values, we can observe some spikes. Also, these spikes result from fact that network facing conditionally new images, which cause more mistakes. Thanks to such "mistakes" the network is becoming more versatile, which would help to handle the real images (never seen by network before) better. So far: more varied data the network getting on input during training (and augmentation allow to get multiple pseudo-unique images from original one), better network optimization we getting, to achieve the best results during real operation.

5 The Conclusion

It is established, that calculation bias of the biomarkers coordinates, based on biomedical images (results of computed tomography), with use of neural network and augmentation, has halved compared with ordinary learning.

References

1. Aggregation Network for Instance Segmentation [Electronic resource]. https://arxiv.org/pdf/1803.01534
2. Keras: Deep Learning for humans [Electronic resource]. https://github.com/keras-team/keras
3. Digital imaging and communications in medicine [Electronic resource]. https://www.dicomstandard.org/current
4. Mask R-CNN for object detection and instance segmentation on Keras and TensorFlow [Electronic resource]. https://github.com/matterport/Mask_RCNN
5. Medvedev, O.S., Dudich, O.N., Krasilkova, V.L., Osipovich, V.S.: Comparative analysis of methods for calculating the system of eye sockets for automated analysis using computed tomography. Bones of the facial skull. In: Seventh International Scientific and Practical Conference "BIG DATA and Advanced Analytics. BIG DATA and High-Level Analysis", Minsk, Republic of Belarus, 19–20 May 2021 (2021)
6. Birillo, A.A., et al.: Improving the accuracy of the neural network for expanding the training of a dataset using augmentation. In: Seventh International Scientific and Practical Conference "BIG DATA and Advanced Analytics. BIG DATA and High-Level Analysis", Minsk, Republic of Belarus, 19–20 May 2021 (2021)
7. warpAffine [Electronic resource]. https://www.kite.com/python/docs/cv2.warpAffine
8. Mean absolute error [Electronic resource]. https://en.wikipedia.org/wiki/Mean_absolute_error

Excitations in Condensed Matter

Switching of Magnetic and Polarizability Characteristics of Dinuclear [CrCo] Complexes via Intramolecular Electron Transfer

Sophia I. Klokishner(✉), O. S. Reu, and M. A. Roman

Institute of Applied Physics, Academy str. 5, 2028 Chisinau, MD, Moldova
sophia.klokishner@ifa.md

Abstract. A model has been developed to describe the valence tautomeric transformation in a crystal containing as a structural element trinuclear complexes with electronic configurations Cr^{3+} –*dhsq*$^{3-}$–*low-spin* Co^{3+} and Cr^{3+} –*dhbq*$^{2-}$–*high-spin* Co^{2+} at low and high temperatures, respectively. The model takes into account the antiferromagnetic exchange coupling between the *dhsq*$^{3-}$–ligand with the itinerant electron and the Cr^{3+} - ion, the ferromagnetic exchange coupling Cr^{3+} – *high-spin* Co^{2+} as well as the interaction of the Co-ion with the full symmetric breathing mode of the nearest crystal surrounding producing a strong localizing effect. In the model there are also accounted for the cooperative dipole-dipole and electron-deformational interactions. The parameters of the main interactions governing the observed phenomena are evaluated through DFT calculations. The interplay between the intracluster exchange coupling and the cooperative interactions has been demonstrated to lead to gradual and abrupt spin transitions as well as to those accompanied by a hysteresis loop. Within the framework of the suggested model a qualitative and quantitative explanation is given of the magnetic susceptibility of the $[(Cr(SS\text{-}cth)(Co(RR\text{-}cth)(\mu-dhbq)](PF_6)_2Cl$ compound.

Keywords: Electron transfer · Polarization · Magnetic susceptibility

1 Introduction

A characteristic feature of valence tautomeric compounds is the ability to transfer electrons between the ligands and metal ions under the action of external stimuli [1]. The accompanying redistribution of electronic density inside the complex is usually associated with large changes in structural, optical and magnetic properties. Meanwhile, recently it has been reported a new valence tautomeric compound [2] $[(Cr(SS\text{-}cth)(Co(RR\text{-}cth)(\mu\text{-}dhbq)](PF_6)_2Cl$ ([CrCo]), which demonstrates the appearance of macroscopic polarization along with the appreciable change of the above mentioned characteristics and especially of the magnetic moment. The polarization originates from the intramolecular electron transfer between the paramagnetic dhsq^{3-}–ligand and the low-spin (ls) Co^{3+}-ion, thus converting the latter ion to the high-spin (hs) Co^{2+} one. In our first attempt [3] to explain the change of the magnetic and polarization characteristics

I. Tiginyanu et al. (Eds.): ICNBME 2021, IFMBE Proceedings 87, pp. 263–270, 2022.
https://doi.org/10.1007/978-3-030-92328-0_35

of the [CrCo]-compound several restricting assumptions concerning the exchange interaction in the ground Cr^{3+} –*dhsq*$^{3-}$–*ls*-Co^{3+} configuration (configuration I) were made, while in the excited Cr^{3+} –*dhbq*$^{2-}$–*hs*-Co^{2+} one (configuration II) of the [CrCo] complex, in which the ligand becomes diamagnetic, the exchange coupling was neglected. The cooperative dipole-dipole interaction, responsible for the redistribution of the electronic density in the [CrCo] complexes caused by ligand-metal electron transfer, has been written in the simplest form, identifying the dipole moments of the cluster in the states of configurations I and II with the dipole moments of the unpaired electrons on the dhsq^{3-}-ligand and hs-Co^{2+} ion, which are oppositely directed and counted off from the middle of the segment, connecting the ligand and the cobalt ion. For the first step the vibronic interaction of the Co-ion with the displacements of the nearest ligand surrounding was not included in the model [3] as well. Therefore, here we generalize the earlier elaborated approach [3] and develop a comprehensive model of the valence tautomeric transformation in the [CrCo] compound, combining the modeling of the observed properties with *ab* initio calculations of the intrinsic parameters of the system under examination.

2 Results and Discussion

The crystal Hamiltonian looks as follows:

$$H = H_0 + V_{dd} + H_{str}, \tag{1}$$

where $H_0 = \sum_n H_e^n$ represents the Hamiltonian of isolated molecules and describes the crystal field splittings, intra- and intercenter interactions inside a single [CrCo]-molecule as well as the Heisenberg exchange interactions H_{ex1}^n and H_{ex2}^n in the pairs Cr^{3+}–*dhsq*$^{3-}$ and *hs*-Co^{2+}–Cr^{3+} which appear in configurations I and II, respectively. The Hamiltonian H_e^n also includes the electron-vibrational interaction H_{eV}^n of the transition metal ions with the full symmetric vibrations of the nearest surrounding and the Hamiltonian of free lattice vibrations. The interaction of the electron located on the *dhsq*$^{3-}$ ligand with the vibrations of the nearest surrounding is supposed to be negligibly small as compared with that for the Co-ion. Otherwise, the excess electron on this ligand cannot transferred to the Co-ion. As to the Cr^{3+}-ion remaining in the orbitally non-degenerate $^4A_2$-state its interaction with the full symmetric vibrations of the nearest surrounding only leads to the shift of all cluster levels on one and the same value. Therefore, further on the model allows for only the interaction of the Co-ion with the breathing full symmetric mode of the nearest surrounding producing the prevalent localization effect. The last two terms in Hamiltonian (1) represent the cooperative dipole-dipole and electron-deformational interactions. These interactions are introduced in the model because they are responsible for the electronic density redistribution inside each molecule and the observed spin transition accompanied by the bond length elongations and deformation of the intermolecular space. The mean field approximation is applied to reduce the problem of interacting clusters to one cluster problem. The energies of the cluster levels look as follows:

$$\begin{aligned}
\varepsilon_1 &= J\overline{\tau} - \overline{d}d_1L + \tfrac{5}{2}j_1 + \hbar\omega\left(k + \tfrac{1}{2}\right), & (g_1 = 3),\\
\varepsilon_2 &= J\overline{\tau} - \overline{d}d_2L - \tfrac{3}{2}j_1 + \hbar\omega\left(k + \tfrac{1}{2}\right), & (g_2 = 5),\\
\varepsilon_3 &= \Delta_1 - J\overline{\tau} - \overline{d}d_{10}L + \tfrac{15}{2}j_2 + \hbar\omega\left(k + \tfrac{1}{2}\right), & (g_3 = 3),\\
\varepsilon_4 &= \Delta_1 - J\overline{\tau} - \overline{d}d_{20}L + \tfrac{11}{2}j_2 + \hbar\omega\left(k + \tfrac{1}{2}\right), & (g_4 = 9),\\
\varepsilon_5 &= \Delta_1 - J\overline{\tau} - \overline{d}d_{30}L + \tfrac{3}{2}j_2 + \hbar\omega\left(k + \tfrac{1}{2}\right), & (g_5 = 15),\\
\varepsilon_6 &= \Delta_1 - J\overline{\tau} - \overline{d}d_{40}L - \tfrac{9}{2}j_2 + \hbar\omega\left(k + \tfrac{1}{2}\right), & (g_6 = 21),
\end{aligned} \tag{2}$$

here the energies $\varepsilon_{1,2}$ and $\varepsilon_{3,4,5,6}$ refer to states arising from configurations I and II, respectively, the numbers g_i indicate the degeneracy of these states, which is only determined by the spin multiplicity $2S + 1$ for states of configuration I with $S = 1$, 2, while for states of configuration II with $S = 0, 1, 2, 3$ g_i account for both the spin multiplicity and different possibilities of placements of the unpaired electrons in *hs*-Co^{2+} [4], j_1 and j_2 characterize the exchange interaction in the pairs Cr^{3+} –*dhsq*$^{3-}$ and Cr^{3+} – *hs*-Co^{2+}, respectively, d_1 and d_2 are the dipole moments of the molecule in the states of configuration I with $S = 1$ and 2, respectively, while the dipole moment values d_{10}, d_{20}, d_{30} and d_{40} refer to states of configuration II with spin values $S = 0, 1, 2, 3$, $\Delta_1 = \Delta + (V_1^2 - V_2^2)/(2\hbar\omega)$, here Δ represents the energy of states arising from the excited configuration II and allows for the crystal field splitting, intra- and intercenter Coulomb interactions, V_1 and V_2 are vibronic coupling constants for states of configurations I and II, respectively, L and J are the parameters of cooperative dipole-dipole and electron-deformational interactions. Finally, $\overline{\tau}$ and $\overline{d}$ are the order parameters, which describe the mean distortion and the mean dipole moment of the [CrCo]-molecule and are determined as

$$\overline{\tau} = \frac{Tr\left[\exp\left(-\frac{\tilde{H}}{kT}\right)\tau^n\right]}{Tr\left[\exp\left(-\frac{\tilde{H}}{kT}\right)\right]}, \quad \overline{d} = \frac{Tr\left[\exp\left(-\frac{\tilde{H}}{kT}\right)d^n\right]}{Tr\left[\exp\left(-\frac{\tilde{H}}{kT}\right)\right]}, \tag{3}$$

here $\tilde{H}$ is the Hamiltonian of the system in the mean field approximation. The determination of the parameters involved in the model has been performed as follows. First the exchange parameter j_1 was obtained, applying the procedure suggested in [5] and based on DFT calculations of the single point energies, corresponding to spin values $S = 1, 2$. The calculations, taking into account all 126 atoms of the Cr^{3+} –*dhsq*$^{3-}$–*ls*-Co^{3+} complex, were performed with the ORCA suite of programs, version 4.2.1 [6].Within the DFT calculations the unrestricted formalism, the B3LYP functional and the def2-TZVP valence triple-basis set with polarization functions together with the auxiliary basis sets def2/J and def2-TZVP/C were employed [6]. By this procedure for the parameter j_1 the value -221 cm^{-1} was obtained. The estimation of the parameter j_1 within the "broken symmetry" formalism [6] gives the value -289 cm^{-1}. Thus, the exchange interaction of the antiferromagnetic type is confirmed by both types of calculation, and the parameters of this interaction fall inside the range of parameters known for the pair Cr^{3+}–sq-ligand [7–9].The DFT calculations also indicate on the ferromagnetic type of exchange interaction in the pair Cr^{3+} –*dhbq*$^{2-}$–*hs*-Co^{2+}. However, the obtained

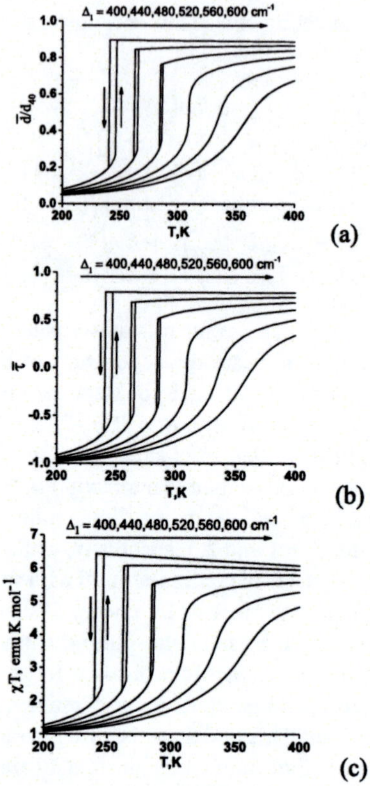

Fig. 1. Temperature dependence of the mean dipole moment $\overline{d}/d_{40}$ (a), mean distortion $\overline{\tau}$ (b) and the χT product (c) calculated with $Ld_{40}^2 = 650\ \mathrm{cm}^{-1}, j_1 = -221\mathrm{cm}^{-1}, j_2 = 70\mathrm{cm}^{-1}, J = 66\mathrm{cm}^{-1}$, $g = 2$, $g_{hs} = 2.2$ and different values of the gap Δ_1 between the states of configurations II and I.

parameters j_2 give such large energy gaps between the states of configurations I and II, which prevent from the thermal population of the states of the latter one. Further on the ferromagnetic type of exchange interaction in the pair Cr^{3+} –*dhbq*$^{2-}$–*hs*-Co^{2+} is preserved. Meanwhile, j_2 is considered as a fitting parameter. The estimated by DFT numerical values of the magnitudes of the dipole moments $d_1, d_2, d_{10}, d_{20}, d_{30}$ and d_{40}, counted from the center of mass of the Cr-ligand-Co molecule, that coincides with the middle of the segment connecting the Cr and Co ions, are $d_1 = 0.44D$, $d_2 = 0.22D$, $d_{10} = -6.5D$, $d_{20} = 9.1D$, $d_{30} = 9.4D$, $d_{40} = 9.9D$. At the same time the dipole moments $d_1, d_{20}, d_{30}, d_{40}$ possess the direction from the Co- ion to the Cr one, d_{10} is oppositely directed, while d_2 is perpendicular to the Co-Cr bond, and it is much smaller in magnitude than d_{10}, d_{20}, d_{30} and d_{40}. In further calculations the obtained values of the latter ones and of d_1 are used. Here it should be mentioned, that the above listed results on the dipole moments of the molecule and, namely, concerning their direction are in agreement with the findings of paper [2], wherein earlier it was

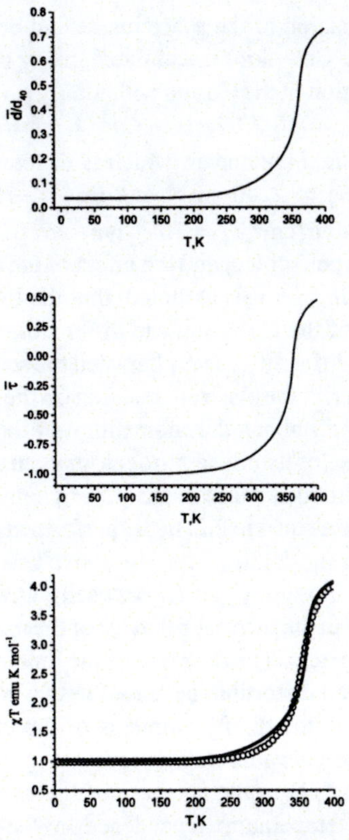

Fig. 2. Temperature dependence of the mean dipole moment $\overline{d}/d_{40}$ (a), mean distortion $\overline{\tau}$ (b) and the χT product (c) for the [(Cr(SS-cth)(Co(RR-cth)(μ − dhbq)](PF_6)$_2$Cl compound calculated with $Ld_{40}^2 = 733$ cm^{-1}, $j_1 = -221$cm^{-1}, $j_2 = 30$ cm^{-1}, $J = 66$cm^{-1}, g = 2, $g_{hs} = 2.1$, $\Delta_1 =$ 530 cm^{-1}. The fit accuracy is $\delta = 5.4\%$.

demonstrated that the polarization magnitude changes under the transition from the low-temperature phase to the high-temperature one, and at the same time the polarization retains its direction from Co to Cr with rising temperature. The difference in numerical values of the dipole moments for spin states $S = 1, 2$ of configuration I as compared with those obtained in [2] can be referred to the utilization of different functionals, sets of monoelectronic functions and options in the DFT calculations. An estimation of the characteristic energy of dipole-dipole interaction for two neighboring clusters with dipole moments equal to $d_{40} = 9.9D$, located at the minimal distance 9.1Å, allowed by the crystal structure, gives a value of 655 cm^{-1}. Therefore, further on the energy of dipole-dipole coupling is varied within the range 655–750 cm^{-1}. Since the observed temperature dependence of the χT product [2] evidences the start of the population of the states arising from configuration II at temperatures higher than room temperature, below the gap Δ_1 is changed in the limits 400–600 cm^{-1}. The parameter of electron-deformational interaction $J = c_2(v_2 - v_1)^2/(4c_1^2\Omega)$ (where c_1 and c_2 are the bulk elastic moduli describing the deformation of the space inside and between the Co-complexes, Ω is the unit cell volume per Co – ion) is estimated, using the following formulae for the constants v_1, v_2 of interaction of the Co-ion with totally symmetric strain in the states of configurations I and II: $v_1 = 40\sqrt{3}Dq(ls - Co^{3+})$, $v_2 = 40\sqrt{3}Dq(hs - Co^{2+})/3$, here Dq is the cubic crystal field parameter, which is different for the *ls*-Co^{3+} and *hs*-Co^{2+} ions. For $Dq(ls - Co^{3+}) = 2240\ \text{cm}^{-1}$ and $Dq(hs - Co^{2+}) = 1300\ \text{cm}^{-1}$ [1], $\Omega = 1236\ \text{Å}^3$ [2], $c_1 = 10^{12}\ \text{dyn/cm}^2$,$c_2 = 10^{11}\ \text{dyn/cm}^2$ the parameter $J = 66\ \text{cm}^{-1}$. First, sample calculations are performed, and the effect of the energy gap Δ_1 on the spin transformation is discussed. In fact, it is assumed, that the ligand, bearing an electron, and the nearest surrounding of the Co^{2+} -ion can differ from those in the [CrCo] compound, and this may result in different gaps Δ_1 between the states of configurations I and II. Further on for the Cr^{3+} ion and the electron, residing on the ligand, the g-factor value, equal to 2, is accepted. The orbital angular momentum of the *hs*-Co^{2+} ion is supposed to be almost quenched because of its mixed nitrogen-oxygen surrounding, strongly distorted from the cubic one, as it takes place in the [CrCo] compound [2], and, therefore, for this ion the g_{hs} value is taken equal to 2.2. In the performed simulations the numerical values for the dipole moments $d_1, d_{10}, d_{20}, d_{30}, d_{40}$ above listed are taken. From Fig. 1 it follows that with decrease of the energy gap Δ_1 between states of configurations I and II significant changes take place in the thermal behavior of the magnetic moment caused by the change of the order parameters. The decrease of the dipole moment is accompanied by the diminishing of the mean distortion, because the two cooperative interactions do not compete in the case, when the dipole moments of the cluster are mainly parallel oriented in the states with different spins arising from configurations I and II. Both interactions lead to a similar effect i.e. to the decrease of the energy gap Δ_1. At the same time with decrease of Δ_1 the temperature dependences of the mean dipole moment and mean distortion change from monotonic ones to those demonstrating a hysteresis loop. From Fig. 1 it is seen that the lower the gap Δ_1 the wider the hysteresis loop predicted in the temperature dependence of the crystal polarization (Fig. 1a), of the mean distortion (Fig. 1b) and of the χT product (Fig. 1c), where χ is the magnetic susceptibility. For certain parameter values the hysteresis loop can be obtained at temperatures close to room temperature.

It has been also obtained that the increase in the strength of electron-deformational interaction also leads to pronounced hysteresis effects in the temperature dependence of the mean dipole moment,mean distortion as well, and as a result in the temperature behavior of the effective magnetic moment.

Finally, the magnetic behavior of the real [CrCo] system is examined (Fig. 2). As in the previous case in the calculations for the g-factor of the unpaired electron residing on the $dhsq^{3-}$ as well as for the Cr^{3+} ion the value $g = 2$ is accepted, while for the hs-*Co*-ion the value $g_{hs} = 2.1$ is taken in these calculations, because the orbital angular of the cobalt ion is almost completely quenched. It is seen that for the parameters given in the caption for Fig. 2 quite a good agreement is obtained between the calculated and experimental curves. The best fit parameters fall within the range of parameters above discussed. From Fig. 2b it also follows that during the spin transformation the change of the mean cluster polarization occurs within the limits $0.04d_{40}$–$0.73d_{40}$. Thus, the change of the mean cluster polarization is equal to $6.8D$. The obtained result looks reasonable, and it is in line with the results of the performed DFT calculations.

3 Conclusions

A crystal containing the heterometalic Cr-ligand-Co cluster with an unpaired electron on the ligand as a structural unit is examined. The elaborated model takes into account the electronic density redistribution caused by the cooperative dipole-dipole coupling, the elongation of the Co-N bonds induced by electron-deformational interaction, the exchange interaction in the pairs Cr^{3+} –$dhsq^{3-}$, Cr^{3+} –hs- Co^{2+} as well as the interaction of the Co-ion with the totally symmetric vibrations of the nearest surrounding. The parameters of the exchange interactions and the cluster dipole moments in different spin states are estimated through DFT calculations. For certain values of the intra- and intermolecular interactions bistability is predicted in the magnetic and polarizability characteristics. An explanation of the magnetic behaviour of the [CrCo]-compound is given.

Acknowledgment. The support of the National Agency for Research and Development of Moldova (project 20.80009.5007.19) is highly appreciated.

Conflict of Interest. The authors declare that they have no conflict of interest.

References

1. Adams, D., Noodleman, L., Hendrickson, D.N.: Density functional study of the valence-tautomeric interconversion low-spin $[Co^{III}(SQ)(Cat)(phen)] \rightleftharpoons$ high-spin$[Co^{II}(SQ)_2(phen)]$. Inorg. Chem. **36**, 3966–3984 (1997). https://doi.org/10.1021/ic9611812
2. Kanegawa, S., Shiota, Y., Kang, S., et al.: Directional electron transfer in crystals of [CrCo] dinuclear complexes achieved by chirality-assisted preparative method. J. Am. Chem. Soc. **138**, 14170–14173 (2016). https://doi.org/10.1021/jacs.6b05089
3. Roman, M.A., Klokishner, S.I.: Modeling of the valence tautomeric transformation in heterometallic [Cr-dhbq-Co] molecules. In: Tiginyanu, I., Sontea, V., Railean, S. (eds.) 4th International Conference on Nanotechnologies and Biomedical Engineering. IP, vol. 77, pp. 67–70. Springer, Cham (2020). https://doi.org/10.1007/978-3-030-31866-6_14
4. Klokishner, S.: Cobalt valence tautomeric compounds: molecular and solid state properties. Chem. Phys. **269**, 411–440 (2001). https://doi.org/10.1016/S0301-0104(01)00349-4
5. Klokishner, S., Reu, O.: Modeling of spin crossover in Iron(II) complexes with N_4S_2 coordination. J. Phys. Chem. C **123**, 19984–19990 (2019). https://doi.org/10.1021/acs.jpcc.9b05064
6. Neese, F., Wennmohs, F.: ORCA-An ab Initio, DFT and semiempirical SCF-MO package-version 4.2.1. Max-Planck-Institut für Kohlenforschung, Müulheim, Germany (2019). https://doi.org/10.1002/wcms.81
7. Fortea-Perez, F.R., Vallejo, J., Pasán, J., et al.: Ferromagnetic coupling through the oxalate bridge in heterobimetallic Cr(III)-M(II) (M=Mn and Co) assemblies. C. R. Chim. **22**, 452–465 (2019). https://doi.org/10.1016/j.crci.2018.10.007

8. Sun, Y.-Q., Zhang, J., Yang, G.-Y.: Molecular self-assemblies of a π-conjugated redox-active bipyridinium cation with magnetic dimetallic oxalate-bridged trimeric clusters. Dalton Trans. **2006**, 1685–1690 (2006). https://doi.org/10.1039/B512695A
9. Vallejo, J., Castro, I., Cañadillas-Delgado, L., et al.: Ferromagnetic coupling and magnetic anisotropy in oxalato-bridged trinuclear chromium(III)-cobalt(II) complexes with aromatic diimine ligands. Dalton Trans. **39**, 2350–2358 (2010). https://doi.org/10.1039/B915546E

Electron Transfer Phenomenon in the Dinuclear {Fe(μ-CN)Co} Complex: Interaction of Molecular Modes with Phonons

S. M. Ostrovsky(✉) and S. I. Klokishner

Institute of Applied Physics, Academiei str. 5, Chisinau, Moldova
serghei.ostrovsky@ifa.md

Abstract. A model for the description of the charge transfer induced spin transition in a crystal containing as a structural element cyanide-bridged binuclear Co-Fe clusters is presented. The cooperative interaction responsible for the spin transformation originates from the coupling of the acoustic crystalline modes with the molecular vibrations of the nearest ligand surroundings of the metal ions. The developed model is applied for the description of the observed magnetic characteristics of the $[\{(Tp)Fe(CN)_3\}\{Co\text{-}(PY5Me_2)\}](CF_3SO_3)\cdot 2DMF$ complex.

Keywords: Spin crossover · Charge transfer induced spin transition (CTIST)

1 Introduction

The study of Co/Fe Prussian Blue analogue compounds is in focus of many papers [1–4]. In these compounds a metal-to-metal charge-transfer transition within Fe-CN-Co units is accompanied by transformation of diamagnetic *ls*-Fe^{II}-CN-*ls*-Co^{III} units into paramagnetic *ls*-Fe^{III}-CN-*hs*-Co^{II} ones (*ls* – low-spin, *hs* – high-spin). From the point of view of nanotechnological applications, homogeneous materials with low dimensionality are more perspective [5] due to the possibility to control the environment around the metal ions by a proper choice of ligands. The synthesis and study of this type of compounds represent a new trend in the chemistry and physics of cyanometalate compounds.

The pentanuclear complex $\{[Co(tmphen)_2]_3[Fe(CN)_6]_2\}$ (tmphen: 3,4,7,8-tetramethyl-1,10-phenanthroline) is the first molecular Fe/Co system in which the electron transfer was demonstrated and proved experimentally [6]. To understand more deeply and transparently the phenomenon of the charge transfer induced spin transition (CTIST), the scientists have turned to complexes with lower nuclearity. The first dinuclear cyanide-bridged $[\{(Tp)Fe(CN)_3\}\{Co\text{-}(PY5Me_2)\}](CF_3SO_3)\cdot 2DMF$ complex (complex **1**·2DMF) has been reported in [7]. The observed thermally induced magnetic and optical bistability for this complex opens new possibilities to design molecular magnetic and optical switches and new multifunctional materials by using this unit as a building block. The present study is focused on the explanation of spin transformation in complex **1**. The new microscopic theoretical approach to the description of spin

I. Tiginyanu et al. (Eds.): ICNBME 2021, IFMBE Proceedings 87, pp. 271–276, 2022.
https://doi.org/10.1007/978-3-030-92328-0_36

crossover phenomena in molecular crystals [8–10] is generalized to the case of molecular crystals containing binuclear structural Fe-Co units demonstrating CTIST under action of temperature. The main electronic mechanisms governing this phenomenon are elucidated.

2 Model

The model for the description of CTIST in a crystal containing Co-Fe clusters as structural units was presented in detail in [11]. Here only the main provisions of this model are discussed. On the basis of the experimental observations, two configurations of the Co-Fe pair are included into consideration, and namely, the diamagnetic *ls*-Co^{III}-*ls*-Fe^{II} (configuration **I**) and the paramagnetic *hs*-Co^{II}-*ls*-Fe^{III} one (configuration **II**). In further examination the approach recently suggested in our papers [8–10] for spin crossover compounds is applied. Within the framework of this approach it is supposed that the ligand surroundings of the metal ions in the binuclear complex participate in two types of vibrations, namely, in the local vibrations and in the vibrations propagating along the crystal (phonons) [12, 13]. Therefore, the ligand displacements of the metal ion surrounding are presented as a linear superposition of the normal coordinates of the local modes and phonons of the same symmetry [8–10]. Further on the interactions of Co and Fe ions with the totally symmetric vibrations of their nearest surroundings are only included into consideration that allows to adapt the general equations presented in [10] to the complex under study and to account for the experimental evidence that the spin transition is not accompanied by the change in the crystal symmetry. At the same time the main result is that such a representation of the ligand displacements leads to the appearance of the effective coupling of the local modes with phonons [8–10], which is responsible for the cooperative spin transition. Along with this cooperative interaction between the dimers the total crystal Hamiltonian includes the Hamiltonian of non-interacting Fe-Co dimers, the Hamiltonian of free phonons, the Zeeman interactions for the cobalt and iron ions when they are in the paramagnetic *hs*-Co^{II} and *ls*-Fe^{III} states. The problem of interacting clusters is solved in the mean field approximation. The role of the order parameters $\overline{\tau^{ls}x^{ls}}$ and $\overline{\tau^{hs}x^{hs}}$ is played by the mean values of the products of electronic diagonal matrices τ^{ls}, τ^{hs} and the coordinates of the local modes x^{ls} and x^{hs} for the high- and low-spin states of the spin crossover complex [8–10].

The transcendental equations for the order parameters $\overline{\tau^{ls}x^{ls}}$ and $\overline{\tau^{hs}x^{hs}}$ have the form:

$$\overline{\tau^{ls}x^{ls}} = -\frac{\left(v_{ls}^{Co^{III}} - J^{ls,ls}\overline{\tau^{ls}x^{ls}} - J^{hs,ls}\overline{\tau^{hs}x^{hs}}\right)}{\hbar\omega^{ls-Co^{III}}}\frac{Z_{ls}\left(\overline{\tau^{ls}x^{ls}}, \overline{\tau^{hs}x^{hs}}\right)}{Z\left(\overline{\tau^{ls}x^{ls}}, \overline{\tau^{hs}x^{hs}}\right)},$$

$$\overline{\tau^{hs}x^{hs}} = -\frac{\left(v_{hs}^{Co^{II}} - J^{hs,hs}\overline{\tau^{hs}x^{hs}} - J^{hs,ls}\overline{\tau^{ls}x^{ls}}\right)}{\hbar\omega^{hs-Co^{II}}}\frac{Z_{hs}\left(\overline{\tau^{ls}x^{ls}}, \overline{\tau^{hs}x^{hs}}\right)}{Z\left(\overline{\tau^{ls}x^{ls}\tau^{hs}x^{hs}}\right)}, \quad (1)$$

with partition functions being

$$Z_{ls}\left(\overline{\tau^{ls}x^{ls}},\overline{\tau^{hs}x^{hs}}\right)=\frac{\exp\left(-\frac{E_{ls}\left(\overline{\tau^{ls}x^{ls}},\overline{\tau^{hs}x^{hs}}\right)}{k_BT}\right)}{2\sinh\left(\frac{\hbar\omega^{ls-\mathrm{Co^{III}}}}{2k_BT}\right)2\sinh\left(\frac{\hbar\omega^{ls-\mathrm{Fe^{II}}}}{2k_BT}\right)},$$

$$Z_{hs}\left(\overline{\tau^{ls}x^{ls}},\overline{\tau^{hs}x^{hs}}\right)=\frac{\sum_i g_{hs}^i\exp\left(-\frac{E_{hs}^i\left(\overline{\tau^{ls}x^{ls}},\overline{\tau^{hs}x^{hs}}\right)}{k_BT}\right)}{2\sinh\left(\frac{\hbar\omega^{hs-\mathrm{Co^{II}}}}{2k_BT}\right)2\sinh\left(\frac{\hbar\omega^{ls-\mathrm{Fe^{III}}}}{2k_BT}\right)},$$

$$Z\left(\overline{\tau^{ls}x^{ls}},\overline{\tau^{hs}x^{hs}}\right)=Z_{ls}\left(\overline{\tau^{ls}x^{ls}},\overline{\tau^{hs}x^{hs}}\right)+Z_{hs}\left(\overline{\tau^{ls}x^{ls}},\overline{\tau^{hs}x^{hs}}\right). \tag{2}$$

In these equations the subscripts *ls* and *hs* correspond to configurations **I** and **II**, respectively, ω and υ with the corresponding sub- and superscripts are the frequencies and the vibronic constants that correspond to the total symmetric molecular vibrations. The values $J^{j,l}$ (j, $l = ls$, hs) represent the parameters of the intercenter cooperative interaction.

Due to spin-orbital interaction, the $^2T_2$ (t_2^5) ground orbital triplet of the *ls*-Fe^{III} ion is split into the ground Kramers doublet and excited quadruplet, while the $^4T_1$ ground orbital triplet of the *hs*-Co^{II} ion is split into the three groups of levels, namely, the ground Kramers doublet, the excited quadruplet and the excited sextet. Since the constants of spin-orbital coupling for free *ls*-Fe^{III} and *hs*-Co^{II} ion are $-486\,\mathrm{cm}^{-1}$ [14] and $-180\,\mathrm{cm}^{-1}$ [15], respectively, the thermal populations of the excited quadruplet of *ls*-Fe^{III} and the excited sextet of *hs*-Co^{II} are negligible. So, in the subsequent consideration the following groups of levels for configuration **II** are included in the model: (i) the group of cluster levels arising from the direct product of two ground Kramers doublets belonging to the *ls*-Fe^{III} and *hs*-Co^{II} ions; (ii) the group of the cluster states, which arises from the ground Kramers doublet of the *ls*-Fe^{III} ion and the quadruplet of the *hs*-Co^{II} ion. For more details of the model see [11].

3 Analysis of the Experimental Data

Let us start this section with the estimation of the values of the key parameters. The vibronic coupling constants $\upsilon_{ls}^{\mathrm{Co^{III}}}$ and $\upsilon_{hs}^{\mathrm{Co^{II}}}$, characterizing the interaction of a single Co - ion in its *ls* $^1A_1$ and *hs* $^4T_1$ states with the local full symmetric breathing mode, can be calculated [16] as the derivatives of the cubic crystal field potential with respect to metal-ligand distances. As a result one finds $\upsilon_{hs}^{\mathrm{Co^{II}}} = 420\ \mathrm{cm}^{-1}$ and $\upsilon_{ls}^{\mathrm{Co^{III}}} = 2408\ \mathrm{cm}^{-1}$ (typical values of the cubic crystal field parameters, averaged metal ligand distances and frequencies of the total symmetric molecular vibrations for Co ion used in these calculations can be found in [17], [7] and [18], respectively).

The parameters $J^{ls,ls}$, $J^{hs,hs}$ and $J^{hs,ls}$ of cooperative interaction are evaluated under the assumption that the long-wave acoustic phonons produce a major contribution to them. The corresponding equation for $J^{j,l}$ parameters (j, $l = ls$, hs) can be found in [8].

One can easily find that for compound **1** the parameters of the cooperative interaction obey the relation $J^{ls,ls}:J^{hs,ls}:J^{hs,hs} = 1.549:1.245:1$ [11]. As a result, only one fitting parameter $J^{ls,ls}$ instead of three independent $J^{j,l}$ parameters is used in the subsequent calculations.

Finally, the frequencies of the total symmetric vibrations of the moieties consisting of the Fe ion and six nearest ligands are $\hbar\omega^{ls-Fe^{III}} = 390\ cm^{-1}$ [19] and $\hbar\omega^{ls-Fe^{II}} = 200\ cm^{-1}$ [20].

Even at very low temperatures a part of binuclear complexes is in configuration **II** and does not participate in the CTIST (see the experimental data in [7] and Fig. 1). The fraction of these complexes is denoted as y_{hs}. The χT product of compound **1** is calculated as $\chi T = (\chi T)_{hs}(n_{hs}[1 - y_{hs}] + y_{hs})$, where y_{hs} is the fraction of dimeric complexes not participating in CTIST, n_{hs} is determined as Z_{hs}/Z and $(\chi T)_{hs}$ describes the magnetic behavior of the Fe-Co complex in configuration **II**. The exchange interaction in the pair Fe-Co is neglected, so the $(\chi T)_{hs}$ represents a sum of susceptibilities of the *ls*-Fe^{III} and *hs*-Co^{II} ions. The magnetic behavior of both *ls*-Fe^{III} and *hs*-Co^{II} ions is calculated using the so called *T-P* isomorphism [21] and includes spin and orbital contributions.

Sample calculations (not presented) demonstrate that the increase of the coupling parameters $J^{j,l}$ shifts the transition temperature to lower values with the transition becoming more abrupt, while the increase of Δ_{hl} makes the transition more gradual and the transition temperature shifts to the high temperature region.

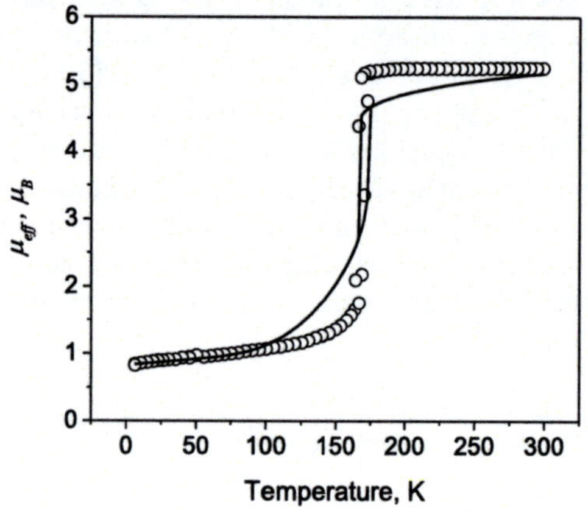

Fig. 1. Experimental μ_{eff} vs. T dependence of complex **1** (open circles [7]) and the theoretical curve calculated with $\Delta_{hl} = 410\ cm^{-1}$, $J^{ls,ls} = 1.435\ cm^{-1}$, $y_{hs} = 4.0\%$.

The result of calculation of the magnetic behavior for complex **1** is presented in Fig. 1 as solid line. The best fit parameters are the part of the figure caption. The corresponding thermal populations of different configurations of the Fe-Co pair are presented in Fig. 2.

The model well reproduces the effective magnetic moment at low temperatures and the observed hysteresis loop (width and position). The account for the real symmetry of the crystal surroundings of the Fe- and Co- ions may improve the coincidence of the experimental data with the calculated curve at high temperatures.

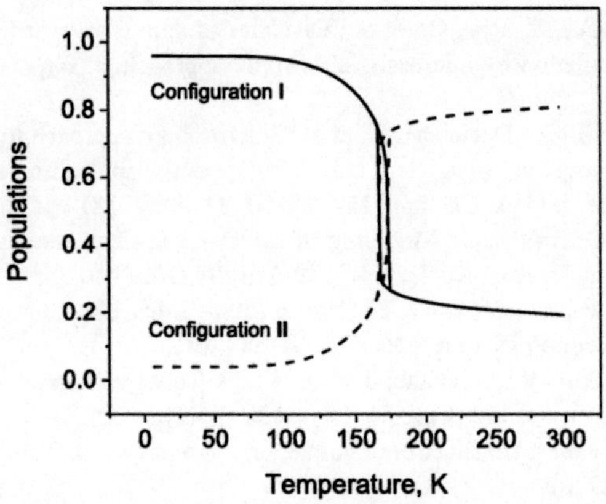

Fig. 2. Thermal variation of the population of different configurations of complex **1** calculated with $\Delta_{hl} = 410\ \text{cm}^{-1}$, $J^{ls,ls} = 1.435\ \text{cm}^{-1}$, $y_{hs} = 4.0\%$.

Acknowledgment. The support of the National Agency for Research and Development of Moldova (project 20.80009.5007.19) is highly appreciated.

Conflict of Interest. The authors declare that they have no conflict of interest.

References

1. Sato, O., Iyoda, T., Fujishima, A., Hashimoto, K.: Photoinduced magnetization of a cobalt-iron cyanide. Science **272**, 704–705 (1996)
2. Escax, V., Bleuzen, A., Cartier dit Moulin, C., et al.: Photoinduced ferrimagnetic systems in Prussian blue analogues $C^{I}_{x}Co_4[Fe(CN)_6]_y$ (C^I = Alkali Cation). 3. Control of the photo- and thermally induced electron transfer by the $[Fe(CN)_6]$ vacancies in cesium derivatives. J. Am. Chem. Soc. **123**, 12536–12543 (2001)
3. Escax, V., Champion, G., Arrio, M.-A., et al.: The Co ligand field: a key parameter in photomagnetic CoFe Prussian blue derivatives. Angew Chem Int Ed **44**, 4798–4801 (2005)
4. Cartier dit Moulin, C., Champion, G., Cafun, J.-D., et al.: Structural rearrangements induced by photoexcitation in a RbCoFe Prussian blue derivative. Angew Chem. Int. Ed. **46**, 1287–1289 (2007)
5. Aguilà, D., Prado, Y., Koumousi, E.S., et al.: Switchable Fe/Co Prussian blue networks and molecular analogues. Chem. Soc. Rev. **45**, 203–224 (2016)
6. Berlinguette, C.P., Dragulescu-Andrasi, A., Sieber, A., et al.: A charge-transfer-induced spin transition in the discrete cyanide-bridged complex $\{[Co(tmphen)_2]_3[Fe(CN)_6]_2\}$. J. Am. Chem. Soc. **126**, 6222–6223 (2004)
7. Koumousi, E.S., Jeon, I.-R., Gao, Q., et al.: Metal-to-metal electron transfer in Co/Fe Prussian blue molecular analogues: the ultimate miniaturization. J. Am. Chem. Soc. **136**, 15461–15464 (2014)
8. Palii, A., Ostrovsky, S., Reu, O., et al.: Microscopic theory of cooperative spin crossover: interaction of molecular modes with phonons. J. Chem. Phys. **143**, 084502 (2015)

9. Palii, A., Ostrovsky, S., Reu, O., et al.: Diversity of spin crossover transitions in binuclear compounds: simulation by microscopic vibronic approach. J. Phys. Chem. C **120**, 14444–14453 (2016)
10. Ostrovsky, S., Palii, A., Decurtins, S., et al.: Microscopic approach to the problem of cooperative spin crossover in polynuclear cluster compounds: application to tetranuclear iron(II) square complexes. J. Phys. Chem. C **122**, 22150–22159 (2018)
11. Klokishner, S., Ostrovsky, S.: Modeling of electron transfer phenomenon in the dinuclear Fe(μ-CN)Co complexes. J. Appl. Phys. **129**, 113901 (2021)
12. Hizhnyakov, V., Pae, K., Vaikjarv, T.: Optical Jahn−Teller effect in the case of local modes and phonons. Chem. Phys. Lett. **525–526**, 64–68 (2012)
13. Pae, K., Hizhnyakov, V.: Nonadiabaticity in a Jahn-Teller system probed by absorption and resonance Raman scattering. J. Chem. Phys. **138**, 104103 (2013)
14. Abragam, A., Bleaney, B.: Electron Paramagnetic Resonance of Transition Ions. Clarendon Press, Oxford (1970)
15. Lines, M.E.: Orbital angular momentum in the theory of paramagnetic clusters. J. Chem. Phys. **55**, 2977–2984 (1971)
16. Klokishner, S., Ostrovsky, S., Palii, A., et al.: Vibronic model for cooperative spin-crossover in pentanuclear $[M^{III}(CN)_6]_2[M'^{II}(tmphen)_2]_3$ (M/M' = Co/Fe, Fe/Fe) compounds. J. Phys. Chem. C **115**, 21666–21677 (2011)
17. Adams, D.M., Noodleman, L., Hendrickson, D.N.: Density functional study of the valence-tautomeric interconversion low-spin $[Co^{III}(SQ)(Cat)(phen)] \rightleftharpoons$ high-spin $[Co^{II}(SQ)_2(phen)]$. Inorg. Chem. **36**, 3966–3984 (1997)
18. Krivokapic, I., Zerara, M., Daku, M.L., et al.: Spin-crossover in cobalt(II) imine complexes. Coord. Chem. Rev. **251**, 364–378 (2007)
19. Rury, A.S., Goodrich, I.E., Galinato, M.G.I., et al.: Ligand recruitment and spin transitions in the solid-state photochemistry of $Fe^{(III)}$TPPCl. J. Phys. Chem. A **116**, 8321–8333 (2012)
20. Jung, J., Spiering, H., Yu, Z., et al.: The debye-waller factor in spincrossover molecular crystals: a Mössbauer study on $[Fe_xZn_{1-x}(ptz)_6](BF_4)_2$. Hyperfine Interact. **95**, 107–128 (1995)
21. Sugano, S., Tanabe, Y., Kamimura, H.: Multiplets of Transition-Metal Ions in Crystals. Academic Press, New York (1970)

Spin Crossover in Trinuclear and Protonated Tetranuclear Iron(II) Complexes: DFT Modelling

S. I. Klokishner and O. S. Reu(✉)

Institute of Applied Physics, Academy str. 5, 2028 Chisinau, Moldova
oleg.reu@ifa.md

Abstract. In the present study the course of spin transformation in the linear trinuclear $[Fe_3(bntrz)_6(tcnset)_6]$ complex, exhibiting a complete one step transition at T = 318 K, and in the tetranuclear protonated $[Fe_4(H_6L_4)]^{6+}$ and $[Fe_4(H_8L)]^{8+}$ complexes is examined with the aid of DFT single point energy calculations. The suggested approach allowed to explain the peculiarities of spin crossover observed in the $[Fe_3(bntrz)_6(tcnset)_6]$ compound. The DFT study of the energy pattern of the tetranuclear $[Fe_4(H_8L)]^{8+}$ complex at low and high temperatures revealed, that upon deprotonation of the complex the number of the Fe^{II} ions participating in the spin crossover transformation is reduced from two Fe^{II} ions to one.

Keywords: Spin crossover · Linear iron(II) trimer · Tetranuclear iron(II) [2x2] grid · DFT calculations · Single point energy

1 Introduction

The search of new spin crossover (SCO) materials with adjustable properties is a very intense field of investigation, since the bistability manifested by these materials promises new routes towards a large panel of potential applications including smart pigments for food status indication, indicators in pharmaceutical and healthcare industry, optical switches, sensors or memory devices and displays [1–3]. The most pronounced effects are observed in Fe^{II} complexes, which show a switching process between the diamagnetic low spin state (*ls*, $S = 0$, ${}^1A_{1g}(t_{2g}^6)$) and paramagnetic high spin state (*hs*, $S = 2$, ${}^5T_{2g}(t_{2g}^4e^2)$). Modulation of the magnetic properties of spin crossover materials by temperature, pressure, light or by chemical processes that affect the field of ligands [4, 5] is very important in the view of their practical applications. Recently, in the field of spin crossover the interest has been switched to polynuclear systems, since an increase in the number of the metal ions of the complex, participating in spin crossover transition, leads to an increase of the critical temperature T_c of the spin transition and offers a possibility to obtain a large difference for the magnetic susceptibility at low and high temperatures, which is a necessary prerequisite for the action of magnetic switches. Currently, two new polynuclear spin crossover compounds have been reported. These are the triazole-based trinuclear $[Fe_3(bntrz)_6(tcnset)_6]$ cluster, which exhibits a complete

I. Tiginyanu et al. (Eds.): ICNBME 2021, IFMBE Proceedings 87, pp. 277–283, 2022.
https://doi.org/10.1007/978-3-030-92328-0_37

one step [*ls-ls-ls*] → [*hs-hs-hs*] SCO transition at T = 318 K [6] and the tetranuclear Fe^{II} [2 × 2] $[Fe_4(H_2L)_4]^{8+}$ grid complex [7], in which the spin crossover phenomena can be controlled by protonation of the two-sided H_2L ligand.

Polynuclear complexes based on 1,2,4-triazole ligands are known as most promising materials, which display wide thermal hysteresis loops around room temperature and could assist the tunable modeling and design of the SCO characteristics of 1D chain polymers [8, 9]. In the previously reported trinuclear complexes the spin crossover transition was manifested only by the central Fe^{II} ion, while the two terminal Fe^{II} ions remained in the *hs* state at all temperatures. In a newly designed, structural and magnetic characterized first linear triazole-based trinuclear $[Fe_3(bntrz)_6(tcnset)_6]$ complex a complete one step [*ls-ls-ls*] → [*hs-hs-hs*] SCO transition is observed above room temperature at T = 318 K [6]. This transition is characterized by the thermal variation of the $\chi_m T$ product from 0 to 9.48 cm^3 K mol^{-1} in the range of 280–345 K and occurs quite abruptly at T = 318 K. The trinuclear $[Fe_3(bntrz)_6(tcnset)_6]$ system and the mechanisms, responsible for the spin transition demonstrated by this system, represent an interesting task for the theoretical investigation undertaken below.

Polymetallic tetranuclear Fe^{II} [2 × 2] grid-like complexes with modulable properties, have also attracted an attention not only due to their abilities to undergo multiple spin state switching under the action of the temperature, light or pressure [4], but also by the reversible ligand deprotonation without destruction of the initial complex [10]. In [7] it has been shown, that the choice of the hydrazine-based ditopic isomeric ligand H_2L offered an opportunity to synthesize and to study physico-chemical properties of the Fe^{II} [2 × 2] grid $[Fe_4(H_2L)_4]^{8+}$ complexes by unic possibiliity of the protonic modulation of their ionizable N–H sites. Under the deprotonation the crystal field created by the two-site H_2L ligand changes its strength from the weak to the strong one. In this case the difference in energies between the t_{2g} and e orbitals of the Fe^{II} ion becomes larger and the low spin $^1A_{1g}(t_{2g}^6)$ state with all six electrons paired in the t_{2g} level becomes more energetically favorable. In [7] the authors showed, that deprotonation of the $[Fe_4(H_2L)_4]^{8+}$ complex reduces the number of the Fe^{II} ions participating in spin crossover transformation, offering a possibility to modulate magnetic properties of the complex.

Thus, the main objective of this work is to explain the peculiarities of the spin crossover transformation in the trinuclear $[Fe_3(bntrz)_6(tcnset)_6]$ and tetranuclear $[Fe_4(H_2L)_4]^{8+}$ complexes with the aid of single point DFT calculations, which allow to obtain the energies, associated to a multi-electron system under the potential created by a certain arrangement of the nuclei in the complex, as functions of the total spin of the complex and temperature. Here, it should be mentioned that single point calculations, performed in our previous work [11], allowed for the description of the spin crossover transformation observed in the $[Fe(ptz_6)](BF_4)_2$ and $[Fe(bpte)(NCSe_2)]$ complexes, containing Fe^{II} in homogeneous and mixed nearest octahedral ligand surrounding, respectively.

2 Computational Details

Single point DFT calculations for the liner trimeric $[Fe_3(bntrz)_6(tcnset)_6]$ [6] complexes and tetranuclear protonated $[Fe^{II}(H_nL)]^{(n+)}$ (n = 6,8) complex [7] have been performed with the aid of the ORCA 4.1.2 program package [12]. The B3LYP hybrid functional together with the Grimme's dispersion correction (D3BJ keyword in the ORCA package) have been applied. In order to increase the accuracy of the calculation the def2-TZVP valence basis set of the triple-zeta quality and corresponding auxiliary basis sets SARC/J and def2-TZVP/C, implemented in ORCA [12], have been selected. The COSX approximation for the exchange terms in conjunction with the RI-J approximation have been accounted for as well with the aid of RIJCOSX keyword and allowed to considerably increase the computation performance. At the same time the precision of the numerical accuracy of the COSX approximation was set with the aid of Gridx5 and FinalGrid6 keywords, while the final target precision of the calculated single point energies for the protonated tetranuclear and linear trimeric iron complexes was set by the VeryTightSCF keyword of the ORCA package and adjusted the convergence of the ground state energies within 10^{-9} Hartree. Detalied description of the ORCA keywords, functionals and basis sets one can find in [12].

3 Results and Discussion

First we start with the evaluation of the linear neutral trimeric $[Fe_3(bntrz)_6(tcnset)_6]$ complex [6]. In the DFT single point calculations for this complex, consisting of 243 atoms, the set of 15 geometrical structures, obtained in the temperature range from T = 250 K to T = 360 K [6], has been employed. Figure 1 shows the plot of the single point energies, calculated in the temperature range 250–360 K, for the set of all possible spin values S = 0, 1, 2, 3, 4, 5 and 6 of the Fe-trimer. From this figure it is clear, that in the temperature range 250–315 K the ground state of the Fe-trimer is the spin singlet, confirming, that up to T = 315 K all three Fe^{II} ions of the trimer are in the $^1A_{1g}$ *ls* – state. (see Fig. 1), which is in accordance with the experimental vanishing value [6] of the magnetic susceptibility $\chi_m T$ for this range of temperatures. This result is also in compliance with the observed characteristic values of the average Fe-N bond lengths for the central Fe1 and the two external Fe2 ions in the *ls*-state, which amount to 1.977–1.979 Å and 1.950–2.002 Å, respectively [6]. The performed DFT calculations also show that, starting from the T = 325 K all three Fe^{II} ions of the trimer are in the *hs*-state, and the total spin of the cluster acquires the value S = 6. The completeness of the spin crossover transition for all three Fe^{II} ions above T = 325 K is also confirmed experimentally [6] by the Fe-N bond lengths magnitude, which fall in this case in the range 2.139–2.175 Å and 2.126–2.158 Å for the central Fe1 and the two external Fe2 ions, respectively. At the same time the calculations show, that the states with S = 4 and S = 6 are close in energy at T = 320 and 325 K, being separated at these temperatures with the energy gaps of 767 and 953 cm^{-1}, respectively, (see Fig. 1), confirming fast one step spin crossover transition [6]. The vicinity of the DFT calculated states of the complex with spin values S = 4 and S = 6 at T = 320 K can be explained on the account of the peculiarities of the octahedral nitrogen surrounding of the two external Fe2 ions, which consists of three Fe2-N bonds of the length 2.094 Å and the other three Fe2-N bonds of the magnitude 2.124 Å. These two Fe2 ions most probably could facilitate a more rapid spin crossover transition as compared with that of the central Fe1 ion with the average length 2.116 Å of the six Fe1-N bonds because of the difference of the strength of the crystal field [6].

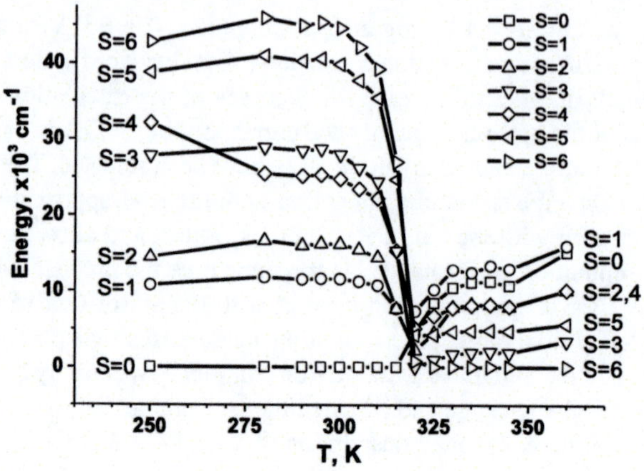

Fig. 1. Spin crossover in the $[Fe_3(bntrz)_6(tcnset)_6]$ complex

Then in the frame of the same computational method the effect of spin crossover observed in the $[Fe_4(H_8L_4)]^{8+}$ complex upon deprotonation of the hydrazine-based N–H sites of the ligand in the complex has been examined. At the first step the results, obtained during single point energy calculations for the $[Fe_4(H_8L_4)]^{8+}$ complex, comprising 196 atoms, are discussed. These results, presented in Fig. 2, show, that at T = 120 K the complex, containing eight attached protons, has the ground state with the total spin S = 0. At this temperature all Fe^{II} ions are in the low-spin state, and the complex is diamagnetic. The ground state of the complex is followed by states with the total spin values equal to S = 2, 4, 6 and 8. At high temperatures the order of the levels changes, and the ground state of the $[Fe_4(H_8L_4)]^{8+}$ complex becomes the state with the total spin S = 4. This state is followed by the other states with spins S = 6,0,2 and 8. The main result of this calculation consists in the following. With the increase of temperature two Fe^{II} ions in the $[Fe_4(H_8L_4)]^{8+}$ complex undergo spin crossover transformation and pass from the diamagnetic state with S = 0 to the high-spin state with S = 2. The obtained result is in agreement with experimental data [7] on the temperature dependence of the magnetic susceptibility. In fact, for the $[Fe_4(H_8L_4)]^{8+}$ complex, the experimental value of the χT product (where χ denotes magnetic susceptibility and T is the temperature) at T = 290 K is equal to 6.2 $cm^3\ Kmol^{-1}$, which is in good agreement with the theoretical value of this product for a pair of two Fe^{II} ions in the high-spin state, calculated with the aid of the formula $\chi T = \mu^2/8 = g^2 S(S+1)/4 = g^2 \cdot 2 \cdot 3/4$ for the Landé factor $g \approx 2.035$, which only slightly differs from the purely electronic one. This value of the Landé g-factor reflects the situation, observed in the fully protonated $[Fe_4(H_8L_4)]^{8+}$ complex. From the structural data for this complex it follows, that the local ligand surrounding possesses a symmetry much lower than the cubic one, which leads to the suppression of the orbital angular momentum of the Fe^{II} ion and a g-factor value, close to the purely electronic one. Thus, the performed DFT calculations show that only two Fe^{II} ions participate in the spin transformation in the $[Fe_4(H_8L_4)]^{8+}$ complex.

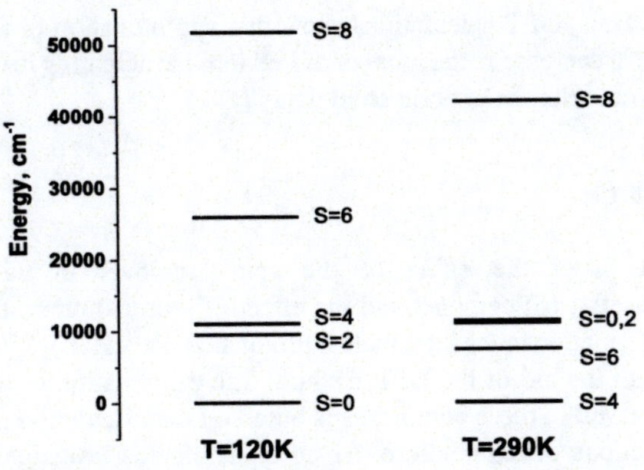

Fig. 2. Temperature dependence of the single point energies corresponding to spin values S = 0, 2, 4, 6, 8 of the $[Fe_4(H_8L_4)]^{8+}$ complex

The same computational procedure has been applied for the identification of the number of the Fe^{II} ions, participating in the spin crossover transformation in the deprotonated $[Fe_4(H_6L_4)]^{6+}$ complex, comprising 194 atoms. The energy scheme, that reflects the order of the single point energy states, calculated for the different values of the total spin of the $[Fe_4(H_6L_4)]^{6+}$ complex is presented in the Fig. 3.

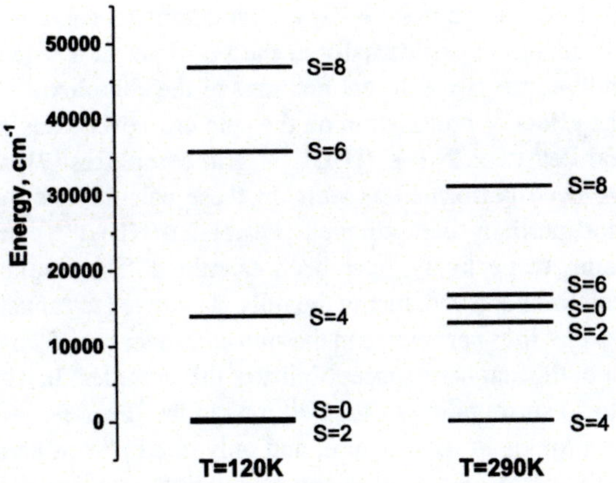

Fig. 3. Temperature dependence of the single point energies corresponding to spin values S = 0, 2, 4, 6, 8 of the $[Fe_4(H_6L_4)]^{6+}$ complex

From Fig. 3 it is clear, that at T = 120 K only one Fe^{II} ion is in the high-spin state with S = 2. The increase of the temperature leads to the spin crossover transformation in another Fe^{II} ion, and at T = 290 K already two Fe^{II} ions are in the high-spin state. The difference of the spin crossover transformation in the $[Fe_4(H_8L_4)]^{8+}$ and $[Fe_4(H_6L_4)]^{6+}$ complexes consists in the following: in the first complex two Fe^{II} ions participate in this transformation, while in the second complex only one of Fe^{II} ions demonstrates spin crossover. Thus, DFT calculations show that deprotonation of the $[Fe_4(H_8L_4)]^{8+}$ complex leads to a decrease in the number of Fe^{II} ions participating in the spin transition, which is consistent with the experimental data [7].

4 Conclusions

In the present paper the effect of the spin crossover in the linear trimeric $[Fe_3(bntrz)_6(tcnset)_6]$ [6] complex and the effect of protonation on the course of spin crossover transformation in the tetranuclear protonated $[Fe^{II}(H_nL)]^{(n+)}$ [7] complex have been studied with the aid of the DFT method. The examination of the spin crossover phenomena observed in these complexes is based on the calculation of the single point energies as functions of temperature. As an input for this investigation the available variable temperature structural data [6, 7] for both complexes have been used.

The course of the spin crossover transformation in the linear trinuclear $[Fe_3(bntrz)_6(tcnset)_6]$ complex has been investigated in details in the range of temperatures from 250 K to 360 K. The plot of the temperature dependence of the single point energy states shows, that up to $T = 315$ K, the ground state of the system is the diamagnetic one with the total spin $S = 0$. The spin crossover transformation occurs abruptly in a very narrow temperature range around $T = 320$ K. In this narrow temperature range each Fe^{II} ion of the trimer passes from the low-spin diamagnetic state with $s = 0$ to the high-spin state with $s = 2$ and the total spin of the $[Fe_3(bntrz)_6(tcnset)_6]$ complex acquires maximal value $S = 6$, confirming a complete spin crossover transformation. The performed single point energy calculations reveal, that the spin crossover transformation, observed experimentally in the Fe-trimer at $T = 318$ K [6], probably occurs more rapidly in the two external Fe^{II} ions of the complex.

To explain the effect of protonation on the spin crossover transformation, observed in the tetranuclear Fe^{II} [2 × 2] $[Fe_4(H_2L)_4]^{8+}$ grid complexes [7], single point energy calculations have been performed as well. In these calculations the fully protonated $[Fe_4(H_8L_4)]^{8+}$ and partially deprotonated $[Fe_4(H_2L)_2(HL)_2]^{6+}$ complexes, possesing 196 and 194 atoms, respectively, have been examined. The results of the performed single point calculations showed, that in the fully protonated tetranuclear $[Fe_4(H_8L_4)]^{8+}$ complex only two Fe^{II} ions participate in the spin crossover transformation, that is in line with the behavior of the magnetic susceptibility of this complex. In its turn deprotonation also affects the local surrounding of the Fe^{II} ions in the $[Fe_4(H_2L)_2(HL)_2]^{6+}$ complex, which includes 6 protons in its structure, and only one Fe^{II} ion demonstrates the spin transition. Thus, the performed single point calculations confirm, that deprotonation of the tetranuclear $[Fe_4(H_8L_4)]^{8+}$ complex suppresses the spin transition and offers a unic possibility to modulate the magnetic properties of the complex in a chemical way [7].

Acknowledgment. The support of the National Agency for Research and Development of Moldova (project 20.80009.5007.19) is highly appreciated.

Conflict of Interest. The authors declare that they have no conflict of interest.

References

1. Sato, O.: Dynamic molecular crystals with switchable physical properties. Nat. Chem. **8**, 644–656 (2016). https://doi.org/10.1038/NCHEM.2547
2. Gamez, P., Costa, J.S., Quesada, M., et al.: Iron spin-crossover compounds: from fundamental studies to practical applications. Dalton Trans. **2009**, 7845–7853 (2009). https://doi.org/10.1039/b908208e
3. Kahn, O., Krober, J., Jay, C.: Spin transition molecular materials for displays and data recording. Adv. Mater. **4**, 718–728 (1992)
4. Breuning, E., Ruben, M., Lehn, J.-M., et al.: Spin crossover in a supramolecular Fe_4^{II} [2×2] grid triggered by temperature, pressure, and light. Ang. Chem. Int. Ed. **39**, 2504–2507 (2000). https://doi.org/10.1002/1521-3773(20000717)39:14%3c2504::AID-ANIE2504%3e3.0.CO;2-B
5. Seredyuk, M., Znovjyak, K.O., Kusz, J., et al.: Control of the spin state by charge and ligand substitution: two-step spin crossover behaviour in a novel neutral iron(II) complex. Dalton Trans. **43**, 16387–16394 (2014). https://doi.org/10.1039/C4DT01885K
6. Pittala, N., Thétiot, F., Charles, C., et al.: An unprecedented trinuclear Fe^{II} triazole-based complex exhibiting a concerted and complete sharp spin transition above room temperature. Chem. Commun. **53**, 8356–8359 (2017). https://doi.org/10.1039/c7cc04112h
7. Dhers, S., Mondal, A., Aguilà, D., et al.: Spin state chemistry: modulation of ligand pK_a by spin state switching in a [2×2] iron(II) grid-type complex. J. Am. Chem. Soc. **140**, 8218–8227 (2018). https://doi.org/10.1021/jacs.8b03735
8. Roubeau, O.: Triazole-based one-dimensional spin-crossover coordination polymers. Chem. Eur. J. **18**, 15230–15244 (2012). https://doi.org/10.1002/chem.201201647
9. Haasnoot, J.: Mononuclear, oligonuclear and polynuclear metal coordination compounds with 1,2,4-triazole derivatives as ligands. Coord. Chem. Rev. **200–202**, 131–185 (2000). https://doi.org/10.1016/S0010-8545(00)00266-6
10. Ruben, M., Breuning, E., Lehn, J.-M., et al.: Supramolecular spintronic devices: spin transitions and magnetostructural correlations in $[Fe_4^{II}L_4]^{8+}$ [2x2]-grid-type complexes. Chem. Eur. J. **9**, 4422–4429 (2003). https://doi.org/10.1002/chem.200304933
11. Klokishner, S., Reu, O.: Modeling of spin crossover in iron(II) complexes with N_4S_2 coordination. J. Phys. Chem. C **123**, 19984–19990 (2019). https://doi.org/10.1021/acs.jpcc.9b05064
12. Neese, F.: The ORCA program system. WIREs Comput. Mol. Sci. **2**, 73–78 (2012). https://doi.org/10.1002/wcms.81

New Ground State of Dipolar Lattice of D_2O@Beryl

Mikhail A. Belyanchikov[1(✉)], M. Savinov[2], V. Thomas[3,4], M. Dressel[5], and B. Gorshunov[1]

[1] Moscow Institute of Physics and Technology, Dolgoprudny, Russia
belyanchikov@phystech.edu
[2] Institute of Physics, Czech Academy of Sciences, Praha, Czech Republic
[3] Institute of Geology and Mineralogy, RAS, Novosibirsk, Russia
[4] Novosibirsk State University, Novosibirsk, Russia
[5] 1.Physikalisches Institut, Universität Stuttgart, Stuttgart, Germany

Abstract. High quality beryl crystals with D_2O molecules in nanocages are synthesized and carefully characterized. IR mapping of the crystals showed drastically different concentration distribution of water-I and water-II molecules. The effect of water concentration on the dielectric properties of D_2O@Beryl was systematically studied. Two areas of the crystal with same water-I concentration and highly different water-II concentrations were studied by temperature-dependent terahertz and impedance spectroscopy. The experiments reveal a strong dependence of the dielectric properties of the crystal on water-II concentration. The sample with low water-II content showed an anomaly at $T = 2$ K in the temperature behavior of radiofrequency permittivity; no saturation in the temperature-dependent behavior of terahertz soft mode is observed. These observations contrast with our previous results on incipient ferroelectricity in H_2O@Beryl. We speculate about a possibility of new ground state developed in dipolar water lattice in beryl.

Keywords: Nanoconfined water · Dipolar lattice · Spectroscopy · Ferroelectricity

1 Introduction

A vibrant frontier of recent condensed-matter science is the study of phenomena occurring on the nanoscales, where qualitatively new properties of matter can emerge not known in the macroscopic bulk state of matter. Understanding the nature of the emerging new phases and their relations to the physical, chemical, geometrical, and morphological characteristics of the environment is of great fundamental and technological interest but is presently still at its infancy. In such studies, special attention is paid to water due to its widespread prevalence and omnipresence on Earth and its critical importance for biological systems and organisms. Although the isolated H_2O molecule seems to be rather simple, bulk water remains one of the least understood liquids. Under the conditions of

I. Tiginyanu et al. (Eds.): ICNBME 2021, IFMBE Proceedings 87, pp. 284–290, 2022.
https://doi.org/10.1007/978-3-030-92328-0_38

nanoconfinement, it acquires an even greater variety of diverse and intriguing properties. Understanding such properties is important for geology, biology, mineralogy, ionic liquids, nanoscience and applications.

Beryl crystals with nanolocalized water represent a unique system to study properties of solitary water molecule in nanoconfinement conditions [1] as well as cooperative behavior of electric dipole lattice composed by water dipoles [2]. Localized in beryl's nanocages (Fig. 1), water molecules form periodic lattices of dipoles with two possible types of localization: freely rotating in the *ab* plane water molecules form water-I dipole subsystem. Another subsystem is built by water-II dipoles formed by water molecules poled along the *c* axis due to coordination by Li or Na atoms at bottlenecks of nanocages channels. In this work, we present the results of our study of influence of presence of water-II molecules on the behavior of water molecular lattice in beryl crystal. Temperature-dependent radio-frequency and terahertz spectra of a sample with low water-II concentration revealed a dielectric anomaly at around $T = 2$ K which might advocate for a new ground state of water dipolar lattice. These observations contrast with our previously observed incipient ferroelectricity ground state in H_2O@Beryl [3]. Elucidation of the nature of the ground state requires additional experiments, but at this stage we can speculate about a possible formation of a quantum dipolar liquid in D_2O dipolar lattice in beryl crystal.

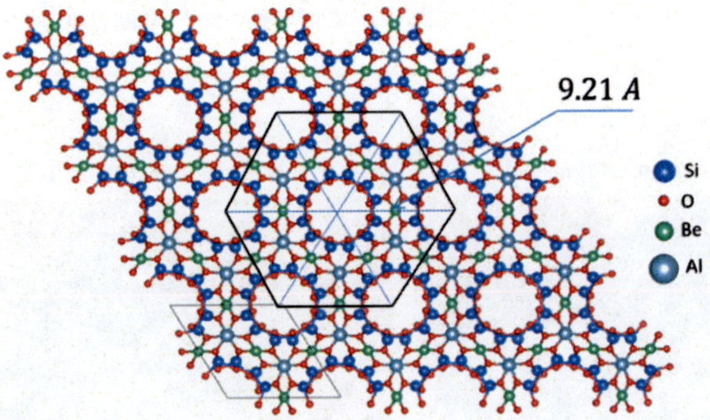

Fig. 1. Crystal structure of beryl crystal. The unit cell is shown in lower left side. The crystal forms triangular lattice of nanocages with period of 9.21 A filled with solitary water molecules (not shown) of first (water-I) or second (water-II) type.

2 Experimental Details

The study was performed with D_2O containing beryl crystal grown with hydrothermal method from oxides, using a complex acidic lithium fluoride mineralizer and growing seed with Miller indices $\{5.5.\bar{1}0.6\}$. The growth process was carried out for 21 days at $T \sim 620$ °C and $P \sim 1.5$ kbar. The sample for spectroscopic measurements was cut from the crystal parallel to the c axis and polished to final thickness of 200 μm. An infrared (IR) mapping was performed with the Bruker Vertex 80v FTIR spectrometer equipped with IR microscope Hyperion 2000. IR maps were obtained from MIR spectra measured on the grid with 100 μm and with the same spot size. At radiofrequencies, we used a Novocontrol Alpha AN High Performance Frequency Analyzer and Andeen-Hagerling 2500 A capacitance bridge. For terahertz measurements, a commercial time-domain TeraView 3000 spectrometer was used. In all experiments, measurements on dehydrated samples allowed us to extract the characteristics determined exclusively by a network of water molecules.

3 Results and Discussion

For a characterization of water distribution in the sample, we obtained IR images of the sample by measuring polarized MIR spectra (Fig. 2). Water-I and water-II molecules have different vibrational modes that are active in E||c polarization [4]. By integration of transmittance spectra in spectral regions of corresponding vibrational D_2O lines we obtained the maps of water-I (Fig. 2, **a**) and water-II (Fig. 2, **b**) molecular distribution. While the water-I map shows homogeneous concentration of water-I molecules over the sample (Fig. 2, a), water-II molecules distribution has highly inhomogeneous character (Fig. 2, b): Water-II concentration is highest near the region of the growing seed and subsequently decreases down to negligible amount at the region of the crystal corresponding to the finished growing process. Such inhomogeneity can be explained by charge compensation process of lattice distortions caused by accretion of subindividuals on the growth front [5].

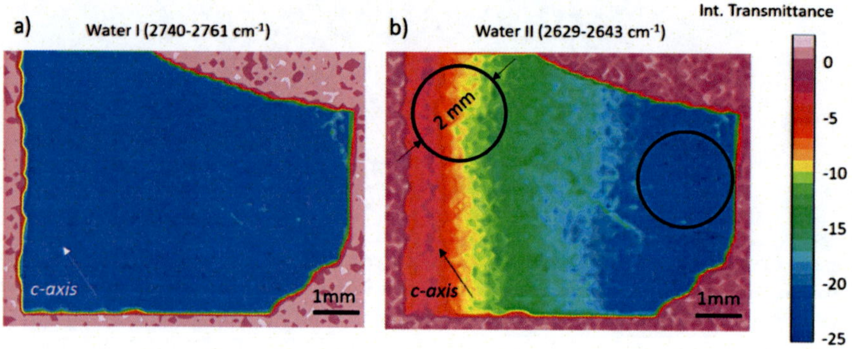

Fig. 2. IR maps of D_2O@Beryl sample. The pictures obtained by integrating transmittance spectra for polarization E||c in the spectral regions 2740–2761 cm^{-1} for water-I (**a**) and 2629–2643 cm^{-1} for water-II (**b**). Black circles show the areas of two different water-II concentrations where THz and impedance spectroscopic experiments were performed.

To investigate the influence of water-II concentration on dielectric properties of dipolar lattice of water-I molecules we choose two regions of the sample, with high and low water-II concentration; they are highlighted by left and right black circles in Fig. 2, **b**, respectively. We performed THz and impedance spectroscopy in selected regions. Figure 3 shows corresponding temperature-dependent dielectric permittivity measured at a fixed frequency 8.4 kHz. Both curves, corresponding to low and high water-II concentration, show the Curie-Weiss behavior of permittivity, typical for paraelectrics. However, the sample with low water-II concentration shows much higher Curie constant (red curve, Fig. 3) compared to sample with high water-II content. The influence of water-II concentration on dielectric permittivity can be caused by polarization of water-I molecular lattice by rigid dipoles of water-II molecules. As a consequence, the polarizability of water-I lattice is decreasing in crystal with high water-II concentration, even if the total water content is higher. Figure 4 Shows low temperature inverse permittivity of the sample with low water-II content. After following the linear Curie-Weiss-like behavior down to $T = 20$ K, the inverse permittivity deviates from straight line at lower temperatures and goes upwards (downwards for permittivity in inset on Fig. 4) below 2 K. Such a behavior contrasts with low temperature saturation of the permittivity both in H_2O@Beryl sample [3] and in high water-II D_2O containing sample (not shown here). The difference in dielectric properties compared with the H_2O@Beryl sample and with the sample with higher water-II (D_2O) content is also seen in the soft mode behavior, Fig. 5. As can be seen from the behavior of imaginary permittivity measured for E||c polarization, in the sample with high water-II content the soft mode of dipolar lattice, that is responsible for paraelectric respond, stabilizes at around 15 cm^{-1} at 5 K, similarly to what was observed in H_2O@Beryl [3], while for low water-II content sample, the soft mode goes below 5 cm^{-1}, that is the low frequency limit of our experimental frequency window.

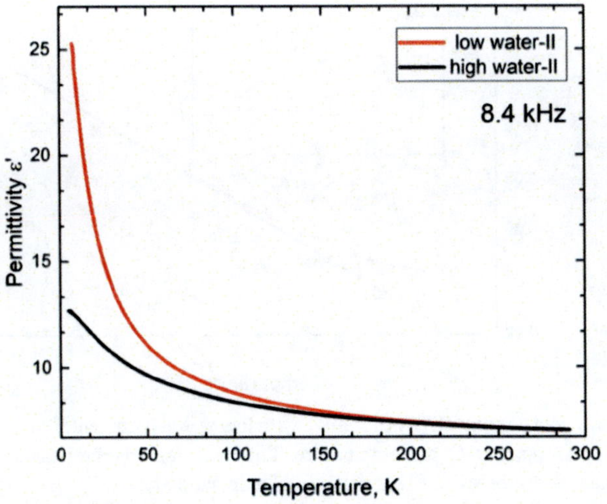

Fig. 3. Dielectric permittivity of samples with low (red line) and high (black line) water-II concentrations at frequency of 8.4 kHz.

The observed difference in the dielectric properties of the samples with low and high water-II concentrations and of previously studied H_2O@Beryl sample [3] tells us the following. First, the presence of rigid dipoles of water-II molecules has dramatic impact on water-I molecular lattice leading to incipient ferroelectric behavior, similar to that in H_2O@Beryl crystal. The precise mechanism of such influence needs to be clarified, but we can speculate about appearing of polarization in water-I lattice around water-II dipoles already at elevated temperature, which leads to suppression of the appearance of order among water-I dipoles at low temperature. The different mechanism suppresses ordering in water-I lattice of H_2O molecules. The higher rotational constant of the light flavor of water leads to higher energy of quantum fluctuations, which can overcome the energy of dipolar interactions and leave such system in highly symmetric disordered state [1]. In contrast to the last two cases, dipolar lattice of heavier D_2O molecules without polarizing centers of water-II dipoles can order due to dipolar interaction. We believe that strong softening of the THz soft mode and especially downward temperature behavior of radiofrequency permittivity are the signs of transition from paraelectric phase to a new ground state, that is different from previously observed quantum paraelectric state. The clarification of exact nature of the new ground state requires further experimental investigation including frequency dependent impedance spectroscopy, specific heat measurements and pyrocurrent measurements, but at this stage we can speculate that intrinsic frustrated nature of triangular lattice along with long range dipolar coupling will suppress the long range order below the transition temperature and might lead to formation of quantum dipolar liquid state [6].

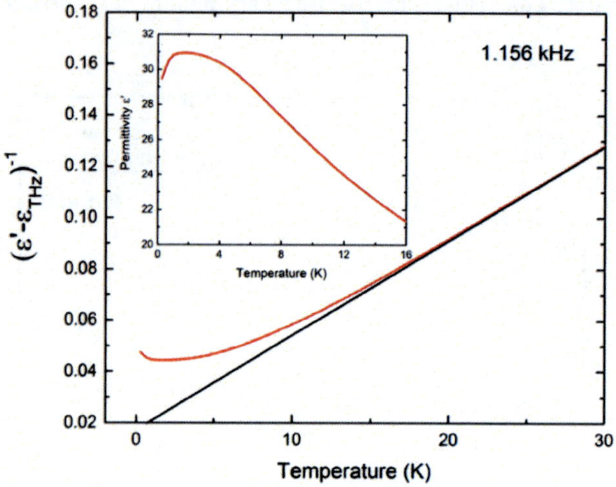

Fig. 4. Inverse dielectric permittivity of sample with low water-II content. Black line shows Curie-Weiss fit with small positive Curie temperature. The inset demonstrates downward permittivity temperature dependence below 2 K indicating dialectic anomaly.

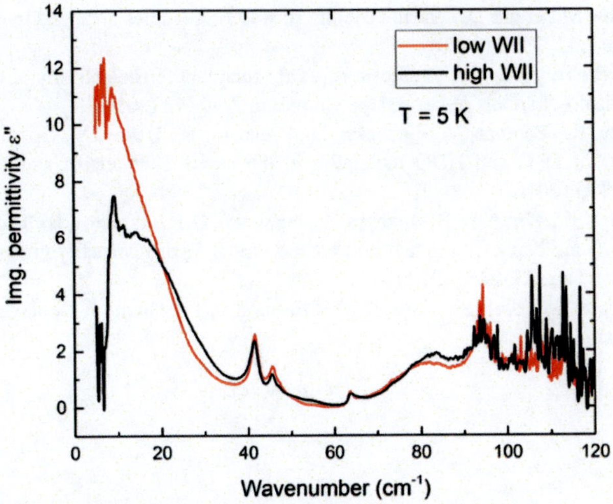

Fig. 5. THz spectra of imaginary permittivity of samples with low (red curve) and high (black curve) water-II concentrations. At 5 K, the soft mode stabilizes at ~15 cm^{-1} in high water-II containing sample, while it goes below 5 cm^{-1} in sample with low content of water-II molecules.

4 Conclusions

Infrared mapping revealed highly inhomogeneous water-II distribution in a synthetic D_2O@beryl crystal, which has a drastic impact on the dielectric properties. The presence of water-II molecules has strong effect on water-I dipolar lattice and leads to incipient ferroelectric behavior previously observed in H_2O@Beryl sample. In contrast, the part of the crystal with almost pure water-I lattice shows signs of transition to a new ground state. Frustration brought in by triangular lattice and long-range character of dipolar interaction can lead to the absence of long-range order in the new ground state which opens a possibility of existence of quantum dipolar liquid state in the dipolar lattice of heavy water molecules.

Acknowledgments. The study was funded by Russian Science Foundation grant 21-72-00026, Czech Science Foundation (Project 20-01527 S) and the Deutsche Forschungsgemeinschaft (DR228/61-1). The research was done in close collaboration with P. Abramov, V. Abalmasov and E. Zhukova.

Conflict of Interest. The authors declare that they have no conflict of interest.

References

1. Kolesnikov, A., Reiter, G., Choudhury, N., et al.: Quantum tunneling of water in beryl: a new state of the water molecule. Phys. Rev. Lett. **116**, 167802 (2016)

2. Belyanchikov, M., et al.: Dielectric ordering of water molecules arranged in a dipolar lattice. Nat. Commun. **11**, 3927 (2020)
3. Gorshunov, B., Torgashev, V., Zhukova, E., et al.: Incipient ferroelectricity of water molecules confined to nano-channels of beryl. Nat. Commun. **7**, 12842 (2016)
4. Mashkovtsev, R., Thomas, V., Fursenko, D., Zhukova, E., Uskov, V., Gorshunov, B.: FTIR spectroscopy of D_2O and HDO molecules in the c-axis channels of synthetic beryl. Am. Miner. **101**, 175 (2016)
5. Belyanchikov, M., Abramov, P., Ragozin, A., Fursenko, D., Gorshunov, B., Thomas, V.: Distribution of D_2O molecules of the first and second type in hydrothermally grown beryl crystals. Cryst. Growth Des. **21**, 2283 (2021)
6. Yao, N., Zaletel, M., Stamper-Kurn, D., Vishwanath, A.: A quantum dipolar spin liquid. Nat. Phys. **14**, 405 (2018)

Excitonic States in Brillouin Zone Center of GaSe Layered Crystals

Victor V. Zalamai[1(✉)], A. V. Tiron[1], E. Cristea[1], and I. G. Stamov[2]

[1] National Center for Materials Study and Testing, Technical University of Moldova, Chisinau, Republic of Moldova
victor.zalamai@cnstm.utm.md

[2] T.G. Shevchenko State University of Pridnestrovie, Tiraspol, Republic of Moldova

Abstract. Optical spectra (absorption and reflection) of GaSe layered crystals were studied at room and low (10 K) temperatures. Contours of measured reflection spectra were fitted by help of dispersion equations. Photoluminescence spectra excited by 448 and 325 nm laser lines were measured at low temperatures. The observed features can be explained in the framework of model of the existence of Frenkel and Wannier-Mott excitons.

Keywords: Gallium monoselenide · Excitonic states · Optical spectroscopy · Absorption · Frenkel and Wannier-Mott excitons

1 Introduction

GaSe belongs to the group of layered crystals and have interesting physical properties. These crystals consist of GaSe layers with strong covalent chemical bonds in layer and weaker Van der Waals forces between layers [1]. Such crystal can be easy cleaved along layers (cleavage plane) or perpendicular to axis ***c***. These weak Van der Waals bonds between layers lead to the absence of dangling bonds. Because of this face it is possibly to receive a very thin plates by cleaving or exfoliation. Thus samples with thicknesses around 100 nm and even less can be received. These samples due to layered structure have a good optical quality and mirror-like surface that it is a good for optical measurements. Also they posses a relatively low absorption coefficient. This fact permits a dip light penetration into samples and thus the better optical spectra as a result [2].

This material can be uses as a detectors of visible and near-infrared light as mentioned in Ref. [3]. The authors of Ref. [4] propose to use GaSe in quantum electronica, as gas sensors, sources of terahertz emission, photovoltaic convertors and in thermoelectric application.

In the present work edge absorption, reflection and photoluminescence spectra of bulk GaSe single crystals were investigated in wide temperature range (10–300 K). Excitonic states of Wannier-Mott type with low binding energy (21.8 meV) were discovered and parameters excitons and bands were estimated. It has been suggested that states with large binding energy (120 meV) can be associated with Frenkel excitons.

I. Tiginyanu et al. (Eds.): ICNBME 2021, IFMBE Proceedings 87, pp. 291–297, 2022.
https://doi.org/10.1007/978-3-030-92328-0_39

2 Experimental Methods

GaSe single crystals were grown by zone melting method in form of ingots with 1 × 1.2 × 15 cm sizes in quartz ampoules. Due to layered structure the grown ingots were easy cleaved and plates with mirror-like surfaces was received. Thus a set of monocrystalline plates with varied thicknesses (from few millimeters to hundred nanometers) was received. The samples with thicknesses of nanometric size were prepared by exfoliation with help of a scotch tape.

The transmission and reflection were measured on high-aperture spectrometer MDR-2. The photoluminescence spectra were recorded by help of double high aperture spectrometer SDL-1. The closed-circuit helium optical cryogenic system LTS-22 C 330 allows to vary sample temperature from 300 down to 10 K. The resolution of our measurements is ~0.5 meV. Received plates surfaces are perpendicular to ***c*** axis and have a high reflection coefficient similar to reflectivity of a polished aluminum mirror.

3 Experimental Results and Discussions

For the layered GaSe crystals there is a large amount of experimental and theoretical data about excitonic and electronic spectra [5]. The conclusion that follows from the analysis of published data is that the properties of excitonic spectra at the absorption edge for layered semiconductors do not differ from the known properties of excitonic spectra in three-dimensional crystals [6].

Figure 1 illustrates absorption spectra in near-edge region measured at low temperature for two extra-thin samples (100 and 300 nm). Due to such thicknesses of samples the interference fringes do not influence sufficient on measured spectra and one can see excitonic features more clearly. One can observe a pronounced maximum at 2.112 eV. This maximum correspond to ground state $\boldsymbol{n} = 1$ of exciton. At higher energies the excited states $\boldsymbol{n} = 2$ and $\boldsymbol{n} = 3$ at 2.127 and 2.130 eV, respectively can be recognized. On the base of energies of ground and excited states the excitonic binging energy or Rydberg ***Ry*** is calculated (***Ry*** = 20.6 meV). Taking into account the binding energy value the observed excitonic state can be associated with Wannier-Mott type exciton.

Detailed the excitonic features can be recognized in reflection spectra measured at low temperatures. Such reflection spectrum measured at 10 K is shown in Fig. 2 as $\boldsymbol{R_{exp}}$ curve. The fitting of experimentally measured in non-polarized light reflection spectrum ($\boldsymbol{R_{exp}}$) with the calculated by dispersion relations ($\boldsymbol{R_{calc}}$) one is shown in Fig. 2. The calculations of reflection spectrum contour R_{calc} were carried out by help of dispersion equations. The method of calculation and used relations are described in Ref. [7]. The observed reflection spectrum R_{exp} is a typical for an exciton. One can see maxima corresponded to ground (n = 1) and excited (n = 2) states of Wannier-Mott exciton. In the reflection spectrum the maximum is corresponds to the transversal exciton (ω_T) and the minimum to the longitudinal exciton (ω_L). The other features observed in the experimentally measured reflection spectrum can be caused by nonideality of sample and appearing an interference between layers of crystal.

The best agreement between theoretical and experimentally measured reflection spectra was achieved by the fitting. The result of this fitting is shown in Fig. 2. The parameters received by the fitting is $\boldsymbol{\varepsilon_b} = 5.0$, $\boldsymbol{\omega_T} = 2.112$ eV, $\boldsymbol{\omega_{LT}} = 3$ meV, $= \gamma 1.4$ meV, and

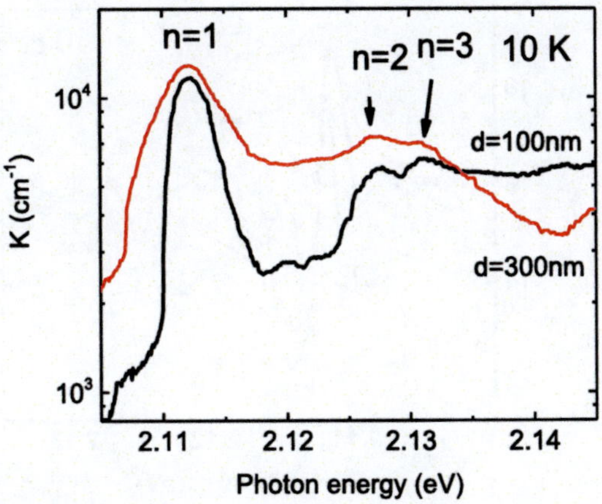

Fig. 1. Absorption spectra of GaSe crystals with thicknesses of 100 and 300 nm measured at temperature 10 K.

$M = 2.5m_0$, where ε_b - background permittivity, ω_T – transversal exciton energy, ω_{LT} – longitudinal-transversal splitting, γ - damping constant of oscillator, and M – exciton mass. The ground ($n = 1$) and excited ($n = 2$) sates of exciton is observed in the reflection spectrum at 2.112 and 2.127 eV, respectively. So the binding energy of Wannier-Mott exciton in this case is equal to 20.4 meV. Taking into account the positon of $n = 1$ ground state (2.1121 eV) and the binding energy (20.4 meV) the minimal direct band gap E_g in Brillouin zone center is equal to 2.132 eV. In Brillouin zone center transitions between valence band V_1 (Γ_1 symmetry) and conduction band C_1 (Γ_6 symmetry) take place. According the selection rules these transition are allowed in E⊥c polarization. On the base of fitting parameters and Rydberg constant the reduced effective mass (μ) of this exciton was estimated ($0.038m_0$). Knowing the reduced effective mass (μ) and translation mass (M) of the exciton one can calculate an effective mass of electron in conduction band. In our case the effective mass of electrons in the bottom conduction band C_1 (Γ_6) is $0.37m_0$.

Measured at room temperature photoluminescence spectra from cleavage face of GaSe plate excited by two different lasers with emissions lines 448 and 325 nm is shown in Fig. 3. One can see an intensive emission maximum E_W at 2.1145 eV observed in the spectrum excited by both lasers. This maximum can be associated with the ground state of Wannier-Mott exciton. Insert of Fig. 3 illustrated the zoomed high-energy part from the intensive line. One can see in the insert the excited states of the Wannier-Mott exciton. It should be mentioned that one can see the excited states up to $n = 4$. So the excited excitonic states $n = 2$, $n = 3$ and $n = 4$ are recognized at energies 2.139, 2.133 and 2.136 eV, respectively. In this case the binding energy of exciton is estimated as 20.5 meV. A weak long-wavelength maximum E_b is observed at 2.088 eV. The authors of Ref. [8] attributed this feature to biexcitonic sates. In the case of 448 nm laser excitation an additional low-energy peak E_F is observed at 2.051 eV. The authors of Ref. [8] were also associated this photoluminescence maximum with bound exciton. The presence of bound excitons in this region is not excluded and even possible.

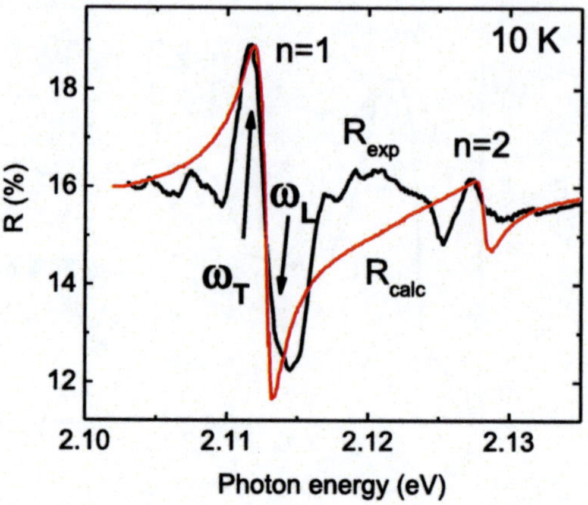

Fig. 2. B - Reflection spectra contours measured in unpolarized light ($\boldsymbol{R_{exp.}}$) and calculated by the dispersion equations ($\boldsymbol{R_{calc.}}$).

On the other hand in the investigated crystals the line $\boldsymbol{E_F}$ can be associated with the ground state n = 1 of transversal mode ($\boldsymbol{\omega_T}$) of Frenkel exciton. In some crystals the weak peak at energy of 2.153 eV is associated more probably with the excited state $\boldsymbol{n}$ = 2 of these excitons.

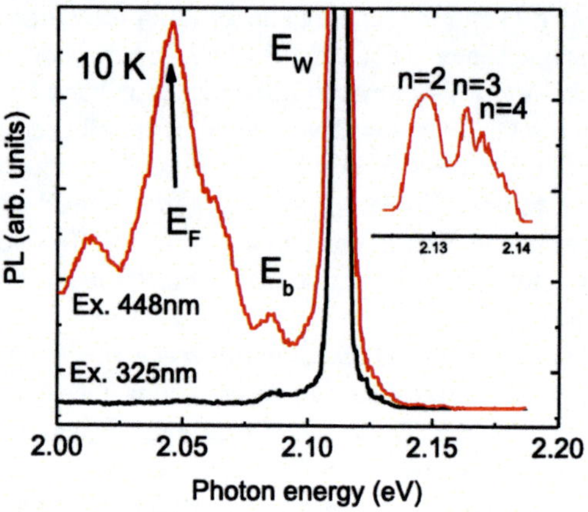

Fig. 3. Photoluminescence spectra received from cleavage surface at excitation by laser line 352 nm measured at temperature 10 K. Insert illustrates the excited states of exciton.

For our opinion the long-wavelength maximum $\boldsymbol{E_F}$ can be attributed with Frenkel exciton ground state ($\boldsymbol{n}$ = 1). The short-wavelength maximum $\boldsymbol{E_W}$ is most probably associated with the ground states of Wannier-Mot exciton. Electrons of the conduction band $\boldsymbol{C_1(\Gamma_6)}$ and holes of the valence band $\boldsymbol{V_1(\Gamma_1)}$ are formed the Wannier-Mott exciton. And in the cals of Frenkel exciton interaction between electrons of the conduction band $\boldsymbol{C_1(\Gamma_6)}$ and holes of the valence band $\boldsymbol{V_2(\Gamma_6)}$ takes place.

Reflection spectra measured from GaSe plate from cleavage surface are shown in Fig. 4. The reflection spectrum measured at 80 K from Ref. [9] was shifted to higher energies on temperature shift coefficient of edge absorption and was marked in Fig. 4 as "***R****". Reflection spectra of our crystals and measured by author of [9] are very similar.

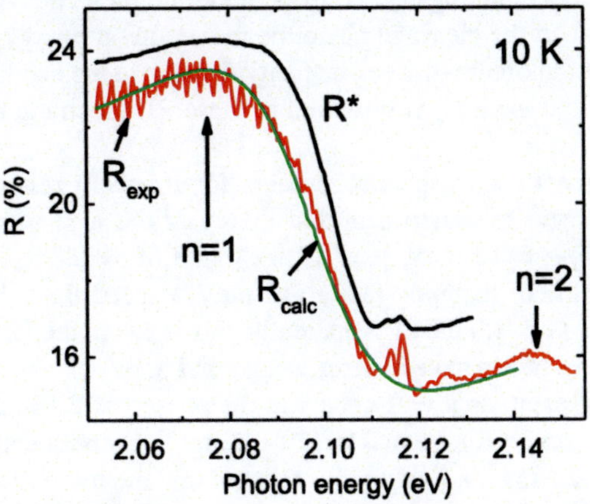

Fig. 4. Reflection spectra of bulk GaSe crystal measured ($R_{exp.}$) at 10 K and calculated ($R_{calc.}$) by dispersion equations. Reflection spectrum (***R****) measured at 80 K form Ref. [9] and shift toward higher energies on the absorption edge temperature shift coefficient.

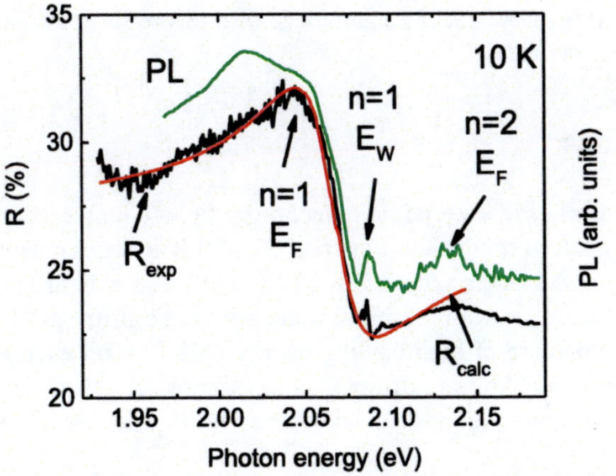

Fig. 5. Measured reflection spectrum ($R_{exp.}$) of bulk GaSe crystal at 10 K (from back side of sample) and calculated by the dispersion equations spectrum ($R_{calc.}$) and photoluminescence spectrum (***PL***) form the face perpendicular to the cleavage face excited by laser line 325 nm.

The maximum at 2.053 eV is attributed to ground state $n = 1$ of exciton is more pronounced in reflection spectra measured from back side of sample plate (Fig. 5). The maximum due to the excited state $n = 2$ of exciton at 2.148 eV is observed at higher energies. At these parameters of the excitonic states the binding energy of exciton is 120–122 meV. Thus, the abovementioned excitonic lines are caused by Frenkel excitons. At the same time, it should be noted that the well-known lines of excitons with low binding energy are at the energy of 2.1 eV [6]. In photoluminescence spectra measured from cleavage plane the sharp maximum at 2.107 eV is more pronounced. These lines are due to excitons with a small binding energy. In the photoluminescence spectra from the side plane perpendicular to the cleavage plane the maximum at energy of 2.05 eV has the most intensity. The maximum of weak intensity E_W due to the exciton with low binding energy (Wannier-Mott exciton) is observed in these PL spectra at energy of 2.107 eV (Fig. 5).

Calculations of reflection spectra contours for Frenkel excitons shown in Fig. 5 give a good agreement of experiment and theory at the next parameters: $\varepsilon_b = 6.0$, $\omega_T = 2.075$ eV, $\omega_{LT} = 13$ meV, $\gamma = 19$ meV and $M = 0.8m_0$. Received form the calculations the value of damping factor (13 meV) exceed the value of longitudinal-transversal splitting (1.6–1.9 meV). Apparently, this is due to the fact that GaSe crystals are layered and these layers have a rather weak bond between. For this excitonic series the excited state of excitons n = 2 appear at the energy of 2.148 eV and the binding energy of Frenkel exciton Ry is equal to 120–122 meV. Consequently, the band gap for these transitions is 2.1751 eV. Taking into account that the translation effective mass of Frenkel exciton is $M = 0.8m_0$ and the effective mass of electrons in the conduction band C_1 with symmetry Γ_6 is equal to $0.37m_0$ the effective mass of holes in the valence band V_2 is $2.13m_0$. Symmetry of the lower zone V_2 is most likely Γ_6. From the obtained data it follows that the band gaps of $V_1(\Gamma_1) - C_1(\Gamma_6)$ is 2.1325 eV and of $V_2(\Gamma_6) - C_1(\Gamma_6)$ is 2.1751 eV. A splitting of valence bands $V_1(\Gamma_1) - V_2(\Gamma_6)$ is 42–43 meV. This model in $\vec{k} = 0$ is in satisfactory agreement with theoretical calculations of the band structure of these crystals.

4 Conclusions

By investigations of edge absorption, reflection and photoluminescence spectra of bulk GaSe single crystals at temperature interval 300–10 K it was shown the simultaneous presence of excitonic states with a low (21.8 meV) and a high (120 meV) binding energies. Excitonic states with low binding energy can be attributed with Wannier-Mott excitons and in the case of high binding energy with Frenkel excitons. The effective masses of electrons and holes in Brillouin zone center were calculated by the fitting the measured reflection excitonic spectra with theoretically calculated ones.

Acknowledgment. The authors acknowledge financial support from the Ministry of Education, Culture and Research of Moldova under Grant No. 20.80009.5007.20.

Conflict of Interest. The authors declare that they have no conflict of interest.

References

1. Belen'kii, G.L., Stopachinskii, V.B.: Electronic and vibrational spectra of III-VI layered semiconductors. Phys. Uspekhi **140**, 234 (1983). https://doi.org/10.1070/PU1983v026n06ABEH004420
2. Hsu, Y.-K., Chen, Ch.g-W., Huang, J.Y., Pan, C.-L.: Erbium doped GaSe crystal for mid-IR applications. Opt. Express **14**(12), 5484 (2006).https://doi.org/10.1364/OE.14.005484
3. Pham, K.D., Phuc, H.V., Hieu, N.N., et al.: Electronic properties of GaSe/MoS2 and GaS/MoSe2 heterojunctions from first principles calculation. AIP Adv. **8**, 075207 (2018). https://doi.org/10.1063/1.5033348
4. Wei, C., Chen, X., Li, D., et al.: Bound exciton and free exciton states in GaSe thin slab. Sci. Rep. **6**, 33890 (2016). https://doi.org/10.1038/srep33890
5. Aziza1, Z.B., Zólyomi, V., Henck, H., et al.: Valence band inversion and spin-orbit effects in the electronic structure of monolayer GaSe. Phys. Rev. B **98**, 115405 (2018).https://doi.org/10.1103/PhysRevB.98.115405
6. Belen'kii, G.L., Godzhaev, M.O., Salaev, E., Aliev, E.T.: High-temperature electron-hole liquid in layered InSe, GaSe and Gas crystals. Sov. Phys. JETP **64**(5), 1886 (1986)
7. Syrbu, N.N., Ursaki, V.V.: Exciton polariton dispersion in multinary compounds. In: Bergin, R.M. (ed.) Exciton Quasiparticles: Theory, Dynamics and Applications, pp. 1–30. Nova Science Publishers Inc., New York (2010)
8. Dey, P., Paul, J., Moody, G., et al.: Biexciton formation and exciton coherent coupling in layered GaSe. J. Chem. Phys. **142**, 212422 (2015). https://doi.org/10.1063/1.4917169
9. Bernier, G., Gagnon, R., Jandl, S.: Thermoreflectance study of direct and indirect excitons near the fundamental edge of GaSe. Solid State Commun. **63**(5), 431 (1987). https://doi.org/10.1016/0038-1098(87)91144-6

Population Dynamics in a Modulated Optomechanical Setup

Victor Ceban(✉) and M. A. Macovei

Institute of Applied Physics, Academiei str. 5, 2028 Chişinău, Moldova
victor.ceban@ifa.md

Abstract. An initially excited two-level quantum-dot embedded on a quantum mechanical resonator is placed in an optical cavity. The quantum-dot has its transition frequency modulated by an off-resonant laser, which leads to a slow-down of the quantum-dot decay within the bad cavity limit. The effect of spontaneous emission control may be enhanced by considering various frequency modulation signals.

Keywords: Spontaneous emission · Slowed-down dynamics

1 Introduction

One of the most prominent issues in the development of modern quantum technologies consists in quantum decoherences that appear due to the interaction of a quantum system with its environment or due to imperfections of the components of the quantum system. The realms of quantum communications and quantum computing often use single or few-photon processes which require a high degree of control and confinement of specific quantum states of the system [1]. The realm of quantum photonics introduces additional engineering requirements in order to implement these processes at micrometer sized on-chip photonic circuits and devices. At this size scale, new interactions appear due to intrinsic mechanical vibrations of the system. The effective optomechanical coupling with the mechanical vibrations is significantly increased when considering optical systems based on emitters made of artificial atoms such as quantum dots [2]. Moreover, imperfections in the band structure of these artificial atoms introduces a new damping phenomenon of dephasing.

A broad gamut of various experimental methods and theoretical investigations have been reported during lasts decades in order to address the problem of decoherences in quantum systems. In this regard, a particular scientific interest have been focused towards the control of the spontaneous emission effect. The first approach of spontaneous emission control have been reported by Purcell [3] and consists in the enhancement of the spontaneous emission decay rate when an atom is placed in resonance with an optical cavity, due to the difference of the density of modes of two environments. The idea of engineering various electromagnetic environments in order to control the spontaneous emission have been further evolved in different optical setups. Some methods rely on

I. Tiginyanu et al. (Eds.): ICNBME 2021, IFMBE Proceedings 87, pp. 298–305, 2022.
https://doi.org/10.1007/978-3-030-92328-0_40

modifying or modulating the reservoir modes or the coupling of the emitter with the reservoir. Enhanced lifetime of the emitter excited state population had been predicted for a two-level atom placed in an optical cavity interacting with modulated environmental reservoir modes [4]. The reservoir modes are engineered by placing the cavity-atom system within a second cavity. In this way, the reservoir modes have their frequency modulated by varying the length of the second cavity.

A different approach in the control of the spontaneous emission consists in modulating the transition frequency of the emitter. The frequency modulation may be obtained by applying on the emitter an off-resonant laser, i.e., via AC Stark effect. A two-level emitter with modulated transition frequency may exhibit longer lifetimes of its excited state population when placed in an optical cavity with a high damping rate [5]. In this case, the spontaneous emission control is achieved due to the interaction of the modulated emitter with the modes of the optical cavity. Enhanced lifetimes of the excited populations are also achieved when modulating both the emitter and the modes of the surrounding reservoir [6, 7] or the coupling strength of the reservoir-emitter interaction [8]. Slow-down or acceleration of the decay dynamics have been predicted for a two-level emitter pumped by a classical strong low-frequency electromagnetic field [9]. While many spontaneous emission control schemes treat pure sinusoidal frequency modulations, considering more diverse types of modulation signals may also affect the dynamics of the population decay and, therefore, the performance of control scheme. Moreover, the implementation of theoretical models able to treat more complex modulation signals would allow a better interpretation of various physical phenomena. For example, considering an emitter-reservoir interaction strength modulated by a signal with a varying amplitude have allowed the description of a model imitating the dephasing effects [8].

In this paper, we investigated the impact of various types of modulation signals, as well as, the influence of the mechanical vibrations, on the quantum dynamics of the spontaneous emission effect of an emitter placed in an optical cavity. The optomechanical setup consists of a two-level emitter placed on a quantum mechanical resonator. The control of the spontaneous emission of the quantum-dot is achieved by placing the emitter in an optical cavity with low quality factor, while modulating its transition frequency by an off-resonant incident laser, as shown in Fig. 1. Quantum-dots or organic molecular crystals are suitable emitters for the considered setup as they are able to couple to the phonons of the substrate on which they are placed. Various types of mechanical resonators such as a nanobeam, vibrational membrane or a phononic crystals, are considered for the current setup as they allow the confinement of single-mode phonon fields with high quality factors [2].

2 The Model

The two-level quantum-dot is characterized by its transition frequency ω_{qd} among its excited state $|e\rangle$ and ground state $|g\rangle$. The atomic quantum-dot operators are $S^+ = |e\rangle\langle g|$, $S^- = |g\rangle\langle e|$ and $S_z = (|e\rangle\langle e| - |g\rangle\langle g|)/2$ and obey the standard commutation relations for SU(2) algebra, i.e., $[S^\pm, S^\mp] = \pm S_z/2$ and $\left[S^\mp, S_z\right] = \pm S^\mp$. The quantum vibrations of the mechanical resonator are described by a single-mode phonon field defined by its oscillation frequency ω_{ph} and the bosonic annihilation and creation operators,

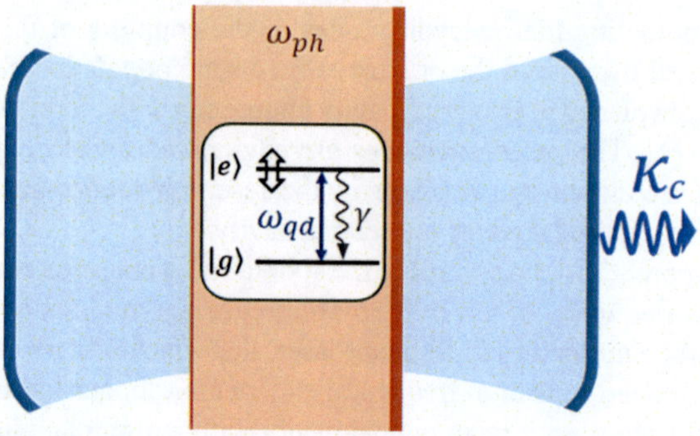

Fig. 1. The optomechanical setup is made of a two-level quantum dot placed on a nanomechanical resonator. The control of the spontaneous emission is achieved by placing the system within an optical cavity and by modulating the transition frequency of the quantum dot.

respectively b and $b^{\dagger}$. The optical cavity of frequency ω_c is described by the phononic annihilation and creation operators, respectively a and $a^{\dagger}$. Bosonic operators obey the commutation rule $[a, a^{\dagger}] = [b, b^{\dagger}] = 1$.

The system Hamiltonian is given as:

$$H = \hbar(\omega_{qd} + s(t))S_z + \hbar\omega_{ph}b^{\dagger}b + \hbar\lambda S^{+}S^{-}\left(b^{\dagger} + b\right) + \hbar\omega_c a^{\dagger}a + \hbar ig\left(S^{-}a^{\dagger} + aS^{+}\right), \quad (1)$$

where the first term represents the free quantum-dot term with its transition frequency modulated by a periodic signal $s(t)$. The second term represents a single-mode phonon field of the free mechanical resonator. The third term represents the interaction of the quantum-dot with the vibrational quanta of the mechanical resonator. This interaction is defined by the coupling constant λ. The fourth term represents the free cavity electromagnetic field term. The last term represents the interaction of the cavity with the quantum-dot, defined by the coupling constant g.

Here, one neglects the spontaneous emission effect which occurs due to the interaction of the quantum-dot with the environmental electromagnetic vacuum. This condition is valid as long as the interaction of the optical cavity with the quantum-dot is strong if comparing to the coupling with the modes of the vacuum.

The system quantum dynamics is defined within the density matrix formalism, where the master equation of the density matrix operator $\rho = |\Psi\rangle\langle\Psi|$ is given as:

$$\frac{\partial\rho}{\partial t} = -\frac{i}{\hbar}[H, \rho] + \kappa(1 + \overline{n})\mathcal{L}(b) + \kappa\overline{n}\mathcal{L}\left(b^{\dagger}\right) + \kappa_c\mathcal{L}(a), \quad (2)$$

where $\mathcal{L}$ representes the Liouville super-operator wich acts on a given system operator Q as follows: $\mathcal{L}(Q) = 2Q\rho Q^{\dagger} - Q^{\dagger}Q\rho - \rho Q^{\dagger}Q$ the first term represents the von-Neumann equation of the system Hamiltonian. The second and the third term represent the damping of the quantum mechanical resonator by the thermal environment, described by the damping and pumping terms, respectively. The phonon damping effect is defined by the damping rate κ and the mean phonon occupation number of the thermal reservoir $\bar{n}$ which is function of environmental temperature. The last terms represent the damping of the cavity photons via the electromagnetic vacuum, defined by the damping rate κ_c.

In its current form of the master equation and system Hamiltonian, the system dynamics cannot be directly solved and further assumptions are required. The main assumption consist in considering an optical cavity with a low quality factor, i.e., $\kappa_c \gg g$. Under this condition, the complexity of the system dynamics may be reduced by eliminating the cavity field operators. This is achieved by expressing the system Hamiltonian and the master equation within the interaction picture, where the equations of motions of the photonic operators are deduced. These equation of motion are solved within the bad cavity limit and their solution are inserted in the expression of the master equation. This operation eliminates the photonic operators from the master equation and reduces the complexity of the system dynamics so that the master equation may be numerically solved [5].

The particularity of the current investigation consists in adapting the previously described theoretic approach for any periodic modulation signal. The system dynamics are solved for a modulation signal $s(t)$ decomposed in a Fourier series, namely:

$$s(t) = \sum_{0}^{\infty} b_j \cos\left(j\omega t + \phi_j\right). \tag{3}$$

Here, for the sake of simplicity one have considered $\phi_j = 0$ and $b_0 = 0$. For numerical calculations, the infinite series of Eq. (3) is truncated and only the first j_{max} are considered. The numerical limitations of the values j_{max} are defined by the complexity of the differential equations of motion of the parameters of interest. For j_{max} terms of the Fourier decomposition, there are j_{max}^2 terms in the differential equation of motion. Therefore, the modulation signals $s(t)$ investigated in following section are chosen according to the numerical limitation of the Fourier decomposition.

Once the previously mentioned assumptions are applied to the system master equation of Eq. (3) and the system Hamiltonian of Eq. (2), the quantum dynamics of the system are solved by building the equation of motion of the parameters of interest by applying to the master equation the property of the density operator ρ:

$$\frac{\partial \langle Q \rangle}{\partial t} = Tr\left[\rho \frac{\partial Q}{\partial t}\right] = Tr\left[Q \frac{\partial \rho}{\partial t}\right] \tag{4}$$

The system dynamics is further investigated by solving the equation of motion of the population of the quantum-dot $\langle S_z \rangle$. The differential equation of $\langle S_z \rangle$ is not affected by the presence of mechanical vibrations or the quantum-dot-phonon coupling to the thermal environment. Therefore, we will focus the following discussion on the influence of the modulation signal $s(t)$ on the spontaneous emission dynamics of the quantum-dot.

3 Results and Discussions

In this section, the impact of various frequency modulation signals on the quantum dynamics of the quantum dot population $\langle S_z \rangle$ is investigated. In what follows, the investigated model is defined by the following parameters: cavity damping rate $\kappa_c/\omega = 2$, optical coupling constant $g/\omega = 0.2$, cavity-quantum-dot detuning $\delta_c = 0$.

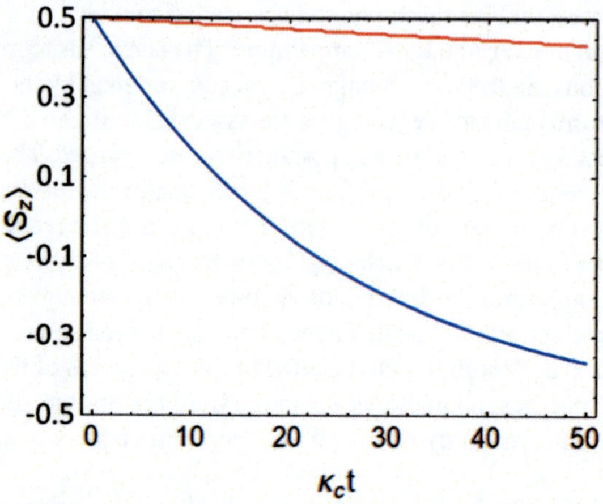

Fig. 2. The population of the quantum-dot as function of time for non-modulated transition frequency s(t) = 0 (blue bottom line) and for a harmonic modulation of the transition frequency s(t) = Acos(ωt) (red top line).

In Fig. 2, the population of the quantum-dot $\langle S_z \rangle$ is represented as a function of time $\kappa_c t$. The case of a quantum-dot with a non-modulated transition frequency, i.e., $s(t) = 0$, is given by the blue bottom line. The case of a quantum-dot with the transition frequency modulated by a sinusoidal signal $s(t) = Acos(\omega t)$, $A/\omega = 60$ is given by the red top line. This case represents the most trivial type of signal used to represent the frequency modulation achieved by AC Stark effect which appears when applying an off-resonant laser on the quantum-dot. The decay of the quantum-dot excited state population is clearly slowed-down when the transition frequency of the quantum-dot is modulated.

In Fig. 3, one compares the previous case of a pure sinusoidal modulation signal with the case of a signal which possess some higher frequency fluctuations (see the inset). These type of fluctuations may appear as a noise-like effect due to the imperfections of the setup. Another source of such fluctuations may appear due to the dephasing effect caused by the imperfection of the quantum-dot energetic levels. The dephasing of the quantum-dot is often represented as a white-noise term added to the transition frequency [10] and is often the source of major decoherences in various quantum setups. With the inclusion of higher frequency terms, an enhancement in the lifetime of the excited state population is predicted.

The current investigation treats only a few terms of the decomposition in Fourier series, due to numerical limitations in the estimation of the quantum dynamics. In Fig. 3, one represents a signal $Acos(\omega t) + A/4cos(7\omega t)$ defined by only one additional high frequency term. However, the addition of more high frequency terms will not change the overall behaviour of the system dynamics.

Another possible deviation from the pure sinusoid signal is represented in Fig. 4. Here, the case of a sinusoid signal $s(t) = A(t)cos(6\omega t)$ with slowly varying amplitude $A(t) = 0.7A + 0.3Acos(\omega t)$ is considered. This case of modulation signal may be

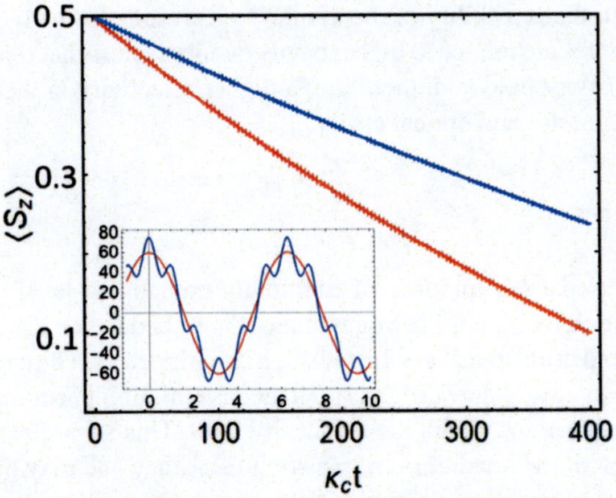

Fig. 3. The population of the quantum-dot as function of time for a pure sinusoidal modulation signal s(t) = Acos(ωt) (red bottom line) and a signal possessing fluctuations s(t) = Acos(ωt) + Bcos(7ωt), B = A/4 (blue top line). The modulation signals s(t) are represented in the inset.

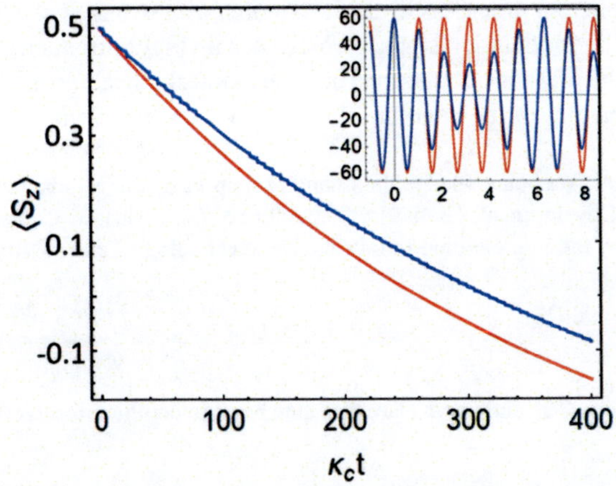

Fig. 4. The population of the quantum-dot as function of time for a pure sinusoidal modulation signal s(t) = Acos(6ωt) (red bottom line) and a sinusoidal signal with slowly varying amplitude s(t) = A(t)cos(6ωt) (blue top line). The modulation signals s(t) are represented in the inset.

achieved by varying the intensity of the applied off-resonance laser. Here, the spontaneous emission effect may be either enhanced or slowed-down depending on the quality factor of the considered optical cavity. In Fig. 4, a clear slow-down of the spontaneous emission for a damping rate $\kappa_c/\omega = 2$ is observed. However, an increase of the cavity damping rate will diminish this effect. For higher enough damping rates, signals with slowly varying amplitudes would lead to the enhancement of the spontaneous emission when compared to a pure sinusoid modulation, e.g., within the current model this condition is achieved for $\kappa_c/\omega > 10$. This difference in the behaviour of the spontaneous emission refers to the sensibility of the system to its resonance with the optical cavity. Lower damping rates are related to higher cavity quality factors that represent a narrower spectrum of the cavity field and, therefore, a higher selectivity in the resonance of the modulated quantum-dot and optical cavity.

4 Conclusions

We have investigated a system made of an initially excited two-level quantum-dot with modulated transition frequency. The quantum dot is embedded on a nanomecanical resonator and is placed in an optical cavity with high damping rates. The cavity-quantum-dot interaction leads to a slow-down of the spontaneous emission phenomenon, when solving the quantum dynamics within the bad cavity limit. This slow-down effect is caused by the modulation of the quantum-dot transition frequency and may be enhanced when considering various modulation signals. Namely, we have demonstrated that sinusoid signals with slowly varying amplitude or sinusoid signals with higher frequency noise-like fluctuations may lead to an enhanced lifetime of the excited quantum-dot population and, therefore, to a decreased effective spontaneous emission rate.

Beyond slowed-down spontaneous emission, this investigation suggests a favorable reduction of decoherent processes in artificial atoms with strong dephasing effects which may be imitated by modulation signals which contain higher frequency terms. We have also shown that the presence of phonons of the nanomechanical resonator does not affect the quantum dynamics of the quantum-dot spontaneous emission.

Acknowledgement. We acknowledge the financial support by the Moldavian National Agency for Research and Development, Grant No. 20.80009.5007.07. Also, V. Ceban is grateful to the financial support by the International Innovative Nanotechnology Centre ININC-CIS, Grant No. 21-116.

Conflict of Interest. The authors declare that they have no conflict of interest.

References

1. Knill, E., Laflamme, R., Milburn, G.J.: A scheme for efficient quantum computation with linear optics. Nature **409**, 46–52 (2000)
2. Aspelmeyer, M., Kippenberg, T.J., Marquardt, F.: Cavity optomechanics. Rev. Mod. Phys. **86**, 1391–1452 (2014)
3. Purcell, E.M.: Spontaneous emission probabilities at radio frequencies. Phys. Rev. **69**, 681 (1946)
4. Linington, I.E., Garraway, B.M.: Dissipation control in cavity QED with oscillating mode structures. Phys. Rev. A **77**, 033831 (2008)
5. Macovei, M., Keitel, C.H.: Quantum dynamics of a two-level emitter with a modulated transition frequency. Phys. Rev. A **90**, 043838 (2014)
6. Janowicz, M.: Non-Markovian decay of an atom coupled to a reservoir: modification by frequency modulation. Phys. Rev. A **61**, 025802 (2000)

7. Agarwal, G.S.: Control of decoherence and relaxation by frequency modulation of a heat bath. Phys. Rev. A **61**, 013809 (1999)
8. Kofman, A.G., Kurizi, G.: Universal dynamical control of quantum mechanical decay: modulation of the coupling to the continuum. Phys. Rev. Lett. **87**, 270405 (2001)
9. Macovei, M., Evers, J., Keitel, C.H.: Spontaneous decay processes in a classical strong low-frequency laser field. Phys. Rev. A **102**, 013718 (2020)
10. Puri, R.R.: Mathematical Methods of Quantum Optics. Springer, Heidelberg (2001). https://doi.org/10.1007/978-3-540-44953-9

Dynamics of Atomic-Molecular Conversion of Alkali Metal Isotopes at Ultralow Temperatures

A. P. Zingan(✉) and O. F. Vasilieva

Dniester State University, Tiraspol, Moldova

Abstract. We study the dynamics of atomic-molecular conversion in ultracold gases of lithium isotopes. It is shown that the time evolution of atoms and molecules in the process of stimulated conversion is largely determined by the initial density of particles and the initial phase difference.

Keywords: Atomic-molecular conversion · Lithium isotopes

In recent years, the study of Bose condensates has become a very promising area of study of the properties of ultracold molecular gases. Currently, an important direction is the production of complex, polyatomic molecules at ultralow temperatures [1–12]. The existence of triatomic Efimov resonant molecules was first observed in ultracold gases in 2006 [13]. Trimeric molecules have been observed experimentally in three-component Fermi gases ^6Li [14, 15], in a Bose gas of atoms ^{39}K [16]. This not only confirms the existence of weakly bound trimeric states, but also opens up new avenues for the study of multiparticle quantum systems. Tetrameric states have recently been realized in an ultracold gas of cesium atoms. In the mean-field approximation, the properties of homonuclear and heteronuclear tetramers were obtained [11, 17, 18]. In [12], the formation of stable homo- and heteronuclear pentamers from ultracold atoms in the process of generalized stimulated Raman adiabatic passage was studied theoretically.

In [19], the generation of a quantum degenerate Fermi-Fermi mixture of two atoms of different types was reported. A quantum-degenerate mixture was realized by cooling fermion gases Li^6 and K^{40} with an evaporatively cooled bosonic gas Rb^{87}. A combination of trapping and cooling methods has been described which have proven to be critical to the successful cooling of the mixture. As for the practical application of Bose condensates, it was shown in [2] how strongly interacting ultracold bosonic gases in periodic potentials can be used as conductors in a circuit and how they can be used to construct atomic analogs of diodes and transistors with a bipolar junction. Hence, the implementation of an atomic amplifier immediately follows. The purpose of this communication is to study the dynamics of atomic-molecular conversion in ultracold gases of lithium isotopes.

Consider a system of condensed particles consisting, for example, of lithium isotopes and their molecules. The formation of homo- and heterodimers of lithium isotopes can be represented as the reaction $^6Li + {}^6Li^7Li \leftrightarrow {}^7Li + {}^6Li_2$. It was shown in [20] that isotope exchange reactions of this type $^nA^mA + {}^nA^mA \rightarrow {}^nA_2 + {}^mA_2$ between the ground states of alkali-halide heteronuclear dimers, consisting of two isotopes of the same

I. Tiginyanu et al. (Eds.): ICNBME 2021, IFMBE Proceedings 87, pp. 306–313, 2022.
https://doi.org/10.1007/978-3-030-92328-0_41

atom, are exothermic with a change in energy in the range of 1–8000 MHz. Therefore, heteronuclear dimers are chemically unstable at ultralow temperatures. For example, two heteromolecules $^6Li^7Li$ decay into two dimers $^6Li_2$ and $^7Li_2$ with an energy release of about 8000 MHz. Let an atom ^6Li and a molecule $^6Li^7Li$ be transformed into a molecule $^6Li_2$ and an atom ^7Li under the action of two pulses of resonant laser radiation with energies $\hbar ck_1$ and $\hbar ck_2$. The process under consideration is optical Raman nutation under conditions of atomic-molecular conversion under the action of two coherent Raman pulses of resonant laser radiation with periodic amplification of one of the pulses and attenuation of the other. The relevance of the topic is based on the fact that one of the important theoretical problems of the physics of condensed matter is the construction of a satisfactory theory describing the process of stimulated Raman atomic-molecular conversion in condensate, taking into account the most significant factors that determine the dynamics of transformations.

Then the interaction Hamiltonian in this case can be represented in the form.

$$H_{int} = -\hbar g\left(a_1 b_1 c_1 a_2^+ b_2^+ c_2^+ + a_1^+ b_1^+ c_1^+ a_2 b_2 c_2\right), \tag{1}$$

where a_1, a_2, b_1 and b_2 are the bosonic annihilation operators of atoms of lithium isotopes $^6Li_2$, $^7Li_2$ and dimers of lithium isotopes $^6Li^7Li$ and $^6Li_2$ with eigenfrequencies ω_1, ω_2, Ω_1 and Ω_2, respectively, c_1 and c_2 are photon annihilation operators, g is the interaction constant. From Hamiltonian (1), a system of Heisenberg equations for the operators of reaction particles was obtained, averaging it and using the mean field approximation [11, 17, 18], one can obtain a system of nonlinear equations for the amplitudes (order parameters) of the material and electromagnetic fields

$$\begin{cases} i\dot{a}_1 = \omega_1 a_1 - g b_1^+ c_1^+ a_2 b_2 c_2 \\ i\dot{b}_1 = \Omega_1 b_1 - g a_1^+ c_1^+ a_2 b_2 c_2 \\ i\dot{c}_1 = ck_1 c_1 - g a_1^+ b_1^+ a_2 b_2 c_2 \\ i\dot{a}_2 = \omega_2 a_2 - g a_1 b_1 c_1 b_2^+ c_2^+ \\ i\dot{b}_2 = \Omega_2 b_2 - g a_1 b_1 c_1 a_2^+ c_2^+ \\ i\dot{c}_2 = ck_2 c_2 - g a_1 b_1 c_1 a_2^+ b_2^+ \end{cases}, \tag{2}$$

We further introduce into consideration the particle density.

$$n_1 = a_1^+ a_1, \tag{3}$$

$$n_2 = a_2^+ a_2, \tag{4}$$

$$N_1 = b_1^+ b_1, \tag{5}$$

$$N_2 = b_2^+ b_2, \tag{6}$$

$$f_1 = c_1^+ c_1, \tag{7}$$

$$f_2 = c_2^+ c_2 \tag{8}$$

and two components of "polarization"

$$Q = i(a_1 b_1 c_1 a_2^+ b_2^+ c_2^+ - a_1^+ b_1^+ c_1^+ a_2 b_2 c_2) \tag{9}$$

and

$$R = a_1 b_1 c_1 a_2^+ b_2^+ c_2^+ + a_1^+ b_1^+ c_1^+ a_2 b_2 c_2. \tag{10}$$

Then it is possible to obtain a closed system of differential equations for the particle densities

$$\dot{n}_1 = -gQ, \tag{11}$$

$$\dot{n}_2 = gQ, \tag{12}$$

$$\dot{N}_1 = -gQ, \tag{13}$$

$$\dot{N}_2 = gQ, \tag{14}$$

$$\dot{f}_1 = -gQ, \tag{15}$$

$$\dot{f}_2 = gQ, \tag{16}$$

$$\dot{R} = -\Delta Q, \tag{17}$$

$$\dot{Q} = \Delta R - 2g\{(N_1 f_1 + n_1 f_1 + N_1 n_1) N_2 n_2 f_2 - (N_2 f_2 + n_2 f_2 + N_2 n_2) N_1 n_1 f_1\} \tag{18}$$

where $\Delta = \omega_1 + \Omega_1 + ck_1 - (\omega_2 + \Omega_2 + ck_2)$ – resonance detuning. Supplementing the system with the initial conditions $n_{1,2}\big|_{t=0} = n_{10,20}$, $N_{1,2}\big|_{t=0} = N_{10,20}$, $f_{1,2}\big|_{t=0} = f_{10,20}$, $Q|_{t=0} = Q_0 = 2\sqrt{n_{10} n_{20} N_{10} N_{20} f_{10} f_{20}} sin\theta_0$, $R|_{t=0} = R_0 = 2\sqrt{n_{10} n_{20} N_{10} N_{20} f_{10} f_{20}} cos\theta_0$, where θ_0 – initial phase difference, from (3) it is possible to obtain the integrals of motion.

$$N_1 + N_2 = N_{10} + N_{20}, \tag{19}$$

$$N_1 + n_2 = N_{10} + n_{20}, \tag{20}$$

$$N_1 + f_2 = N_{10} + f_{20}, \tag{21}$$

$$n_1 - N_1 = n_{10} - N_{10}, \tag{22}$$

$$f_1 - N_1 = f_{10} - N_{10}, \tag{23}$$

$$R = \frac{\Delta}{g}(N_1 - N_{10}) + 2\sqrt{n_{10}n_{20}N_{10}N_{20}f_{10}f_{20}} cos\theta_0, \tag{24}$$

$$\begin{aligned} Q^2 &= 4N_1(n_{10} - N_{10} + N_1)(f_{10} - N_{10} + N_1) \\ &\quad \times (N_{10} + N_{20} - N_1)(N_{10} + n_{20} - N_1)(N_{10} + f_{20} - N_1) \\ &\quad - \left(\frac{\Delta}{g}(N_1 - N_{10}) + 2\sqrt{n_{10}n_{20}N_{10}N_{20}f_{10}f_{20}}\, cos\,\theta_0\right)^2 \end{aligned} \tag{25}$$

and reduce the problem to a single nonlinear differential equation describing the time evolution $N_1(t)$ of the density of dimers of lithium isotopes $^6Li^7Li$. Under conditions of exact resonance, at $\Delta = 0$, the equation of dynamics of atomic-molecular conversion can be represented as $\dot{N}_1^2 + W(N_1) = E_0$, where $\dot{N}_1^2$, $W(N_1)$ and E_0 play the role of kinetic, potential and the total energies of the nonlinear oscillator, respectively, whose oscillations occur in the range of values N_1, in which $W(N_1) \leq E_0$. Here

$$\begin{aligned} W(N_1) &= -4N_1(n_{10} - N_{10} + N_1)(f_{10} - N_{10} + N_1) \\ &\quad \cdot (N_{10} + N_{20} - N_1)(N_{10} + n_{20} - N_1)(N_{10} + f_{20} - N_1) \\ E_0 &= -4n_{10}n_{20}N_{10}N_{20}f_{10}f_{20}\, cos^2\theta_0. \end{aligned} \tag{26}$$

Considering the dependence of the potential energy of the nonlinear oscillator $W(N_1)$, one can judge the qualitative nature of the dynamics of $N_1(t)$. In Fig. 1 and Fig. 2 shows that the evolution of the system at nonzero initial densities of all particles is periodic and consists in a cyclic change in the density of dimers of atoms bonding from isotopes. Even with the pairwise equality of the densities $n_{10} = f_{10}$ and $n_{20} = f_{20}$, the dynamics remains periodic (Fig. 1, (4)). However, the features of the evolution of the system are determined not only by the values of the initial particle densities, but also by the relationships between them. For example, if the initial particle densities are equal $N_{10} = n_{10} = f_{10}$ and $N_{20} = n_{20} = f_{20}$, the periodic evolution mode turns into aperiodic (Fig. 1, (5)). Analytical solutions in the general case cannot be obtained in the known algebraic functions, therefore we use approximations of the given initial densities of the particles participating in the reaction. In this case, both periodic transformations of atoms into molecules and irreversible processes of disintegration and binding of molecules are possible (Fig. 2).

As for the practical application of condensates, it was shown in [20] how strongly interacting ultracold bosonic gases in periodic potentials can be used as conductors in a circuit and how they can be used to construct atomic analogs of diodes and transistors with a bipolar junction. Hence, the implementation of an atomic amplifier follows - a device that allows you to control a large atomic current. The transistor presented in [20] serves this purpose directly, since small changes in base current lead to large changes in collector current. From here, it is easy to imagine more complex bistable devices that use cross-negative feedback between two transistors. First, in the case of an atom, the band gap is the result of interactions, not statistics, as in electronics. Secondly, atomic currents are superfluid. As a consequence, the relationship between voltage and current has the value of non-dissipative resistance. Further differences arise in both diodes and transistors. In this case, the atomic diode does not have a depletion layer, that is, it does not have an energy barrier depending on the voltage across the junction. Asymmetry in the current-voltage curve results from voltage-sign dependent transitions into the insulating phase. As a consequence, atoms move from P to N in forward bias, rather than, as in electronics, from N to P. Due to differences in the behavior of the diode, the atomic collector current in the transistor flows from the collector to the base, and the emitter current flows from the base to emitter, that is, opposite to the electron flow in the NPN transistor. A significant difference in the qualitative behavior of electronic and atomic transistors is due to the negative gain in the atomic case. The collector current increases with decreasing base current. It is expected that this will not affect the functionality of devices based on the operation of bipolar junction transistors. The data presented in [20] were obtained from calculations for a one-dimensional lattice. This choice is purely practical. The basic ideas are also valid for two-dimensional and three-dimensional cubic lattices and are extended to other lattice geometries that undergo transitions between the superfluid and insulating phases when the chemical potential changes. Thus, the model provides an excellent description of current experiments with ultracold bosonic atoms in optical lattices. Other Hamiltonians may offer alternative ways of drawing the analogy with electronics. Hamiltonians describing bosons locally and Hamiltonians for fermionic gases are natural choices for further study.

Thus, it can be concluded that the dynamics of atomic-molecular conversion of isotopes, for example, lithium, and their molecules in the general case, without using approximations of given initial particle densities, is periodic and consists in the cyclic decay and binding of atoms into dimers.

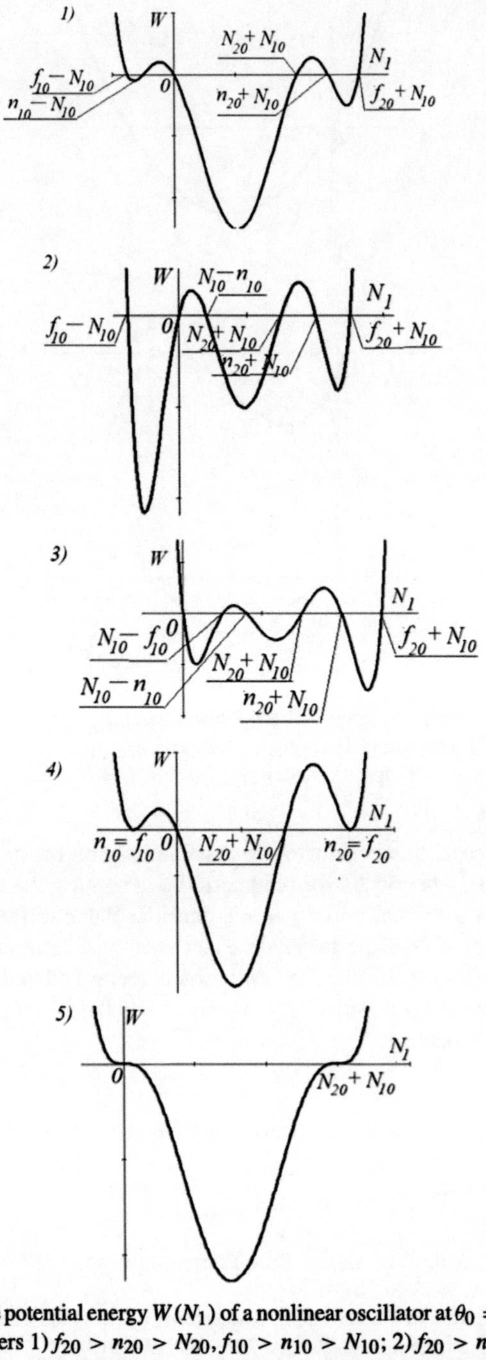

Fig. 1. Graphs of the potential energy $W(N_1)$ of a nonlinear oscillator at $\theta_0 = \frac{\pi}{2}$ and various ratios between the parameters 1) $f_{20} > n_{20} > N_{20}, f_{10} > n_{10} > N_{10}$; 2) $f_{20} > n_{20} > N_{20}, f_{10} > N_{10}, N_{10} > n_{10}$; 3) $f_{20} > n_{20} > N_{20}, N_{10} > f_{10} > n_{10}$; 4) $f_{10} = n_{10} > N_{10}, f_{20} = n_{20} > N_{20}$; 5) $f_{20} = n_{20} = N_{20}, N_{10} = f_{10} = n_{10}$.

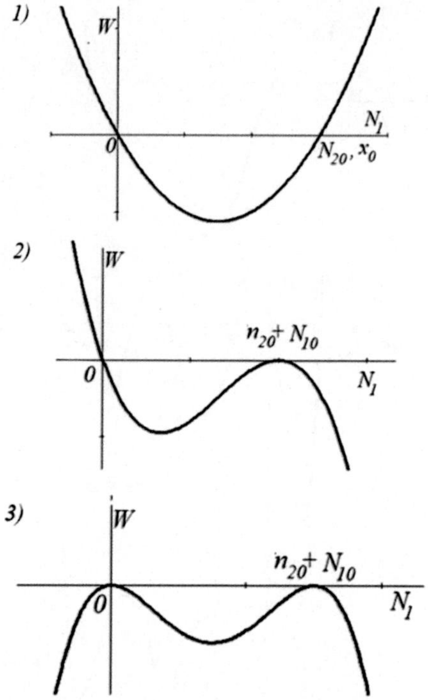

Fig. 2. Graphs of the potential energy $W(N_1)$ of a nonlinear oscillator at $\theta_0 = \frac{\pi}{2}$ and various ratios between the parameters 1) approximation $n_{10}, n_{20}, f_{10}, f_{20} \gg N_{20}$; 2) approximation $n_{10}, f_{10}, f_{20} \gg n_{20} = N_{20}$; 3) approximation $f_{10}, f_{20} \gg n_{10} = N_{10}, n_{20} = N_{20}$.

In the general case, this conclusion is possible only on the basis of a study of the behavior of potential energy. As for the aperiodic dynamics, then on is possible only in the case of pairwise equal initial particle densities using different approximations, and in these cases it is possible to obtain exact analytical solutions. From the results presented above, it follows that the time evolution of atoms and molecules in the process of stimulated conversion is substantially determined by the initial particle densities and the initial phase difference.

Conflict of Interest. The authors declare that they have no conflict of interest.

References

1. Burchianti, A.C., et al.: Dual-species Bose-Einstein condensate of ^{41}K and ^{87}Rb in a hybrid trap. Phys. Rev. A. **98**, 063616 (2018)
2. Wang, K.: Preparation of a heteronuclear two-atom system in the three-dimensional ground state in an optical tweezer. Phys. Rev. A. **100**, 063429 (2019)
3. Voges, K.: Formation of ultracold weakly bound dimers of bosonic ^{23}Na^{39}K. Phys. Rev. A **101**, 042704 (2020)

4. Hood, J.D.: Multichannel interactions of two atoms in an optical tweezer. Phys. Rev. Res. **2**, 023108 (2020)
5. Gregory, P.: Loss of ultracold $^{87}Rb^{133}Cs$ molecules via optical excitation of long-lived two-body collision complexes. Phys. Rev. Lett. **124**, 163402 (2020)
6. Wang, X.: Observation of state-to-state hyperfine-changing collisions in a Bose-Fermi mixture of ^{6}Li and ^{41}K atoms. Phys. Rev. A. **101**, 041601 (2020)
7. Wang, F.: Observation of resonant scattering between ultracold heteronuclear Feshbach molecules. Phys. Rev. A. **100**, 042706 (2019)
8. Liu, L.: Observation of interference between resonant and detuned stirap in the adiabatic creation of $^{23}Na^{40}K$ molecules. Phys. Rev. Lett. **122**, 253201 (2019)
9. Guijarro, G.: Few-body bound states of two-dimensional bosons. Phys. Rev. A **101**, 041602 (2020)
10. Dey, A.: Interaction-induced instability and chaos in the photoassociative stimulated Raman adiabatic passage from atomic to molecular Bose-Einstein condensates. Phys. Rev. A **101**, 053627 (2020)
11. Meng, S.-Y., Chen, X.-H., Ning, S.-N., Wen, J.-M., Fu, L.-B.: Instability, adiabaticity and controlling effects of external fields for the dark state in a heteronuclear atom–tetramer conversion system. J. Phys. B **47**, 185303 (2014)
12. Dou, F.-Q., Li, S.-C., Cao, H., Fu, L.-B.: Creating pentamer molecules by generalized stimulated Raman adiabatic passage. Phys. Rev. A **85**, 023629 (2012)
13. Kraemer, T., et al.: Evidence for Efimov quantum states in an ultracold gas of caesium atoms. Nature **440**, 315–318 (2006)
14. Ottenstein, T.B., Lompe, T., Kohnen, M., Wenz, A.N., Jochim, S.: Collisional stability of a three-component degenerate fermi gas. Phys. Rev. Lett. **101**, 203202 (2008)
15. Hucknaus, J.H., Williams, J.R., Hazlett, E.L., Stites, R.W., O'Hara, K.M.: Three-body recombination in a three-state fermi gas with widely tunable interactions. Phys. Rev. Lett. **102**, 165302 (2009)
16. Zaccanti, M., et al.: Observation of an Efimov spectrum in an atomic system. Nat. Phys. **5**, 586–591 (2009)
17. Jing, H., Jiang, Y.: Surface breathers in discrete magnetic metamaterials. Phys. Rev. E **77**, 065601 (2008)
18. Li, G.Q., Peng, P.: Formation of a heteronuclear tetramer A3B via Efimov-resonance-assisted stimulated Raman adiabatic passage. Phys. Rev. A **83**, 043605 (2011)
19. Taglieber, M.: Quantum degenerate two-species fermi-fermi mixture coexisting with a Bose-Einstein condensate. Phys. Rev. Lett. **100**, 010401 (2008)
20. Seaman, B.T.: Atomtronics: ultracold atom analogs of electronic devices. Phys. Rev. A **75**, 023615 (2007)

Photoinduced Anisotropy in Azopolymer Studied by Spectroscopic and Polarimetric Parameters

C. Losmanschii(✉), E. Achimova, V. Abashkin, V. Botnari, and A. Meshalkin

Institute of Applied Physics, Chisinau, Moldova
constantin.losmanschii@ifa.md

Abstract. We discuss methods for synthesis and fabrication of photosensitive thin azopolymer films. Spectroscopy of films presents information as about nonactinic and recording wavelengths as a for refraction index calculations by Swanepoil method. Photoinducing of anisotropy was done by blue laser. The resulting polarimetric light parameters produced by photoinduced anisotropy measured at nonactinic He-Ne red wavelength. For this aim a polarimeter was applied. Kinetics of a value of photoinduced anisotropy, azimuth and ellipticity of light passed through films were investigated.

Keywords: Azopolymer · Photoinduced birefringence · Spectroscopy · Polarimetry

1 Introduction

The interaction of laser radiation with matter is of great practical importance and is used effectively in various fields, including laser processing and material structuring, optical capture and microparticle manipulation, developing optical recording and storage media [1–3].

In laser processing and structuring, in addition to the beam intensity distribution, the polarization state of the laser beam plays an important role.

The laser radiation force components, both gradient and non-gradient nature, were considered in various works. These optical force components have a significant effect on the formed microstructures under the laser action. Models describing light-induced processes based on gradient forces are applicable to amorphous compounds, while azopolymer materials are more susceptible to the non-gradient forces due to the anisotropy of the molecules of the azo-containing polymer. Undergoing cyclic photoinduced isomerization from *trans* to *cis* form, the polymer matrix is copolymerized with azodye. It acquires molecules mobility and the ability to form anisotropy structure under the action of optical forces.

The azopolymers undergo *trans-cis-trans* isomerization cycles, while they are irradiated by linearly polarized light and are accompanied by molecular orientation. After several of these cycles, an excess of chromophores is aligned perpendicularly to the

I. Tiginyanu et al. (Eds.): ICNBME 2021, IFMBE Proceedings 87, pp. 314–321, 2022.
https://doi.org/10.1007/978-3-030-92328-0_42

laser polarization, resulting in a macroscopic photoinduced birefringence in the sample. When the excitation laser is turned off, some relaxation of the orientation occurs in the sample [4]. Nevertheless, some percentages of molecules remain oriented. Such optically induced birefringence can be observed by measuring the transmittance of a probe beam that passes through the sample.

The aim of this work is to study photoinduced birefringence in azopolymer by measuring alteration of polarimetric parameters, such as ellipticity and azimuth of probe polarized laser beam that is passing through azopolymer thin film illuminated from its active optical range.

2 Experimental Procedure

2.1 Obtaining Azopolymer Thin Films

Following reagents were used for the synthesis without additional purification: toluene (solvent), Solvent Yellow 3 (SY3, chromophore) both manufactured by Sigma Aldrich. Azopolymers were obtained by refluxing 1 g of poly-N-(2,3-epoxypropyl) carbazole (PEPC) and SY3, with 0.3 g, 0.4 g, and 0.5 g weights, in 8 ml of toluene for 8 h. The resulting solution was cooled and filtered. Thin azopolymer films were obtained by spin-coating at 800 rpm for 25 s, followed by drying at room temperature for 24 h. The chemical route used for this synthesis is shown in Fig. 1.

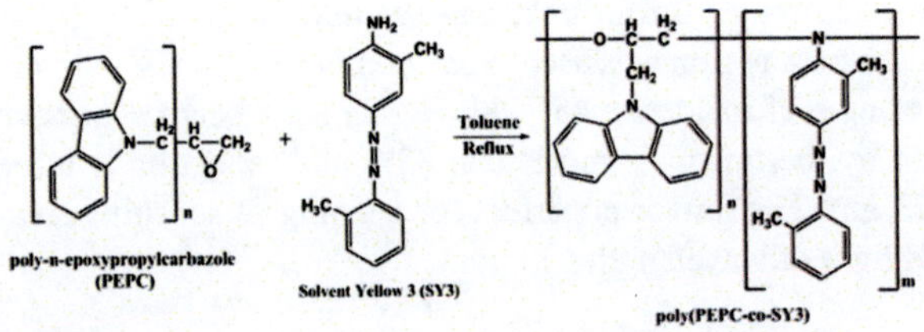

Fig. 1. Chemical route of PEPC-co-SY3 azopolymer synthesis

2.2 Polarimetric Measurement

Optical setup of photoinduced anisotropy measurements in polymer thin films is pictured in Fig. 2. Two lasers were used in our setup. Non actinic multimode unpolarized probe He-Ne laser beam at wavelength 632 nm was polarized by Glan-Thompson calcite polarizer (extinction ratio - 100 000:1). The continuously variable zero-order retarder of Soleil-Babinet compensator was applied as wave plate. It was adjusted in position as λ/2 waveplate with orientation as pointed on Fig. 3, (orientation 45° ± 0,05°, ellipticity ±0.05°). It should be noted that compensator maintains a uniform retardance across the beam aperture at given setting. This beam passes normally through the sample surface in direction to the polarimeter. Second DPSS single mode laser beam at wavelength 473 nm was used for inducing anisotropy in azopolymer films. A wave plate λ/2 for wavelength 473 nm was mounted inside stepper motor rotative mount K10CR1, resulting in polarization state rotation step by step. This laser beam was expanded on sample surface illuminated by the probe beam mounting the optical expander BE5x to achieve quasi uniform intensities distribution.

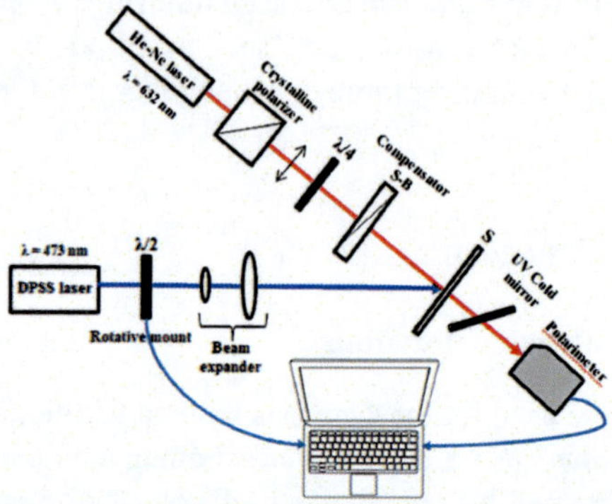

Fig. 2. Optical set-up for anizotropic control: He-Ne laser (power-1 mW multimode), wave plate λ/4, Soleil-Babinet Compensators, UV cold mirror FM04 Thorlabs, polarimeter PAX1000VIS Thorlabs, DPSS laser (473 nm, power – 75 mW), wave plate λ/2 inside rotative mount K10CR1/M Thorlabs, beam expander BE5x Thorlabs.

To prevent polarimeter from direct and scattered light of DPSS laser a dichroic mirror with a high pass region was used. Such component as UV cold mirror FM04 was mounted at the angle of incidence 45° with He-Ne laser beam. Optical transmission of this wavelength by the mirror is more than 85% and reflection is more than 90% for wavelength 473 nm. The mirror is perfect for keeping of sensitive area of polarimeter from unwanted light of wavelength 473 nm.

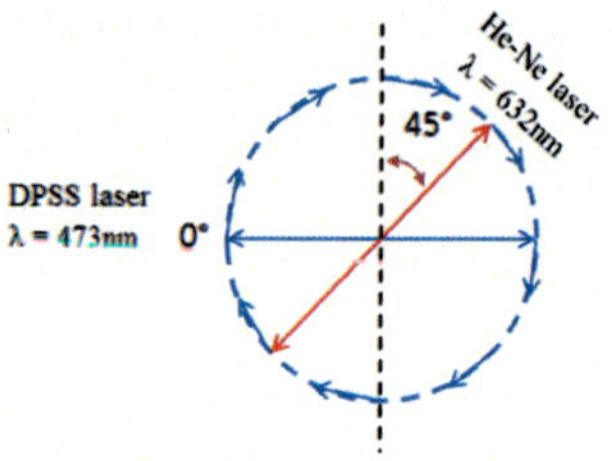

Fig. 3. Polarization states: initial polarization state of blue laser 0° to final polarization state 180°, He-Ne laser constant polarization state 135°.

3 Results and Discussions

3.1 Spectroscopic Investigation of Azopolymer Films

To determine the refractive index of the obtained films, the transmission spectra were recorded, before and after irradiation, in the range of 333–800 nm. The transmission spectra and spectral dependences of the refractive index are shown in Fig. 4. From the transmission spectra, it can be noted that the absorption edge has changed after irradiation. The shift in the transmission spectra of azopolymer films after irradiation indicates the presence of the cis-isomers of SY3 due to the photoisomerization process.

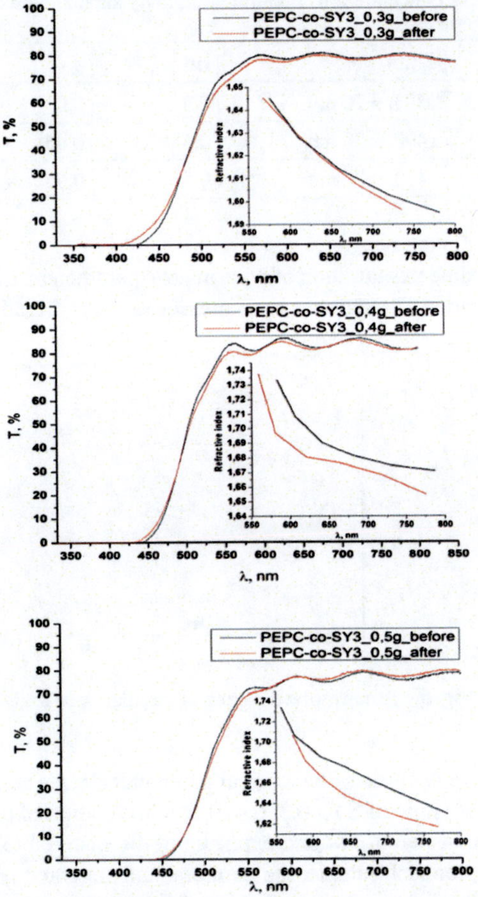

Fig. 4. Transmission spectra and spectral dependences of the refractive index for films with different concentrations of the azo dye SY3.

From the curves of the spectral dependences of the refractive index, it is noticeable that $\Delta n = n_{\text{before ill.}} - n_{\text{after ill.}}$ increases with increasing mass ratio of SY3 from 0.3 g to 0.5 g. Increase in the concentration of the azo dye in the azo polymer leads to rise in the probability of the interaction of the incident light with the azo dye molecules, which manifests to a more pronounced change in the refractive index [5].

3.2 Polarimetric Investigations of Dependence of Exposure Angle and Chromophore Concentration

To study the photoinduced changes in azimuth(ψ) and ellipticity(χ), films with the parameters presented in Table 1 were used.

Table 1. Thickness, concentration and initial ellipticity and azimuth of azopolymer films

Sample name	d, nm	rm	χ	ψ
PEPC_SY03	1393, ±20 nm	0,3	0,01°	44,99°
PEPC_SY04	1490, ±20 nm	0,4	0,05°	45,05°
PEPC_SY05	1470, ±20 nm	0,5	0,05°	44,95°

To estimate the time of saturation of polarimetric parameters, the kinetics of changes in degree of polarization (DOCP) has been measured. Thus, based on Fig. 5 an optimal exposure time of 6 min was selected.

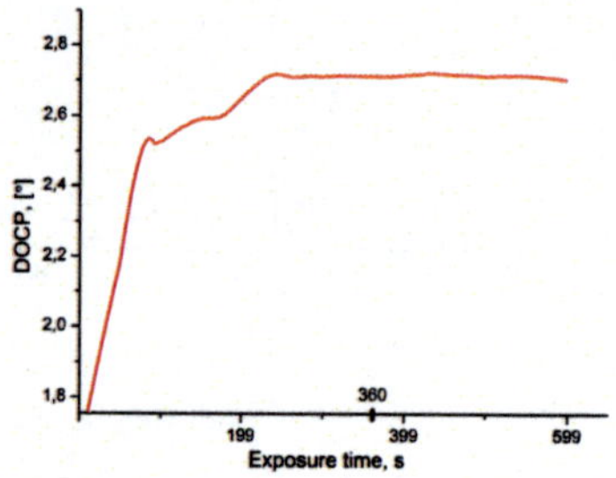

Fig. 5. Time-resolved degree of circular polarization

For a comparative analysis of the optical parameters, three series of measurements were performed (mass ratio of SY3 0.3, 0.4, 0.5) with rotation of the polarization plane from 0° to 180° with a step of 10° and exposure of the material to 473 nm radiation in within 6 min. Thus, for each sample, the temporal and angular changes of azimuth and ellipticity were recorded depending on the polarization plane angle of the 473 nm laser. The temporal changes in the azimuth angle (ψ) for each the angle of the polarization plane from 0° to 180° is shown in Fig. 6.

From the curves, one can see a cyclical change in the azimuth angle, which directly indicates the reversibility of the process associated with the photoinduced changes in the position of molecules in the volume of the azopolymer. At the same time, a relationship is

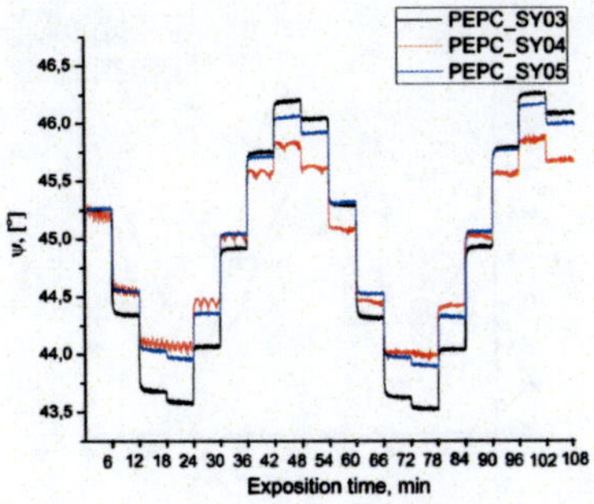

Fig. 6. The temporal changes in the azimuth angle for each the angle of the polarization plane from 0° to 180°

observed between the maximum angle of the azimuth and the thickness of the azopolymer film. Based on Table 1 and Fig. 6, it can be assumed that increasing the film thickness leads to a decrease in the azimuth angle of the reading laser beam.

The temporal variation of the ellipticity (χ) for each the angle of the polarization plane from 0° to 180° is shown in Fig. 7. Like azimuth, ellipticity changes cyclically over time. Analysis of Table 1 and plot in Fig. 7, it can be assumed that the maximum value of ellipticity depends on the concentration of the chromophore in the azopolymer.

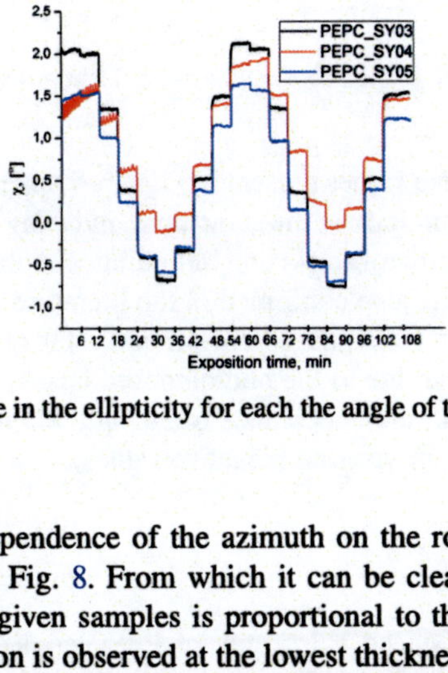

Fig. 7. The temporal change in the ellipticity for each the angle of the polarization plane from 0° to 180°

The graphs of the dependence of the azimuth on the rotation angle of the probe laser beam are shown in Fig. 8. From which it can be clearly distinguished that the azimuth average for the given samples is proportional to the film thickness, where a deeper azimuth modulation is observed at the lowest thickness.

The plots of the dependence of ellipticity on the rotation angle of the probe laser beam are shown in Fig. 9. From this dependence, it can be noted that the average of ellipticity depends on the concentration of the azo dye in the azo dye and decreases with an increase in the concentration of the azo dye.

Analyzing Fig. 8 and Fig. 9 it can be observed that with the increase of the elliptic value the decrease of the azimuth angle takes place, which is also confirmed in the literature [6]. This is because $\chi = 0$ means linear polarization which induces a maximum

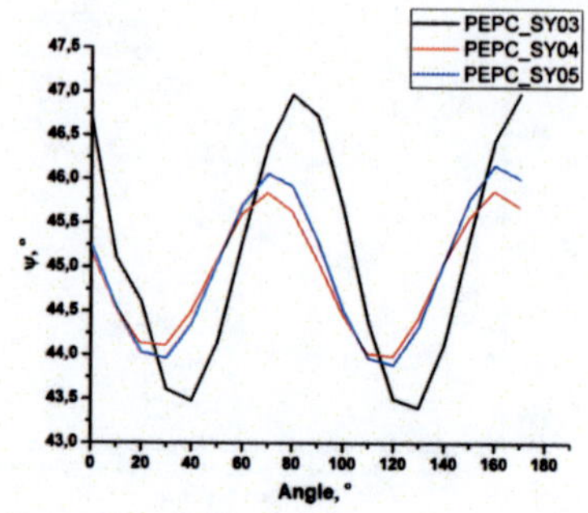

Fig. 8. Dependence of the azimuth angle of the incident probe beam polarization.

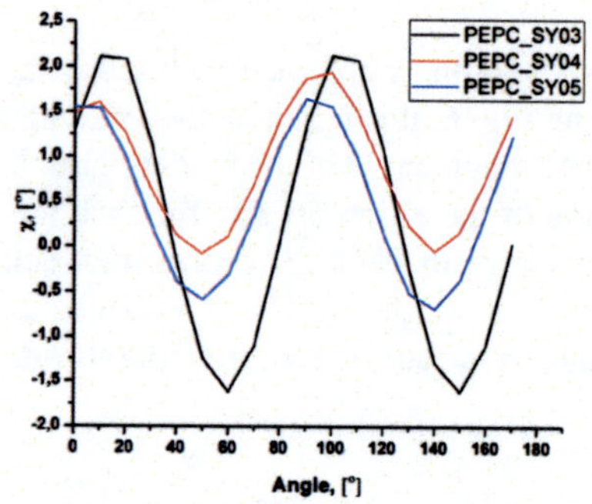

Fig. 9. Dependence of the ellipticity of the incident probe beam polarization.

value of the ψ. The negative values present in Figs. 6–9 appear due to the photoinduced change of azimuth and ellipticity at the same time, inducing the appearance of an right elliptical polarization (positive values) and left elliptical polarization (negative values). Angle takes place, which is also confirmed in the literature [7]. This is because $\chi = 0$ means linear polarization which induces a maximum value of the ψ. The negative values present in Figs. 6–9 appear due to the photoinduced change of azimuth and ellipticity at the same time, inducing the appearance of an right elliptical polarization (positive values) and left elliptical polarization (negative values).

4 Conclusions

In summary, the photo-induced anisotropy of PEPC-co-SY3 films, irradiated with a linearly polarized laser with $\lambda = 473$ nm was investigated. It was determined that the refractive index of azopolymer thin films depends on the thickness and the difference in values increases with increasing film thickness and at 632.8nm $\Delta n_{PEPC\_SY03} = 0.005$, $\Delta n_{PEPC\_SY04} = 0.009$ and $\Delta n_{PEPC\_SY05} = 0.028$. Also, the maximum azimuth change was achieved in a sample with a mass ratio of 0.3, $\Delta\psi$ reaching 3.49. It was shown that an increase in the azimuth angle increases with an increase in the thickness of the azopolymer film. It was shown that an increase in ellipticity occurs as the concentration of the chromophore in the azopolymer decreases, and for the sample with mass ratio 0.3, the maximum value of ellipticity reached 1.3 for the right azimuth and -1.62 for the left azimuth.

Acknowledgment. The work was partly supported by the project of National Agency for Research and Development of Moldova (ANCD 20.80009.5007.03) and International Project of Program ERA.Net Rus Plus (ANCD 21.80013.5007.1M).

Conflict of Interest. The authors declare that they have no conflict of interest.

References

1. Keller, W.J., Shen, N., Rubenchik, A.M., et al.: Physics of picosecond pulse laser ablation. J. Appl. Phys. **125**, 085103–085111 (2019)
2. Khonina, S.N., Kotlyar, V.V., Skidanov, R.V., et al.: Rotation of microparticles with Bessel beams generated by diffractive elements. J. Modern Optics **51**(14), 2167–2184 (2004)
3. Подлипнов, В.В., Ивлиев, Н.А., Хонина, С.Н., et al.: Компьютерная оптика **42**(5), 779–785 (2018)
4. Hore, D.K., Natansohn, A.L., Rochon, P.L.: Anomalous cis isomer orientation in a liquid crystalline azo polymer on irradiation with linearly-polarized light. J. Phys. Chem. B **107**(10), 2197–2204 (2003). https://doi.org/10.1021/jp026295f
5. Nikolova, L., et al.: Photoinduced circular anisotropy in side-chain azobenzene polyesters. Opt. Mater. **8**(4), 255–258 (1997). https://doi.org/10.1016/s0925-3467(97)00046-3
6. Matsuoka, T., et al.: Optically inscribed grating in azo-carbazole dye: concentration dependence. e-J. Surface Sci. Nanotechnol. **13**, 69–74 (2015). https://doi.org/10.1380/ejssnt.2015.69
7. Mateev, G., et al.: J. Phys.: Conf. Ser. **1859**, 012010 (2021)

Molecular, Cellular and Tissue Engineering

The Isolation of Fibroblasts by Volumetric Regulation Cycles

Mariana Jian(✉), V. Cobzac, and V. Nacu

Laboratory of Tissue Engineering and Cells Culture, State University of Medicine and Pharmacy "Nicolae Testemitanu", Chisinau, Republic of Moldova
mariana.jian@usmf.md

Abstract. The fibroblasts are the most used cells for *in vitro* testing of various substances. The explant modality of fibroblast isolation is a widely used method. In order to fix the explant to the cell culture surface are used various substances or are performed various manipulations, such as applying the mechanical force or explant sticking to cell culture surface after a short period of drying. This paper presents another way of fibroblast isolation using the explant modality, which consists in manipulation with the volume of cell culture media in a well of a 12-well plate during several cycles of cell isolation. From 3 domestic rabbits, under general anesthesia, pieces of dermis were harvested and cutted to 32 ± 8 mm^3 (n = 7), and placed by one per well, in a 12-well plate. For 3 days the explants were incubated in 3 ml of cell culture media to ensure the cellular multiplication on the explants. After removing the cell culture media, a small volume of medium was added to maintain the explants moist but fixed to the cell culture surface. When the explants self-attached to the cell culture surface, in the wells were poured 2 ml of cell culture media. After cellular colonies formation, the explants were transferred to another well in which the previous procedure was repeated, starting with the addition of a small volume of medium. The isolated cells from the wells, were cultured to a 80–90% confluence and subcultured in a 75 cm^2 flask. As a result, at 16 ± 2 days, from the attached explants (n = 6), after isolated cells subculture to a confluency of 70–80%, were obtained $2.9 \times 10^6 \pm 1.6 \times 10^5$ cells per flask, which were identified as fibroblasts by Hematoxylin-Eosin and Masson Trichrome stainings.

Keywords: Fibroblast · Cycles · Volumetric regulation · Explant attachment

1 Introduction

Utilisation of various techniques of cellular isolation for therapeutic purposes or *in vitro* tests are widespread nowadays [1–3, 12, 13], and the most commonly used cells for the *in vitro* tests are the fibroblasts. Most of existing protocols describe the isolation of fibroblasts from skin taken from different regions of the donor's body [3, 4], but they can be extracted also from lung tissue [4, 5], Tenon's capsule of the eye sclera [6] and heart [7]. To isolate the fibroblasts from the explant are used various enzymatic solutions [2, 4, 5], basal cell culture media to which are added fibroblast growth factors or manufactured

I. Tiginyanu et al. (Eds.): ICNBME 2021, IFMBE Proceedings 87, pp. 325–333, 2022.
https://doi.org/10.1007/978-3-030-92328-0_43

complete culture media special for fibroblasts [6]. The protocols according to which the cells are isolated directly from tissue use various enzymes for complete tissue digestion, and fibroblasts are separated from tissue debris and other cells by filtration and flow-cytometry [7], which is wery expensive. The disadvantage of isolating fibroblasts from the explant is poor adhesion of the explant to the cell culture surface. In order to increase explant adhesion to the cell culture surface, different tricks are used, such as explant disection to pieces from 2–3 mm^2 to 1 cm^2 with their further application to the cell culture surface to which they will adhere after a short period of drying [2, 6], or tissue mincing followed by enzymatic digestion to a very small pieces of ~1 mm [4]. Also, to increase explant attachment to the cell culture surface are used various substances, such as type I collagen, gelatin, fibronectin [8] and fetal bovine serum (FBS) [9].

The paper describes the implementation of a simplified protocol, according to which, the isolation of fibroblasts by explant does not require enzymatic treatment of the tissue or utilisation of any additional substances for attachment. Due to the good attachment of the explant to the cell culture surface, fibroblasts migrate from the explant, and the whole isolation process is performed only by using various volumes of cell culture media at certain stages.

2 Materials and Methods

2.1 Tissue Sample Harvest for Explant Preparation

From 3 domestic rabbits under general anesthesia with xylla solution 5 mg/kg (De Adelaar, Netherlands) and ketamine 35 mg/kg (Farmako, Moldova) [3, 12], from the lumbosacral region, after removing the fur with a trimmer, the operating field was sanitized and delimited with sterile sheets. With a scalpel, a cutaneous incision was made in the lumbosacral region and a piece of dermis was taken and placed in a 15 ml tube with cell culture media composed of DMEM/F-12 (Sigma, Germany), 10% FBS (Lonza, Belgium) and 100 U/ml penicillin, 100 μg/ml streptomycin and 25 μg/ml amphotericin B (HiMedia, India) [1, 3, 12] (Fig. 1). After wound suture the animal was taken to the vivarium, and the harvested derma was taken to the sterile chamber and placed in the laminar flow hood (LN 090, Nuve). Under the hood the harvested tissue was poured into a 10 cm Petri dish and washed 3 times with HBSS without calcium, magnesium and red phenol (Sigma, Germany), then it was sectioned with a scalpel in 2–3 pieces with a size of 32 ± 8 mm^3 ($n = 7$). The dermis pieces were suspended in cell culture media and centrifuged at 500 rpm for 5 min at 25 °C (Universal 32R, Hettich Zentrifugen). After removing the supernatant, the dermal pieces were individually transferred into the wells of a 12-well plates (Sofra, China).

2.2 The Fibroblasts Isolation by Cycles of Volumetric Regulation

After explant introduction into a well of the 12-well plate, 3 ml of cell culture media was poured instead of recommended 2 ml, to suspend the explant in a large volume of cell culture media. The plate was incubated for 3 days at 37 °C, 5%CO_2 in a humid environment (Smart Cell, Heal Force), to ensure cellular multiplication on the explant.

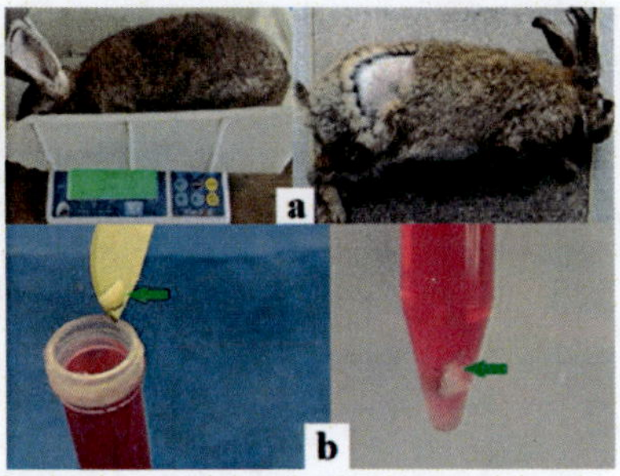

Fig. 1. Preparation of animal for a short term anesthesia (a) and harvesting of small piece of derma (green arrow) from rabbit (b).

After complete removal of the cell culture media, to explant was added a small volume of media sufficient to maintain the explant moist, but fixed to the cell culture surface. Thus was ensured migration of cells from the explant to the cell culture surface of the well, after which the plate was incubated repeatedly for 1–2 days. When changing the cell culture media, to the wells in which the explants were attached, 2 ml of culture medium were added, the working volume specific to used vessel [11] (Fig. 2). The cell culture media was completely changed every 2 days until formation of cellular colonies capable of normal growing. Then with a sterile tweezer, or a micropipette the explants were transferred to another wells, where a small volume of cell culture media was poured again, to ensure their repeated adhesion (Fig. 3). The explants which did not adhered to the cell culture surface after 6 days of culture in a small volume of medium and explants from which the cells did not migrated within 7 days, were removed from the experiment. After cellular confluence to 80–90% in the wells, the cells were trypsinized [1, 12] and transferred to a 75 cm^2 cell culture flasks (Nunc, Denmark) by 5.2×10^3 cells/cm^2, in which they were cultivated to a 70–80% confluence, with complete medium changement every 2 days. Only cells isolated in the first cycle were subcultured in 75 cm^2 flasks. The cells isolated during other cycles were conserved in FBS (Lonza, Belgium) with 10% DMSO (OriGen Biomedical, Germany) and stored at −84 °C (Uluf 450-2m, Artctiko) [1, 3, 10, 12].

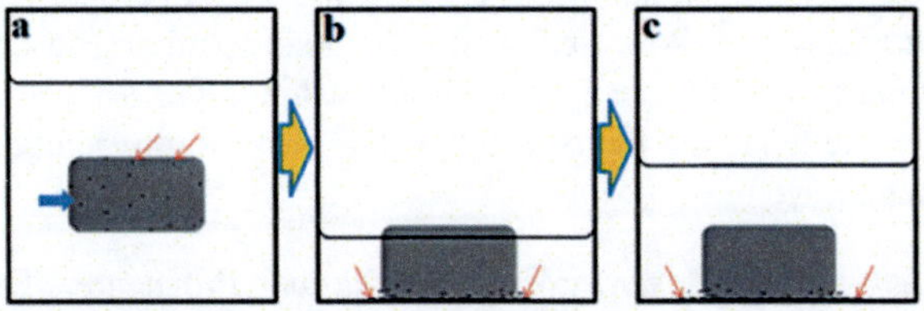

Fig. 2. Representation of a cycle of volumetric regulation during fibroblast isolation from the explant (blue arrow). Germination of the explant with fibroblasts (red arrow) in a high quantity of cell culture media (a), decreasing the amount of cell culture medium to ensure explant fixation to the cell culture surface and cellular migration from it (b) and increasing the amount of cell culture media to normal values (c).

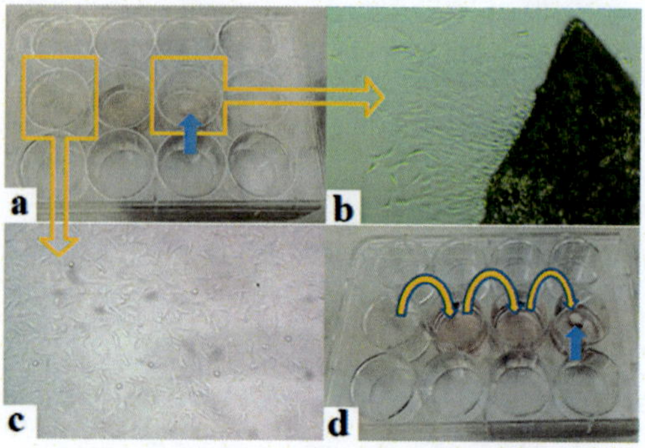

Fig. 3. The explant (blue arrow) transfer from a well of 12-well plate to another (yellow inflected arrow) during fibroblast isolation by cycles of volumetric regulation. The explant adherence in 3rd cycle of volumetric regulation for fibroblast isolation (a), the adhered explant and the fibroblasts migrated from it (b), 70–80% cellular confluency after explant extraction, (d) initiation of 4th cycle for fibroblasts isolation

2.3 The Fibroblasts Identification

In order to identify the isolated cells, they were cultured in two 25 cm^2 cell culture flasks (Nunc, Denmark), in which by 2.5×10^5 cells were seeded. As a control were used mesenchymal stem cells (MSC) obtained from bone marrow of a rabbit [1, 3, 10, 12], which were seeded in cell culture flasks in the same number as the isolated cells. The cells were cultured in over-confluence during 21 days, with the cell culture media changement every 1–2 days (Fig. 4). On the 21st day, the cell culture media was removed from the flasks and stained in monolayer with Hematoxylin-Eosin (Bio-Optika, Italy) and Masson Trichrome (DDK, Italy), by one flask with isolated cells and one with MSC.

Staining of a Cellular Monolayer with Hematoxylin and Eosin: After preparing the staining solutions, the cells in the flasks were washed 3 times with ddH_2O, then 2 times by 3 min with 96% alcohol (Luxfarmaol, Moldova) and tap water for 2 times, after which 3 ml of Hematoxylin (Bio-Optika, Italy) were added in each flask. The cells were washed repeatedly with tap water 3 times, which was kept in the flask for 3 min. After washing for one minute with 96% alcohol, 0.25% Eosin solution (Bio-Optika, Italy) was added for one minute, with further 3 consecutive washings with 96% alcohol. Then, 5 ml of tap water were added in the flasks, and the cells were examined under a microscope (MT-2, Olympus).

Staining of a Cellular Monolayer by Masson Trichrome Procedure: The staining is made according to the protocol of the commercial coloring kit Masson Trichrome (DDK, Italy) for fibrous tissue. After washing the cells from the flasks for 3 times with ddH_2O, Weigert Ferric Hematoxylin was added for 20 min, which was washed with tap water. The water was kept in the flask for 10 min. After washing 3 times with ddH_2O, in flasks was added Biebrich fusidic acid for 15 min and washed again with ddH_2O for 3 times. Then, in the flasks for 15 min phosphotungstic acid was added, which was washed 3 times

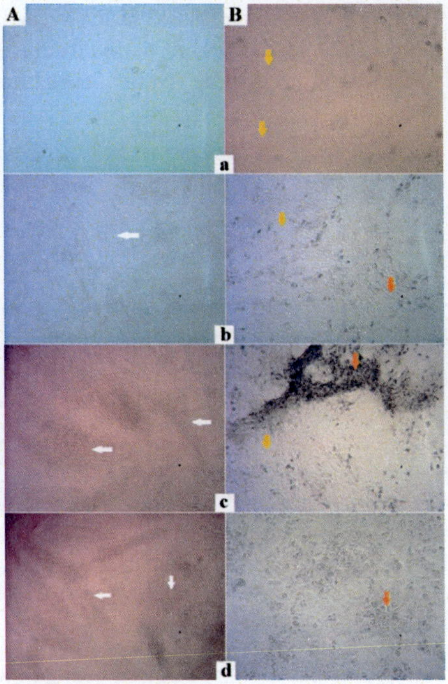

Fig. 4. Culture in over confluence of isolated cells (A) and MSC (B) during 21 days for fibroblast identification. At 2nd (a) and 7th (b) days of culture, in flask with isolated cells (A) we see formation of organised cellular structures similar to shoals of fish (white arrow), when in the MCS (B) the aspect is disorganised (orange arrow) with formation of voids of different dimensions with a radial inside structure (yellow arrow). At 15th (c) and 21st (d) days the aspect of shoals of fish (white arrow) formed by isolated cells (A) is thicker and more pronounced, and in the MSC (B) flask the aspect is more disorganised and hypertrophied (orange arrow).

with ddH_2O. The counter-staining was done with aniline blue for 2 min, after washing with ddH_2O, the cells were washed twice every 2 min with 96% alcohol (Luxfarmol, Moldova).

3 Results

From the 7 prepared explants for fibroblasts isolation, 6 explants attached to the cell culture surface, or 85.71%. The process of explants attachment the to the cell culture surface lasted 2 ± 1 days. The confluence to 80–90% of isolated cells after attachment of explants to the cell culture surface lasted 6 ± 1 days, and the transfer of the explants to other wells was at 3 ± 1 days after attachment. The 6 attached explants for fibroblast isolation were used during 23 cycles, in average by 4 ± 1 cycles per explant, which lasted 31 ± 7 days. During those 23 isolation cycles in the wells, were isolated 9.04×10^6 cells, and from each well were obtained by $3.93 \times 10^5 \pm 2.2 \times 10^4$ cells. The subculture of isolated cells from the first cycle in 75 cm^2 flasks lasted in each case 5 days, obtaining by $2.9 \times 10^6 \pm 1.6 \times 10^5$ cells per flask. Thus, the period from the initiation of fibroblast isolation to 70–80% cellular confluence in the 75 cm^2 flasks lasted 16 ± 2 days.

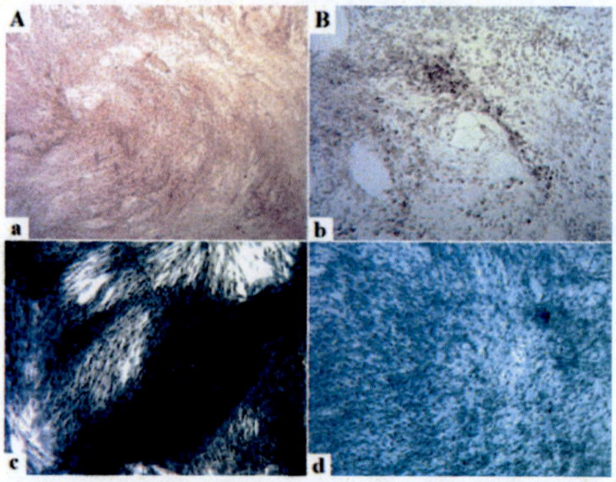

Fig. 5. Fibroblasts identification of isolated cells (A) cultured in over confluence for 21 days, MSC used as control (B). The isolated cells are stained in red with Hematoxylin-Eosin (a) and deep blue by Masson Trichrome (c). The specific staining of control is absent (b and d).

The monolayer staining of isolated cells, cultured over 21 days in over-confluence confirmed that the isolated cells are fibroblasts, because at staining with Hematoxylin and Eosin the monolayer of isolated cells turned red, and at staining by Masson Trichrome, the cellular monolayer turned to deep blue. MSC cultures used as controls did not secreted extracellular matrix specific to fibrous tissue, that is why they did not retained the specific coloring agent (Fig. 5).

4 Discussion

The studies in which fibroblasts are isolated by explant method use different protocols for processing and attaching the explants to the cell culture surfaces [2, 4–6, 8, 9]. In the research we have showed that the practical need for the enzymatic processing of the explant does not exist, because cells migrate by themselves from the explant during cellular multiplication. Selective utilisation of dermis as an explant, theoretically allowed isolation of fibroblasts pure cultures, without their contamination with keratinocytes [2, 4]. According to an research, obtaining of a pure culture of fibroblasts when using as a explant whole skin is possible after the explant was transfer from one culture vessel to another at least 2 times after obtaining a cell migration from the explant in a radius of 20 mm [2]. Contamination of fibroblastic cultures can be done with many other types of cells present in lung tissue and airways [4, 5], heart, liver, kidney, ovary, testicle [8] and other tissues from which the explants were obtained.

A major importance in the process of fibroblasts isolation by explant plays the size of the explants. Utilisation of a 1 cm^2 explant for exemple, allows migration of cells from it during several cycles, and its preservation ensure fibroblasts isolation when it is needed [2, 9]. When using small sized explants, the ability of cellular multiplication and migration from them is greatly reduced [1, 4–6, 8, 12], leading during few cycles to explant depletion. In addition, the transfer of small sized explants to another vessel is possible only after releasing the explants by cells trypsinization from this vessel, thus leading to detachment of the cells present on the explant surface which are necessary to obtain cellular colonies in the next cycle, or after their mechanical detachment, thus leading to a slower cellular confluence.

Detachment of explants from the cell culture surface is a common phenomenon, but poorly described in the literature. From our experiment, those 6 explants attached to the cell culture surface represent 85.71% of used explants, with a volume of 32 ± 8 mm^3, while in another study were tested the adhesion capacities of different substances on the explants from various organs and tissues. In which 1–2 mm^3 explants were used, those were introduced by 30 pieces in 25 cm^2 flasks. The adhesion rate of which depended directly by explant tissue origin, and by used adhesive substance, the yield varying from 15% up to 65% and more [8].

The period for fibroblasts and fibroblast-like cells isolation in various sources is different. At utilisation of a explant obtained from 1 cm^2 of skin and fixed to the cell culture surface with FBS, a confluence of 70–80% in a 75 cm^2 cell culture flask was obtained after 30 days of culture [9], while in our case the process lasted 16 ± 2 days and without utilisation of additional adhesive substances ($p < 0.001$). In another study, in 10 cm Petri dishes, with the area of 56.7 cm^2, the cells reached to 70–80% confluence in 14 days, without taking any measures for adhesion of the explants obtained from skin and lung tissue, with the size of ~1 mm. The maximum period of these explants utilisation was 14 days [4]. In other study, the period necessary to obtain a culture of fibroblasts from lung explants lasted 3–4 weeks [5]. When using as explant a fragment of $2 - 3 \times 1$ mm of eye Tenon's capsule, which adherence to the cell culture surface was carried out by simple pressing, the formation only of one monolayer in a well of a 12-well plate lasted 15 ± 2 days, which in our case was 6 ± 1 days ($p < 0.001$), and the subculture in a 25 cm^2 flask lasted another 7 ± 1 days [6]. Another study shows that the period needed to obtain the fibroblast-like cells from different types of tissue of a fish, depends directly by the type of used tissues as explants, thus, obtaining of a cellular monolayer in a 25 cm^2 cell culture flask is faster from explants obtained from the testicles and liver, about 7–9 days, while the slowest isolation was from the fin, about 20–21 days [8].

The identification of isolated cells from various tissues is often used by our team [1, 3, 12, 13], because at their culture in over-confluence takes place the synthesis of the extracellular matrix specific to the origin tissue [1, 12], and the MSC used as control, are devoid of such capacity [3]. Thus, the isolation of specific cells we need, allows us to perform targeted *in vitro* tests on certain types of cells or use them for therapeutic purposes *in vivo*.

5 Conclusions

The utilisation of volumetric regulation cycles for fibroblasts isolation, ensure a high level of explant adherencc and a faster fibroblasts isolation from a dermal tissue. Also the size of the explants plays an important role in the process of fibroblasts isolation, because the large-sized explants can be transferred several times to other vessels to continue cell isolation. The described method of fibroblasts isolation is faster, cheaper and much simpler in execution compared to others described in the literature ($p < 0.001$).

Acknowledgment. The study was supported by the State Program Project: "GaN-based nanoarchitectures and three-dimensional matrices from biological materials for applications in microfluidics and tissue engineering", # 20.80009.5007.20.

Ethical Requirements. The surgical procedures related to animals and cells isolation from different animal tissues was approved by the Ethics Committee of Research from "N. Testemitanu" SUMPh with No. 1 from 22.05.2020.

Conflict of Interest. The authors declare no conflict of interests.

References

1. Jian, M., Cobzac, V., Mostovei, A., Nacu, V.: The procedure of bone cells obtaining, culture and identification. In: Tiginyanu, I., Sontea, V., Railean, S. (eds.) ICNBME 2019. IP, vol. 77, pp. 595–599. Springer, Cham (2020). https://doi.org/10.1007/978-3-030-31866-6_106
2. Huschtscha, L.I., Napier, C.E., Noble, J.R., Bower, K., et al.: Enhanced isolation of fibroblasts from human skin explants. Biotechniques **53**(4), 239–244 (2012). https://doi.org/10.2144/0000113939
3. Cobzac, V., Mostovei, A., Jian, M., Nacu, V.: An efficient procedure of isolation, cultivation and identification of bone marrow mesenchymal stem cells. Mold. Med. J. **62**(1), 35–41 (2019). https://doi.org/10.5281/zenodo.2590011
4. Seluanov, A., Vaidya, A., Gorbunova, V.: Establishing primary adult fibroblast cultures from rodents. J. Vis. Exp. **44**, e2033 (2010). https://doi.org/10.3791/2033
5. Hutton, A.J., Polak, M.E., Spalluto, C.M., Wallington Joshua, C.: Human lung fibroblasts present bacterial antigens to autologous lung Th cells. J. Immunol. **198**(1), 110–118 (2017). https://doi.org/10.4049/jimmunol.1600602
6. Przekora, A., Zarnowski, T., Ginalska, G.: A simple and effective protocol for fast isolation of human Tenon's fibroblasts from a single trabeculectomy biopsy – a comparison of cell behaviour in different culture media. Cell Mol. Biol. Lett. **22**, 5 (2017). https://doi.org/10.1186/s11658-017-0034-4
7. Stellato, M., Czepiel, V., Distler, O., Błyszczuk, P.: Identification and isolation of cardiac fibroblasts from the adult mouse heart using two-color flow cytometry. Front Cardiovasc. Med. **6**, 105 (2019). https://doi.org/10.3389/fcvm.2019.00105
8. Nanda, P.K., Swain, P., Nayak, S.K., Behera, T., et al.: Evaluation of different coating factorsto establish cell culture from tissue explants of Indian major carp, cirrhinus mrigala. Asian J. Anim. Vet. Adv. **9**, 395–404 (2014). https://doi.org/10.3923/ajava.2014.395.404
9. Butnaru, M., Luca, A.: Cultura de celule animale: tehnici uzuale şi tehnici speciale. Editura PIM, Iasi (2014). ISBN 978-606-13-2199-5
10. Braniste, T., Cobzac, V., Ababii, P., Plesco, I., et al.: Mesenchymal stem cells proliferation and remote manipulation upon exposure to magnetic semiconductor nanoparticles. Biotechnol. Rep. **25**, e00435 (2020). https://doi.org/10.1016/j.btre.2020.e00435

11. https://www.thermofisher.com/md/en/home/references/gibco-cell-culture-basics/cell-culture-protocols/cell-culture-useful-numbers.html
12. Cobzac, V., Verestiuc, L., Jian, M., Nacu, V.: Chondrocytes isolation from hyaline cartilage by continuous monitoring method. Mold. Med. J. **64**(4), (2021)
13. Jian, M., Cobzac, V., Vartic, V., Nacu, V.: Hepatocytes isolation from adult rats for liver recellularization. Mold. Med. J. **62**(1), 13–16 (2019). https://doi.org/10.5281/zenodo.2589998

The Cartilaginous Tissue Regeneration on Weight Bearing and Non-weight Bearing Surfaces of the Knee

Vitalie Cobzac(✉), M. Jian, T. Globa, and V. Nacu

Laboratory of Tissue Engineering and Cells Culture, State University of Medicine and Pharmacy "Nicolae Testemitanu", Chisinau, Republic of Moldova
vitalie.cobzac@usmf.md

Abstract. Regeneration of articular cartilage is a major problem in the field of orthopedic surgery and regenerative medicine. Most research on cartilage regeneration performed on rabbits, tests the possibilities of cartilage regeneration on the non-weight bearing surfaces of the distal femur - in the trochlear groove of the femur, giving less importance to the weight bearing surfaces, the areas where the joint surface is subjected to various high forces. The study was performed on 4 month old domestic rabbits in which type I collagen sponges combined with bone marrow mesenchymal stem cells (MSC) were transplanted. For comparison, the combined grafts were transplanted at 2 levels - in a defect on the weight bearing articular surface of the medial femoral condyle ($n = 3$) and in 2 defects made in the trochlear groove of the femur ($n = 3$). As a negative control served the experimental defects ($n = 2$) made in the the trochlear groove of the rabbits femurs ($n = 6$). The results were evaluated using the Unified Histological Score of Regenerated Cartilage (UHSRC) after removing the animals from the experiment at 12 weeks, and the last 3 from the control group at 52 weeks. As a result, at 12 weeks no difference was determined between the control group and the combined grafts transplanted into the defects from trochlear groove ($p > 0.8$), but, a significant difference was found between results obtained after transplantation of combined grafts in the weight-bearing and non-weight-bearing areas ($p = 0.002$). The control group removed from the experiment from the experiment at 52 weeks has better results comparing to those removed at 12 weeks ($p < 0.1$).

Keywords: Cartilage · Collagen sponge · Weight-bearing area · Non-weight-bearing area · Trochlear groove

1 Introduction

Articular cartilage injury are detected in 61–66% of all performed arthroscopies [1]. Experimental methods of cartilage lesions treatment, performed on the laboratory animals, especially rabbits, are mostly created in the non-weight bearing area - the trochlear groove of the femur [3–5, 7] and less in the weight-bearing surfaces, as the medial condyle of the femur [6]. The present study was performed on domestic rabbits in order to evaluate the results obtained after transplantation of combined grafts [2, 8] with autologous

I. Tiginyanu et al. (Eds.): ICNBME 2021, IFMBE Proceedings 87, pp. 334–341, 2022.
https://doi.org/10.1007/978-3-030-92328-0_44

mesenchymal stem cells (MSC) in experimental osteochondral defects made in the non-weight bearing region surface of the knee joint - the trochlear groove of the femur and the weight-bearingsurface - the articular surface of the femoral medial condyle.

2 Materials and Methods

2.1 The Graft Obtaining for Combination with the MSC

The graft that was combined with MSC was obtained from collagen extracted from bovine tendons taken from the slaughterhouse. After harvesting, the tendons were washed from blood and frozen at −25 °C. To extract the collagen, after the tendons thawing, they were minced with scissors to 1–3 mm^3. Then, the tendon pieces were placed in 0.05M Na_2HPO_4 solution (Sigma, USA) on magnetic stirrer (MS-3000, Biosan) for 4 days. Then, the tendons were dissolved in a 0.5 M of glacial acetic acid solution (Chem-Lab, Germany) with 5 mM EDTA (Sigma, Germany) for another 4 days. After removing the undissolved tendon pieces by straining, the collagen was sedimented with 2M NaCl (Sigma, Germany) solution. After separating the collagen by centrifugation at 4000 rpm for 10 min at 4 °C (Universal 32R, Hettich Zentrifugen), it was dissolved with 1 M glacial acetic acid solution (Chem-Lab, Germany) to a concentration of 1%.

The 10 ml of 1% collagen suspension was frozen at −25 °C and lyophilized at 0.05 mbar for 16 h (VaCo 2, Zirbus). The obtained collagen sponge was sectioned with a circular knife into pieces of 3.5 mm in diameter. The collagen sponge pieces were conserved and sterilized using the buffered solution of 0.25% formaldehyde with pH-7.4 and sterilized using 0.45 μm filtration systems (Biosigma, Italy). The preservation solution was changed daily for 10 days in sterile conditions, under the laminar air flow hood (LN 090, Nuve) [8]. Then the pieces of collagen sponge were introduced for 3 days in HBSS without calcium, magnesium and red phenol (Sigma, Germany) with changing the solution 1–2 times a day. The sponges were sterilized with Sabouraud and Thioglycolic medium (HiMedia, India) by incubation at 37 °C for 14 days (TC-80M-2).

To determine the cytotoxicity of the obtained collagen sponges, was performed the MTT assay with MSC at 24, 48 and 72 h.

2.2 Isolation and Culture of Autologous MSC

MSC were isolated from rabbit bone marrow. Under general anesthesia with 5 mg/kg of xylla solution (De Adelaar, The Netherlands), 2 mg/kg of 0.5% diazepeks solution (Grindex, Latvia) and 35 mg/kg of ketamine solution (Farmako, Moldova), after removing the fur from the pelvis, asepticization and delimitation of the operating field, with a 18G needle accompanied by a trocar, parallel to the plane of the iliac bone wing, by rotary movements was perforated first cortex of the iliac bone [10]. After removing the trocar from the needle, a syringe with 1000 U of heparin (BalkanPharma, Moldova) was connected and the bone marrow was aspirated. The obtained bone marrow was introduced into the laminar flow air hood (LN 090, Nuve), where it was diluted 1:1 with PBS (HiMedia, India) and then slowly poured onto the same amount of HiSep 1077 concentration gradient (HiMedia, India) in a 15 ml tube. After centrifugation at 2000 rpm for 15 min at 25 °C (Universal 32R, Hettich Zentrifugen), the bone marrow separated into layers. The mononuclear cell layer was collected and placed in another 15 ml tube where it was washed with 10 ml of PBS (HiMedia, India) and centrifuged at 1000 rpm for 10 min. The cells were then washed repeatedly with HiMesoXL MSC expansion media (HiMedia, India) and centrifuged again. After resuspension of the cell pellets in culture media, they were introduced in a 25 cm^2 cell culture flask (Nunc, Denmark) and incubated at 37 °C, 5% CO_2 in humidified environment (Smart Cell, Heal Force) [13]. MSC were cultured to a 70–80% confluence.

2.3 Combination of MSC with Collagen Sponges

The day when the MSC reached 70–80% confluence, the collagen sponges were placed in 15 ml tubes on sterile gauze and dessicated by centrifugation at 4000 rpm for 10 min [2]. The sponges were then removed from the tube and placed in a plastic Petri dish under the hood. After trypsinization, the MSC were washed and resuspended in 1 ml of cell culture medium and then transferred to a 1.5 ml Eppendorf tube and centrifuged at 3500 rpm for 3 min (Combi-Spin FVL-2400N, Boeco). The entire culture medium was extracted from the Eppendorf tube except 50–60 µl, in which the cells were resuspended. With a micropipette (LightDrop 10–100 µl, ThermoFisher) the cells were poured on the collagen sponges placed in a Petri dish [12]. In 3 cases the MSC suspension was divided in 2 equal volumes and poured on 6 sponges by $5.9 \times 10^5 \pm 1 \times 10^5$ cells/sponge - for the defects made in the non-weight bearing region. In other 3 cases the cell suspension was poured in total volume on 3 sponges, by $1.17 \times 10^6 \pm 3 \times 10^5$ cells/sponge, for the defect in the weight bearing region. The sponges were incubated at 37 °C, 5%CO2 in humidified environment (Smart Cell, Heal Force) until transplantation, for approximately 70 ± 10 min. (Fig. 1).

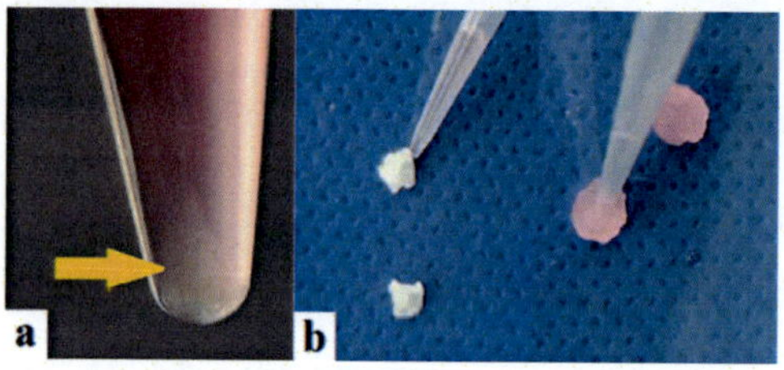

Fig. 1. Cellular population of collagen sponges: (a) pellets of MSC in a Eppendorf tube (orange arrow) and (b) pouring of MSC suspension on prepared 3.5 mm collagen sponges.

2.4 The Fibrin Glue Obtaining

From the National Blood Transfusion Center of the Republic of Moldova were obtained bags with 50 ml of fresh frozen plasma (FFP) and vials with 250U of human thrombin. One day before surgery, a 50 ml bag with FFP was thawed in the refrigerator at 4 °C. Then the bag with thawed plasma was placed in a laminar flow hood where the plasma was trasnfered in a 50 ml tube and centrifuged at 4000 rpm, 2 °C for 20 min (Universal 32R, Hettich Zentrifugen) to precipitate the fibrinogen. After centrifugation the plasma was removed and the fibrinogen tube was placed in a water bath at 37 °C (BAE-2, Raypa). After solubilization the fibrinogen was collected in a 2 ml syringe.

Few hours before transplantation, in sterile conditions, under the laminar air flow hood (LN 090, Nuve), the thrombin solubilisation solution was prepared by mixing 1.7 ml of 10% CaCl (Darnita, Ukraine) and 3.3 ml of 0.5% aminocapronic acid (Iuria-Farm, Ukraine) in a 5 ml syringe. Then, the 250 U of thrombin were solubilised in 1 ml of prepared solution.

After preparation, the fibrin glue components were placed in a thermostat at 37 °C (TC-80M-2) until use.

2.5 Transplantation of Collagen Sponges Combined with Autologous MSC

After rabbits anesthesia from which bone marrow was harvested to isolate the MSC, 25 mg/kg cefazoline was injected intramuscularly, and the fur was removed from the left knee, then treated with antiseptics and the operating field was dressed with sterile sheets. Through a medial parapatellar approach to the knee, a 4–5 cm incision was made. After opening of knee joint the patella was laterally dislocated and the knee flexed. With a 3.5 mm drill bit, to 3 rabbits in the femoral trochlear groove two defects were made, and to other 3 rabbits only one defect was made on the bearing surface of the femoral medial condyle. The osteochondral defects were created with caution not to open the medullary canal, they were about 2–3 mm deep. Collagen sponges combined with MSC were taken from the incubator and brought to the operating room. The grafts were placed on sterile gauze to absorb the fluid excess. With a syringe in each created experimental defects a drop of thrombin was injected to stop bleeding from the bone. After remouving the clots from the defects, they were washed with saline and dried, after which a drop of thrombin was poured in each defect. At the same time, to the sponge combined with the MSC a drop of fibrinogen was applied. With a tweezer, the sponges were inserted into the defects with the fibrinogen drop facing down to come in contact with the thrombin. Sponges with $5.9 \times 10^5 \pm 1 \times 10^5$ SCM ($n = 6$) were introduced by two in the defects created in the trochlear groove of the femur of 3 rabbits, and sponges combined with $1.17 \times 10^6 \pm 3 \times 10^5$ MSC ($n = 3$) were introduced in a single defect created in the weight bearing surface to other 3 rabbits (Fig. 2). After suturing the wound, a long sterile dressing was applied from the pelvis to the foot. The antibiotics were injected for another 2 days, and the wounds were treated with a 10% povidone-iodine solution (EGIS, Hungary). The rabbits were removed from the experiment at 12 weeks postoperatively.

2.6 The Control Group

As a negative control group served the experimental defects created by two in the femoral trochlear groove to 6 rabbits. The postoperative approach to all rabbits was the same. Three rabbits from control group were removed from the experiment at 12 weeks and the other 3 at 52 weeks postoperatively.

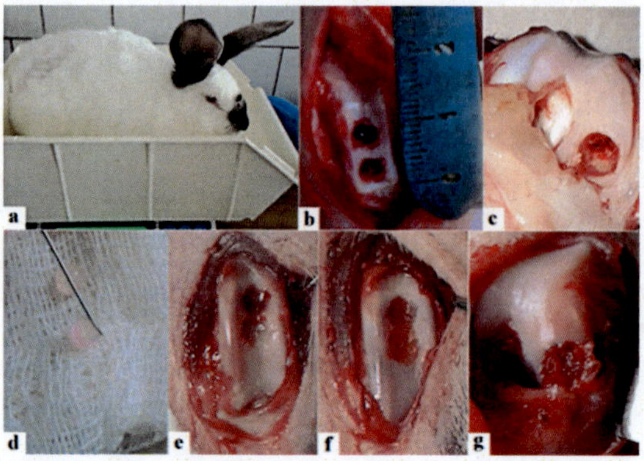

Fig. 2. Transplantation of collagen sponges combined with MSC: (a) rabbit preparation for surgery, (b) two 3.5 mm experimental osteochondral defects performed in the femoral trochlea, (c) one 3.5 mm experimental defect performed on the medial condyle surface, (d) fibrinogen drop application on the combined grafts surfaces, (e) thrombin application in the experimental defects, grafts implantation in defects from (f) trochlear groove and (g) medial femoral condyle.

2.7 The Results Evaluation

In order to evaluate the results after removing the animals from the experiment, the distal femurs were collected and fixed in 10% buffered formaldehyde. For histological examination, was performed stainings with Hematoxylin-Eosin and Toluidine Blue with Fast Green. For objective histological evaluation of obtained results, three similar, popular histological scores were unified in one score were used. Those are the histological scores of Sellers and coauthor. (1997), Wakitani and coauthor. (1994) and O'Driscoll et al. (1986) [11], which became the Unified Histological Score of Regenerated Cartilage (UHSRC). Thus, the obtained results are evaluated according to 13 criteria, most of which are found in all histological evaluation scores variants presented above, and the individual criteria specific to each of those scores were simply added into the unified one. The unified score scale has the values oriented in descending order, the higher is the final score, the worse is the result. The maximum score is 43 points.

3 Results

For combination of collagen sponges with cells in total were obtained $1.18 \times 106 \pm 2.3 \times 105$ MSC per culture, without any difference between the number of cells used in each group ($p > 0.5$), and the mean period for cell culture was 9 ± 1 days.

MTT cytotoxicity test performed before transplantation of collagen sponges combined with MSC showed a high cellular viability (Fig. 3).

During the postoperative period, complications related to septic processes, knee misalignment or stiffness did not occurred. Following the application of UHSCR for the objective evaluation of histological results, we obtained a score of 13.33 ± 3.07 for the control group removed from the experiment at 12 weeks and a score of 9 ± 4.24 for the control group removed from the experiment at 52 weeks, indicating a better result in the

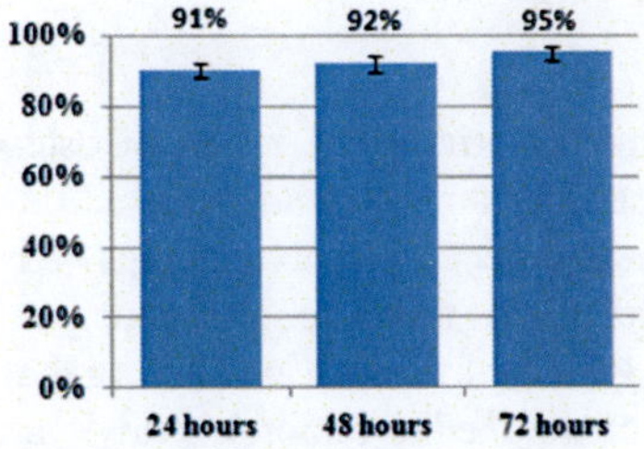

Fig. 3. MTT assay with MSC.

regeneration of untreated defects in a long term period ($p < 0.1$). For the two defects from the trochlear groove of 3 rabbits treated with grafts combined with MSC at 12 weeks distance a score of 13.86 ± 4.62 was obtained, while in the single defects from the medial condyle of the femur in 3 other rabbits we obtained a score of 39.66 ± 2.08, this being a very poor result ($p = 0.002$). However, at compariation of the scores obtained at 12 weeks in the control group and the experimental group with defects defects in the femoral trochlea treated with grafts combined with MSC, were obtained similar results ($p > 0.8$) (Fig. 4).

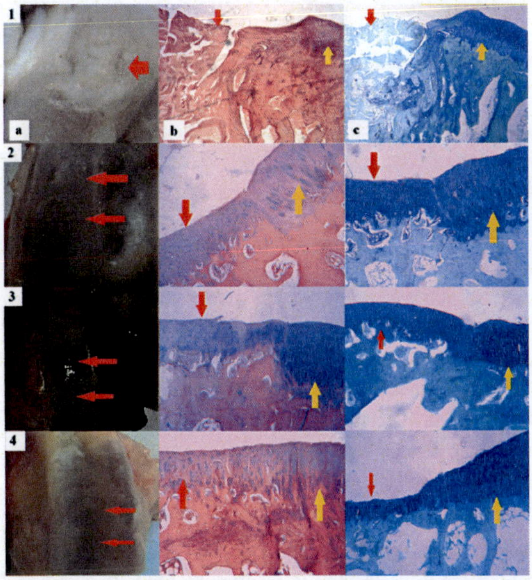

Fig. 4. The experimental defects regeneration made on the weight bearing surface of the femoral condyle with cellularised graft (1), on the non weight bearing surface of the femoral groove with cellularised grafts (2) the negative defect filling from the femoral groove (3) at 12 weeks post-operation, and the negative defect filling from the femoral groove at 52 weeks post-op. On the images are seen the macroscopic view (a), H-E staining (b), Toluidine blue with Fast green staining (c) (MT-2, Olympus). With the red arrow is indicated the regenerated tissue, with the orange arrow is indicated the normal cartilage.

4 Discussions

The results of the experiments oriented to repair of osteochondral defects, are generally focused on demonstrating the efficiency of used three-dimensional structure in the restoration of concerned defect [3–7]. No studies were been found in which the same three-dimensional structure is tested simultaneously in the weight-bearing and non weight-bearing region of the distal femur. Considering that at the level of the weight-bearing articular surface of the medial femoral condyle are acting shear, tensile, and compressive forces [9] which occur at contact of the articular surface of the tibia during movement, it is unlikely that the graft fixed with fibrin glue could properly withstand. That we cannot say about the grafts transplanted in the trochlear groove, where the same forces act much weaker or at all, and as a consequence the results are much better. Also, the poor result obtained at the level of the weight-bearing surface is also due to the fact that the tested grafts do not have the mechanical resistance corresponding to the forces acting at this level, while three-dimensional structures resistant to high forces show good results [6, 7]. As a result, there are doubts about the practicability of critical experimental defects in non weight-bearing regions of the distal femur [7].

The very poor preliminary results in the treatment of osteochondral defects in the weight-bearing articular surface in the first 3 rabbits treated with collagen sponges combined with MSC and the results from the literature, determined us not to create the negative control group with a single defect at the level of weight-bearing surface, because it is known that these defects do not regenerate [6]. The fact that two critical defects in the trochlear groove of the femur treated with collagen sponges containing 2 times less MSC per sponge have much better results compared to a single defect in the weight-bearing surface of the femoral medial condyle treated with a single sponge with double MSC content, it requires that *in vivo* testing of three-dimensional structures obtained for the regeneration of articular cartilage to be realised certainly at this level.

5 Conclusions

The term of 12 weeks is insufficient for the definitive and objective assessment of the obtained results, this term can only serve as an intermediate stage in the evaluation of the cartilage regeneration process.

The UHSCR is the unification of three scores with several similar criteria, which allows a more objective assessment of cartilaginous tissue regeneration results at the preclinical stage.

The presence of a significant difference ($p = 0.002$) between the scores obtained as a result of utilisation the same graft for the treatment of two defects on a non weight-bearing surface and of a single defect on the weight-bearing surface, leads to conclusion that all grafts used for the regeneration of articular cartilage must have a corresponding resistance to the mechanical forces present at the joint level or to take supplementary measures to limit the action of these forces on the grafts after transplantation.

Acknowledgment. The study was supported by the State Program Project: "GaN-based nanoarchitectures and three-dimensional matrices from biological materials for applications in microfluidics and tissue engineering", # 20.80009.5007.20.

Ethical Requirements. The exposed research on the domestic rabbits received the permission from Ethics Committee of Research from "N. Testemitanu" SUMPh with No. 31 from 14.12.2016.

Conflict of Interest. The authors declare no conflict of interests.

References

1. Arøen, A., Løken, S., Heir, S., et al.: Articular cartilage lesions in 993 consecutive knee arthroscopies. Am. J. Sports Med. **32**(1), 211–215 (2004). https://doi.org/10.1177/0363546503259345
2. Cobzac, V., Verestiuc, L., Jian, M., Nacu, V.: Assesment of ionic and anionic surfactants effect on demineralized osteochondral tissue. IOP Conf. Ser.: Mater. Sci. Eng. **572**, 012084 (2019). https://doi.org/10.1088/1757-899X/572/1/012084
3. Boopalan, P.R.J.V.C., Sathishkumar, S., Kumar, S.: Rabbit articular cartilage defects treated by allogenic chondrocyte transplantation. Int Orthop **30**(5), 357–361 (2006). https://doi.org/10.1007/s00264-006-0120-0
4. Zhao, M., Chen, Z., Liu, K., Wan, Y., et al.: Repair of articular cartilage defects in rabbits through tissue-engineered cartilage constructed with chitosan hydrogel and chondrocytes. J. Zhejiang Univ. Sci. B **16**(11), 914–923 (2015). https://doi.org/10.1631/jzus.B1500036
5. Zhang, J., Ming, D., Ji, Q., Liu, A., et al.: Repair of osteochondral defect using icariin-conditioned serum combined with chitosan in rabbit knees. BMC Complement Med. Ther. **20**, 193 (2020). https://doi.org/10.1186/s12906-020-02996-3
6. Aulin, C., Jensen-Waern, M., Ekman, S., Hagglund, M., et al.: Cartilage repair of experimentally 11 induced osteochondral defects in New Zealand White rabbits. Lab. Anim. **47**(1), 58–65 (2013). https://doi.org/10.1177/0023677212473716
7. Meng, X., Ziadlou, R., Grad, S., Alini, M., et al.: Animal models of osteochondral defect for testing biomaterials. Biochem. Res. Int. **2020**, 12 (2020). https://doi.org/10.1155/2020/9659412
8. Nacu, V.: Grefe tisulare în optimizarea regenerării osoase posttraumatice dereglate. N Testemitanu SUMPh, Chisinau C.Z.U.: 616.12-001.5-089.84+615.7 (2010)
9. Alice, J., Fox, S., Bedi, A., Rodeo, S.A.: The basic science of articular cartilage: structure, composition and function. Sports Health Orthop. **1**(6), 461–468 (2009). https://doi.org/10.1177/1941738109350438
10. Cobzac, V., Mostovei, A., Jian, M., Nacu, V.: An efficient procedure of isolation, cultivation and identification of bone marrow mesenchymal stem cells. Mold. Med. J. **62**(1), 35–41 (2019). https://doi.org/10.5281/zenodo.2590011
11. Orth, P., Madry, H.: Complex and elementary histological scoring systems for articular cartilage repair. Histol. Histopathol. **30**, 911–919 (2015). https://doi.org/10.14670/HH-11-620
12. Cobzac, V., Jian, M., Nacu, V.: Cellularization of small sized grafts from biological material using the gravitational modality principle. J. Phys. Conf. Ser. **1960**, 012004 (2021). https://doi.org/10.1088/1742-6596/1960/1/012004
13. Braniste, T., et al.: The influence of semiconductor nanoparticles upon the activity of mesenchymal stem cells. In: Tiginyanu, I., Sontea, V., Railean, S. (eds.) ICNBME 2019. IP, vol. 77, pp. 607–611. Springer, Cham (2020). https://doi.org/10.1007/978-3-030-31866-6_108

Composite Scaffolds with Inclusion of Magnetite Nanoparticles for Bone Tissue Engineering

F. D. Cojocaru[1], A. S. Mihai[1], V. Balan[1], C. A. Peptu[2], M. Butnaru[1], and Liliana Verestiuc[1](✉)

[1] Department of Biomedical Sciences, Faculty of Medical Bioengineering, Grigore T. Popa University of Medicine and Pharmacy of Iasi, Iasi, Romania
liliana.verestiuc@bioinginerie.ro

[2] Department of Natural and Synthetic Polymers, Faculty of Chemical Engineering and Environmental Protection, Gheorghe Asachi Technical University of Iasi, Iasi, Romania

Abstract. More than 200 bones provide vital functions to the human body. The unique regenerative capacity of bone allows the healing without structural or functional impairment in case of minor defects, while in case of major defects, bone tissue engineering (based on a scaffold, cells and bioactive factors) can be seen as an alternative to the conventional methods. Composite scaffolds for bone tissue engineering based on biopolymers and ceramics, the main components of human bone, successfully combined the key properties of the two biomaterials, as is reported in the present paper. The inclusion of magnetite in the scaffolds brings also great advantages, and moreover, functionalized with drugs, will make possible its use as targeted drug delivery systems.

Keywords: Composite scaffolds · Calcium phosphate · Biopolymers · Magnetite · Bone tissue engineering

1 Introduction

In the human body are found more than 200 bones with various dimensions and shapes. Their main functions are: movement and locomotion of the body, protection of some vital organs (brain, spinal cord, heart, lungs, most of the digestive system organs, kidneys, bladder, reproductive organs), contain the microenvironments for adult haematopoiesis to occur, stores minerals, and regulates hormones. Regarding its physiology, very important to mention is the fact that bone has an amazing feature, unique in the human body: a regenerative capacity that allows its heal without structural or functional impairment, mainly attributed to the balance of osteoblasts and osteoclasts activity [1–3].

Unluckily, this ability is not steady for long-term healing approaches of large bone defects [4], the reconstruction of the damaged bone being a real challenge in the clinical practice [5]. Because the conventional treatment methods (autografts, metallic implants, and so on) have proved to be only satisfactory and not great, in the last two decades bone tissue engineering has been seen as an alternative therapeutic approach [5]. The aim of

I. Tiginyanu et al. (Eds.): ICNBME 2021, IFMBE Proceedings 87, pp. 342–349, 2022.
https://doi.org/10.1007/978-3-030-92328-0_45

this technique involves the replacement of the affected tissue and functions reestablishment by bio-miming its native regeneration ability. It is based on the combination of a scaffold (the main component), cells and bioactive factors [5]. The scaffold is often fabricated from different biopolymers (collagen, hyaluronic acid, glycosaminoglycans, cellulose, chitosan, silk fibroin and so on) due to their bioactivity and unique biochemical functions *in vivo* and in the human body [6]. Considering the poor mechanical strength of biopolymers, they are frequently used together with ceramics: hydroxyapatite, bioactive glass, and tricalcium phosphate. Used alone, ceramic biomaterials are brittle and negatively affect cell growth and proliferation, that's why combining biopolymers and ceramics, will lead to an advanced composite scaffold with complex properties [7]. Magnetite, an iron oxide often used for hyperthermia and drug delivery, has gained attention also for bone tissue engineering, due to the fact that can remarkably increases the number of osteoblasts [8].

Based on these theoretical aspects, the aim of this study was the preparation and characterization of composite scaffolds based on three biopolymers (collagen, chitosan, hyaluronic acid), calcium phophates and magnetite for bone tissue engineering.

2 Materials and Methods

2.1 Materials

The scaffolds have been prepared using: chitosan – Cs (Mw = 309.900 Da, Vanson Chemicals, Redmond, WA, USA), collagen – Col (Type I, bovine tendon), hyaluronic acid, sodium, salt – Hya (5 mg/mL solubility in H_2O), magnetite nanoparticles – MNs colloidal suspension previously prepared in our group [9], calcium chloride – $CaCl_2 \cdot 2H_2O$, aqueous ammonia solution – NH_4OH (25% solution) and $NaH_2PO_4 \cdot H_2O$. The last reagent has been also used for the retention of simulated body fluids, together with Na_2HPO_4. *In vitro* cytotoxicity assays have been performed using: Hank's Balanced Salt Solution – HBSS, Dulbecco's Modified Eagle's Medium – DMEM (high glucose with L-glutamate and pyruvate), Bovine Fetal Serum – BFS (heat inactivated, sterile-filtered), Penicillin- Streptomycin-Neomycin – P/N/S (5000 units P, 5 mg S and 10 mg N/mL, sterile-filtered) and Thiazolyl Blue Tetrazolium Bromide – MTT. With the exception of chitosan, all of the reagents were purchased from Sigma-Aldrich, Germany.

2.2 Scaffolds Preparation

Three composite scaffolds based on chitosan, collagen, hyaluronic acid, calcium phosphates and magnetite have been prepared according to a protocol previously described [10], all the steps being represented in Fig. 1. They have the some composition in terms of biopolymers: 50% Col, 50% CS, and 5% Hya (wt%, reported to Col and Cs), but different Ca/P theoretical ratios: 1.579 (codified as CS1), 1.65 (codified as CS2) and 1.721 (codified as CS3).

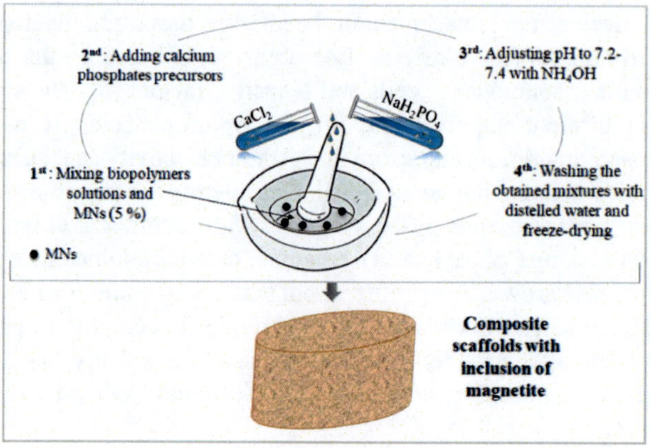

Fig. 1. Composite scaffolds preparation

2.3 Morphology and Chemical Structure

The cross-sectional morphology of the scaffolds has been studied using scanning electron microscopy – SEM (Tescan Vega SBH II, Czech Republic) at an accelerating voltage of 30 kV, while their chemical structure has been scanned within the range of 400–4000 cm^{-1} by an Fourier transform infrared spectroscope – FTIR (Bio-Rad Win-IR instrument, USA).

2.4 Mechanical Properties

Axial compression assays have been performed with a texture analyzer device TA-XT2 Plus (Stable Microsystems, UK), Young modulus being calculated according to Hooke's law (Eqs. 1 and 2, where F is the force, A the surface aria, σ is the stress and ε is the strain or deformation) [11]. Composite scaffold samples (4–6 mm height and 5–6 mm diameter) have been tested for 60 s at an initial fast deformation of 20% [10].

$$\sigma = \frac{F}{A}\left(\frac{N}{m^2}\right) \tag{1}$$

$$E = \frac{\sigma}{\varepsilon}\left(\frac{N}{m^2}\right) \tag{2}$$

2.5 In Vitro Behavior

Three samples from each composite scaffold (around 20 mg) have been completely submerged in a QIA quick VR Spin Column 50, Ø 10 mm, connected to a 1 mL syringe containing phosphate buffered solution – PBS (with pH = 7.2 and 0.01 M), prepared from $NaH_2PO_4 \cdot H_2O$ and Na_2HPO_4. The systems have been incubated at 37 °C for 24 h.

The volume of retained PBS has been measured and the maximum retention degree – RD (%) has been calculated using Eq. 3:

$$RD\,(\%) = \frac{we - w0}{w0} \times 100\,(\%) \tag{3}$$

w_0 is the initial weight of the scaffold, we the equilibrium weight: $w_e = w_0 + w_{abs}$, w_{abs} is the product between the PBS volume retained by the scaffold and the PBS density (1.02 g/mL) [12].

In vitro cytotoxicity of composite scaffolds has been analyzed in indirect contact with the human osteosarcoma MG-63 cell line, using porous membrane-based inserts with ø 4 μm pores, for 120 h. In addition, for a high accuracy of the results, the morphology of the cell was analyzed using an inverted microscope (Leica, Germany).

3 Results and Discussions

3.1 Morphology, Chemical Structure and Composition

SEM is one of the most used microscopy technique for the study of bone tissue engineering scaffolds, being an accurate tool that can image bio-nanostructures. It provides valuable data about scaffolds morphology from their general architecture to fiber surface, cross-section and material microstructure, based on its high resolution (as far as the nanometer range), high depth of field, and a significant range of accessible magnifications [13, 14].

A three-dimensional porous structure with calcium phosphate crystals and MNs uniformly distributed within the biopolymers matrix can be observed in Fig. 2, these characteristics being essential for an ideal bone tissue engineering scaffold.

For many years, FTIR has been extensively used for the characterization of engineered tissue composition and chemical structure because delivers key data about molecular vibrations within functional groups, resulting characteristic absorptions bands in the spectra that can be associated with the composition [15].

Fig. 2. Composite scaffolds morphology

For many years, FTIR has been extensively used for the Regarding our study, the most representative bands in the FTIR spectra are showed in Table 1, the results confirming the presence of all the functional groups specific to the three biopolymers, CPs and MNs.

Table 1. FTIR bands

Sample	Typical bands (cm^{-1})	Functional groups	Ref.
CS1	3422	OH	[16]
CS2	3415		
CS3	3419		
CS1	1651	Amide I, C = O	[16, 17]
CS2	1650		
CS3	1653		
CS1	1239	Amide III, N-H	[18]
CS2	1237		
CS3	1241		
CS1	1029	C-O-C	[16]
CS2	1027		
CS3	1031		
CS1	601	PO_3^{-4}	[19]
CS2	599		
CS3	603		
CS1	562	Fe-O	[16]
CS2	561		
CS3	564		

3.2 Mechanical Properties

Using Eqs. 1 and 2, mentioned in Sect. 2. *Materials and methods*, Young modulus has been calculated. The study has been performed in triplicate for each composite scaffold, the results being reported as the average ± the standard deviation (Fig. 3).

The values of E obtained for all three scaffolds can be framed in the range of those for the natural human spongy bone: 4–40 MPa [20].

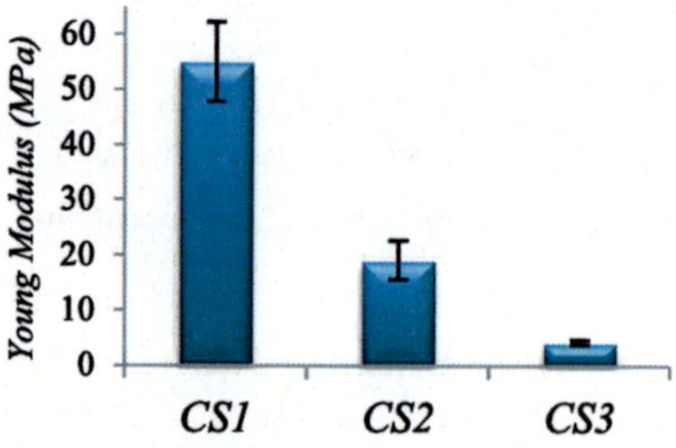

Fig. 3. E (MPa) for the three composite scaffolds

3.3 In Vitro Behavior

Before performing *in vitro* cytotoxicity assays is very important to study *in vitro behavior* of the scaffolds in presence of simulated body fluids (e.g. PBS). For example, the determination of the retention degree of PBS will offer a perspective about the swelling behavior of the scaffolds; a high value of RD, involves an increase of the pore size, which facilitate cell attachment and growth in three dimensions [21]. In this paper, the interaction of the sca-ffolds with PBS has been studied using a volumetric method. The values obtained, detailed in Fig. 4 (minimum: around 800% for CS1 and maximum around 1700% for CS2), can be explained by the porous structure of the scaffolds and their biopolymeric composition.

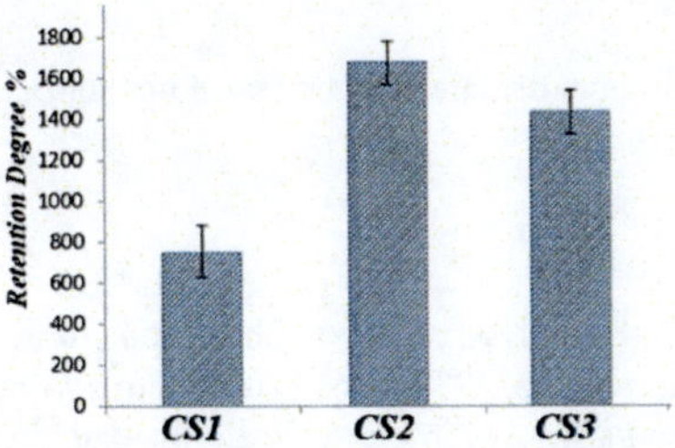

Fig. 4. RD (%) of the composite scaffolds at 24 h

Concerning *in vitro* cytotoxicity of the three composite scaffolds, MG-63 viability was higher that 85% even at 120 h of contact (Fig. 5).

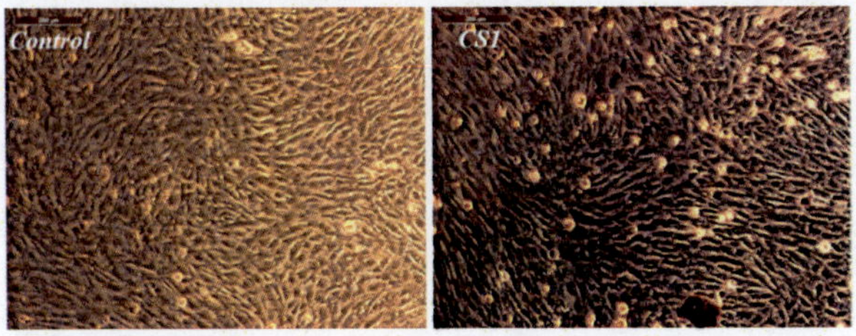

Fig. 5. Cell morphology after 120 h of indirect contact with composite scaffolds aliquots

Moreover, the cell morphology with or without scaffold (control) revealed, in both cases, a normal cell spreading and shape confirming once again the fact that the prepared composite scaffolds are not cytotoxic *in vitro.*

4 Conclusions

Based on the key characteristics of chitosan, collagen, hyaluronic acid (same quantities) and ceramics (different Ca/P ratios), three composite scaffolds have been prepared and analyzed. Magnetite was also incorporated in their structure, with the aim to be functionalized with bioactive molecules in a future study. The morphology and chemical structure are adequate for the intended application, while the values obtained for Young Modulus are comparable with those of natural human spongy bone. An average PBS retention degree around 1300% will allow the increase of the pore size, which can facilitate cell attachment and growth. Preliminary *in vitro* cytotoxicity assays showed a non-cytotoxic behavior of the scaffolds.

Acknowledgment. This work was supported by a grant of the Ministry of Research, Innovation and Digitization, CNCS/CCCDI – UEFISCDI, project number PN-III-P2-2.1-PED-2019-4524, within PNCDI III.

Conflict of Interest. The authors declare that they have no conflict of interest.

References

1. Noori, A., Ashrafi, S.J., et al.: A review of fibrin and fibrin composites for bone tissue engineering. Int. J. Nanomed. **12**, 4937–4961 (2017). https://doi.org/10.2147/IJN.S124671
2. Yong, K.W., et al.: Recent advances in mechanically loaded human mesenchymal stem cells for bone tissue engineering. Int. J. Mol. Sci. **21**, 5816 (2020). https://doi.org/10.3390/ijms21165816
3. Salhotra, A., Shah, H.N., et al.: Mechanisms of bone development and repair. Nat. Rev. Mol. Cell Biol. **21**, 696–711 (2020). https://doi.org/10.1038/s41580-020-00279-w
4. Verrier, S., et al.: Tissue engineering and regenerative approaches to improving the healing of large bone defects. Eur. Cell Mater. **32**, 87–110 (2016)
5. Safari, B., et al.: Osteogenic effects of the bioactive small molecules and minerals in the scaffold-based bone tissue engineering. Colloids Surf. B Biointerfaces **198**, 111462 (2021). https://doi.org/10.1016/j.colsurfb.2020.111462
6. Liu, J.Y., et al.: Hemostatic porous sponges of cross-linked hyaluronic acid/cationized dextran by one self-foaming process. Mater. Sci. Eng. C **83**, 160–168 (2018). https://doi.org/10.1016/j.msec.2017.10.007
7. Valtanen, R.S., et al.: Synthetic and bone tissue engineering graft substitutes: what is the future? Injury **52**, S72–S77 (2021). https://doi.org/10.1016/j.injury.2020.07.040
8. Pistone, A., et al.: Engineering of chitosan-hydroxyapatite-magnetite hierarchical scaffolds for guided bone growth. Materials **12**, 2321 (2019). https://doi.org/10.3390/ma12142321
9. Hritcu, D., Popa, M.I., et al.: Preparation and characterization of magnetic chitosan nanospheres. Turk. J. Chem. **33**, 785–796 (2009)
10. Cojocaru, F.D., et al.: Development and characterisation of microporous biomimetic scaffolds loaded with magnetic nanoparticles as bone repairing material. Ceram. Int. **47**, 11209–11219 (2021). https://doi.org/10.1016/j.ceramint.2020.12.246
11. https://www.engineeringtoolbox.com/stress-strain-d_950.html
12. Cojocaru, F.D., Balan, V., et al.: Biopolymers – calcium phosphates composites with inclusions of magnetic nanoparticles for bone tissue engineering. Int. J. Biol. Macromol. **125**, 612–620 (2019). https://doi.org/10.1016/j.ijbiomac.2018.12.083
13. Koch, M., Włodarczyk-Biegun, M.K.: Faithful scanning electron microscopic (SEM) visualization of 3D printed alginate-based scaffolds. Bioprinting (2020). https://doi.org/10.1101/2020.03.18.997668

14. Gashti, M.P., et al.: Microscopic methods to study the structure of scaffolds in bone tissue engineering: a brief review. In: Current Microscopy Contributions to Advances in Science and Technology FORMATEX Microscopy Series, Badajoz, Spain, no. 5, vol. 2, pp. 625–638 (2012)
15. Querido, W., et al.: Vibrational spectroscopy and imaging: applications for tissue engineering. Analyst **142**, 4005–4017 (2017). https://doi.org/10.1039/c7an01055a
16. Mohseni-Bandpi, A., Kakavandi, B., Kalantary, R.R., Azari, A., Keramati, A.: Development of a novel magnetite–chitosan composite for the removal of fluoride from drinking water: adsorption modeling and optimization. RSC Adv. **5**(89), 73279–73289 (2015). https://doi.org/10.1039/c5ra11294j
17. Khanarian, N.T., et al.: FTIR-I compositional mapping of the cartilage-to-bone interface as a function of tissue region and age. J. Bone Miner. Res. **29**, 2643–2652 (2014). https://doi.org/10.1002/jbmr.2284
18. Stani, C.: FTIR investigation of the secondary structure of type I collagen: new insight into the amide III band. Spectrochim. Acta Part A Mol. Biomol. Spectrosc. **229**, 118006 (2020). https://doi.org/10.1016/j.saa.2019.118006
19. Berzina-Cimdina, L., Borodajenko, N.: Research of calcium phosphates using Fourier transform infrared spectroscopy. Infrared Spectrosc. Mater. Sci. Eng. Technol. **12**, 251–263 (2012). https://doi.org/10.5772/36942
20. Edward, X.G., Yizhong, J.H., Dinescu, T.A.: Bone structure and function. Encycl. Bone Biol. **30**, 233–246 (2020)
21. Thein-Han, W.W., Misra, R.D.K.: Biomimetic chitosan–nanohydroxyapatite composite scaffolds for bone tissue engineering. Acta Biomater. **5**(4), 1182–1197 (2009). https://doi.org/10.1016/j.actbio.2008.11.025

Check for updates

Evaluation of Ultrasound Application for the Decellularization of Small Caliber Vessels

Tatiana Malcova[1,2](✉), V. Nacu[1], Gh. Rojnoveanu[2], B. Andrée[3], and A. Hilfiker[3]

[1] Laboratory of Tissue Engineering and Cell Cultures, Nicolae Testemitanu State University of Medicine and Pharmacy, Chisinau, Republic of Moldova
[2] Department of Surgery No. 1 "Nicolae Anestiadi", Nicolae Testemitanu State University of Medicine and Pharmacy, Chisinau, Republic of Moldova
[3] Leibniz Research Laboratories for Biotechnology and Artificial Organs (LEBAO), Department of Cardiothoracic, Transplantation and Vascular Surgery, Hannover Medical School, Hannover, Germany

Abstract. Decellularized matrices for tissue engineering seem to be an attractive material for providing biological vascular grafts for patients with advanced peripheral arterial disease who require bypass surgery, but do not have suitable autologous small-caliber vessels (<6 mm diameter). Currently, a variety of decellularization (DC) techniques have been proposed such as physical, chemical, and/or enzymatic methods; however, identifying an optimal protocol resulting in preservation of favorable physiochemical properties of the vascular scaffold is still elusive.

The goal of the proposed study was to examine the capacity of sonication to completely decellularize small-diameter blood vessels when applied alone, to test the effect of waves' parameters on the processing quality and matrix microarchitecture preservation, and to evaluate the possibility to reduce the time required for cell removal when ultrasound is used in combination with non-ionic detergents.

Contradictory to other DC protocols reported previously, we were not able to record completely or even partially cell removal in all studied groups. Interestingly, the combination of conventional chemicals, as Triton X-100, with physical method did not result in improving the DC efficiency and did not offer tissue permeabilization and easier chemicals access towards deeper tissue layers. In addition, when high sonication power was applied, significant destruction of the vessel matrix was determined. In summary, the use of sonication had no beneficial effect on DC in this study.

Keywords: Decellularization · Small caliber vascular grafts · Ultrasound · Detergent · Histological assessment

1 Introduction

Increased interest for tissue engineering and regenerative medicine is determined by the scarcity of available organs and the gap between demand and supply [1].

Decellularization (DC) techniques seem to be a promising method for the preparation of natural scaffolds for regenerative medicine, because the retrieved matrices maintain

I. Tiginyanu et al. (Eds.): ICNBME 2021, IFMBE Proceedings 87, pp. 350–357, 2022.
https://doi.org/10.1007/978-3-030-92328-0_46

biochemical and biomechanical properties, are biodegradable, biocompatible, stimulate cell migration and proliferation, and do not stimulate immune response, if allogeneic material is used [2, 3].

Taking the existing advantages of acellular materials into consideration, they represent an attractive option for patients with advanced end-stage cardiovascular disease, as peripheral arterial disease or coronary artery disease, who require bypass surgery, but do not have suitable autologous small-caliber vessel (<6 mm diameter) [1, 4, 5].

DC can be realized by a variety of methods. The majority of the publications in the field refer to application of chemical and biological agents or combined approaches; physical methods being assessed inadequately so far [4–7].

Till now, the contributions of physical methods in DC, such as rapid freeze-thaw (thermal shock), perfusion, supercritical carbon dioxide, immersion and agitation, hydrostatic pressure and sonication [2] when used alone or in combination with chemical and biological approaches are insufficiently evaluated.

For instance, so far, sonication treatment has been used for DC of tissues such as larynx [8, 9], trachea [10], kidney [11–13], meniscal tissue [14–17], aorta [3, 18–21], skeletal muscle [22], or osteochondral tissue [23]. It is assumed that sonication may have the potential to decrease the time required for complete tissue decellularization by inducing cell membrane perforation and enhancing detergent (Det) molecules uptake. This effect is determined by its ability to reversible loosen the junctions sealing the cells together [21, 24, 25].

The effects induced from the ultrasonic waves, including homogenization, dispersing, deagglomeration, emulsification, altering cell membrane integrity, debris removal, disintegration, loosening cells junctions, and sonochemical effects, are caused by cavitation. Cavitation phenomenon consists in the creation of vapor cavities (micro bubbles) in a fluid determined by altering high-pressure and low-pressure cycles, with rates depending on the frequency (the number of bubbles increasing with increasing sonication power). During the low-pressure cycle (the rarefying phase of sound wave), these bubbles grow. During this event the gas inside the bubble reaches very high temperature [26]. When the bubbles cannot longer absorb energy, they collapse violently during a high-pressure cycle (the compression phase), inducing significant local temperature, pressure increase, and generating a violent shock wave through the medium. In this way the sonic energy is transformed into mechanical energy responsible for the majority of biological effects caused by the ultrasound [14, 27–29].

The sonication process can be carried out by different types of ultrasonic systems, as probe-type ultrasonic homogenizer (direct sonication) or an ultrasonic bath (indirect sonication).

However, the cavitation process generated in an ultrasonic bath seems to be uncontrollable due to by the uneven distribution of the ultrasonic irradiation field in the ultrasonic tanks; thus, the repeatability and scalability of the process is very poor. Alternatively, a probe-type device offers full control over the sonication parameters ensuring reproducibility of the experimental results [12, 23].

This study was carried out to evaluate the capacity of sonication treatment to complete cell removal from small-diameter blood vessels when used alone or in combination with a non-ionic Det, as Triton X-100.

2 Materials and Methods

2.1 Decellularization System

Porcine vessels were collected from German Landrace pigs (3–5 months old) in the animal facility of the Medizinische Hochschule Hannover (MHH). Vessels were cleaned from fat and adjacent tissue using forceps and scissors. Vessels were stored at −80 °C.

DC of the vascular tissue has been done by physical (sonication and osmotic shock) and combined (sonication with Det) methods.

For sample processing a direct sonication method (direct sonicator UP200S Hielscher, Germany and Sonotrode S1 for samples from 0.1 to 5 mL) was used. Taking into consideration that collapsing bubbles may produce significant thermal loads emitted in the surrounding liquid [8], the experiments were performed in a cold room (+4 °C). In addition, the samples were placed in an ice bath in order to prevent tissue over-heating.

Physical Treatment: For decellularization, the vessels (1 cm long segment, internal diameter 4 mm) were flushed with PBS and submerged in 2.0 mL Eppendorf tubes containing 1.5 mL hypotonic lysis buffer (0.3% NaCl in distilled water). The samples were exposed to sonication with a frequency of 24 kHz, 200 W, control mode "1" (permanent acoustic irradiation). Two different amplitude values and two different exposure times for DC were applied: 20% *vs* 100% and 3 h *vs* 12 h, respectively.

Combined Method: The samples from this group were placed in a 2.0 mL Eppendorf tubes containing 1.5 mL 1% Triton X-100 and exposed to sonication with a frequency of 24 kHz, 200 W, amplitude 20%, control mode "1" for 48 h. Distance of the sample to the tip of the ultrasound probe was set at 1 cm. As control, samples were treated with the same solution under continuous rotation (50 rpm speed, Biometra WT 17).

2.2 Histological Analysis

The retrieved matrices were evaluated qualitatively for remaining cellular components and DNA through H&E and DAPI staining, respectively. Native porcine vessels were used as controls.

H&E Staining. Frozen samples were cut into 7 μm sections. Sections were transferred on slides and stored overnight at −80 °C. Samples were fixed in acetone at −20 °C for 8 min. The fixed slides were placed in a staining rack and moved through hemotoxylin (8 min), eosin (20 s), and a series of alcoholic solutions (namely, 5 min in 95% ethanol and 5 min in 100% ethanol). Finally, after clearing the samples in xylene (10 min), they were covered with cover slips using corbit balsam and left to dry overnight.

DAPI Staining. Sections from snap frozen samples were used for DAPI staining. The sections were washed in PBS and fixed in 4% paraformaldehyde for 20 min in a wet chamber. Sections were washed for 15 min with PBS, and incubated for 15 min with DAPI (0.33 μg/mL in PBS). After three washing steps in PBS, 5 min each, the samples were covered with cover slips using Shandon Immu-Mount fluorescent mounting medium. The stained sections were then analyzed with a fluorescence microscope.

3 Results

The pictures of the histological stainings of samples treated by sonication without supplementary Det exposure for 3 and 12 h-treatments are presented in Fig. 1.

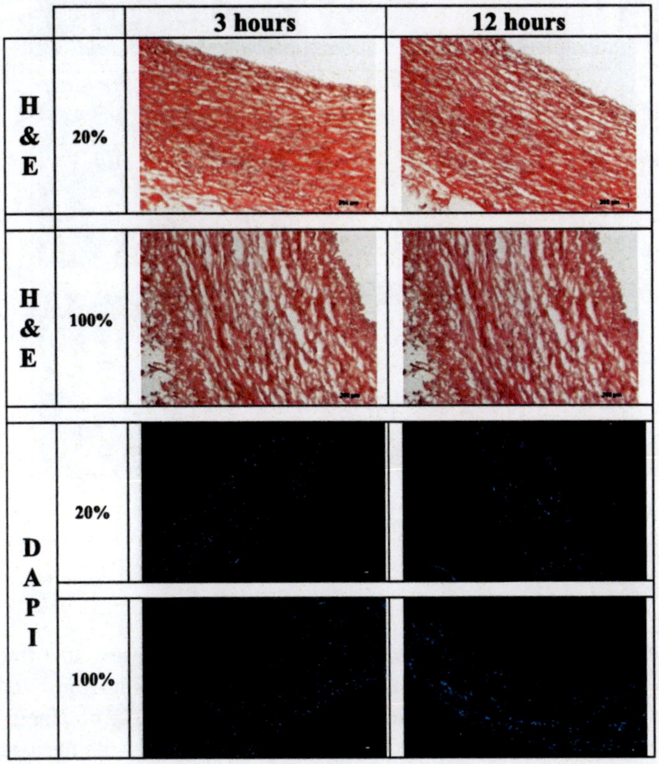

Fig. 1. Ultrasound application in carotid artery decellularization (histological assessment)

Interestingly, H&E and DAPI staining revealed the presence of huge amounts of intact cells under all investigated conditions demonstrating that sonication was not able to remove cells efficiently from the tissue even with high amplitude setting and prolonged exposure. In addition, H&E staining analysis showed that the tissue structures were significantly affected by high amplitude waves (100%).

Results from combined treatment with Det and ultrasound are shown in Fig. 2. Native carotid artery was used as control. H&E and DAPI staining demonstrated the presence of intact cells and nuclei in treated samples (Triton X-100 in combination with sonication or rotation) to the same extent as in the control. Nuclei within the vascular wall remained intact in treated tissue, suggesting that the reagent was not able to solubilize the membranes and to induce cell removal even with sonication.

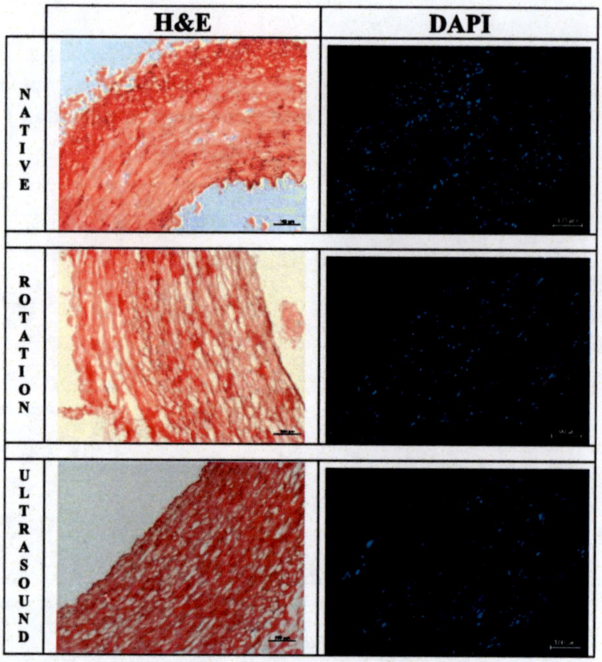

Fig. 2. Combined approach in carotid artery decellularization (histological assessment)

4 Discussion

Azhim A *et al.* (2011) demonstrated for the first time that using low frequency (20 kHz), high power (15 or 30 W) ultrasound in combination with strong ionic Det (2% sodium dodecyl sulfate (SDS) complete DC of aortic tissue can be obtained. The results were compared to a typical DC process (immersion or shaking with the same chemical solution). A novel DC system consisting of a commercially available ultrasonic horn and a roller pump was used by the group. They stated that adding sonication significantly improved DC depth with cells being completely removed from the scaffold, which did not happen in immersion treatment [20].

In a second study, the recellularization potential of such DC matrix with vascular smooth muscle cells (VSMCs) was evaluated. In this case, DC was carried out at 36 ± 1 °C temperature in continuous oscillation with ultrasonic power of approximately 500 mVrms and ultrasonic frequency of 170 kHz for 10 h in 2% SDS solution. The study confirmed successful VSMCs adherence and its infiltration into the tissues [21]. Further, the same working group reported in 2019 the development of a closed sonication DC system used together with 0.1% or 2% SDS for preparing acellular aortic scaffolds [3].

Also other reports demonstrated the efficiency of ultrasound application in combination with other chemicals in larynx (probe-type ultrasonic homogenizer Sonifier 250, output control setting 6, 6% DMSO and 1% Triton X-100) [9], meniscus (closed sonication system, 40 kHz, 0.1% SDS) [16, 17], and kidney (bath-type sonicator, sonicator power

0 W/60 W/120 W, 0.25%/0.625%/1% SDS, Triton X-100 OR 150 W/200 W/250 W, frequency 20 kHz, perfusion 10 mL/min, 1% SDS) DC [11–13].

Only the report byTchoukalova YD *et al.* (2018) did not confirm the efficiency of sonication combined with Det (4% sodium deoxycholate, SDC) in washout of cells from the cartilaginous tissue [10].

In the current study, we were not able to obtain DC by treating porcine carotides with ultrasonic waves even in combination with a Det; however, using Det and rotation together was not resulting in DC as well. As Triton X-100 is considered a weak Det this might explain the inefficiency of the treatment compared to other studies. To investigate if sonication may offer permeabilization of the matrix and easier access of the chemicals to the deep layers, we suggest the application of stronger detergents such as SDS or SDC.

In addition, the tissue structure seemed to be significantly affected by high amplitude waves (Fig. 1). The negative impact of long exposure to high sonicator settings can be explained by prolonged subjection to the cavitation effect and microbubbles action that can puncture and destroy the matrix. The same results were reported previously [11, 14, 18, 19, 22, 30].

5 Conclusions

No study thus far has provided results for small-diameter blood vessel decellularization, as porcine carotid artery, when using sonication alone or combination of chemical immersion and ultrasound. Nevertheless, using ultrasound in DC did not appear to be an efficient strategy to remove cells from the tissue. The same results were obtained when using ultrasonic waves in combination with a non-ionic detergent.

In addition, the lack of DNA reduction when applying Triton X-100 suggests the necessity of using sonication in combination with stronger chemicals, such as sodium dodecyl sulfate or sodium deoxycholate, for small caliber vessel DC.

Acknowledgment. The authors are grateful to the programme Horizon 2020 for its finance support through a NanoMedTwin project "Promoting smart specialization at the Technical University of Moldova by developing the field of Novel Nanomaterials for BioMedical Applications through excellence in research and twinning" (810652) and the project "GaN-based nanoarchitectures and three-dimensional matrices from biological materials for applications in microfluidics and tissue engineering" (20.80009.5007.20, State Program 2020–2023).

Conflict of Interest. The authors declare that they have no conflict of interest.

References

1. Gilpin, A., Yang, Y.: Decellularization strategies for regenerative medicine: from processing techniques to applications. Biomed. Res. Int. **2017**, 9831534 (2017). https://doi.org/10.1155/2017/9831534

2. Rabbani, M., Zakian, N., Alimoradi, N.: Contribution of physical methods in decellularization of animal tissues. J. Med. Signals Sens. **11**(1), 1–11 (2021). https://doi.org/10.4103/jmss.JMSS_2_20
3. Hazwani, A., Sha'Ban, M., Azhim, A.: Characterization and in vivo study of decellularized aortic scaffolds using closed sonication system. Organogenesis **15**(4), 120–136 (2019). https://doi.org/10.1080/15476278.2019.1656997
4. Pashneh-Tala, S., MacNeil, S., Claeyssens, F.: The tissue-engineered vascular graft-past, present, and future. Tissue Eng. Part B Rev. **22**(1), 68–100 (2016). https://doi.org/10.1089/ten.teb.2015.0100
5. Simsa, R., Padma, A.M., Heher, P., et al.: Systematic in vitro comparison of decellularization protocols for blood vessels. PLoS ONE **13**(12), e0209269 (2018). https://doi.org/10.1371/journal.pone.0209269
6. Starnecker, F., König, F., Hagl, C., et al.: Tissue-engineering acellular scaffolds-the significant influence of physical and procedural decellularization factors. J. Biomed. Mater. Res. B Appl. Biomater. **106**(1), 153–162 (2018). https://doi.org/10.1002/jbm.b.33816
7. Lin, C.H., Hsia, K., Ma, H., et al.: In vivo performance of decellularized vascular grafts: a review article. Int. J. Mol. Sci. **19**(7), 2101 (2018). https://doi.org/10.3390/ijms19072101
8. Hung, S.H., Su, C.H., Lee, F.P., et al.: Larynx decellularization: combining freeze-drying and sonication as an effective method. J. Voice **27**(3), 289–294 (2013). https://doi.org/10.1016/j.jvoice.2013.01.018
9. Nayakawde, N.B., Methe, K., Premaratne, G.U., et al.: Combined use of detergents and ultrasonication for generation of an acellular pig larynx. Tissue Eng. Part A **27**(5–6), 362–371 (2021). https://doi.org/10.1089/ten.TEA.2020.0054
10. Tchoukalova, Y.D., Hintze, J.M., Hayden, R.E., et al.: Tracheal decellularization using a combination of chemical, physical and bioreactor methods. Int. J. Artif. Organ. **41**(2), 100–107 (2018). https://doi.org/10.5301/ijao.5000648
11. Say, S., Dugos, N., Roces, S., et al.: Effect of sonication power on perfusion decellularization of cadaveric porcine kidney. In: MATEC Web of Conference (RSCE 2018), vol. 268, p. 01009 (2019). https://doi.org/10.1051/matecconf/201926801009
12. Manalastas, T.M., Dugos, N., Ramos, G., Mondragon, J.M.: Effect of Decellularization parameters on the efficient production of kidney bioscaffolds. Appl. Biochem. Biotechnol. **193**(5), 1239–1251 (2020). https://doi.org/10.1007/s12010-020-03338-2
13. Say, S., Dugos, N.P., Roces, S.A., et al.: Sonication-assisted perfusion decellularization of whole porcine kidney. Int. J. Biol. Biomed. **13**, 78–81 (2019)
14. Azhim, A., Ono, T., Fukui, Y., et al.: Preparation of decellularized meniscal scaffolds using sonication treatment for tissue engineering. In: Annual International Conference of the IEEE Engineering in Medicine and Biology Society 2013, pp. 6953–6956 (2013). https://doi.org/10.1109/EMBC.2013.6611157
15. Mardhiyah, A., Sha'ban, M., Azhim, A.: Evaluation of histological and biomechanical properties on engineered meniscus tissues using sonication decellularization. In: Annual International Conference of the IEEE Engineering in Medicine and Biology Society 2017, pp. 2064–2067 (2017). https://doi.org/10.1109/EMBC.2017.8037259
16. Yusof, F., Sha'ban, M., Azhim, A.: Development of decellularized meniscus using closed sonication treatment system: potential scaffolds for orthopedics tissue engineering applications. Int. J. Nanomed. **14**, 5491–5502 (2019). https://doi.org/10.2147/IJN.S207270
17. Yusof, F., Sha'ban, M., Azhim, A.: Engineered meniscus scaffolds using sonication decellularization treatment system. Int. J. Integr. Eng. **11**(3), 81–89 (2019)
18. Azhim, A., Syazwani, N., Morimoto, Y., et al.: The use of sonication treatment to decellularize aortic tissues for preparation of bioscaffolds. J. Biomater. Appl. **29**(1), 130–141 (2014). https://doi.org/10.1177/0885328213517579

19. Syazwani, N., Azhim, A., Morimoto, Y., Furukawa, K., Ushida, T.: Immune response of implanted aortic scaffolds decellularized by sonication treatment. In: Goh, J. (ed.) The 15th International Conference on Biomedical Engineering. IP, vol. 43, pp. 275–278. Springer, Cham (2014). https://doi.org/10.1007/978-3-319-02913-9_70
20. Azhim, A., Yamagami, K., Muramatsu, K., et al.: The use of sonication treatment to completely decellularize blood arteries: a pilot study. In: Annual International Conference of the IEEE Engineering in Medicine and Biology Society 2011, pp. 2468–2471 (2011). https://doi.org/10.1109/IEMBS.2011.6090685
21. Fitriatul, N., Sha'ban, M., Azhim, A.: Evaluation of recellularization on decellularized aorta scaffolds engineered by ultrasonication treatment. In: Annual International Conference of the IEEE Engineering in Medicine and Biology Society 2017, pp. 2072–2075 (2017). https://doi.org/10.1109/EMBC.2017.8037261
22. Lin, C.H., Yang, J.R., Chiang, N.J., et al.: Evaluation of decellularized extracellular matrix of skeletal muscle for tissue engineering. Int. J. Artif. Organ. **37**(7), 546–555 (2014). https://doi.org/10.5301/ijao.5000344
23. Forouzesh, F., Rabbani, M., Bonakdar, S.: A Comparison between ultrasonic bath and direct sonicator on osteochondral tissue Decellularization. J. Med. Signals Sens. **9**(4), 227–233 (2019). https://doi.org/10.4103/jmss.JMSS_64_18
24. Peruzzi, G., Sinibaldi, G., Silvani, G., et al.: Perspectives on cavitation enhanced endothelial layer permeability. Colloids Surf. B Biointerfaces **168**, 83–93 (2018). https://doi.org/10.1016/j.colsurfb.2018.02.027
25. Ross, J.P., Cai, X., Chiu, J.F., et al.: Optical and atomic force microscopic studies on sonoporation. J. Acoust. Soc. Am. **111**(3), 1161–1164 (2002). https://doi.org/10.1121/1.1448340
26. Magaletti, F., Marino, L., Casciola, C.M.: Shock wave formation in the collapse of a vapor nanobubble. Phys. Rev. Lett. **114**(6), 064501 (2015). https://doi.org/10.1103/PhysRevLett.114.064501
27. Saranya, N., Devi, P., Nithiyanantham, S., Jeyalaxmi, R.: Cells disruption by ultrasonication. BioNanoScience **4**(4), 335–337 (2014). https://doi.org/10.1007/s12668-014-0149-2
28. Probe-Type Sonication vs. Ultrasonic Bath: An Efficiency Comparison. https://www.hielscher.com/probe-type-sonication-vs-ultrasonic-bath-an-efficiency-comparison.htm. Accessed 25 May 2021
29. Hesson, J.R.: High energy sound ultrasonics (2014). http://www.hessonic.com/fundamentals-of-ultrasonic-cleaning.html. Accessed 20 May 2021
30. Wang, Y., Bao, J., Wu, Q., et al.: Method for perfusion decellularization of porcine whole liver and kidney for use as a scaffold for clinical-scale bioengineering engrafts. Xenotransplantation **22**, 48–61 (2015). https://doi.org/10.1111/xen.12141

Mimicking *In Vivo* Tissue Microenvironment for *In Vitro* Testing – Hydrogels for Cell Encapsulation

A. Luca, T. R. Craescu, L. Verestiuc, and Maria Butnaru(✉)

Faculty of Medical Bioengineering, Grigore T. Popa University of Medicine and Pharmacy, Iasi, Romania
maria.butnaru@umfiasi.ro

Abstract. Cell-encapsulating hydrogels must be obtained in rigorous and completely cytocompatible conditions, offering this way tissue engineered products that mimic the *in vivo* microenvironment. These hydrophilic structures bring *in vitro* testing to a new level of complexity by offering a tissue-like architecture. Various polymeric structures have been studied in the aim to build hydrogels with physical and mechanical integrity as cell-supporting medium. In this study, collagen-based hydrogels crosslinked with oxidized polysaccharides (chitosan and hyaluronic acid) were used for encapsulating two different types of cells, primary *Albino rabbit* fibroblasts and human epidermoid carcinoma A-431 cells. The cells were able to proliferate into hydrogels and maintained their viability for at least eight days, for both cell lines and initiate tumor formation in the case of A-431 cells. It was observed a better cell integration into oxidized hyaluronic acid crosslinked hydrogels which allowed cell-cell interactions and the formation of spheroids.

Keywords: Collagen · Tissue-like architecture · Cell encapsulation · Oxidized chitosan · Oxidized hyaluronic acid

1 Introduction

Cell encapsulation technique have been developed for tissue engineering applications from the need of offering a closer medium to native environment, which permit the cells growing and developing in structures that will eventually replace the damaged tissue. This environment is represented *in vivo* by the extracellular matrix (ECM), which is a very complex network of fibrous proteins (such as collagen), proteoglycans, glycosaminoglycans with an important role in influencing cell behavior and functions. There is an inter-dependent relation between cells and ECM, the cells being able to remodel the ECM, while signals from ECM greatly influence cell characteristics and functioning such as proliferation, adhesion, migration, polarity, differentiation, and apoptosis. It offers mechanical support especially through the protein component, cues for cell adhesion, while the proteoglycans offer hydration and compression resistance [1–3].

Creating an accurate extracellular matrix that will mimic the complexity of the native one involves the use of components that resemble in structure and properties with those

I. Tiginyanu et al. (Eds.): ICNBME 2021, IFMBE Proceedings 87, pp. 358–364, 2022.
https://doi.org/10.1007/978-3-030-92328-0_47

that are found *in vivo*. Unlike culturing cells on rigid surfaces, in monolayer, three-dimensional (3D) cell cultures offer to the cells the possibility of developing and behaving closer to their native state, allowing more accurate *in vitro* studies in the field of cell response to drugs or cancer cell biology, the resulting tissue engineered products being one step closer to practical use [4, 5].

Cell encapsulation in hydrogels offers these 3D conditions and necessitates completely cytocompatible working parameters regarding the pH, temperature and chemical reagents used in hydrogel formation, conditions imposed by the presence of cells in the crosslinking process [6, 7]. Various materials have been considered for 3D scaffolding in the presence of cells. Natural polymers have several advantages including biocompatibility and the ability of promoting cell functions; however, some of their poor mechanical properties imposed crosslinking processes [8–10].

Considering the need for tissue engineered products that will mimic the *in vivo* conditions, the purpose of this study was to create an environment for encapsulated cells closer to the native ECM, using collagen-based hydrogels, the most abundant protein found in ECM. The structure of the collagen was stabilized by crosslinking with oxidized chitosan and oxidized hyaluronic acid, respectively. These hydrogels were used for encapsulating primary fibroblasts or A-431 cells (human epidermoid carcinoma cell line) to offer a 3D growth environment for observing cell behavior, with the potential of being used as *in vitro* models for tissue engineering constructs and drug effectiveness studies.

2 Materials and Methods

2.1 Materials

The collagen used for hydrogel formation was purchased for Stem Cell Technologies, in a concentration of 6 mg/ml, pH 2.0. Chitosan (medium molecular weight, 76 000 Da) was purchased from Sigma and was oxidized at a concentration of 1% and used for crosslinking at a concentration of 10 mg/ml. Hyaluronic acid was also purchased from Sigma and prepared at a concentration of 1 mg/ml for oxidation, while in the hydrogels was used at 10 mg/ml. Sodium meta-periodate ($NaIO_4$) used for oxidation was purchased from Merck. Calcein AM, DAPI (4′,6-diamidino-2-phenylindole) and Rhodamine-Phalloidin were purchased from Life Technologies. *Albino rabbit* fibroblasts were isolated in the Training and Research Center in Tissue Engineering, Artificial Organs and Regenerative Medicine of Medical Bioengineering Faculty from UMF "Gr. T. Popa", Iasi, while human A-431 cells were purchased from ATCC Cell Bank. All the culture media and other reagents used were purchased from Sigma.

2.2 Methods

Chitosan and Hyaluronic Acid Oxidation
The oxidation protocol was described in literature and in our previous work [11, 12]. Briefly, a solution of 1% chitosan was prepared by dissolving the polysaccharide in HCl (1% in double distilled water) and adjusting the pH at 4 for complete dissolution. After

this was achieved, the sodium meta-periodate solution was obtained in double distilled water, at 4 °C, and the oxidation process was initiated by adding $NaIO_4$ dropwise to the chitosan solution, away from light exposure, under continuous stirring for 24 h. The reaction was stopped by adding ethylene glycol into the mixture and leaving it overnight, under stirring.

The same protocol was used for hyaluronic acid oxidation, after preparing a solution of 1% hyaluronic acid in double distilled water, at room temperature.

After finalizing the oxidation process, the solutions were purified by placing them in dialysis membranes and dialyzed against double distilled water, for three days (changing the double distilled twice every day), after which they were lyophilized (−54 °C, 0.05 mBa) and kept away from direct light exposure, at 4 °C. The oxidation was confirmed by infrared spectroscopy – FTIR and ^{1}H NMR.

Cell Culture Conditions
The culture media used for primary fibroblasts culture was Dulbecco's Modified Eagle's Medium/Nutrient mixture F12 Ham (DMEM/F12) with L-glutamine, supplemented with 10% fetal bovine serum (FBS) and 1% penicillin/streptomycin/neomycin antibiotic mixture. A-431 cells were cultured in DMEM High Glucose (4500 mg/ml), in incubation conditions at 37 °C, 6% CO_2 and 96% humidity. After reaching a monolayer, the cells were detached by trypsinization, counted and the vitality was determined by trypan blue staining.

Hydrogel Cell Encapsulation
The encapsulation followed the protocol that was described in our previous studies [11, 12]. Briefly, for hydrogel formation, collagen solution was neutralized and brought to a concentration of 3 mg/ml by adding culture media (specific for the type of cells encapsulated) and 10 x PBS. The stabilization of the collagen fiber was obtained by adding the oxidized component, either chitosan or hyaluronic acid as a crosslinking agent and as a structural component, in a ratio of 4.5:1 v/v to collagen. All the components of the hydrogels were mixed in 60 mm center-well organ culture dish (Falcon), in triplicate. Cell suspension was added as a final step of the mixture, fibroblasts being used at a final concentration of 200.000 cells/mL, while A-431 cells, 400.000 cells/mL. The hydrogels with encapsulated cells were allowed to crosslink in incubation conditions, at 37 °C, for 20 min, after which culture media was added.

Fluorescence Microscopy
For analyzing cell viability and shape after encapsulation, these were visualized using fluorescence microscopy and specific markers: Calcein AM for evaluating cell viability, DAPI and Rhodamine-Phalloidin for evaluating cell nucleus and cytoskeleton. The staining procedure was performed according to producer instructions and the cells were observed using an inverted Leica DMI3000 fluorescence microscope.

3 Results and Discussions

3.1 Hydrogel Formation

Collagen is the preferred protein when it comes to applications which involve direct cell contact, but its low mechanical properties must be improved in order to obtain a proper support for tissue engineering applications [13]. This can be done by physical or chemical crosslinking the collagen, even though the presence of cells in the hydrogel formation process limits greatly the possible choice of substances. Therefore, in this study, it was chosen the use of oxidized biocompatible polysaccharides as crosslinking agents. The cytocompatibility of the oxidized polysaccharides used as crosslinkers was tested in our previous studies [11, 12]. The hydrogels studied in this paper are obtained based on the reaction between the amino groups present in collagen and the aldehydes resulted from the oxidation process of chitosan and hyaluronic acid, respectively. The behavior of encapsulated cells in the two types of hydrogels obtained after collagen crosslinking with the oxidized macromolecules was evaluated and compared. Fibroblasts were chosen because of their important role in tissue regeneration *in vivo*, their role as a cell source for tissue engineered products and because of their implications in pathological conditions [14, 15]. Epidermoid carcinoma cells, A-431 were chosen as model tumor cells because of their ability to initiate tumors, 3D culture conditions promoting interactions between cells and with the extracellular matrix, which better replicates natural environment [16].

3.2 Cell Encapsulation

Fibroblasts were encapsulated in collagen-based hydrogels crosslinked with oxidized chitosan and oxidized hyaluronic acid, respectively. In this study, cell development was monitored comparatively at 48 h after encapsulation, on day 4 and on day 8. It was observed that initially round-shaped cells started to adhere to the hydrogels and gain an elongated shape. This morphology was seen especially in hyaluronic acid crosslinked hydrogels, as it can be observed in Fig. 1(A). Eight days after encapsulation, oxidized hyaluronic acid crosslinked hydrogels presented more elongated shaped fibroblasts, compared with oxidized chitosan crosslinked hydrogels which presented round-shaped cells (Fig. 1(B)). In our previous work, similar hydrogels crosslinked with oxidized polysaccharides were able to maintain cell viability for at least 12 days, with an exponential cell proliferation until day 3 of culture, continuing until day 5 of culture, as evaluated with Alamar Blue assay [11].

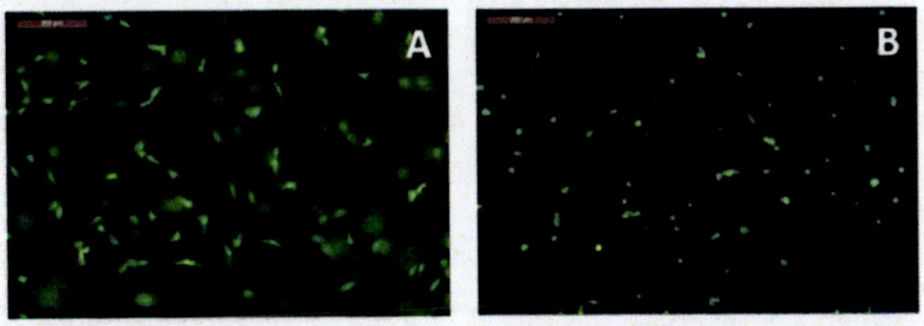

Fig. 1. Encapsulated fibroblasts morphology and viability, eight days after encapsulation in (A) oxidized hyaluronic acid crosslinked hydrogel; (B) oxidized chitosan crosslinked hydrogel. Green – Calcein AM, fluorescent live staining for viable cells.

Calcein AM was used as a viability marker in order to assess which hydrogel performed better in offering a more cytocompatible support for cellular development, eight days after encapsulation. It can be observed in Fig. 1 that hyaluronic acid crosslinked hydrogel presented more viable cells, compared with chitosan crosslinked one, indicating that the use of another important component of the native extracellular matrix, along with its most abundant protein, collagen, offers a proper support for cells developing. This type of crosslinking proves to be optimal as it eliminates the use of cytotoxic compounds that, even though might form stable supports, cannot be used in direct contact with the cells and without supplementary purification steps [6].

For this reason, finding a completely cytocompatible method for hydrogel crosslinking raises many obstacles and one possibility is the stabilization of the structure with oxidized ECM components, in this case hyaluronic acid.

Considering the results obtained for fibroblasts encapsulated in oxidized hyaluronic acid crosslinked hydrogel, it was decided to continue the study by also encapsulating A-431 cells into the same type of hydrogel for observing the ability of the obtained 3D support to allow cell-cell interaction and the initiation of spheroid formation (Fig. 2). The cells were observed for eight days, with the daily changing of half of the culture medium. Cell morphology was observed with the use of an inverted microscope, under brightfield, but also with the use of fluorescent markers, DAPI and Rhodamine-Phalloidin. Cell viability eight days after encapsulation was determined using Calcein AM.

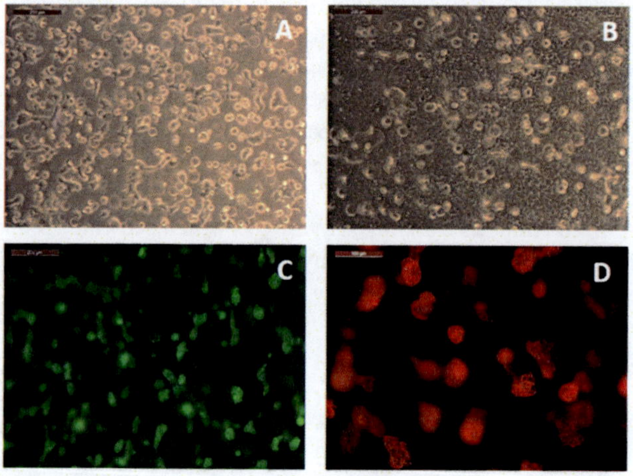

Fig. 2. A-431 encapsulated cells morphology after 48 h (A) and on day 8 (B); cell viability on day 8 (C); spheroid formation, day 8 (D). Green – Calcein AM, fluorescent live staining for viable cells, red – Rhodamine-Phalloidin staining for cell cytoskeleton.

In Fig. 2 it can be observed the initial cell adhesion to the support represented by the collagen-based hydrogel and the tendency of the cells to cluster, with the formation of spheroids. Figure 3(A) presents the interactions between cells, as the hydrogel support enables cell-cell communication favoring the formation of a tissue specific microenvironment.

The formation of spheroids by tumor cells can be observed in Fig. 3(B), the obtained hydrogel proving to be an excellent matrix for studying tumor cell biology. This method of study shortens the testing time for tumor agents that might present a different response

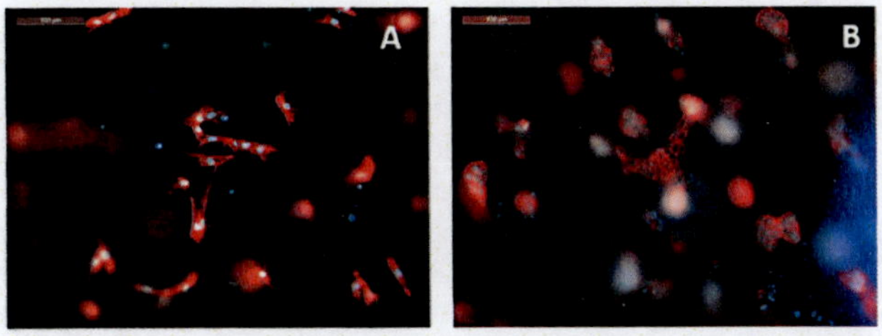

Fig. 3. Cell-cell interaction between encapsulated cells (A) encapsulated fibroblasts; (B) spheroid formation of A-431 encapsulated cells. Blue – DAPI, fluorescent staining for nuclear DNA, red – Rhodamine-Phalloidin staining for cell cytoskeleton

when studied on cell monolayers. Spheroids can recreate more accurate gene expression patterns, like those found on native tissues [16].

The results encourage us in considering the collagen-based hydrogel, crosslinked with oxidized hyaluronic acid as a proper support for cell proliferation and future *in vitro* studies in which therapeutic agents can be tested for their effects on cells organized in structures that offer more complex conditions than those offered by a two-dimensional cell culture.

4 Conclusions

In this study two different hydrogels were obtained by crosslinking collagen in cytocompatible conditions, with oxidized chitosan and oxidized hyaluronic acid, respectively. The hydrogels were used for encapsulating primary mammalian fibroblasts and human A-431 tumor cells in order to obtain three-dimensional structures that mimic the *in vivo* tissue architecture. Oxidized hyaluronic acid crosslinked hydrogel performed better in terms of cell proliferation, viability and cell-cell interactions and it was chosen as a support for tumor cell encapsulation. The obtained structure allowed spheroid formation observed with fluorescent markers indicating that this type of support can be used as a model for future *in vitro* studies like drug effects on cells found in an environment as close as possible as the one given by native tissues.

Acknowledgment. This work was supported by a grant of the Ministry of Research, Innovation and Digitization, CNCS/CCCDI – UEFISCDI, project number PN-III-P2-2.1-PED-2019-4524, within PNCDI III.

Conflict of Interest. The authors declare that they have no conflict of interest.

References

1. Assuncao, M., Dehghan-Baniani, D., Yiu, C., et al.: Cell-derived extracellular matrix for tissue engineering and regenerative medicine. Front. Bioeng. Biotechnol. **8**, 1378 (2020). https://doi.org/10.3389/fbioe.2020.602009

2. Frantz, C., Stewart, K.M., Weaver, V.M.: The extracellular matrix at a glance. J. Cell Sci. **123**(Pt 24), 4195–4200 (2010). https://doi.org/10.1242/jcs.023820
3. Kim, Y., Ko, H., Kwon, I.K., et al.: Extracellular matrix revisited: roles in tissue engineering. Int. Neurourol. J. **20**(Suppl 1), S23–S29 (2016). https://doi.org/10.5213/inj.1632600.318
4. Park, Y., Huh, K.M., Kang, S.W.: Applications of biomaterials in 3D cell culture and contributions of 3D cell culture to drug development and basic biomedical research. Int. J. Mol. Sci. **22**(5), 2491 (2021). https://doi.org/10.3390/ijms22052491
5. Bassi, G., Panseri, S., Dozio, S.M.: Scaffold-based 3D cellular models mimicking the heterogeneity of osteosarcoma stem cell niche. Sci. Rep. **10**(1), 22294 (2020). https://doi.org/10.1038/s41598-020-79448-y
6. Nicodemus, G.D., Bryant, S.J.: Cell encapsulation in biodegradable hydrogels for tissue engineering applications. Tissue Eng. Part B Rev. **14**(2), 149–165 (2008). https://doi.org/10.1089/ten.teb.2007.0332
7. Echalier, C., Valot, L., Martinez, J., et al.: Chemical cross-linking methods for cell encapsulation in hydrogels. Mater. Today Commun. **20**, 100536 (2019). https://doi.org/10.1016/j.mtcomm.2019.05.012
8. Pupkaite, J., Rosenquist, J., Hilborn, J., Samanta, A.: Injectable shape-holding collagen hydrogel for cell encapsulation and delivery cross-linked using thiol-michael addition click reaction. Biomacromol **20**(9), 3475–3484 (2019). https://doi.org/10.1021/acs.biomac.9b00769
9. Somo, S.I., Brown, J.M., Brey, E.M.: Dual crosslinking of alginate outer layer increases stability of encapsulation system. Front. Chem. **8**, 575278 (2020). https://doi.org/10.3389/fchem.2020.5752782
10. Han, C., Zhang, H., Wu, Y., et al.: Dual-crosslinked hyaluronan hydrogels with rapid gelation and high injectability for stem cell protection. Sci. Rep. **10**, 14997 (2020). https://doi.org/10.1038/s41598-020-71462-4
11. Luca, A., Butnaru, M., Maier, S.S., et al.: Atelocollagen-based hydrogels crosslinked with oxidised polysaccharides as cell encapsulation matrix for engineered bioactive stromal tissue. Tissue Eng. Regen. Med. **14**(5), 539–556 (2017). https://doi.org/10.1007/s13770-017-0063-0
12. Luca, A., Maier, V., Maier, S.S., et al.: Biomacromolecular-based ionic-covalent hydrogels for cell encapsulation: the atelocollagen - oxidized polysaccharides couples. Carbohydr. Polym. **169**, 366–375 (2017). https://doi.org/10.1016/j.carbpol.2017.04.046
13. Brovold, M., et al.: Naturally-derived biomaterials for tissue engineering applications. In: Chun, H.J., Park, K., Kim, C.-H., Khang, G. (eds.) Novel Biomaterials for Regenerative Medicine. AEMB, vol. 1077, pp. 421–449. Springer, Singapore (2018). https://doi.org/10.1007/978-981-13-0947-2_23
14. Costa-Almeida, R., Soares, R., Granja, P.L.: Fibroblasts as maestros orchestrating tissue regeneration. J. Tissue Eng. Regen. Med. **12**(1), 240–251 (2018). https://doi.org/10.1002/term.2405
15. Wong, T., McGrath, J.A., Navsaria, H.: The role of fibroblasts in tissue engineering and regeneration. Br. J. Dermatol. **156**(6), 1149–1155 (2007). https://doi.org/10.1111/j.1365-2133.2007.07914.x
16. Durbin, K.R., Nottoli, M.S., Jenkins, G.J.: Effects of microtubule-inhibiting small molecule and antibody-drug conjugate treatment on differentially-sized A431 squamous carcinoma spheroids. Sci. Rep. **10**, 907 (2020). https://doi.org/10.1038/s41598-020-57789-y

New Hydrogels Based on Methacrylated Collagen and Hyaluronic Acid for Soft Tissue Engineering

A. Raicu[1], I. Cobzariu[1], A. L. Vasilache[1], C. A. Peptu[2], M. Butnaru[1], and Liliana Verestiuc[1](✉)

[1] Faculty of Medical Bioengineering, Grigore T. Popa University of Medicine and Pharmacy, Iasi, Romania
liliana.verestiuc@umfiasi.ro
[2] Cristofor Simionescu Faculty of Chemical Engineering and Environmental Protection, Gheorghe Asachi Technical University of Iaşi, Iaşi, Romania

Abstract. Polymeric hydrogels are soft and hydrophilic materials of great interest for the engineering of soft tissues, as mimicking structures of the extracellular matrix; because of their 3D architecture and biological properties constitute an adequate environment for cells function. Biopolymers like collagen and hyaluronic acid have a good biological response but the mechanical characteristics make necessary the crosslinking of the networks in order to build resistant and easily workable structures. The paper presents preparation of new hydrogels based on methacrylated collagen and methacrylated hyaluronic acid/hyaluronic acid as natural-based networks with controlled morphology and biointeractivity. FTIR spectra confirmed the proposed structure and the scanning electron microscopy data indicated a porous morphology dependent of the ratio between polymers. The prepared hydrogels are highly swellable, bioadhesive and degradable by enzymes. *In vitro* cytotoxicity tests revealed a very good material response in contact with cells (fibroblasts).

Keywords: Artificial ECM · Hydrogel · Collagen · Hyaluronic acid · Soft tissue engineering

1 Introduction

Collagen (Col) is the major component of the extracellular matrix that performs a number of functions in tissues, such as: maintaining the biologic and structural integrity; binding, storage and release of growth factors and cytokines, having an important role in tissue development and regeneration. Based on its excellent biocompatibility collagen is one of the most used biomaterials in soft tissue applications, as matrices for tissue regeneration, supports for the controlled release of active principles, implants etc. [1–3].

Along with collagen, hyaluronic acid (HA) is the basic substance of connective tissue, the extracellular matrix and the cytoskeleton. The chemical and stereometric

I. Tiginyanu et al. (Eds.): ICNBME 2021, IFMBE Proceedings 87, pp. 365–372, 2022.
https://doi.org/10.1007/978-3-030-92328-0_48

structure of the molecule, high molecular weight, high viscosity of the solution, elasticity, hydrophilicity, the presence on the cell membrane of specific receptors determine the biological qualities of hyaluronic acid. HA has various functions, participating in the embryogenesis and morphogenesis, intercellular relationships and communication, mechanical resistance of tissues, water retention, transport and metabolism of substances, permeability of biological membranes [4, 5]. HA is used for skin protection, the formation of semipermeable barriers in the treatment of eczema and trophic ulcers. HA is widely used in wound healing, cosmetology - as an emollient in the composition of creams, lotions, gels. Various 3D networking scaffolds based on collagen and hyaluronic acids have been developed and the characteristics of the both biopolymers were combined in order to form adequate artificial ECM. The lower mechanical properties and fast degradability of the obtained structures required methods for crosslinking and physical or chemical processes were involved for preparing three-dimensional matrices with both physical-chemical and biological performances (support for cells infiltration into scaffold, their attachment, proliferation, and differentiation) [6, 7]. The product resulting from these interactions has a better ability to adhere to mucosal surfaces, an ideal feature for the treatment of an affected area.

The paper presents the obtaining and characterization of 3D scaffolds based on methacrylated collagen (ColMa) and methacrylate hyaluronic acid (HAMa)/hyaluronic acid and the materials architecture and properties have been compared in order to form EMC mimicking architectures.

2 Materials and Methods

2.1 Materials

The hydrogels were obtained by using solution 0.5% (wt/vol, double distilled water) of methacrylate collagen (Cellink, Switzerland) and methacrylated hyaluronic acid/hyaluronic acid solution (0.5%, wt/vol, double distilled water). Hyaluronic acid was modified with methacrylic anhydride using the protocol described in [8]. Methacrylic anhydride (MA), initiating system (ammonium persulfate-APS, *N, N, N'*, *N'*-tetramethylethylene diamine - TEMED) were purchased from Sigma Aldrich. APS was purified by recrystallization from a mixture methanol/water (70/30, vol/vol).

For the citocompatibility tests, the following materials were used (Sigma Aldrich): MTT assay (3-(4,5-dimethylthiazol-2-yl)-2,5-diphenyltetrazolium bromide); Hank's Balanced Salt Solution (HBSS); Dulbecco's Modified Eagle Medium (DMEM) as culture medium; Penicillin-Streptomycin-Neomycin (PSN) and fetal bovine serum (FBS), sterile, suitable for cell culture, May-Grünwald and Giemsa staining solutions. All other chemical have been used as received.

2.2 Preparation of Hydrogels

Two types of hydrogels based on collagen modified with methacrylic anhydride (MA) have been prepared; in the first one modified hyaluronic acid (HA) with methacrylic anhydride was added and, respectively, in the second one the pure hyaluronic acid was used for scaffolds preparation. Briefly, a homogeneous mixture of methacrylated collagen and methacrylated hyaluronic acid/hyaluronic acid was adjusted to pH = 8, by adding 1N NaOH, and then an initiating system was added (ammonium persulphate (APS) and tetramethylethylenediamine (TEMED)), in the appropriate proportions, (1% mol/mol, reported to the methacrylate group) and stirred until a homogenous system was obtained. The mixtures were places in culture plates and exposed to ultraviolet radiation (UV-365 nm) for 10 min, washed with double distilled water and finally freeze-dried (−51 °C, 0.08 mBa). The hydrogels composition is presented in the Table 1.

Table 1. Hydrogels composition

Hydrogel	ColMa (ml)	HAMa (ml)	HA (ml)	SD_{eq}* (%)	SD_{eq}** (%)
ColMaHAMa1	30	5	0	5530.3 ± 24	15075 ± 75
ColMaHAMa2	30	10	0	8030.5 ± 132	9201.7 ± 13
ColMaHAMa3	30	15	0	15719.7 ± 255	2443.4 ± 83
ColMaHA1	30	0	5	6052.2 ± 16	6871.9 ± 24
ColMaHA2	30	0	10	6660.3 ± 42	14705.8 ± 93
ColMaHA3	30	0	15	8186.9 ± 112	8761.9 ± 117

*PBS; **SWE

2.3 Structural and Morphological Characterization

The Fourier transform infrared spectra (FT-IR) were registered on dried hydrogels using a spectrophotometer (Bio-Rad Win-IR instrument, USA) by scanning within the range of 400–4000 cm^{-1}, 16 scans at a resolution of 4 cm^{-1}. Gold coated cross-sections of hydrogels were examined using scanning electron microscopy – SEM (Tescan Vega SBH II, Czech Republic) at an accelerating voltage of 30 kV.

2.4 Swelling Properties, Bioadhesion and in Vitro Degradability

The swelling study was performed in phosphate buffer solution (PBS, pH = 7.4) and simulated wound exudate (SWE, pH = 7.4), at 37 °C, using the volumetric method; the swelling degree (SD, %) was calculated with the Eq. 1:

$$SD = \frac{W_t - W_0}{W_0} \times 100 \tag{1}$$

where W_o is the weight of the dry hydrogel, W_t is the weight of the swollen hydrogel and SD_{eq} is the swelling degree at equilibrium. The *in vitro* degradation properties were evaluated by incubating the hydrogels with a collagenase solution (1×10^{-2}%, in PBS solution, pH = 7.4, at 37 °C) and the spectrophotometrically monitoring (reaction with ninhydrin reagent - 420nm). The bioadhesion tests were performed on a TA.XT Plus® analyser (Stable Micro Systems - Godalming, UK) on a hydrated dialysis tubing membrane (cellulose, Visking DTV14000), in PBS (pH 7.4), at 37 °C.

2.5 In Vitro Cytotoxicity Tests

In vitro evaluation of citocompatibility (normal fibroblasts from albino rabbit) was also performed by the direct contact method, supported by morphological analysis of cell culture [9]. The cells were cultured in DMEM with 10% FBS and a mixture of penicillin, neomycin and streptomycin (1%), at 37 °C, in a humidified atmosphere of 5% CO_2. A number of 1×10^4 cells per well in a 24-well culture plate was maintained in contact with each hydrogel (sterilized by exposure to UV light for 20 min and then incubated for 24 h in DMEM). Four parallel samples from each hydrogel were put in direct contact with the cell culture in the wells. For reference purpose, cells were seeded in the same condition. At 24, 48 and 72 h the MTT assay was performed. The formazan resulted was detected by a Tecan Sun-rise Plate Reader (570 nm). The cell viability was calculated by the following equation:

$$Cell\ viability = \frac{A_h \times 100}{A_c}[\%], \quad (2)$$

where A_h is absorbance of samples and A_c is control absorbance. For cells morphology analysis a double staining with May-Grünwald solution and Giemsa solution was performed.

3 Results and Discussions

3.1 Hydrogel Structure and Morphology

The main markers for collagen identification are amides A, I and II (Fig. 1). Amide A can be observed in the absorption zone with wavelengths between 2800 and 3500 cm^{-1}, with the peak located at the value of 3437.3 cm^{-1} which indicates the presence of N-H type bonds. Another peak at a wavelength of 1647.0 cm^{-1} and 1643.9 cm^{-1}, respectively, indicates the presence of the C = O double bonds representing the absorption zone of amide I. Amide II can also be found near this zone, having the peaks at 1558.7 cm^{-1} and 1560.1 cm^{-1}.

Regarding the peaks 1030.2 cm^{-1} and 1042.4 cm^{-1}, respectively, which are relatively isolated from the areas mentioned above, it can be noted that their wavelengths correspond to the simple C-O and C-O-C connections. The band at 1413 cm^{-1} is attributed to the stretching of COO –, which refers to the acid group of molecule HA [10].

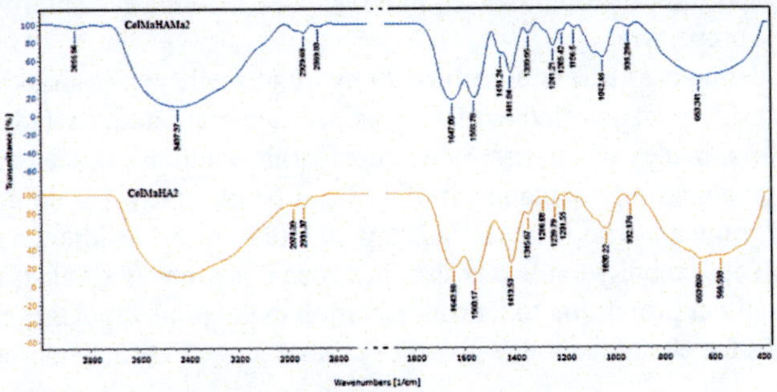

Fig. 1. FTIR spectra for ColMa-HAMa and ColMa-HA hydrogels

Following the SEM analysis, it was possible to observe the structural differences of the prepared hydrogels: both hydrogels, with methacrylated HA (Fig. 2a) and pure HA (Fig. 2b) have a porous morphology and the predominant appearance was the nanofiber. The difference between the two is given by the accentuated uniform distribution of pores in the ColMa-HA material, while the ColMa-HAMa material did not show this homogeneity, being mostly fibrous.

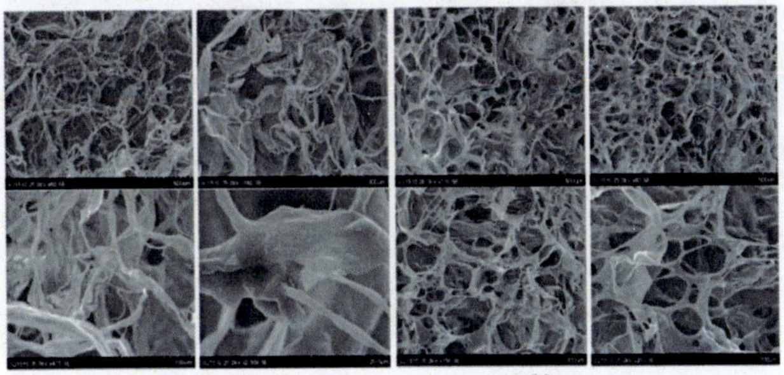

a) b)

Fig. 2. SEM images for ColMa-HAMa and ColMa-HA hydrogels

3.2 Swelling Characteristics, Bioadhesion and in Vitro Degradability

The prepared hydrogels swell extensively in PBS and the highest values for the swelling degree at equilibrium were recorded for the material containing collagen-Ma and HA (modified or not with Ma) with an increased content of polysaccharide; as was expected the hydrophilic groups from HA improved water interactions with in the 3D network. In

SWF, the highest value for SD was measure for the hydrogel ColMaHA2. The swelling behavior is an important characteristic for hydrogels and their capacity to absorb fluids is due to morphology, such as the network porosity, and due to chemical structure and the presence of hydrophilic groups such as carboxyl (-COOH), hydroxyl (-OH), amine(NH_2) amides (-CO-NH-, -CO-NH_2) from collagen and hyaluronic acid [10]. The bioadhesion tests revealed that all hydrogels are bioadhesive and hyaluronic acid improved this property.

Biodegradation is an essential feature for hydrogels with applications in tissue engineering. In vitro degradation studies of the obtained supports demonstrated their tendency to lose structural integrity in the prepared degradation solution (0.01% collagenase, pH 7.2). In the graphical representation in Fig. 3 can be observed that the highest values were found for the material ColMa-HAMa with a ratio of 6:1. Although collagen provides strength and stability to the mixtures in which it is found, the values of these traits increasing in direct proportion to its concentration in the product, in this case, the main difference was the change with the presence or absence of MA in hyaluronic acid.

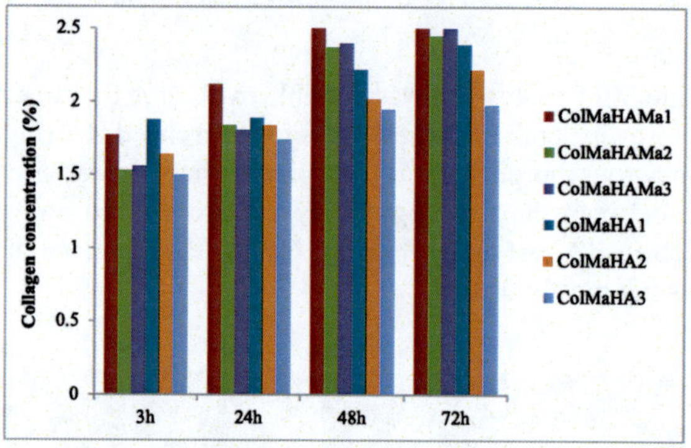

F

Fig. 3. Degradation behaviour of ColMa-HAMa and ColMa-HA hydrogels

3.3 In vitro citocompatibility

Biocompatibility is an important feature for materials designed to be used in various engineering applications. In this sense, the evaluation of cell viability by MTT method was performed in order to determine the cytotoxicity of the obtained hydrogels. The study resulted in a value of cell viability that met current standards, both materials having viability of over 95%, both at 48 and 72 h of culture (Fig. 4). They were also close to the value of the control, at 48 h still showing a higher viability. These results pointed out that hydrogels are not cytotoxic, but they stimulate cell growth.

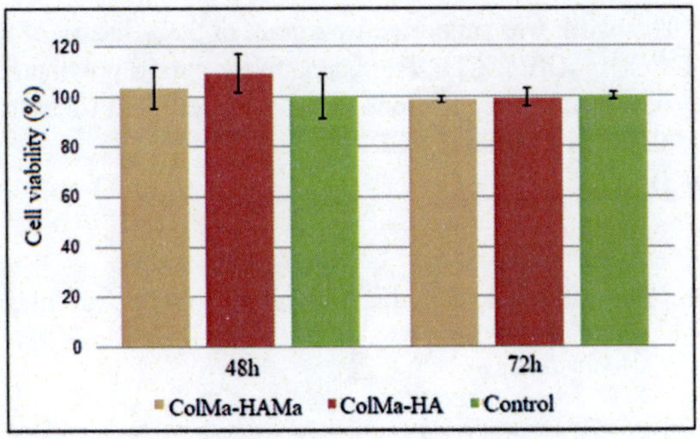

Fig. 4. Cell viability at 48 and 72 h

After staining the samples (Fig. 5), it was found that the cells had a normal morphology, specific to fibroblasts, some areas showing a tendency to populate the material.

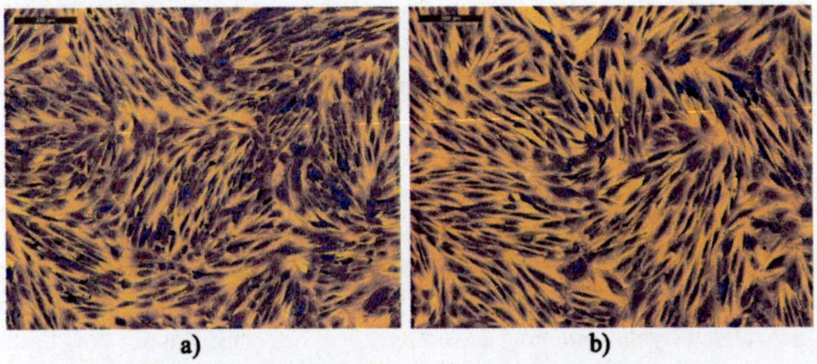

Fig. 5. Microscopic images of ColMa-HAMa (a) and ColMa-HA (b) hydrogels after staining (x100)

It can be noticed that this tendency of the cells led to overpopulation, hence the pronounced shape and their high density, the materials representing an environment favorable to cell proliferation, comparatively better than control. Finally, the results of the microscopic analysis confirm those of the cytotoxicity tests by MTT method, both types of supports having a high citocompatibility with fibroblast cells.

4 Conclusions

In the present paper, methacrylated collagen and methacrylated hyaluronic acid were UV crosslinked and high swellable hydrogels in PBS and SWF fluids were obtained. The hydrogels are bioadhesive and degradable in presence of collagenase these properties being strongly dependent of the content of hyaluronic acid. The hydrogels are citocompatible for fibroblasts and stimulate their growth and proliferation.

Acknowledgment. This work was supported by a grant of the Ministry of Research, Innovation and Digitalization, CCCDI - UEFISCDI, Romania, "New hybrid polymer/peptide hydrogels as innovative platforms designed for applications in cell cultures (HYPCELGEL)", project number PN-III-P2-2.1-PED-2019-2743, within PNCDI III.

Conflict of Interest. The authors declare that they have no conflict of interest.

References

1. Lai, V.K., Nedrelow, D.S., Lake, S.P., et al.: Swelling of collagen-hyaluronic acid co-gels: an in vitro residual stress model. Ann. Biomed. Eng. **44**(10), 2984–2993 (2016)
2. Abdeen, A.A., Saha, K.: Manufacturing cell therapies using engineered biomaterials. Trends Biotechnol. **35**, 971–982 (2017)
3. Dai, C., Shawn, S., Amor, K.: Skin substitutes for acute and chronic wound healing: an updated review. J. Dermatol. Treat. **31**(6), 639–648 (2020)
4. Davidenko, N., Campbell, J.J., Thian, E.S., Watson, C.J., Cameron, R.E.: Collagen-hyaluronic acid scaffolds for adipose tissue engineering. Acta Biomater. **6**, 3957–3968 (2010)
5. Gonzalez, S.R., Wolter, K.G., Yuen, J.C.: Infectious complications associated with the use of integra: a systematic review of the literature. Plast Reconstr Surg Glob Open. **8**(7), e2869 (2020)
6. Shahrokhi, S., Arno, A., Jeschke, M.G.: The use of dermal substitutes in burn surgery: acute phase. Wound Repair Regen. **22**(1), 14–22 (2014)
7. Vig, K., Chaudhari, A., Tripathi, S., et al.: Advances in skin regeneration using tissue engineering. Int. J. Mol. Sci. **18**(4), 789 (2017)
8. Spearman, B.S., Agrawal, N.K., Rubiano, A., Simmons, C.S., Mobini, S., Schmidt, C.E.: Tunable methacrylated hyaluronic acid-based hydrogels as scaffolds for soft tissue engineering applications. J. Biomed. Mater. Res. **108A**, 279–291 (2020)
9. Stockert, J.C., Horobin, R.W., Colombo, L.L., Blázquez-Castro, A.: Tetrazolium salts and formazan products in cell biology: viability assessment, fluorescence imaging, and labeling perspectives. Acta Histochem. **120**(3), 159–167 (2018)
10. Yousefi, F., Kandel, S., Pleshko, N.: Infrared spectroscopic quantification of methacrylation of hyaluronic acid: a scaffold for tissue engineering applications. Appl. Spectrosc. **72**, 1455–1466 (2018)

Zinc Oxide and Gallium Nitride Nanoparticles Application in Biomedicine: A Review

Ştefan Cojocari[1](✉), O. Ignatov[1], M. Jian[1], V. Cobzac[1], T. Branişte[2], E. V. Monaico[2], A. Taran[3], and V. Nacu[1]

[1] Laboratory of Tissue Engineering and Cell Cultures, Nicolae Testemitanu State University of Medicine and Pharmacy, Chişinău, Republic of Moldova
stefan.cojocari@usmf.md

[2] National Center for Materials Study and Testing, Technical University of Moldova, Chişinău, Republic of Moldova

[3] Department of Orthopedic and Traumatology, Nicolae Testemitanu State University of Medicine and Pharmacy, Chişinău, Republic of Moldova

Abstract. Currently available data have a major impact on widening the applications area of zinc oxide (ZnO) and gallium nitride (GaN) nanoparticles (NPs). Being a new medical domain, nanomedicine shows a spectacular growth of published works. Thus, in this paper we aimed to provide comprehensive current information on the implementation of inoffensive synthetized ZnO and GaN nanoparticles. The articles in the PubMed database, Bethesda (MD): US National Library of Medicine, "PubMed.gov", Google Scholar Academic containing the keywords "nanoparticles, zinc oxide, gallium nitride, cytotoxicity, adhesion" were selected. From these articles it was collected and processed the information related to the applicability of ZnO and GaN NPs. Nanoparticles based on ZnO and GaN currently have a wide range of implementation in the field of oncology, antibacterial, antifungal domains. The combination of ZnO and GaN nanoparticles as adjuvants in target factor treatments shows an increased efficacy of the active substance obtained by ecological methods. The application of ZnO GaN NPs requires innovative methods to obtain beneficial results in biomedicine. Possession of clinical nomenclature for use of ZnO and GaN NPs would reduce their cytotoxic effects in practical applications.

Keywords: Nanoparticles · Zinc oxide · Gallium nitride · Cytotoxicity · Adhesion

1 Introduction

The purpose of the study is to update information on the implementation of zinc oxide and gallium nitride nanoparticles in the medical field. The exponential growth of studies in the field of medicine, make these changes unnoticed by specialists in this field.

The paper highlighted the current data on the peculiarities of zinc oxide and gallium nitride nanoparticles, including in relation to the living organism which will facilitate the perception of information by medical specialists, which would increase the proposal/implementation of new nanotechnologies in the treatment of human diseases.

I. Tiginyanu et al. (Eds.): ICNBME 2021, IFMBE Proceedings 87, pp. 373–381, 2022.
https://doi.org/10.1007/978-3-030-92328-0_49

2 Materials and Methods

The source of the information was represented by the articles in the online database PubMed (service of the National Library of Medicine of the National Institute of Health of the United States; US National Library of Medicine, National Institute of Health) PubMed [Internet]. Bethesda (MD): US National Library of Medicine, "PubMed.gov" (available at: https://www.ncbi.nlm.nih.gov/pubmed), Google Scholar Academic, where the keyword search was performed: nanoparticles, zinc oxide, gallium nitride, cytotoxicity, adhesion. The particularities of zinc oxide and gallium nitride nanoparticles, definition and description of nanoparticles with subcategories were included: risk factors, toxicity, and their application in the treatment of diseases. The results of recent studies have taken precedence over old hypotheses. The results, reproduced in several studies, or supported by data from other tests or experiments, took precedence over single studies with unconfirmed or contradictory results. The conclusions of existing literature journals were critically examined.

Subsequently, the information was systematized with the presentation of the main aspects of the contemporary vision on ZnO and GaN NPs.

3 Results and Discussion

The term "nano" is a Greek word meaning "extremely small or dwarf" with a size of one to one hundred nanometers. Nanoparticles are designated as a fixed extension between the raw material and the atomic or molecular compositions. The piezoelectric properties (a phenomenon discovered in 1880 by Jacques and Pierre Curie) of nanoparticles are of great interest to be widely implemented in medicine. The first attention to ZnO was shown in 1960 by Hutson, who remained unique until 1990. The first successful application of converting mechanical energy into electricity using ZnO NPs belongs to Wang and Song in 2006 [1–3].

Nanotechnologies bring about key changes by introducing innovative methods in electronics, biomedicine and materials science [4].

According to literature, nanoparticles are polymorphic being classified by classes, property, shape or size. The group includes fullerenes, metallic NPs, ceramic NPs and polymeric NPs. Nanoparticles have unique physical and chemical properties due to their active surface and nano size [5]. Metal oxide NPs such as ZnO have anti-inflammatory effects (through inhibitory mechanisms → of the enzyme for the expression of nitric oxide synthesis, → release of proinflammatory cytokines, → myeloperoxidase, → NF-κβ pathway, → mast cell degranulation [6]), anticancer, antimicrobial, the property of transporting the active drug substance being synthesized by green methods [7, 8].

A special interest in the field of biomedicine is given to zinc oxide by widely discussed biocompatibility and cyto-toxicity, which is easily available in a large variety of shapes at a relatively low price [9]. The morphology of ZnO NPs depends on the synthesis process and at the nano scale can have the following shapes - rods, plates, spheres, box, hexagon, tripods, tetrapods, wires, tubes, rings, cages and flowers [10]. Zinc is an indispensable universal inorganic element used in medicine, biology and industry. The daily intake in an adult is 8–15 mg/day, of which approximately 5–6 mg/day is lost through urine and perspiration [10].

The methods that allow the production of ZnO nanoparticles with different shapes and size, are as follows [11]:

→ chemical vapor deposition;

→ precipitation in aqueous solution: hydrolysis of a Zn (II) solution, under conditions which limit the uncontrolled growth of particles, finally followed by a heat treatment to improve crystallinity;

→ hydrothermal synthesis: heat treatment of aqueous Zn (II) solutions under automatically generated pressure, using an autoclave as a reaction vessel;

→ gel solution method;

→ synthesis by microemulsions;

→ mechano-chemical processes: dry grinding with high energy [11].

Gallium nitride is a semiconductor material, which exhibit high chemical and physical stability, moreover, being biocompatible and possessing piezoelectric properties. Thus, GaN can be used in a wide range of bio-medical applications like biosensors, transducers for direct stimulation of living cells and tissues, as well as in the theragnostic of skeletal diseases, calcium metabolism and cancer[12–15].

GaN nanoparticles with shapes of rods, spheres, wires, tubes, disks, orthorhombic can be synthesized using methods as: chemical vapor deposition, hydride vapor phase epitaxy, molecular beam epitaxy, nitriding, solvothermal technique and grinding, etc. [12, 13]. GaN-based antibacterial therapy is also "Trojan Horse" due to siderophore ligands [16]. The inverse proportional relationship has been established where with the decrease of the GaN NPs diameter and the second order nonlinear optical sensitivity increases [17]. GaN NPs can be widely used in the treatment of cancer. A low concentration of GaN nanoparticles in the blood plasma has a therapeutic effect by facilitating membrane penetration, playing a crucial role in the distribution and immunity of the host [13].

Studies show (according to Table 1) that GaN NPs in combination with gold NP with in situ application at a size between 5 and 200 nm has the functionality of a thermo-transporter biosensor, electrochemical immune sensitization in cancer cells, photoelectrochemical aptasensor as a cancer biomarker [18–20]. In vitro GaN NPs activity with a size between 1 and 100 nm has a functionality to stimulate the proliferation of specific tissue cells (neurotypical, endothelial, mesenchymal stem) [14, 15, 21, 22].

ZnO NPs (from Table 2), according to performed studies (with an average size of 45.49 nm, maximum 350.00 nm, minimum 3.50 nm)) are most often determined in hexagonal form (with an average size of 38.84 nm, maximum 97.50 nm, minimum 10.00 nm) and spheres (with an average size of 39.09 nm, maximum 300.00 nm, minimum 3.50 nm). Other shapes or combinations of them are also determined, such as rods, quasi-sphere, spheres and hexagon, heterostructure, spheres and sheets, stems.

ZnO NPs (from Table 3) show different effects that are studied in situ, in vitro and/or in vivo, and especially the most frequent results are determined to be antibacterial (with an average size 45.49 nm, maximum 350.00 nm, minimum 3.50 nm) and antitumor (with an average size 60.06 nm, maximum 350.00, minimum 3.50 nm), followed by the antifungal effect (with an average size 35.94 nm, maximum 61.00 nm, minimum 20.00 nm) and antioxidant (with an average size 45.58 nm). The antileishmanial (with an average size of 25.63 nm), anti-inflammatory (with an average size of 190.00 nm)

Table 1. Application of GaN NPs in situ and in vitro

Size	Studies	Functionality	Method of synthesis	References
200 nm	In vitro	Mesenchymal stem cell proliferation	Epitaxial growth of vapor phase hydride (HVPE)	[21]
1–14 nm	In vitro	Stimulation of neuro-typical cells (PC12)	Chemical vapor storage	[22]
20–100 nm	In vitro	Proliferation of porcine aortic endothelial cells	Vapor phase epitaxial growth (HVPE)	[14]
50–100 nm	In vitro	Porcine aortic endothelial cell proliferation	Vapor phase epitaxial growth (HVPE)	[15]
50–10 nm	In situ	Thermo-transporter biosensor in the detection of cancer cells	– // –	[19]
200 nm	In situ	Electrochemical immune sensitization in the detection of cancer cells	Chemical vapor deposition (hydrogen vapor deposition)	[18]
20–5 nm	In situ	Photoelectrochemical aptasensor as a cancer biomarker	– // –	[20]

Table 2. ZnO shapes depending on NP size

Form	Average size (nm)	Maximum size (nm)	Minimum size (nm)	References
Rods	22.50	22.50	22.50	[23]
Quasi-sphere	22.50	22.50	22.50	[24]
Spheres and hexagon	24.00	24.00	24.00	[25]
Hetorostructure	33.25	33.25	33.25	[26]
Spheres sheets	350.00	350.00	350.00	[27]
Stems	36.00	36.00	36.00	[28]
Hexagon	38.84	97.50	10.00	[1, 3, 11, 29–42],
Spheres	39.09	300.00	3.50	[43–59]
Overall	**45.49**	**350.00**	**3.50**	

and in particular antidiabetic (with a size of 26.55 nm), antiangiogenesis (with a size of 30.00 nm) were also studied.

Table 3. ZnO effect depending on NP size

Effect	Average size (nm)	Maximum size (nm)	Minimum size (nm)	References
Cytotoxic	18.00	18.00	18.00	[1]
Anti-leishmanial	25.63	33.25	18.00	[1, 26]
Antidiabetic	26.55	26.55	26.55	[41]
Anti-angiogenesis	30.00	30.00	30.00	[53]
Antifungal	35.94	61.00	20.00	[3, 11, 30, 39]
Antibacterial	45.49	350.00	3.50	[1, 11, 24, 26, 27, 29, 30, 35–37, 39–41, 44–46, 49, 54, 55]
Antioxidant	45.58	66.25	18.00	[1, 34, 55]
Antitumor	60.06	350.00	3.50	[25, 27, 31, 33, 34, 36, 38–40, 42, 43, 46, 47, 51–53, 55, 57–61]
Anti -inflammatory	190.00	350.00	30.00	[27, 53]
Overall	**45.49**	**350.00**	**3.50**	

Zinc oxide and gallium nitride nanoparticles currently have a wide scope both in the field of oncology (anticancer, anti-inflammatory effect), as well as in the production of antibacterial and antifungal (antimicrobial effect) products. The combination of ZnO or GaN NPs as an adjuvant in target factor treatments shows an increase in the efficacy of the active substances. Determination of the NPs administration dose of ZnO and GaN can obtain a low toxicity by using green methods of theragnostic (in vivo, in situ).

4 Conclusions

The application of ZnO and GaN NPs requires innovative methods to obtain beneficial results in biomedicine. Spherical and hexagonal forms of NPs are the most used, and as their size decreases, the effects are determined as antibacterial, antitumoral, antifungal, and antioxidant. Possession of detailed databases of the cyto-effects of NPs depending their shape size and surface treatment will definitely advance the practical side of nanomedicine.

Acknowledgment. This research was funded by the European Commission under the Grant #810652 "NanoMedTwin" and by the National Agency for Research and Development of the Republic of Moldova under the Grant #20.80009.50007.20.

Conflict of Interest. The authors declare that they have no conflict of interest.

References

1. Abbasi, B.A., et al.: Bioactivities of geranium wallichianum leaf extracts conjugated with zinc oxide nanoparticles. Biomolecules **10** (2019). https://doi.org/10.3390/biom10010038
2. Agarwal, H., Nakara, A., Shanmugam, V.K.: Anti-inflammatory mechanism of various metal and metal oxide nanoparticles synthesized using plant extracts: a review. Biomed. Pharmacother. **109**, 2561–2572 (2018)
3. Agarwal, H., Shanmugam, V.: A review on anti-inflammatory activity of green synthesized zinc oxide nanoparticle: mechanism-based approach. Bioorg. Chem. **94**, 103423 (2019)
4. Ahmad, H., et al.: Green synthesis and characterization of zinc oxide nanoparticles using eucalyptus globules and their fungicidal ability against pathogenic fungi of apple orchards. Biomolecules **10** (2020). https://doi.org/10.3390/biom10030425
5. Akbarian, M., Mahjoub, S., Elahi, S.M., Zabihi, E., Tashakkorian, H.: Green synthesis, formulation and biological evaluation of a novel ZnO nanocarrier loaded with paclitaxel as drug delivery system on MCF-7 cell line. Colloids Surf. B Biointerfaces **186**, 110686 (2019)
6. Alves, M.M., Andrade, S.M., Grenho, L., Fernandes, M.H., Santos, C., Montemor, M.F.: Influence of apple phytochemicals in ZnO nanoparticles formation, photoluminescence and biocompatibility for biomedical applications. Mater. Sci. Eng. C Mater. Biol. Appl. **101**, 76–87 (2019)
7. Ansari, M.A., et al.: Cinnamomum verum bark extract mediated green synthesis of ZnO nanoparticles and their antibacterial potentiality. Biomolecules **10** (2020). https://doi.org/10.3390/biom10020336
8. Awwad, A.M., Amer, M.W., Salem, N.M.: Green synthesis of zinc oxide nanoparticles (ZnO-NPs) using Ailanthus altissima fruit extracts and antibacterial activity. Chemistry (2020)
9. Banerjee, S., et al.: Antibacterial, anti-biofilm activity and mechanism of action of pancreatin doped zinc oxide nanoparticles against methicillin resistant Staphylococcus aureus. Colloids Surf. B Biointerfaces **190**, 110921 (2020)
10. Beegam, A., et al.: Environmental fate of zinc oxide nanoparticles: risks and benefits. In: Soloneski, S., Larramendy, M. (eds.) Toxicology - New Aspects to this Scientific Conundrum. InTech (2016)
11. Braniste, T., et al.: Mesenchymal stem cells proliferation and remote manipulation upon exposure to magnetic semiconductor nanoparticles. Biotechnol. Rep. (Amst) **25**, e00435 (2020)
12. Braniste, T., et al.: Targeting endothelial cells with multifunctional GaN/Fe nanoparticles. Nanoscale Res. Lett. **12**(1), 1–6 (2017). https://doi.org/10.1186/s11671-017-2262-y
13. Braniste, T., et al.: Viability and proliferation of endothelial cells upon exposure to GaN nanoparticles. Beilstein J. Nanotechnol. **7**, 1330–1337 (2016)
14. Buşilă, M., et al.: Size-dependent biological activities of fluorescent organosilane-modified zinc oxide nanoparticles. J. Biomed. Nanotechnol. **16**, 137–152 (2020)
15. Chauhan, A., et al.: Photocatalytic dye degradation and antimicrobial activities of pure and Ag-doped ZnO using Cannabis sativa leaf extract. Sci. Rep. **10**, 7881 (2020)
16. Cheng, J., et al.: Green synthesized zinc oxide nanoparticles regulates the apoptotic expression in bone cancer cells MG-63 cells. J. Photochem. Photobiol. B **202**, 111644 (2019)
17. Cruz, D.M., et al.: Green nanotechnology-based zinc oxide (ZnO) nanomaterials for biomedical applications: a review. J Phys. Mater. **3**, 034005 (2020)
18. Duan, X., et al.: Zinc oxide nanoparticles synthesized from Cardiospermum halicacabum and its anticancer activity in human melanoma cells (A375) through the modulation of apoptosis pathway. J. Photochem. Photobiol. B **202**, 111718 (2019)
19. Eixenberger, J.E., et al.: Defect engineering of ZnO nanoparticles for bioimaging applications. ACS Appl. Mater. Interfaces **11**, 24933–24944 (2019)

20. ElAfandy, R.T., et al.: Nanomembrane-based, thermal-transport biosensor for living cells. Small **13** (2016). https://doi.org/10.1002/smll.201603080
21. Fouda, A., El-Din Hassan, S., Salem, S.S., Shaheen, T.I.: In-Vitro cytotoxicity, antibacterial, and UV protection properties of the biosynthesized zinc oxide nanoparticles for medical textile applications. Microb. Pathog. **125**, 252–261 (2018)
22. Hameed, S., et al.: Greener synthesis of ZnO and Ag-ZnO nanoparticles using Silybum marianum for diverse biomedical applications. Nanomedicine **14**, 655–673 (2019)
23. Hu, D., et al.: Highly sensitive and selective photoelectrochemical aptasensor for cancer biomarker CA125 based on AuNPs/GaN Schottky junction. Anal. Chem. **92**, 10114–10120 (2020)
24. Hu, D., et al.: Cucurbita pepo leaf extract induced synthesis of zinc oxide nanoparticles, characterization for the treatment of femoral fracture. J. Photochem. Photobiol. B **195**, 12–16 (2019)
25. Jevapatarakul, D., Jiraroj, T., Payungporn, S., Chavalit, T., Khamwut, A., T-Thienprasert, N.P.: Utilization of Cratoxylum formosum crude extract for synthesis of ZnO nanosheets: characterization, biological activities and effects on gene expression of nonmelanoma skin cancer cell. Biomed. Pharmacother. **130**, 110552 (2020)
26. Jiang, J., Pi, J., Cai, J.: The advancing of zinc oxide nanoparticles for biomedical applications. Bioinorg. Chem. Appl. **2018**, 1062562 (2018)
27. Khan, I., Saeed, K., Khan, I.: Nanoparticles: properties, applications and toxicities. Arab. J. Chem. **12**, 908–931 (2019)
28. Kircheva, N., Dudev, T.: Gallium as an antibacterial agent: a DFT/SMD study of the Ga3+/Fe3+ competition for binding bacterial siderophores. Inorg. Chem. **59**, 6242–6254 (2020)
29. Kuang, Y., et al.: Poly(amino acid)/ZnO/mesoporous silica nanoparticle based complex drug delivery system with a charge-reversal property for cancer therapy. Colloids Surf. B Biointerfaces **181**, 461–469 (2019)
30. Kulkarni, S., Pandey, A., Mutalik, S.: Liquid metal based theranostic nanoplatforms: application in cancer therapy, imaging and biosensing. Nanomedicine **26**, 102175 (2020)
31. Lan, Y., Li, J., Wong-Ng, W., Derbeshi, R.M., Li, J., Lisfi, A.: Free-standing self-assemblies of gallium nitride nanoparticles: a review. Micromach. (Basel) **7** (2016). https://doi.org/10.3390/mi7090121
32. Li, F., et al.: Anticancer and genotoxicity effect of (Clausena lansium (Lour.) Skeels) peel ZnONPs on neuroblastoma (SH-SY5Y) cells through the modulation of autophagy mechanism. J. Photochem. Photobiol. B **203**, 111748 (2019)
33. Liu, Q., et al.: A highly sensitive label-free electrochemical immunosensor based on an aligned GaN nanowires array/polydopamine heterointerface modified with Au nanoparticles. J. Mater. Chem. B Mater. Biol. Med. **7**, 1442–1449 (2019)
34. Mahdizadeh, R., Homayouni-Tabrizi, M., Neamati, A., Seyedi, S.M.R., Tavakkol Afshari, H.S.: Green synthesized-zinc oxide nanoparticles, the strong apoptosis inducer as an exclusive antitumor agent in murine breast tumor model and human breast cancer cell lines (MCF7). J. Cell Biochem. **120**, 17984–17993 (2019)
35. Miri, A., Mahdinejad, N., Ebrahimy, O., Khatami, M., Sarani, M.: Zinc oxide nanoparticles: biosynthesis, characterization, antifungal and cytotoxic activity. Mater. Sci. Eng. C Mater. Biol. Appl. **104**, 109981 (2019)
36. Mohammad, G.R.K.S., Tabrizi, M.H., Ardalan, T., Yadamani, S., Safavi, E.: Green synthesis of zinc oxide nanoparticles and evaluation of anti-angiogenesis, anti-inflammatory and cytotoxicity properties. J. Biosci. **44**(2), 1–9 (2019). https://doi.org/10.1007/s12038-019-9845-y

37. Pandiyan, N., Murugesan, B., Arumugam, M., Sonamuthu, J., Samayanan, S., Mahalingam, S.: Ionic liquid - a greener templating agent with Justicia adhatoda plant extract assisted green synthesis of morphologically improved Ag-Au/ZnO nanostructure and it's antibacterial and anticancer activities. J. Photochem. Photobiol. B **198**, 111559 (2019)
38. Rahimi Kalateh Shah Mohammad G., Karimi, E., Oskoueian, E., Homayouni-Tabrizi, M.: Anticancer properties of green-synthesised zinc oxide nanoparticles using Hyssopus officinalis extract on prostate carcinoma cells and its effects on testicular damage and spermatogenesis in Balb/C mice. Andrologia **52**, e13450 (2019)
39. Rajan, S.A., Khan, A., Asrar, S., Raza, H., Das, R.K., Sahu, N.K.: Synthesis of ZnO/Fe(3)O(4)/rGO nanocomposites and evaluation of antibacterial activities towards E. coli and S. aureus. IET Nanobiotechnol. **13**, 682–687 (2019)
40. Rajeshkumar, S., Kumar, S.V., Ramaiah, A., Agarwal, H., Lakshmi, T., Roopan, S.M.: Biosynthesis of zinc oxide nanoparticles usingMangifera indica leaves and evaluation of their antioxidant and cytotoxic properties in lung cancer (A549) cells. Enzyme Microb. Technol. **117**, 91–95 (2018)
41. Rajeshkumar, S., Sandhiya, D.: Biomedical applications of zinc oxide nanoparticles synthesized using eco-friendly method. In: Shukla, A.K. (ed.) Nanoparticles and their Biomedical Applications, pp. 65–93. Springer, Singapore (2020). https://doi.org/10.1007/978-981-15-0391-7_3
42. Ruddaraju, L.K., Pammi, S.V.N., Pallela, P.N.V.K., Padavala, V.S., Kolapalli, V.R.M.: Antibiotic potentiation and anti-cancer competence through bio-mediated ZnO nanoparticles. Mater. Sci. Eng. C Mater. Biol. Appl. **103**, 109756 (2019)
43. Sana, S.S., et al.: Crotalaria verrucosa leaf extract mediated synthesis of zinc oxide nanoparticles: assessment of antimicrobial and anticancer activity. Molecules **25** (2020). https://doi.org/10.3390/molecules25214896
44. Saravanan, M., Gopinath, V., Chaurasia, M.K., Syed, A., Ameen, F., Purushothaman, N.: Green synthesis of anisotropic zinc oxide nanoparticles with antibacterial and cytofriendly properties. Microb. Pathog. **115**, 57–63 (2017)
45. Selim, Y.A., Azb, M.A., Ragab, I., Abd El-Azim, M.H.M.: Green synthesis of zinc oxide nanoparticles using aqueous extract of deverra tortuosa and their cytotoxic activities. Sci. Rep. **10**, 3445 (2020)
46. Shobha, N., et al.: Synthesis and characterization of Zinc oxide nanoparticles utilizing seed source of Ricinus communis and study of its antioxidant, antifungal and anticancer activity. Mater. Sci. Eng. C Mater. Biol. Appl. **97**, 842–850 (2018)
47. Siddiqi, K.S., Ur Rahman, A., Tajuddin, H.A.: Properties of zinc oxide nanoparticles and their activity against microbes. Nanoscale Res. Lett. **13**, 141 (2018)
48. Sisubalan, N., et al.: ROS-mediated cytotoxic activity of ZnO and CeO(2) nanoparticles synthesized using the Rubia cordifolia L. leaf extract on MG-63 human osteosarcoma cell lines. Environ. Sci. Pollut. Res. Int. **25**, 10482–10492 (2017)
49. Snyder, P.J., Reddy, P., Kirste, R., LaJeunesse, D.R., Collazo, R., Ivanisevic, A.: Noninvasive stimulation of neurotypic cells using persistent photoconductivity of gallium nitride. ACS Omega **3**, 615–621 (2018)
50. Steffy, K., Shanthi, G., Maroky, A.S., Selvakumar, S.: Enhanced antibacterial effects of green synthesized ZnO NPs using aristolochia indica against multi-drug resistant bacterial pathogens from diabetic foot ulcer. J. Infect. Public Health **11**, 463–471 (2017)
51. Suriyaprabha, R., et al.: A sensitive refining of in vitro and in vivo toxicological behavior of green synthesized ZnO nanoparticles from the shells of Jatropha curcas for multifunctional biomaterials development. Ecotoxicol. Environ. Saf. **184**, 109621 (2019)
52. Tang, Q., Xia, H., Liang, W., Huo, X., Wei, X.: Synthesis and characterization of zinc oxide nanoparticles from Morus nigra and its anticancer activity of AGS gastric cancer cells. J. Photochem. Photobiol. B **202**, 111698 (2019)

53. Umar, H., Kavaz, D., Rizaner, N.: Biosynthesis of zinc oxide nanoparticles using Albizia lebbeck stem bark, and evaluation of its antimicrobial, antioxidant, and cytotoxic activities on human breast cancer cell lines. Int. J. Nanomed. **14**, 87–100 (2018)
54. Vinotha, V., et al.: Synthesis of ZnO nanoparticles using insulin-rich leaf extract: anti-diabetic, antibiofilm and anti-oxidant properties. J. Photochem. Photobiol. B **197**, 111541 (2019)
55. Wang, D., Cui, L., Chang, X., Guan, D.: Biosynthesis and characterization of zinc oxide nanoparticles from Artemisia annua and investigate their effect on proliferation, osteogenic differentiation and mineralization in human osteoblast-like MG-63 Cells. J. Photochem. Photobiol. B **202**, 111652 (2019)
56. Wang, K., Qian, H., Liu, Z., Yu, P.K.L.: Second-order nonlinear susceptibility enhancement in gallium nitride nanowires. Prog. Electromagn. Res. B Pier B (2020)
57. Wang, Y., et al.: Synthesis of zinc oxide nanoparticles from Marsdenia tenacissima inhibits the cell proliferation and induces apoptosis in laryngeal cancer cells (Hep-2). J. Photochem. Photobiol. B **201**, 111624 (2019)
58. Xue, Y., Yu, G., Shan, Z., Li, Z.: Phyto-mediated synthesized multifunctional Zn/CuO NPs hybrid nanoparticles for enhanced activity for kidney cancer therapy: a complete physical and biological analysis. J. Photochem. Photobiol. B **186**, 131–136 (2018)
59. Yadi, M., et al.: Current developments in green synthesis of metallic nanoparticles using plant extracts: a review. Artif. Cells Nanomed. Biotechnol. **46**, S336–S343 (2018)
60. Zhao, C., Zhang, X., Zheng, Y.: Biosynthesis of polyphenols functionalized ZnO nanoparticles: characterization and their effect on human pancreatic cancer cell line. J. Photochem. Photobiol. B **183**, 142–146 (2018)
61. Zheng, M., Wang, S., Liu, Z., Xie, L., Deng, Y.: Development of temozolomide coated nano zinc oxide for reversing the resistance of malignant glioma stem cells. Mater. Sci. Eng. C Mater. Biol. Appl. **83**, 44–50 (2017)

Cellular Lifesaving Flexible Device

S. Meulesteen[1], Andriy O. Semenov[2](✉), O. Semenova[2], K. Koval[2,3], D. Datsiuk[3], and H. Fomenko[3]

[1] Montr B.V., The Hague, The Netherlands
[2] Faculty for Infocommunications, Radioelectronics and Nanosystems, Vinnytsia National Technical University, Vinnytsia, Ukraine
semenov.a.o@vntu.edu.ua
[3] Droid Technologies, Kyiv, Ukraine

Abstract. The paper proposes a new device for remote healthcare. This device is intended for elderly patients with cardiovascular disease who are living alone. The advantage of the device is that it can be used 24/7. The device has simple and understandable interface. The design of the device is based on two modern technologies, they are modern infocommunication technologies of LTE networks and technologies of flexible electronics. Also, this device can be used for remote healthcare during the COVID-19 pandemic.

Keywords: e-health · Mobile · Wearable · Remote healthcare · Flexible electronic

1 Introduction

The rapid growth of mobile communication systems stimulates development, manufacture and application of new types of medical devices and development of technologies for diagnosis, treatment and rehabilitation of patients [1, 2]. Nowadays, the new generation of cellular communication systems NB IoT, LTE m1 are widely used [3]. They all have a broadband unified radio access system that allows information transmission in real time.

Vinnytsia National Technical University (VNTU) has received a request from vascular surgeons upon a device for measuring physical activity in a reliable and scientific validated way. The point is, for the best treatment of vascular diseases, exercise therapy must become evidence based. Some patients tend to overestimate their physical activity. Multiple studies conclude that 72% patients did not meet the physical activity recommendations [4]. Governments also promote monitoring physical activity with activity meters [5]. However, current technology is rather designed as fitness-applications on smartphones than a solution for medical applications. They are not designed for elderly and generate uncalibrated results that are inadmissible for medical purposes.

A fifth of the population over the age of 65 has intermittent claudication [6]. 20.6% of the EU population is over 65 [7]. Yearly 20 million EU-citizens suffer symptoms related to intermittent claudication, with an annual growth of 5%. Vascular surgeons estimate

I. Tiginyanu et al. (Eds.): ICNBME 2021, IFMBE Proceedings 87, pp. 382–389, 2022.
https://doi.org/10.1007/978-3-030-92328-0_50

that half of these patients do not need invasive therapy if they intensify physical activities [8]. The target market is the EU cardiovascular medical sector applicable to an estimated yearly amount of 10 million patients with current associated treatment costs of approx. €3000 per patient [8]. Extrapolation results in a possible annual market turnover is 15 billion euros.

Patients usually are insufficiently aware of their inactivity or physical condition. The device will address two essential needs for evidence-based therapy for elderly: calibrated, reliable results are collected by the device which sends data over cellular IoT. A comfortable, rugged elastomer wristband helps patients to monitor and reflect on their individual physical activity. It will be the first device on a global market that requires zero user set up and can transmit data directly over cellular IoT to any medical cloud solution for additional processing. Another use is emergency messaging. Moreover, no user interaction is needed, all sensors are integrated, and the cellular performance are guaranteed as well as a decent battery performance.

Current state-of-the-art products that are available within the market don't have wearable and water-proof flexible electronics, or inclusiveness for all elderly users. Clear market demand and integration of flexible electronics, cellular IoT and calibrated accelerometers put the focus on a development of breakthrough technology for traditional industries.

The aim of the study was to develop a key-enabling technology, a cellular IoT wearable device on flexible PCB equipped with multiple sensors like an accelerometer and thermometer. Also, alarm messaging must be enabled. The device will send its data through cellular IoT (NB IoT/ LTE m1) straight to a (cloud) service, to support medical specialists in evidence-based therapy for physical activity.

2 Design of the Wearable Device

There is a strong trend towards incorporating more connectivity in medical devices in order to enable new services and valuable propositions. However, there are a number of barriers, including resistance to changes among end-users (patients) and healthcare organizations. Virtually all of the current available wearables are using regular LTE (2/3/4G) or BLE technology which is either expensive and very energy hungry (LTE) or requires regular user interaction to setup (BLE). The proposed cellular (NB IoT/LTEm1) solution is low in energy consumption, cheap to use, requires no user setup and can be utilized indoors and outdoors. The proposed solution is also sourced from European components, is GDPR compliant, and is secure because of the use of cellular networks, used communication protocol (TCP) and the ability to easily update the firmware. According to the overall description and general requirements the next block diagram was suggested (Fig. 1). The components were chosen according to the availability on market.

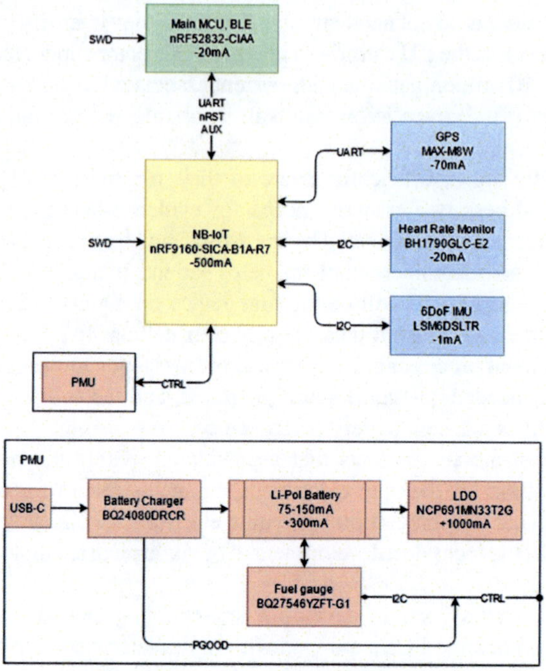

Fig. 1. General block diagram of the device.

The device allows monitoring physical activity of elderly people with cardiovascular diseases. The device has sensors to measure parameters of the body (pulse, temperature etc.). It is designed for medical applications and transfers data via cellular IoT (NB IoT/LTE m1) from any place on the Earth. Moreover, the device has GPS and Bluetooth 5.0. The device can produce an alarm signal and transfer required data to a doctor's computer or smartphone automatically. New technical solutions from the theory of antennas and feeders are applied for the device, as well as the original microwave circuitry to match IoT antennas. The device differs from the well known fitness trackers because of the easy set up (plug and play) for the user and the 24/7 connectivity from the LTEm1/NB IoT networks. It is small and needs low power supply. It is easy in use, requires no adjustment and no user interaction (that is extremely important for elderly people). It can be used 24 h a day, also under the shower (where a lot of accidents happen), able to transmit an emergency message from the wristband.

Main components of the device (Fig. 1) are: nrf9160, MAX-M8W, LSM6DSLTR, BH1790GLC-E2, nRF52832-CIAA, nRF9160-SICA-B1A-R7, BH1790GLC-E2 (nRF9160 is used as main controller, nRF52832 is used as BLE peripheral). The wristband is semi-flexible with a rigid section. Hardware is placed on the rigid section. The wristband has a waterproof USB-charger. Its lifetime between charges is about a few days.

3 LTE Flexible Antenna

Electronic devices in a flexible form is essential for remote healthcare. An important task that arises when designing broadband antennas for wearable medical devices is the development of an antenna shape that will fit into the device shell and provide a required gain. Strict requirements are imposed on a shape of the radiation pattern of such antenna. The antenna must provide stable receiving from any direction in urban environment. Also, the shape of radiation pattern must meet the requirements of electromagnetic compatibility when using a mobile medical device. The last requirement for the shape of the antenna radiation pattern is due to the presence of reflective surfaces that affect matching of the radio device in an operating frequency band.

VNTU has developed an LTE Flexible antenna for the frequency ranges B3 1800 MHz (DCS, Uplink 1710–1785 MHz, Downlink 1805–1880 MHz) and B20 800 MHz (Digital Dividend, Uplink 832–862 MHz, Downlink 791–821 MHz). The linear dimensions of the antenna are 22 × 38 mm. Material of the antenna is 0.3 mm thick polyimide. The peak power is 23 dBm. The radiation pattern of the LTE flexible antenna is shown in Fig. 2.

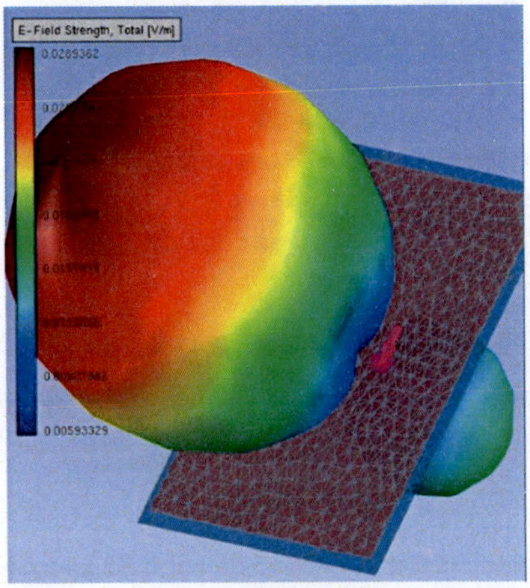

Fig. 2. Radiation pattern of the LTE flexible antenna in EMCoS Studio.

The original design of this antenna and its optimal coordination with the wearable device allow reducing the effect of capacitance and resistance of the antenna when putting the wristband on the hand and providing a reliable communication of the wearable device in the LTE (or NB IoT) network.

4 Operation

The authors have developed the software for the device. The block diagram of the device algorithm is shown in Fig. 3. Below, there is a description of operations.

Main Thread

1. After the setup, the device downloads settings from the Non Volatile Random Access Memory. If the settings are invalid, the settings from the firmware are applied.
2. The device established connection with the Montr server via NB-IoT, downloads settings, checks for updates.

   a. If there is an update, the device tries to download, to check, and to apply the update.
   b. In case of errors, depending on the system settings, the device tries to connect to the server again.
   c. If the connection to the server failed, the device skips the update stage and proceeds to the normal operation mode (point 3).
   d. The action can be skipped, the update request can be sent via Bluetooth.

3. The device initializes Bluetooth low energy (BLE). The device adds services: BatteryService, LocationService, AccelerometerService, HeartRateService, BuzzerService, ButtonService, DeviceSettingsService. Any device connected via BLE will be able to monitor data, as well as configure the device itself.
4. Initialization of drivers and threads:

   a. Button initialization
   b. GPS initialization
   c. Heart rate initialization
   d. Accelerometer initialization

5. The thread is responsible for transferring data via BLE and NB-IoT. The data transfer is activated on demand or periodically, depending on the settings. When a request for sending data is received, the system collects data from the sensors, having converted it previously into a required form.
6. The collected data is transferred via LTEm1/NB IoT. Also, some data monitored depending on the settings can be updated via BLE without transferring data via NB-IoT.
7. The device tries to connect to the Montr server.

   a. If the connection fails, it tries to connect again. The number of times and the pause between attempts depend on the settings.
   b. In case of successful connection, the device transfers the collected data to the server.
   c. In case of data transfer error, it tries to transfer again.

d. If the event is an emergency (pressing the button, exceeding the heart rate, falling, etc.), this event will be transmitted urgently, even if some data have not been collected. If there was no GPS position, the system will try to get the correct GPS signal and send it.

8. After sending the data, the system goes into energy-saving NB-IoT mode with or without disconnection from the network.
9. Requests are expected (item 5).

Button Thread

1. The event of pressing the button is expected
2. After pressing the button, a request is sent to the main thread to send the data to the server. The corresponding indication is set.

Accelerometer Thread

1. An event from the accelerometer is expected. A falling can be an emergency event.
2. The event/data is/are processed.
3. A request is sent to the main thread to send data to the server. The corresponding indication is set.

Heart Rate Sensor Thread

1. A heart rate is expected/monitored.
2. The data is processed.
3. In case of a critical heart rate, a request is sent to the main thread to send the data to the server. The corresponding indication is set.

GPS Thread

1. A request for geo-location data is expected. Also, the data can be monitored (depending on settings).
2. The data is processed.
3. The data are saved to the internal memory.

5 Conclusions

The authors have proposed the new device for a remote control of physical activity of elderly people with cardiovascular disease, which monitors such health parameters as temperature, pulse and fall detection. One of the key features of the device is that it will always work, even in degraded circumstances with poor LTE coverage on lower bandwidth and/or degraded circumstances due to contact with water (fall in a shower for example).

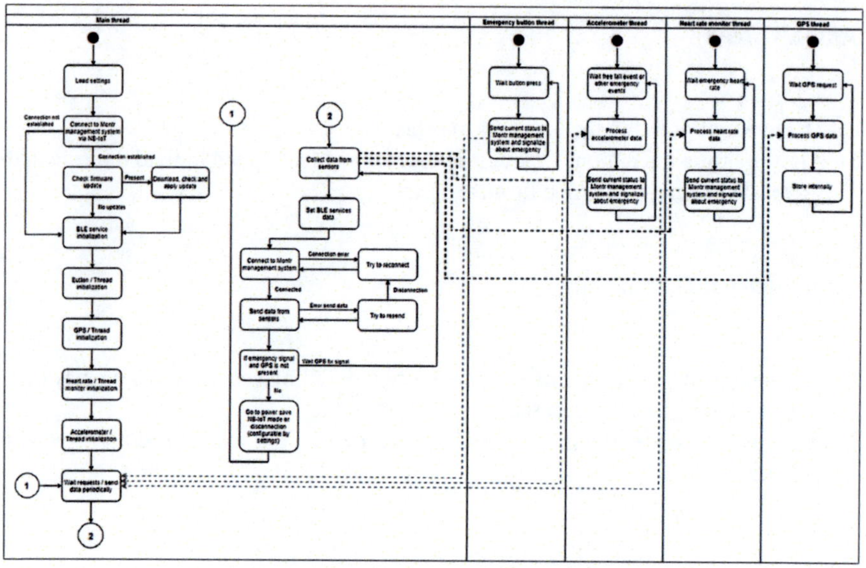

Fig. 3. Block diagram of the algorithm.

Acknowledgment. This work was supported by the European Union, represented by the European Commission, Grant Agreement no. 872614 for the implementation of the Innovation Action "SMART4ALL" within the framework of the Horizon 2020 Program, Call Reference no. H2020-DT-2018-2020.

Conflict of Interest. The authors declare that they have no conflict of interest.

References

1. Soliman, M., Chowdhury, M., Khandakar, A., et al.: Review on medical implantable antenna technology and imminent research challenges. Sensors **21**(9), 3163 (2021). https://doi.org/10.3390/s21093163

2. Flores-Cuadras, J., Medina-Monroy, J., Chavez-Perez, R., Lobato-Morales, H.: Novel ultra-wideband flexible antenna for wearable wrist worn devices with 4G LTE communications. Micr. Opt. Tech. Let. **59**(4), 777–783 (2017). https://doi.org/10.1002/mop.30393
3. Ali Khan, M., Raad, R., Tubbal, F., et al.: Bending analysis of polymer-based flexible antennas for wearable, general IoT applications: a review. Polymers **13**(3), 357 (2021). https://doi.org/10.3390/polym13030357
4. Loyen, A., et al.: Sedentary Time and physical activity surveillance through accelerometer pooling in four European countries. Sports Med. **47**(7), 1421–1435 (2016). https://doi.org/10.1007/s40279-016-0658-y
5. Dutch National Health Council, no. 2017/08. https://www.gezondheidsraad.nl/documenten/adviezen/2017/08/22/beweegrichtlijnen-2017
6. Cassar, K.: Intermittent claudication. BMJ **333**(7576), 1002–1005 (2006). https://doi.org/10.1136/bmj.39001.562813.DE
7. Eurostat: Population structure and ageing (2020). https://ec.europa.eu/eurostat/statistics-explained/index.php?title=Population_structure_and_ageing
8. Smolderena, K., Wanga, K., de Pouvourville, G., et al.: Two-year vascular hospitalisation rates and associated costs in patients at risk of atherothrombosis in France and Germany: highest burden for peripheral arterial disease. EJVES **43**(2), 198–207 (2012). https://doi.org/10.1016/j.ejvs.2011.09.016

Assessment of Gold Nanoparticles Uptake in Tissues of Female Mice and Their Offspring Using Neutron Activation Analysis

A. Ivlieva[1], Inga Zinicovscaia[2,3,4](✉), E. Petritskaya[1], N. Yushin[2], D. Rogatkin[1], and A. Peshkova[2]

[1] M.F. Vladimirskiy Moscow Regional Research and Clinical Institute, Moscow, Russia
[2] Joint Institute for Nuclear Research, Dubna, Moscow Region, Russia
inga@jinr.ru
[3] Horia Hulubei National Institute for R&D in Physics and Nuclear Engineering, Bucharest - Magurele, Romania
[4] Institute of Chemistry, Chisinau, Moldova

Abstract. The uptake of gold in different tissues of female mice and their offspring after prolonged oral administration of gold nanoparticles to the females during pregnancy and lactation was investigated. Gold content in different organs was determined using neutron activation analysis. The highest content of gold was accumulated in kidneys followed by liver, lungs, blood and brain. Accumulation of gold in brain confirms ability of nanoparticles to pass through placental and blood-brain barriers. The average specific mass content of gold which crossed the blood-brain barrier was 3.8 ng/g (for female) and 1.1 ng/g (for offspring). The results are important for research on developmental and reproductive toxicity and safety of nanomaterials.

Keywords: Gold nanoparticles · Neutron activation analysis · Mice · Brain · Reproductive function

1 Introduction

Gold nanoparticles (AuNPs) already have wide applications in medicine – in optical visualization, target photosensitization, and they are used as carriers for drug molecules [1, 2]. AuNPs have been also extensively used as nanomaterials for theranostic applications due to their multifunctional characteristics in therapeutics, imaging, and surface modification [3]. The potential of AuNPs' use in medicine is enormous, therefore the questions arise about their safety for patients, especially for pregnant women and children.

Several biological barriers prevent uncontrollable exposure of tissues to substances carried by bloodstream, including potentially harmful: blood-brain barrier, blood-testis in males or blood–follicles, and the blood–placenta barrier in females. The latest of them controls exchange of substances between a fetus and its mother [4]. However, it is known that the permeability of biological barriers for small nanoparticles is higher, thus AuNPs designed for medical use can be transferred through to the brain, and from mother to

I. Tiginyanu et al. (Eds.): ICNBME 2021, IFMBE Proceedings 87, pp. 390–395, 2022.
https://doi.org/10.1007/978-3-030-92328-0_51

fetus, or to the infant during lactation, with milk. In De Jong et al. [5] AuNPs with size of 10 nm were found in various organs including the liver, spleen, kidney, heart, lungs, testis, brain, and thymus, whereas nanoparticles with size 250 nm were accumulated mainly in the liver and spleen.

The aim of the present study was to examine the distribution of AuNPs in different tissues of female mice and their offspring in comparison to control animals after prolonged oral exposure of experimental females to gold nanoparticles during pregnancy and lactation. Neutron activation analysis (NAA) was used to determine the total gold content in blood, brain, kidney, liver, and lungs samples. The data obtained were compared with previously obtained results, describing the distribution of another potentially toxic but widely used nanoparticles – silver [6].

2 Materials and Methods

Small, 3–5 nm in diameter, AuNPs were selected, since accumulation of smaller particles is usually higher and more widespread than that of larger particles [5]. Polyethylene glycol coated gold nanoparticles (PEG-AuNPs) were purchased from the M9 Company (Tolyatti, Russia). The preparation of experimental solution with gold concentration of 25 μg/ml was prepared according to procedure described in our previous study [7].

Outbred white mice (SHK) 15 females and 5 males, which are used to assess the safety of medical and cosmetic products, and dietary supplements, were purchased from the farm "Krolinfo" (Moscow region, Russia). The animals were maintained in the vivarium of M.F. Vladimirskiy Moscow Regional Research and Clinical Institute. The experiments and animals' maintenance in the vivarium were performed according to the principles of the Directive 2010/63/EU of the European Parliament and of the Council of 22 September 2010 on the protection of animals used for scientific purposes.

To study uptake of AuNPs by mice, experimental females were administrated with solution with gold concentration of 25 μg/ml starting one week before pregnancy and to the end of lactation. The total daily intake per animal was 125 μg [8, 9]. Data for control females, which drank pure water and their offspring were taken from the similar experiment performed previously with silver nanoparticles [6].

At the end of the lactation females and a part of offspring were euthanized by intraperitoneal injection of urethane solution (1.2 g dry urethane/kg). Selected tissues (brain, liver, lungs, kidneys and blood) were extracted from each mouse, weighted and dried freeze dried.

The gold content in the tissues was determined using neutron activation analysis at the IBR-2 reactor in Dubna, Russia. The description of irradiation process can be found elsewhere [10].

For irradiation samples were packed in aluminum foil cups and irradiated for 3 days at a neutron flux of $1.2 \cdot 10^{11}$ cm^{-2} s^{-1}. The samples were irradiated simultaneously with two reference materials: SRM 2710 (Montana Soil, Highly Elevated Trace Element Concentrations (NIST, USA) and liquid gold standard (Merck, Germany). Reference materials and blanks were placed in each container [8]. After irradiation samples were repacked and measured using HPGe detectors with an efficiency of 40–55% and resolution of 1.8–2.0 keV for total-absorption peak 1332 keV of the isotope ^{60}Co for 30 min.

The analysis of the spectra was performed using the Genie2000 software from Canberra company. The calculation of concentration was carried out using software "Concentration" developed in the Frank Laboratory of Neutron Physics (Dubna, Russia). The calculation of concentration was carried out using software "Concentration" developed in FLNP. The software automatically proposes to create the group standard from reference materials irradiated simultaneously with samples for calculation of concentrations of elements in the analyzed samples. Using the group standard, it is possible to check the quality of analysis by calculating the concentrations in each of irradiated reference materials and automatically comparing the obtained values with the certified ones. In the present study with lower net inaccuracy the isotope ^{198}Au from reference material liquid standard was used as a calibrator, and SRM 2710 was used for quality control. The obtained values for concentrations for SRM 2710 in all irradiated containers differed from the certified values in the range (0.2–5.5%).

3 Results and Discussion

After oral intake AuNPs were transported to different organs via the bloodstream, however, the content of gold in the blood of females and offspring was lower than in analyzed organs (Table 1). This suggests the rapid transfer of AuNPs from blood that confirms the data about high permeability of tissues for very small (<20 nm) nanoparticles [7]. Slightly more gold was detected in females' blood than in offspring'. In contrast, in previously performed research with the same design content of silver was four times higher in offspring compared to females. In control mice the content of gold in all the samples was below the detection threshold of the NAA method.

Accumulation of gold in different organs and blood was similar in females and offspring: the lowest content was accumulated in the blood, followed by lungs, liver, and the highest – in kidneys (Table 1), which may indicate active excretion of gold. The results contradict the data, obtained for silver nanoparticles, described in [6], where the content of silver in kidneys was the lowest, lower than in blood, and the highest content of silver was found in lungs.

The authors attributed this to the high density of blood vessels in the lungs. Taking into account present data it can be assumed that gold content in the lungs most likely correlates with the content of the metal in blood. Lipka and co-authors [7] showed that at intratracheal injection the majority of AuNPs were accumulated in the lungs.

The liver also has a high density of blood vessels, and the phagocytosis of nanoparticles by Kupffer cells was described [11]. Thus, the high accumulation of gold in the liver of mothers was expected. However, in offspring the content of gold in the liver was ten times lower than in females, and it differs from the data reporting silver content in the liver [6], which did not show the difference between its content in mothers and offspring. Since the content of gold in offspring' kidneys was two times lower than in females', it can be suggested that in case of indirect contact with AuNPs, i.e. through females' blood and/or milk, the excretion of AuNPs by kidneys became the main route of AuNPs elimination from organism. This can also correlate with the known chemical inertness of gold, in contrast to the high chemical activity of nanosilver [12]: AuNPs can be excreted inert.

Table 1. Content of gold in the organ and tissue samples from experimental females and their offspring.

Sample	Organ or tissue	Mean ± SD (μg/g dry weight)	Range (μg/g dry weight)
Females (5)	Blood	0.012 ± 0.007	0.006–0.023
	Liver	0.239 ± 0.117	0.131–0.410
	Kidneys	0.323 ± 0.128	0.182–0.516
	Lungs	0.079 ± 0.049	0.038–0.145
Offspring (10)	Blood	0.007 ± 0.005	0.002–0.018
	Liver	0.028 ± 0.022	0.008–0.083
	Kidneys	0.161 ± 0.116	0.053–0.418
	Lungs	0.017 ± 0.010	0.005–0.036

The brain is one of the most complex and important organs of living organisms. Thus, it is important to protect it from contamination with foreign substances which could affect brain functions. Protection of the brain is provided by the blood-brain barrier, which act as a local gateway against foreign compounds [13].

The amount of gold accumulated in brain of females and their offspring (Table 2) was significantly lower than during the accumulation of silver nanoparticles [6]. The chemical inertness of gold could be responsible for the low accumulation of AuNPs – despite the high permeability of the blood-brain barrier for very small nanoparticles. It indicates the safe use of AuNPs in new treatments in medicine. Sokolova et al. showed that AuNPs (core diameter 2 nm; hydrodynamic diameter 3–4 nm) were able to cross the blood brain barrier [14].

Table 2. Gold accumulated in the brain of female mice and their offspring. without the amount of gold in blood vessels in brain.

Sample	Specific mass in a sample (ng)		Gold content (ng/g dry weight)
	Mean ± SD	Range	
Females (5)	0.25 ± 0.10	0.14–0.36	3.79
Offspring (10)	0.08 ± 0.03	0.04–0.13	1.10

4 Conclusions

Gold was detected in all experimental samples, and its content in organs of control mice was below the detection limit of the NAA. The content of gold in the liver, kidneys, lungs, and brain of offspring was lower than in organs of females and the content of gold in blood was almost similar. The highest content of gold was detected in the kidneys of females and offspring, which indicates the high rate of its excretion, especially in offspring, where kidneys were the major deposition site. Accumulation of gold in the brain was the lowest among analyzed organs, pointing to the relative safety of AuNPs for their application in medical treatment.

Acknowledgments. This work was supported by the Russian Foundation for Basic Research (RFBR) [Grant No. 19-015-00145 A].

Conflict of Interest. The authors declare that they have no conflict of interest.

References

1. Sukhorukov, G., Fery, B., Möhwald, H.: Intelligent micro- and nanocapsules. Prog. Polym. Sci. **30**, 885–897 (2005). https://doi.org/10.1016/j.progpolymsci.2005.06.008
2. Kurapov, Y., Litvin, S., Belyavina, N.: Synthesis of pure (ligandless) titanium nanoparticles by EB-PVD method. J. Nanopart. Res. **23**, 20 (2021). https://doi.org/10.1007/s11051-020-05110-3
3. Khongkow, M., Yata, T., Boonrungsiman, S., Ruktanonchai, U.R., Graham, D., Namdee, K.: Surface modification of gold nanoparticles with neuron-targeted exosome for enhanced blood–brain barrier penetration. Sci. Rep. (2019) https://doi.org/10.1038/s41598-019-44569-6
4. Takeda, K., Suzuki, K., Ishihara, A., et al.: Nanoparticles transferred from pregnant mice to their offspring can damage the genital and cranial nerve systems. J. Health Sci. **55**, 95–102 (2009). https://doi.org/10.1248/jhs.55.95
5. De Jong, W.H., Hagens, I.W., Krystek, P., Burgera, M.C., Sips, A.J.A.M., Geertsmad, R.E.: Particle size-dependent organ distribution of gold nanoparticles after intravenous administration. Biomaterials **29**, 1912–1919 (2008). https://doi.org/10.1016/j.biomaterials.2007.12.037
6. Zinicovscaia, I., Grozdov, D., Yushin, N., et al.: Neutron activation analysis as a tool for tracing the accumulation of silver nanoparticles in tissues of female mice and their offspring. J. Radioanal. Nucl. Chem. **322**(2), 1079–1083 (2019). https://doi.org/10.1007/s10967-019-06746-9
7. Lewinski, N., Colvin, V., Drezek, R.: Cytotoxicity of Nanoparticles. Small **4**, 26–49 (2008). https://doi.org/10.1002/smll.200700595
8. Lipka, J., et al.: Biodistribution of PEG-modified gold nanoparticles following intratracheal instillation and intravenous injection. Biomaterials **31**, 6574–6581 (2010)
9. Lopatina, M., Petritskaya, E., Ivlieva, A.: Dynamics of body mass of laboratory mice in dependence of feed type and regimen of feeding. Lab. Anim. Sci. Res. **2** (2019). (In Russian). https://doi.org/10.29296/2618723X-2019-02-08
10. Frontasieva, M.: Neutron activation analysis in the life sciences. PEPAN **42**, 332–378 (2011). https://doi.org/10.1134/S1063779611020043
11. Valentini, X., Rugira, P., Frau, A., et al.: Hepatic and renal toxicity induced by TiO2 nanoparticles in rats: a morphological and metabonomic study. J. Toxicol. 5767012 (2019). https://doi.org/10.1155/2019/5767012
12. Behra, R., Sigg, L., Clift, M.J.D., et al.: Bioavailability of silver nanoparticles and ions: from a chemical and biochemical perspective. J. R. Soc. Interface **10**, 20130396 (2013). https://doi.org/10.1098/rsif.2013.0396

13. Teleanu, D.M., Chircov, C., Grumezescu, A.M., Volceanov, A., Teleanu, R.I.: Blood-brain delivery methods using nanotechnology. Pharmaceutics **10**(4), 269 (2018). https://doi.org/10.3390/pharmaceutics10040269
14. Sokolova, V., Mekky, G., van der Meer, S.B., Seeds, M.C., Atala, A.J., Epple, M.: Transport of ultrasmall gold nanoparticles (2 nm) across the blood–brain barrier in a six-cell brain spheroid model. Sci. Rep. **10**, 18033 (2020). https://doi.org/10.1038/s41598-020-75125-2

Modern Devices and Tools for the Cornee Collection and Processing. Synthesis of Literature

Adrian Cociug[1](✉), O. Macagonova[2], V. Cusnir Jr.[3], V. Cusnir[3], and V. Nacu[2]

[1] Human Tissue Bank, Chisinau, Republic of Moldova
[2] Laboratory of Tissue Engineering and Cells Culture, State University of Medicine and Pharmacy "Nicolae Testemitanu", Chisinau, Republic of Moldova
[3] Department of Ophthalmology and Optometry, "NicolaeTestemitanu" State University of Medicine and Pharmacy, Chisinau, Republic of Moldova

Abstract. The devices and tools used in the cornea sampling and processing is and will be one of the requirements used by the Eye Bank. Technological and informational progress tends to modernize all types of the devices that lead to increased quality of the cornea giving a longer storage with the of the viability of all epithelial corneal layers, stromal and endothelial. Bioreactors are complex devices that maintain the storage of the cornea *ex vivo* that are closer to the physiological norm. Many studies show that the materials that are made of these devices directly or indirectly influence the activity of the cells.

Keywords: Endothelial cells · Culture medium · Intraocular pressure · Epithelial cells · Stroma · Bioreactor · Eye bank

1 Introduction

The cornea is the window of our eye, allowing the light reach to the sensory cells that increase us to see the world around us. In humans, it has a thickness of 0.5 mm, a diameter of 10–12 mm, has a spherical shape that forms a refractive power of the eye of 43 diopters. Due to corneal lesions or regenerative, infectious, dystrophic and inflammatory disorders, many corneal diseases lead to the opacity and blindness. Currently, 5.2 million people have impaired visibility. The corneas are among the most successful transplanted tissues [1]. However, the lack of a sufficient number of corneal grafts imposes queues, especially in the developing countries [2]. Moreover, the number of donors is declining due to the demographic changes and the increasing frequency of the surgery on the cornea. Increased corneal transplantation requires more donors to be recruited and corneal tissue should be stored for more than 30 days [3].

Studying the cornea, which is based on the maintaining of the cell viability with the keeping of the transparency, is one of the challenges for the researchers. They always tend to implement the various methods and devices necessary for the various stages of the cornea sampling and processing that make the given steps easier, short-lived that maximize the lifespan of the cornea cells [4].

I. Tiginyanu et al. (Eds.): ICNBME 2021, IFMBE Proceedings 87, pp. 396–403, 2022.
https://doi.org/10.1007/978-3-030-92328-0_52

2 Materials and Methods

At the sampling stage, the various devices were invented in the Laboratory of Tissue Engineering and Cell Cultures, one of them being the "Corneal sectioning device" (Fig. 1) [5], which allow the cornea sectioning in one stage being made of the stainless steel. This device is efficient, portable, easy to apply and sterilized.

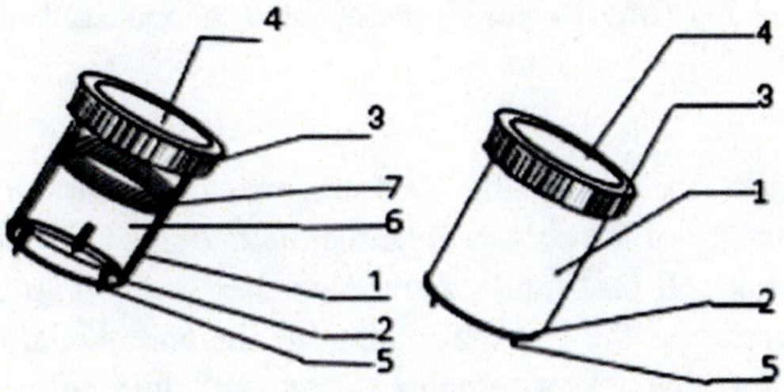

Fig. 1. Patent. (MD) 1048. Device for the cutting of the cornea [5].

At the stage of the cornea processing are used a number of the devices used in the Bank of Human Tissue in the Republic of Moldova, which aims to shape the cornea and remove debris that allows us to increase the resistance to the bacterial and fungal infections. One of them being the cornea fixation device (Figs. 2) [6].

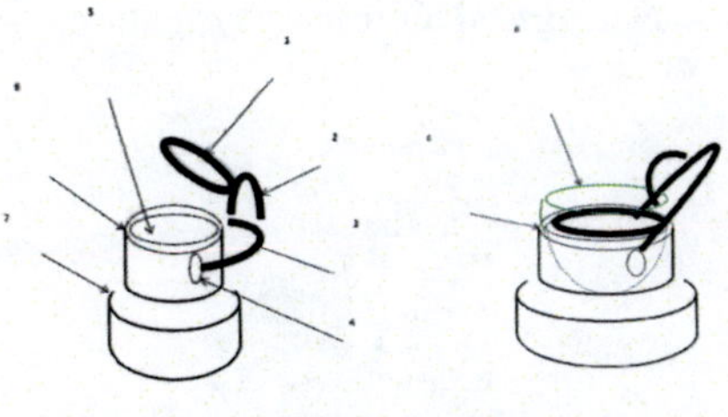

Fig. 2. Patent number 1504 MD. Cornea fixation device [6].

After the cornea is processed in the clean room of the Human Tissue Bank (BȚU), the stage of the evaluating of the cornea follows, which morphologically determines the integrity of the epithelial layer, the opacity and the transparency of the cornea. The number of the endothelial cells is performed with the help of the inverted phase contrast myocroscope in which the cornea is placed in the "Vessel for counting the endothelial cells of the cornea" (Fig. 3) [7], which is filled with Tissue culture medium. "C". After the immersing the cornea in the recently highlighted vessel, it returns with the downward concavity to which it is applied on the table of the inverted phase contrast microscope, easily highlighting the endothelial cells in which they can be quantified in the various quadrants and establish the endothelial cell average.

The storage of the corneas in the various culture media with the increased of the cell viability of more than 30 days remains one of the unresolved issues to date.

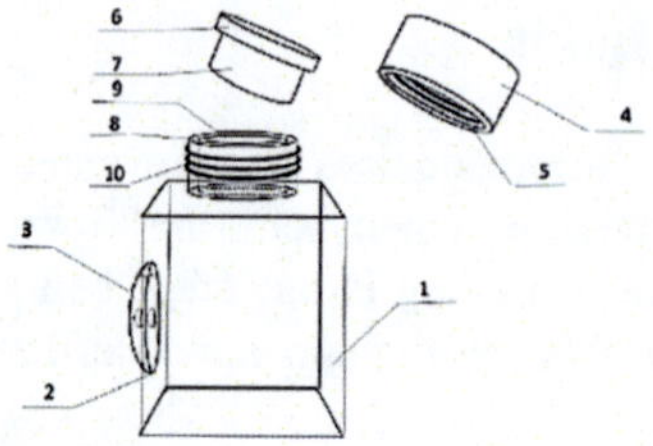

Fig. 3. Patent number 1311 MD. Device for preservation of cornea and corneal endothelial cells examination [7].

Currently, devices are known that perform a function similar to the eyeball that allows the upper surface of the cornea to be maintained in a CO_2-rich environment that induces the epithelial cell proliferation and the inner layer in an environment similar to the anterior chamber of the eye that maintains the endothelial cell viability. These devices are bioreactors that have a complex system with the automated systems that permanently change the nutritional environments of the cornea cells. One of the bioreactors is designed in SolidWorks (Dassault Systemes Soli dWorks Corp, Waltham, USA). The bioreactor crowns were milled from the stainless steel and the body was milled from polysulfone, by 3D printing, using the digital light processing technology. The material for 3D printing was a biocompatible acrylate-ba sed polymer (Technical University of Munich, Germany) (Fig. 4). It was established that culture configuration, autoclaving and surface preparation of the materials were the important factors influencing the cell viability. Poly (ether tert-ketone) and titanium-6Al-4V have proven to be the most suitable materials for use in a corneal bioreactor system.

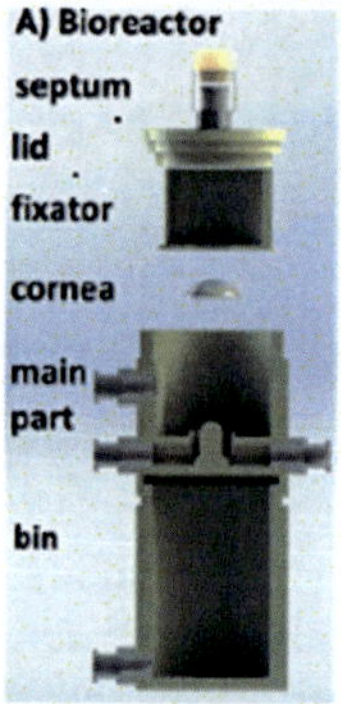

Fig. 4. Dynamic culture of native cornea. A) A cornea was fixed in a bioreactor using a custom-made fixator [27].

The bioreactor system consisted of two separate circuits. Circuit 1 with the pump P1 provided medium to the endothelial side of the cornea; in circuit 2 the pump P2 delivered medium to the epithelial side of the cornea to mimic tear fluid. The sensor PI measured the intraocular pressure.

After 7 days of culture in different media combinations (III, IV, V, and VI), the corneal tissue was investigated by (B) (H&E) staining, and C) fluorescein staining. D) The assessment of transparency, E) the integrated optical density, and F) the electrical resistance allowed the comparison to (I) native corneas and (II) standard submerse culture used in eye banks.

This bioreactor is used only in the scientific study as biological material being Pig eyes that have similar characteristics in size and structure of the stratum corneum to human.

The eyes were transferred directly in DMEM (Invitrogen, Darmstadt, Germany) with 10% fetal bovine serum (Lonza, Colog ne, Germany) and 1% Penicillin/Streptomycin (PAA, Coe lbe, Germany).

The cornea placed with the concave face on the surface of the crown (Fig. 5) is fixed hermetically in which the convex surface allows to form a high pressure that is similar to the anterior chamber of the eyeball, and the convex surface forms a chamber containing CO_2 and drops similar to the tear content (patent for eye drops (US8648057 B2) containing the high concentrations of calcium. It has positive effect on epithelial growth. In the body of the bioreactor are various culture media, the most suitable being biochrome 1 medium (MerckBiochrom GmbH, Berlin, Germany). It provides the optimal environmental conditions for epithelium and endothelium.

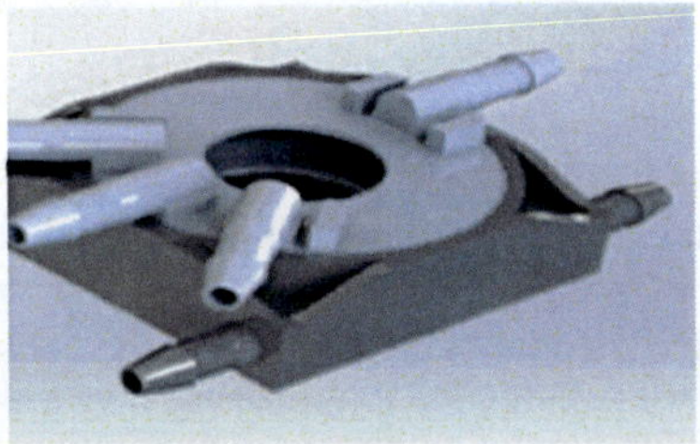

Fig. 5. Application domain of the bioreactor device [27].

Another bioreactor that studies the cornea is from France, Quantel Medical Laboratory (Fig. 6), in addition to studying the three layers, it was also studied the pressure inside the anterior chamber of the eyeball and determining the biochemical composition of the layers after swelling of the cornea in various medium.

It has been determined that control of the corneal shape, transparency, and thickness requires an endothelial pump; epithelial integrity, especially with tight, continuous junctions between the most superficial cells; and normal intraocular pressure (IOP) [8]. After donor death and even more so after the cornea excision, loss of IOP and concomitant gradual loss of the endothelial pumping function explain the rapid swelling of the cornea. Moreover, the preservation of corneal epithelial cells and lingual epithelial stem cells requires a specific culture medium, intermittent exposure to air, [9] and possibly shear stress [10].

BR applies the principles of the infusion chambers described above and of the anterior artificial chambers, [11–17], (Fig. 6). It has been designed to maintain a sterile closed environment that allows for long-term storage. The cornea is tightly fixed on the base of the BR, using the scleral rim as a tight seal to separate the epithelial and endothelial chambers; each chamber is connected to a separate control system for the culture medium. The complete system, except for its control panels, was placed in a dry incubator with 5% CO_2 at 31 °C.

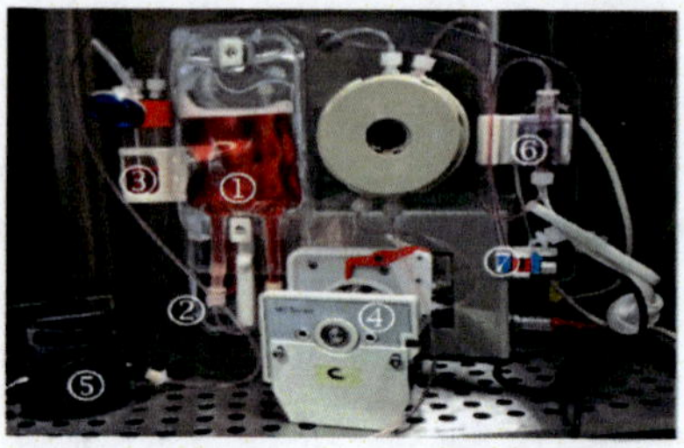

Fig. 6. Main characteristics of the corneal bioreactor (BR) in a CO2 incubator. (A) Overview of the BR in its cassette including the bag for fresh endothelial medium (1) and its waste (2), flask for epithelial medium (3), peristaltic pumps (4, 5), pressure sensor (6), and micro solenoid valve (7) [28].

3 Discussion

The devices and instruments used in the cornea sampling and processing is and will be one of the necessities used by the Eye Bank. Technological and informational progress tends to modernize all types of the devices that lead to increase the cornea quality giving a longer storage maintenance keeping of the viability of the corneal layers at the epithelial, stromal and endothelial. Bioreactors are complex devices that maintain *ex vivo* corneal storage that are closer to the physiological norm. Many studies show that the materials that are made of these devices directly or indirectly influence the activity of the cells. The closest to the cornea of the human eyeball parameters are those of 6-month-old pig that have the same sizes and cellular structures. One difference is only the architectural structure of the stroma which is richer reticular fibers in humans that shape the shape of the eyeball [19], in pigs they are smaller for which there is an age-related non-enzymatic reticulation that affects stromal collagen fibrils. [20]. Only when maintained in culture media with hydration of the layers do they return to normal shape. In addition, the cornea swelling begins immediately after scleral cutting (+70 μm in the present study), suggesting that the shrinkage of naturally tensed elastic collagen fibers plays a role in increasing thickness.

Porcine corneas stored for 7 days in BR with 20 mm Hg in the endothelial chamber, continuous renewal of the culture medium and lifting of epithelial air keep characteristics close to those of the postmortem cornea for each of the three layers: epithelial, stromal, and endothelial.

Organ culture is clearly inadequate for the storage of the porcine cornea. Loss of IOP and epithelial barrier partially explains the dramatic stromal edema and subsequent loss of EC in the present study. The pig corneas available from slaughterhouses are on average 6 months old and are much less rigid than human corneas [18] after the corneal excision. The porcine cornea loses its shape, unlike the human cornea.

Massive stromal edema seen in BR with cornea without iris creates an abnormal pathway for water when IOP is restored. In particular, this phenomenon was not observed in the OC because a pressure gradient was missing. On the contrary, when the anatomy of the iridocorneal angle is preserved, the pressure restored in the BR partially compensates for the SP and thus significantly limits the stromal swelling; corneal swelling is also limited by the integrity of the epithelial barrier related to tight junctions in the most superficial epithelial cells. The stored corneas thus remain thin without the addition of macromolecules, commonly used by eye banks in conventional OCs, to increase the average osmolality. This result may have applications for the Eye Banks, as the hyperosmolar disintegration media used during corneal storage may be toxic to the endothelium [21, 22]. In addition, reduction of endothelial folds related to corneal edema is likely to reduce EC losses, given the relationship between endothelial folds and EC apoptosis [23, 24]. In addition, the continuous renewal of MC in BR could also improve EC survival.

In addition, the air lifting system in the epithelial chamber could induce a shear stress on the limbus that could be involved in the regulation of epithelial stem cells. reporting the indirect interactions between these two layers described in the literature. Epithelial cell differentiation is indeed influenced by the presence of endothelial cells [25]. Cases of limbic stem cell deficiency secondary to severe endothelial dysfunction have been reported, which also supports the hypothesis of endothelial-epithelial interactions [26].

In addition, the present study shows that BR is an innovative experimental platform. It allows the storage of the cornea in the sterile conditions, but also the direct visualization of the tissues through transparent windows, without interrupting the experiment. The pressure inside the endothelial chamber can be adjusted if necessary, and the medium that bathes the endothelium and epithelium may be different and may have separately adjustable flow rates, as in the present study. In addition, media samples can be easily collected at various points in the circuit for the specific analysis.

By restoring of the physiological environment close to the cornea (IOP, flow, air epithelial lift, CM specific for each part), this BR version keeps the porcine corneas for 1 week in conditions not previously achieved, with excellent EC survival, thickness reduced, forms the maintenance and integrity of the epithelium and limbus. Although still improving, this model could be used for preclinical experiments. Research is underway to extend the shelf life and to create a pathological test for the healing of the corneal lesions.

4 Conclusions

1. The devices used at the sampling stages as well as at the processing stages have one of the guarding elements that lighten and due to the morphological quality of the cornea;
2. The use of the corneal calculation vessel allows the sterility of the cornea to be maintained until the transplant stage;
3. Bioreactors are modern devices that allow the cornea to be kept in the environment closer to the physiological norm;

4. Modernization of the bioreactors is one of the main ways to maintain the viability of endothelial cells for longer;
5. There is currently no method or device that would maintain the corneas for more than 30 days with a number of the viable endothelial cells per mm^2;
6. Culture media used in ocular banks were developed more than 30 years ago and should be reviewed to benefit from advances in cell biology and physiology.

Acknowledgment. The study was supported by the State Program Project: "GaN-based nanoarchitectures and three-dimensional matrices from biological materials for applications in microfluidics and tissue engineering", # 20.80009.5007.20.

Conflict of Interest. The authors declare no conflict of interests.

References

1. Resnikoff, D., et al.: World Health Organ. **82**, 844 (2004)
2. Armitage, W.J.: Transfus. Med. Hemoth. **38**, 143 (2011)
3. Rock, T., et al.: Arch. Clin. Exp. **254**, 135 (2016)
4. Cociug, A., Macagonova, O., Cusnir Jr., V., Cusnir, V., Nacu, V. Evaluation of the endothelial cell regenerative proprieties of the cornea in the culture media. Results and prospects. In: 4th International Conference on Nanotechnologies and Biomedical Engineering Proceedings of ICNBME-2019, 18–21 September 2019, pp. 601–605 (2019)
5. Cociug, A., Macagonova, O., Nacu, V.: Patent. (MD) 1048. Device for cutting of the cornea
6. Cociug, A., Macagonova, O., Nacu, V.: Patent number 1504 MD. Cornea fixation device
7. Cociug, A., Macagonova, O., Nacu, V.: 1311 MD Device for corneal preservation and examination of endothelial cells of the cornea
8. Bryant, M.R., McDonnell, P.J.: A triphasic analysis of corneal swelling and hydration control. J. Biomech. Eng. **120**, 370–381 (1998)
9. Massie, I., Levis, H.J., Daniels, J.T.: Response of human limbal epithelial cells to wounding on 3D RAFT tissue equivalents: effect of airlifting and human limbal fibroblasts. Exp. Eye Res. **127**, 196–205 (2014)
10. Kang, Y.G., Shin, J.W., Park, S.H., et al.: Effects of flow-induced shear stress on limbal epithelial stem cell growth and enrichment. PLoS ONE **9**, e93023 (2014)
11. Piehl, M., Gilotti, A., Donovan, A., DeGeorge, G., Cerven, D.: Novel cultured porcine corneal irritancy assay with reversibility endpoint. Toxicol. Vitro **24**, 231–239 (2010)
12. Dikstein, S., Maurice, D.M.: The metabolic basis to the fluid pump in the cornea. J. Physiol. **221**, 29–41 (1972)
13. Hull, D., Green, K., Boyd, M., Wynn, H.: Corneal endothelium bicarbonate transport and the effect of carbonic anhydrase inhibitors on endothelial permeability and fluxes and corneal thickness. Invest. Ophthalmol. Vis. Sci. **16**, 883–892 (1977)
14. Brunette, I., Nelson, L.R., Bourne, W.M.: A system for long-term corneal perfusion. Invest. Ophthalmol. Vis. Sci. **30**, 1813–1822 (1989)
15. Tanelian, D.L., Bisla, K.: A new in vitro corneal preparation to study epithelial wound healing. Invest. Ophthalmol. Vis. Sci. **33**, 3024–3028 (1992)
16. Thiel, M.A., Morlet, N., Schulz, D., et al.: A simple corneal perfusion chamber for drug penetration and toxicity studies. Br. J. Ophthalmol. **85**, 450–453 (2001)

17. Zhao, B., Cooper, L.J., Brahma, A., et al.: Development of a three-dimensional organ culture model for corneal wound healing and corneal transplantation. Invest. Ophthalmol. Vis. Sci. **47**, 2840–2846 (2006)
18. Elsheikh, A., Alhasso, D., Rama, P.: Biomechanical properties of human and porcine corneas. Exp. Eye Res. **86**, 783–790 (2008)
19. Hatami-Marbini, H., Etebu, E., Rahimi, A.: Swelling pressure and hydration behavior of porcine corneal stroma. Curr. Eye Res. **38**, 1124–1132 (2013)
20. Elsheikh, A., Wang, D., Brown, M., et al.: Assessment of corneal biomechanical properties and their variation with age. Curr. Eye Res. **32**, 11–19 (2007)
21. Borderie, V.M., Baudrimont, M., Lopez, M., Carvajal, S., Laroche, L.: Evaluation of the deswelling period in dextran-containing medium after corneal organ culture. Cornea **16**, 215–223 (1997)
22. Thuret, G., Manissolle, C., Campos-Guyotat, L., Guyotat, D., Gain, P.: Animal compound-free medium and poloxamer for human corneal organ culture and deswelling. Invest. Ophthalmol. Vis. Sci. **46**, 816–822 (2005)
23. Albon, J., Tullo, A.B., Aktar, S., Boulton, M.E.: Apoptosis in the endothelium of human corneas for transplantation. Invest. Ophthalmol. Vis. Sci. **41**, 2887–2893 (2000)
24. Gain, P., Thuret, G., Chiquet, C., et al.: Value of two mortality assessment techniques for organ cultured corneal endothelium: trypan blue versus TUNEL technique. Br. J. Ophthalmol. **86**, 306–310 (2002)
25. Pai, V.C., Glasgow, B.J.: MUC16 as a sensitive and specific marker for epithelial downgrowth. Arch. Ophthalmol. **128**, 1407–1412 (2010)
26. Uchino, Y., Goto, E., Takano, Y., et al.: Long-standing bullous keratopathy is associated with peripheral conjunctivalization and limbal deficiency. Ophthalmology **113**, 1098–1101 (2006)
27. Schmid, R., et al.: In vivo-like culture conditions in a bioreactor facilitate improved tissue quality in corneal storage. Biotechnol. J. **13**, 202 (2018)
28. Guindolet, D., et al.: Storage of porcine cornea in an innovative bioreactor. Invest. Ophthalmol. Vis. Sci. **58**(13), 5907–5917 (2017)

Techniques of Dental Pulp Decellularization

Stella Samson(✉) and V. Nacu

Laboratory of Tissue Engineering and Cells Culture SMPhU "Nicolae Testemiţanu", Chisinau, Republic of Moldova

Abstract. Regeneration of the endodontic tissues shows promise in treating dental pulp diseases; however, no suitable scaffolds exist for pulp regeneration. Different natural, biological extracellular matrix could be used as scaffolds for tissue regeneration. This study is dedicated for investigation of the characteristics of decellularized dental pulp from extracted wisdom teeth and evaluated whether it could mediate pulp regeneration. Decellularized dental pulp represents a suitable scaffold for improving clinical outcomes and functions of teeth with dental pulp diseases.

Keywords: Decellularization · Recellularization · Dental pulp

1 Introduction

The dental pulp is a soft, vascualrized tissue located in the pulp cavity. It contain endothelial cells, fibroblasts, lymphocytes, odontoblasts, etc. The sensorial, and protective functions of the healthy pulp-dentin complex is very important for normal teeth activity during the hole life. The pulp provide nutrient supply to the teeths, dentin formation, sensing, defense, etc. Pulpitis or pulp necrosis is a common disease caused by trauma and dental caries [1, 11]. Because of the small root canal volume and narrow apical foramen, in case of the dental pulp inflammation, blood supply to the pulp become insufficient and often causes unrecoverable pulp necrosis. In our days, development of regenerative medicine, provide new ideas to regenerat dental pulp, recovering it biological functions for maintaining dental normal activity.

By transplanting stem cells, growth factors with biological scaffold into the dental cavity, it could be achieve functional pulp regeneration [9, 15]. Biological, ECM, scaffolds play an important role in teeth tissue engineering by providing a temporary, three-dimensional spatial structure, and extracellular matrix (ECM) component to maintain the regenerative environment and stem cell function [3, 4, 6]. The ECM is essential for cell differentiation, proliferation, survival, migration and promote tissue and organ regeneration [7–9]. Various scaffolds have been used in dental pulp regeneration, including polylactic acid and collagen [7], but the major problem is lack of dental pulp extracellular matrix (ECM), thus limiting the ability to form dentin [9, 12, 14]. Therefore, it is essential to construct a dental pulp regeneration scaffold containing dental pulp ECM to promote odontoblast differentiation and dentin formation. Acellular natural ECM is used as biological scaffold for preserving the anatomical structure of the natural organ

I. Tiginyanu et al. (Eds.): ICNBME 2021, IFMBE Proceedings 87, pp. 404–409, 2022.
https://doi.org/10.1007/978-3-030-92328-0_53

and ECM. Acellular natural ECM has been used for skin [13], diaphragm [3], blood vessels [5], liver [10], regeneration. ECM used for tissue regeneration is basically derived from biological tissues [17]. Therefore, the present study investigated the structure and composition of decellularized dental pulp.

2 Materials and Methods

This study was conducted in accordance with the requirements of the ethics committee based on an informative agreement or an informative agreement close to each patient. In order to achieve the objectives outlined in this study, research was performed according to the clinical investigation protocol, which includes the collection and interpretation of data: anamnestic, objective clinical examination and paraclinical examination results. We used information methods of data collection: interviewing patients according to the special questionnaire developed, analysis of medical documentation to complete the data, assessment of the results of clinical and paraclinical investigations, dynamic evaluation of clinical manifestations. The teeth were extracted for orthodontic and prophylactic reasons.

After processing the oral cavity with betadine solution, local anesthesia is performed with Septanest 4% solution - 1.7 ml. After performing the syndesmotomy, the actual extraction of the tooth takes place, then the suture is applied. The extracted teeth are processed with a 70% ethanol solution and then placed in sterile test tubes containing DMEM (HIMEDIA) nutrient medium, antifungals, antibiotics, to be transported to laboratories for decellularization procedures. The excised dental pulp was mounted to the perfusion system and washed with 0.1–0.3 L of distilled water and after decellularized by two protocols: first - 0.25% Sodium Dodecyl Sulfate (SDS);the second – after washing the dental pulp with citrate phosphate dextrose the decellularization was done with 0.25% SDS solution. Subsequently, the dental pulp was washed three times in 1% of PBS solution. After the segments of the native and decellularized dental pulp were fixed in 10% formalin, late was performed morphological investigation HE staining. Nucleic acids counting was performed according to the QIAamp Blood Mini Kit extraction protocol (2003). Each sample was quantified in the "Nano Drop 200 C Thermo Scientific" spectrophotometer. The hydroxyproline content was determined on a spectrophotometer based on the oxidation of hydroxyproline in pyrrole under the action of chloramine B.

3 Results

After washing the dental pulp with citrate dextrose phosphate (second method) the dental pulp became to be paler in comparison to that were the samples were washed only with PBS (Fig. 1, 2).

After, decellularization the dental pulp had a whitish color in the second experimental lot compared to the first one, were the yellow color indicates the presence of the cellular residue.

Was observed that in case of decellularization of the dental pulp without using the anticoagulant solution, morphologically was discovered necrotic detritus and lyzed cells, sometimes the vascular wall has autolyzed cellular elements (Fig. 3).

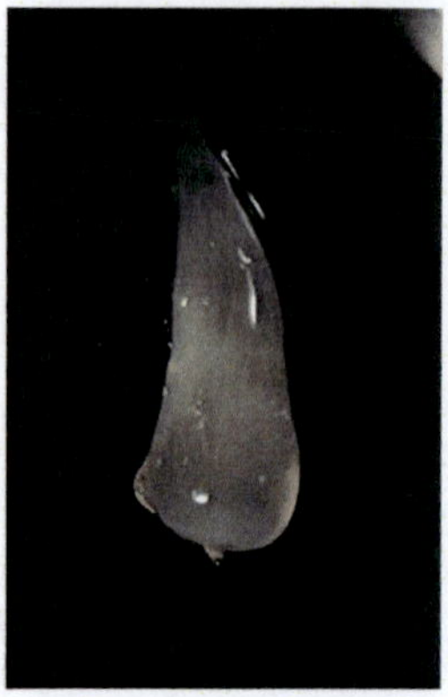

Fig. 1. Dental pulp perfusion with distilled water.

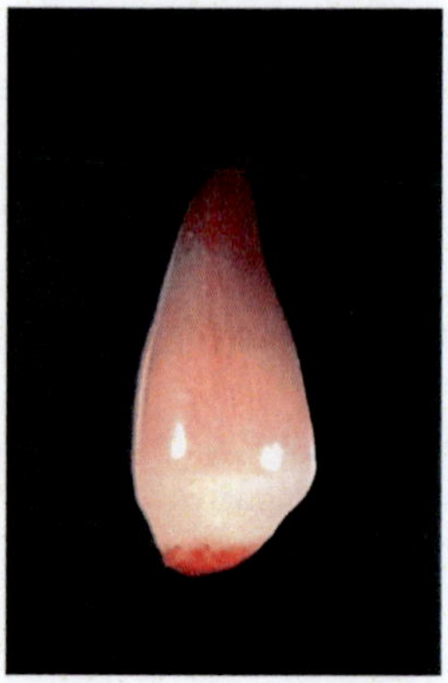

Fig. 2. Perfuzed dental pulp with distilled water and anticoagulant solution.

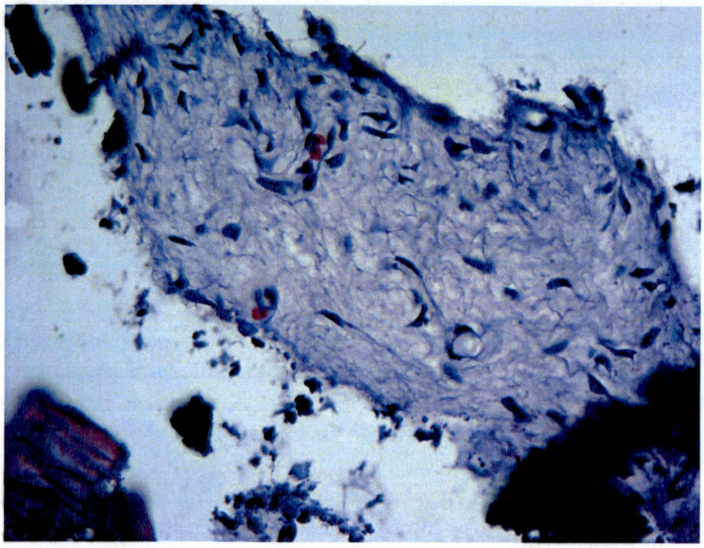

Fig. 3. Decellularized dental pulp by the SDS without anticoagulant washing. H-E, x 90.

The decellularized dental pulp scaffolds initially washed with anticoagulant solution showed the presence of collagen fibers which are in good condition (Fig. 4, 5).

The DNA concentration obtained from the samples of the dental pulps from both experimental lots were lower regarding native one. In first group was 1.01 ± 0.16 ng/μl, which constituted 4 times less DNA than in only decellularized with SDS samples - $4{,}2 \pm 1{,}4$ ng/μl. The date shows a more efficient decellularization in case of previous anticoagulant treatment.

Also was evaluated the free hydroxyproline content that must be proportional to the collagen content in the samples. Was obtained an amount of hydroxyproline of $0.23 \pm$

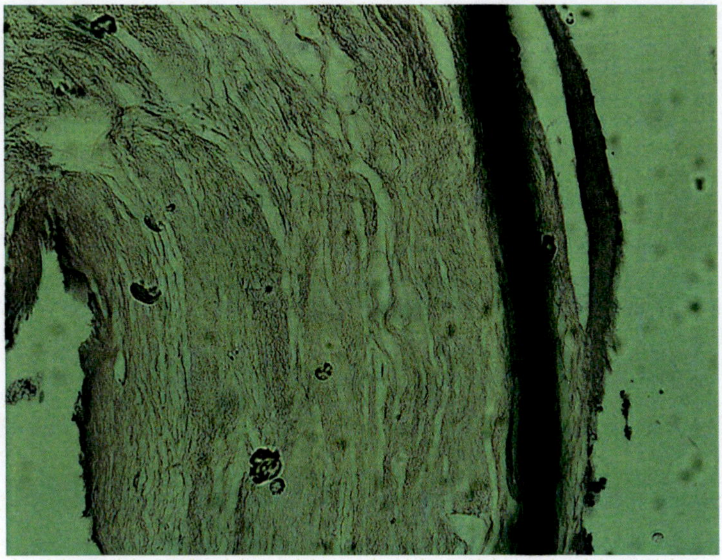

Fig. 4. Decellularized dental pulp by the SDS with anticoagulant washing. H-E, x 90.

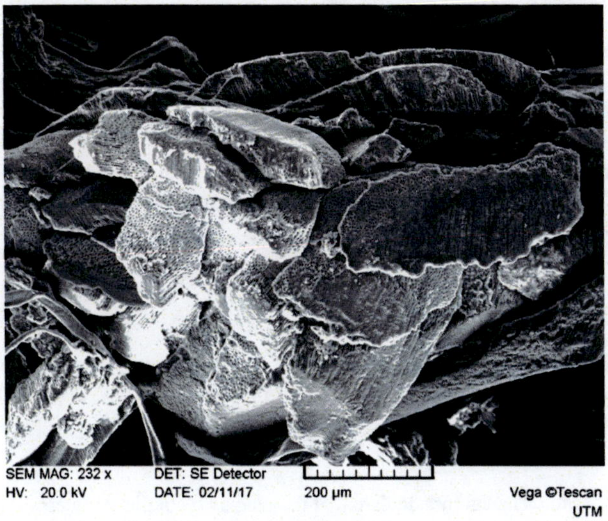

Fig. 5. Decellularized dental pulp.

0.011 mM/g of dried tissue from native dental pulp. In the same time the hydroxyproline in decellularized samples of the first experimental group was 1.93 ± 0.6 mM/g; in the second experimental group (used phosphate citrate dextrose) was 3,94 ± 0.23 mM/g.

It could be explained that the using of anticoagulant provid better decellularization that results in higher amount of hydroxyproline.

4 Discussion

An correct selected method of decellularization allows to preserv the normal structure of the extracellular scaffold. This will have a high influence on seded cells differentiation, proliferation after scaffold repopulation [6, 18].

The received three-dimensional matrices contain a small amount of residual cellular material, so the matrices are less immunogenic, which would not lead to posttransplant immunological rejection. Some authors shows that extracellular matrices preserve their structural components such as collagen and laminin [12, 14], and in our researche the hydroxyproline content revealed the preservation of architecture of the obtained matrix. It is very important the evaluation of the detergent influence on the extracellular matrix structure [15, 16].

We have use SDS ionic detergent that complete cell removal, but also it may be help to obtain a matrix with improved chemical, physical, biological properties [1, 3]. What is also very important – it to preserv vascular network structures for efficient supply of oxygen and nutrients [2, 5].

5 Conclusions

The DNA content after dental pulp decellularization shows that decellularization with SDS has better results.

The used decellularization methods maintains the connective architecture and form of the dental pulp. Obtained ECM probably could be used for the tissue bioengineering in dental pulp reconstruction.

Acknowledgment. The study was supported by the State Program Project: "GaN-based nanoarchitectures and three-dimensional matrices from biological materials for applications in microfluidics and tissue engineering", # 20.80009.5007.20.

Ethical Requirements The research was approved by the Research Ethics Committee of *Nicolae Teste*miţanu State University of Medicine and Pharmacy (protocol No 70/75 of 21.05.2018).

Conflict of Interest. The authors declare that they have no conflict of interest.

References

1. Baptista, P.M., Orlando, G., Mir Malek-Sani, S.H., Siddiqui, M., Atala, A., Soker, S.: Whole organ decellularization: a tool for bioscaffold fabrication and organ bioengineering. In: Conference on Proceedings of IEEE Engineering in Medicine and Biology Society, vol. 65, pp. 26–29 (2009)

2. Khan, A.A., Vishwakarma, S.K., Bardia, A., Venkateshwarulu, J.: Repopulation of decellularized whole organ scaffold using stem cells: an emerging technology for the development of neo-organ. J. Artif. Organs **17**(4), 291–300 (2014). https://doi.org/10.1007/s10047-014-0780-2
3. Eremia, V., Babuci, S., Petrovici, V., Negru, I., Nacu, V., Cobzac, V.: Reconstrucţia chirurgicală experimentală a defectelor diafragmatice prin utilizarea grefelor biologice decelularizate. Rezultate preliminare. In: Arta Medica. Chişinău, vol. 3, no. 68, pp. 4–8 (2018). ISSN 1810-1852
4. Jazayeri, H.E., Lee, S.-M., Kuhn, L., Fahimipour, F., et al.: Polymeric scaffolds for dental pulp tissue engineering: a review. Dent. Mater. **36**(2), e47–e58 (2020). https://doi.org/10.1016/j.dental.2019.11.005. Epub 30 Nov 2019
5. Malcova, T., Globa, L., Vascan, A., Ţugui, E., Stoian, A., Nacu, V.: Mechanical and morphological characterization of decellularized umbilical vessels as tissue engineering scaffolds. In: Tiginyanu, I., Sontea, V., Railean, S. (eds.) ICNBME 2019. IFMBE Proceedings, vol. 77, pp. 589–594. Springer, Cham (2020). https://doi.org/10.1007/978-3-030-31866-6_105. ISBN 978-3-030-31865-9
6. Howard, D., Buttery, L.D., Shakesheff, K.M., Roberts, S.J.: Tissue engineering: strategies stem cells and scaffolds. J Anat. **213**(1), 66–72 (2008). https://doi.org/10.1111/j.1469-7580.2008.00878.x
7. Dong, J., Mo, X., Li, Y., Chen, D.: Recent research progress of decellularization of native tissues. Sheng Wu Yi Xue Gong Cheng Xue Za Zhi **29**(5), 1007–1013 (2012)
8. Gilbert, T.W.: Strategies for tissue and organ decellularization. J. Cell Biochem. **113**(7), 2217–2222 (2012). https://doi.org/10.1002/jcb.24130
9. Jazayeri, H.E., Fahmy, M.D., Razavi, M., et al.: Dental applications of natural-origin polymers in hard and soft tissue engineering. J. Prosthodont. **25**(6), 510–517 (2016). https://doi.org/10.1111/jopr.12465. Epub 21 Mar 2016. PMID: 27003096
10. Jian, M., Cobzac, V., Moghildea, I., Popescu, V., Nacu, V.: Techniques of liver decellularization. Moldmedjournal. **61**(4), 21–24 (2018). ISSN 2537-6373
11. Jarvelainen, H., Sainio, A., Koulu, M., Wight, T.N., Penttinen, R.: Extracellular matrix molecules: potential targets in pharmacotherapy. Pharmacol. Rev. **61**(2), 198–223 (2009)
12. Keane, T.J., Swinehart, I.T., Badylak, S.F.: Methods of tissue decellularization used for preparation of biologic scaffolds and in vivo relevance. Methods **84**, 25–34 (2015). https://doi.org/10.1016/j.ymeth.2015.03.005. Epub 16 March 2015
13. Macagonova, O., Risnic, D., Cociug, A., Nacu, V.: Comparative analysis of the skin decellularization methods. Moldovan Med. J. **64**(2), 79–86 (2021). ISSN 2537-6373/ISSNE 2537-6381
14. Keane, T.J., Londono, R., Turner, N.J., Badylak, S.F.: Consequences of ineffective decellularization of biologic scaffolds on the host response. Biomaterials **33**(6), 1771–1781 (2012). https://doi.org/10.1016/j.biomaterials.2011.10.054. Epub 2 Dec 2011
15. Shupe, T., Williams, M., Brown, A., Willenberg, B., Petersen, B.E.: Method for the decellularization of intact rat liver. Organogenesis **6**(2), 134–136 (2010)
16. Samson, S., Nacu, V.: The dental pulp chamber evaluation by using cone-beam computed tomography. Moldovan Med. J. **64**(2), 49–52 (2021). ISSN 2537-6373/ISSNE 2537-6381
17. Zheng, C.X., Sui, B.D., Hu, C.H., Qiu, X.Y., Zhao, P., Jin, Y.: Reconstruction of structure and function in tissue engineering of solid organs: toward simulation of natural development based on decellularization. J. Tissue Eng. Regen. Med. **12**(6), 1432–1447 (2018). https://doi.org/10.1002/term.2676. Epub 15 May 2018
18. Gudumac, E., Babuci, S., Petrovici, V., Nacu, V., Tica, C., Negru, I.: Histological evaluation of the efficiency of human amniotic membrane used in experimental reconstruction of the anterior abdominal wall defects. J. Pediatrului – Rev. Societăţii Române Chirurgie Pediatrică **XIX**(75–76), 69–78 (2016)

Innovation, Development and Interdisciplinary Research

Breathing Pattern in Subjects with Borderline Personality Disorder in Pain Test

S. Lozovanu[1(✉)], I. Moldovanu[1,2], V. Vovc[1], T. Besleaga[1], A. Ganenco[1], and I. Tabirta[1]

[1] "Nicolae Testemitanu" State University of Medicine and Pharmacy, Chisinau, Republic of Moldova
svetlana.lozovanu@usmf.md

[2] Department of Headache and Autonomic Disorders within Institute of Neurology and Neurosurgery, Chisinau, Republic of Moldova

Abstract. Individuals with borderline personality disorder (BPD) are more likely to experience pain and assess the pain as more severe than individuals with other personality disorders. Actual study researched psychophysiological reactivity of subjects with BPD by studying the breathing pattern and pain test. The study proceeded in two stages: first stage - psychometric testing using Personality Inventory for DSM-5 (PID-5; DSM-5 - Diagnostic and Statistical Manual of Mental Disorders); second stage included recording the parameters of the breathing pattern in resting state and pain test by compression of the middle third of the left shoulder. Time parameters of breathing pattern were significantly different between groups, unlike volume parameters. Pain test produces changes in breathing parameters in both groups, but in different way. Pain test appears to be a valuable method to detect psychophysiological changes in patients with BPD.

Keywords: Borderline personality disorder · PID-5 · Breathing pattern · Pain test

1 Introduction

The study of the variations of the components of the respiratory pattern under the influence of physiological or pathological factors allow the deepening of knowledge in the field of physiology of systemic interaction at the level of the whole organism as well as in the field of pathophysiology of functional, psychosomatic diseases. Taking into account the prevalence of borderline personality disorder (BPD) between 0.2 and 1.8% in the general community, 6% in patients in primary care, about 10–15% in patients in outpatient mental health centers, and 20% [1] to 50% [12] in specialized psychiatric centers, the study of the respiratory pattern could offer doctors, especially those at the primary level, an alternative to pharmacological treatment, by correcting the psychophysiological mechanisms of development of systemic dysfunctions.

Simultaneous evaluation of respiratory parameters and respiratory functional disorders can be applied in the analysis of the role of different brain and psychological

I. Tiginyanu et al. (Eds.): ICNBME 2021, IFMBE Proceedings 87, pp. 413–421, 2022.
https://doi.org/10.1007/978-3-030-92328-0_54

components in establishing the respiratory pattern in healthy individuals and patients with psychofunctional respiratory dysfunction. Highlighting the patients with personality disorders (PD) and affective disorders (anxiety, depression, obsessive-compulsive disorder, etc.) with pronounced respiratory symptoms may have an impact on the creation of models of types of psychorespiratory personality in healthy individuals and patients with respiratory dysfunction, and to choose specific methods of treatment, such as bio-feedback techniques of respiratory reeducation.

Patients with BPD are frequently seen by primary care physicians and specialists with pain problems. They are more likely to experience pain and assess the pain as more severe than patients with other personality disorders [8], they report significant pain that interferes with their lives. An emphasis on the management of medical and psychiatric comorbidities can improve their long-term functionality [1, 9].

Significant headaches are a common complaint in patients with BPD [3, 11]. In a previously published article [5], we presented the results of the analysis of the relationship between BPD and headache. These results are confirmed by other authors who suggest that depression, suicidal tendencies, impulsivity, violence and migraines that have been associated with BPD [7], are based on a central biochemical disorder, and especially related to serotonergic disorders. Namely that serotonergic mechanisms and other biochemical mechanisms are common for BPD and certain disorders associated with pain and headache [2, 10, 11].

Based on these considerations, the research on the influence of some factors on the psychophysiological reactivity of the subjects in the examination of the respiratory pattern is of interest. These data will allow the deepening of knowledge in the field of respiratory psychophysiology for the validity and fidelity of methods, diagnostic criteria, analysis techniques and data interpretation.

2 Subjects and Methods

The current study was designed to determine the relationship between personality disorders and nervous control of breathing and was performed on a group of 237 people who signed an informed agreement, at the Department of Human Physiology and Biophysics, USMF "Nicolae Testemitanu", Republic of Moldova, from March 2017 till February 2020.

In the current article will be presented some of the data of this comprehensive study, as a continuation of the results presented in previous publications [4, 5]. Thus, 96 people were selected for this study. The average age of the people included in the study was 33.5 years (19–60 years). The study excluded subjects with acute and chronic pulmonary or cardiac pathology, as well as subjects with recent diseases or interventions on the rib cage or abdominal wall.

The study proceeded in two stages:

First stage - Psychometric testing:

The Personality Inventory for DSM-5 (PID-5; DSM-5 - Diagnostic and Statistical Manual of Mental Disorders) is a tool for self-assessment of personality traits, developed by the American Psychiatric Association (APA) in 2012.

The Personality Inventory for DSM-5 (PID-5) was translated and validated by a working group consisting of researchers of the Department of Human Physiology and Biophysics USMF „Nicolae Testemitanu" and the Department of Headache and Vegetative Disorders within the Institute of Neurology and Neurosurgery with respect to the norms of translation, adaptation and validation of International Test Commission (ITC) and with the agreement of the authors. PID-5 is a questionnaire, which contains 220 elements of self-report of personality (items), used to measure the maladaptive personality traits that are described in DSM-5. Answers are selected on a four-point scale, from 0 ("very false or often false") to 3 ("very true or often true"). Thus PID-5 offers scores evaluated on a scale of 4 points for 25 facets (features). Each facet / feature includes 4 to 14 elements. These facets correspond to the maladaptive personality traits described in section III of DSM-5 and are included in the five domains of higher order, also described in section III: negative affectivity, detachment, antagonism, disinhibition, and psychoticism.. The score higher than 2 is of a certain number of facets is a quantitative index of one of the 6 types of PD: Antisocial, Borderline, Schizotypal, Avoidant, Obsessive-compulsive or Narcissistic [1].

Finally, based on the results obtained from the PID-5 test, it is possible to distribute the numerical results of the scores for each facet from which the profile of personality disorders is formed. For borderline personality disorder, the distribution of numerical results of 7 facets is studied. For a positive result, it is necessary that 4 of the 7 facets on the ordinate axis to have a score higher than 2 (Fig. 1).

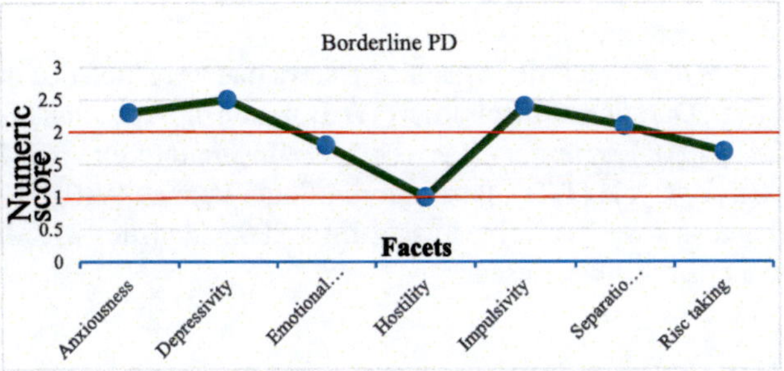

Fig. 1. The profile of a subject with BPD. Subject with BPD, with numeric score more than 2 in 4 facets.

Further analysis, namely the distribution of facets and their numerical score (higher than 2), made possible to divide the subjects into two groups:

- I group – PID score 0–1.99 in all 7 facets – control group;
- II group – PID score more than 2.0 in at least 4 facets - borderline personality disorder group (BPD).

Thus, the control group included 64 people (N = 64), aged between 19 and 60 years, average age 35.5 years, 34 women and 30 men.

The group with borderline personality disorder (BPD) - 32 people (N = 32) aged between 19 and 60 years, average age 34.7 years, of which 20 women and 12 men.

The second stage of the study included recording the parameters of the breathing pattern by respiratory inductive plethysmography, using VISURESP (RBI instruments, Meylan, France; http://www.visuresp.com), as well as end-tidal CO_2 using Capnostream™ 20p bedside monitor (Medtronic, Minneapolis, USA). The experimental protocol included recording the parameters of the respiratory pattern in people lying on their backs, in a quiet room, with moderate light, at a comfortable temperature, in 3 functional tests. During the recording, subjects were asked to breathe quietly, not to speak, and to avoid further movements.

1. Breathing at rest (R) - for 5 min in conditions of physical, mental and emotional rest (the first minute was excluded from the calculations to exclude artifacts obtained from the movements of application and accommodation of subjects in the vest).
2. Pain test (PT) - for 3 min, the pain was induced through compression of the middle third of the left shoulder by applying the cuff of the sphygmomanometer at a constant pressure of 200 mm Hg. Respiratory cycles were recorded for 3 min.
3. The period of recovery after pain test (post-PT) - examiners were asked again to breathe normally for 5 min, to see if the parameters of the respiratory pattern returned to rest values, and how fast it occurred.

Prior to recording the respiratory parameters, the plethysmograph was calibrated, for each person, by inhaling a standard volume of air from a graded syringe (1 L) and recording changes to the chest corresponding to this volume, from which all calculations were subsequently performed.

The parameters of the recorded breathing pattern that were included in the study are: Tidal Volume (Vt); Duration of Inspiration (Ti), Duration of Expiration (Te), Duration of Respiratory Cycle (Tt), Respiratory Rate (F), Mean Inspiratory Flow (VtTi), Inspiration Ratio in Respiratory Cycle (TiTt), Respiratory Minute Volume (VtF).

Statistical analysis was performed using IMB SPSS Statistics 23; Mann –Whitney U test and Student's t-test were used.

3 Results and Discussions

In functionally resting adult human subjects, there is a variety of breathing patterns not only in terms of tidal volume, duration of inspiration, expiration, and derived variables (T, VtTi, and TiTt), but also in the airflow profile. Each individual seems to select a certain pattern from the infinite number of possible combinations of respiratory variables and the airflow profile. This variability is not random and can be explained either by a central neural mechanism or by instability in the chemical feedback loops that control respiration. Moreover, the unrepeatability and individuality of the respiratory pattern led some scientists to launch the idea of the so-called "ventilatory personality", an idea that marked the beginning of a whole direction in contemporary research on the psychophysiology of respiration [6].

Thus, the results of the current study show that there are no significant differences in the parameters of volume, time and flow of the respiratory pattern between control group and BPD group.

Significant differences were recorded in the time parameters between the study groups during test with pain stimulation, namely a shorter Te by 20.8% in the BPD group (1.72 ± 0.3 s) compared to control group (2.17 ± 1.0 s), $p < 0.05$ (Fig. 2a) and a shorter Tt by 15.6% in the BPD group (3.03 ± 0.44 s) compared to control group (3.61 ± 1.21 s), $p < 0.05$ (Fig. 2c).

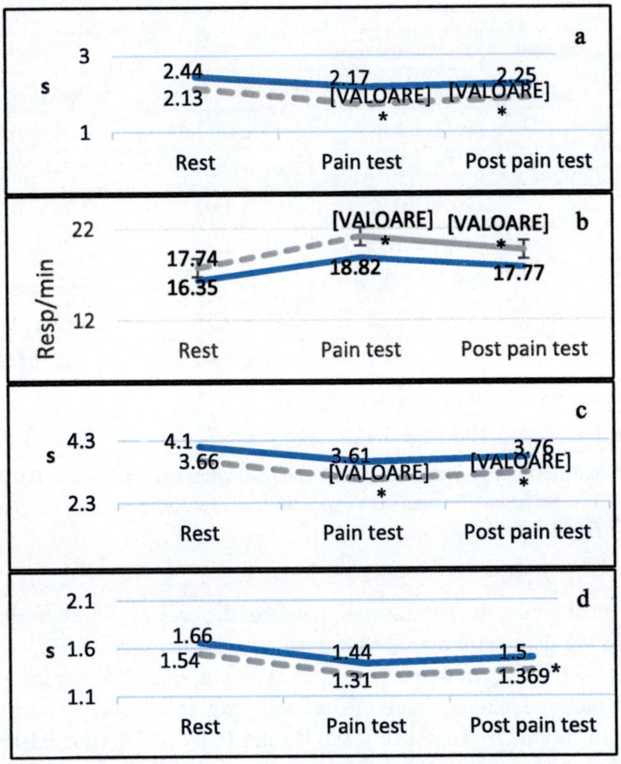

Fig. 2. Time parameters in control and BPD group. Duration of expiration Te (a), respiratory rate F (b), duration of respiratory cycle Tt (c) and duration of inspiration Ti (d) in control group and BPD group. Note: * indicates $p < 0.05$, continuous line indicates the values in control group, dashed line indicates the values in BPD group

Respectively, in healthy subjects during pain test, a significant lower respiratory rate, 18.82 resp/min was observed, compared to BPD group with 21.15 resp/min ($p < 0.05$) (Fig. 2b).

Compared to healthy subjects, subjects with BPD after the pain test continue to breathe with a higher rate by 10.7% (19.67 resp/min in BPD group compared to 17.77 resp/min in the control group, $p < 0.05$) (Fig. 2b); because Tt is lower in BPD group

(3.27 ± 0.43 s) than in the control group (3.76 ± 0.95 s), $p < 0.05$ (Fig. 2c), Ti is 9% lower in BPD group (1.37 ± 0.19 s) compared to the control group (1.5 ± 0.27 s), $p < 0.05$, (Fig. 2d), and Te is 15.6% lower in BPD group (1.9 ± 0.29 s) compared to the control group (2.25 ± 0.76), $p < 0.05$, (Fig. 2a).

The analysis of the parameters of breathing pattern within the study groups attests considerable and significant changes in these parameters during the pain test and after pain test compared to rest.

Table 1. The parameters of breathing pattern in subjects of control group

Respiratory parameter	Functional test		
	Rest	Pain test	Post pain test
Vt (l)	0.43 ± 0.2	0,37 ± 0,18 ***	0.39 ± 0.21*
Ti (s)	1.66 ± 0.35	1,44 ± 0,31 ***	1.5 ± 0.26*** •
Te (s)	2.44 ± 0.92	2,17 ± 1,01 **	2.25 ± 0.76*
Tt (s)	4.10 ± 1.23	3,61 ± 1,21 ***	3.76 ± 0,95**
F (resp/min)	16.35 ± 3.78	18,82 ± 3,47 ***	17.77 ± 3.17 *** •••
TiTt	0.42 ± 0.25	0,43 ± 0,05	0.42 ± 0.05
ViTi (l/s)	0.25 ± 0.09	0,24 ± 0,10	0.25 ± 0.11
VtF (l/min)	6.41 ± 2.40	6,28 ± 2,85	6.27 ± 3.31
$etCO_2$ (mm Hg)	35.73 ± 3.43	36,3 ± 3,57	35.92 ± 3.39 ••

Note: significant differences between pain test and resting condition, * – $p < 0.05$, ** – $p < 0.01$, *** - $p < 0.001$; significant changes between pain test and post pain test, • – $p < 0.05$; •• – $p < 0.01$, ••• – $p < 0.001$

In the control group, the volume parameters of the breathing pattern show significant difference during pain test compared to rest; Vt decreases by 15.6%, during pain test being 0.37 ± 0.18 l compared to 0.43 ± 0.2 l at rest ($p < 0.001$). Similarly, we recorded a decrease in the time parameters of the breathing pattern during the pain test compared to resting: Ti decreased by 13.4% ($p < 0.001$) and Te by 10.9% ($p < 0.01$); Tt shortened to 3.61 ± 1.21 s compared to 4.1 ± 1.23 s at rest ($p < 0.001$). Healthy subjects during pain test breathed with a frequency of 18.82 ± 3.47 resp/min compared to 16.35 ± 3.78 resp/min at rest ($p < 0.001$) (Table 1).

In post pain test period, the pattern parameters behaved identically as in the pain test pain. In the control group, significant differences were observed in following parameters: Vt lowered to 0.39 ± 0.21 l compared to 0.43 ± 0.2 ml at rest ($p < 0.05$); Ti in post pain test shortened to 1.5 ± 0.26 s from 1.66 ± 0.35 s at rest ($p < 0.001$); Te was 2.25 ± 0.76 s compared to 2.44 ± 0.92 s ($p < 0.05$); Tt was 3.76 ± 0.95 s compared to 4.1 ± 1.23 s at rest ($p < 0.01$); and F was higher by 8.7%, 17.77 ± 3.17 resp/min compared to 16.35 ± 3.78 resp/ min at rest ($p < 0.001$) (Table 1).

Significant differences were observed in the following parameters of the breathing pattern during post pain test compared to the pain test in the control group: Ti was 4.6%

higher in post-pain, 1.5 ± 0.26 s compared to 1.44 ± 0.31 s in pain test ($p < 0.05$); F in post-pain decreased to 17.77 ± 3.17 resp/min compared to 18.82 ± 3.42 resp/min in pain ($p < 0.001$). $EtCO_2$ was also lower in post-pain, 35.92 ± 3.39 mm Hg compared to 36.3 ± 3.52 mm Hg in pain test ($p < 0.01$) (Table 1).

In the BPD group, significant differences in several parameters of the breathing pattern were observed during the pain test compared to the rest: Vt decreased from 0.39 ± 0.16 l at rest to 0.34 ± 0.13 l in the pain test ($p < 0.01$); Te shortened from 2.12 ± 0.58 s to 1.72 ± 0.29 s ($p < 0.01$) and Tt changed from 3.66 ± 0.79 s at rest to 3.03 ± 0.43 s in pain test ($p < 0.01$); F in these subjects increased from 17.74 ± 3.27 resp/min at rest to 21.15 ± 3.08 resp/min in the pain test ($p < 0.01$). Also, an increase in the $etCO_2$ during pain test is observed, 35.43 ± 2.38 mm Hg compared to 34.48 ± 2.42 mm Hg at rest ($p < 0.01$). (Table 2).

Table 2. The parameters of breathing pattern in subjects of BPD group

Respiratory parameter	Functional test		
	Rest	Pain test	Post pain test
Vt (l)	0.39 ± 0.16	0.34 ± 0.13 **	0.36 ± 0.18
Ti (s)	1.54 ± 0.25	1.31 ± 0.17 ***	1.37 ± 0.18 ** •••
Te (s)	2.12 ± 0.58	1.72 ± 0.29 **	1.90 ± 0.28 ••
Tt (s)	3.66 ± 0.79	3.03 ± 0.43 **	3.27 ± 0.42 •••
F (resp/min)	17.74 ± 3.27	21.15 ± 3.08 **	19.67 ± 2.72 * •••
TiTt	0.43 ± 0.03	0.44 ± 0.03	0.43 ± 0.03 •
VtTi (l/s)	0.26 ± 0.11	0.16 ± 0.11	0.26 ± 0.13
VtF (l/min)	6.7 ± 3.20	6.95 ± 2.95	6.7 ± 3.41
$etCO_2$ (mmHg)	34.48 ± 2.42	35.43 ± 2.38 **	34.96 ± 2.75 •

Note: significant differences between pain test and resting condition, * – $p < 0.05$, ** – $p < 0.01$, *** – $p < 0.001$; significant changes between pain test and post pain test, • – $p < 0.05$; •• – $p < 0.01$, ••• – $p < 0.001$.

During post pain test, F was higher by 10.8%, 19.67 ± 2.72 resp/min compared to 17.74 ± 3.27 resp/min at rest ($p < 0.05$), while Ti decreased by 11%, from 1.54 ± 0.25 s to 1.37 ± 0.18 s ($p < 0.01$) (Table 2).

The most significant differences in studied parameters were observed when comparing the parameters of the breathing pattern during pain test and after pain test in BPD group. The time parameters of the breathing pattern in post-pain test changed as following: Ti increased by 4.8% ($p < 0.001$), Te by 10.4% ($p < 0.01$), and Tt by 8.4% ($p < 0.001$). F was lower by 7.1% in post pain test, 19.67 ± 2.72 resp/min compared to 21.15 ± 3.08 resp/min in pain test ($p < 0.001$). TiTt was lower in post pain period, 0.43 ± 0.03 compared to 0.44 ± 0.3 in pain test ($p < 0.05$). $EtCO_2$ also decreased in post pain period, 34.96 ± 2.75 mm Hg compared to 35.43 ± 2.38 mmHg in pain test ($p < 0.05$) (Table 2).

Pain causes changes in all functional systems of the human body. Breathing as a vital function of the body is one of the most sensitive to any change in homeostasis.

The current data regarding the mechanisms of respiratory rhythmogenesis as well as the mechanisms responsible the appearance of borderline personality disorder, denote changes in the serotonergic system important in generating the breathing pattern.

4 Conclusions

Pain test appears to be a valuable method to detect psychophysiological changes in patients with BPD. Time parameters of breathing pattern were significantly different between groups, unlike volume parameters. The respiratory response to induced pain and psychological traits may be linked with dysfunctional activity of prefrontal and limbic areas in BPD although exact cause is not yet clarified.

Conflict of Interest. The authors declare that they have no conflict of interest.

References

1. Al-Dajani, N., Gralnick, T.M., Bagby, R.M.: A psychometric review of the personality inventory for DSM–5 (PID–5): current status and future directions. J. Pers. Assess **98**(1), 62–81 (2016). https://doi.org/10.1080/00223891.2015.1107572
2. Amad, A., Ramos, N., Thomas, P., et al.: Genetics of borderline personality disorder: systematic review and proposal of an integrative model. Neurosci. Biobehav. Rev. **40**, 6–19 (2014). https://doi.org/10.1016/j.neubiorev.2014.01.003
3. Biskin, R.S., Frankenburg, F.R., Fitzmaurice, G.M., Zanarini, M.C.: Pain in patients with borderline personality disorder. Pers. Ment. Health **8**(3), 218–227 (2014). https://doi.org/10.1002/pmh.1265
4. Lozovanu, S., Moldovanu, I., Vovc, V., Besleaga, T., Ganenco, A.: Recording of the breathing pattern in the test with controlled hyperventilation in subjects with a borderline type personality disorder. In: Tiginyanu, I., Sontea, V., Railean, S. (eds.) ICNBME 2019. IP, vol. 77, pp. 475–479. Springer, Cham (2020). https://doi.org/10.1007/978-3-030-31866-6_85
5. Lozovanu, S., Moldovanu, I., Vovc, V., et al.: Relationship between personality disorders and headaches using PID for DSM-5. Moldovan Med. J. **61**(4), 29–35 (2018). https://doi.org/10.5281/zenodo.2222309
6. Moldovanu, I., Vovc, V.: Conceptul neurologiei funcţionale. Bul. Acad. Ştiinţe Moldovei Ştiinţe Med. **8**(4), 16–21 (2006)
7. Niedtfeld, I., Schulze, L., Kirsch, P., et al.: Affect regulation and pain in borderline personality disorder: a possible link to the understanding of self-injury. Biol. Psychiatry **68**(4), 383–391 (2010). https://doi.org/10.1016/j.biopsych.2010.04.015
8. Pavony, M.T., Lenzenweger, M.F.: Somatosensory processing and borderline personality disorder: pain perception and a signal detection analysis of proprioception and exteroceptive sensitivity. Pers. Disord. **5**(2), 164–171 (2014). https://doi.org/10.1037/per0000017
9. Sansone, R.A., Whitecar, P., Meier, B.P., Murry, A.: The prevalence of borderline personality among primary care patients with chronic pain. Gen. Hosp. Psychiatry **23**(4), 193–197 (2001). https://doi.org/10.1016/s0163-8343(01)00148-7

10. Saper, J.R., Lake III, A.E.: Borderline personality disorder and the chronic headache patient: review and management recommendations. Headache **42**(7), 663–674 (2002). https://doi.org/10.1046/j.1526-4610.2002.02156.x
11. Schmahl, C., Baumgärtner, U.: Pain in borderline personality disorder. Mod. Trends Pharmacopsychiatry **30**, 166–175 (2015). https://doi.org/10.1159/000435940
12. Widiger, T.A., Weissman, M.M.: Epidemiology of borderline personality disorder. Hosp. Commun. Psychiatry **42**(10), 1015–1021 (1991). https://doi.org/10.1176/ps.42.10.1015

Cathodoluminescence and X-Ray Luminescence of $ZnIn_2S_4$ and $CdGa_2S_4$ Single Crystals

E. Arama[1], V. Pîntea[2], and T. Shemyakova[3](✉)

[1] Nicolae Testemitanu University of Medicine and Pharmacy, Chisinau, Moldova
[2] Technical University of Moldova, Chisinau, Moldova
[3] Institute of Applied Physics, Chisinau, Moldova
tatiana.shemyakova@ifa.md

Abstract. Zinc thioindate and gallium thiogallate single crystals were grown by a chemical vapor transport method. The cathodoluminescence and X-ray luminescence spectra of $ZnIn_2S_4$ and $CdGa_2S_4$ single crystals were studied. From cathodoluminescence spectra of $ZnIn_2S_4$ at low temperatures the forbidden gap width of (2.96 ± 0.02) eV at 80 K and optical depth of the deep acceptor level $E_A = (E_V + 0.30)$ eV were determined. In the X-ray luminescence spectra of $CdGa_2S_4$ a single emission band is observed with an energy maximum at 2.14 eV and a slope within the high-energy range at approximately 2.34 eV identified as optical transitions of donor-acceptor type.

Keywords: Ternary semiconductor · Zinc thioindate · Cadmium thiogallate · Cathodoluminescence · X-ray luminescence

1 Introduction

Ternary semiconductor compounds - thioindate of zinc $ZnIn_2S_4$, thiogallate of cadmium $CdGa_2S_4$, and related compounds - are interesting objects for investigations since they exhibit specific properties: a wide forbidden gap ≈3 eV, low sensitivity to the introduction of impurities, high stability to ionization radiation. High photosensitivity, intensive photoluminescence [1], and cathodoluminescence [2] in visible and ultraviolet spectral range determine wide possibilities for their applications in optics and optoelectronics [3], as well as for medical applications. Technology of synthesis and growth of single crystals of these semiconductors is well developed. Therefore, they are of high scientific and practical interest.

As for many ternary $A^{II}B_2{}^{III}C_4{}^{VI}$ compounds, $ZnIn_2S_4$ and $CdGa_2S_4$ crystals can be grown from a vapor phase [4] or by the vertical Bridgman method [5]. These materials are also suitable for many applications, for example, as acousto-optic and electro-optic modulators and switches, optical filters, frequency doublers, optical parametric oscillators, and photovoltaic solar cells [6]. Cadmium thiogallate is promising for application in X-ray dosimenty [7], as photoresistors and optical narrow-band filters [8, 9].

A number of research works were accomplished using thin film samples of $CdGa_2S_4$. El-Nahass et al. studied the dark electrical conductivity and dielectric properties of cadmium thiogallate films [10]; the structural and optical properties of thermally evaporated $CdGa_2S_4$ nanostructure films are reported in [11].

I. Tiginyanu et al. (Eds.): ICNBME 2021, IFMBE Proceedings 87, pp. 422–428, 2022.
https://doi.org/10.1007/978-3-030-92328-0_55

In recent decades semiconductors with bright cathodoluminescence attracted a special attention of researchers since the works on the development of new types of light sources – cathodoluminescent lamps (CLLs) – have been in progress [12–14]. The necessity for these works is determined primarily by the fact that all now existing light sources possess serious limitations [12]. Therefore, it is expedient to test a great variety of semiconducting materials for their applicability as screens for CLLs.

To the best of our knowledge, only a limited number of published papers have been devoted to the study of cathodoluminescence (CL) and X-ray luminescence (RL) in $ZnIn_2S_4$ and $CdGa_2S_4$ [2, 15–17].

2 Cathodoluminescence of Zinc Thioindate and Cadmium Thiogallate. Experimental Results and Discussion

In this work the experimental rezults are presented of the investigations of luminescent properties of thioindate of zinc and thiogallate of cadmium. A complex investigation of cathodoluminescence (CL) and X-ray luminescence (RL) of the studied compounds was carried out with the aim to widen substantially the range of excitation energy using the accelerated electrons and X-rays. The investigations were accomplished at temperatures of 80 and 296 K in a wide range of excitation energies.

For experimental investigations of energy spectra variations of charge carriers in the samples under the action of X-rays an installation described earlier [17] was used.

The studied single crystals were grown using a chemical vapor transport (CVT) method with iodine as a transportation agent. The luminescent radiation was excited with an electron flux with the energy of 60 keV and electric current density in the flux of $2.5 \cdot 10^{-4}$ $A \cdot cm^{-2}$.

The single crystals of the studied materials were thin platelets with the area of the smooth surface of ≈ 0.25 cm^2 and the thickness of ≈ 250 μm. The electron flux was modulated and transformed into pulses with a duration of (1–10) μs. The signal was registered in a modulated way with optical resolution of the installation approximately of 0.02 eV. At a temperature of 296 K in the CL spectrum of $ZnIn_2S_4$ a band with the energy maximum of (1.79 ± 0.02) eV and steep increasing in the low energy region is observed (Fig. 1, curve 1). When the temperature decreases to 80 K the emission spectrum changes (Fig. 1, curve 2); the maximum of the emission band weakly shifts into the high energy region up to (1.88 ± 0.02) eV, and $dE/dT = 4.1 \cdot 10^{-4}$eV/K.

In the high-energy wing a weak peculiarity is observed at the energy of 2.13 eV. The spectral maximum at 296 K coincides with the spectral peculiarity in the low energy wing at 80 K. The half-width of the emission band amounts to 0.30 eV. The position of the energy maximum weakly shifts dependent on the temperature; the absolute intensity in the maximum approximately doubles. The temperature decreasing leads to the increasing of the luminescence intensity of the band by a factor of two.

Investigations of CL of $ZnIn_2S_4$ at low temperatures allowed us to determine more precisely the forbidden gap width of (2.96 ± 0.02) eV at 80 K and optical depth of the deep acceptor level $E_A = (E_V + 0.30)$ eV.

Taking into account the properties of CL of cadmium thiogallate we demonstrated that the spectral distribution of the luminescence intensity strongly depends on the excitation conditions. These conditions are determined by the acceleration voltage of electrons, surface density of the anode current in the electron gun, and by the temperature variation. For investigations crystals with a surface area of 0.2 cm^2 and medium thickness of 250 μm were used.

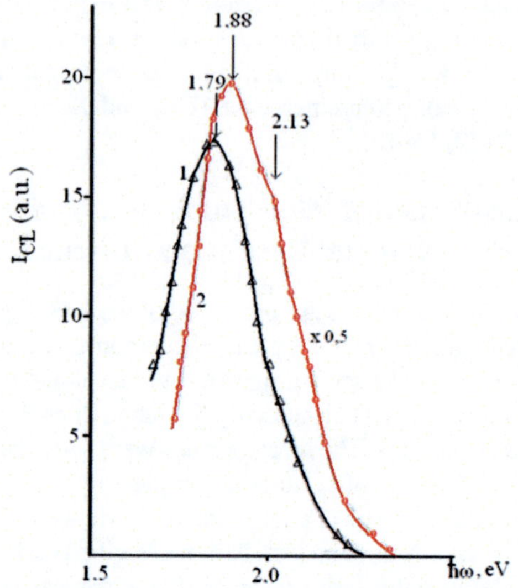

Fig. 1. CL spectra of $ZnIn_2S_4$. Excitation density $j = 2.5 \cdot 10^{-4}$ $A \cdot cm^{-2}$; $U = 60$ keV; T, K: (1) 296 and (2) 80 K. Penetration depth of electrons amounts to 11.4 μm.

The CL spectra of $CdGa_2S_4$ single crystals are presented in Fig. 2. The luminescent radiation was excited by an electron beam with acceleration energy of 60 keV and the current flux density of the order of $j = 10^{-4}$ $A \cdot cm^{-2}$. In the CL spectra at a temperature of 296 K an emission band with a half-width of 0.4 eV and energy maximum at 2.13 eV was registered (Fig. 2, curve 1). In the high-energy wing a peculiarity is observed – a pronounced plateau at an energy of 2.85 eV.

When the temperature decreases (80 K) it was observed that at the same excitation conditions the energy maximum of the fundamental band shifts to the high energy range and is located at 2.33 eV; the major part of the radiated energy is within the high-energy spectral range. With decreasing of temperature a considerable increasing of the integrated radiation approximately by a factor of ten is observed (Fig. 2, curve 2). The analysis of CL spectra of $CdGa_2S_4$ single crystals at 296 K and 80 K shows that the energy maxima change their positions. The high-energy wing reaches up to 3.5 eV. It is interesting that in the CL spectrum at 80 K the energy peculiarity registered at 2.13 eV in the low-energy wing corresponds to the energy maximum at a temperature of 296 K.

The luminescence bands in the studied range are identified as donor-acceptor and band-acceptor transitions. The energy levels in the forbidden gap are determined by structural defects. We can suppose that they can be most probably caused by the cadmium excess or by the mutual substitution of the metal atoms Cd_{Ga} and Ga_{Cd}.

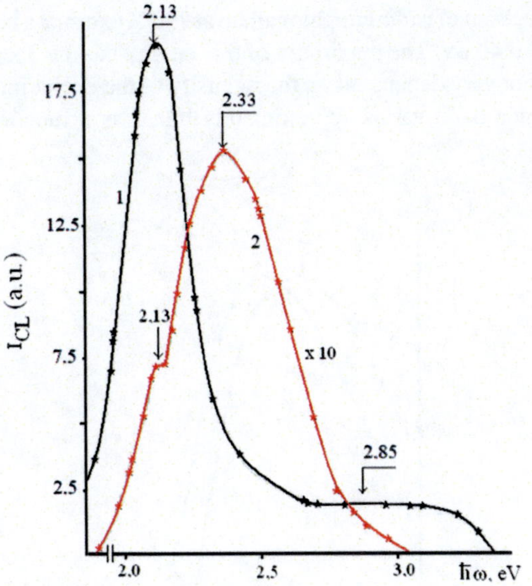

Fig. 2. CL spectra of $CdGa_2S_4$ at 296 K (1) and 80 K (2). $U = 60$ keV; $j = 10^{-4}$ A·cm^{-2}. Penetration depth of electrons amounts to 13.6 μm

3 X-Ray Luminescence of Zinc Thioindate and Cadmium Thiogallate

The comparison of the emission spectra of single crystals at various methods of excitation demonstrates that the positions of the absolute maxima at X-ray excitation shift to the lower energy range, though the half-widths of the emission energy bands increase. This difference between the spectra excited by X-rays and those excited by accelerated electrons can be explained by the fact that the excitation by X-rays occurs within the entire volume of the investigated sample; therefore, the reabsorption of the energy of emitted photons is more intensive than in the case of accelerated electrons for which the excitation is strictly localized.

A specific feature of the X-ray excitation in comparison with cathode excitation of semiconductors is the fact that in the case of X-rays the energy of photons is considerably higher. A specific feature is also that the efficiency of the luminescence excitation increases with the increasing of the absorption coefficient of X-ray radiation by the emitting substance.

The emission spectrum of $ZnIn_2S_4$ (polytype II) at a temperature of 296 and 80 K (Fig. 3) exhibits a bell shape where the low-energy and high-energy wings are symmetrical relative the vertical axis, which passes through the energy maxima of the emission bands at 1.54 and 1.65 eV, respectively. The calculations have demonstrated that the inclination of the curve in the high-energy range amounts to $\Delta I/\Delta h\nu = 4$, though in the low-energy range it amounts to 1.7. This evident difference is conditioned by the increasing of the absorption coefficient $\alpha(h\nu)$, which within the specified spectral range increases exponentially.

The X-ray spectrum of cadmium thiogallate at 296 K exhibits a band with a relatively large half-width ~0.40 eV. The maximum of this energy band is located at 2.14 eV, and its position does not vary dependent on the intensity of the excitation flux. The intensity of the luminescence band varies approximately linear as a function of the excitation level.

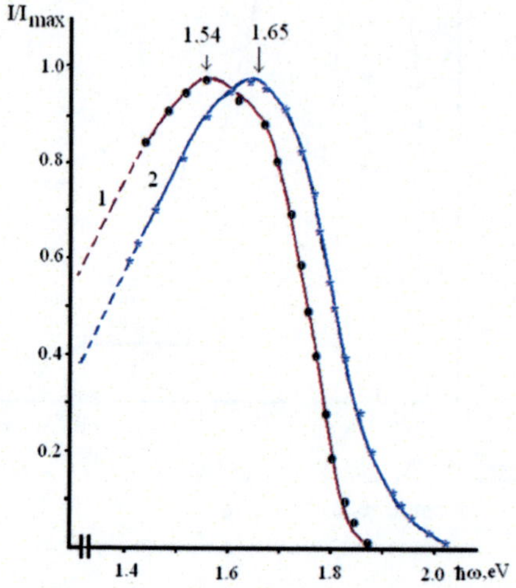

Fig. 3. RL spectra of $ZnIn_2S_4$ (polytype II) at a temperature T, K: (1) 296; (2) 80

The X-ray spectrum of $CdGa_2S_4$ at 296 K is presented in Fig. 4. As it differs from the X-ray spectrum of $ZnIn_2S_4$ single crystals, for $CdGa_2S_4$ a more pronounced energy maximum at 2.14 eV and a plateau at 2.34 eV are observed. However, comparatively with the CL spectrum of the same compound, in the X-ray spectrum only an energy maximum at 2.14 eV and a plateau at 2.34 eV are observed. The energy range of the RL band (1.84–2.50) eV is much narrower than the range of (2.00–3.00) eV for the case of CL spectrum of the same single crystal. This is conditioned by the peculiarities of the excitation method.

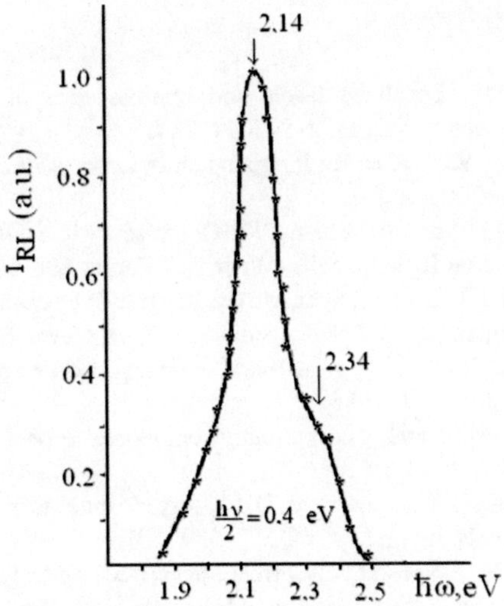

Fig. 4. RL spectrum of $CdGa_2S_4$ at a temperature of 296 K

4 Conclusions

One can conclude that an effective mechanism exists of the energy transmission from the areas where energy is absorbed to the luminescence centers.

According to the experimental studies, the possibility of transmission of the exciting energy depends on the crystal structure of the substance, though the efficiency of the process depends on the presence of impurities and defects that can capture and transmit the excitation energy to the luminescent center.

We can suggest that in the case of X-ray radiation and accelerated electrons a phenomenon of considerable interaction of the incident radiation with the crystal lattice of the sample occurs. This leads to a relative increasing of the half-width of the luminescent emission band.

Investigations of CL of $ZnIn_2S_4$ at low temperatures allowed us to determine more precisely the forbidden gap width of (2.96 ± 0.02) eV at 80 K and optical depth of the deep acceptor level $E_A = (E_V + 0.30)$ eV. The structure of CL spectrum of $ZnIn_2S_4$ single crystals has shown that the red band with the maximum at 1.79 eV correlates with the donor-acceptor optical transitions.

The CL spectra of $CdGa_2S_4$ single crystals with stoichiometric composition contain within the energy range of (1.50–3.50) eV the following bands: a red-orange band (1.50–2.10) eV, a green band (2.33 eV), and a blue band (2.78–2.85) eV. In the RL spectra a single emission band is observed with an energy maximum at 2.14 eV and a slope within the high-energy range at approximately 2.34 eV identified as optical transitions of donor-acceptor type.

Conflict of Interest. The authors declare that they have no conflict of interest.

References

1. Guzzi, M., Grillii, E.: Localized levels and luminescence of AB_2X_4 semiconducting compounds. Mater. Chem. Phys. **11**, 295–304 (1984)
2. Machuga, A., Zhitar', V., Arama, E.: Interband emission of cadmium thiogallate. Semiconductors **34**, 626–628 (2000)
3. Kumar, P., Sahariya, J., Soni, A., Bhamu, K.: Optoelectronic analysis of $CdGa_2X_4$ (X = S, Se): a promising material for solar cells. Mater. Sci. Forum **900**, 69–73 (2017)
4. Nitsche, R., Bolsterli, H., Lichtensteiger, M.: Crystal growth by chemical transport reactions-I: binary, ternary, and mixed-crystal chalcogenides. J. Phys. Chem. Solids **21**, 199–205 (1961)
5. Feigelson, R., Route, R.: Crystal growth and optical properties of $CdGa_2S_4$. Prog. Cryst. Growth Charact. **10**, 189–197 (1984)
6. Gentile, A.: Devices using ternary or multinary compounds. Prog. Cryst. Growth Charact. **10**, 241–256 (1984)
7. Mustafaeva, S., Asadov, M., Guseinov, D.: X-Ray dosimetric properties of vapor grown $CdGa_2S_4$ single crystals. Inorg. Mat. **46**, 587–589 (2010)
8. Isaev, V., Serezhkin, V., Avanesov, A.: Correlation between optical and crystal-structure characteristics of single crystals of triple and mixed chalcogenides. Ekol. Vestn. Nauchn. Tsentrov. CHES **2**, 61–68 (2008)
9. Stamov, I., Syrbu, N., Parvan, V., Zalamai, V., Tiginyanu, I.: The band structure of birefractive $CdGa_2S_4$ crystals. Opt. Commun. **309**, 205–211 (2013)
10. El-Nahass, M., El-Shazly, E., El-Barry, A., Omar, H.: J. Mater. Sci. **46**, 5743–5750 (2011)
11. El-Nahass, M., El-Barry, A., El-Shazly, E., Omar, H.: Structural and optical properties of thermally evaporated cadmium thiogallate $CdGa_2S_4$ nanostructure films. Eur. Phys. J. Appl. Phys. **52**, 10502 (2010)
12. Sheshin, E., Kolodyazhnyj, A., Chadaev, N., Getman, A., Danilkin, M., Ozol, D.: Prototype of cathodoluminescent lamp for general lighting using carbon fiber field emission cathode. J. Vac. Sci. Technol. B **37**, 031213 (2019)
13. Ozol, D.: Preliminary study of cathode ray tube phosphors on the basis of nanocrystal quantum dots. In: 29th International Vacuum Nanoelectronics Conference (IVNC), Vancouver, BC, 15–16 July 2016, pp. 1–2 (2016)
14. Ozol, D.I., Sheshin, E.P., Danilkin, M.I., Vereschagina, N.Y.: Cathodoluminescent UV sources for biomedical applications. In: Tiginyanu, I., Sontea, V., Railean, S. (eds.) ICNBME 2019. IP, vol. 77, pp. 313–317. Springer, Cham (2020). https://doi.org/10.1007/978-3-030-31866-6_60
15. Venger, E., Ermolovich, I., Milenin, V., Papusha, V.: The effect of composition on the properties and defect structure of the $CdS–Ga_2S_3$ solid solution. Semiconductors **36**, 763–771 (2002)
16. Park, H., et al.: Anomalous dependence of photoluminescence properties on composition x in $Cd_{1-x}Mn_xGa_2S_4$ mixed crystals. Phys. Stat. Sol. C **3**, 2915–2918 (2006)
17. Machuga, A., Radu, R., Pintea, V., Arama, E., Zhitar, V., Shemyakova, T.: X-ray luminescence in $ZnIn_2S_4$, $CdGa_2S_4$ and $Zn_3In_2S_6$. In: Proceedings of International Semiconductor Conference, CAS 2007, 15 October–17 September 2007, Sinaia, Romania, vol. 2, pp. 377–380 (2007)

Molecular Modeling of the Interaction of Taxifolin with Quorum Sensing Regulator LasR of *Pseudomonas aeruginosa*

Hovakim Grabski[1(✉)], Siranuysh Ginosyan[1], and Susanna Tiratsuyan[2]

[1] Department of Medical Biochemistry and Biotechnology, Institute of Biomedicine and Pharmacy, Russian-Armenian University, Yerevan, Armenia
hovakim.grabski@rau.am

[2] Department of Bioengineering, Bioinformatics and Molecular Biology, Institute of Biomedicine and Pharmacy, Russian-Armenian University, Yerevan, Armenia

Abstract. *Pseudomonas aeruginosa* is one of the most dangerous superbugs and is responsible for both acute and chronic infection. Current therapies are not effective because of biofilms that increase antibiotic resistance. Biofilm formation is regulated through a system called quorum sensing, and includes transcriptional regulators LasR and RhlR. These regulators are activated by their own natural autoinducers. Targeting quorum sensing is a promising strategy to combat bacterial pathogenicity. Flavonoids are very well known for their antimicrobial activity and inhibit *Pseudomonas aeruginosa* biofilm formation, but the mechanism of action is unknown. In the present study, we analyze the mode of interactions of LasR with taxifolin. We use a combination of molecular docking, molecular dynamics simulations to study the interaction of LasR with taxifolin. We show that taxifolin has two binding modes. One binding mode is the interaction with ligand-binding domain. The second mode is the interaction with the "bridge", which is a cryptic site. Biochemical studies show hydroxyl group of ring A in flavonoids is necessary for inhibition. In our model the hydroxyl group ensures the formation of hydrogen bonds during the second binding mode. This study may offer insights on how taxifolin inhibits LasR and the quorum sensing circuitry.

Keywords: Flavonoids · Taxifolin · Molecular modeling · *Pseudomonas aeruginosa* · Quorum sensing

1 Introduction

Antibiotic resistance problem is worsening, so urgent action has to be taken to fix this problem. *Pseudomonas aeruginosa* is one of the ESKAPE pathogens and affects cystic fibrosis sufferers, burn victims and patients with implanted medical devices [1]. One potential strategy is to target the quorum sensing system, which is responsible for the cell-cell communication of the bacteria. Quorum sensing (QS) system regulates gene

I. Tiginyanu et al. (Eds.): ICNBME 2021, IFMBE Proceedings 87, pp. 429–438, 2022.
https://doi.org/10.1007/978-3-030-92328-0_56

expression depending on the cell population density [1]. Bacteria and fungi produce signaling molecules, termed autoinducers (AI), which regulate the QS system [1]. Autoinducers regulate expression of virulence factors, coordination between microorganisms and the microorganism and the host [1]. Quorum sensing plays an essential role in *P. aeruginosa's* virulence [2] and it also activates the CRISPR-Cas immunity system, thus suppressing it would make *P. aeruginosa* more susceptible to phage therapy and antibiotics [2]. Four main QS systems have been identified so far, specifically the LasI/LasR and RhlI/RhlR [3], the PqsABCDE/PqsR [3] and AmbBCDE/IqsR [3]. Each QS system has its own AI' and these different QS systems adjust and control each other [3]. So LasI/LasR regulate all the other three systems. Transcriptional regulator LasR is one of the main quorum sensing regulatory protein of *P. aeruginosa* and it is activated by its autoinducer, the N-3-oxododecanoyl homoserine lactone (3OC12-HSL). Receptor LasR is also activated by other homoserine lactones, which form part of the interbacterial communication. LasR protein has two domains: ligand binding domain (LBD) and DNA binding domain (DBD). One issue with this protein is that the whole structure is not available, because it is insoluble [4]. The inhibition of LasR will prevent the activation of virulence genes and stop biofilm formation. Flavonoids are a group of natural products and are known as Quorum quenching (QQ) agents [2]. There are reports that flavonoids and their derivatives inhibit biofilm formation of *P. aeruginosa* [1]. Studies show that flavonoids inhibit transcriptional regulator LasR, but the mechanism of action is not known [1]. Flavonoids can bind to LasR LBD when AI is bound. Paczkowski et al. [1] deduce that flavonoids do not use the canonical AI binding site and also note that at low concentrations and only in the presence of the AIs, the flavonoids modestly activate the reporter strains. Suneby et al. [5] reveal that multiple antagonists bind to LasR, stabilize it and provoke an unnatural fold, which does not allow the binding of LasR to response elements, which also does not affect the dimerization. In our previous work we modelled the interaction of 3OC12-HSL with LasR [6]. It appears there could be a possible second binding site. In another work we show that quercetin has two binding modes: one site with LBD and the other with the "bridge" [7], which encompasses ligand-binding domain and DNA binding domain. Another research study shows that the LBD is capable of having multiple ligands in the binding pocket [8], but yet flavonoids do not inhibit in the presence of 3OC12-HSL [1]. The inhibitors do not affect dimerization interface [1, 5], so there has to be another mechanism for inhibition. Taxifolin is also a flavonoid, and it is not mutagenic, and it is less toxic than quercetin [9]. We show that taxifolin also has two binding modes, one with the LBD and the "bridge", which includes amino acids from LBD and DBD, like quercetin [5, 7]. The results of Paczkowski et al. [1] indicate that hydroxyl group at position 7 in flavonoids is essential for inhibition [1], so we remove the hydroxyl group and run another set of independent simulations. In this work we use a combination of molecular docking, molecular dynamics (an aggregate of 5.8 ms of simulations) techniques. The purpose is to understand how taxifolin inhibits transcriptional regulator LasR on molecular level, and give explanation to the biochemical studies [1, 5].

2 Methods

2.1 Protein Model and Ligand Molecule

The protein model is taken from our previous research [6, 7]. The molecule of the ligand is obtained from PubChem [CID: 439533] and then parameterized using the acpype [10] tool. It has been shown that the OH group at position 7 is necessary for inhibition [1], we create a new ligand model by removing OH group and replacing it with hydrogen.

2.2 Clustering Metrics

Clusterization is an unsupervised machine learning technique, but the determination of the number of clusters is not an easy task. For this purpose, different metrics exist to solve the problem of cluster selection. In this work we use various metrics such as the David-Bouldin Index [11], Dunn Index [12], Silhouette score [13] and Calinski Harabasz [14], because they are widely used in the machine learning community. These metrics give indication of the quality of clustering. The final number of clusters is selected using these 4 criteria by selecting the 3 best scores for each respective metric. For each place a score is assigned, the first place gets a score of 1, 2nd - 0.85 and 3rd – 0.5. In the end we select the number of clusters with the highest score.

2.3 Ligand Docking Using Multiple Docking Programs

We perform blind docking by using multiple docking programs to identify binding sites of taxifolin to LasR protein. Each program has its own bias, thus using multiple docking programs allow to compensate for their individual deficiencies, so to arrive at a consensus using Autodock Vina [15], rDock [16], LeDock [17] and FlexAid [18]. A total of 40 docking independent trials are performed, 10 for each docking program. Each docking run is set to generate 20 poses. Later the coordinates of the center of mass of docking poses are subjugated to principal component and cluster analysis using Scikit-learn library [19]. The first two principal components were used for cluster analysis, which was performed using the DBSCAN algorithm [20]. After that only common clusters were selected, where all programs predicted a binding conformation. The centroid conformations were extracted from each cluster using equation [21].

2.4 Molecular Dynamics Simulations Using the Centroid Docked Poses

Gromacs 2018 [22] with AMBER 99SB-ILDN [23] force field is used for molecular dynamics (MD) simulations. Steepest descent is used for minimization and it is set F_{max} of no greater than 1000 $kJmol^{-1}nm^{-1}$. Temperature is set to 300 K. In all experiments, the timestep is set to 2 fs. Explicit water solvation model is used with TIP3P [24] water molecules. Periodic boundary conditions are used for all the simulations. In all simulations, the concentration of NaCl is 100 mM. In all cases, the short range non bonded interactions are truncated to 1.2 nm. PME [25] is used for long range electrostatics. Ligand parameters are generated using the acpype [10] tool. After minimization, the system is equilibrated in 2 stages: the first stage involved simulating for 200 ps under a constant volume (NVT) ensemble, the second stage involved simulating for 200 ps under a constant-pressure (NPT) for maintaining pressure isotropically at 1.0 bar. Agglomerative clustering is used for the cluster analysis of MD trajectories. We use a value of 12 for the stride, which means every snapshot apart 120 ps are used for the analysis. Hydrogen bonds are calculated using the Wernet-Nilson criteria, which is an available module in MDTraj [26]. Final conformation is obtained by calculating the average structure using the last 500 frames. Binding energy is calculated using g_mmpbsa [27], using the last 20 ns from the trajectory. MMPBSA [27] approach is used because it provides good balance between accuracy and speed. It should be noted that MMPBSA approach can calculate the relative binding energy and not the absolute binding energy and for each interaction 167 snapshots are used for MMPBSA calculation.

3 Results

3.1 Docking of Taxifolin with LasR

The methodology is inspired by our previous works [6, 7, 21]. Molecular docking is performed using multiple docking programs: Autodock Vina, rDock, LeDock, FlexAid. Multiple docking programs are used for obtaining a consensus. For each program 10 simulation runs are performed. Later Principal Component Analysis (PCA) is performed using scikit-learn on the center of mass coordinates (COM) of the docking results. The first two principal components explain 86.17% variance in data. The identification of binding sites is performed in two stages. The first stage involved clusterization using the DBSCAN algorithm (Fig. 1b). The second stage involved selecting clusters where all 4 docking programs are featured for a consensus (Fig. 1c). It appears there are 2 potential binding sites for taxifolin with and without the hydroxyl group. For taxifolin these 2 clusters feature 80.4% of all the docking poses (Fig. 1c).

For the taxifolin without the hydroxyl group these 2 clusters amount 73.8% of all the docking poses. These two clusters correspond to ligand binding domain, which corresponds to the experimental binding site and the other is the "bridge" region. Then we extract the "centroid" (representative structure) from each cluster of the docking simulations using Eq. 2 for molecular dynamics.

3.2 Binding Modes of Taxifolin with LasR by MD Simulations

The extracted centroid poses are used for MDs. Overall, 3.4 μs (DQC) and 2.4 μs (DHO) of simulation data are collected. The obtained data are used for the analysis of the interaction of DQC and DHO with LasR monomer. An aggregate of 5.8 μs of simulation data are used for the analysis of the interactions.

The structure of the LasR protein in all of the interactions appears to be stable and converged based on the RMSD plot (Fig. 2). The high RMS value is due to the very high mobility of the DBD region.

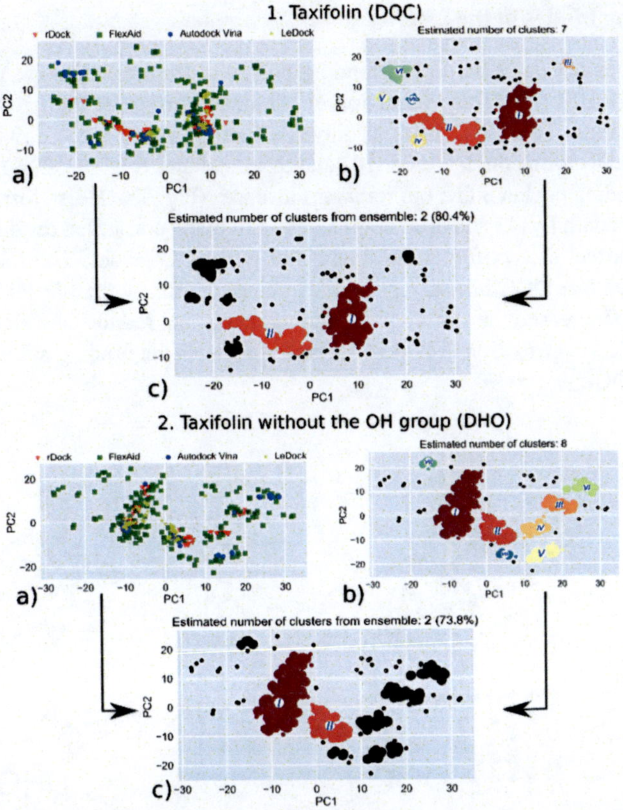

Fig. 1. Analysis of blind docking of 1. taxifolin (DQC) and 2. taxifolin without OH group (DHO) with LasR. a) Visualization of the first two PCs of COM data from docking poses. b) Number of clusters c) Clusters, where all docking programs converged

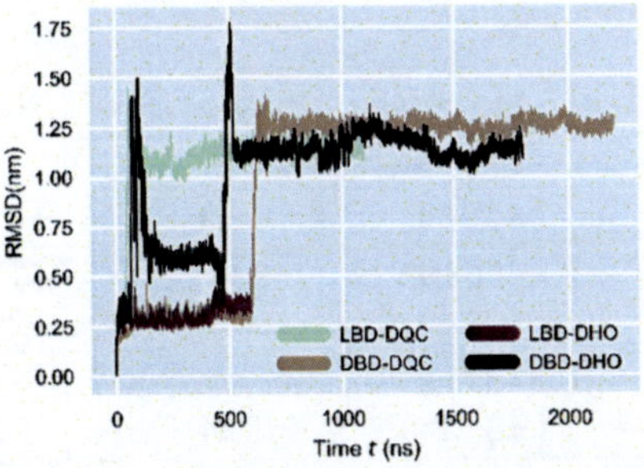

Fig. 2. RMSD evolution through time for LasR monomer structure

3.3 Binding Modes in the LBD

Taxifolin does not enter the ligand binding pocket of the LasR like the native autoinducer 3OC12-HSL [9]. This interaction on the outer surface of the LBD provokes an allosteric change of the DBD. It forms only one hydrogen bond with Gln94 and interacts hydrophobically with conserved Glu103 and other residues (Fig. 3a). The DHO does not enter the binding pocket either but remains close to it (Fig. 3b). It also forms one hydrogen bond but with Lys42 and does not interact with many conserved residues like DQC. It does not affect the conformation dramatically like DQC does. The relative binding energy shows that DQC has stronger relative binding affinity than DHO. The removal of OH group increased the van der Waals components, but reduced the electrostatic and polar solvation contribution (Table 1). Overall it appears the binding affinity of DHO is lower than DQC.

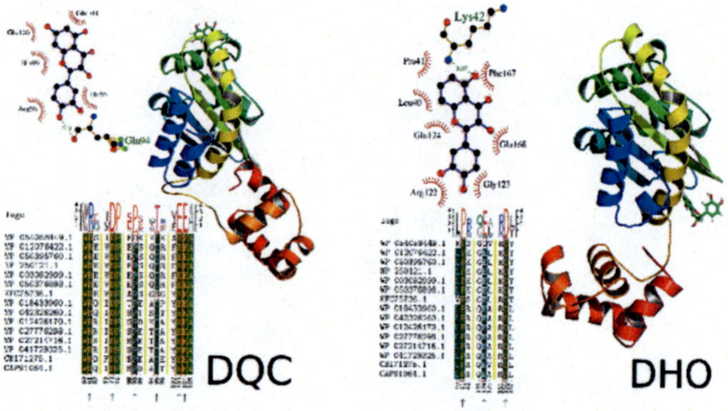

(a) Binding modes of DQC (b) Binding modes of DHO

Fig. 3. Binding modes of taxifolin (DQC) and taxifolin without OH group (DHO) group to the LBD of transcriptional regulator LasR after MD simulations. It also presents the respective hydrogen bonds and interactions with the conserved residues

Table 1. Relative binding energy calculations using MMPBSA approach for LBD interaction using g_mmpbsa (kJ/mol)

Binding sites	Van der waals	Electrostatic	Polar salvation	Non-polar	Binding Energy
HSL-LBD	−215.67 ±8.007	−39.59 ±18.82	147.52 ±13.99	−20.84 ±0.79	−128.58 ±18.76
DQC-LBD	−90.95 ±12.55	−163.92 ±39.43	148.55 ±24.70	−10.17 ±0.64	−116.51 ±18.92
DHO-LBD	−130.62 ±9.401	−51.66 ±33.186	100.64 ±18.34	−13.26 ±0.64	−94.90 ±17.32

3.4 Binding Modes in the "Bridge" DBD

Taxifolin provokes major changes of the DBD, while DHO does not. Taxifolin forms 3 hydrogen bonds and interacts hydrophobically with Leu234, Glu168, Gln160, His169 and Val176 (Fig. 4a), while DHO forms 2 hydrogen bonds with Lys42 and Leu39 and interacts hydrophobically with Leu40, Pro41, Val171, Arg122 and Glu124 (Fig. 4b).

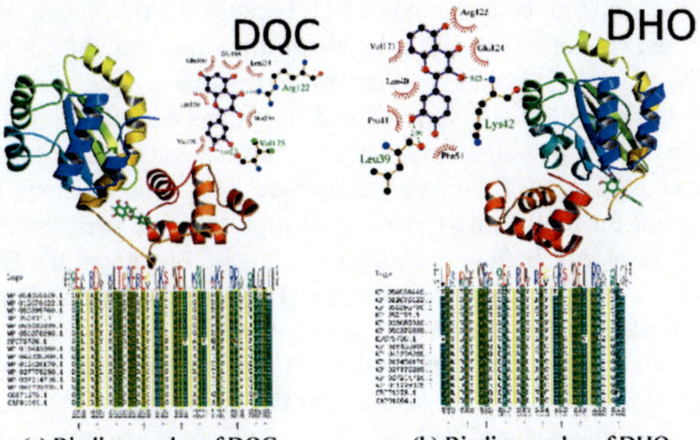

(a) Binding modes of DQC (b) Binding modes of DHO

Fig. 4. Binding modes of taxifolin (DQC) and taxifolin without OH (DHO) group to the LBD of transcriptional regulator LasR after MD simulations. It also presents the respective hydrogen bonds and interactions with the conserved residues

It should be noted that DQC interacts with many conserved amino acid residues, including Leu177 and Leu236. The native autoinducer [6] and quercetin [7] also interacts with Leu177 and Leu236, while DHO does not. The relative binding energy shows that DQC has stronger binding affinity than DHO. The removal of OH group decreased the electrostatic contribution (Table 2). Overall, it appears the binding affinity of DHO is lower than DQC. The binding affinity of DQC is similar to the native autoinducer.

Table 2. Relative binding energy calculations using MMPBSA approach for the "bridge interaction using g_ mmpbsa (kJ/mol)

Binding sites	Van der waals	Electrostatic	Polar salvation	Non-polar	Binding Energy
HSL-LBD	−205.14 ± 11.80	−142.97 ±29.03	201.89 ±20.12	−20.23 ±0.84	−166.46 ±20.49
DQC-LBD	−128.78 ±13.89	−172.99 ±22.98	153.19 ±12.14	−14.61 ±0.81	−163.19 ±16.39
DHO-LBD	−125.20 ±12.63	−147.17 ±20.77	152.25 ±18.01	−13.64 ±0.97	−133.76 ±14.66

4 Conclusions

Pseudomonas aeruginosa is a difficult bacterium to deal with and antibiotics are not effective against them. Biofilms are huge problem as well, affecting from tooth plaques up to food processing and health issues. One potential way to counter is targeting of the quorum sensing system. In *P. aeruginosa* the LasR protein is one of the main regulators of biofilm formation [1]. LasR protein is cytoplasmic protein, but the currrent crystallographic structure contains only LBD, because the full length of LasR protein is not soluble and cannot be crystallized without the native autoinducer [4]. Because of the insolubility of LasR, it makes very hard to study LasR using experimental methods. Biochemical studies indicate the flavonoids do not affect dimerization [1] and the inhibition mechanism is non-competetive. They neither interact directly with the DNA nor through the canonical site of the native autoinducer [1]. The hydroxyl group at position 7 is essential for inhibition [1]. So, for that purpose we perform multiple sets of independent simulations, both with taxifolin and taxifolin without the OH group, with the intention to imitate the biochemical study [1]. The direct interaction of DQC with the "bridge" of LasR is more suitable to the experimental data, because it appears that 3OC12-HSL can also interact with this region [9], so it is more stable. Longer simulations demonstrate that the structures are stable. The OH group ensures the interaction with conserved amino acid residues like Leu177 and Leu236. It appears these two residues are essential for inhibitory activity. The allosteric interaction which leads to the change of the DBD conformation does not explain fully the experimental data. The biochemical study shows that in the presence of 3OC12-HSL, flavonoids do not inhibit, but can even activate it [1]. The direct interaction of taxifolin with the "bridge", which is also not competitive, could explain the experimental studies of flavonoids and inhibitors [1, 5]. All these interactions show that they are not competitive and these results match the biochemical studies [1, 5]. This research could give a potential answer on how flavonoids inhibit biofilm, virulence gene expression and CRISPR-Cas adaptive immune system of *P. aeruginosa*.

Acknowledgment. The work was supported by the Russian-Armenian University within the subsidy of the Ministry of Science and Higher Education of Russian Federation. This work was made possible by a research grant from the Armenian National Science and Education Fund (ANSEF) based in New York, USA. The research was done within the Ministry of Education and Science of the Republic of Armenia, State Committee of Science: 10-2-1-4, Government budget financing.

Conflict of Interest. The authors declare that they have no conflict of interest.

References

1. Paczkowski, J.E., Mukherjee, S., McCready, A.R., et al.: Flavonoids suppress *Pseudomonas aeruginosa* virulence through allosteric inhibition of quorum-sensing receptors. J. Biol. Chem. **292**, 4064–4076 (2017). https://doi.org/10.1074/jbc.M116.770552

2. Høyland-Kroghsbo, N.M., Paczkowski, J., Mukherjee, S., et al.: Quorum sensing controls the *Pseudomonas aeruginosa* CRISPR-Cas adaptive immune system. Proc. Natl. Acad. Sci. **114**, 131–135 (2017). https://doi.org/10.1073/pnas.1617415113
3. Pérez-Pérez, M., Jorge, P., Pérez Rodríguez, G., et al.: Quorum sensing inhibition in *Pseudomonas aeruginosa* biofilms: new insights through network mining. Biofouling **33**, 128–142 (2017). https://doi.org/10.1080/08927014.2016.1272104
4. Bottomley, M.J., Muraglia, E., Bazzo, R., Carfì, A.: Molecular insights into quorum sensing in the human pathogen *Pseudomonas aeruginosa* from the structure of the virulence regulator LasR bound to its autoinducer. J. Biol. Chem. **282**, 13592–13600 (2007). https://doi.org/10.1074/jbc.M700556200
5. Suneby, E.G., Herndon, L.R., Schneider, T.L.: Pseudomonas aeruginosa LasR·DNA binding is directly inhibited by quorum sensing antagonists. ACS Infect. Dis. **3**, 183–189 (2017). https://doi.org/10.1021/acsinfecdis.6b00163
6. Grabski, H., Hunanyan, L., Tiratsuyan, S., Vardapetyan, H.: Interaction of N-3-oxododecanoyl homoserine lactone with transcriptional regulator LasR of Pseudomonas aeruginosa: insights from molecular docking and dynamics simulations. F1000Research **8**, 324 (2019). https://doi.org/10.12688/f1000research.18435.1
7. Grabski, H.V., Tiratsuyan, S.G.: Interaction of quercetin with LasR of Pseudomonas aeruginosa: mechanistic insights of the inhibition of virulence through quorum sensing. In: Tiginyanu, I., Sontea, V., Railean, S. (eds.) 4th International Conference on Nanotechnologies and Biomedical Engineering, vol. 77, pp. 585–588. Springer, Cham (2020)
8. Bassler, B.L., McCready, A.R., Paczkowski, J.E., Henke, B.R.: The Pseudomonas aeruginosa LasR quorum-sensing receptor balances ligand selectivity and sensitivity. bioRxiv (2018). https://doi.org/10.1101/269001
9. Makena, P.S., Pierce, S.C., Chung, K.-T., Sinclair, S.E.: Comparative mutagenic effects of structurally similar flavonoids quercetin and taxifolin on tester strains *Salmonella typhimurium* TA102 and *Escherichia coli* WP-2 *uvrA*. Environ. Mol. Mutagen. **50**, 451–459 (2009). https://doi.org/10.1002/em.20487
10. Sousa da Silva, A.W., Vranken, W.F.: ACPYPE - AnteChamber PYthon parser interfacE. BMC Res. Notes **5**, 367 (2012). https://doi.org/10.1186/1756-0500-5-367
11. Davies, D.L., Bouldin, D.W.: A cluster separation measure. IEEE Trans. Pattern Anal. Mach. Intell. **PAMI-1**, 224–227 (1979). https://doi.org/10.1109/TPAMI.1979.4766909
12. Dunn, J.C.: A fuzzy relative of the ISODATA process and its use in detecting compact well-separated clusters. J. Cybern. **3**, 32–57 (1973). https://doi.org/10.1080/01969727308546046
13. Rousseeuw, P.J.: Silhouettes: a graphical aid to the interpretation and validation of cluster analysis. J. Comput. Appl. Math. **20**, 53–65 (1987). https://doi.org/10.1016/0377-0427(87)90125-7
14. Calinski, T., Harabasz, J.: A dendrite method for cluster analysis. Commun. Stat. - Theory Methods **3**, 1–27 (1974). https://doi.org/10.1080/03610927408827101
15. Trott, O., Olson, A.J.: AutoDock Vina: improving the speed and accuracy of docking with a new scoring function, efficient optimization, and multithreading. J. Comput. Chem. 455–461 (2009). https://doi.org/10.1002/jcc.21334
16. Ruiz-Carmona, S., Alvarez-Garcia, D., Foloppe, N., et al.: rDock: a fast, versatile and open source program for docking ligands to proteins and nucleic acids. PLoS Comput. Biol. **10**, e1003571 (2014). https://doi.org/10.1371/journal.pcbi.1003571
17. Wang, Z., Sun, H., Yao, X., et al.: Comprehensive evaluation of ten docking programs on a diverse set of protein–ligand complexes: the prediction accuracy of sampling power and scoring power. Phys. Chem. Chem. Phys. **18**, 12964–12975 (2016). https://doi.org/10.1039/C6CP01555G

18. Gaudreault, F., Najmanovich, R.J.: FlexAID: revisiting docking on non-native-complex structures. J. Chem. Inf. Model. **55**, 1323–1336 (2015). https://doi.org/10.1021/acs.jcim.5b00078
19. Varoquaux, G., Buitinck, L., Louppe, G., et al.: Scikit-learn: machine learning without learning the machinery. GetMobile Mob. Comput. Commun. **19**, 29–33 (2015). https://doi.org/10.1145/2786984.2786995
20. Schubert, E., Sander, J., Ester, M., et al.: DBSCAN revisited, revisited: why and how you should (still) use DBSCAN. ACM Trans. Database Syst. **42**, 1–21 (2017). https://doi.org/10.1145/3068335
21. Grabski, H., Ginosyan, S., Tiratsuyan, S.: Molecular simulations and Markov state modeling reveal inactive form of quorum sensing regulator SdiA of Escherichia coli. IEEE/ACM Trans. Comput. Biol. Bioinform. **PP** (2021). https://doi.org/10.1109/TCBB.2021.3074567
22. Abraham, M.J., Murtola, T., Schulz, R., et al.: GROMACS: high performance molecular simulations through multi-level parallelism from laptops to supercomputers. SoftwareX **1–2**, 19–25 (2015). https://doi.org/10.1016/j.softx.2015.06.001
23. Lindorff-Larsen, K., Piana, S., Palmo, K., et al.: Improved side-chain torsion potentials for the Amber ff99SB protein force field. Proteins Struct. Funct. Bioinform. 1950–1958 (2010). https://doi.org/10.1002/prot.22711
24. Jorgensen, W.L., Chandrasekhar, J., Madura, J.D., et al.: Comparison of simple potential functions for simulating liquid water. J. Chem. Phys. **79**, 926–935 (1983). https://doi.org/10.1063/1.445869
25. Darden, T., York, D., Pedersen, L.: Particle mesh Ewald: An $N \cdot \log(N)$ method for Ewald sums in large systems. J. Chem. Phys. **98**, 10089–10092 (1993). https://doi.org/10.1063/1.464397
26. McGibbon, R.T., Beauchamp, K.A., Harrigan, M.P., et al.: MDTraj: a modern open library for the analysis of molecular dynamics trajectories. Biophys. J. **109**, 1528–1532 (2015). https://doi.org/10.1016/j.bpj.2015.08.015
27. Kumari, R., Kumar, R., Open Source Drug Discovery Consortium, Lynn, A.: *G_mmpbsa*—a GROMACS tool for high-throughput MM-PBSA calculations. J. Chem. Inf. Model. **54**, 1951–1962 (2014). https://doi.org/10.1021/ci500020m

Investigation of Dynamical Properties of a Laser with Incorporated DBR Section Under the Influence of External Optical Feedback

Eugeniu Grigoriev(✉) and V. Tronciu

Department of Physics, Technical University of Moldova, Chisinau, Republic of Moldova
eugeniu.grigoriev@fiz.utm.md

Abstract. We report in this paper the results of theoretical investigations of the dynamical properties of a laser with incorporated Distributed Brag Reflector (DBR) section under the influence of external optical feedback. The adapted Lang-Kobayashi model was used to simulate and analyze the dynamics of the considered laser device. We have identified the nature of the bifurcations that occur in such a system. We plotted the Hopf bifurcation, responsible for instabilities, in the plane of different parameters. The conditions that are necessary for stable laser operation are identified. We also demonstrate the influence of the length of active region on the stability of devise emission, and show how this property is changed by variation of detuning of the mode of solitary laser.

Keywords: DBR laser · Bifurcations · Stability · External cavity modes

1 Introduction

DBR lasers are good candidacy as a laser source with single mode operation and small spectral linewidth. Such devices are applicable in spectroscopy, quantum metrology and optical communication [1–4]. Many currently used devices need an isolator in order to reduce the effect of feedback. Recently the feedback sensitivity of DBR type lasers are investigated theoretically [5, 6]. It is well known that the unwanted feedback can cause instabilities in laser operation [7]. To investigate the dynamic properties of Fabry Perot laser under the influence of feedback widely the Lang Kobayashi model [8] is used.

In this paper we report the investigation of stability analysis of laser with an incorporated DBR under the influence of external feedback based on adopted Lang Kobayashi model. The paper is structured as follows. In Sect. 2 we present the laser setup, the adapted to DBR laser Lang Kobayashi model and the stationary solutions. Section 3 is devoted to results obtained. The conclusions are given in Sect. 4.

I. Tiginyanu et al. (Eds.): ICNBME 2021, IFMBE Proceedings 87, pp. 439–447, 2022.
https://doi.org/10.1007/978-3-030-92328-0_57

2 Setup, Model and Equations

Figure 1 shows the configuration setup considered in this paper. The laser emission from front facet of laser is passed from a waveguide and reflected from a distant mirror with reflectivity R. Phase shifts of coated mirrors and facets are taken into account. φ is phase of the external mirror. τ is the delay of feedback arm. The solitary laser wavelength λ_1 is 1.12 μm. The optical losses in DBR section is 200 m^{-1}. The central wavelength λ_K of DBR is 1.12 μm. The front facet reflectivity of active region is $R_a = 0.1$.

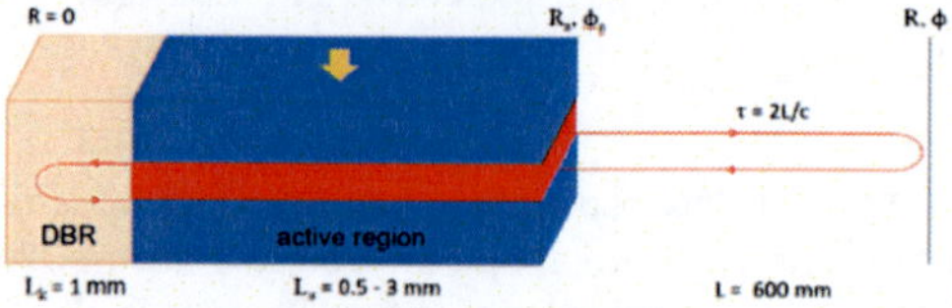

Fig. 1. Setup of DBR laser with external feedback.

Let first look for the properties of DBR. Figure 2 shows the spectrum of intensity reflection. In the case considered here the maximal field intensity $R_{\max} = 0.34$ was obtained.

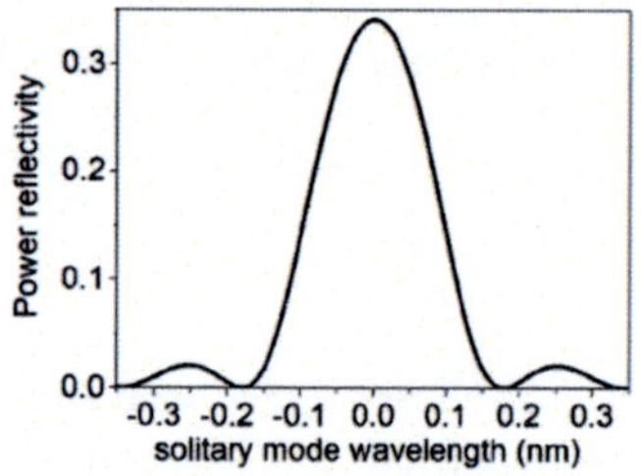

Fig. 2. Reflectivity spectra of the DBR laser.

We start our analysis from the well-known standard Lang-Kobayashi model [8] for a semiconductor laser subject to the delayed external optical feedback

$$\dot{E}(t) = \left[i\omega_1 + (1 + i\tilde{\alpha}_H)\frac{N(t)}{\tilde{\tau}_p} \right] E(t) + \eta e^{-i\phi} E(t - \tau) \quad (1)$$

$$\dot{N} = \frac{1}{\tau_N}\left[N_0 - N(t) - (1 + 2N(t))|E(t)|^2) \right] \quad (2)$$

These equations have been widely studied in literature. The model is simple and well understood for Fabry-Perot lasers with conventional optical feedback However, the situation is more complicated here. The main parameters depend on the solitary mode wavelength i.e. on the detuning of the mode of solitary laser. In Eq. (1) and (2) $N_0 = 0.5(I - I_{th})/(I_{th} - I_{tr})$.

The rescaled feedback strength has the following form

$$\eta = \frac{\sqrt{R/R_a}}{\left|\partial_{\omega_s}\tilde{q}\right|}. \tag{3}$$

The phase are written as.

$$\phi = \varphi + \arg(\partial_{\omega_s}\tilde{q}) - \frac{\pi}{2} - \arg(r). \tag{4}$$

The Henry factor can be written in form

$$\tilde{\alpha}_H = -\frac{\mathrm{Re}(\partial_g\tilde{q}/\partial_\omega\tilde{q})}{\mathrm{Im}(\partial_g\tilde{q}/\partial_\omega\tilde{q})}. \tag{5}$$

Rescaled photon life time

$$\tilde{\tau}_p = \left(\tilde{v}_g g_l\right)^{-1} \tag{6}$$

with the group velocity

$$\tilde{v}_g = +2\mathrm{Im}(\partial_g\tilde{q}/\partial_\omega\tilde{q}). \tag{7}$$

The threshold condition are as follows

$$g_l = \alpha_a - \frac{1}{L_a}\ln\left(|r_k(\omega_l)|\sqrt{R_a}\right). \tag{8}$$

The other parameters and notations are as in [5].

In what follows, we consider the stationary lasing states of systems (1)–(2). They are given by rotating wave solutions so called External Cavity Modes (ECMs)

$$E = E_s e^{i\omega_s t} \tag{9}$$

We obtain a transcendental equation for ωs

$$i\omega_s = i\omega_1 + (1 + i\tilde{\alpha}_H)\frac{N_s}{\tilde{\tau}_p} + \eta e^{-(\phi + \tau\omega_s)}. \tag{10}$$

The real part yields the stationary carrier density

$$\frac{N_s}{\tilde{\tau}_p} = -\eta\cos(\tau\omega_s + \phi) \tag{11}$$

and the imaginary part to a transcendental equation

$$\omega_s - \omega_1 = -\eta\left[\tilde{\alpha}_H\cos(\tau\omega_s + \phi) + \sin(\tau\omega_s + \phi)\right]. \tag{12}$$

3 Results and Discussions

We study the stability of solitary laser modes in dependence on their spectral position relative to the center of the DBR stop band for fixed other parameters. We begin to consider how the Henry factor α_H depends on solitary mode wavelength so called detuning. By dotted line in Fig. 3 we show the Henry Factor for Fabry Pertot laser being a constant equal to 1.2. One can observe in this figure a strong dependence of Henry factor on the changes of solitary mode wavelength. For laser with 3 mm active medium the variation of Henry factor is smooth (see blue line). When we decrease the length of active region to 1 mm the changes in the dependence of α_H on solitary mode wavelength is more pronounced and for negative detuning reaches value 3.2 (see red line). When active region of laser is reduced to 0.5 mm the value of α_H reaches value of 7 for negative detuning. Thus, for short cavity length and negative detuning due to an increase of Henry factor the instabilities can occur. On the other hand, the positive detuning the stable operation is present.

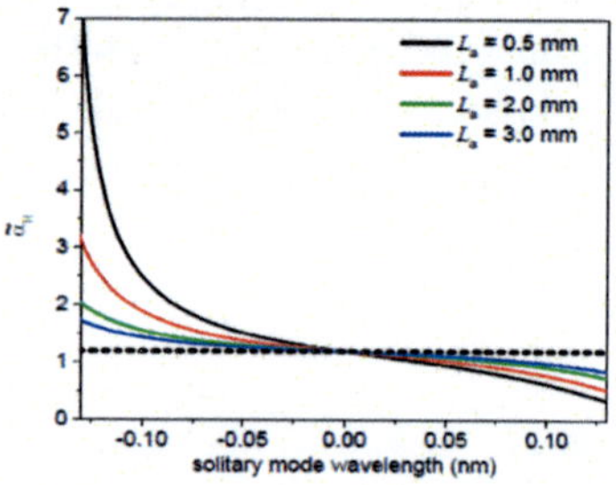

Fig. 3. The dependence of alpha factor on the solitary mode wavelength.

In what follows, we look for stationary states of system (1)–(2) obtained from Eqs. (8)–(12). Figure 4 shows the dependence of stationary mode gain on stationary mode wavelength for feedback reflectivity $R = 10^4$.

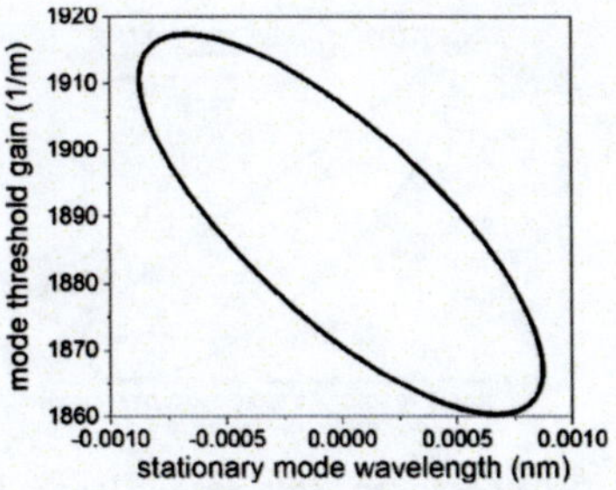

Fig. 4. The dependence of mode threshold gain on stationary mode wavelength i.e. the distribution of ECMs

We observe that the ECMs are distributed along to an ellipse. The situation is similar to that of well-known convention optical feedback. We mention that, this dependence strongly depends on g_l (Eq. 8). As long as the length of active region is increase the size of ellipse is reduced.

Let consider other dependencies. Figure 5 shows the dependence of stationary on solitary mode wavelength. One can distinguish in this figure the phenomena of bistability and multistability in this dependence.

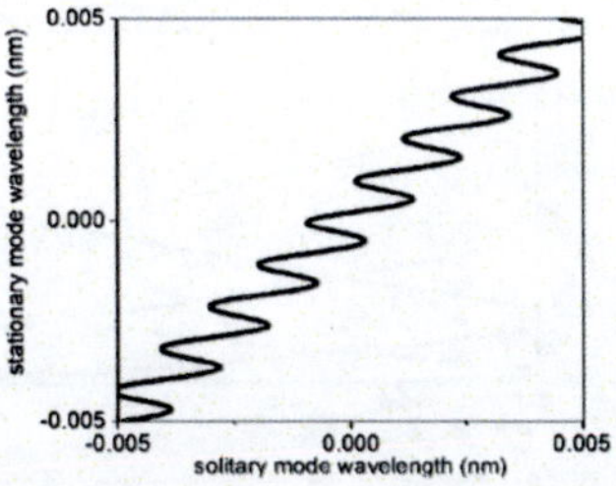

Fig. 5. Dependence of stationary vs solitary mode wavelengths.

An important issue is to investigate the stability of stationary states. Figure 6 shows the stability of external cavity modes in the plane of stationary mode gain – stationary mode wavelength.

Red solid lines in these figures show the anti-modes, that not present interest for our investigations. We use continuation method DDE-biftool [9] to analyze the stability of modes. This figure shows that for laser with cavity active length of 1.0 mm the region of instability is wide (see blue line). By green solid line we show stable stationary states. These state are separated by Hopf bifurcation marked by black square. We mention that for laser with higher length the region with instabilities increases. By gray line we show the saddle –node bifurcation for any values of external feedback reflectivity *R*. The intersection points between line and ellipse shown by gray circle represent the saddle –node bifurcation for fixed value of reflectivity $R = 10^{-4}$. Thus, it should be interesting to look for saddle –node bifurcation in the plane of two parameters. Figure 7 shows the lines of saddle node bifurcation for different length of active region La obtained

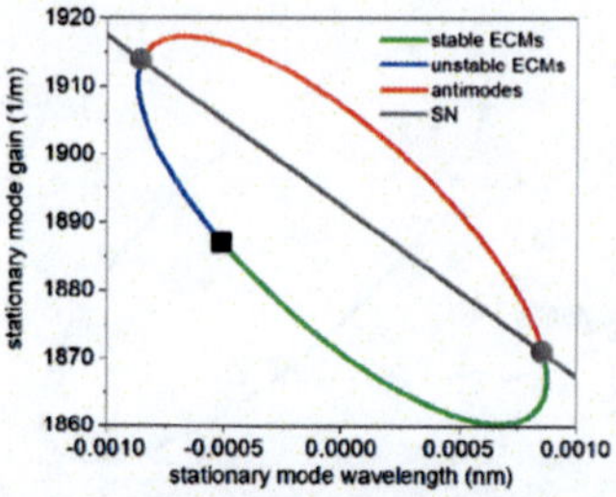

Fig. 6. Curve of external cavity modes for an active section length $L_a = 1$ mm ($I_{tr} = 0.013$ A, $I_{th} = 0.0262$ A and $I = 0.0524$ A), and external facet reflectivity of $R = 10^{-4}$. Solid green line – stable stationary states. Blue line – unstable stationary states. Gray circle Saddle Node bifurcation for $R = 10^{-4}$. Gray line saddle node bifurcation for any reflectivity. Square – Hopf bifurcation.

derivating Eq. (12). It is well known, that for lower feedback strength a single mode is observed. When crossing a line of SN bifurcation, the number of modes in increased by pare (mode and antimode). We plotted SN bifurcation in Fig. 7 for external facet reflectivity of $R < 10^{-3}$ where the LK model is suitable. One can observe that for laser with active region length $L_a = 3$ mm the number of ECMs is much lower than for that of $L_a = 0.5$ mm. This is an indication that for laser with 3 mm less mode is involved.

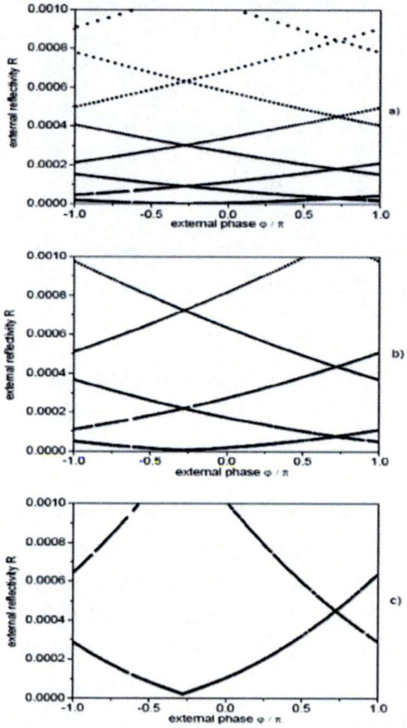

Fig. 7. Saddle-node bifurcation for different length of cavity of active region a) $L_a = 0.5$ mm, b) $L_a = 1.0$ mm c) $L_a = 3.0$ mm.

Figure 8 shows the Hopf – bifurcation in the plane of two parameters (external reflectivity R – external phase) for different length of active cavity L_a.

The Hopf line separate the stable region by unstable one. This Figure confirms the results of previous figures. Laser with long active region ($L_a = 3$ mm) are more stable that those with short length of active region. Thus, for laser with $L_a = 0.5$ mm the region of instability is very wide (see Fig. 8a). An increase of active region cavity length leads to a narrowing the unstable region and wider stable region. A further increase of cavity length reduces the unstable region (see Fig. 8c).

In what follows, we consider the influence of the detuning of solitary mode wavelength on stability of states for different length of active region cavity.

One can observe that for lasers with $L_a = 0.5$ mm the region of instability (blue lines) increases (see Fig. 9a). An increase of active region cavity length to $L_a = 3$ mm leads to disappearance of instability regions and reduction of bistability regions (see Fig. 9c). The instability appears due to the Hopf-bifurcation marked by square. Finally, we conclude that long DBR lasers with feedback are more stable in the operations that those with short active regions.

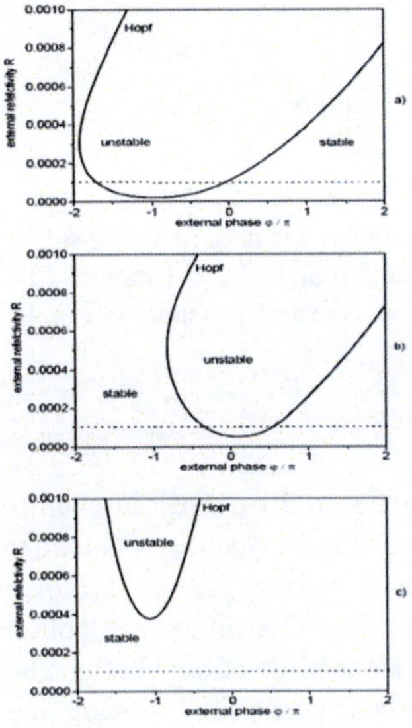

Fig. 8. Hopf bifurcation in the plane of two parameters (external reflectivity R - external phase φ/π. a) $L_a = 0.5$ mm, b) $L_a = 1.0$ mm c) $L_a = 3.0$ mm).

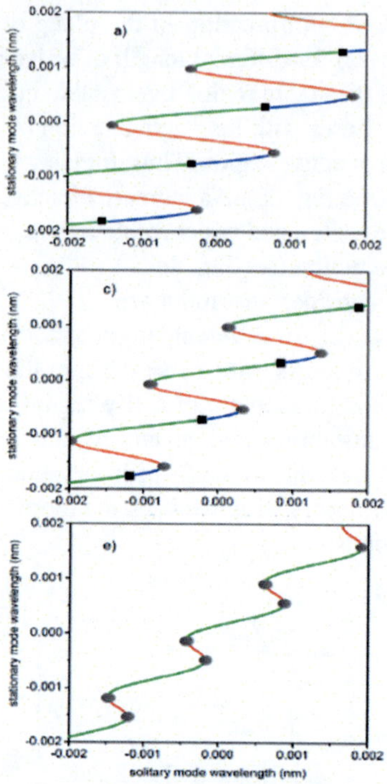

Fig. 9. Stationary mode wavelength in the dependence of solitary mode wavelength for different length of active region a) $L_a = 0{:}5$ mm, b) $L_a = 1$ mm, c) $L_a = 3$ mm. Other parameters: $R = 10^{-4}$. The symbols and line colors are similar to those of Fig. 8.

4 Conclusions

Theoretical investigation on the stability of DBR laser emission is presented. An adopted Lang Kobayashi model is considered. Stability of stationary states strongly depend on the detuning between the lasing wavelength and the Bragg wavelength. For a negative detuning the Henry factor is increased leading to unstable stationary state. The instability comes due to appearance of Hopf bifurcation. On the other hand, the positive detuning reduces the Henry factor and make stable the stationary states. We found that, DBR lasers with long active region are more stable. We believe that our work provides a good basis for future studies and, in particular, provides some hints for more detailed experimental investigations of DBR lasers with feedback.

Acknowledgment. This work was supported by the National Agency for Research and Development of Moldova within the project 20.80009.5007.08 "Study of optoelectronic structures and thermoelectric devices with high efficiency".

Conflict of Interest. The authors declare that they have no conflict of interest.

References

1. Herschbach, N., et al.: Photoassociation spectroscopy of cold He (23S) Atoms Phys. Rev. Lett. **84**, 1874–1878 (2000)
2. Sodnik, Z., Furch, B., Lutz, H.: Optical intersatellite communication IEEE. J. Sel. Top. Quantum Electron. **1**, 1051 (2010)
3. Matsui, Y., Schatz, R., Di CheKhan, F., Kwakernaak, M., Sudo, T.: Low-chirp isolator-free 65-GHz-bandwidth directly modulated lasers. Nat. Photon. **15**, 59–63 (2021)
4. Hofmann, J., Sahm, A., John, W., Bugge, F., Paschke, K.: Opt. Laser Technol. **83**, 55 (2016)
5. Tronciu, V., Werner, N., Wenzel, H., Wünsche, H.-J.: Feedback sensitivity of detuned DBR semiconductor lasers. IEEE J. Quantum Electron. **57**(5), 1–7 (2021)
6. Morthier, G.: Feedback sensitivity of DBR-type laser diodes. IEEE Photon. J. **13**(4), 1–8 (2021)
7. Krauskopf, B.: Unlocking Dynamical Diversity: Optical Feedback Effects on Semiconductor Lasers. In: Kane, D.M., Shore, K. A. (eds.) pp. 147–183. Wiley (2005)
8. Lang, R., Kobayashi, K.: External optical feedback effects on semiconductor injection laser properties. IEEE J. Quantum Electron. 347 (1980)
9. Sieber, J., Engelborghs, K., Luzyanina, T., Samaey, G., Roose, D.: DDE-BIFTOOL manual - bifurcation analysis of delay differential equations (2016)

Preclinical Stage of Building a Machine Learning System for Sepsis Prediction: A Comparative Study of Four Algorithms

Victor Iapăscurtă[1(✉)] and A. Belîi[1,2]

[1] Department of Anesthesia and Intensive Care, N. Testemitanu Sate University of Medicine and Pharmacy, Chişinău, Moldova
victor.iapascurta@usmf.md

[2] Institute of Emergency Medicine, Chişinău, Moldova

Abstract. From the point of view of data engineering, the application of machine learning methods in the case of data represented by multivariate multimodal time series is a difficult task, but with a possible practical potential, which is insufficiently explored.

From a medical point of view, the conditions described by this type of data are current problems in medicine, especially in anesthesia and intensive care and particularly in case of sepsis. The early diagnosis of life-threatening conditions can lead to increased treatment success, reduced mortality and reduced cost of care for this group of patients, especially in complex cases.

This paper describes a research focused on identifying the most appropriate machine learning algorithm to be used for further investigations. It presents some results concerning the preclinical stage in developing of a machine learning system for early sepsis prediction with an anticipated potential of improving the clinical management of patients with sepsis.

Keywords: Machine learning · Algorithmic complexity · Machine learning algorithms · Sepsis management

1 Introduction

A time series (TS) is a sequence of data points (measurements) that has a natural time order. Many important real-world model recognition tasks deal with TS analysis. In medicine and especially in anesthesia and intensive care (AIC), TSs are ubiquitous reflecting the continuous recording of biological signals: pulse, plethysmogram, blood pressure, ECG, EEG, etc. Unlike static data, the application of machine learning (ML) to TS is more difficult (many data points, many predictors, large volumes of initial data, missing values, etc.)

The original purpose of the study was the evaluation of the use of ML technologies in medicine and especially in AIC, in particular for the management of continuous physiological data and especially in patients who may develop sepsis [1, 2] with the

I. Tiginyanu et al. (Eds.): ICNBME 2021, IFMBE Proceedings 87, pp. 448–455, 2022.
https://doi.org/10.1007/978-3-030-92328-0_58

creation of a system with discriminatory abilities (sepsis vs non-sepsis), which allows early prediction of sepsis development.

Before being passed to the ML module the data generated by clinical monitors need to be preprocessed and there are a number of data processing techniques. In this research we use a combination of preprocessing methods including the Block Decomposition Method (BDM), a novel approach coming from the field of *Algorithmic Information Dynamics,* which is a new type of discrete calculus based on computer programming to study causation by generating mechanistic models to help find first principles of physical phenomena building up the next generation of machine learning [3].

The final goal of the current research is the identification of the most suitable ML algorithm to be subsequently used for the creation of an ML-based system for early sepsis prediction.

2 Matherials and Methods

2.1 Data

The data used in this research represent time series with hourly sampling that include 40 parameters (vital signs, laboratory indices and logistics). The data sets are public access data from the 'Early Prediction of Sepsis from Clinical Data - the PhysioNet Computing in Cardiology Challenge 2019' [4] and come from two geographically distinct US hospital systems: Beth Israel Deaconess Medical Center (set A) and Emory University Hospital (set B). These data were collected in the last decade with the approval of the institutional review boards, were de-identified and labeled using Sepsis-3 clinical criteria [5].

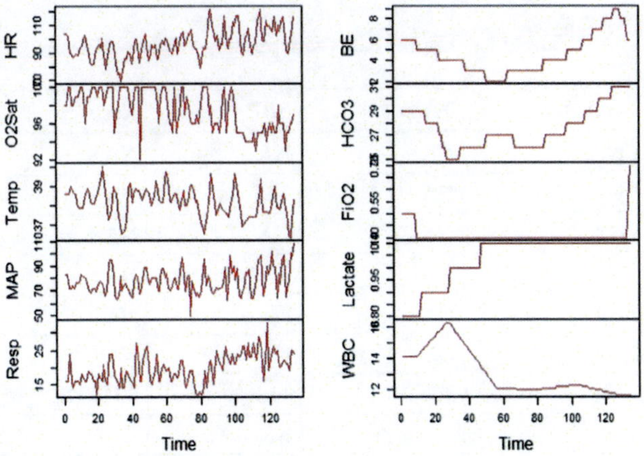

Fig. 1. Appearance of original data

It includes over 1.5 million time-windows and over 10.4 million data points (non-missing physiological and laboratory variables).

The Fig. 1 represents the dynamics of several parameters (heart rate (HR), oxygen saturation (SaO_2), temperature (Temp), mean arterial pressure (MAP), respiratory rate (RR), base excess (BE), bicarbonate (HCO3), fraction of inspired oxygen (FiO2), lactate, white blood count (WBC)) over time.

In total there are 40336 patients/clinical cases from Intensive Care Units. Since the set A includes fewer missing values it was selected for the further processing.

2.2 Data Preprocessing

Initial data are explored using standard statistical methods: mean, standard deviation, median, percentage ratio. After experimenting with different parameters/biomedical signals available in the dataset, six physiological parameters were selected for further exploratory data analysis and processing: heart rate (HR), oxygen saturation (SaO_2), systolic blood pressure (SBP), diastolic blood pressure (DBP), respiratory rate (RR) and temperature (Temp).

Original set comes with up to 80.9% missing data. Restoring missing values can be done with specialized packages (e.g., the "mclust" package in R and others), but most of the specialized software is not time-oriented, in fact ignoring the "continuous" nature of the physiological processes and data that they generate. To restore the data in this set, a recovery algorithm has been developed which consists of the following steps: (1) each column (containing the hourly values of a parameter) is evaluated for missing data (NA); (2) The first and last value in a column that describe a certain parameter is "restored" according to the value in the nearest cell; (3) finally the interpolation is performed, with the calculation of the values between two present values in compliance with the trend (increase or decrease). For interpolation, the function {na.approx ()} is used (the "zoo" package in R). Tables 1 and 2 show data before and after restoring missing values (NA):

Table 1. Original data (with missing values – NA)

HR	O2Sat	Temp	SBP	Resp	...
NA	NA	NA	NA	NA	...
117	99	NA	116	20	...
NA	NA	NA	NA	NA	...
NA	NA	NA	NA	NA	...
NA	NA	NA	NA	NA	...
NA	NA	NA	NA	NA	
120	100	36	118	30	...
...	...	...	...	...	...

For further processing data was grouped in "Circulatory" (HR, SBP, DBP) and "Respiratory-Metabolic" (SaO_2, RR, Temp) and reshaped to 3 × 3 matrices. For the analysis and processing of data, including statistics, and their subsequent processing,

Table 2. Restored data

HR	O2Sat	Temp	SBP	Resp	...
117	99	36	116	20	...
117	99	36	116	20	...
118	99	36	116	22	...
119	99	36	117	24	...
119	100	36	117	26	...
120	100	36	118	28	...
120	100	36	118	30	...
...	...	...	...	...	...

the RStudio integrated development environment was used based on the R (R) programming language, which was also the language of interaction with the H2O platform (https://www.h2o.ai/) for machine learning itself.

To evaluate the performance of the created models and report the results, the TRIPOD (Transparent Reporting of a multivariable prediction model for Individual Prognosis or Diagnosis) checklist [6] and the PROBAST (Risk Prediction Model of Bias Assessment Tool) [7] principles were used.

Algorithmic complexity estimation (via BDM) was performed by a special tool for providing reliable estimations to uncomputable functions, namely the online algorithmic complexity calculator (OACC) [8], which provides estimations of algorithmic complexity and algorithmic probability for short and long strings and for two-dimensional arrays better than any other traditional tool, none of which can capture any algorithmic content beyond simple statistical patterns.

In order to solve the classification/forecasting task, the method of transforming the time series into vectors, consisting of (a) the value of algorithmic complexity (BDM) calculated on the matrix, representing certain physiological systems/processes and (b) the dynamics of predictor values of interest over time (V1 to V12), represented by the difference of the values of these parameters in consecutive samples. Table 1 represents three such vectors/cases.

Table 3. Format of data to be passed to ML.

BDM1	BDM2	V1	V2	V3	V4	...	V12
16.956	17.491	−4.5	1.5	0	−1.5	...	0.5
18.328	14.815	−8.5	0	0.5	−10	...	0
18.456	15.942	1	0	0.2	28	...	0

A more detailed description of BDM calculation is presented in [9].

2.3 Machine Learning

Data in the format described in Table 3 with an additional column for labeling (0 – for non-septic cases and 1 – for sepsis) was passed to a number of ML algorithms: (a) Generalized linear model – GLM, (b) Gradient boosting machine – GBM, (c) Distributed random forest – DRF, (d) Deep neural network – DNN for a binomial classification task (sepsis vs non-sepsis).

Optimal hyperparameters (number of predictors for GLM, learning rate, number of trees, depth, number of branches for GBM and DRF, number of layers, activation functions for DNN as a multilayer perceptron) were initially identified by automated machine learning (AutoML) using the H2O preset platform. These hyperparameters were then justified in the context of the present data by a 10-fold cross-validation grid search.

The best set of main hyperparameters for a particular algorithm/model is presented in Table 4.

Table 4. Best main hyperparameters: 4 ML-algorithms/models

Algorithm	Hyperparameters
Generalized linear model (GLM)	Type "binomial" with 14 predictors
Gradient boosting machine (GBM)	Number of trees – 116, Average depth – 6, Average number of leaves – 55.36
Distributed random forest (DRF)	Number of trees – 120, Average depth – 20, Average number of leaves – 532.4
Deep neural network/Multilayer perceptron (DNN/MLP)	4 layers, 2 hidden with 200 units each, Activation – ReLu and Softmax – for output

3 Results

3.1 ML Algorithms Performance

The process of creating machine learning models in this study was more or less standard, including: data identification/collection, data preprocessing, selection of the most successful algorithms for the data set of interest, training, validation and testing, evaluation of model performance. A crucial moment is the creation from the very beginning of the test set, which was 10% of the total data, with 90% data for training and validation.

After the exploratory analysis, 6096 patients were selected, of which 1384 - with sepsis. The selection criteria included: (a) the presence of at least 7 hourly observation windows (4 h - forecast horizon, 3 h - to assess the dynamics of the parameters describing the patient) and (b) the maximum possible data density (for non-septic patients - the presence of 7 consecutive values for each of the parameters of interest). The data of septic patients with missing values were reconstructed according to the algorithm described in II_B above. The test set consisted of 609 subsets/cases.

The same data set was fed to the four algorithms/models (i.e., GLM, GBM, DRF, DNN) with hyperparameters presented in Table 4. The performance of these algorithms is shown by Fig. 2 and Table 5 below:

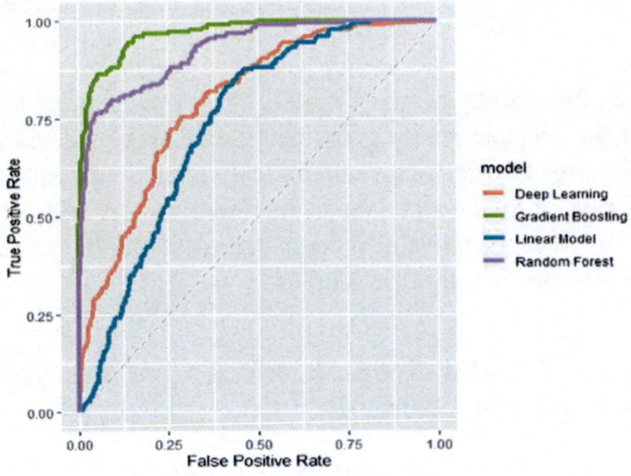

Fig. 2. ROC curves of four algorithms (DNN, GBM, GLM, DRF)

Table 5. Performance of the 4 ML-algorithms/models

Metrics/Algorithm	GLM	GBM	DRF	DNN
Area under the curve (AUC)	0.73	**0.95**	0.89	0.79
Recall	0.79	**0.84**	0.75	0.77
Specificity	0.61	**0.93**	0.80	0.71
Accuracy	0.65	**0.91**	0.89	0.72

As an orientation point for the creation of the sepsis prediction system in this research was the InSight system, reported in the literature as one of the best in the field [2]. Using fairly general information from [2] this system was replicated in the R language and tested for performance on the data set in the current study. Results were obtained that correspond to those reported by the authors. These results further served as milestones in the creation and evaluation of our own system. Table 6 presents the comparative performance metric of these two systems.

The metrics used are traditional for such systems. The mean values and the standard deviation are indicated. AUC is area under the receiver operating characteristic curve.

The performance of these two systems is quite close. A higher "Recall" in the case of the created system would indicate a more balanced system, which detects sepsis cases

Table 6. Comparative metrics of two sepsis prediction systems (InSight vs the GBM system created in the current research).

Metrics	InSight	Our system
AUC	0.95 (0.01)	0.95 (0.008)
Accuracy	0.91 (0.01)	0.91 (0.007)
Recall	0.82 (0.01)	0.84 (0.03)
Specificity	0.94 (0.02)	0.93 (0.01)

a little better, than the InSight system. The resulting system differs from InSight in the initial data used for machine learning, the data processing methods and the final sets data format, which are ultimately delivered for learning and validation. For example, the final data in the case of the created system represent vectors with length 14 (numeric strings made up of 14 numbers/values) compared to 30, in the case of InSight, i.e., a reduction in dimensionality more than 2 times.

3.2 Incorporation of the ML-System Based on GBM Algorithm into an Application for Clinical Use

For the clinical use of the system, it has been embedded in a Shiny application (R programming language) which is supposed to be easy to use by AIC physicians. A demo-version of the application is available at: https://viapascurta.shinyapps.io/ISAAC_sepsis_demo/.

The application can be used as follows: (1) entering the data describing the case/patient (the 6 parameters); (2) the data can be viewed graphically in the form of time series; (3) with the accumulation of 3 time-windows, the risk of sepsis in the targeted patient is automatically determined ("High risk" vs "Low risk"); (4) as the time progresses, after each hour the patient's condition (in relation to the risk of sepsis) is re-evaluated with the display of the risk for each evaluation. The risk dynamics can be viewed in a separate window.

4 Conclusions

The phenomenon of the penetration of ML technologies in medicine/AIC is a relatively a new one [1], but of proportions and perspectives difficult to predict. It seems that the medical practitioner has a lot to gain from this. But this interaction requires from the human specialist visions and abilities, which until recently were in the field of exoticism. The current study, which is an excerpt from a series of ongoing research, shows that this is possible and probably even useful.

The sufficiently high performance of the sepsis prediction system based on GBM creates premises for its use in clinical conditions in the Republic of Moldova and can create a potential benefit for patients, as a result of alerting the AIC doctor about an imminent risk of sepsis, which involves primarily an early start of antibiotic treatment with all favorable consequences, extensively researched and described in the literature [1, 2, 5].

Acknowledgment. This research is part of the activity within "Algorithmic Information Dynamics" research group headed by Hector Zenil (Oxford Immune Algorithmics, Oxford University Innovation, Department of computer science, University of Oxford/Alan Turing Institute, London, UK/Algorithmic Dynamic Lab, Karolinska Institute, Stockholm, Sweden).

Conflict of Interest. The authors declare that they have no conflict of interest.

References

1. Hashimoto, D., et al.: Artificial intelligence in anesthesiology: current techniques, clinical applications, and limitations. Anesthesiology **132**(2), 379–394 (2020)
2. Mao, Q., et al.: Multicentre validation of a sepsis prediction algorithm using only vital sign data in the emergency department, general ward and ICU. BMJ Open **8**, e017833 (2018). https://doi.org/10.1136/bmjopen-2017-017833
3. Zenil, H., Kiani, N., et al.: Algorithmic information dynamics: a computational approach to causality and living systems from networks to cells (2018). www.complexityexplorer.org/courses/63-algorithmic-informationdynamics-a-computational-approach-to-causality-and-living systems-from-networks-to-cells-2018
4. Reyna, M., et al.: Early prediction of sepsis from clinical data -- the PhysioNet computing in cardiology challenge 2019. (Version 1.0.0) PhysioNet (2019). https://doi.org/10.13026/v64v-d857
5. Shankar-Hari, M., Phillips, G., Levy, M., et al.: Developing a new definition and assessing new clinical criteria for septic shock: for the third International consensus definitions for sepsis and septic shock (Sepsis-3). JAMA **315**, 775–787 (2016)
6. Collins, G., et al.: Transparent reporting of a multivariable prediction model for individual prognosis or diagnosis (TRIPOD): the TRIPOD statement. BMC Med. **13**, 1 (2015). https://doi.org/10.1186/s12916-014-0241-z
7. Wolff, R., et al.: PROBAST: a tool to assess the risk of bias and applicability of prediction model. Ann. Intern. Med. **170**, 51–58 (2019). https://doi.org/10.7326/M18-1376
8. Zenil, H., et al.: A decomposition method for global evaluation of shannon entropy and local estimations of algorithmic complexity. Entropy **20**(8), 605 (2018). https://doi.org/10.3390/e20080605
9. Iapăscurtă, V.: Detection of movement toward randomness by applying the block decomposition method to a simple model of the circulatory system. Complex Syst. J. **28**(3), 59–77 (2019)

Impact of the Covid-19 Pandemic on the Use of Microsoft 365 and Learning Outcomes at the Technical University of Moldova

D. Țurcanu[1], Rodica Siminiuc[2(✉)], V. Bostan[3], and T. Țurcanu[4]

[1] Informatization, Partnerships, Institutional Image and Communication Office, Technical University of Moldova, Chisinau, Republic of Moldova
[2] PhD and Postdoctoral Department, Technical University of Moldova, Chisinau, Republic of Moldova
rodica.siminiuc@adm.utm.md
[3] Rector's Office, Technical University of Moldova, Chisinau, Republic of Moldova
[4] Department of Software Engineering and Automatics, Technical University of Moldova, Chisinau, Republic of Moldova

Abstract. Faced with the restrictions imposed by the Covid-19 pandemic, universities, as providers of higher education, have been forced to provide their students exclusively through a digital interface. For this, the institutions were challenged to select resources and platforms to support online education. The World Health Organization (WHO) advocates for research focusing on the impact and prediction of post-Covid evolution on education, in order to obtain adequate explanations for these challenges. The Technical University of Moldova (UTM) is the only technical higher education institution in the Republic of Moldova. TUM invests continuously in information technology infrastructure, as well as in the digital education of both students and academics. The aim of the study is to evaluate the university potential, from the perspective of digital technologies, in providing now, but also in the future, e-Learning. The most dominant form of providing learning content to TUM, during the pandemic, was online lectures via real-time video conferencing, which is confirmed by the continuous increase in the number of active MTeams users. And e-Learning did not negatively influence students' learning. The obtained results show that TUM has demonstrated its potential to ensure the continuity and good development of the teaching process during the pandemic, maintaining the good quality of studies.

Keywords: Information technologies · Higher education · e-learning · Students performance · Covid-19

1 Introduction

Education is the basis of development of every nation. Respectively, its sustainability is paramount for the growth and development of all. Over the centuries, education has been hit by several challenges. The Covid 19 pandemic has affected education systems around

I. Tiginyanu et al. (Eds.): ICNBME 2021, IFMBE Proceedings 87, pp. 456–462, 2022.
https://doi.org/10.1007/978-3-030-92328-0_59

the world, leading to the widespread closure of educational institutions [1], which has affected not only students, teachers and families, but the economy and society. in general [2, 3]. In this context, universities, as providers of higher education, have been forced to provide their students exclusively through a digital interface [4].

Used to support both teaching and learning, digital technologies have transformed the way we teach, by introducing a new connection-based model. A model that links teachers to their students and professional content, resources and systems to help them improve their own instruction and personalize learning. The impressive acceleration of the digitalisation of society in recent years, thanks to European policies to encourage technological development and, more recently, the Covid-19 pandemic, has reconfigured educational practices from "face-to-face" interaction to the online environment. UNESCO noted that about one billion students were forced to stay at home due to the emergency, and the scale and speed with which educational institutions were closed was an unprecedented challenge [5–7]. UNESCO also recommends distance learning programs and open educational applications during school closure caused by COVID-19 [8].

In order to face a different daily life from the one that characterized our days until the pandemic, the educational institutions were encouraged to develop pandemic management strategies and to face digital transformations, being challenged to select resources and platforms that could be used to support online education [9, 10]. As the crisis developed, the nature of the rapid change in communication offered to students and academics could have added additional sources of pressure. To date, multiple researches and case studies have been conducted and published focusing on the impact of COVID-19 on educational institutions and the prediction of post-COVID-19 developments in the field of education. The World Health Organization advocates for rapid analyzes to obtain adequate explanations for such challenges [11].

2 Materials and Methods

2.1 TUM-Digital Portrait

The Technical University of Moldova (TUM) is the only technical higher education institution in the Republic of Moldova. Its community includes about 9520 students from full-time and part-time sections, being provided by about 733 teachers. The university hosts 9 respective faculties and departments. It offers courses for 60 specialties and specializations in the first cycle, 43 master's specialties and 45 doctoral specialties, covering the needs of the national economy in engineering. TUM is an innovative university, which promotes a multi and interdisciplinary approach to higher education and research and which aims to broaden the professional and personal horizons of its students [12].

TUM offers complex digital services to staff and students, as well as to stakeholders, through several departments (IT, continuing education, communication, image, etc.). In order to maintain its relevance and remain competitive in today's digital age, TUM is continuously investing in information technology infrastructure, as well as in literacy among academics, students and administrative staff, which contributes to the reshaping of digital culture and strengthens confidence in new technologies [13].

Based on the partnership with Microsoft, UTM has fully implemented, since 2016, the Office 365 for Education platform, which offered a new process of learning, communication and virtual collaboration for both students and employees, using personalized e-mail, portals, online document editing and storage, instant messaging and web video conferencing - tools already used in the classical training process and which have helped a faster online migration. And, following the precise organizational interventions, anchored on needs and, in accordance with the development strategy of the information technology field, TUM managed to quickly adapt its mode of operation to ensure the continuity of its activities and to accompany its students in the process. study in an effective way. In the created situation, predictable or not, but the digital transformation, from opportunity and alternative has become, without much care, an imperative necessity [10, 14].

Within TUM, research has already been carried out aimed at evaluating the university's potential to provide e-Learning from the perspective of digital technologies. The results of that study, which included a period of 180 days, starting with the installation of lockdown, allowed the shaping of organizational, procedural, infrastructural and pedagogical strategies to ensure the university education process.

The use of digital technologies has helped to reconfigure attitudes towards learning, seen as a process that pursues the individual only throughout the educational process, in a process throughout life. However, research on the impact of ICT on e-Learning is current and required.

2.2 The Purpose of the Research

Through this research we wanted to achieve the following objectives:

- Assessment of the impact of the Covid - 19 pandemic on the dynamics of the use of Microsoft 365 services at TUM, which would allow the evaluation of the university potential in providing now, but also in the future, e-Learning, from the perspective of digital technologies implemented within TUM.
- Evaluating the impact of e-Learning on learning outcomes, by analyzing the grades obtained by students in the years of study until the pandemic and during the pandemic.

2.3 Theoretical Framework and Exploratory Procedures

In order to carry out the research, data on the use of the Microsoft 365 suite of applications were taken and compared in the time intervals 15.02.2020 - 15.08.2020 (at the beginning of the installation of the state of emergency in the Republic of Moldova) and 24.03.2021–19.09.2021 (current period). The data was exported from the M365 User Report for a period of 180 days.

In order to identify the impact of e-Learning on the quality of learning, the students' skills were evaluated during the years 2016–2021. In order to carry out the study, the average values of the students' characteristics per faculty in the winter session and the summer session were analyzed and calculated. Those calculations included all grades from current student assessments, individual assignments, tests and examinations. Subsequently, the average values of the property per university were determined.

3 Results and Discussion

Focusing on the first goal set, we note that the rapid transition from physical presence to distance learning has contributed to the increase in the total number of active users, which is currently 11, 6k. During 24.03.2021–19.09.2021 the total number of users increased by 32%, compared to the period since the beginning of the pandemic. Also, the number of active users for One Drive services increased by 40%, Share Point - 30%.

The means of providing learning content to TUM during the pandemic were mainly online lectures through real-time video conferencing. This approach is also confirmed by the results obtained: until the outbreak of the pandemic, the MTeams application was used to a relatively small extent by the TUM community, the average number of users per day being about 50 people. The number of active MTeams users has increased compared to the period since the beginning of the pandemic by 57% (Fig. 1). Other research that relied on the means of providing learning content between different higher education institutions confirmed / found that the most dominant form, in lockdown, was synchronous video activities, which were supplemented with other asynchronous forms, such as video recording, written communication, sending presentations to students, etc. [6, 15].

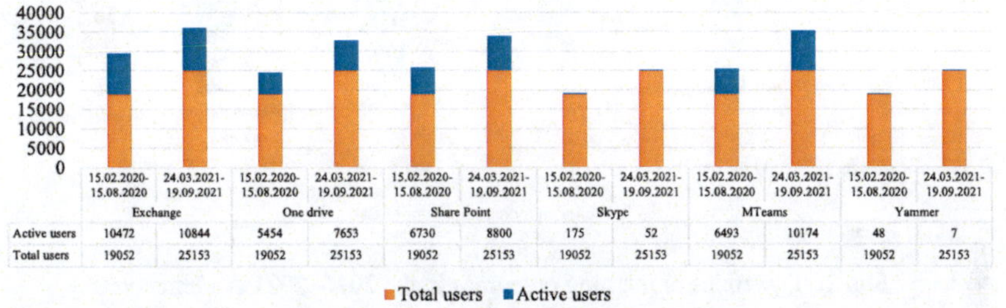

	Exchange		One drive		Share Point		Skype		MTeams		Yammer	
	15.02.2020-15.08.2020	24.03.2021-19.09.2021	15.02.2020-15.08.2020	24.03.2021-19.09.2021	15.02.2020-15.08.2020	24.03.2021-19.09.2021	15.02.2020-15.08.2020	24.03.2021-19.09.2021	15.02.2020-15.08.2020	24.03.2021-19.09.2021	15.02.2020-15.08.2020	24.03.2021-19.09.2021
Active users	10472	10844	5454	7653	6730	8800	175	52	6493	10174	48	7
Total users	19052	25153	19052	25153	19052	25153	19052	25153	19052	25153	19052	25153

Fig. 1. Dynamics of Microsoft service users

The rapid transition to online education has been an adaptation to the requirements of social distancing, but it has also come with challenges: for many universities, especially for tum - as an engineering institution, distance education could not be impervious to the impact of physical restrictions because it is based on some activities that require a physical location. Activities that involve high levels of trust and authentication. It is believed that teaching and learning approaches (e-Learning, mixed learning and mobile learning) that use digital technologies improve the teaching and learning process and can lead to better learning outcomes for students. However, regardless of these benefits, some research reports that the degree of absorption and use is low [16]. Student performance is considered an important indicator of quality, being defined as a multidimensional and multilayered concept, which depends on the requirements set by different actors. Grades are widely perceived as what students "earn" for their achievement and have a pervasive influence on students and the institution [17]. Student assessment is of the utmost importance, as examining variation helps the administration to fully understand the actual performance of learning and its processes. In that study we focused on one of the criteria for assessing performance - student evaluation. The results, calculated for a period of about 6 years, in order to identify how the covid - 19 pandemic influences and, respectively, e-learning on learning outcomes, are presented in Fig. 2. The average values of the students' grades report a constant ascent with about 15% and 18% in 2017 and, respectively, in 2018, after which, the results achieved, practically, remain at the same values during the years 2019–2021. The dynamics in the summer session is slower, amounting to about 4–6% each year, including during the pandemic. The average values of student performance being 85.44% in the summer session (about 4 months after the pandemic), and in the winter session 2020–2021-88.83%. It seems that online education has not negatively influenced learning outcomes. Of course, any assessment is not exempt from the degree of error, as there is a risk that teachers will not properly interpret the learning outcomes. There is research that shows that even teachers who explicitly decide to score only based on the achievement of learning standards sometimes mix effort, improvement and other academic facilities in setting grades [18]. These components could form the basis of new research.

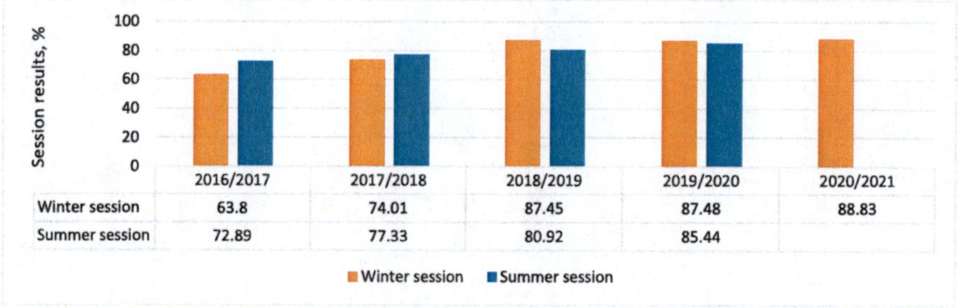

Fig. 2. Dynamics of learning outcomes in the 2016–2021 academic years

4 Conclusions

Probably anticipating the current situation and following precise organizational interventions, but also in accordance with the development strategy of the information technology field, the Technical University of Moldova has demonstrated its potential to ensure the continuity and proper conduct of the teaching process during the pandemic. There are still many issues to be elucidatedregarding the negative impact of the covid-19 pandemic on students' study habits, life and mental health. Additional studies are needed that would investigate the effect of the pandemic on the workload of learning, the interactions between students, student-teachers. Studies that can provide a better understanding of how to proactively support students during forced elearning.

It is our duty, of the academic community, to analyze the possible future scenarios and to look, with seriousness and reflection, for strategies that can be followed, regardless of the circumstances [7].

Conflict of Interest. The authors declare that they have no conflict of interest.

References

1. Owusu-Fordjour, C., Koomson, C. K., Hanson, D.: The impact of covid-19 on learning - the perspective of the Ghanaian student (2020). https://doi.org/10.5281/ZENODO.3753586
2. Khlaif, Z.N., Salha, S., Affouneh, S., et al.: The Covid-19 epidemic: teachers' responses to school closure in developing countries. Technol. Pedagog. Educ. **30**, 95–109 (2021). https://doi.org/10.1080/1475939X.2020.1851752
3. Siminiuc, R., Turcanu, D.: The impact of the pandemic on the agri-food system. JSS **3**, 85–94 (2020). https://doi.org/10.5281/zenodo.3971973
4. Watermeyer, R., Crick, T., Knight, C., Goodall, J.: COVID-19 and digital disruption in UK universities: afflictions and affordances of emergency online migration. High. Educ. **81**(3), 623–641 (2020). https://doi.org/10.1007/s10734-020-00561-y
5. Tømte, C.E., Fossland, T., Aamodt, P.O., Degn, L.: Digitalisation in higher education: mapping institutional approaches for teaching and learning. Qual. High. Educ. **25**, 98–114 (2019). https://doi.org/10.1080/13538322.2019.1603611
6. Aristeidou, M., Cross, S.: Disrupted distance learning: the impact of Covid-19 on study habits of distance learning university students. Open Learn.: J. Open Dist. e-Learn. 1–20 (2021). https://doi.org/10.1080/02680513.2021.1973400
7. Giancola, O., Piromalli, L.: Apprendimenti a distanza a più velocità. L'impatto del COVID-19 sul sistema educativo italiano. Il Mulino (2020). https://doi.org/10.12828/97097
8. Gopal, R., Singh, V., Aggarwal, A.: Impact of online classes on the satisfaction and performance of students during the pandemic period of COVID 19. Educ. Inf. Technol. **26**(6), 6923–6947 (2021). https://doi.org/10.1007/s10639-021-10523-1
9. Kim, C.L.: E-learning, information technology, and student success in higher education. ORE (2017). https://doi.org/10.1093/acrefore/9780190224851.013.78
10. Ţurcanu, D., Siminiuc, R., Bostan, V.: The impact of the COVID-19 pandemic on the use of digital technologies in ensuring the efficient e-learning process at the Technical University of Moldova. CE **11**, 2116–2132 (2020). https://doi.org/10.4236/ce.2020.1110154
11. Khan, M.A.: COVID-19's impact on higher education: a rapid review of early reactive literature. Educ. Sci. **11**, 421 (2021). https://doi.org/10.3390/educsci11080421
12. Technical University of Moldova Brief history of TUM. In: Technical University of Moldova. https://utm.md/en/about-us/brief-history-of-tum/
13. Grosseck, G., Maliţa, L., Bunoiu, M.: Higher education institutions towards digital transformation—the WUT case. In: Curaj, A., Deca, L., Pricopie, R. (eds.) European Higher Education Area: Challenges for a New Decade, pp. 565–581. Springer, Cham (2020). https://doi.org/10.1007/978-3-030-56316-5_35
14. Technical University of Moldova: UTM is preparing for several scenarios for the 2020/2021 academic year. In: Technical University of Moldova (2020). https://utm.md/en/blog/2020/08/05/utm-is-preparing-for-several-scenarios-for-the-2020-2021-academic-year/
15. Aucejo, E.M., French, J., Ugalde Araya, M.P., Zafar, B.: The impact of COVID-19 on student experiences and expectations: evidence from a survey. J. Public Econ. **191**, 104271 (2020). https://doi.org/10.1016/j.jpubeco.2020.104271
16. Lwoga, E.: Making learning and Web 2.0 technologies work for higher learning institutions in Africa. Campus-Wide Info Syst. **29**, 90–107 (2012). https://doi.org/10.1108/10650741211212359

17. Brookhart, S.M., Guskey, T.R., Bowers, A.J., et al.: A century of grading research: meaning and value in the most common educational measure. Rev. Educ. Res. **86**, 803–848 (2016). https://doi.org/10.3102/0034654316672069
18. Hay, P.J., Macdonald, D.: (Mis)appropriations of criteria and standards-referenced assessment in a performance-based subject. Assess. Educ.: Principles Policy Pract. **15**, 153–168 (2008). https://doi.org/10.1080/09695940802164184

Nanotechnology and Nonproliferation

Artur Buzdugan[1,2](✉), S. Railean[1], and Au. Buzdugan[3]

[1] Technical University of Moldova, Chisinau, Moldova
[2] National Nuclear Security Support Centre, Chisinau, Moldova
[3] Moldova State University, Chisinau, Moldova

Abstract. Nanotechnology has demonstrated performance in top technologies of the 50s-60s of the XX- century such as nuclear, cosmic and other domain. Although the first results were promising, the crucial point for further development took place later in 2000 with the adoption by the US of the National Nanotechnology Initiative (NNI) [1]. This served as an impetus for third countries that have adopted similar programs or strategies (EU member states, the Russian Federation, etc.) [2, 3]. Today, nanotechnologies are an indispensable component of the 4th industrial revolution, which radically changes the world. Nanotechnology is considered as a new frontier of engineering and research with an unimaginable long-term potential impact, giving rise to new interdisciplinary fields of research, engineering and education. Unfortunately, like any cutting-edge technology, it can also lead to the development of a new destructive technologies, especially of weapons of mass destruction (WMD). Therefore, risk assessment becomes critical in the context of dual-use technologies.

This can be achieved only through joint efforts and consensus of the governments, researchers and large producers, and requires long negotiations. We find that a promising solution would be by regulating this domain at a national level, as well as via a Codes of Ethics for professional societies. The implementation can be initiated with a new University course covering the responsibilities of engineers and researchers in the preventing proliferation of WMD via nanotechnology and nanoscale science. This path has been followed by the Technical University of Moldova by developing a new curriculum in engineering and non-proliferation culture for master degree study.

Keywords: Nanotechnology · Nanoscale science · Dual use risk · WMD nonproliferation

1 Introduction

About 4000 years ago, in ancient civilization such as Assyria, Babylon, Egypt, and films of various substances were obtained on the surface of the water. Ancient alchemists described how adding tiny Au and Ag particles to the composition of glass essentially changed its color, depending on the size of the particles (e.g. the Lycurgus cup is recognized as one of the oldest obtained nanomaterials) [4].

The first to use the concept of "*nanotechnology*" was the physicist R. Feynman (1959, the speech "There's Plenty of Room at the Bottom" [5]) describing a process

I. Tiginyanu et al. (Eds.): ICNBME 2021, IFMBE Proceedings 87, pp. 463–469, 2022.
https://doi.org/10.1007/978-3-030-92328-0_60

capable of managing with discrete atoms and molecules. The widespread use of the term "*nanotechnology*" was attributed in 1974 to Professor N. Taniguchi, defined as "nanotechnology mainly consists of the processing of separation, consolidation, and deformation of materials by one atom or one molecule". Later, in 1986 Eric Drexler introduced a world readership to the concept of nanotechnology via his book "Engines of Creation".

However, the practical use of nanomaterials took place much earlier, especially in top technologies for special purposes, without mentioning that they are nanomaterials or nanotechnologies. Thus, we recall that modern industrial nanotechnology had its first applications in the 1930s, in processes used to create silver coatings for photographic film or when chemists have been making polymers. During the 1950s, research on nanoscale materials was called research into the physics of ultra-disperse materials. It was not clear what nanostate was, it was the period of accumulation of theoretical and experimental results. Even biologists, who studied small particles (e.g. *viruses*), did not mention that they were actually dealing with a nanostate of matter.

As another example we mention the first nanoscale materials in former Soviet Union - metal powders specifically designed and successfully used behind closed doors for solving "uranium project" in the 1950s [6]. Applications of nanotechnologies for the nuclear field has been demonstrated also from the 70s–80s regarding increasing the efficiency of nuclear fuel combustion by adding nanoadditives, hardening steel by adding 0.5% by mass of Yttrium Oxide nanoparticles, or 0.37% by mass of F_2O_3 nanoparticles. The usefulness of nanomaterials has also been demonstrated in the manufacturing of radiation protection clothing with a four times higher radiation absorption efficiency. Nanoscale TiO_2 and ZnO powders are now commonly used by cosmetics manufacturers, whereas nanoscale iron oxide powder is now being used as a base material for rouge and lipstick. Paints with reflective properties are also being manufactured using nanoscale TiO_2 particles. Nanostructured cemented carbide coatings are used on some ships for their increased durability, etc. [7].

As mentioned, a decisive step in the development of nanotechnologies was the approval in 2000 by the US of NNI. This helped the realization of the program for which 20 departments and agencies worked together [4]. According to the NNI "nanoscale science and engineering research will focus on fundamental understanding and synthesis of nanometer-size building blocks aimed at potential breakthroughs in areas such as materials and manufacturing, nanoelectronics, medicine and health care, environment and energy, the chemical and pharmaceuticals industries, biotechnology and agriculture, computation and information technology, and national security" [1]. An important fact of NNI is that it defined nanotechnology *as a new state of manifestation of matter,* with properties very different from the same macrostructure.

2 Current Status

Nanotechnologies are considered one of the four pillars (nanotechnologies, biotechnologies, technologies of a new materials and advanced digital production technologies) of the IVth industrial revolution.

What are the risks of nanotechnologies? If we analyze this technology we identify the following risks (Fig. 1):

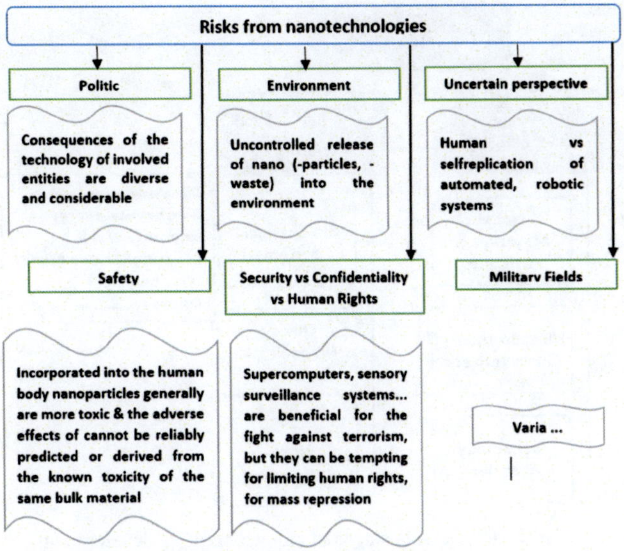

Fig. 1. The risks expected from the development of nanotechnologies

We note that nanoscale science & nanotechnology is triggering of a new era and opportunity. However, its gravity depends on the degree of control over the fundamental elements of matter and of the advanced goal. The challenge is so great that it is hard to imagine what awaits us on the horizon. In this context, we remember the God Janus, the God of Doors, with the duality of reality that awaits us when we pass through that door - the large-scale hostile use of nanotechnology is comparable to the benefits expected by society.

The scientific community and policy-makers are focused for decades on non-proliferation of chemical, biological and nuclear WMD. The attention is drawn to means of transport to the destination, which have often shown functionality and destructive power. But nanotechnologies and nanoscale science in the contemporary phase are the most convincing technologies for WMD processing and delivery methods. In addition, nanotechnologies are so powerful and versatile that they potentially can give rise to new technologies of WMD.

The analysis of the nanotechnology roadmap currently shows that virtually all stages of development can contribute to the development of WMD (see Fig. 2).

In addition to the dangerous effects on the upgrade of WMD, the manufacture of nanotechnologies also presents problems of surveillance, terrorism, irreversible damage to the environment or security issues in various fields.

The availability of nanotechnologies is already shown for nuclear WMD [8], but it is becoming faster and more widespread than nuclear proliferation. The destruction caused by weapons based on nanotechnologies can be more catastrophic, but more limited in space than nuclear explosions, thus being more demanding.

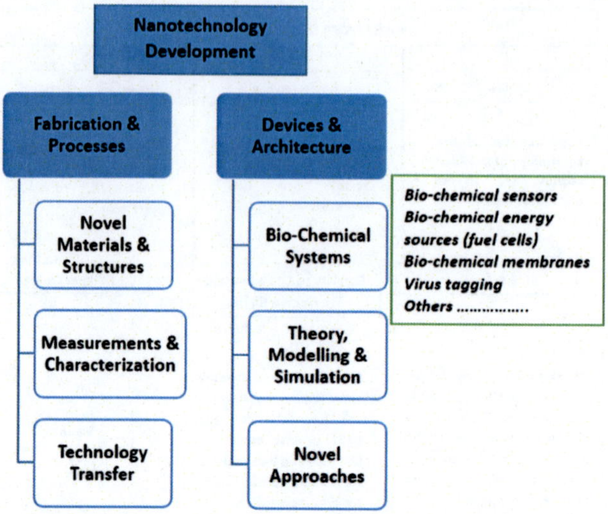

Fig. 2. Predictable stages of nanotechnology development

If we refer to the development of biological or chemical WMD, nanoparticles are characterized as much more reactive (and currently unpredictable due to lack of knowledge) compared to biological and chemical agents. Nanotechnologies are the basis for the production and delivery of chemical and biological agents. Thanks to nanotechnologies, the required quantities of chemical and biological agents will be much smaller, generating small production facilities that are difficult to detect by existing methods. But the effects of the new materials at the nanoscale are much stronger, because it increases the contact surface, which makes them more toxic and viral. The convergence of nanotechnology with chemical and biological technologies creates unique premises such as for editing the DNA of bacteria, thus creating new species or new chemicals with targeted properties. New methods of targeted delivery of nanotoxins make them virtually invisible to medical diagnostic and countermeasures.

For these reasons, the existing challenges due to nanotechnologies are not yet officially recognized.

Nanotechnologies not only allow the increase the existing threats related to WMD, but also create new threats that we are barely aware of. The process of identifying and assigning nanoparticles is little known when referring to WMD. Thus, the counteractions of these "nanocontains" ADMs are extremely limited at present.

We currently assume the distinction of the following threats caused by nanotechnologies in the development of WMD.

a) Extension or dissemination of nanotechnologies due to easy access to information sources through software technologies, which are accessible in terms of building a nanotechnology laboratory. For example, the guide for building such a technological laboratory for DNA is about $500. Nowadays, many harmless chemical ingredients can be found over the counter. Unfortunately, the result of chemical reactions of these ingredients can become extremely dangerous. The relatively easy accessibility and acceptable price of nanotechnologies are attracting more and more states to develop their own WMD.

b) The temptation of state actors to develop their own WMD, initially for defensive purposes, is to discourage potential aggressors. Nanotechnologies make these products substantially cheaper, as well as making them less visible, so they are easier to transport. These things consequently generate the temptation to deliver them (proliferate) to ill-willed states. Thus, they become more accessible to non-state actors, who are less influenced by state or international non-proliferation regulations. These conditions of proliferation are fueled by the drastic lag behind departmental, national and international regulations in the field of nanotechnologies.

Present legal and scientific uncertainty regarding nanotechnologies makes it difficult to amend the arms limitation and control conventions [10]. It is only one step (decisional) from harmless nanotechnologies, beneficial to humans to technologies of repression and aggression - the technology remaining the same, but the purpose is different. However, it is now clear that nanotechnologies are dual-use, so regulation is required in this field [8–11]. However, the risks must be assessed and managed in such a way as not to suppress scientific and therefore economic progress.

The educational institutions, indispensable for research, are one of the cornerstones of the prosperity and security of future generations. The present and inevitable diversification of research fields and information technologies impose new ethical attitudes in the development and dissemination of research results. Responsible management of research in high school also requires active commitments from relevant institutions, including the following security considerations.

First of all, researchers are those who provide or generate, volens-nolens, stimuli for the irresponsible use of technical-scientific achievements and innovations.

Secondly, the nanotechnology is the future of information processing, whether that processing is based on a nanoscale silicon transistor manufactured to tolerances partially determined by processes of chemical self-assembly or on one or more of the new developed molecular devices.

In the third, the progress of electronic communications in free access to information shows an increase in the role of information security in scientific and innovative databases, especially in unclassified information for various reasons for dual-use components, materials and technologies. For these reasons, information security becomes an important and ascending player, indispensable for the non-proliferation of CBRN materials, because some research may generate knowledge, materials, methods or technologies that could also be used in unethical ways. Although such research is argued and conducted with healthy intentions, the perpetrators may harm humans, animals or the environment by using the results of research.

So, researchers and engineers play a key and responsible role in non-proliferation.

How to counteract these trends? Researchers are the most knowledgeable and best placed professionally to assess the nature and seriousness of the potential for misuse of knowledge, products or technologies. In this context, they should be responsible for evaluating and reporting these findings to the research institution or national bodies.

Research is most vulnerable to unethical uses because:

- generates and provides knowledge, materials, methods and technologies (including in the form of CBRN products) that could be channeled into crime or terrorism;
- leads to the general free dissemination of information for useful exchange of information but also guided by the support of human rights to information access;
- does not ensure the awareness of future researchers in terms of innovations that could represent dual-use;
- even if there are databases of scientific results and innovations, their vulnerability is not periodically assessed and security measures are exceeded for various reasons.

We recommend the following steps for mitigating or at minimizing the undesirable proliferation:

- conducting self-assessment of research risk during or before research, including by updating regularly the assessment models for high-risk areas [12];
- adjusting safety and security measures to counteract safety risks (at the end of and after the completion of the research);
- introducing at the bachelor's or master's degree disciplines in the non-proliferation of WMD, their interference with nanotechnologies and nanoengineering in liaisons with the growing role of information security in these areas.

We note that such a course is currently being developed in the National Nuclear Safety Support Center of the Technical University of Moldova [13] with the support of development partners.

3 Conclusions

Nanotechnology requires an enormous responsibility from researchers and engineers. Beneficial and abusive use are two indispensable facets of the outcome. It is everyone's responsibility is to assess the potential risk for dual use, and if necessary to comply with ethical and regulatory regulations to counteract it.

Governments have an obligation to use commercial potential, sociological and environmental impacts of nanotechnology in a controlled manner and avoid risks. For this it is necessary to harmonize the definitions, terminology, and characterization compatible with the international ones. They also require aspects of standardization and metrology [14, 15] in nanotechnology research to improve in establishing the framework conditions for the successful transfer of research and development results into innovative products and services.

The development of nanotechnologies is impossible without an interdisciplinary culture necessary to make progress in nanoscale science and technology, including for the assessment of the risk of dual-use. For this scope, certain models have already been developed and proposed. One of the models looks at the risk implication of nuclear and radiological domain in the context of safety and security of an institution, or even at the industrial branch level [12].

An interdisciplinary interactions must be encouraged and sustained in the long term. The interdisciplinary nature of nanoscale science and nanotechnology requires that we implement new paradigms for educating scientists and engineers.

Conflict of Interest. The authors declare that they have no conflict of interest.

References

1. National Nanotechnology Initiative (NNI). https://www.nano.gov/about-nni
2. Action Plan. Nanotechnology 2020. An inter-departmental strategy of the Federal Government. https://www.bam.de/_SharedDocs/EN/Downloads/as-nano-action-plan-nanotechnology-2020.pdf?__blob=publicationFile
3. Federal Law of July 19, 2007 N 139- ФЗ "On the Russian Corporation of Nanotechnologies". (In Russian). https://rg.ru/2007/07/25/nano-korporacia-dok.html
4. Nanotechnology Timeline. https://www.nano.gov/timeline
5. Feynman, R.P.: There's plenty of room at the bottom. http://www.nanoparticles.org/pdf/Feynman.pdf
6. Petrunin, V.F.: Development of nanomaterials for nuclear energetics. Phys. Proc. **72**, 536–539 (2015). Conference of Physics of Nonequilibrium Atomic Systems and Composites, PNASC 2015, 18-20 February 2015 and the Conference of Heterostructures for microwave, power and optoelectronics: physics, technology and devices (Heterostructures), 19 February 2015
7. Tabassum, N.: An empirical exploration of the nanotechnology. Int. J. Adv. Res. **8**, 885–915 (2020). https://doi.org/10.21474/IJAR01/11352
8. Gsponer, A.: From the lab to the battlefield? Nanotechnology and fourth-generation nuclear weapons. In: Disarmament Diplomacy, Issue No. 67, October – November 2002
9. Howard, S.: Nanotechnology and mass destruction: the need for an inner space treaty. In: Disarmament Diplomacy, Issue No. 65, July - August 2002
10. Brehm, M.: Article 36, Nanoweapons. Discussion paper for the Convention on Certain Conventional Weapons (CCW). Discussion paper, Geneva, November 2017. https://article36.org/wp-content/uploads/2020/12/nanoweapons.pdf
11. Marchant, G.E., Sylvester, D.J.: Transnational models for regulation of nanotechnology. J Law Med Ethics Winter **34**(4), 714–725 (2006). https://doi.org/10.1111/j.1748-720X.2006.00091.x. Nanotechnology. Winter
12. Buzdugan, A.: Model for cyber security maturity assessment in critical infrastructures. Official Catalogue "Cadet INOVA", pp. 154–157, June 2021. ISSN 2501-3157
13. Buzdugan, Ar., Buzdugan, A.: Knowledge and role of technology in university's nonproliferation culture. In: SnT 2021. CTBT Science and Technology Conference, 28 June–02 July 2021, Vienna, Austria (2021). https://conferences.ctbto.org/event/7/contributions/1345/contribution.pdf
14. Buzdugan, A.: Nanotehnology and nanometrology. Metrologia **2**(8), 6–10 (2013). (In Romanian). https://inm.md/static-files/img/revista/R_M_8.pdf
15. Allan, J., Ferreri, A., Grande, S., Giannantonio, R., Matteucci, F.: Technology transfer in nanotechnology, EUR 29686 EN, Publication Office of the European Union, Luxembourg (2019). https://doi.org/10.2760/183529. ISBN 978-92-76-00671-8, JRC115968

Sorbents Obtained from Cellulose-Containing Waste for Water Purification

T. Marsagishvili, G. Tatishvili(✉), N. Ananiashvili, E. Tskhakaia, N. Giorgadze, M. Gachechiladze, M. Matchavariani, and L. Kvinikadze

R. Agladze Institute of Inorganic Chemistry and Electrochemistry, Iv. Javakhishvili Tbilisi State University, Mindeli Str. #11, Tbilisi, Georgia
tati@iice.ge

Abstract. As a rule, ions of some heavy metals are always found in natural and waste waters. The purpose of this work was to study the sorption properties of carbon material obtained from secondary raw materials (hazelnut and walnut shells) according to the technology we have developed; determination of its applicability in the purification of drinking and waste water, in which ions of several heavy metals are simultaneously present. The salts of lead, cobalt, iron and cadmium, as the most important pollutants of the water basin, were the adsorbates.

Studies of solutions of various concentrations of metal ions showed that in the presence of all four ions, the maximum adsorption rate and the degree of extraction are achieved at a solution concentration of 0.0025 M Me^{+2} in 30 min, as for the effect of pH on the degree of sorption, the experimental results showed that the best values are achieved at pH 3 for cobalt ions and pH 3 ÷ 5 for copper, lead and iron ions.

It was found that the nature of sorption by these absorbers is the same: for a given metal, hydrated ions are adsorbed in the same way on different absorbers.

The larger the crystal radius of an ion with the same charge, the better it is adsorbed (sorption capacity for Pb^{++} ions is maximum, and minimum for Co^{++} ions).

Keywords: Adsorption · Heavy metals · Carbon · Waste water · Secondary raw materials

1 Introduction

Environmental pollution by industrial emissions is currently an acute environmental problem. A significant part of wastewater of various origins enters the environment without treatment, which negatively affects human health, the environment and the economy [1].

Among the most dangerous carcinogenic toxicants contained in industrial waste water are heavy metals belonging to I and II hazard classes. They rank second in terms of toxicological effects on the human body, behind pesticides.

Heavy metals include lead (Pb), cadmium (Cd), arsenic (As), chromium (Cr), copper (Cu), nickel (Ni), cobalt (Co), zinc (Zn) and others. It is necessary to constantly monitor

I. Tiginyanu et al. (Eds.): ICNBME 2021, IFMBE Proceedings 87, pp. 470–474, 2022.
https://doi.org/10.1007/978-3-030-92328-0_61

their concentration in water to detect and assess contamination. Large amounts of heavy metals in wastewater can be caused by waste accumulation, increased human activity or industrial accidents. They accumulate in drinking water and can cause various diseases, as well as biological and genetic disorders [2].

To purify wastewater from water-soluble salts of heavy metals, chemical, physico-chemical, biological and combined methods of water purification are applicable. However, these technologies are either often ineffective, produce secondary products, or are too expensive. In fact, when choosing a specific wastewater treatment technology, in addition to its effectiveness, environmental and economic aspects must be taken into account.

Among the physicochemical methods of detoxification of both gas and liquid media, adsorption stands out, as a universal process that allows you to almost complete removing of toxic impurities. The successful solution of many practical problems is determined by the choice of sorbents with optimal physicochemical properties and cost for these purposes [3].

Currently, the most versatile sorbents are materials made from chemically activated carbon [4]. These are materials with the best sorption properties, high porosity, developed surface area and mechanical strength.

2 Materials and Methods

At the Institute of Inorganic Chemistry and Electrochemistry of the Ivane Javakhishvili Tbilisi State University, a technology has been developed [5, 6] for obtaining carbon materials with a high surface area. This technology was used for many types of raw materials (used tires, hazelnut and walnut shells, nectarine kernel, sawdust, plastics, etc.).

The process is one-stage and does not require preliminary processing of raw materials. Carbon dioxide, water and various gases generated in the process are used as a reagents and energy sources. In this case, the reactor vessel (stainless steel) acts as a catalyst. Using this technology, carbon materials were obtained from recycled organic waste, which are cheap and efficient.

The physicochemical properties of the above materials were measured, namely, BET surfaces, areas and volumes of micropores (by capillary nitrogen condensation, on **Micromeritics'** GEMINI VII 2390T surface area analyzer), chemical composition (by a **Bruker's** QUANTAX energy dispersive X-ray spectrometry system (EDS) to **HITACHI** TM 3030 Plus scanning electron microscope (SEM)), ash content. For comparison, commercial activated carbon (BAU A) was used. The adsorption capabilities of these materials have been determined. The measurement results are published in works [7–9].

Studies have shown that these materials are characterized by a large specific surface area, porous structure, high adsorption capacity, and can be successfully used to extract various heavy metal ions from water separately. It was found that carbon materials obtained from hazelnut and walnut shells are characterized by the best adsorption properties with respect to ions of cobalt, lead, copper and iron. The maximum adsorption of ions such as cobalt, copper and iron, separately, is achieved in 30–45 min, and for lead ions, 5–10 min is enough.

As a rule, ions of some heavy metals are always found in natural and waste waters. The purpose of this work was to study the sorption properties of carbon material obtained from secondary raw materials (hazelnut and walnut shells) according to the technology we have developed; determination of its applicability in the purification of drinking and waste water, in which ions of several heavy metals are simultaneously present.

Studies of solutions of various concentrations of metal ions have shown that in the presence of all four ions, the maximum adsorption rate and the degree of extraction are achieved at a solution concentration of 0.0025 M Me^{+2} in 30 min. As for the effect of pH on the degree of sorption, the results of experiments have shown that the best values are achieved at pH 3 for cobalt ions and pH 3 ÷ 5 for copper, lead and iron ions. The adsorbates were aqueous solutions of nitrates of lead, cobalt, iron, and cadmium.

Table 1. The values of the sorption capacity (A, mg/g) and the degree of extraction (α, %) of metal ions (Pb^{++}, Cd^{++}, Co^{++}, Fe^{++} - 0,01 M), by sorbents obtained from various carbon materials (hazelnut and walnut shells) and factory activated carbon (fractional size 40 microns) after holding (stirring) in a solution (t = 25–270 C, T = 30 min. $V_{solution}$ = 100 ml)

Metal ions	Hazelnut shells		Walnut shells		Activated carbon BAU A	
	Sorption capacity (A, mg/g)	Degree of extraction (α, %)	Sorption capacity (A, mg/g)	Degree of extraction (α, %)	Sorption capacity (A, mg/g)	Degree of extraction (α, %)
pb^{++}	83,5	80,7	88,0	85,0	84,0	90,2
Cd^{++}	22,5	80,0	22,8	81,3	22,5	80,0
Fe^{++}	19,8	53,6	21,3	45,4	20,1	57,0
Co^{++}	13,2	89,5	12,6	85,6	12,2	83

The adsorption on various carbonaceous materials was studied in the joint presence of the indicated metal ions, in particular, from mixed model solutions containing equal amounts of lead (0.01 M), iron, cobalt and cadmium.

The carbon materials obtained by us in terms of sorption activity were compared with activated carbon of the BAU A brand. The results are presented in the Table 1.

3 Results and Discussion

As the results obtained show (Table 1), the parameters characterizing the sorption by these absorbers are approximately the same. This is due to the fact that the molecules of a substance, dissolving in water, decompose into ions, which are in a hydrated state. In this case, the solute is adsorbed on the surface of the adsorbent in the form of hydrated ions. Therefore, for a given metal, hydrated ions are adsorbed in the same way on different absorbers.

The sorption capacity in relation to Pb^{++} ions is maximum and minimum in relation to Co^{++} ions. In this case, the adsorption of metal ions is greatly influenced by the value of their ionic radius. The larger the radius of an ion with the same charge, the better it is adsorbed, since with an increase in the radius, its polarizability increases, and, consequently, the ability to be attracted to the polar surface of the sorbents, i.e. its adsorbability, increases. At the same time, an increase in the ionic radius leads to a decrease in the hydration of the ion, which also contributes to an increase in the adsorption value.

Thus, the ions of heavy metals, which were used as adsorptives in our work, can be arranged in the following row in terms of increasing adsorption capacity, called the lyotropic series or the Hofmeister series:

$$A(Pb{+}{+}) > A(Cd{+}{+}) > A(Fe{+}{+}) > A(Co{+}{+}),$$

where, A is adsorption value.
Correspondingly [10],

$$r(Pb{+}{+})=1{,}21\ \text{Å} > r(Cd{+}{+})=0{,}99\ \text{Å} > r(Fe{+}{+})=0{,}8\ \text{Å} > r(Co{+}{+})=0{,}78\ \text{Å}$$

where, r is ionic radius of metal.

4 Conclusions

1. It was found that the nature of sorption by these absorbers is the same: for a given metal, hydrated
2. Ions are adsorbed in the same way on different absorbers.
3. The larger the crystal radius of an ion with the same charge, the better it is adsorbed (sorption capacity for Pb^{++} ions is maximum and minimum for Co^{++} ions).

Conflict of Interest. The authors declare that they have no conflict of interest.

References

1. Sadeek, S.A., Negm, N.A., Hefni, H.H., Wahab, M.M.: Metal adsorption by agricultural biosorbents: adsorption isotherm, kinetic and biosorbents chemical structures. Int. J. Biol. Macromol. **81**, 400–409 (2015). https://doi.org/10.1016/j.ijbiomac.2015.08.031. Epub 2015 Aug 15 PMID: 26282929
2. Maystrenko, V.N., Khamitov, R.Z.: Ecological and Analytical Monitoring of Supertoxicants. Khimia, Moscow, p. 105 (1996)
3. Kinetic modeling of multiple adsorption of heavy metal ions using activated carbon from Nigerian bamboo for design of adsorbers. Am. J. Chem. Eng. **4**(5), 105–113 (2016)
4. Uzun, I., Güzel, F.: Adsorption of some heavy metal ions from aqueous solution by activated carbon and comparison of percent adsorption results of activated carbon with those of some other adsorbents. Turkish J. Chem. **24**, 291–297 (2020)
5. United States Patent and Trademark Office, patent US 9,663,662 B1
6. Patent AP 2019 15030/1Positive decision on the patent application

7. Marsagishvili, T., et al.: Adsorption of lead ions on carbonaceous sorbents of nutshell obtained from secondary raw material. In: Tiginyanu, I., Sontea, V., Railean, S. (eds.) 4th International Conference on Nanotechnologies and Biomedical Engineering. ICNBME 2019. IFMBE Proceedings, vol. 77. Springer, Cham (2020). https://doi.org/10.1007/978-3-030-31866-6_21
8. Giorgadze, N.V., et al.: Adsorption of cobalt ions on carbonaceous sorbents obtained from secondary raw materials. Int. J. Green Herbal Chem., IJGHC **10**(1), 101–108 (2020). https://doi.org/10.24214/IJGHC/GC/10/1/07077, E-ISSN: 2278-3229
9. Marsagishvili, T., et al.: Adsorption of lead ions on carbonaceous sorbents of nutshell obtained from secondary raw material. SCIREA J. Electr. Eng. **6**(1), 23–41 (2021)
10. Handbook of a chemist. Ionic radii according to Belov and Bokiy. Inorganic chemistry, Appendix 5, Tomsk (1997)

Measuring and Information System for Monitoring Microwave Contamination of Urban Environment

A. Simakov, I. Vodokhlebov, Yuri Bocharov(✉), V. Butuzov, and M. Simakov

National Research Nuclear University, Kashirskoe Shosse, 31, Moscow, Russian Federation
YIBocharov@mephi.ru

Abstract. There is a lot of data confirming the effect of microwave radiation on psychophysiology and human health. The paper considers a method and a measuring and information system for dynamic monitoring of microwave environmental pollution in cities. It is assumed that a fully implemented system will consist of thousands of portable personal devices – mobile terminals held by volunteers who move around the city in their usual life and, possibly, a number of city vehicles will be equipped with such terminals. The structure and features of the technical implementation of the proposed distributed, non-deterministic system, which allows real-time mapping of the microwave dose load of the city population, are considered. The personal terminal records the geographic position using a GPS navigator as well as the basic parameters of microwave dosimetry. The data are transferred to a web server where they are stored in a cloud database. From there, they can be transferred to the environmental and medical services of the city, and they can also be read and displayed in a convenient form on any device with access to the corresponding Internet resource. The results of testing a prototype of the system when acquiring data on one of the routes in the city of Moscow are presented.

Keywords: Microwave pollution · Ecological monitoring · Microwave dosimetry · Measuring and information system

1 Introduction

Currently, a lot of experimental data has been collected and systematized, confirming the effect of microwave fields on biological objects, including psychophysiology and human health [1–3]. The maximum permissible exposure levels are defined in the national standards for electromagnetic safety [4–6]. However, there are significant contradictions in the analysis and interpretation of these data. Modern foreign safety standards are based on the thermal effects of absorption of microwave radiation in biological tissues [6].

This approach leads, in our opinion, to an artificial overestimation of safe absorption levels. In particular, in the American standard [5], the maximum incident microwave power density (PD) is in the range from 2 to 10 W/m^2 (depending on the parts of the human body), and the Specific Absorption Rate (SAR) should not exceed 0.08 W/kg

I. Tiginyanu et al. (Eds.): ICNBME 2021, IFMBE Proceedings 87, pp. 475–482, 2022.
https://doi.org/10.1007/978-3-030-92328-0_62

(for the body as a whole). With such irradiation parameters, a significant heating of tissues occurs, and heating by 1 °C already irreversibly damages the brain cells [7, 8]. In Russian sanitary regulations [6], similar parameters are almost an order of magnitude less. They are based on the genetic principle of hazard assessment. Dangerous are the levels of irradiation of the parent of the experimental animal, which lead to the death of its offspring. Then, these values are approximated for humans.

Currently, numerous studies show that even two orders of magnitude lower microwave radiation levels have significant destabilizing and often dangerous effects on humans [2, 9]. Moreover, the current standards practically do not consider the individual sensitivity of a person to microwave radiation. Individual hypersensitivity can lead to severe consequences and even damage to the main functional systems [10–12].

Another factor is the accumulation of the absorbed dose. In [13, 14] it is shown that the sensitivity of biological objects to microwave radiation increases if they were previously exposed to such radiation. And in [15], the phenomenon of dose accumulation was reported, in which biological effects appear as multiple nonlinear summation of the absorbed dose. The same effect can occur both with short, but powerful irradiation, and with weak, but prolonged or repeated exposure.

The problems and uncertainties that have arisen in this scientific area of a technical, biological and philosophical nature have been analyzed in reviews of the International Commission on Non-Ionizing Radiation Protection (ICNIRP) [4, 16]. However, so far, the conclusions of the reports are only advisory in nature. Work on the analysis and revision of modern standards is being carried out in Russia, in particular, in the Russian National Committee for Protection against Non-Ionizing Radiation [17].

It should be added that over the past two decades, the level of artificial microwave pollution and its territorial density have increased thousands of times in cities. However, the effects of microwave exposure on human communities, for example, residents of a separate metropolitan area, have not been practically studied. First of all, this refers to the urban ecological situation, the influence of microwave radiation on the statistical indicators of the health and diseases of the townspeople.

The International Agency for Research on Cancer of the World Health Organization in 2011 classified the electromagnetic field of mobile phones as a class 2B carcinogen.

The basis for this decision was, first of all, data on the growth of malignant tumors in cell phone users, especially in teenagers [18].

All this creates the preconditions for the development of systems for continuous operational monitoring of the electromagnetic situation in the city and, if necessary, for the formation of warning signals about a dangerous situation. It should be noted that the first attempts to create such control systems are already being undertaken in a number of cities around the world, for example, in Stockholm [19]. Typically, these systems consist of several stationary microwave monitors that serve small local areas.

The paper considers a method as well as a measuring and information system named GEO-DOSE for dynamic monitoring of microwave environmental pollution in cities. It presents the results of the next step of research, the preliminary results of which were discussed in [20]. It is assumed that a fully implemented system will consist of thousands of portable personal devices (mobile terminals named GEO-DEMI) held by volunteers who move around the city in their usual life and. A number of city vehicles may be equipped with such terminals.

A method for acquiring data on the electromagnetic environment in a city is proposed, based on the deployment of a dynamic distributed non-deterministic system in it, which allows real-time mapping of the microwave field levels and determining the dose load in a specific place and along routes of movement for everyone who has a personal mobile terminal.

The personal terminal records the geographic position using a GPS navigator as well as the basic parameters of microwave dosimetry. The data are transferred to a web server where they are stored in a cloud database. From there, they can be transferred to the environmental and medical services of the city for processing and visualizing the electromagnetic environment, assessing the level of microwave environmental hazard and generating alarms. These data can also be used by district health centers to statistically analyze the health effects of microwave radiation on community residents. They can also be read and displayed in a convenient form on any device with access to the corresponding Internet resource.

The results of testing a prototype of the system when acquiring data on one of the routes in the city of Moscow are presented.

2 System Prototype Implementation

The proposed system provides for the free behavior and movement of a person or vehicles across the territory with a mobile terminal for data acquisition. There are two options for implementing the system. The first one involves recording the measurement results into the memory card of the terminal with subsequent loading data into local computer and transferring them to a central server for processing and visualization. The second involves the use of telecommunication means of wireless data transmission from mobile terminals to a cloud web server for processing and storing data and transferring it to city medical centers and client applications on computers and smartphones of personal users of the system. Figure 1 illustrates this implementation of the system. Both of these options are combined in the GEO-DOSE system. Combining long-term data on the electromagnetic environment in an area with data on the distribution of diseases will allow medical professionals to consider the impact of microwave radiation on public health.

The characteristics of the GEO-DOSE system are largely determined by the technical characteristics and functional capabilities of the GEO-DEMI mobile terminal, in particular, its microwave dosimetry unit. A large number of portable microwave dosimetry instruments are produced. As a rule, they measure the power density of electromagnetic radiation, which is then converted into other dosimetry parameters.

The authors of this work proposed a new method for assessing dosimetry characteristics [21]. The two microwave sensors are separated by a material that mimics biological tissue. In the presence of radiation, due to absorption in this material, a difference signal appears, which serves as a measure of the absorbed dose. As a phantom of biological tissue, a special conductive rubber is used, the electrophysical properties of which are selected close to the average characteristics of human tissues. This principle is the

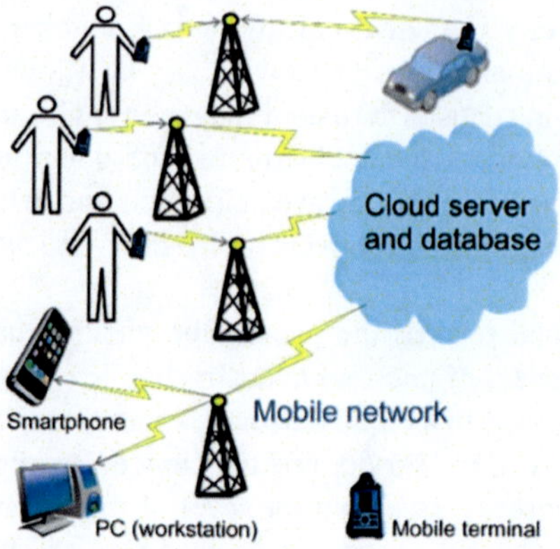

Fig. 1. The implementation of proposed system

basis for the operation of the dosimetry unit of the GEO-DEMI mobile terminal. The block-scheme of the device is shown in Fig. 2, and its photo is shown in Fig. 3.

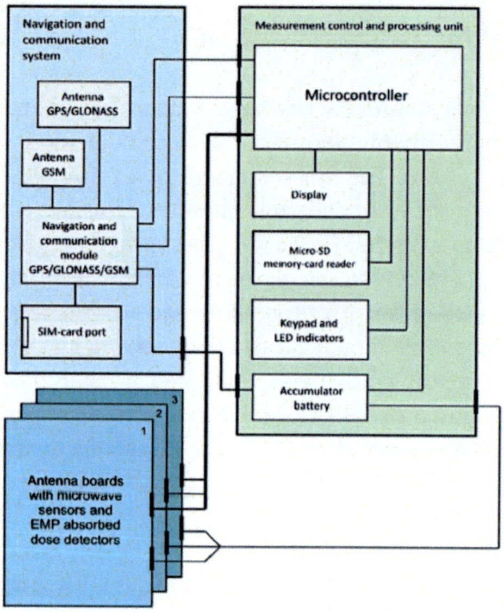

Fig. 2. Block-scheme of the personal mobile terminal

The software of the GEO-DOSE system is designed to transfer data from mobile terminals to a cloud server, process and store data in the cloud and transfer them upon request to client devices. Data transmission is carried out via a GSM mobile communication channel. The system software includes embedded mobile terminal software, cloud server application software, and client applications.

Fig. 3. Photo of personal mobile terminals

The microcontroller of the mobile terminal receives data from the detector and navigation units, generates HTTPPOST requests containing the registered data along with service information, and transmits them through the GSM module to the web server.

The client and server parts of the system software are located in the Heroku cloud server. It is a web service that is accessed via the HTTP protocol. The object-relational system PostgreSQL was chosen as the database management system. Client applications are used to obtain data by the system users. Through HTTPGET requests to the server applications receive the data required for visualization.

3 System Prototype Testing

Technical tests of the prototype of the GEO-DOSE system were carried out using two samples of GEO-DEMI personal mobile terminals. To measure the characteristics of the device, continuous, unmodulated electromagnetic radiation was used as a measuring signal with frequencies most often used for various types of communication (GSM, 3G, 4G, Wi-Fi) in the range from 0.8 GHz to 3 GHz. The main characteristics of the mobile terminal based on the test results are shown in Table 1.

Field tests of the mobile terminal prototype were carried out in the Southern District of Moscow. The results of these tests are illustrated in Fig. 4. It shows a fragment of a map with the route of movement of experimenters equipped with a mobile terminal (points on the map). This map in a time scale close to real is displayed on the display of the central server of the system, as well as on any device with access to the Internet if its owner has access rights to the system. The experimenters traveled from the entrance to NRNU MEPhI (test point 1) by car to the Kolomenskoye museum (test point 2), and then by metro to the Tsaritsyno museum (test point 3).

Clicking the computer mouse on a route point opens a window with data on the microwave environment at this point, as well as on the human radiation dose during the

Table 1. Main characteristics of the personal mobile terminal

Measured values	Ranges of values
Power density (PD)	10 μW/cm^2–1.6 mW/cm^2
Specific absorption rate (SAR)	10 nW/kg–1.3 W/kg
Dose of microwave exposure	10 nJ/kg–465 J/kg
Mean position measurement error	7.6 m
Frequency range of measured EM-fields	**0.8–3 GHz**

time spent in this location. The color of the point corresponds to the degree of radiation hazard. In the GEO-DOZE system the color indication is based on the Russian standard: green (safe) indicates PD values not exceeding 10 μW/cm^2; red (dangerous) indicates PD values exceeding 100 μW/cm^2; PD values from 10 μW/cm^2 to 100 μW/cm^2 are indicated in yellow (normal).

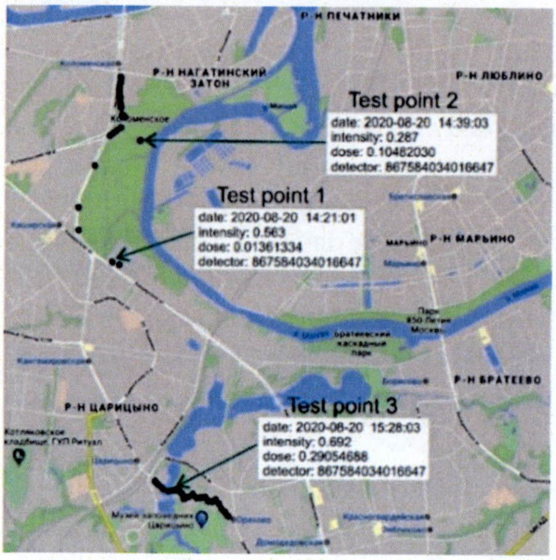

Fig. 4. Fragment of map with route information and measured data

4 Conclusions

The paper presents methods and a prototype of the system for monitoring of electromagnetic environmental pollution in large metropolitan areas and other regions and territories. The system can serve as a source of information that may be used for medicine and biological studies of microwave radiation effects on people's health.

Acknowledgment. The work was financially supported by the Russian Foundation for the Development of Small Innovative Enterprises in the Scientific and Technical Sphere.

Conflict of Interest. The authors declare that they have no conflict of interest.

References

1. Elwood, J.M.: Epidemiological studies of radio frequency exposures and human cancer. Bioelectromagnetics **24**, S63–S73 (2003)
2. Markov, M.S.: Electromagnetic Field in Biology and Medicine. CRC Press, Boca Raton (2015)
3. Lai, Y.F., Wang, H.Y., Peng, R.Y.: Establishment of injury models in studies of biological effects induced by microwave radiation. Mil. Med. Res. **8**, 1–18 (2021). https://doi.org/10.1186/s40779-021-00303-w
4. ICNIRP: Guidelines for limiting exposure to electromagnetic fields (100 kHz to 300 GHz). Health Phys. **118**, 483–524 (2020). https://doi.org/10.1097/HP.0000000000001210
5. IEEE C95.1–2019: IEEE Standard for Safety Levels with Respect to Human Exposure to Electric, Magnetic, and Electromagnetic Fields, 0 Hz to 300 GHz. IEEE, Inc., New York (2019). https://doi.org/10.1109/IEEESTD.2020.9238523
6. Russian Sanitary Regulations: SANPIN 2.1.8/2.2.4.1190-03. Hygienic Requirements for Placement and Operation of Land Mobile Satellite Radio Communication Facilities (2003). https://meganorm.ru/Index2/1/4294817/4294817554.htm
7. Foster, K.R.: Thermal mechanisms of interaction of radiofrequency energy with biological systems with relevance to exposure guidelines. Health Phys. **92**, 609–620 (2007)
8. Leitgeb, N., Danker-Hopfe, H., Auvinen, A., et al.: Potential health effects of exposure to electromagnetic fields (EMF). Sci. Comm. Emerg. Newly Identified Health Risks (SCENIHR) (2015). https://doi.org/10.2772/75635
9. Vecchia, P., Matthes, R., Ziegelberger, G., et al.: Exposure to high frequency electromagnetic fields, biological effects and health consequences (100 kHz–300 GHz). International Commission on Non-Ionizing Radiation Protection (2009). https://ocpm.qc.ca/sites/ocpm.qc.ca/files/pdf/P52/5z.pdf
10. Hallberg, O., Oberfeld, G.: Letter to the editor: will we all become electro sensitive? Electromagn. Biol. Med. **25**, 189–191 (2006). https://doi.org/10.1080/15368370600873377
11. Eltiti, S., Wallace, D., Zougkou, K., et al.: Development and evaluation of the electromagnetic hypersensitivity questionnaire. Bioelectromagnetics **28**, 137–151 (2007)
12. Balroop, M.: Electromagnetic Hypersensitivity – An issue of today's IoT (2018). https://www.justis.com/electromagnetic-hypersensitivity.
13. D'Andrea, J.A., DeWitt, J.R., Emmerson, R.Y., et al.: Intermittent exposure of rat to 2450 MHz microwaves at 2.5 mW/cm^2: behavioral and physiological effects. Bioelectromagnetics **7**, 315–328 (1986)
14. Schrot, J., Thomas, J.R., Banvard, R.A.: Modification of the repeated acquisition of response sequences in rats by low-level microwave exposure. Bioelectromagnetics **1**, 89–99 (1980)
15. Magras, I.N., Xenos, T.D.: RF radiation-induced changes in the prenatal development of mice. Bioelectromagnetics **18**, 455–461 (1997)
16. ICNIRP: General approach to protection against non-ionizing radiation. Health Phys. **82**, 540–546 (2002)
17. Grigoriev, O.A.: Actual issues of radiobiology and hygiene of non-ionizing radiation for new electromagnetic technologies, Proc Actual problems of radiobiology and hygiene of non-ionizing radiation, Moscow, Russia (2019). http://www.bioemf.ru/conf/conf2/grigoriev_oleg.pdf. (In Russian)
18. Non-ionizing radiation, Part 2: Radiofrequency electromagnetic fields: IARC monographs on the evaluation of carcinogenic risks to humans, vol. 102, Lyon, France (2011)

19. Hardell, L., Koppel, T., Carlberg, M., et al.: Radiofrequency radiation at Stockholm central railway station in Sweden and some medical aspects on public exposure to RF fields. Int. J. Oncol. **49**, 1315–1324 (2016). https://doi.org/10.3892/ijo.2016.3657
20. Simakov, A., Onishchenko, E., Gurkovskiy, B., et al.: Microwave dosimeters for domestic use. In: Proceedings of the IEEE Workshop on Environmental, Energy, and Structural Monitoring Systems, EESMS, Milan, Italy, pp. 1–4 (2017)
21. Gurkovskiy, B., Vodohlebov, I., Onishchenko, E., Simakov, A.: The personal microwave dosimeter. In: Proceedings of the lst URSI Atlantic Radio Science Conference (URSI-RASC), Las Palmas, Spain, p. 1 (2015). https://doi.org/10.1109/URSI-AT-RASC.2015.7303218

The Effects of Terahertz Radiation on the Development of Biological Organisms I: Wheat Seeds

Robin-Cristian Bucur-Portase(✉)

National Institute for Research and Development in Electrical Engineering, Splaiul Unirii 313, Bucharest, Romania
robin.bucur-portase@cnmv.ro

Abstract. The uses of Terahertz (THz) radiation are broad, ranging from scanning packages to identifying various chemical compounds. Its effects on living matter have been neglect. The scientific literature available to date describes various effects on living matter depending on the type of cells studied. No studies thus far have investigated the effects of THz radiation on dormant seeds, whose animal-like metabolism creates a unique opportunity to assess the full extent of THz radiation as well as further analyze the mechanism of action behind this type of non-ionizing radiation. The aim of this study is to assess the radiation's effects on the sprouting of wheat seeds. Previous studies have shown a positive effect on the growth and proliferation of plants when placed under the effects of THz radiation; however, due to the aforementioned unique metabolism of seeds' cells, the ongoing hypothesis is that the seeds will suffer under the effects of the THz radiation. The seeds have been placed into two distinct batches, depending on whether they were exposed dry or wet. Multiple time-frames were assessed to different sub-groups, having one control sub-group in each batch. The exposed seeds saw a significant decrease in the success of germination, albeit the fact that those that survived saw a significant increase in their growth rate without any subsequent exposure. More research is needed to assess the effects of THz radiation on seeds, but the technology seems promising for use in agriculture, especially seeing the daunting effects of climate change on crop yields.

Keywords: Terahertz · Plant metabolism · Plants · Seeds · EM radiation

1 Introduction

Terahertz radiation represents a special type of electromagnetic (EM) radiation whose frequency lies between those of microwaves and infrared waves, or ranging from 0.1 to 10 THz, where one THz equals 10^{12} Hz. It is a non-ionizing type of EM radiation [1].

The literature on THz radiation's mechanism of effect on biological organisms is limited, although various cellular changes have been observed, mostly due to the thermal effect of THz radiation on cellular machinery. These changes range from increased oxidative stress [2] to disruption of the properties of the cellular membrane, such as

I. Tiginyanu et al. (Eds.): ICNBME 2021, IFMBE Proceedings 87, pp. 483–488, 2022.
https://doi.org/10.1007/978-3-030-92328-0_63

changes in permeability [3]. ATP production is also known to be affected [2]. THz radiation seems to usually have a detrimental effect on the structure and metabolism of animal cells [4–7] (though notable exceptions apply [8, 9]), a positive effect on the growth of plant cells [10, 11], and mixed results on bacteria and fungi depending on the species used [8, 12, 13].

Seeing that the metabolism of seeds after their dormancy period ends more closely resembles that of animal cells, consuming oxygen for burning through their stores of starch until they are ready to photosynthesize [14], we predict that, if THz radiation acts solely on the seeds' cells metabolism, then it will have detrimental effects on their development post-exposure. However, if the results are mixed as to indicate both damage and proliferation, then the THz radiation does not act solely on the metabolism of the seeds' cells but rather on various other structures and mechanisms as well.

2 Materials and Methods

600 wheat seeds have been purchased from a local farmer's market. They were then placed in groups of 50 seeds each. In total, there were 12 groups that were categorized depending on whether the seeds were exposed dry or wet as well as the time spent subjected to THz radiation. The timeframes were as follows: 0 (control sub-group), 1, 5, 10, 15 and 30 min.

The seeds (10 at a time) were placed in a rectangular capsule (Fig. 1) made of polyethylene which has a minimal absorbance coefficient for THz radiation (Fig. 2).

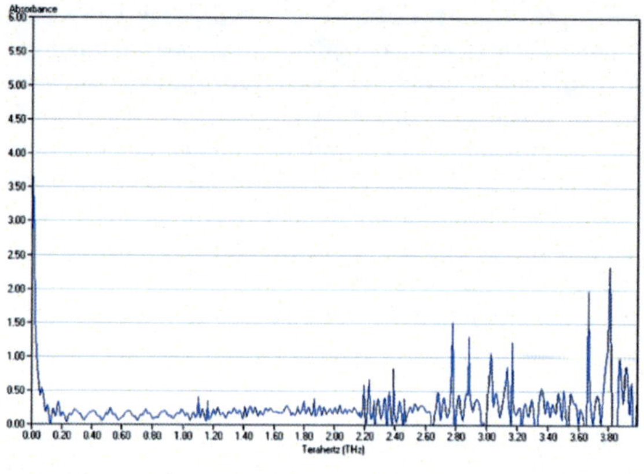

Fig. 1. Absorbance coefficient of polyethylene at various values of THz, peaking close to 0, 3.8 and 3.65 THz, with values of 3.5, 2.4 and 2 absorbance units respectively

The aforementioned capsule was placed in normal atmospheric conditions, at 22 °C, with ambient air as the medium between the device's output window and the analyzed sample. After the exposure, the seeds were left to germinate for seven days during which all the seeds' growth was stunted due to improper access to water.

The seeds suffered no subsequent exposure to THz radiation after the initial one.

Fig. 2. Rectangular capsule with polyethylene windows used as a temporary storage of the exposed sample

The spectrometer used, TeraView (TPS Spectra 3000), has an optical tunable Ti: Sapphire ultrashort pulsed laser with a spectral range of 0.1–3 THz. For the reported experiments, it was set to 150 mW, half its original power.

3 Results

The absorbance coefficient of dry wheat seeds (Fig. 3) is highest around 0.2 THz and 3.18 THz.

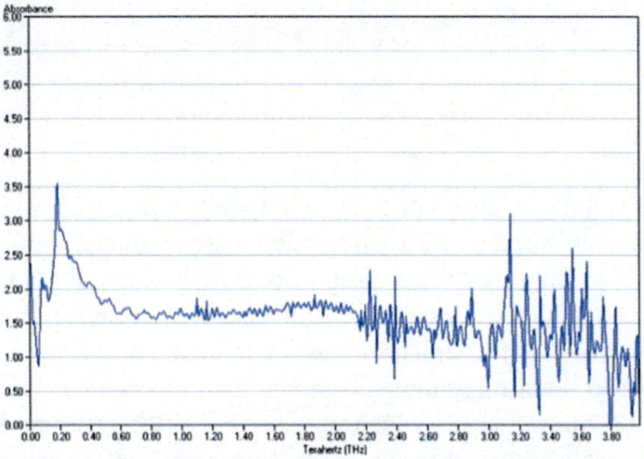

Fig. 3. Absorbance coefficient of dry wheat seeds at various values of THz, peaking at 0.2 and 3.18 THz, with values of 3.5 and 3.1 absorbance units respectively

The absorbance coefficient of wet wheat seeds (Fig. 4) peaks around 0.18 THz, 0.23 THz, and 3.49 THz. Significantly higher values have been reached by the wet seeds, suggesting that water had a significant role in shielding the seeds' cells' organic compounds from the THz radiation.

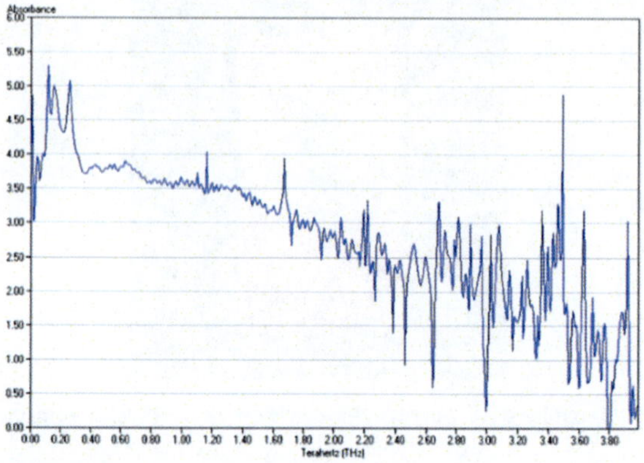

Fig. 4. Absorbance coefficient of wet wheat seeds at various values of THz, peaking at 0.18, 0.23 and 3.4 THz, with values of 5.28, 5.06 and 4.9 absorbance units respectively

Table 1. Influence of exposure time on the percentage of success of germination

Exposure min	Control sample (0 min.)	1 min.	5 min.	10 min.	15 min.	30 min.
Germination (%) Dry	76	19	17	16	19	31
Wet	52	52	44	38	26	26

4 Discussion

As seen in Table 1, the percentage of germinated seeds was highest in the control group out of both the dry and wet sets. This observation matches our hypothesis. However, the growth rate of the seedlings is directly proportional to the amount of time spent exposed to the THz radiation, with the dry batch's 30 min. sub-group having grown about 4 times faster than the control group (rough estimate due to their growth speed not being part of the initially intended set of measurements and monitored changes). This does not seem to be the case with the wet seeds which may be due to the water molecules shielding the biological molecules present in the seeds. Water molecules are known to have the highest absorbance coefficient for THz radiation of all molecules. The wet seeds also had a slower rate of germination than the dried batch.

It is worth noting that at the time of the exposure, the dry seeds were dormant with no cellular activity to speak of, at least according to the common scientific consensus on wheat seeds' suspended metabolism prior to any contact with water. The wet seeds, on the other hand, would have had minimal levels of cellular activity because water exposure would have led to the production of gibberellins by the seeds' embryo [14]. This class of compounds, besides having other effects as well, diffuse into the aleurone layer and stimulate the cells to synthesize amylase whose role is that of converting starch into maltose which can then be further broken down into glucose to be used as a source of energy [14]. After the release of gibberellins, seeds can start sprouting though without the use of any photosynthetic activity, relying solely on their reserves for energy and growth [14]. This is in many ways similar to the metabolism of an animal's cell.

An observation for which we could not find an explanation yet was the odd percentages of the number of germinated seeds, through which we cannot properly assess which length of time had the most detrimental effects on the seeds. We attribute these inconsistencies to the low number of seeds present in each group which made statistical improbabilities more pronounced.

5 Conclusions

Seeing the accelerated growth rate of the seeds, the author believes that exposure to Terahertz radiation could be used to enhance the speed at which seeds grow prior to their selling to farmers. Such a technique could bolster agriculture all while helping to tackle famine and the effects of climate change on crop yields. However, more research is needed to thoroughly assess the significance of THz radiation's effects on dry seeds as well as its side effects, especially those on the rate of success of germination.

Acknowledgments. The research has not been subject to any monetary funding from any institution. The materials and tools used for the purpose of this study have been lent by ICPE-CA Romania's National Institute of Research – Development for Electrical Engineering. The author also acknowledges and is grateful for the tutoring provided by biology teacher with a teaching degree of level 1 (top level) Claudia Negut, biologist Nicoleta Oana Nicula, Mihai Badic Ph.D. and Mircea Ignat Ph.D.

Conflict of Interest. The author declares that they have no conflict of interest.

References

1. Sun, Q., et al.: Recent advances in terahertz technology for biomedical applications. Quant. Imaging Med. Surg. **7**, 345 (2017)
2. Macouillard-Poulletier de Gannes, F., et al.: Mitochondrial impairment and recovery after heat shock treatment in a human microglial cell line. Neurochem. Int. **36**, 233–241 (2000)
3. Dewey, W.C., et al.: Cellular responses to combinations of hyperthermia and radiation. Radiology **123**, 463–474 (1977)
4. Olshevskaya, J.S., et al.: Effect of terahertz electromagnetic waves on neurons systems. In: Computational Technologies in Electrical and Electronics Engineering (2008)
5. Wilmink, G.J., et al.: Quantitative investigation of the bioeffects associated with terahertz radiation. In: Optical Interactions with Tissues and Cells XXI (2010)

6. Berns, M., et al.: Free electron laser irradiation at 200 microns affects DNA synthesis in living cells. In: Proceedings of the National Academy of Sciences of the United States of America (1990)
7. Zalyubovskaya, N.P., et al.: To biological activity of radiation in millimeter and submillimeter ranges. Eksperimental'noy i Klinicheskoy Radiologii **6**, 202–205 (1970)
8. Hadjiloucas, S., Chahal, M., Bowen, J.: Preliminary results on the non-thermal effects of 200–350 GHz radiation on the growth rate of S. cerevisiae cells in microcolonies. Phys. Med. Biol. **47**, 3831 (2002)
9. Clothier, R.H., Bourne, N.: Effects of THz exposure on human primary keratinocyte differentiation and viability. J. Biol. Phys. **29**, 179–185 (2003)
10. Xiong, S., Shaomin, P.: Influence of submillimeter laser radiation on the growth of paddy rice. Appl. Laser **6**, 33–37 (1986)
11. Xu, M., Xiong, S.: FIR laser irradiation in wheat. Appl. Infrared Optoelectron. **4**, 30 (1988)
12. Webb, S.J., Dodds, D.D.: Inhibition of bacterial cell growth by 136 GC microwaves. Nature **218**, 374–375 (1968)
13. Giles, J.P., et al.: Investigating the effects of terahertz radiation on bacillus subtilis. SPIE (2012)
14. Rosental, L., Nonogaki, H., Fait, A.: Activation and Regulation of Primary Metabolism During Seeds Germination. Cambridge University Press, Cambridge (2014)

Microbiological Decontamination of Air and Surfaces Due to Nanosecond Discharges

Iurie Bosneaga(✉), M. Bologa, and E. Agarwal

Institute of Applied Physics, Chisinau, Republic of Moldova

Abstract. Presented observational data indicate that a significant number of infections with the *SARS-CoV-2* coronavirus occur by air without direct contact with the source, in addition, in a tangibly long time interval. It is noticed that atmospheric precipitations help to cleanse the air from pollution and at the same time from viruses, reducing non-contact infections. These facts additionally actualize the problem of optimal microbiological decontamination of air and surfaces. In order to optimize microbiological sterilization, a thermodynamic approach is applied. It is shown that irreversible chemical oxidation reactions are the shortest way to achieve sterility, they being capable of providing one hundred percent reliability of decontamination. It is established that oxygen is optimal as an oxidant, including ecologically, because it and all of its reactive forms harmoniously fit into natural exchange cycles. The optimal way to obtain reactive oxygen species for disinfection is the use of low-temperature ("cold") plasma, which provides energy-efficient generation of oxidative reactive forms - atomic oxygen (O), ozone (O_3), hydroxyl radical ($\cdot OH$), hydrogen peroxide (H_2O_2), superoxide (O_2^-), singlet oxygen $O_2(a^1\Delta_g)$. Due to the short lifetime for most of the above forms outside the plasma applicator, remoted from the plasma generator objects should be sterilized with ozone (O_3), the minimum lifetime of which is quite long (several minutes). It is substantiated that microwave method of generating oxygen plasma is optimal for energy efficient ozone production. A modular principle of generation is proposed for varying the productivity of ozone generating units over a wide range. The module is developed on the basis of an adapted serial microwave oven, in which a non-self-sustaining microwave discharge is maintained due to ionizations produced by radionuclides-emitters.

Keywords: Coronavirus · Optimization of microbiological sterilization · Thermodynamic approach · Reactive oxygen species · Low-temperature plasma · Ozone · Microwaves

With the growth of the Earth's population, its accelerated urbanization combined with unprecedented mobility, the risk of the emergence and explosive spread of infections increases sharply. Anthropogenic environmental degradation also contributes to the continuous renewal (actualization) of microbiological threats. The economic and social impact of the *SARS-CoV-2* (*COVID-19*) coronavirus has surpassed all previous epidemics of dangerous diseases of the 21st century (*SARS, Ebola, swine and avian flu*). The lightning-fast (according to epidemiological criteria) transformation of a separate

I. Tiginyanu et al. (Eds.): ICNBME 2021, IFMBE Proceedings 87, pp. 489–504, 2022.
https://doi.org/10.1007/978-3-030-92328-0_64

focus into a terrifying pandemic was facilitated, in our opinion, not only by the underestimation of the virulence of the "fresh" coronavirus *SARS-CoV-2*, but also the interpretation of this coronavirus infection mainly as a "disease of dirty hands," or the result of direct contacts. Accordingly, the main tools for preventing the spread of infection are personal hygiene, avoiding direct contact and wearing a mask, as well as spraying liquid disinfectants.

Experience, however, has shown that these (very expensive) measures cannot stop the infection with the *SARS-CoV-2* coronavirus. The reason seems to be that a significant number of infections occur through the air, usually from an undetermined source, rather than through direct contact. This is confirmed by numerous data - as a rule, the sick cannot establish when or where they were infected, although they respected the rules of personal hygiene.

The *SARS-CoV-2* virus has a protective lipid layer (its capsid protein coat functions as a protection for the virus's genetic material - *RNA* - from mechanical, physical and chemical damage), which allows to survive better in difficult conditions, including low relative humidity. We believe that coronavirus particles in the airborne aerosol coming from the source of infection are electrostatically attached to dust particles and other colloidal particles contained in the air and (after probable dehydration and actual transformation into biopolymers) can be transferred in such a "canned" form for significant distance. Additionally, this version is confirmed by a decrease in the levels of contamination after atmospheric precipitations [1], which cleanse the air from fine impurities. Note that dehydrated (dried) viruses pose a particular danger in the context of their unhindered migration over long distances. The persistence of such dried viruses can be explained by their reduced sensitivity to daytime ultraviolet radiation: photoabsorption of ultraviolet radiation by dry viruses does not lead to the formation of oxidizing radicals (*OH* and *O*) derived from water - due to the absence of nearby water molecules. As a result, the effect of ultraviolet radiation on *RNA* and other components of the coronavirus is minuscule without oxidative fixing. This explains the unhindered spread of coronavirus infection in warm, dry and sunny weather - in India, Brazil, the United States in 2020, as well as in Russia - in the summer of 2021. The problem can be solved by cleaning the air from pollutions. For open spaces, this is possible with the help of atmospheric precipitation, and for closed spaces - applying adequate oxidation of contaminating agents (see below).

The *COVID-19* pandemic is the root cause of multifaceted socio-economic shocks, the largest global recession since the *Great Depression* of the last century. Taking into account the inevitable prospect of new epidemic threats emergence, the *optimization of microbiological decontamination* (especially of the air and surfaces), is becoming a priority civilizational task today.

1 Optimization of Microbiological Decontamination

It is crucial to propose and implement methods and means for microbiological decontamination that are optimal in terms of critical parameters - fast, irreversible, universal sterilization; unlimited resource; maximum energy efficiency; simplicity, availability, acceptable final cost, etc. Thermodynamic ("energy") approach is applicable and effective for microbiological security optimization. All processes near the earth's surface occur at constant (average) pressure and temperature. Therefore, it is the isobaric-isothermal potential, or *Gibbs free energy* $G = H - TS$ that characterizes the total potential energy of a thermodynamic system (an organism, or a simpler formation). The supply of free (i.e. high potential, high quality) energy from the environment is a prerequisite for life sustaining, the Sun being practically the only significant primary source of free energy on Earth.

In the absence of an influx of free energy from the outside, all systems inevitably spontaneously tend to a minimum of free energy G - due to the consumption of enthalpy H and increase of disorder S. Thus, the course of the isobaric-isothermal process is determined by two factors: enthalpy, associated with a change in the enthalpy of the system (ΔH), and entropy ($T\Delta S$), which reflects a change in the order of the system. The difference between these thermodynamic factors is a function of the state of the system. The dynamics of growth (or degradation) of an organism at a given temperature and pressure is characterized by a change in the accumulated free energy:

$$\Delta G = \Delta H - T\Delta S \tag{1}$$

An increase in free energy in the body (due to the assimilation of energy from the environment) is a prerequisite for its existence (growth and multiplication); accordingly, a decrease in free energy means degradation and, ultimately, the death of a living formation (see Fig. 1).

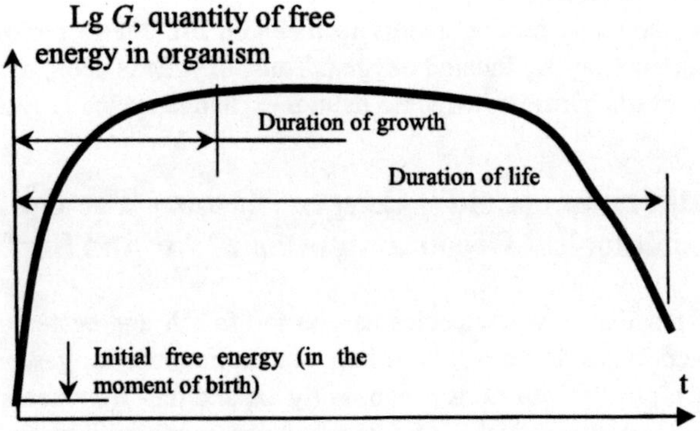

Fig. 1. Typical free energy variation in the organism during life

Thus, in order to achieve decontamination, one should strive to reduce the free energy of the living formation (up to a certain limit, when the guaranteed non-viability occurs).

For the realization of life-supporting reversible non-spontaneous (endothermic) reactions, the change in free energy must necessarily be positive ($\Delta G > 0$), which means the need of an external energy supply. An illustrative example of a non-spontaneous endothermic reaction that symbolizes life is the reaction of photosynthesis, for which the absorption of high-potential solar energy is a prerequisite for the accomplishment:

$$6CO_2 + 6H_2O \xrightarrow{h\nu} C_6H_{12}O_6 + 6O_2 \tag{2}$$

The reverse reaction (3) – of oxidation - on the contrary, is a spontaneous exothermic reaction and means the destruction of organic life. The release of low-potential energy *Q* ("dead" energy) makes this reaction irreversible (respectively, $\Delta G < 0$ for it):

$$C_6H_{12}O_6 + 6O_2 \rightarrow 6CO_2 + 6H_2O + Q \tag{3}$$

In such spontaneous exothermic oxidation reactions, a complete decrease in free energy ($\Delta G < 0$) occurs due to an increase in entropy (degree of disorder) ΔS and a decrease in enthalpy (heat content) ΔH - see formula (1).

Thus, *irreversible chemical oxidation reactions are the shortest way to achieve sterility. They are able to provide one hundred percent reliability of decontamination - if the free energy contained in virulent forms is sufficiently reduced ("burned") by oxidation and, ultimately, respective organic compounds conversed into the gas phase.*

The best oxidizing agent can be chosen using the data from Table 1, which show that the fluorine atom is the strongest of the naturally occurring oxidizing agents (fluorine is capable of oxidizing even oxygen - with the formation of oxygen difluoride OF_2). However, fluorine is so active that any manipulation with it is extremely dangerous (with respect to fluorine, only diamond and some types of glassy carbon are stable). In addition, the large-scale introduction of fluorine as an oxidizing agent would compromise the existing natural equilibrium. Therefore, atomic oxygen, the second element in terms of oxidative activity (see Table 1), is the most acceptable (optimal) oxidant. The choice of oxygen as the optimal oxidizer is logical and natural - oxygen is the most abundant (by weight) element of the earth's crust. Living organisms are about 70% by weight of oxygen, and they also receive energy to maintain life due to biological oxidation of organic accumulations by inhaled oxygen. Thus, oxygen is ecologically ideal as an oxidizing agent, it fits harmoniously into natural exchange cycles.

2 Non-equilibrium ("Cold") Oxygen Plasma - The Optimal Means for Microbiological Decontamination of Air and Surfaces

The source of reactive oxygen species can be the fourth aggregate state of matter - plasma generated in gas discharges. The range of applications for gaseous plasma generated by electro-physical methods is constantly expanding - from aerospace technologies to medicine and food industry. Low-temperature (non-equilibrium, "cold" plasma) *at atmospheric pressure* is of particular interest because of its energy efficiency and relative ease of widespread adoption.

Gas plasma is a high-energy quasi-neutral state of a gas (with a certain amount of ionization of components, as well as dissociation of molecules - with the formation of radicals of high chemical activity). For creation and maintaining of unstable plasma state of a gas, a continuous supply of free (high-potential) energy from the outside is required ($\Delta G > 0$) - in this, the condition for the existence of plasma coincides with the condition for the existence of life.

For the purposes of microbiological decontamination, only low-temperature ("cold") oxygen plasma is suitable - for reasons of need thermolabile surfaces disinfection, as well as high energy efficiency during plasma generation. Such a plasma is essentially

Table 1. Constants of atoms (molecules), characterizing the oxidation capacity

Molecule type	F_2	Cl_2	O_2	O_3	N_2
Bond (dissoc.) mol. energy, *eV* *(kJ/mol)* &respec.rad.λ (λ_{max} dissociation limit)	1,61 (155,0) (λ_{max} = 772,0 nm) $F_2 + h\nu \to F + F$	2,52 (242,6) (λ_{max} = 493,3 nm) $Cl_2 + h\nu \to Cl + Cl$	5,12 (493,6) (λ_{max} = 242,4 nm) $O_2 + h\nu \to O + O$	1,11 (107,0) (λ_{max} = 1117,3 nm) $O_3 + h\nu \to O_2 + O$	9,81 (945,3) (λ_{max} = 126,6 nm) $N_2 + h\nu \to N + N$
Electron affinity, *eV*	3,45 (for atom F)	3,61 (for atom Cl)	1,47 (for atom O) 0,44 (for mol.O_2)	2,26 (for mol. O_3)	−0,21 (for atom N)
Ionization energy (for atoms, molecules), *eV*	17,42 ($F + h\nu \to F^+ + e$) 15,70 ($F_2 + h\nu \to F_2^+ + e$)	12,96 ($Cl + h\nu \to Cl^+ + e$) 11,48 ($Cl_2 + h\nu \to Cl_2^+ + e$)	13,62 ($O + h\nu \to O^+ + e$) 12,08 ($O_2 + h\nu \to O_2^+ + e$)	12,52 ($O_3 + h\nu \to O_3^+ + e$)	14,53 ($N + h\nu \to N^+ + e$) 15,58 ($N_2 + h\nu \to N_2^+ + e$)
Electronegativity, atoms (by L.Pauling)	**3,98**	**3,16**	**3,50**	**3,50**	**3,05**
Covalent radius of atoms, *nm*	0,064	0,099	0,066	0,066	0,074

non-equilibrium - in it the average kinetic energy of electrons is much higher than that of gas molecules (respectively, $T_e \gg T_g$). It is impossible to achieve a non-equilibrium plasma by the thermal method - kinetic energy is effectively transferred only to massive molecules, and only then - to electrons. The problem of the selective transmission of kinetic energy to electrons can be solved using electric fields. Due to significant difference in masses (for example, $M_{O2}/M_e \sim 6 \times 10^4$) electrons are accelerated in the same electric field ~6×10^4 times more intensely than mono-charged ions.

In air gas discharge (when the ionization threshold is exceeded), collisional ionization by electron impacts occurs with the formation of molecular ions and electrons multiplication:

$$O_2 + e^- \to O_2^+ + 2e^-;\ \ N_2 + e^- \to N_2^+ + 2e^-; \tag{4}$$

The dynamic ionization balance is maintained by electron volume recombination, which at low temperatures occurs mainly in a dissociative way:

$$O_2^+ + e^- \to O + O;\ \ N_2^+ + e^- \to N + N; \tag{5}$$

Recombination of electrons with molecular ions occurs predominantly dissociatively - since in this case the participation of a third particle is not required.

There is also a mechanism of dissociative attachment of electrons to oxygen molecules, for which there is an energy threshold of ~4 eV (this estimate follows from the data given in Table 1):

$$O_2 + e^- \to O^- + O \tag{6}$$

The factor of predominantly dissociative recombination is crucial for obtaining the sterilizing qualities of plasma - as a result, atomic oxygen (O) is generated, which is a highly active oxidizing agent. Reverse associative reactions (reduction of O_2 and N_2 molecules) are possible, but their course - in an undesirable - reverse direction is hampered by the need of contact with a third particle that absorbs the energy of association:

$$O + O + M \rightarrow O_2 + M^*;\ N + N + M \rightarrow N_2 + M^* \tag{7}$$

Any molecule or atom of a gas mixture (including molecular oxygen O_2 and nitrogen N_2) can act as the third particle. Generated as a result of reactions (5–6) highly active - and, as a consequence, short-lived - atomic oxygen (O) is of the greatest interest for the implementation of disinfection. It enters into a three-particle reaction with molecular oxygen to form ozone O_3:

$$O + O_2 + M \rightarrow O_3 + M^* \tag{8}$$

The reaction takes place with the obligatory participation of the third particle, which perceives the relatively low binding energy of the oxygen atom in the ozone molecule (Table 1: $E = 1.11$ eV, or 107.0 kJ/mol). Such a low binding energy predetermines the relative instability of ozone molecule, which grows with the increase of temperature. On the other hand, this low separation energy of the O atom from the O_3 molecule, coupled with the high electron affinity for O_3 (Table 1: $E = 2.26$ eV, or 217.9 kJ/mol), results in the high oxidizing capacity of ozone, which is required for decontamination. The lifetime of ozone under favorable conditions is measured in tens of hours - due to this, the decrease of its concentration at night in the stratosphere protective ozone layer is insignificant, which provides conditions for life on Earth (by absorbing harmful daytime solar ultraviolet radiation by ozone in the range $\lambda = 200$–310 nm). Sufficient stability of gaseous ozone should be considered as an invaluable gift of nature, guaranteeing organic life, as well as allowing accumulating and delivering the chemical oxidation energy necessary for decontamination to the right place. (However, unfortunately, the issue of long-term safe storage (accumulation) of highly concentrated ozone does not lend itself to solution - even for the case of a liquid or solid (cryogenic) state of ozone - because of the danger of a spontaneous avalanche-like heat release according to the reaction $2O_3 \rightarrow 3O_2$, with an explosive release of $\Delta H = -285$ kJ/mol).

The optimal composition of the plasma-forming gas should contain molecular oxygen, dissociation of which in the discharge provides, ultimately, the maximum oxidation of microbiological forms.

Equations (4–5) also include molecular nitrogen (N_2), which implies the use of air as the most convenient (cheap) plasma-forming agent. The highly active atomic nitrogen N formed according to Eq. (5) enters into the reactions of formation of nitrogen oxides NO_X (mainly monoxide NO and dioxide NO_2). Both nitrogen oxides - directly and indirectly - contribute to a decrease in the production of ozone in the discharge:

$$O_3 + NO \rightarrow O_2 + NO_2;\quad O + NO_2 \rightarrow O_2 + NO \tag{9}$$

Other reactions of violation of ozone formation in the discharge plasma with the participation of nitrogen oxides are also possible, for example:

$$NO_2 + O_3 \rightarrow NO_3 + O_2;\quad O + NO_3 \rightarrow O_2 + NO_2 \tag{10}$$

By the sum of the negative effects from the presence of nitrogen in the plasma-forming gas, it follows that nitrogen should be excluded from the gas mixture, despite the increase in the cost of the process due to the need for preliminary separation of oxygen. (At the same time, it should be noted the positive dynamics of improvement and widespread introduction of affordable oxygen concentrators, up to the household level).

Note that even in the absence of reducing ozone formation nitrogen, simultaneously with the target reaction of ozone synthesis (8), an undesirable reaction $O + O + M \rightarrow O_2 + M^*$ - takes place in the plasma, with the obligatory participation of the third particle M, which absorbs significant association energy (Table 1: 5.12 eV, or 493.6 kJ/mol) and turns it into useless heat. However, fortunately, given a high concentration of molecular oxygen in the plasma, atomic oxygen is much more likely to form ozone than to recombine, and the energy efficiency of the process remains high. Let us consider a variant of humidification (saturation) of oxygen supplied to the plasmatron with water vapor - in order to obtain at the outlet, along with ozone, hydrogen peroxide (H_2O_2). As a result of impact ionization by electrons (the ionization energy of the H_2O molecule is 12.61 eV, or 1217.1 kJ/mol), molecular ions of water arise:

$$H_2O + e^- \rightarrow H_2O^+ + 2e^- \quad (11)$$

The competing process of recombination of electrons with water ions occurs, for the most part, through their dissociation, because the probability of collision with a suitable third body precisely during the recombination process is small:

$$H_2O^+ + e^- \rightarrow H + OH \quad (12)$$

The energy expended for this dissociation ($\Delta H = 498.7$ kJ/mol) is close to that for the oxygen ion O_2^+ ($\Delta H = 493.6$ kJ/mol in accordance with Table 1); the rest - kinetic energy - is carried away by the radicals formed as a result of dissociation. Highly reactive OH radicals enter into a three-particle reaction, forming hydrogen peroxide (H_2O_2):

$$OH + OH + M \rightarrow H_2O_2 + M^* \quad (13)$$

The kinetic energy of association carried away by the particle M ($\Delta H = 214.2$ kJ/mol) significantly exceeds that in the case of ozone generation ($\Delta H = 107.1$ kJ/mol, in accordance with Table 1), which indicates higher losses for useless gas heating. The more durable - in comparison with the ozone O_3 molecule - the hydrogen peroxide H_2O_2 molecule enters into oxidation reactions more difficult, requiring higher activation energy. Accordingly, there is better efficiency of ozone application for disinfection of air and surfaces in comparison with hydrogen peroxide. Nevertheless, hydrogen peroxide has its own areas of priority application - this is, first of all, the reduction of surface contaminations. At the same time, the effectiveness of gas disinfectants depends on the concentration of the infection (the degree of surface contamination). Their preliminary cleaning before gas disinfection with hydrogen peroxide (with an appropriate aqueous solution) for reduction of the viral load is the optimal, environmentally friendly option. The residues of hydrogen peroxide H_2O_2, and ozone O_3, - unlike other chemical reagents - do not leave any traces, except water H_2O, oxygen O_2, and heat release, in accordance with the reactions:

$$2O_3 \rightarrow 3O_2 (\Delta H = -285\,\text{kJ/mol});$$

$$2H_2O_2 \rightarrow 2H_2O + O_2(\Delta H = -196\,\text{kJ/mol}) \tag{14}$$

An important advantage of hydrogen peroxide is the possibility of its relatively simple (in comparison with ozone) accumulation and long-term storage due to the greater stability of the H_2O_2 molecule. For the production of hydrogen peroxide, methods other than gas discharge plasma are optimal and are currently used. The conclusion follows from the above data: non-equilibrium ("cold") plasma generating ozone in pure oxygen is an optimal microbiological energy-saving means for decontamination of air environment and extended surfaces.

3 Optimal Methods and Devices for Generation and Application of Non-equilibrium Oxygen Plasma

Essentially non-equilibrium plasma (with $T_e \gg T_g$) can be generated in sufficiently short (nanosecond) discharges. It is clear *a priori* that in order to generate a non-equilibrium plasma, the gas should be exposed to an alternating electric field of high (maximal) frequency and intensity (naturally, exceeding the breakdown strength). The development time of an electron avalanche in gas discharges is ~10^{-7}–10^{-3} s. Therefore, the frequency of the microwave field f ~ 2.45 GHz (respectively, with period T ~ 0.4 x10^{-9} s), used for industrial applications, is optimal in terms of a number of features (significant plasma non-equilibrium, availability of generation sources, relative ease of breakdown realization and of prevention discharge contraction).

Plasma state is characterized by the fact that a significant part of the gas molecules (atoms) is ionized. From the Table 1 follows, that ionization energies of molecules (atoms) significantly exceed the dissociation energies of molecules. As a result of dissociative recombination, in plasma of molecular gases are necessarily present chemically active radicals (neutral, excited, and ionized) required for decontamination.

The mean free path of electrons in air at atmospheric pressure is fractions of a micrometer. The minimum energy required for dissociation of the target oxygen molecule (O_2) is 5.12 eV, and for its ionization - 12.08 eV (Table 1). Hence it follows that the minimum electric field strength E required for plasma generation in air at atmospheric pressure is several kilovolts per centimeter (E ~ N × kV/cm), i.e. it is necessary to realize a *high-voltage discharge*. However, when trying to create a high-voltage discharge at atmospheric pressure of a significant duration (comparable to the development time of an electron avalanche in gas discharges) between flat conducting electrodes, one inevitably has to deal with discharge contraction, i.e. with its rapid transition into a spark, and then into a destructive arc discharge. Overcoming contraction is possible with the help of a corona discharge (in which the current is limited by a small area of corona electrodes), or barrier discharge (in which stabilization is achieved by installing dielectrics on the electrodes, with an alternating voltage applied to them). It is advisable to use the listed types of discharges as independent ozone generators of *relatively low productivity* (due to their low energy-efficiency and high cost).

In our opinion, the optimal solution to the problem of ozone generation is the use of an electrodeless microwave discharge [2]. The absence of electrodes means not only the convenience in the design of the applicator, but also the exclusion of heat losses on the electrodes, as well as their possible sputtering. In a microwave discharge, the coefficient of conversion of electrical energy into a plasma state reaches large values (over 80%), unattainable for other types of discharge. This is due to the fact that thousands of times more massive positive ions practically do not have time to gain significant kinetic energy in a short half-period of oscillations (less than ~10^{-9} s), and all the kinetic energy goes to incomparably lighter electrons. Fast, but light electrons poorly transfer energy to massive particles, which leads to a significant non-equilibrium of the microwave plasma with $T_e \gg T_g$. The risk of contraction is reduced due to the volumetric nature of the microwave discharge and the short oscillation period ($T \sim 10^{-9}$ s - small compared to the time of development of an electron avalanche: ~10^{-7}–10^{-3} s). The maximum density and, accordingly, the reactivity of the microwave plasma is also unavailable for other types of discharge. The relatively high density of the microwave discharge plasma (correspondingly, the increased concentration of free electrons) ensures efficient absorption of microwave energy [3]. The listed advantages of non-equilibrium microwave plasma are especially important for decontamination of large volumes of air and extended surfaces.

The highest reactivity is observed directly in the microwave plasma torch and depends, *inter alia*, on the composition of the working mixture. Naturally, the concentration of reactive particles sharply decreases with distance from the torch. Therefore, the entire reaction potential of microwave plasma can be fully realized only with respect to relatively small surfaces in direct contact with the plasma.

The choice of the composition of the working mixture is determined by the goals that are set during plasma exposure. For example, the presence of monoatomic helium (He) in plasma-forming gas makes it possible to count on the generation of ultraviolet radiation of the highest severity (since the energy of the first ionization for inert He atom - the largest of all atoms - is 24.59 eV). The presence of water vapor (H_2O) in oxygen ensures the generation of all oxidative reactive oxygen species - atomic oxygen (O), ozone (O_3), hydroxyl radical (OH), hydrogen peroxide (H_2O_2), superoxide (ion of oxygen molecule with unpaired electron O_2^-), singlet oxygen - an excited state of oxygen $O_2(a^1\Delta_g)$, with a half-life under normal conditions of 72 min). Due to the generation of the listed reactive oxygen species, non-equilibrium ("cold") plasma is the optimal means for local microbiological decontamination. Note that there is no information that any microorganisms can withstand the destructive effects of oxidative low-temperature plasma.

In fairness, it should be noted that the "patent" for the use of optimal reactive oxygen species belongs to nature: in support of this, it is shown in [4] that, if necessary, even such strong oxidants as ozone (O_3) and hydrogen trioxide (H_2O_3) are synthesized in the human body - from singlet oxygen and water.

To implement the decontamination of large volumes of air and extended remote surfaces, the optimal, in our opinion, is the use of ozone generated *in situ* in microwave oxygen plasma. An additional argument for this decision is the widespread, close to ubiquitous, prevalence of microwave heating technology in everyday life and at work, and the use of generators of sufficient power (in general, many millions of resonator-type microwave ovens have been produced and are in operation in the world). To provide an additional option (function) - a plasma generator of reactive oxygen species, including ozone - only relatively simple modifications are required, while maintaining the basic purpose of a microwave oven. It is preferable - but not necessary - to use a microwave oven with a dissector that provides the best uniformity of microwave field distribution throughout the volume, as well as with a variable frequency (*variable frequency microwave oven)*. This is desirable because in resonators (even in multimode ones) electromagnetic field of standing waves is inhomogeneous - neighboring nodes and antinodes are located at a distance of quarter wavelength Λ.

To exclude the formation of toxic nitrogen oxides (NO_X) in the plasma, it is advisable to purify the air from nitrogen as much as possible, or to use medical oxygen.

Initiation of a microwave discharge in the microwave oven is possible with the help of various types of ignitors - flames of various origins (gas flame, gas lighter, spirit lamp, etc.), auxiliary electrical discharges (for example, corona), as well as using passive electric field concentrators of a special design. After ignition, such a discharge can function as an independent one (Fig. 2).

In order to ensure the most reliable ignition, as well as to guarantee the stable operation of the generator of reactive oxygen species at variable load, it is necessary to generate in the microwave oven a non-self-sustaining microwave discharge (i.e., which functions with the use of an additional ionizer), continuously supported by specially selected radionuclides-emitters. Additional advantage of this solution is the maximum energy efficiency of generation. Radionuclides used for generation of cold (non-thermal) plasma must have a number of properties: the half-life ($t_{1/2}$) of the emitter must be long enough to ensure the release of energy for ionization at a relatively constant rate, over a reasonable period of time (during several months - at least). In addition, the specific power (measured in *W/g*) is important - the amount of energy released (and spent on ionization) per unit time by unit mass of a radionuclide. The power density (specific power) is inversely proportional to the half-life ($t_{1/2}$) and is directly proportional to the energy released during each emission of a given radionuclide.

Table 2 lists the most suitable radionuclides for plasma generation. Based on the sum of the above characteristics, the polonium isotope *Po210* should be recognized as the preferred radionuclide for generating cold plasma. Particularly attractive is its extremely high specific power (Table 2: P_s ~ 140 W/g), which guarantees the generation of a sufficiently concentrated (dense) plasma. The application of a thin (if possible - monomolecular) layer of *Po210* as an emitting surface is a prerequisite for ensuring

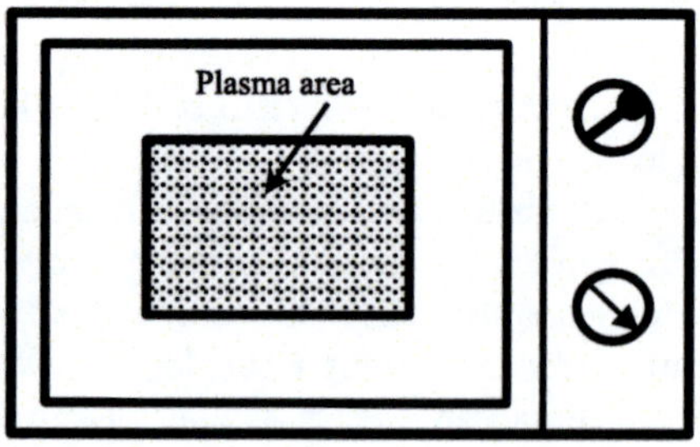

Fig. 2. Plasma area configuration (foreground of the microwave oven, converted into plasma generator)

high energy efficiency by reducing useless volumetric heat release. (This is due to the fact that the specific power of *Po210* is extremely high, and the compact *Po210* capsule weighing only 0.5 g under normal conditions reaches temperatures above 500 °C - due to the kinetic energy of α-particles absorbed in the volume of the radionuclide). At normal atmospheric pressure in air, the alpha particles emitted by *Po210* completely lose their energy (~5.49 meV) on a path of ~4.14 cm (as established by *William H. Bragg*). This characteristic α-particle path length is important for optimizing the design of the respective applicators (see caption of the Fig. 3).

Table 2. The most expectant radionuclides for cold plasma generation

Radio-nuclide	Half-life ($t_{1/2}$)	Power density	Emitted energy structure	Eventual shielding	Provenience
Po210	138,376 *days*	140 W/g	Practically pure α at 5,49 meV (non-significant γ or *X*-ray)	Low	Sub-terrain (incl. CH_4 extraction)
Pu238	87,6 *years*	0,568 W/g	Practically pure α (low γ and n levels)	Low	SNF re-processing
Am241	432,2 *years*	0,114 W/g	α-decay and γ	Medium	SNF re-processing
Sr90	28,8 *years*	0,46 W/g	Practically pure β, but -second. bremsstrahlung	High	SNF re-processing
Cs137	30,17 *years*	~0,8 W/g	β-decay and γ (not pure β)	Very high	SNF re-processing

On the basis of α-radiation from *Po210*, it is easy to obtain plasma of the desired density - by simple increasing the area of the emitting surfaces. The cost (availability) of radionuclides for generating cold plasma is a decisive parameter for their wide, economically justified application. *Po210* is usually produced artificially - by bombarding the practically stable ($t_{1/2}$ ~ 1.9×10^{19} *years*) bismuth isotope *Bi209* with neutrons in nuclear reactors:

$$_{83}Bi^{209} + {_0n^1} = {_{83}Bi^{210}} \tag{15}$$

Then radioactive *Bi210*, as a result of spontaneous β-decay (with a half-life of only $t_{1/2}$ ~ 5 *days*), turns into target *Po210*:

$$_{83}Bi^{210} = {_{84}Po^{210}} + e^- \tag{16}$$

This method of obtaining and practical use of *Po210* for plasma generation has already been implemented, but, apparently, it can be surpassed. At the research stage,

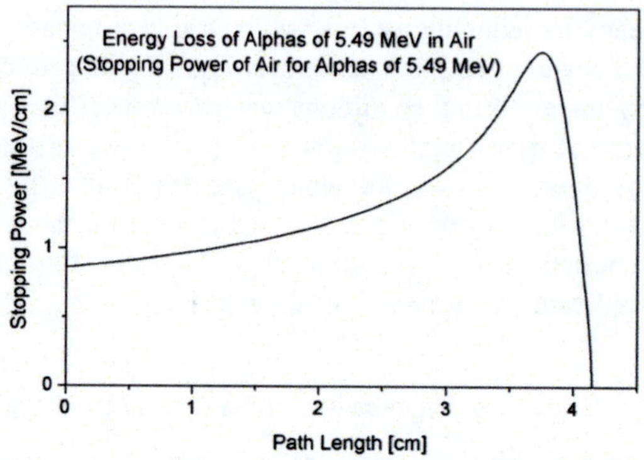

Fig. 3. Bragg's curve for alpha-particles (of 5,49 meV), which shows energy transfer in air (at normal atmospheric pressure). Stopping power experiences the "*Bragg peak*" before the full stop (at path length ~4,14 cm). The peak occurs because the interaction cross section increases as the charged particle's energy decreases. Energy lost by charged particles is inversely proportional to the square of their velocity, which explains the peak occurring just before the particle comes to a complete stop

there is a competing method based on the possibility of utilizing the radioactive isotope *Rn222* that comes naturally from the bowels, with its subsequent decay to *Po210* (this topic, which also has relevant environmental aspects, is beyond the scope of this paper). The use of radionuclides to generate cold plasma is an example of efficient direct conversion of nuclear energy into high-potential energy of a non-equilibrium plasma.

To vary the productivity of installations for generating microwave plasma (ozone) within a wide range, the modular principle of generation is convenient (with the increase, if required, the number of microwave generators in parallel). In Fig. 2 plasma area is limited by a rectangular applicator. However, if necessary, the entire useful volume of the oven can be used to generate plasma, including for sterilization of thermolabile dielectric materials in plasma, along with the simultaneous production of (relatively) stable ozone at the outlet - for its use on site or subsequent transportation to other places of consumption. The effectiveness of a microwave oven as a plasma sterilizer can be demonstrated by the example of disinfecting masks and respirators. Synthetic filter materials are thermolabile, plus detergents and disinfectants enter into chemical reactions with polypropylene, which damages filters and leads to the passage of hazardous fine aerosols. The same effect is observed when using boiling, hard washing, detergents and disinfectants. Therefore, the use of low-temperature microwave plasma for sterilizing such materials is the best option to achieve 100% sterility in a minimum time. (Just remember that the metal nose clip cannot be placed in a microwave oven).

For the generation of the non-equilibrium (non-thermal) plasma it is not obligatory to use *harmonic* high frequency oscillations. The same effect can be obtained as a result of application of the short (nanosecond) high voltage impulses.

Periodic short impulses could be presented as the sum of simple oscillating functions, namely *sines* and *cosines* (i.e., decomposed in *Fourier series*) - of "microwave" range of frequencies. In this case plasma parameters can be controlled through the amplitude, duration and duty ratio of impulses. The main differences come from the hardware design, which appears to be more complex in the case of microwaves (see Fig. 4).

At first glance, the "impulse" option is simpler.

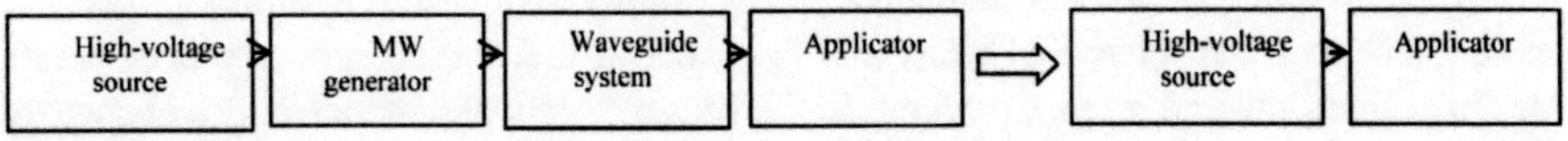

Fig. 4. The scheme of microwave plasma installation as compared with high voltage impulses variant

However, today the massively perfected "microwave" version is superior to the "pulsed" one - because of the complexity and high cost of powerful pulsed high-voltage sources (which does not exclude a change in the situation in the future).

Ozone, as the optimal oxidizing agent for decontamination, must certainly be used indoors - *in the absence of people* - in concentrations and for durations that guarantee 100% decontamination. In these cases, restrictions on the concentration and duration of ozone use are usually associated with the damageability of materials sensitive to oxidation (the permissible mole fraction should be determined experimentally, as the first approximation it is about 25–30 ppm). The same harsh regimes are applicable when disinfecting masks, clothing and other potentially infected objects. Repeated, safe and, if necessary, impersonal use of expensive components has an important economic effect, which is easily achieved due to ozone oxidation. Microorganisms do not have protection against ozone oxidation. Unlike all living creatures unarmed against ozone, *homo sapiens* can - with the proper desire - create such protection. It is thanks to abstract thinking - a person is able to provide the increase in free energy ΔG necessary to maintain life (which in this particular case means the creation of effective technologies of combating a pandemic).

To prevent the harmful effects of ozone, its strong side - high reactivity - should be used.

Unlike the classic gas mask, it is proposed separate (comprehensive) protection for ozone-sensitive organs - eyes, nose and mouth.

To protect the eyes, it is sufficient to use insulating glasses or special contact lenses.

A protective respirator mask should undergo the greatest degree of modernization. The full face protective respirator mask equipped with carbon filters is in most cases functionally redundant and, accordingly, too expensive for mass use. Following the comprehensive principle of mass protection, it is proposed to provide the possibility of independent (separate) protection for the nose and mouth - for less vulnerable intake of drinks, food, etc. Further, the respirator mask for breathing through the nose and mouth must be aggregate, i.e. consist of several removable mask layers. An outer layer-mask containing effective neutralizers of ozone (and, in combination, other toxic gases) based on activated carbons, should be worn when entering areas with a high ozone content. In a critical epidemiological situation, such zones can be not only specialized *COVID* centers, but also retail spaces, transport, etc. Outside the ozone rich zones, it is advisable not to use the ozone neutralizing mask and to maintain its consumption potential with the help of a sealed package.

Wearing a mask causes not only an increase in breathing resistance, but also deterioration in the quality of the inhaled air (a certain part of the used air with an increased content of carbon dioxide CO_2 remains in the intra-mask space during exhalation). In addition, moisture condensation may occur on the mask at low temperatures. The fundamental solution to these issues works optimally at low temperatures (since it is accompanied by the release of heat) and is based on the unique property of superoxides to accumulate and release oxygen. In the presence of water, superoxides absorb the exhaled carbon dioxide CO_2 and liberate oxygen O_2 abundantly. Using the example of potassium superoxide KO_2, the total reaction of replacing carbon dioxide CO_2 with a larger number of oxygen molecules O_2 has the form:

$$4KO_2 + 2CO_2 + H_2O \rightarrow H_2O + 2K_2CO_3 + 3O_2 \quad (17)$$

The introduction of this method of enriching the inhaled air with oxygen as applied to mask respirators is facilitated by the fact that this principle of converting carbon dioxide into oxygen is already used in orbital stations and submarines.

The improvement of masks (respirators) is the subject of increased attention of researchers - largely due to the *SARS-CoV-2* coronavirus pandemic. The development and implementation of an intelligent mask [5] - a kind of *i-mask* - has been launched to determine the presence of *SARS-CoV-2* coronavirus in the host's body. The functions of the smart mask will certainly be expanded. For example, there is an acute issue of prompt express diagnostics of the presence and concentration of coronavirus in the air - both in critical rooms and in free spaces.

The astronomical number of produced masks requires adequate disposal. In our opinion, the optimal way to dispose of spent masks is their high-temperature pyrolysis, with further utilization of activated carbon (including for the manufacture of new masks), as well as residual hydrocarbons (see Fig. 5). High temperatures guarantee the microbiological safety of pyrolysis products - regardless of the nature and degree of their contamination.

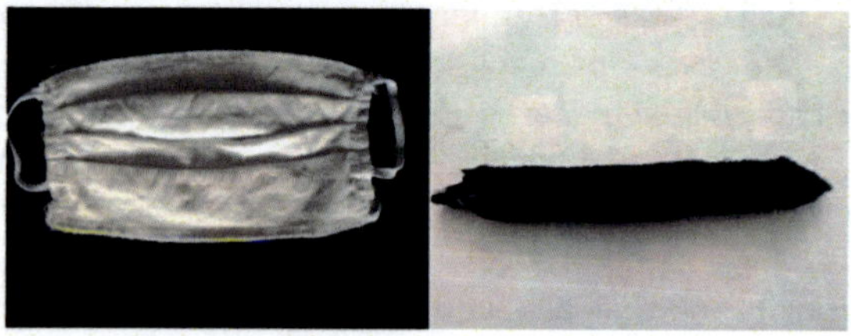

Fig. 5. On the left - a used cotton mask; on the right - the same mask pyrolyzed at t ~ 900 °C in a cylindrical applicator

The coronavirus pandemic has highlighted the importance of modernizing indoor air conditioning systems. To ensure microbiological safety, widely used centralized air conditioning systems must be supplemented with means of reliable decontamination - both for fresh supply air and for polluted exhaust air discharged into the atmosphere. At the same time, it should be noted that centralized air conditioning systems are inertial and are not able to respond quickly to sudden changes in weather, contingent and viral load in specific critical areas. Therefore, such places (operating rooms, hospital reception rooms, waiting rooms, etc.) should have a maneuverable local air conditioning system combined with controlled decontamination. In the optimal variant, in our opinion, at large facilities, only the base load should be centrally provided, with the obligatory possibility of correcting the conditioning by local means.

Obviously, in local air conditioning systems, it is easier to take into account the viral load (along with other local factors) and, if necessary, vary the air conditioning parameters and the intensity of disinfection. But the term "local system" does not mean that its closure is permissible (*split-systems* are practically closed, without significant air renewal). The inherent disadvantage of *split-systems* is recirculation - the lack of a proper inflow of fresh outside air (as well as the removal of polluted air). Local *split* air conditioning systems - even those equipped with filters and cold plasma - are categorically contraindicated in institutions with a large contingent (hospitals, schools, libraries, etc.). For such facilities, energy-optimized air conditioning systems are required with an adequately high air exchange rate and controlled decontamination of the supply and exhaust air. The creation of such systems is, without exaggeration, one of the most vital challenges, closely related to energy and ecology and additionally stimulated by the intensification of microbiological threats. Note that - within the framework of the formulated complex problem - the problem (component) of disinfection can be solved, as noted above, by using low-temperature ("cold") plasma for energy-efficient generation of reactive oxygen species.

4 Conclusions

Optimizing microbiological decontamination, taking into account the inevitable prospect of new epidemic threats, appears to be a priority task. Irreversible chemical oxidation reactions are the shortest way to achieve sterility with 100% decontamination reliability. At the same time, oxygen is optimal as an oxidizing agent, including ecologically, since it and all of its reactive forms harmoniously fit into natural exchange cycles. The optimal way to obtain reactive oxygen species for disinfection is the use of low-temperature ("cold") plasma, which provides energy-efficient generation of oxidative reactive forms - atomic oxygen (O), ozone (O_3), hydroxyl radical ($\cdot OH$), hydrogen peroxide (H_2O_2), superoxide (O_2^-), singlet oxygen $O_2(a^1\Delta_g)$. Due to the short lifetime of most of the above forms outside the plasma applicator, objects remoted from the plasma generator should be sterilized with ozone (O_3) - its minimum lifetime is long enough (measured in minutes). The microwave method of generating oxygen plasma is optimal for energy efficient ozone production. It should be emphasized that no microorganisms can withstand the destructive effects of ozone and oxidizing low-temperature plasma. To exclude any damage to the human body when using ozone, the issue of finalizing and certifying a modern multifunctional respirator mask is relevant. Energy efficient air conditioning systems with adequate decontamination of aspirated and used air are extremely relevant.

Conflict of Interest. The authors declare that they have no conflict of interest.

References

1. Menebo, M.M.: Temperature and precipitation associate with Covid-19 new daily cases: a correlation study between weather and Covid-19 pandemic in Oslo, Norway. Sci. Total Envir. **737** (2020). https://doi.org/10.1016/j.scitotenv.2020.139659
2. Azharonok, V., Filatova, I., Bosneaga, I., Bologa, M., Shedikova, O.: Non-thermal plasma sterilization in RF and MW discharges. Rom. J. Phys. **56**(Suppl.), 62–68 (2011)
3. Bosneaga, Iu., Bologa, M., Agarwal, E.: Intensification of electro-magneto-hydrodynamic effects using radionuclides. In: Proceedings of 12th International Conference on Modern Problems of Electrophysics and Electrohydrodynamics (MPEE 2019), Peterhof, Russia, pp. 189–193, 24 June 2019. ISBN 978-5-4386-1740-2
4. Babior, B.M., Takeuchi, C., Ruedi, J., Gutierrez, A., Wentworth, P.: Investigating antibody-catalyzed ozone generation by human neutrophils. Proc. Natl. Acad. Sci. U.S.A. (PNAS) **100**(6), 3031–3034 (2003). https://doi.org/10.1073/pnas.0530251100
5. Nguyen, P.Q., et al.: Wearable materials with embedded synthetic biology sensors for biomolecule detection. Nat. Biotechnol. (2020). https://doi.org/10.1038/s41587-021-00950-3

Biomedical Devices and Sensors

PEG-Ylated Phenothiazine Derivatives. Synthesis and Antitumor Activity

Sandu Cibotaru[1(✉)], Valentin Nastasa[2], Andreea-Isabela Sandu[1], Andra-Cristina Bostanaru[2], Mihai Mares[2], and Luminita Marin[1]

[1] "Petru Poni" Institute of Macromolecular Chemistry of Romanian Academy, Grigore Ghica Voda Alee, 41A, Iasi, Romania
cibotaru.sandu@icmpp.ro
[2] Laboratory of Antimicrobial Chemotherapy, "Ion Ionescu de la Brad" University of Life Sciences, Iasi, Romania

Abstract. Phenothiazine based compounds are well known for their successful application in bio-medicine. Used for many years for the synthesis of many classes of drugs, in the last two decades the phenothiazine derivatives proved a promising potential for the cancer treatment. Taking into account phenothiazine properties and poly(ethylene glycol) biocompatibility, a series of three new PEGylated phenothiazine derivatives were prepared by grafting PEG chains to the phenothiazine core. The structure of the targeted molecules was confirmed by FTIR and NMR spectroscopy. The capacity of the synthetized compounds to self-assembly in water was studied by DLS and UV-vis techniques. Their biocompatibility was assessed on normal human dermal fibroblasts and five human cancer cell lines. The synthetized compounds proved excellent biocompatibility on normal cells. A concentration dependent cytotoxicity against cancer cell lines was noticed for two of synthesis PEGylated phenothiazine derivatives. In vivo anti-tumor investigations presented high tumor inhibition comparable to traditional drugs.

Keywords: Phenothiazine · Poly(ethylene glycol) · Tumour growth inhibition

1 Introduction

Phenothiazine (PTZ) is a fused tricyclic system which has a bent shape known as 'butterfly geometry' due to its heteroatoms on median axe of the central ring. This compound is an important building block for the synthesis of different derivatives used in medicine or for optoelectronic devices [1, 2]. In medicine, phenothiazine derivatives proved their activity as antihistaminic, neuroleptic bactericides, fungicides, antiviral, anti-inflammatory, antifilarial, trypanocidal, anticonvulsant and analgesic drugs [3]. Moreover, recent studies demonstrated that some phenothiazine derivatives had antitumor activity on several cell lines [2]. On the other hand, phenothiazine has a degree of cytotoxicity on the normal cells too. In order to improve the antitumor potential of phenothiazine, the paper proposed to modify its structure by PEGylation, envisaging that PEG presence will selectively improve the phenothiazine cytotoxicity. To this aim,

I. Tiginyanu et al. (Eds.): ICNBME 2021, IFMBE Proceedings 87, pp. 507–514, 2022.
https://doi.org/10.1007/978-3-030-92328-0_65

three different phenothiazine derivatives were synthetized, by grafting a PEG chain to the nitrogen atom via three different functional groups, i.e. ether, ester and amide. The structure of the compounds was confirmed by FTIR and NMR spectroscopy, and their antitumor activity was investigated on normal and tumor cells lines *in vitro* and *in vivo*.

2 Results and Discussions

A. Synthesis

The compounds under study were synthesized by grafting a polyethylene glycol chain to the nitrogen atom via three different functional groups, ether (PP), ester (PPO) and amide (PPN). The PP compound was obtained by grafting a tosyl-ated polyethylene glycol chain to the nitrogen atom of the phenothiazine core in a strong basic DMF solution. To obtain the other two compounds, firstly it was obtained a phenothiazine activated ester which was transformed into the PPO compound by a transesterification reaction with methoxy poly(ethylene glycol), and into the PPN compound by reaction with methoxy poly(ethylene glycol) amine. A schematic representation of their synthesis is presented in Fig. 1 [4].

Fig. 1. Synthesis of the PEGylated derivatives.

The structure of the synthesized compounds was confirmed by FTIR and NMR spectroscopy. FTIR spectra displayed characteristic vibrations of the C-H and C-O-C bonds of PEG, at 2900 cm^{-1} and 1100 cm^{-1}, and characteristic vibrations from the phenothiazine aromatic ring, such as C-H at 3060 cm^{-1} and C = C at 1570 cm^{-1}. Specific vibrations bands of C = O bond in amide and ester groups from PPN and PPO were noticed at 1690 cm^{-1} and 1742 cm^{-1} [5].

The ^{1}H NMR spectra displayed the chemical shift characteristic to the aromatic protons from phenothiazine core between 6.6 and 7.4 ppm and those characteristic to the aliphatic ones from PEG between 3.4 and 4.4 ppm. The disappearance of the chemical shift characteristic for the proton linked to the nitrogen atom of phenothiazine indicated the chemical modification of the compounds and in the case of PPN derivative, the occurrence of the chemical shift of amide proton at 8.32 ppm indicated the amide formation. The ratio of the integrated intensities of the signals characteristic to the aromatic and aliphatic protons, gave the right value for the pure compounds, indicating no side products or unreacted reagents.

B. Compounds solubility

The obtained PEG-ylated phenothiazine derivatives were viscous liquids with high solubility in water and in other organic solvents such as ethanol (EtOH), dimethyl sulfoxide (DMSO) or dichloromethane (DCM). A lower solubility was noticed in case of hexane due to its strong nonpolar character (Fig. 2).

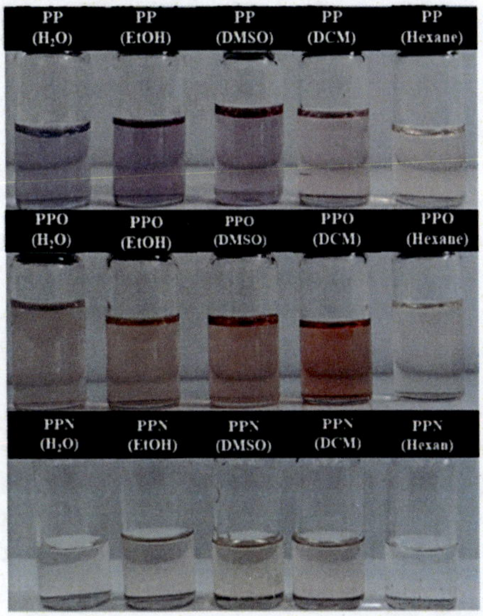

Fig. 2. Compound solutions in different solvents

C. Photophysical properties

Phenothiazine derivatives are known for their electron-donator properties that can be tuned by different structural modifications [6]. Due to this, the water solutions of the PEG-ylated compounds were investigated by UV-vis spectroscopy in comparison with water solution of pristine phenothiazine. The absorption spectra of PTZ presented two absorption bands, a sharp one at 250 nm characteristic to π-π^* electronic transitions and a broad absorption band at 279 nm characteristic to n-π^* electronic transitions [7]. For all three PEG-ylated compounds (PP, PPN and PPO), the second absorption band was bathochromically shifted approximately with approx. 25 nm, at about 305 nm (Fig. 3), in line with the appearance of aggregates due to the self-assembly properties of the PEG chain [8]. The bathocromic shift appeared to rise due to the intermolecular forces between phenothiazine units, which favored the electron delocalization and consequently the decrease of HOMO-LUMO band gap [9].

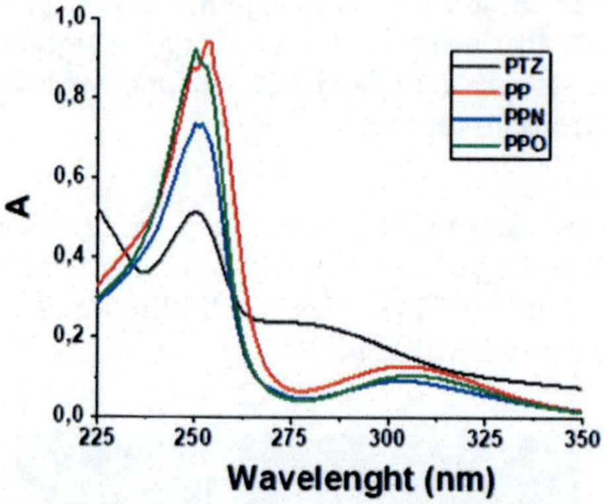

Fig. 3. Representative absorption spectra of the water solutions of the compounds

D. Self-assembling behavior

To confirm the hypothesis about aggregate formation indicated by the absorption spectra, the compound solutions for three different concentrations were investigated by dynamic light scattering (DLS) technique at different times during 60 days. The results confirmed the formation of aggregates and displayed their hydrodynamic stability in time (Fig. 4). PPO and PP presented a regular increasing of the hydrodynamic diameter in time in agreement with a continuous aggregation of the compounds. PPO displayed the lowest hydrodynamic diameter, around 63 nm, while PPN presented the biggest aggregates. Moreover, PPN aggregates were stable over time. It was assumed that the higher stability of the aggregates formed by PPN is prompted by the amide linker through the developing strong H-bonds and also by the higher length of the PEG chain leading to a higher surface density, less susceptible to dissolution [10, 11].

E. In vitro investigation

Because all three compounds were designed as building blocks for bioaplications, their biocompatibility was investigated on normal cells. All of them presented a good biocompatibility, but only two of them (PP and PPO) presented a cytotoxic effect when in contact with HeLa tumor cells.

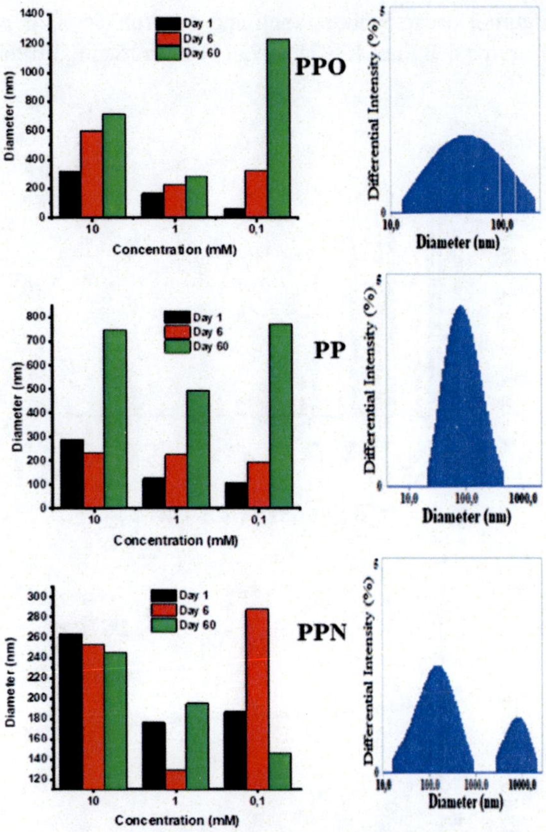

Fig. 4. DLS representative data

The deeper *in vitro* investigation of the antitumor activity for these two compounds was further done on five human cancer cell lines: cervical carcinoma (HeLa); malignant melanoma (MeWo); osteosarcoma (HOS), breast cancer (MCF7) and liver cancer (HepG2) versus a normal cell line (NHDF). For further *in vivo* investigations, the antitumor activity of the compounds was also investigated on mouse colon carcinoma cell line (CT26). The measurements were done for solutions of different concentration (from 0.01 to 1 mM) in order to determine the half maximal inhibitory concentration (IC_{50}).

The best results were obtained for PPO compound tested on MCF7 cancer human cell line (Fig. 4) with an $IC_{50} = 0.13$ mM and a selectivity index SI = 2.17. Cytotoxic investigation on mouse tumor cell line CT26 presented better results for both compounds with an IC_{50} of low concentration, 0.024 mM for PPO (Fig. 5) and 0.047 mM for PP. In comparison with normal cells (NHDF), the IC_{50} values are 12 and 6 times lower which indicated a good selectivity index. These results were comparable with those of traditional antitumor drugs 5-fluorouracil and doxorubicin (IC_{50} = 0,039 mM and 0,035 mM) on the same cell lines (CT26) [12, 13], encouraging further *in vivo* tests.

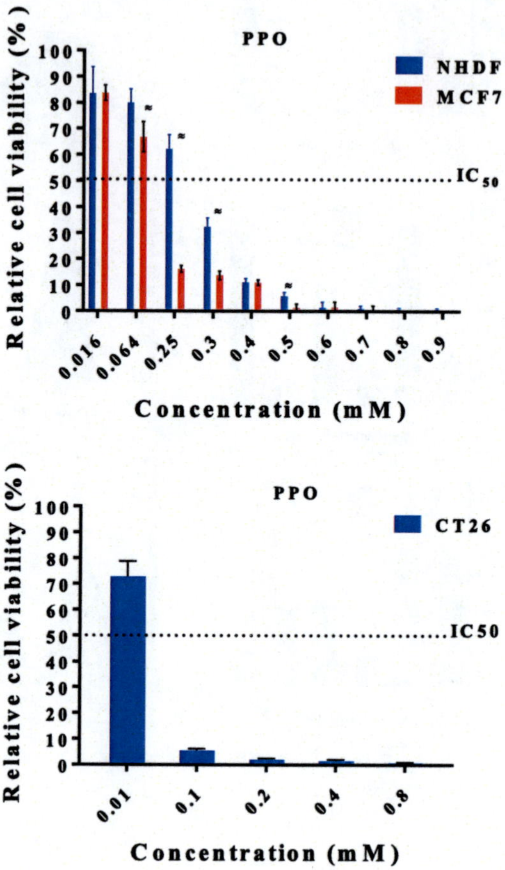

Fig. 5. Graphical representation of relative viability of human tumor cell line MCF7 and mouse tumor cell line CT26 when in contact with PPO

F. In vivo investigations

In order to establish the doses to be used for the *in vivo* antitumor tests, the LD_{50}, (a dose that causes death in 50% of the studied experimental animals) was measured firstly. The test was made by intraperitoneal administration of a single dose of PP and PPO in comparison with PTZ. The LD_{50} doses for PTZ, PP and PPO were 1450 mg/kg, 1300 mg/kg, and respectively 952.38 mg/kg in phenothiazine units.

The in vivo antitumor activity was investigated on mice with induced tumors. The mice were divided in four groups; the first received PBS solution, and the others received PTZ, PP and PPO, in doses lower than LD_{50} administrated daily during 10 days. Tumor size was measured every day in order to monitor tumor evolution. The results indicated the tumor inhibition TI = 92% for both studied compounds (PP and PPO), in comparison with PBS control (Fig. 6). Comparing these results with those reported for 5-fluorouracil and doxorubicin on the same tumor line (TI = 85% and 27%) [14, 15], it can be cocluded that these two phenothiazine PEG-ylated derivatives results are very promising candidates for developing of antitumor drugs.

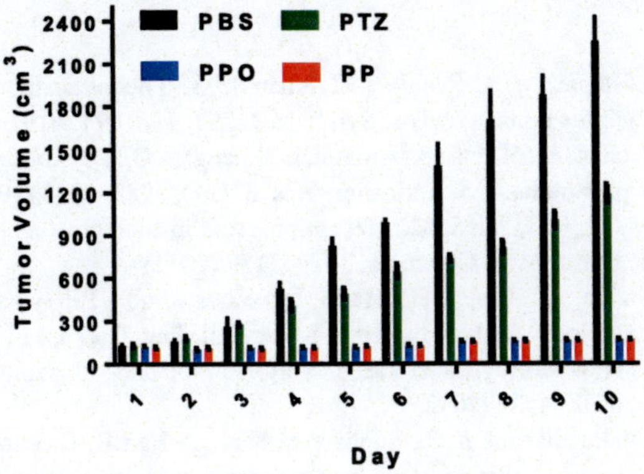

Fig. 6. Graphical representation of the tumour volume (mm^3) during the treatment

3 Compliance with Ethical Requirements

A. Ethical implications
The study was conducted in accordance with national and international regulations on animal welfare, identification, control and elimination of factors causing physiological and behavioural disorders: Directive EC86/609 EU; Government Ordinance no. 37/2002, approved by Law no. 471/2002; Law 205/2004 on animal protection, amended and supplemented by Law no. 9/2008; Joint Order of ANSVSA and of the Ministry of Interior and Administrative Reform no. 523/2008 for the approval of the Methodological Norms for the application of Law 205/2004 on animal protection.

B. Conflict of Interest
The authors declare that they have no conflict of interest conflict of interest.

4 Conclusion

A series of three PEG-ylated phenothiazine derivatives were synthesized and their right structure was confirmed by FT-IR and ^{1}H-NMR spectroscopy. They presented a good solubility in common organic solvents and water, and concentration dependent formation of nanoaggregates. *In vitro* tests demonstrated good cytotoxicity against five different human tumor lines and *in vivo* tests on mice demonstrated that two compounds were good inhibitors of tumor growth.

Acknowledgment. The paper was supported by a project financed through a Romanian National Authority for Scientific Research MEN – UEFISCDI, grant project PN-III-P4-ID-PCCF-2016-0050.

References

1. Al-Busaidi, J., Haque, A., Al Rasbi, N.K., Khan, M.S.: Phenothiazine-based derivatives for optoelectronic applications: a review. Synth. Met. **257**, 116189 (2019)
2. Varga, C.Á., Csonka, A., Molnár, J., Amaral, L., Spengler, G.: Possible biological and clinical applications of phenothiazines. Anticancer Res. **37**(11), 5983–5993 (2017)
3. Morak-Mlodawska, K.P., Jeleń, M.: Recent progress in biological activities of synthesized phenothiazines. Eur. J. Med. Chem. **46**, 3179–3189 (2011)
4. Cibotaru, S., Sandu, A.I., Belei, D., Marin, L.: Water soluble PEGylated phenothiazines as valuable building blocks for bio-materials. Mater. Sci. Eng. C **116**, 111216 (2020)
5. Belei, D., et al.: New farnesyltransferase inhibitors in the phenothiazine series. Bioorg. Med. Chem. Lett. **22**, 4517–4522 (2012)
6. Hart, S.A., Chandra Bikram, K.C., Subbaiyan, K.N., A Karr.P., D'Souza F.: Phenothiazine-sensitized organic solar cells: effect of dye anchor group positioning on the cell performance. J. Am. Chem. Soc. **4**, 5813–5820 (2012)
7. Marin, L., Bejan, A., Ailincai, D., Belei, D.: Poly(azomethine-phenothiazine)s with efficient emission in solid state. Eur. Polym. J. **95**, 127–137 (2017)
8. Risteen, B.E., et al.: Enhanced alignment of water-soluble Polythiophene using cellulose nanocrystals as a liquid crystal template. Biomacromol **8**, 1556–1562 (2017)
9. Bejan, A., Marin, L.: Phenothiazine based nanocrystals with enhanced solid state emission. J. Mol. Liq. **265**, 299–306 (2018)
10. Johansson, A., Kollman, P., Rothenbergv, S., McKelvey, J.: Hydrogen bonding ability of the amide group. J. Am. Chem. Soc. **96**, 3794–3800 (1974)
11. Owena, S.C., Chana, D.P.Y., Shoichet, M.S.: Polymeric micelle stability. Nano Today **7**, 53–65 (2012)
12. Moghimipour, E., et al.: Folic acid-modified liposomal drug delivery strategy for tumor targeting of 5-fluorouracil. Eur. J. Pharm. Sci. **114**, 166174 (2017)
13. Danhier, F., et al.: Vitamin E-based micelles enhance the anticancer activity of doxorubicin. Int. J. Pharm. **476**, 9–15 (2014)
14. Lee, S.J., et al.: Enzyme-responsive doxorubicin release from dendrimer nanoparticles for anticancer drug. Int. J. Nanomed. **10**, 5489–5503 (2015)
15. Pangeni, R., Choi, S.W., Jeon, O.C., Byun, Y., Park, J.W.: Multiple nanoemulsion system for an oral combinational delivery of oxaliplatin and 5-fluorouracil: preparation and in vivo evaluation. Int. J. Nanomed. **11**, 6379–6399 (2016)

Analysis of Melanin Properties in Radio-Frequency Range Based on Distribution of Relaxation Times

P. A. Abramov, S. S. Zhukov, Z. V. Bedran, B. P. Gorshunov, and Konstantim A. Motovilov(✉)

Moscow Institute of Physics and Technology, Dolgoprudny, Russia
k.a.motovilov@protonmail.com

Abstract. Being a family of biodegradable materials with natural origin, melanins are widely used for development of model bioelectronic devices. However, the mechanism of their electric conductivity is still a matter of discussions.

Current study is devoted to the room temperature impedance measurements of pure and copper-doped synthetic eumelanin at different values of humidity in frequency range 0.1–$5 \cdot 10^6$ Hz. To analyze the obtained impedance spectra, we utilize density relaxation times (DRT) methodology. The performed analysis demonstrates an absence of significant difference in relaxation times in the studied materials. At the lowest frequencies, the doped material has about 30 times lower conductance than pure material. Possible origins of the observed phenomena are discussed in terms of copper ions activity as complexing agent for water molecules and semiquinone groups of melanin.

Keywords: Melanin · DRT · Electrochemical impedance spectroscopy

1 Introduction

Broad use of electric and electronical devices resulted in the huge increase of corresponding waste. According to the forecast [1], this year world economy is going to produce 52.2 million metric tons of e-waste hazardous to environment. Therefore, the development of techniques that will allow production of biodegradable "green electronics" [2] is the matter of urgent need for modern society. Melanins are one of the perspective families of biodegradable materials that are able to partially substitute conventional semiconductors used in devices with low energy input [3, 4]. Melanins are pigments that are widely spread in nature in the tissues of cellular organisms [5]. Despite the intensive investigations during more than last 70 years, their structural and conductive properties remain controversial to the scientific community. One of distinctive features of melanin is its hydration-dependent conductance [6, 7].

On the basis of melanin, organic transistors have already been built [8, 9], but their properties are greatly inferior analogs. The purpose of the present study was to investigate mechanisms of conduction of melanin in radio-frequency (RF) range, as well as the evolution of its properties change with humidity.

I. Tiginyanu et al. (Eds.): ICNBME 2021, IFMBE Proceedings 87, pp. 515–521, 2022.
https://doi.org/10.1007/978-3-030-92328-0_66

Properties of melanins in the RF range have been investigated by many scientific groups [10–12], however, the results were analyzed exclusively within the framework of a phenomenological approach, and the physical interpretation of them was either not made or raises a lot of questions. Most studies were devoted to investigations of the dielectric properties of melanin films, whose properties are definitely different from those of bulk melanin, due to the different topology and, probably, different oxidation states as a result of film growth.

2 Samples Preparation

Two types of synthetic melanin were studied: pure DOPA eumelanin (M-pure) and doped with Cu^{2+} ions (M-Cu). Copper content in M-Cu was estimated to be 40491 μg/g according to GC-MS element analysis. The synthesis was described earlier in [9].

Powders of melanin were pressed into 5 mm in diameter plane-parallel pellets with a thickness of 700 μm under a pressure of the 800MPa and kept for 60 min. Gold electrodes (4 mm in diameter) were deposited on both sides of the pellets. Relative humidity of prepared samples was reached by keeping the pellets in the moist air over saturated salts solutions (Table 1).

Table 1. Salts and corresponding RH values at 27 °C

Cu (II) melanin		Pure melanin	
Salt	RH, %	Salt	RH, %
–	–	10^{-6} mbar vacuum	0
$LiCl$	12	$LiCl$	12
$MgCl_2$	33	$MgCl_2$	33
$NaNO_2$	64	$Na_2Cr_2O_7$	54
–	–	$NaCl$	75
$(NH_4)_2SO_4$	81.7	KCl	84

3 Experimental Setup

For the experiments, a special hygrostatic system was developed. In a vessel with saturated salt solution, a fan was mounted that stirred the humidified air over the solution. We found that permanent air mixing is required to achieve the needed relative humidity. With a membrane pump, the humid air was blown into a sealed sample chamber with the studied sample. The container with the sample was enclosed in an electromagnetically shielded aluminum box. The entire hygrostatic system was located in a specially designed thermostat. This complex system was designed because the properties of melanin samples are known to be extremely sensitive to moisture and temperature: we found that a change in temperature by only 1 degree leads to a change in the impedance of the sample up to 10%, in accordance with the literature data [7]. In addition, the RH of air above the salt solution also depends on the temperature.

The samples were kept at a fixed humidity for 3 days, resulting in a complete establishment of humidity within the sample. This time interval was determined empirically from the assessment of changes in the impedance of the system.

Radio-frequency measurements were conducted using an impedance analyzer MFIA Zurich Instruments 4-contact scheme in broadband radio-frequency range 0.1 Hz to 5 MHz at a temperature of $27 \pm 0.5\,^{\circ}\mathrm{C}$ with voltage amplitude 100 mV.

4 Data Processing

The experimental data were processed in two ways: a) by using an approach that utilizes a distribution of relaxation times (DRT) analysis [13] and b) by fitting the Nyquist plots with equivalent circuits. Using the two methods was necessary due to the variation of the DRT distributions depending on the regularization parameter, noise and number of points [14]. In the DRT model, melanin dielectric impedance spectra were modeled by the Voight model that considers a serial chain of parallel-connected resistor and capacitor. One element of such chain gives a delta function in the time domain. The integral impedance of this method is represented as

$$Z_{DRT} = R_{\infty} + \int_{0}^{\infty} \frac{g(\tau)}{1 + i2\pi f \tau} d\tau,$$

with R_{∞}-resistance at a cut off frequency, $g(\tau)$ - probability density function of relaxation times.

All data for DRT analysis were noise filtered using a moving average filter (window size = 5) and then cubically interpolated from 1 MHz up to 5 MHz at 100 points logarithmically at even spacing.

When fitting with an equivalent circuit, instead of capacitors, constant phase elements (CPE) were used that are model elements with an impedance given by the formula $1/Q(i\omega)^{\varphi}$, $\varphi \in [0, 1]$ [14]. Depending on φ, the CPE can represent a capacitor, a resistor, and a Warburg element. In terms of the distribution of relaxation times, a resistor connected to CPE in parallel (R-CPE element) gives a broadened peak at $\tau = (RQ)^{1/\varphi}$. At low frequencies, the behavior of the melanin pellet was characterized by a Warburg element (from ≈10 Hz to ≈0.3 MHz, the Nyquist plot has 45 ° slope with good accuracy). The high-frequency response is represented by an inductance $L \approx 10^{-5} H$ with an ohmic resistor R_{∞}. The remained frequency range corresponds to a series of six R-CPE elements connected in parallel.

Thus, with an equivalent circuit fitting for preliminary estimates of the samples characteristics, their final values were refined according to the DRT analysis.

5 Result and Discussion

From the frequency dependence of the real part of the impedance (Z') we found that the conductance of both melanin samples monotonically grows with increase of their relative humidity over the frequency range from 100 MHz up to approximately 1 kHz for pure melanin, and up to 3 kHz for M-Cu melanin. At frequency 10 Hz, it rises thousand times when humidity changes from 12% to 84% for pure melanin, and from 12% to 81% for M-Cu. In the high-frequency region, the dependency of the real part of impedance reveals a local minimum.

For RH values of 12%, 33%, and 81%, the conductance of the pure melanin was found to be at least 30 times larger than that for M-Cu melanin, at a frequency of 10 Hz. At lower frequencies, copper tends to reduce the conductivity of melanin. At higher frequencies, the situation is reversed. Interestingly, the position of the critical point on the frequency axis, where corresponding conductivity curves intercept each other, tends to increase with growing water content in the samples. The location of the critical point increases with the humidity: it is 5 kHz for 12% RH and 60 kHz for 33% RH; for 81% RH this point is not observable and probably is above 5 MHz.

To further analyze the influence of absorbed water on melanin conductance, we used the approach that is standard in the electrochemical impedance spectroscopy. Figure 1,2 shows the Nyquist plot of the measured spectra obtained in the frequency range from 100 MHz to 5 MHz for the pure melanin pellet at the 0%, 12%, 33%, 54%, 75%, and 84% RH, and for the M-Cu melanin at the 12%, 33%, 64% and 81% RH.

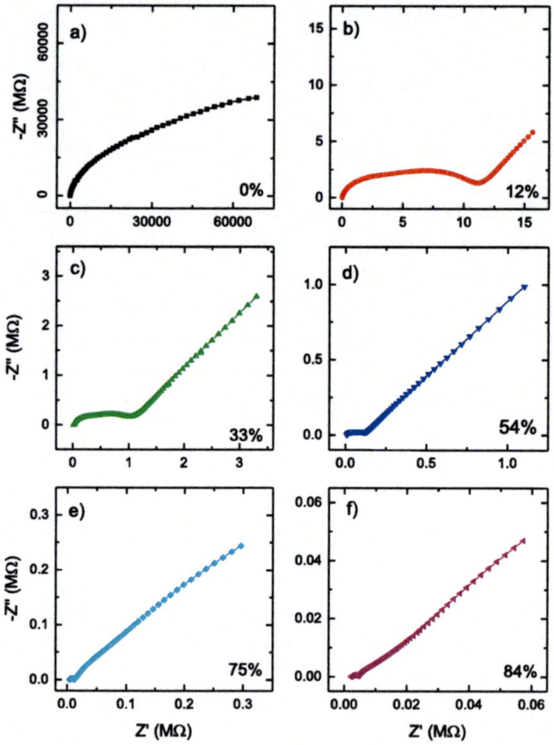

Fig. 1. Nyquist plot of the impedance spectra of the melanin pellet measured at the (a) 0%, (b) 12%, (c) 33%, (d) 54%, (e) 75% and (f) 84% RH.

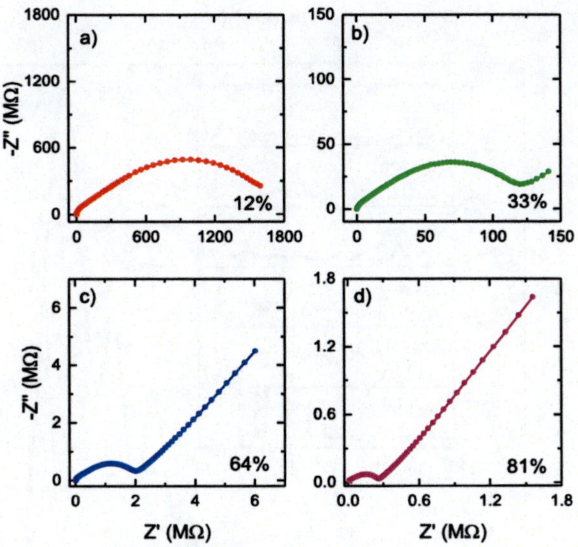

Fig. 2. Nyquist plots of the impedance spectra of the M-Cu 1.0 mmol/g (Cu-melanin) pellet measured at the (a) 12%, (b) 33%, (c) 64% and (d) 81% RH.

It is seen that the evolution of the impedance with humidity growth is basically the same in both samples. However, the impedance spectra of 12% M-Cu sample (Fig. 2, a) look similar to that of the 0% M-pure (Fig. 1, a) sample, the spectra of the 33% M-Cu (Fig. 2, b) look similar to that of the 12% M-pure (Fig. 1.b), etc. Since copper is a strong complexing agent, it can effectively bind within its coordination sphere both the water molecules and the nucleophilic groups of the melanin units, including semiquinone anions formed in the course of comproportionation reaction [9].

It is clearly seen that the low-frequency tails of the spectra are well approximated by the finite-length Warburg element, except for the sample with the lowest RH of 0% pure, and for the 12%-RH M-Cu sample.

When modeling the spectra, we found that the part that is well approximated by the finite-length Warburg spectra broadens and shifts towards high frequencies with humidity increase. We thus assume that for the 0%-RH pure sample, and for the 12%-RH M-Cu sample, the Warburg-type tail is located below 100 MHz. If so, then at the lowest frequencies we should see a series of the decaying peak in the relaxation times distribution. The results obtained for pure melanin using the DRT approach are shown in Fig. 3. For humidity changing from 12% to 84%, well distinguishable three decaying peaks are seen that are located close to 10, 1, and 0.1 s. For 0%-RH sample, we can also see a series of the merged decaying peaks.

Applying the DRT treatment allows to distinguish a number of relaxation times, and their variation can be traced with a change in humidity. Most peaks remain at practically the same positions and only vary their amplitude and width. All peaks with a relaxation time exceeding two milliseconds decrease monotonically in amplitude with an increase of the amount of absorbed water. In the region with times less than two microseconds, we can observe peaks that are totally vanishing at 12% and 0% RH. The most intriguing

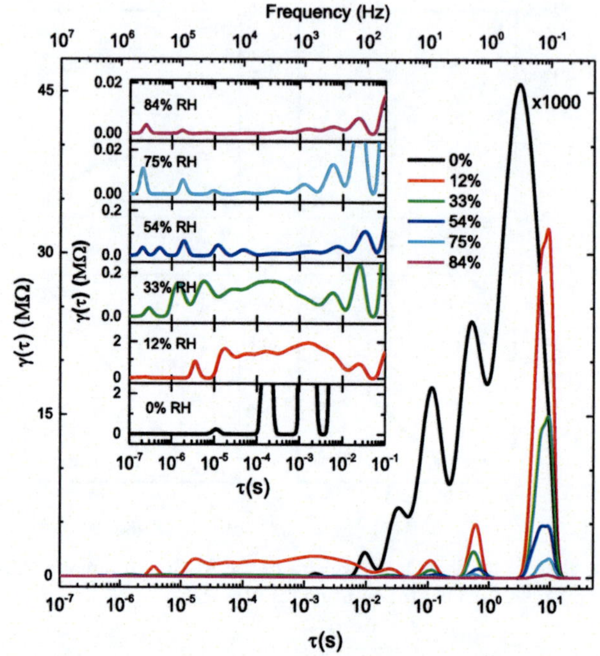

Fig. 3. Distribution of the relaxation times for the pure melanin pellet measured at the 0%, 12%, 33%, 54%, 75% and 84% RH. Plot for 0% RH has been reduced in amplitude by 10^3 for convenience. The inset shows zoomed region for relaxation time smaller than 0.1 s.

process is connected with the relaxation times ranging from two microseconds up to ten milliseconds. In this region, with an increase of humidity from 0% to 12%, the amplitude of relaxation times also increases. A further increase in humidity from 12% to 84% leads to reduced amplitude of the peaks.

6 Conclusions

We have measured the RF impedance spectra of pellets of pure and Cu melanins in a wide range of humidity values at a fixed temperature. It was shown that conductance of the pure melanin is at least 30 times larger than that of M-Cu melanin, at frequencies <1 kHz.

We found that the presence of Cu leads to a lag in the evolution of the impedance with humidity growth, in contrast with pure melanin. This effect is a subject of further research.

We have shown that the frequency window, within which the spectra are well approximated by the finite-length Warburg element, shifts towards high frequencies with the increase of absorbed water content.

The results of DRT analysis allow us to claim that the amplitudes of time peaks (10^{-2}–10^{-6} s) increase with humidity growth from 0% to 12% and that process is reverse with further increase of humidity from 12% to 84% RH.

Acknowledgment. The work was supported by the Russian Science Foundation under grant 19-73-10154. We thank Dr. Bernard Mostert for providing characterized melanin samples for this study.

Conflict of Interest. The authors declare that they have no conflict of interest.

References

1. http://collections.unu.edu/view/unu:6341
2. Irimia-Vladu, M.: Green electronics: biodegradable and biocompatible materials and devices for sustainable future. Chem. Soc. Rev. **43**(2), 588–610 (2013). https://doi.org/10.1039/C3CS60235D
3. Mostert, A.B.: Melanin, the what, the why and the how: an introductory review for materials scientists interested in flexible and versatile polymers. Polymers **13**(10) (2021). https://doi.org/10.3390/polym13101670. Art. no. 10
4. d'Ischia, M., et al.: Melanins and melanogenesis: from pigment cells to human health and technological applications. Pigment Cell Melanoma Res. **28**(5), 520–544 (2015). https://doi.org/10.1111/pcmr.12393
5. D'Alba, L., Shawkey, M.D.: Melanosomes: biogenesis, properties, and evolution of an ancient organelle. Physiol. Rev. **99**(1), 1–19 (2018). https://doi.org/10.1152/physrev.00059.2017
6. Mostert, A.B., et al.: Role of semiconductivity and ion transport in the electrical conduction of melanin. Proc. Natl. Acad. Sci. U. S. A. **109**(23), 8943–8947 (2012). https://doi.org/10.1073/pnas.1119948109
7. Jastrzebska, M.M., Isotalo, H., Paloheimo, J., Stubb, H.: Electrical conductivity of synthetic DOPA-melanin polymer for different hydration states and temperatures. J. Biomater. Sci. Polym. Ed. **7**(7), 577–586 (1996). https://doi.org/10.1163/156856295X00490
8. Sheliakina, M., Mostert, A.B., Meredith, P.: An all-solid-state biocompatible ion-to-electron transducer for bioelectronics. Mater. Horiz. **5**(2), 256–263 (2018). https://doi.org/10.1039/C7MH00831G
9. Mostert, A.B., et al.: Engineering proton conductivity in melanin using metal doping. J. Mater. Chem. B **8**(35), 8050–8060 (2020). https://doi.org/10.1039/D0TB01390K
10. Sheliakina, M., Mostert, A.B., Meredith, P.: Decoupling ionic and electronic currents in Melanin. Adv. Funct. Mater. **28**(46), 1805514 (2018). https://doi.org/10.1002/adfm.201805514
11. Reali, M., et al.: Electronic transport in the biopigment sepia Melanin. ACS Appl. Bio Mater. **3**(8), 5244–5252 (2020). https://doi.org/10.1021/acsabm.0c00373
12. Paulin, J.V., et al.: Melanin thin-films: a perspective on optical and electrical properties. J. Mater. Chem. C **9**(26), 8345–8358 (2021). https://doi.org/10.1039/D1TC01440D
13. Wan, T.H., Saccoccio, M., Chen, C., Ciucci, F.: Influence of the discretization methods on the distribution of relaxation times deconvolution: implementing radial basis functions with DRTtools. Electrochim. Acta **184**, 483–499 (2015). https://doi.org/10.1016/j.electacta.2015.09.097
14. Saccoccio, M., Wan, T.H., Chen, C., Ciucci, F.: Optimal regularization in distribution of relaxation times applied to electrochemical impedance spectroscopy: ridge and lasso regression methods - a theoretical and experimental study. Electrochim. Acta **147**, 470–482 (2014). https://doi.org/10.1016/j.electacta.2014.09.058

Nanostructuring of Protein Systems by Electroactivation

E. G. Vrabie[1](✉), M. K. Bologa[1], I. V. Paladii[1], V. G. Vrabie[2], A. A. Policarpov[1], V. P. Gonciaruc[1], C. Gh. Sprincean[1], and T. G. Stepurina[1]

[1] Institute of Applied Physics, Chisinau, Republic of Moldova
elvira.vrabie@ifa.md
[2] Institute of Physiology and Sanocreatology, Chisinau, Republic of Moldova

Abstract. A possibility of nanostructuring protein systems is demonstrated based on whey electroactivation. Whey as a raw material is not only a valuable by-product available in most of the cheese-production industry areas, it is also relatively cheap, with high nutritional value and several functionalities, with unique properties, which allows the study of nano-sized structures of whey proteins. Various non-uniform isolations of protein fractions in the protein-mineral complexes (PMCs) at the electroactivation of the studied whey is determined, first of all, by the properties of each fraction separately and by their behavior at the electrochemical action. The experimental data on the isolation of Beta-Lactoglobulin (β-Lg), the most abundant whey protein, are presented; the aggregation of β-Lg in the PMCs via various mechanisms and modes has been presented by the authors earlier. The isolation of Alpha-Lactalbumin (α-La) in the PMCs upon the electroactivation of whey differs from the isolation of β-Lg in the PMCs. Electroactivation of whey allowed the formation of high molecular weight protein compounds. Three fractions of casein: –α-CSN, β CSN, and κ-CSN, with a molecular weight of 37, 33, and 46 kDa, respectively, were identified. The electroactivation of whey and the extractions in the PMCs ennobled with certain protein fractions allowed to specify different sedimentation and isolation mechanisms of the whey proteins: the formation of a calcium phosphate caseinate complex; salinization of proteins; and sedimentation of proteins in their isoelectric point.

Keywords: Electroactivation processing · Protein mineral concentrates · Electrolyzer · Whey · Whey proteins

1 Introduction

The development of nano-sized ingredients, supplements and additives, as well as nanostructures as carriers of bioactive compounds or for incorporation in food packaging (to improve their barrier and mechanical properties) has been receiving increasing attention from the scientific community bringing new applications of nanotechnologies in the food industry. Properly designed and controlled processes can form nanostructures of whey proteins useful as carriers of bioactive compounds (e.g., antimicrobials, antioxidants, and

I. Tiginyanu et al. (Eds.): ICNBME 2021, IFMBE Proceedings 87, pp. 522–529, 2022.
https://doi.org/10.1007/978-3-030-92328-0_67

nutraceuticals) [1, 2]. An emerging approach presents the application of moderate electric fields in combination with conventional heating, which can inhibit/prevent excessive denaturation and aggregation of whey proteins, thus opening new perspectives for the development of innovative protein nanostructures. Whey is a valuable by-product in most of the cheese-production industry areas, being relatively cheap, with a high nutritional value and several functionalities with unique properties; all this allows the development of nanostructured systems of whey proteins [3]. During primary milk processing and the production of different types of cheese, large quantities of various secondary dairy products are also obtained: whey, buttermilk, skim milk, caseinate, etc., with various solid contents. Global whey production is estimated at about $180–190 \cdot 10^6$ tons/year, of this quantity only 50% is processed [4]. The gelling property of whey proteins is of particular importance for the development of nanostructures, which is affected by the amino acid composition, proteins distribution and polarity, and environmental conditions affecting molecular interactions between them (e.g., ionic crosslinking between amino and carboxyl groups, the presence of hydrogen bonds and intramolecular and disulphide bonds) [5]. The development of wasteless technologies at electroactivation permits the fractionation of whey, having as the main objective the desired isolation of whey proteins and of amino acids in concentrates, in a special combination with minerals, which will make it possible to manufacture a specific range of very healthy food additives and dietetic supplements [6]. Electroactivation of whey allows the control of the content of whey proteins in the obtained concentrates, depending on the processing mode.

The protein composition of the initial whey (IW) is represented by the major protein fractions: β-lactoglobulins, α-lactalbumins, immunoglobulins, and bovine serum albumin (BSA) [7, 8]. Among whey proteins, there are also minor protein fractions, such as lactoferrin, for example, which are in an insignificant quantity; those proteins do not have a particular nutritional value but perform other very important functions. The amino acids composition of IW proteins is mostly similar to that of the human muscular tissue; as for the content of essential amino acids (lysine, tryptophan, methionine, and threonine), they are superior to many proteins of the animal and plant origin. The assimilability of IW proteins is extremely high [9]. There are different known methods for processing IW with the view of obtaining protein concentrates used as different supplements, including biologically active ones [10].

2 Materials and Methods

The technical requirements to ensure the management and control of the technological process that takes place at the electroactivation of whey with the isolation of the protein mineral complexes (PMCs) and the simultaneous isomerization of lactose into lactulose, reveal the need of certain conditions. In the framework of the experiments of electroactivation of whey, the whey after the manufacture of the "Cottage cheese" with a 2% fat content provided by the "JLC" Joint Stock Company, Chisinau, RM, was used.

The electroactivation of whey was performed in the membrane electrolyzer EDP-4 (with a short distance between the electrodes; the ratio of whey volumes to the surface of electrodes is $V/S = 1.0$), in a stationary regime, at current densities of 10 and 20 mA/cm^2, remaining constant during processing, and also in a stationary regime of the discharge of the working liquid (whey discharged in the cathode cell) and of the secondary liquid (2% $CaCl_2$ solution discharged in the anode cell) [11, 12]. The PMCs at certain treatment periods were collected, and the results of experiments were reported by the authors earlier [12]. The soluble fractions extracted from the PMCs with 0.05 M buffer solution 0.5 M Tris-HCl buffer NaCl, 0.5 mM EDTA (0.04% NaN_3), pH 8.0, were analyzed by the sodium dodecyl sulfate electrophoresis (SDS-PAGE) and by the non-denaturing polyacrylamide gel concentration of 15% according to [13]. The obtained results were scanned through the HP Scanget 3800 with the software Microsoft Photo Editor and analyzed with the Phoretix 1D Advans in order to determine the quantity of the major fractions in the PMCs.

3 Experimental

The purpose of the paper was the investigation of the degree of the isolation of each fraction, depending on the current density, active acidity, and the duration of the processing, as well as on the mechanisms that influence the process of isolation of whey proteins from concentrates during electroactivation.

Various non-uniform isolations of protein fractions in the PMCs at the electroactivation of the studied type of whey were determined, first of all, by the properties of each fraction separately and by their behavior at the electrochemical action.

The formation of the PMCs from whey depends on the electroactivation regime and on such parameters as: electric current density, energy consumption per unit volume, variations in pH values, variations in temperature and Joule heat. Under the action of an electric current on the solid content of whey of about 6% and over 200 ingredients, multiple inter- and intra-molecular processes take place that cause variations in the isolation of protein fractions in the PMCs.

Whey proteins have a high solubility in a wide range of pH values due to the composition of amino acids and the arrangement of their free radicals during the formation of tertiary structures. Increasing the pH values is also compulsive due to the processes generated by electroactivation. It is due to the arrangement of hydrophilic radicals and a large number of disulfide bridges on the surfaces of the globular structures of whey proteins, which confirms the information of other researchers. In addition, the globular structure induces the resistance of these proteins to proteolysis.

The protein fractions extracted from whey in the PMCs were divided into four groups:

- High weight proteins (HWPs), in which 2–5 fractions with variations of the molecular weight (MW) of 54–249 kDa, containing BSA with a MW of 66 kDa, lactoperoxidase with a MW of 70 kDa, and lactoferrin with a MW of 80 kDa, as well as protein complexes with a MW of approximately 200–249 kDa;
- Casein (CSN), in which 3 fractions were identified: α-CSN, β CSN, and κ-CSN, with a MW of 37, 33, and 46 kDa, respectively;
- Beta-Lactoglobulin (β-Lg) with a MW of 18.4 kDa whose isolation is significant in all collected samples;
- Alpha-Lactalbumin (α-La) with a MW of 14.2 kDa which had an extracted maximum at the end of processing.

Electroactivation of whey in the electrolyzer EDP-4, in the stationary regime, at j = 10 mA/cm^2, vividly demonstrates different degrees of isolation of protein whey fractions in the PMCs depending not only on the processing regime, which is mostly based on the energy consumption, the volume of the processed whey, the duration of processing, and the variations of pH, but also on the type of the whey being processed that, in turn, depends on the primary processing of milk (Fig. 1).

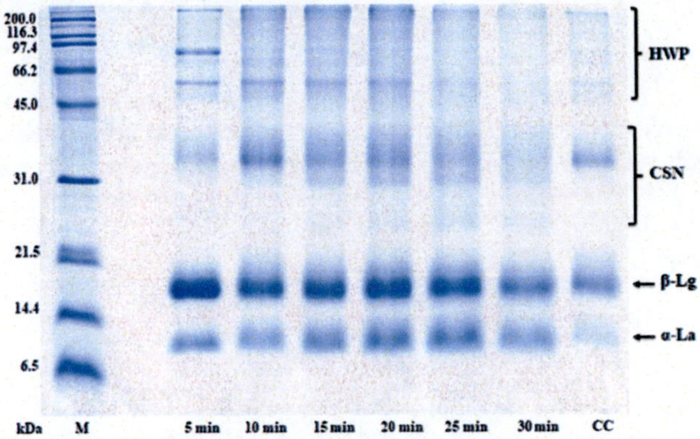

Fig. 1. SDS-PAGE 15% of soluble protein concentrates isolated with 0.05 M Tris-HCl buffer 0.5 M NaCl, 0.5 mM EDTA (0.04% NaN_3), pH 8.0, of whey, at processing at j = 10 mA/cm^2 (M - marker, 5–30 min – duration of processing, CC – content of cathode cell).

The maximum quantitative recovery (Q_f, mg) of β-Lg was recorded in the first 5 min and decreases during processing. Subsequent processing of the studied whey allows a more intense quantitative recovery of α-La (20, 25 min).

Both CSN and HWP fractions recover quantitatively more intensely at 10–20 min of processing (Fig. 2).

Electroactivation of whey at j = 20 mA/cm^2 allowed for the electrofractionation of the PMCs at 5 min pace, thus making it possible to register the protein fractions isolated in the PMCs. A larger quantity of β-Lg was registered in the first 5 min, which is the largest value under similar processing conditions, however, it was descending during processing. As to α-La fractions, they were accumulated in the PMCs in an ascending mode during processing, reaching the maximum amount at 20 min of processing. Both CSN and HWP recovered quantitatively more intensively in the first 5–10 min of processing. (Figs. 3 and 4).

Experimental data on the isolation of β-Lg (about 60%), the most abundant whey protein, in the first 5–10 min, when the pH is maintained in the neutral range (pH 6.0–8.0), at a low temperature, were recorded in the cathode cell volume. The mechanism of β-Lg aggregation has been described in more detail in [6].

The isolation of α-La (about 40%) in the PMCs upon the electroactivation of whey differs from the isolation of β-Lg in the PMCs. This is revealed through the data accumulated during processing of whey at j = 20 mA/cm^2 when a high quantity is registered at the end of the process (after 15 min of processing).

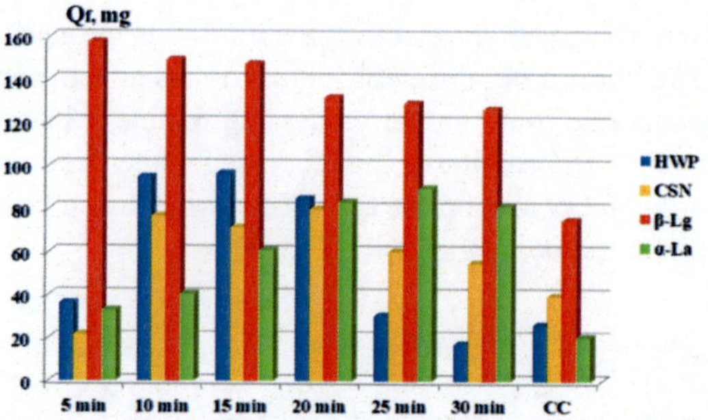

Fig. 2. Variations of content (Q_f, mg) of major soluble protein fractions isolated in 0.05 M Tris-HCl buffer 0.5 M NaCl, 0.5 mM EDTA (0.04% NaN_3), pH 8.0, of whey analyzed with SDS-PAGE 15%, at processing at $j = 10$ mA/cm^2 (5–30 min – duration of processing, CC – content of cathode cell).

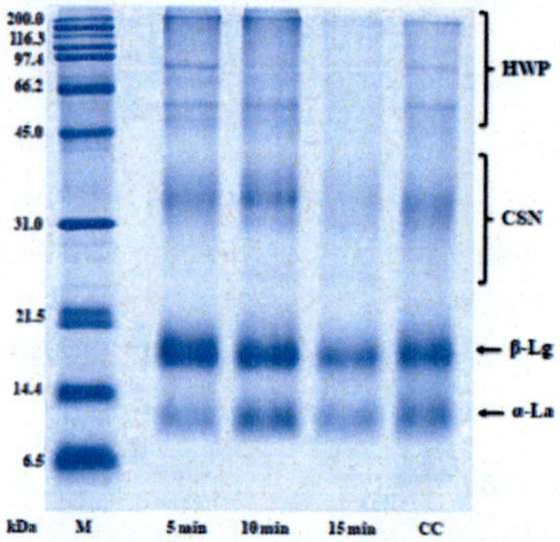

Fig. 3. SDS-PAGE 15% of soluble protein concentrates isolated with 0.05 M Tris-HCl buffer 0.5 M NaCl, 0.5 mM EDTA (0.04% NaN_3), pH 8.0, of whey at processing at $j = 20$ mA/cm^2 (M - marker; 5–15 min – duration of processing, CC – content of cathode cell).

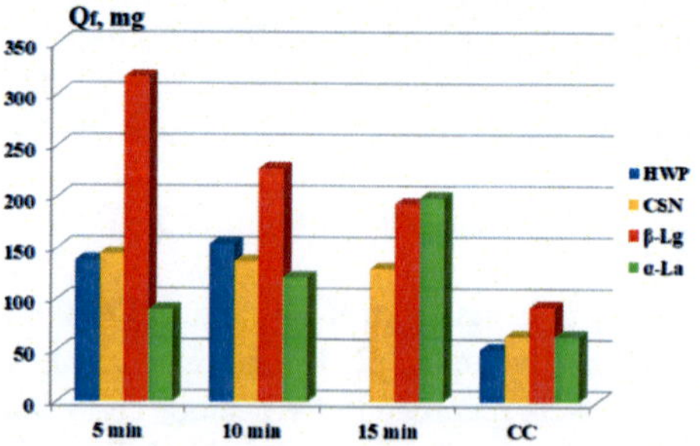

Fig. 4. Variations of content (Q_f, mg) of major soluble protein fractions isolated in 0.05 M Tris-HCl buffer 0.5 M NaCl, 0.5 mM EDTA (0.04% NaN_3), pH 8.0, of whey analyzed with SDS-PAGE 15%, at processing at $j = 20$mA/cm^2 (5–15 min – duration of processing, CC – content of cathode cell).

The whey under study here had, initially, a high content of calcium ions because of the primary processing of milk with calcium chloride supplemented by the Ca ions from the anode cell, which resulted in the formation of the PMCs with a high level of α-La that makes up 20% of the whey proteins of the bovine milk. It is a small protein with a MW of 14.2 kDa, with an increased affinity to the ions of Ca^{2+} and to other bivalent cations; this takes place at the level of the aspartic acid that has two groups –COOH.

Upon electroactivation of the whey studied here, HWP fractions (20–30%) were identified. HWPs contain 2–5 fractions with variations of MW of 54–240 kDa: BSA with a MW of 66 kDa, lactoperoxidase with a MW of 70 kDa, and lactoferrin with a MW of 80 kDa, as well as protein complexes with a MW of about 200–240 kDa. Electroactivation at $j = 20$ mA/cm^2 of the studied whey allowed the formation of high molecular weight protein compounds that cannot be identified with 15% SDS-PAGE.

Three fractions of CSN were identified (see above). Their isolation varied from 9% to 24% and depended on the processing regime and the chemical properties of each fraction separately. The availability of CSN confirms the aggregation of protein compounds revealed by the formation of a calcium phosphate caseinate complex, which occurs due to the presence of colloidal calcium phosphate complexed with calcium caseinate. In cow's milk (pH 6.47–6.67), calcium salts are mostly phosphates that have a low solubility and a low degree of dissociation. Only a small part can be formed as a pure solution, while the larger part is in the form of a colloidal solution. The colloidal calcium phosphate content of a calcium phosphate caseinate complex, as well as the case of complexation with casein, are not yet elucidated; however, it is clear that calcium caseinate is formed at the interaction of Ca ions with carboxylic groups and casein serophosphates.

Isolation of whey proteins in the PMCs by the processing used in this research depends on the action of several complexing and coagulation mechanisms. For protein sedimentation, the salinization reaction is widely used which is based on the decrease of protein solubility under an increase in salt concentration. When whey is electroactivated, the concentration of ions in the electrode area can be several times higher (at an order of magnitude) than in the initial solution, thus creating the conditions for protein salinization, which is indirectly confirmed by intensive foaming in the cathodic area and by their subsequent isolation sustained by the ionic flotation due to electroactivation.

Another mechanism that allows the formation of protein compounds is the sedimentation in their isoelectric point (pI). The pH variations in the CC are not homogeneous and depend both on the constructive/geometric parameters of the electrolyzer (including the V/S ratio) and on the flux of the initial solution, which, during processing, increases, reaching the pI of the protein fractions recovered in the PMCs, therefore, their sedimentation in pI.

4 Conclusions

Whey - a valuable by-product of cheese making, was studied with the aim of demonstrating a possibility to develop nanostructured systems of whey proteins.

Isolation of protein fractions in the PMCs at the electroactivation of the studied whey was determined, first of all, by the properties of each fraction separately and also by their behavior at the electrochemical action.

The experimental data on the isolation of β-Lg (about 60%), the most abundant whey protein within the first 5–10 min, revealed the aggregation of β-Lg in the PMCs via various mechanisms and modes as has been presented by the authors earlier. The isolation of α-La (about 40%) in the PMCs upon the electroactivation of whey differs from the isolation of β-Lg in the PMCs. A high quantity of α-La is registered at the end of the process (after 15–25 min of processing). At the electroactivation of the whey after the production of the "Cottage cheese" with a 2% fat content, HWP fractions (20–30%) were identified. Electroactivation of whey at $j = 20 mA/cm^2$ allowed the formation of high molecular weight protein compounds which cannot be identified with SDS-PAGE 15%. Three fractions of CSN were identified: –α-CSN, β CSN, and κ-CSN, with a MW of 37, 33, and 46 kDa, respectively. Their isolation varied from 9 to 24% and depended on the processing regime and the chemical properties of each fraction separately.

At the electroactivation of the studied whey and the extraction of the PMCs ennobled with certain protein fractions, different sedimentation and isolation mechanisms of the whey proteins were identified: the formation of a calcium phosphate caseinate complex, salinization of proteins, and sedimentation of proteins in their pI.

Acknowledgment. This research was carried out within the state program project ANCD 20.80009.5007.06 (2020-2023).

The authors express their gratitude to Dr. Olga Iliasenco for assistance in preparing this paper.

Conflict of Interest. The authors declare that they have no conflict of interest.

References

1. Chaudhry, Q., Scotter, M., Blackburn, J., et al.: Applications and implications of nanotechnologies for the food sector. Food Addit. Contam. **25**(3), 241–258 (2008). https://doi.org/10.1080/02652030701744538
2. Kane, R.S., Stroock, A.D.: Nanobiotechnology: protein–nanomaterial interactions. Biotechnol. Prog. **23**(2), 316–319 (2007). https://doi.org/10.1021/Bp060388n
3. Acosta, E.: Bioavailability of nanoparticles in nutrient and nutraceutical delivery. Curr. Opin. Colloid Interface Sci. **14**(1), 3–15 (2009). https://doi.org/10.1016/j.cocis.2008.01.002
4. El-Tanboly, E.S., El-Hofi, M.K.: Recovery of cheese whey, a by-product from the dairy industry for use as an animal Feed. J. Nutr. Health Food Eng. **6**(5), 148–154 (2017). https://doi.org/10.15406/jnhfe.2017.06.00215
5. Ramos, O., Pereira, R., Rodrigues, R.M., et al.: Physical effects upon whey protein aggregation for nano-coating production. Food Res. Int. **66**, 344–355 (2014). https://doi.org/10.1016/j.foodres.2014.09.036
6. Bologa, M., Vrabie, E., Paladii, I., et al.: Peculiarities of extraction of β-lactoglobulin in protein mineral concentrates at electroactivation of whey. One Health Risc Manag. **2**(1), 52–68 (2021). https://doi.org/10.38045/ohrm.2021.1.06
7. Sprincean, E.: The optimization of protein extraction during electrophysical whey processing. Surf. Eng. Appl. Electrochem. **46**, 605–611 (2010). https://doi.org/10.3103/S1068375510060116
8. Santos, M.J., Teixeira, J., Rodrigues, L.: Fractionation of the major whey proteins and isolation of β-Lactoglobulin variants by anion exchange chromatography. Sep. Purif. Technol. **90**, 133–139 (2012). https://doi.org/10.1016/j.seppur.2012.02.030

9. Naidoo, K., Naidoo, R., Bangalee, V.: Understanding the amino acid profile of whey protein products. Global J. Health Sci. **10**(9), 45–53 (2018). https://doi.org/10.5539/gjhs.v10n9p45
10. Paladii, I.V., et al.: Whey: the state of the art. Part I. classification, composition, properties, derivatives, applications. Электронная обработка материалов **1**(57), 52–69 (2021). https://doi.org/10.52577/eom.2021.57.3.83
11. Vrabie, E.G., et al.: Nanotechnological aspects at electro-activation of secondary dairy products. In: Tiginyanu, I., Sontea, V., Railean, S. (eds.) ICNBME 2019. IP, vol. 77, pp. 799–803. Springer, Cham (2020). https://doi.org/10.1007/978-3-030-31866-6_142
12. Vrabie, E.G., et al.: Electrical processing of whey. Role of construction, technological and energy characteristics of reactors. Surf. Eng. Appl. Electrochem. **55**(2), 197—209 (2019). https://doi.org/10.3103/S1068375519020145
13. Laemmli, U.K.: Cleavage of structural proteins during the assembly of the head of bacteriophage T4. Nature **227**(5259), 680–685 (1970). https://doi.org/10.1038/227680a0

Silver Nanoparticles as Stimulators in Biotechnology of *Porphyridium cruentum*

Liliana Cepoi[1](✉), L. Rudi[1], T. Chiriac[1], A.Valuta[1], I. Zinicovscaia[2,3], V. Miscu[1], and V. Rudic[1]

[1] Institute of Microbiology and Biotechnology, Ministry of Education and Research, Chisinau, Republic of Moldova

[2] Joint Institute of Nuclear Research, Dubna, Russia

[3] Horia Hulubei National Institute for Physics and Nuclear Engineering (IFIN-HH), Bucharest, Magurele, Romania

Abstract. Silver nanoparticles (AgNPs) are one of the most actively used nanomaterials. Therefore, their release into the environment becomes inevitable. Consequently, studies regarding the harmful effects of AgNPs towards organisms are very common. In the process of investigating the effects of silver nanoparticles on living cells, other effects besides the toxic ones were revealed. Thus, different nanoparticles can stimulate biosynthetic processes, which will significantly expand their application in biotechnology. The aim of this paper was to elucidate the possibility of using small silver nanoparticles (10–20 nm) stabilized with polyethylene glycol and citrate as stimulators in the biotechnology of microalga Porphyridium cruentum. This red microalga is recognized as a producer of polysaccharides, proteins, polyunsaturated fatty acids. Silver nanoparticles stabilized with polyethylene glycol of 12 nm in size, as well as nanoparticles stabilized with citrate of 10 and 20 nm in size, were used in concentrations ranging from 0.01 to 1.0 μM. Concentrations were determined for each type of particles, where an increased accumulation of polysaccharides and proteins was observed in Porphyridium cruentum biomass. Moreover, the lack of toxicity of these nanoparticles was confirmed by maintaining a high level of productivity.

Keywords: Silver nanoparticles · Stabilizing agents · *Porphyridium cruentum* · Biomass · Proteins · Carbohydrates

1 Introduction

Silver nanoparticles are widely applied in various fields, such as optoelectronics, biomedicine, cosmetology, food packaging, and other fields, while the problem of their release into the environment and toxicity to living organisms is one of the most intensively studied [1]. Algae, as primary producers in the food chain, are the most commonly used organisms for assessing the toxicity of nanoparticles [2], and their toxic effects depend on shape, size, chemical composition, concentration, solubility, degree of dispersion of nanoparticles, characteristics of microalgae cell and, equally important, the environment [3].

I. Tiginyanu et al. (Eds.): ICNBME 2021, IFMBE Proceedings 87, pp. 530–536, 2022.
https://doi.org/10.1007/978-3-030-92328-0_68

At the same time, silver nanoparticles can stimulate different processes in the cells of microalgae and cyanobacteria [4]. For instance, polyethylene glycol-coated small-sized nanoparticles enhanced the productivity of *Dunaliella salina* by 15% and carotene content in biomass by 20–36% [5]. Small nanoparticles in polyethylene glycol coating were non-toxic or showed low toxicity towards cyanobacterium *Arthrospira platensis* grown in a closed system [6].

Porphyridium cruentum is a marine red microalga known as a producer of omega-3 polyunsaturated fatty acids and sulfated polysaccharides known to inhibit retrovirus replication [7]. Meanwhile, various substances have been proposed as stimulators of the biosynthetic processes of porphyridium -organic compounds, mineral compounds and *3d* transition metal complexes [8, 9]. *Porphyridium cruentum* can serve as a matrix for the biosynthesis of metal nanoparticles [10]. It has also been shown that CdSe nanoparticles in concentrations of 4.0–8.0 mg/L ensured an increase in the amount of biomass by 18.4–47%. This stimulating effect can probably be explained by the formation of a hybrid NP-phycoerythrin system, which captured very efficient light energy [11].

In this paper, we aimed to elucidate the possibility of applying small silver nanoparticles (10–20 nm) as stimulators in the biotechnology of microalga *Porphyridium cruentum.*

2 Material and Methods

In this study, 12 nm silver nanoparticles stabilized with polyethylene glycol (AgNP(PEG)) manufactured by M9 Company (Tolyatti, Russia), and 10 and 20 nm stabilized with citrate (AgNP(CYT)) (Sigma) were used. The strain *Porphyridium cruentum* CNMN-AR-01 grown on Brody's nutrient medium was used. The cultivation conditions were as follows: the quantity of inoculums - 0.55–0.6 g/L absolutely dry biomass (ADB); temperature of 25–28°C , pH of the medium 6.8–7.2, and continuous illumination of 50–57 μmol photons m^{-2} s^{-1}. Cultivation was carried out in 100 ml Erlenmeyer flasks with a working volume of 50 ml. The cultivation cycle lasted 14 days.

The quantity of biomass was determined by recording the absorbance of the cell suspension at a wavelength of 565 nm, and the quantitative calculation was based on the calibration curve. Estimation of carbohydrate content in algal biomass was determined by the Anthrone method [12], and protein content – by the Lowry method [13].

3 Results and Discussions

Silver nanoparticles were added in concentrations between 0.01 and 1.0 μM prior to the inoculum was added. Microalgal cells were in contact with nanoparticles throughout the entire cultivation cycle. At the end, the biomass was collected, its quantity and the content of proteins and carbohydrates were determined. The amount of biomass varied in the experimental samples within ranges very close to control samples (Fig. 1).

An increase of 10–11% was found for concentrations of 0.01, 0.025 and 0.25 μM AgNP(PEG) and 0.25 μM AgNP(CYT) of 10 nm in size. When applying 20 nm AgNP(CYT) in concentration of 0.5 μM, a slight decrease in the amount of *Porphyridium cruentum* biomass was observed. There was no significant dependence of this response

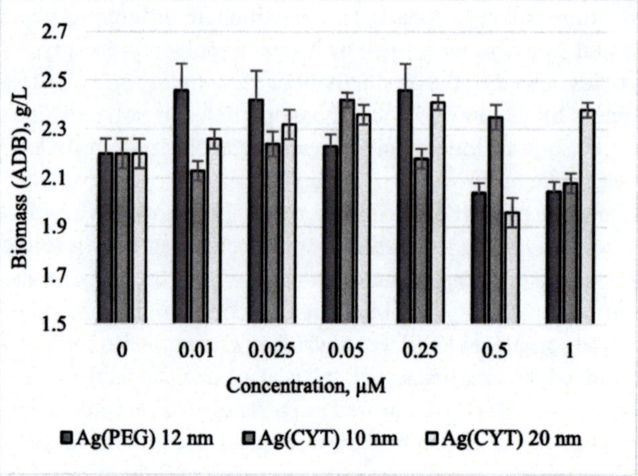

Fig. 1. Changes in the amount of *P. cruentum* biomass grown in the presence of AgNPs in concentrations of 0.01–1.0 µM

on the concentration and size of nanoparticles. In addition, the stabilizer did not appear to alter the culture response either.

Organic or bioorganic stabilizing compounds used to coat the metal cores of nanoparticles can affect cultures of microalgae and cyanobacteria in different ways [14]. Some authors argued that stabilizers reduce toxicity, while others showed reverse effects. For instance, the toxicity of 5 nm AgNPs stabilized with poly(N-vinyl-2-pyrrolidone)/polyethyleneimine (PVP/PEI) towards microalga *Phaeodactylum tricornutum* exceeded the toxicity of large nanoparticles - 47 nm without a coating. It is assumed that shell components are involved in the formation of small conglomerates with a higher affinity for the algal cell membrane [14, 15].

For the most part, studies of the effects of nanoparticles on aquatic microorganisms aim to identify their toxic effects. The contact of AgNPs 2 nm and 15 nm in size at concentrations of 10–300 µg/L with microalgae *Chlamydomonas reinhardtii* and *Phaeodactylum tricornutum* led to a sharp decrease in the amount of biomass [16]. At the same time, silver nanoparticles are less toxic than ions [17].

Recently, there have been reports upon the stimulatory effect of silver nanoparticles on microalgae. For example, 10–150 µg/L of 30–50 nm AgNPs enhanced the growth of microalgae *Phaeodactylum tricornutum* [16]. Silver nanoparticles 12 nm in size in a polyethylene glycol coating in concentrations from 0.025 to 0.5 µM favoured the growth of *Spirulina platensis* biomass by 24–36% [6].

AgNP(PEG) added to the mineral medium in concentrations of 0.054–0.108 mg/L stimulated the accumulation of *Dunaliella salina* biomass by 12–33% [5], and for other AgNPs at the same concentrations, the inhibitory effect of *D. salina* growth was found [18]. At the same time, quite high concentrations of AgNPs used in the cultivation of *Chlamydomonas reinhardtii* (10–300 µg/L) for 24 h did not alter the cell density [16].

In this study, silver nanoparticles stabilized with polyethylene glycol in concentrations of 0.01–0.25 µM enhanced the growth of *P. cruentum*, while in the concentration range of 0.01–1.0 μM, there was a tendency toward a decrease in the amount of biomass. In the case of citrate-stabilized nanoparticles, no specific pattern of action was revealed, but it can be argued that silver nanoparticles in the tested concentrations were not toxic to *P. cruentum* culture. One of the responses of porphyridium culture to stress was accelerated synthesis and accumulation of carbohydrates in biomass. Starting from the biological role of sulfated polysaccharides, which constitute an important fraction of carbohydrates, and the fact that *P. cruentum* is grown on an industrial scale to obtain these substances, this parameter was determined in the resulting biomass (Fig. 2).

Concentrations of AgNP(PEG) that stimulated biomass production also resulted in an increase in carbohydrate content by more than 30%.

For a concentration of 0.25 μM AgNP(CYT) of 10 nm in size, there was recorded a 20% increase in carbohydrate content. The carbohydrate content also increased by 20% in biomass obtained in the presence of 0.01 and 0.05 μM AgNP(CYT) of 20 nm. Thus, studied silver nanoparticles stimulated the accumulation of carbohydrates in biomass, and the effect depended on the concentration.

Carbohydrates are the main photosynthetic products that make up the carbon supply. The increase in carbohydrate production may be associated with a defense mechanism in response to stress caused by changes in environmental conditions. The protective action of carbohydrates against oxidative stress is based on the annihilation of reactive oxygen species. An increase in carbohydrate content by 20, 22 and 80% was established for concentrations of 360, 720 and 1440 μg/L AgNPs with a size of 140 nm upon 24-h contact with green microalga *Chlorella vulgaris*. It is assumed that the culture response was caused by severe stress, since the applied concentrations significantly reduced the number of algal cells [19].

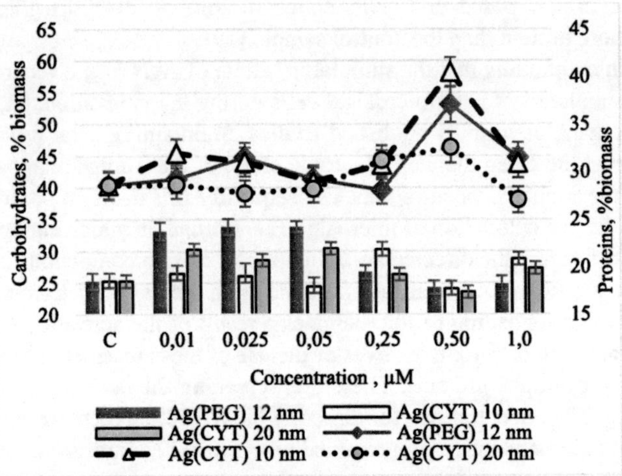

Fig. 2. Carbohydrate and protein content in *P. cruentum* biomass grown in the presence of AgNPs in concentrations of 0.01–1.0 μM. (Columns – carbohydrates; lines – proteins)

At the same time, when microalga *Skeletonema costatum* was exposed to oleyamine-capped 10 nm AgNPs in concentrations from 0.05 to 5.0 mg/L, no changes in the carbohydrate content were observed, while the stress intensity was very high [20]. Thus, in some cases, this may be evidence for the involvement of nanoparticles in cellular biosynthetic activity with a stimulating effect that does not necessarily lead to oxidative stress in the cell. In the case of cultivation of microalga *Porphyridium cruentum* in the presence of low concentrations of AgNPs, a stimulatory effect can be revealed, which led to an increase in the carbohydrate content. An obvious stimulatory effect was determined for silver nanoparticles stabilized with polyethylene glycol.

Protein content is one of the most important biochemical parameters characterizing the biomass quality. In most experimental variants, the amount of proteins in porphyridium biomass was at the control level (Fig. 2). The concentrations were determined for all types of used nanoparticles, which acted as stimulators of protein accumulation in biomass. Thus, for AgNPs(PEG), a concentration of 0.5 μM was established, which increased the protein content by 39%. The same concentration of AgNP(CYT) of 10 nm provided an increase in the amount of proteins by 42%, and AgNP(CYT) of 20 nm in size - by 14%. The nanoparticle concentrations of 0.01–0.025 μM and 1.0 μM did not change the protein content in biomass.

In the presence of 0.5 μM AgNPs, carbohydrate content in *P. cruentum* biomass was below the limit of control values, which indicated their intensive consumption. Some authors suggest that protein accumulation in the presence of low concentrations of silver nanoparticles as a result of the activation of detoxification mechanisms. This strategy may be one of the ways in which algae can eliminate the toxic effects of metals, using carbohydrates in favour of protein accumulation [21, 22]. Silver is known to have an affinity for sulfur, resulting in an increase in the content of sulfated polysaccharides and sulfur-containing amino acids [23]. The increase in the amount of proteins in microalgal biomass under the action of AgNPs has been observed by other authors. Thus, microalga *Chlorella vulgaris*, exposed to AgNPs concentrations of 90–720 μg/L, accumulated 110–293% more protein than the control sample [19].

It is worth mentioning that the stimulatory effect of AgNPs was recorded due to the contact of nanomaterials with microalgal cells during the cultivation, and not during a short-term contact, as in most published studies. Maintaining algal productivity at an acceptable level indicated the lack of a toxic effect of the studied nanoparticles within the limits of the applied concentrations. Consequently, this makes it possible to develop technologies for the cultivation of microalga *P. cruentum*, in which nanoparticles act as stimulators that ensure the directed accumulation of certain compounds in biomass.

The concentration of nanoparticles was the main factor that determined changes in the biochemical composition of microalga as a result of the activation of mechanisms for the accumulation of carbon reserves or the use of these reserves for other groups of substances, for example, proteins. Therefore, screening the doses of nanoparticles can reveal the optimal concentrations for directing biosynthetic processes in microalgae in order to obtain biomass with a programmed biochemical composition.

4 Conclusions

Silver nanoparticles coated with polyethylene glycol and citrate 10–20 nm in size at certain concentrations can stimulate some biosynthetic processes in red microalga *Porphyridium cruentum*, which can serve as a basis for their use in biotechnology.

Silver nanoparticles stabilized with polyethylene glycol at concentrations of 0.01–0.25 μM stimulated the accumulation of biomass and increase polysaccharide content by more than 30%.

Silver nanoparticles stabilized with citrate, 10 and 20 nm in size, at a concentration of 0.01 and 0.05 μM, also increased the carbohydrate content by 20% compared to control sample.

It was found that the concentration of 0.5 μM for all types of studied nanoparticles had a stimulating effect on the accumulation of proteins in *Porphyridium cruentum* biomass. In the case of silver nanoparticles coated with polyethylene glycol, the protein content in algal biomass increased by 39%, in the case of AgNP (CYT) 10 nm - by 42%, and AgNP (CYT) 20 nm - by 14%.

The concentration of nanoparticles was the main factor determining changes in the biochemical composition of microalgae due to the activation of mechanisms for accumulating carbon reserves or using these reserves for other groups of substances, for example, proteins.

Acknowlegment. The results were obtained within the project 20.80009.5007.05, NARD, Republic of Moldova.

Conflict of Interest. The authors declare that they have no conflict of interest.

References

1. Rudramurthy, G.R., Swamy, M.K.: Potential applications of engineered nanoparticles in medicine and biology: an update. J. Biol. Inorg Chem. **23**, 1185–1204 (2018)
2. Wang, F., Guan, W., Xu, L., et al.: Effects of nanoparticles on algae: adsorption, distribution ecotoxicity and fate. Appl. Sci. **9**, 1534 (2019)
3. Bundschuh, M., Seitz, F., Rosenfeldt, R.R., et al.: Effects of nanoparticles in fresh waters: risks, mechanisms and interactions. Freshwater Biol. **61**, 2185–2196 (2016)
4. Vargas-Estrada, L., Torres-Arellano, S., Longoria, A., et al.: Role of nanoparticles on microalgal cultivation: a review. Fuel **280**, 118598 (2020). https://doi.org/10.1016/j.fuel.2020.118598
5. Maftei, E., Rudic, V.: The use of gold and silver nanoparticles in cultivation of microalga Dunaliella salina. Buletinul AŞM. Ştiinţele vieţii **3**(336), 159–165 (2018)
6. Cepoi, L., Zinicovscaia, I., Rudi, L., et al.: Effect of PEG-coated silver and gold nanoparticles on spirulina platensis biomass during its growth in a closed system. Coatings **10**, 717 (2020). https://doi.org/10.3390/coatings10080717
7. Lutzu, G.A., Zgang, L., Zgang, Z., et al.: Feasibility of attached cultivation for polysacharides production by Porphyridium cruentum. Bioprocess Biosyst. Eng. **40**, 73–83 (2017)
8. Coropceanu, E., Rudic, V., Cepoi, L., et al.: Synthesis and cristal structure of $[Co(DmgH)_2(Thio)_2]_2F[PF_6]$. The effect of fluorine-containing Co(III) dioximates on the physiological proceses of themicroalga porphyridium cruentum. Russ. J. Coord. Chem. **45**, 200–207 (2019)
9. Lu, X., Nan, F., Feng, J., et al.: Effects of different environmental factors on the growth and bioactive substance accumulation of porphyridium purpureum. Int. J. Environ. Res. Public Health **17**(7), 2221 (2020)

10. Cepoi, L., Rudi, L., Zinicovscaia, I., et al.: Biocemical changes in microalga porphyridium cruentum associated with silver nanoparticles biosynthesis. Arch. Microbiol. **203**, 1547–1554 (2021)
11. Rudic, V., Cepoi, L., Gutsul, T., et al.: Red algae porphyridium cruentum growth stimulated by CdSe Qantum Dots covered with thioglycerol. J. Nanoelectron. Optoelectron. **7**(7), 681–687 (2012)
12. Hedge, J., Hofreiter, B.: Methods of Estimating Starch and Carbohydrates, Carbohydrate Chemistry, pp. 163–201. Academic Press, New York, NY. (1962)
13. Lowry, O., et al.: Protein measurement with the folin phenol reagent. J. Biol. Chem. **193**(1), 265–275 (1951)
14. Tsiola, A., Pitta, P., Callol, A.J., et al.: The impact of silver nanoparticles on marine plankton dynamics: dependence on coating, size and concentration. Sci. Total Environ. **601–602**, 1838–1848 (2017)
15. Schiavo, S., Duroudier, N., Bilbao, E., et al.: Effects of PVP/PEI coated and uncoated silver NPs and PVP/PEI coating agent on three species of marine microalgae. Sci. Total Environ. **577**, 45–53 (2016)
16. Sendra, M., Yeste, M.P., Gatica, J.M.: Direct and indirect effects of silver nanoparticles on freshwater and marine microalgae (Chlamydomonas reinhardtii and Phaeodactylum tricornutum). Chemosphere **179**, 279–289 (2017)
17. Di, H., Dorantes-Aranda, J.J., Waite, T.D.: Silver nanoparticle algae interactions: oxidative dissolution, reactive oxygen species generation and synergistic toxic effects. Environ. Sci. Technol. **46**, 8731–8738 (2012)
18. Johari, S.A., Sarkheil, M., Tayemeh, M.B., et al.: Influence of salinity on the toxicity of silver nanoparticles (AgNPs) and silver nitrate (AgNO3) in halophilic microalgae Dunaliella salina. Chemosphere **209**, 156–162 (2018)
19. Romero, N., et al.: Physiological and morphological responses of green microalgae Chlorella vulgaris to silver nanoparticles. Environ. Res. **189**, 109857 (2020)
20. Huang, J., Cheng, J., Yi, J.: Impact of silver nanoparticles on marine diatom Skeletonema costatum: Silver nanoparticles are phototoxic to marine diatom. J. Appl. Toxicol. **36**, 1343–1354 (2016)
21. Marchello, A.E., Barreto, D.M., Lombardi, A.T.: Effects of titanium dioxide nanoparticles in different metabolic pathways in the freshwater microalgae Chlorella sorokiniana (Trebouxiophyceae). Water Air Soil Pollut. **229**(48) (2018). https://doi.org/10.1007/s11270-018-3705-5
22. Miao, A.J., Schwehr, K.A., Xu, C., et al.: The algal toxicity of silver engineered nanoparticles and detoxification by exopolymeric substances. Environ. Pollut. **157**, 3034–3041 (2009)
23. McShan, D., Ray, P.C., Yu, H.: Molecular toxicity mechanism of nanosilver. J. Food Drug Anal. **22**(1), 116–127 (2014)

Study the Effect of UVC Radiation on Specific Regions of the SARS-CoV-2 Coronavirus Genome Encoding the Synthesis of Structural Proteins

Iurie Nica(✉), L. Pogorelischi, S. Zavrajny, V. Dimitriu, L. Peev, and A. Sidorenko

Institute of Electronic Engineering and Nanotechnologies "D. Ghitu", Chişinău, Republic of Moldova
tehmed@nanotech.md

Abstract. The COVID19 pandemic that terrorizes the world with terrible aggression calls for the urgent finding of tools that would quickly inactivate viruses in the environment to reduce the chance of aerosol infection and contact transmission. For the inactivation of SARS – CoV – 2 viruses we used LEDs with maximum emission on the wavelength 255 ± 5 nm. All sources of UVC bactericidal radiation have emission peaks close to the center of the DNA and RNA absorption spectrum and may underlie devices and therapies to disrupt the spread of infection.

We studied the inactivation of SARS – CoV – 2 viruses by annihilating the structural components of the viruses that manifest in the amplification process in the PCR procedure.

Keywords: SARS-CoV-2 · COVID-19 · N,E – genes · UVC radiation

1 Introduction

Hundreds of studies and many applications (decontamination of enclosed spaces, water, easily alterable products, etc.) have demonstrated the ability of UVC bactericidal ultraviolet light (200–280 nm) to kill bacteria, viruses and fungi by inducing photodimerization in the genomes of microorganisms or the destruction of the protein membrane. In the last 20 years, especially after the acute respiratory syndrome coronavirus (SARS – CoV) epidemic from 2002–2003, many experiments have been performed to determine the energy doses capable of inactivating coronaviruses. For the 90% inactivation rate of viruses, the range of energy dose variation is estimated to be between 6.6 and 2410 J/m^2 [1]. Almost all experiments were performed with mercury vapor lamps with peak emission at 254 nm (UVC), which is close to the RNA uptake peak. Individual investigations were performed with peak wavelengths at 222 nm (UVC), or 365 nm (UVA), or even in daylight [2]. With the apparition of the SARS – CoV – 2 virus and the spread of the COVID19 pandemic, research was intensified to determine the radiation doses capable of annihilating the new virus. In published studies, lethal doses for SARS – CoV – 2

I. Tiginyanu et al. (Eds.): ICNBME 2021, IFMBE Proceedings 87, pp. 537–543, 2022.
https://doi.org/10.1007/978-3-030-92328-0_69

viruses range from 34 J/m^2 (in wet droplets) [3] to 10476 J/m^2(in the SARS-CoV-2 culture isolated from a nasopharyngealswab of a patient suffering from COVID-19 disease) [4]. Intermediate lethal doses were reported in [5–7].

To study the influence of UVC bactericidal radiation on SARS-CoV-2 viruses, which causes COVID-19, the impact of UVC on the structural components of viruses that manifest in the amplification process in the PCR procedure can be investigated. Figure 1 schematically shows the structure of the SARS-CoV-2 virus and the organization of genes in the COVID - 19 genome. This type of virus contains single-stranded ribonucleic acid (RNA) genetic material with positive polarity [(+) ssRNA]. The virus consists of a nucleocapsid, which in turn is composed of (+) ssRNA and nucleoprotein (N gene). This structure is covered by a lipid layer. Here are other structural proteins of the coronavirus, such as the spike protein (S gene), which covers this virus particle, the extremely hydrophobic shell protein (E gene) and the membrane protein (M gene).

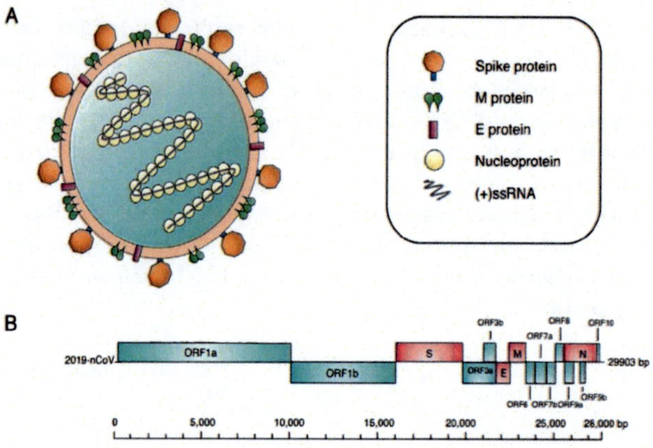

Fig. 1. A) Coronavirus particle. (B) Organization of genes in the 2019-nCoV genome. Structural genes are shown in pink and non-structural genes are displayed in blue [8].

The advantage of molecular genetic methods for detecting a new coronavirus, including the reverse transcription and real-time polymerase chain reaction (RT qPCR) method, compared to other virus detection methods is its high sensitivity, speed of results, standardization and technological effectiveness of the study. It is important that no manipulation with live viral cultures is required, which serves to prevent the spread and circulation of viruses in medical and laboratory diagnostic facilities. The method is designed to detect coronavirus RNA SARS-CoV-2, which causes COVID-19 coronavirus infection, in human biological material (nasopharyngeal, oropharyngeal, bronchoalveolarlavage, endotracheal, nasopharyngeal aspirate, sputum) by RT qPCR). It consists of repetitive cycles of thermal denaturation of cDNA, annealing of primers with complementary sequences and subsequent expansion of the polynucleotide chains in these primers with Taq polymerase and is performed with the Amplifier that detects DTprime [9].

The RT qPCR multiplex detection kit contains in the reaction mixture for amplification deoxyribonucleic acid (DNA) indicators, each of which contains a certain fluorescent label and a fluorescence extinguishing agent. When a specific product is formed, the DNA indicator is destroyed, the effect of the extinguisher on the fluorescent label is completed, which leads to an increase in the level of fluorescence. The number of indicators destroyed (and therefore the level of fluorescence) increases in proportion to the number of specific amplicons formed and is measured at each amplification cycle.

SARS-CoV-2/SARS-CoV Multiplex REAL TIME PCR Detection Kit includes fluorescent dye (Hex) for internal control (RNA-IC), which is intended to evaluate the quality of RNA extraction and polymerase chain reaction, fluorescent dye (Fam) used to detect SARS-CoV-like coronavirus amplification product, fluorescent dye (Rox) to detect SARS-CoV-2 amplification (gene E) and fluorescent dye (Cy5) to detect SARS-CoV-2 amplification (N gene).

2 Materials and Methods

To clarify the degree of influence of the UVC on SARS-COV-2, we conducted a number of experiments with the irradiation of biomaterial containing this virus.The biomaterial, which was irradiated, was chosen in the process of PCR-testing of the population and contained a SARS-COV-2. The selected material was applied with a thin layer on the slide glasses or the surface inside the Petri dish and exposed to various doses of UVC. In order for the material under study to form a uniform thin layer on the surface, the slide glasses were pre-exposed to special processing, which provided the maximum adhesion of the virus-containing substrate to the glass.

The scientific community has discovered and confirmed that the radiation spectrum that causes a noticeable effect of deactivation of microorganisms corresponds to the absorption spectrum of nucleic acids with maximum absorption at the wavelengths of about 258 nm [10, 11]. In this regard, we have manufactured and applied a LED matrix emitting UVC with a power of 190 mW in a narrow wavelength range near 255 nm.

For the irradiation of the material containing, an installation was developed and manufactured, which includes:

1. the table for placing slides or Petri dishes with samples, a LED matrix holder and an electromagnetic shutter mounted as a single module;
2. the control unit, which includes a timer for controlling the shutter, a current stabilizer through a LED matrix;
3. the device is powered by AC ~20 V and the stabilization of the operating modes of internal electrical circuits is provided by the AC/DC 220 V/30 V 2A adapter and stabilizing DC/DC converters.

Each individual portion of the biomaterial (sample) was irradiated with a certain dose. The dose was determined by the duration of irradiation with a constant power density on the sample surface of 4 mW/cm^2. One of the samples was not exposed to radiation. Then, all samples, including the non-irradiated one, were analyzed using real-time PCR. The effect of irradiation on the virus was evaluated by comparing the data of PCR analysis of irradiated samples with data for non-irradiated sample.

In accordance with the diagnostic requirements of the SARS-CoV-2 coronavirus, the effect of radiation on three characteristic regions of the viral genome was studied [12, 13]:

- SARS-CoV-2-specific sections of gene E and gene N;
- conservative section of the E gene common to a group of coronaviruses similar to SARS-CoV (including SARS-CoV and SARS-CoV-2).

The degree of influence of irradiation on the target fragments of the virus RNA was determined by evaluating the ratio of the concentration No*i* of target fragments in irradiated samples to their concentration No in a non-irradiated sample.

3 Results

The logarithm values of the ratio of the concentrations in irradiated samples to the concentration in non-irradiated samples obtained in a series of experiments performed in the order described above are presented in Table 1. The change in the content of target genome fragments in irradiated samples is shown in the graphs of the logarithm of the concentration ratio (Fig. 2 and Fig. 3). The data obtained during the experiments showed that irradiation of samples with 255 nm ultraviolet radiation with a power density of 4 mW/cm^2 for 70 s (280 mJ/cm^2) reduces the content of the gene capable of translating the E – protein by 100 times. To reduce the amount of this gene by ~1000 times, an exposure with a duration of 320 s (1.28 J/cm^2) is necessary and exposure to radiation with the above characteristics for a time of about 400 s (~1.6 J/cm^2) will reduce the E – gene content by 10^4 times.

Table 1. The logarithm values of the ratio of the concentrations of the N – gene and E – gene of the SARS-CoV-2 in irradiated samples to the concentration in non-irradiated samples

n.n	Exposure duration, sec	Dose, mJ/cm^2	$\lg\left(\frac{No_i}{No}\right)$	
			E-gene	N-gene
1	0	0	0	0
2	30	120	1.053	1.223
3	50	200	1.249	0.644
4	70	280	2.094	1.526
5	100	400	2.249	1.361
6	150	600	2.683	1.501
7	200	800	2.864	2.324
8	320	1280	2.939	1.434
9	340	1360	3.542	1.788
10	360	1440	3.748	2.034
11	400	1600		3,041
12	600	2400		2,706
13	650	2600		4,176
14	700	2800		3,57

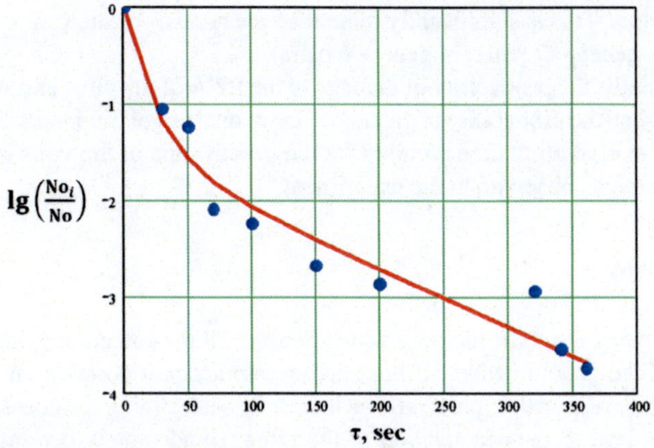

Fig. 2. The change in the concentration of the target fragment of the E – gene of the SARS-CoV-2 virus in the samples during UV radiation exposure (relative to the concentration in the unirradiated sample).

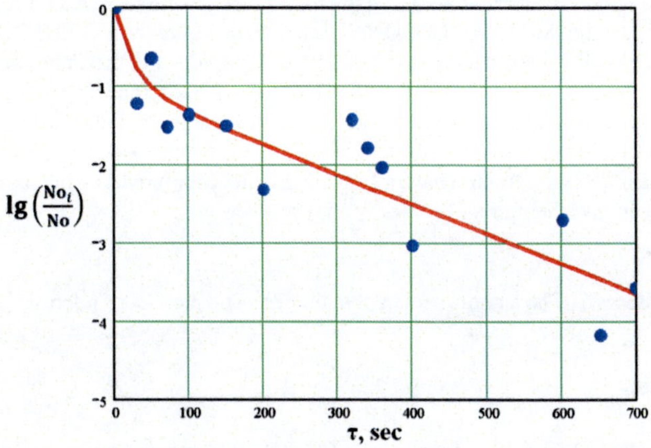

Fig. 3. The change in the concentration of the target fragment of the N – gene of the SARS-CoV-2 virus in the samples during UV radiation exposure (relative to the concentration in the unirradiated sample).

4 Discussion

It follows from Fig. 3 that irradiation of the gene encoding the structure of the N-protein with an ultraviolet light of 255 nm with a power density of 4 mW/cm^2 for ~270 s leads to a 100-fold decrease in the content of the analyzed code fragment. To reduce the number of the target code fragment by 1000 times, irradiation is necessary for ~530 s. The SARS-CoV-2 genome contains ~2700 pairs of sequentially arranged uracil nucleotides, each of which, under the influence of quanta of characteristic ultraviolet radiation, is able to form a dimer with a certain probability, which leads to a violation of the genetic code of the virus. This is significantly more than the number of such pairs in the studied fragments (E-gene: ~17 pairs; N-gene: ~6 pairs).

Consequently, the probability of damage to the RNA of the virus as a whole is many times higher than the probability of damage to individual target fragments. Thus, it can be argued that the irradiation time required for the deactivation of the virus is significantly less than the values observed in the experiment.

5 Conclusions

The peculiarity of the PCR method, which consists in the forced copying of specified fragments of the genome code, without taking into account possible violations of the genome in other responsible parts, allows indirectly determining the doses of ultraviolet radiation that lead to the deactivation of the virus. To accurately determine the doses that completely suppress the activity of the SARS-CoV-2 virus, studies of changes in the population of the irradiated virus in Vero cells cultured in a nutrient medium are required.

Acknowledgments. The current work, TEHMED COVID 20.70086.07/COV(70105), was supported by a grant from the "ERAvsCORONA" Action Plan part of the CG-ENI/2017/386-980 EC "Financial Support for the participation of Republic of Moldova in the Framework Program of the European Union for research and innovation HORIZON 2020", and also of the Moldova State Program, Project «Nanostructuri şi nanomateriale funcţionale pentru industrie şi agricultură» No. 20.80009.5007.11.

We are thankful to Ion Rotari (Invitro Diagnostics SRL) for providing his support and access to PCR testing in his laboratory.

Conflict of Interest. The authors declare that they have no conflict of interest.

References

1. Kowalski, W.J., Walsh, T.J., Petraitis, V.: COVID-19 Coronavirus Ultraviolet Susceptibility (2020). https://www.researchgate.net/publication/339887436
2. Heßling, M., Hönes, K., Vatter, P., Lingenfelder, C.: Ultraviolet irradiation doses for coronavirus inactivation - review and analysis of coronavirus photoinactivation studies. GMS Hyg. Infect. Control (2020). https://doi.org/10.3205/dgkh000343, https://pubmed.ncbi.nlm.nih.gov/32547908/
3. Storm, N., et al.: Rapid and complete inactivation of SARS-CoV-2 by ultraviolet-C irradiation (2020). https://www.nature.com/articles/s41598-020-79600-8#Fig1
4. Heilingloh, C.S., et al.: Susceptibility of SARS-CoV-2 to UV irradiation. Am. J. Infect. Control **48**, 1273–1275 (2020). https://doi.org/10.1016/j.ajic.2020.07.031
5. Bianco, A., et al.: UV-C irradiation is highly effective in inactivating and inhibiting SARS-CoV-2 replication (2020). https://www.medrxiv.org/content/, https://doi.org/10.1101/2020.06.05.20123463v2.full.pdf
6. Inagaki, I., et al.: Rapid inactivation of SARS-CoV-2 with deep-UV LED irradiation. Emerg. Microbes Infect. **9**, 1744–1747 (2020). https://doi.org/10.1080/22221751.2020.1796529

7. Caetano, P.S., et al.: UV-C (254 nm) lethal doses for SARS-CoV-2. Photodiagnosis Photodyn. Ther. **32**, 101995 (2020). https://www.ncbi.nlm.nih.gov/pmc/articles/PMC7477605/
8. Palacios Cruz, M., Santos, E., Velázquez Cervantes, M.A., León Juárez, M.: COVID-19 unaemergencia de saludpúblicamundial. Rev. Clin. Esp. **221**, 55–61 (2021). https://www.ncbi.nlm.nih.gov/pmc/articles/PMC7173827/
9. DNA-Technology Research & Production, LLC: https://www.dna-technology.ru
10. Jagger, J.: Introduction to research in ultraviolet photobiology. Photochem. Photobiol. **7**, 413 (1968). https://doi.org/10.1111/j.1751-1097.1968.tb08029.x
11. Kowalski, W.: Ultraviolet Germicidal Irradiation Handbook. Springer, Berlin, Heidelberg (2009)
12. Гончарова Е.В. и др: Диагностика вируса, вызывающего COVID-19, методом ПЦР в реальном времени ФАРМАКОЭКОНОМИКА, No1, стр. 56,57 (2020)
13. Corman, V.M., et al.: Detection of 2019 novel coronavirus (2019-nCoV) by real-time RT-PCR. Eurosurveillance **25**(3) (2020). https://doi.org/10.2807/1560-7917.es.2020.25.3.2000045

Organic Crystals of *p* - Type TTT_2I_3 and *n* - Type $TTT(TCNQ)_2$ as Prospective Thermoelectric Materials for Biomedical Sensors

I. I. Sanduleac(✉) and S. C. Andronic

Department Electronics and Telecommunications, Technical University of Moldova, Av. Stefan cel Mare 168, 2004 Chisinau, Republic of Moldova
ionel.sanduleac@mt.utm.md

Abstract. In this paper, the prospective of using organic nanostructured crystals of p – type TTT_2I_3 (tetrathiotetracene-iodide) and n – type $TTT(TCNQ)_2$ (tetrathiotetracene-iodide-tetracyanoquinodimethan) as components of thermoelectric biosensors is investigated. A thermoelectric biosensor consists of a p-n module, specially designed to be used as power generator, converting human body heat into small electrical signals, or as local cooler, able to create low temperatures (up to −20°C) on small surfaces. In biomedical applications, the temperature gradients are low and, in order to obtain as much as possible high electrical signal, materials with enhanced thermoelectric properties are required. Organic crystals of TTT_2I_3 and $TTT(TCNQ)_2$ were investigated earlier and it was established that these organic compounds are prospective thermoelectric materials if an appropriate optimization of carrier concentration with further purification of the crystal is performed during synthesis. In the following, the electrical conductivity, thermopower (Seebeck coefficient) and the delivered voltage from a p-n module constructed from the mentioned crystals are calculated for different crystals parameters at room temperature. It is established that a single p-n module made of organic crystals can deliver up to 5 mV under a temperature difference of 20 K around the room temperature if optimization procedures are applied.

Keywords: Nanostructured organic crystal · TTT_2I_3 · $TTT(TCNQ)_2$ · Thermoelectric module · Biosensor

1 Introduction

The growing number of advanced medical devices and sensors has significantly changed the approach of using physics and electronics in healthcare. The recent years have come with a tight interpenetration of medicine with physics, resulting a very promising area of science – bioengineering. In a specific sense, the recent achievements in thermoelectricity are beginning to meet more and more needs of bioengineering. Thermoelectric (TE) devices have no moving parts, are reliable, eco-friendly and provide an accurate temperature control, with fast response time (that happens at the speed of electrons). Medical applications of thermoelectric devices can be divided in two categories: cooling/heating systems and power generation.

I. Tiginyanu et al. (Eds.): ICNBME 2021, IFMBE Proceedings 87, pp. 544–551, 2022.
https://doi.org/10.1007/978-3-030-92328-0_70

Thermal cooling systems are used in medical refrigerators and chiller/heater and wrap systems. Precise temperature control is necessary for safe transportation of vaccines, biological samples or drugs, for performing medical procedures or to control PCR (Polymerase Chain Reaction) [1, 2].

Thermoelectric generators (TEG) can be used to convert body heat into electricity. The low delivered current is sufficient to power the electronic parts of other biosensors. Wearable TEGs are prime candidates to power wireless sensors for continuous remote patient monitoring, while implantable devices cand act as pacemakers, drug pumps or neurological stimulators [3].

Thermal imaging systems use TEGs that convert small amount of infrared radiation, coming from human bodies, into electrical signals. In airports, train stations or other agglomerated places, these devices can quickly identify ill people by temperature monitoring.

Organic thermoelectric materials (OTE) are promising candidates for a wide range of thermoelectric applications. Such compounds have the advantage of a diverse molecular design, good solution processability, flexibility and eco-friendly production technology, abundant raw material. Moreover, with an accurate tuning of carrier concentration and precise synthesis control, organic thermoelectric materials can provide high values of thermopower. In particular, it was found theoretically that in nanostructured organic crystals of TTT_2I_3 type the TE properties can be significantly enhances by simple chemical techniques [4].

In the recent years, very promising results were reported for different polymeric TE compounds [5]. High values of electrical conductivity (1415.7 ± 21.2 S cm^{-1}) and of power factor (61.4 μW m^{-1} K^{-2}) were reported in PEDOT:PSS self-supporting thick films [6]. Simultaneous enhancement of electrical conductivity and Seebeck coefficient was obtained in organic thermoelectric SWNT/PEDOT:PSS nanocomposites (1701 S cm^{-1} and 55.6 μV K^{-1}), resulting in a high power factor of ~526 μW m^{-1} K^{-2} [7]. In the recent years, mixed inorganic-organic TEGs were reported with full fabrication guidelines [8].

In the domain of inorganic thermoelectric materials, values of $P \sim 108$ μW m^{-1} K^{-2} and an output power of 23.6 μW at a temperature gradient of 70 K in a TE devide made of 10 couples of legs were reported in [9]. In carbon fiber thermoelectric generator an output power of 0.87 μW at $\Delta T = 75$ K with a voltage of V 19.56 mV was reported [10]. In (PEDOT:PSS)/tellurium nanowires (TeNWs) composites a power factor (78.1 μW m^{-1} K^{-2}) was reported [11].

In this paper, we investigate the thermoelectric properties of organic crystals of TTT_2I_3 and $TTT(TCNQ)_2$ as reliable compounds for thermoelectric sensors. Earlier, it was established that thermoelectric properties can be enhanced by simple adjustment of charge carrier concentration and further crystal purification. Also, it was demonstrated that for near-to room temperatures, the electrical conductivity and Seebeck coefficient remains practically constant. The output voltage of a p-n organic thermoelectric module made of the mentioned materials is calculated for different parameters.

2 Physical Model of the Crystal

Internal structure of TTT_2I_3 and $TTT(TCNQ)_2$ crystals is consisting of stack-like packaging of molecules. In perpendicular to stacks direction, the molecules form linear chains. The electrical properties are very anisotropic, and the crystals are belonging to the class of quasi one-dimensional (Q1D) structures. In the following, the chain direction will be considered as x – direction of a cartesian coordinate system. In TTT_2I_3 crystal, the main charge carriers are holes on TTT chains, iodine chains having a low, negligible electrical conductivity. In the direction of TTT chains, the overlap of electronic wavefunctions opens a narrow conduction band of ~ 0.64 eV [12]. The lattice constants are $b = 4.96$ Å, $a = 18.35$ Å and $c = 18.46$ Å for x, y and z directions. The electrical conductivity of $TTT(TCNQ)_2$ crystals is provided by electrons, travelling along TCNQ chains. In this case, the conduction band has a width of ~ 0.5 eV. The lattice constants are $b = 3.75$ Å, $a = 12.97$ Å and $c = 19.15$ Å, for x, y and z directions [13].

The energy of charge carriers is obtained in the tight binding and nearest neighbor approximations:

$$E(\boldsymbol{k}) = \mp 2w_1[1 - \cos(k_x b)] \mp 2w_2[1 - \cos(k_y a)] \mp 2w_3[1 - \cos(k_z c)] \tag{1}$$

where the sign "–" is for p – type crystal and "+" is for n – type crystal. The energy is renormalized in order to be measured from the top (or from the bottom) of the conduction band. w_1, w_2 and w_3 are the transfer energies of a charge carrier from the given molecule to the nearest ones along lattice vectors $\boldsymbol{b}$, $\boldsymbol{a}$ and $\boldsymbol{c}$, (k_x, k_y, k_z) are the projections of the 3D quasi-wave vector $\boldsymbol{k}$. Considering the pronounced quasi-one dimensionality of the crystal, it was considered that w_1 is much bigger than w_2 and w_3.

As can be shown, the charge carriers interact mainly with the longitudinal acoustic phonon modes of frequency $\omega_{\boldsymbol{q}}$, which has the form:

$$\omega_{\boldsymbol{q}}^2 = \omega_1^2 \sin^2(bq_x b/2) + \omega_2^2 \sin^2(q_y a/2) + \omega_3^2 \sin^2(q_z c/2) \tag{2}$$

here ω_1, ω_2 and ω_3 are the cut-off frequencies along x, y and z directions. Because of quasi-one dimensionality $\omega_1 \gg \omega_2 \approx \omega_3$ and we considered only the first term in (2).

The mean polarizabilities of TTT and TCNQ molecules are $\alpha_0 = 46$ Å^{-3} and $\alpha_0 = 10.2$ Å^{-3}, respectively. The acoustic vibrations of molecules surrounding the conduction hole (or electron in case of n – type) induces the variation in polarization energy. A supplementary electron-phonon interaction, of the polaron-type, appears in addition to the mechanism of the deformation potential type. In order to describe the ratio of amplitudes of the polaron-type to the deformation type interaction mechanisms, the dimensionless parameter was defined [14]:

$$\gamma_1 = 2e^2\alpha_0/b^5|w_1'|, \gamma_2 = 2e^2\alpha_0/a^5|w_2'| \\ \gamma_3 = 2e^2\alpha_0/c^5|w_3'| \tag{3}$$

here $|w_1'|$, $|w_2'|$, $|w_3'|$ are the derivatives of charge carrier transfer energy with respect to the intermolecular distances for x, y and z directions.

The interaction of charge carriers with point-like, electrically neutral impurities and lattice defects:

$$D_0 = \left(n_i I_i^2 V_{0i}^2 + n_d I_d^2 V_{0d}^2 e^{-\frac{2E_0}{k_0 T}} \right) M v_{s1}^2 / 4b^3 acw_1'^2 k_0 T \tag{4}$$

where n_i is the concentration of impurity centers, and I_i^2, V_{0i}^2- the energy of interaction of the electron with an impurity within region of volume V_{0i}^2. The second term with n_d, I_d^2, V_{0d}^2 describe the lattice structural defects. E_0 is the activation energy of a lattice dislocation, v_{s1} is the sound velocity in the direction of molecular chains.

3 Transport Properties

The charge transport along molecular chains is described in the approximations of a weak electric field and low temperature gradient. The kinetic equation of a Boltzmann type was obtained by the method of retarded, two-particle Green functions. Analytical expressions for electrical conductivity σ_{xx}, the thermopower S_{xx} and the power factor P_{xx} in the direction of molecular chains have the form:

$$\sigma_{xx} = \sigma_0 R_0; S_{xx} = \left(\pm \frac{2|w_1|}{eT} \right) \frac{R_1}{R_0}; P_{xx} = \sigma_{xx} S_{xx}^2 \tag{5}$$

$$\sigma_0 = \frac{2e^2 M v_{s1}^2 |w_1|^3 z}{\pi^3 \hbar abc (k_0 T)^2 w_1'^2} \left(\frac{T_0}{T} \right)^2$$

where k_0 is the Boltzmann constant, e – elementary charge, z is the number of molecular chains per elementary cell. The upper sign in S_{xx} is for p – type crystals and the lower - for n – type. R_0 and R_1 are the transport integrals as follows:

$$\begin{aligned} R_n = & \iiint_{-\pi}^{\pi} dxdydz |\sin x|^3 [1 - \cos x + d_1(1 - \cos y) \\ & + d_2(1 - \cos z) - (1 + d_1 + d_2)\varepsilon_F]^n \times n_k(1 - n_k) / \\ & + \{ \left(1 \mp |\gamma_1| [1 + \beta_{xx}(T - T_0)]^5 \cos x \right)^2 \\ & + \frac{d_1^2}{8 \sin^2 x} \left(1 + 2\sin^2 y \mp 2|\gamma_2| \cos y + |\gamma_2|^2 \right) \\ & D_0 + D_1 \exp(-E_0/k_0 T) \}. \end{aligned} \tag{6}$$

where the upper sign in "$\mp$" is for p – type crystals and the lower for n – type. n_k is the Fermi distribution function in terms of dimensionless quasi-momentum $\boldsymbol{k} = (k_x b, k_y a, k_z c)$ and $\varepsilon_F = E_F/2w_1$ is the dimensionless Fermi energy. The parameters $d_1 = w_2/w_1$ and $d_2 = w_3/w_1$ describe the ratio of transfer energies. D_0 describes the scattering on point-like impurities and D_1 – on thermally activated lattice dislocations. The thermal expansion with a coefficient β_{xx} was considered only in the direction of molecular chains. In transversal directions it is negligible.

4 Numerical Modeling of the Results

The transport integrals (6) were calculated numerically, and the electrical conductivity σ_{xx}, the Seebeck coefficient S_{xx} and the thermoelectric power factor P_{xx} were modeled for different values of dimensionless Fermi energy ε_F with the following crystals parameters:

TTT_2I_3: $M = 6.5 \cdot 10^5 m^e$ (m_e is the mass of the free electron), $w_1 = 0.16$ eV, $w_1' = 0.26$ eVÅ^{-1}, $v_{s1} = 1.5 \cdot 10^3$ m/s, $z = 4$, $d_1 = d_2 = 0.015$, $D_0 = 0.001$, $D_1 = 0.25$, $E_0 = 0.024eV$, The parameters D_0 and D_1 were determined by comparison with experimentally-reported data for electric conductivity of crystals synthesized from gaseous phase [15]. The stoichiometric concentration of conducting holes in TTT_2I_3 crystals was estimated to $n_h = 1.2\cdot 10^{21}$ cm^{-3}, which corresponds to $\varepsilon_F = 0.35$. In this case the reported electrical conductivity $\sigma_{xx} \sim 10^4$ Ω^{-1}cm^{-1} [15].

TTT(TCNQ)$_2$: $M = 3.72\cdot 10^5$ m_e, $w_1 = 0.125$ eV, $w_1' = 0.22$ eVÅ^{-1}, $v_{s1} = 2.8$ 10^3 m/s, $z = 2$, $d_1 = 0.01$, $d_2 = 0.015$, $D_0 = 0.001$, $D_1 = 0.25$, $E_0 = 0.024eV$. Due to the similar internal structure, the parameters D_0 and D_1 were considered as for TTT_2I_3. The stoichiometric concentration of conducting electrons in TTT(TCNQ)$_2$ crystals was estimated to $n_e = 1.1\cdot 10^{21}$ cm^{-3}, which corresponds to $\varepsilon_F = 0.36$. The experimentally reported electrical conductivity at room temperature is $\sigma_{xx} = 0.48\cdot 10^3$ Ω^{-1}cm^{-1} [14].

Electrical conductivity and thermopower of TTT_2I_3 crystals are presented in Fig. 1 as a function of conduction hole concentration at room temperature. For stoichiometric carrier concentration ($\varepsilon_F = 0.35$), the calculated electrical conductivity $\sigma_{xx} = 8.8\cdot 10^3$ Ω^{-1}cm^{-1}, and Seebeck coefficient $S_{xx} = 60$ μV/K near to the experimentally reported values [15].

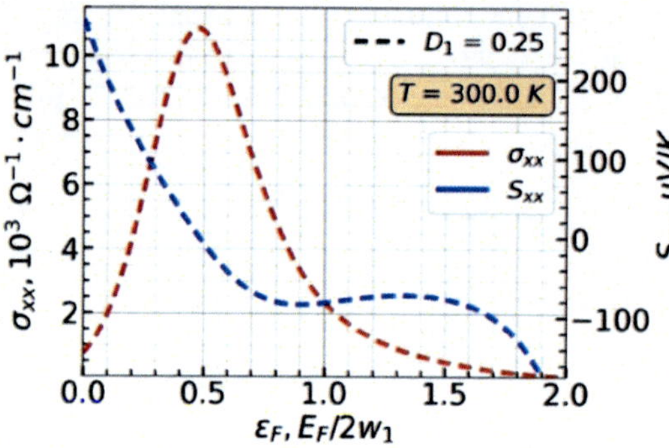

Fig. 1. Electrical conductivity and Seebeck coefficient of TTT_2I_3 crystals as function of dimensionless Fermi energy

From Fig. 1 it is observed that the diminution of ε_F down to 0.18 (wich corresponds to the diminution of carrier concentration by two times), electrical conductivity decreases to $\sigma_{xx} = 3.6\cdot 10^3$ Ω^{-1}cm^{-1} bud Seebeck coefficient and the power factor increases up to $S_{xx} = 150$ μV/K and $P_{xx} = 80$ μV/K^2. This procedure can be performed by evaporating iodine from structure.

In Fig. 2 the electrical conductivity and Seebeck coefficient for TTT(TCNQ)$_2$ crystal are presented. In stoichiometric compounds ($\varepsilon_F = 0.36$), $\sigma_{xx} = 0.48 \cdot 10^3\ \Omega^{-1}\text{cm}^{-1}$, $S_{xx} = -115\ \mu\text{V/K}$ and $P_{xx} = 80\ \mu\text{V/K}^2$.

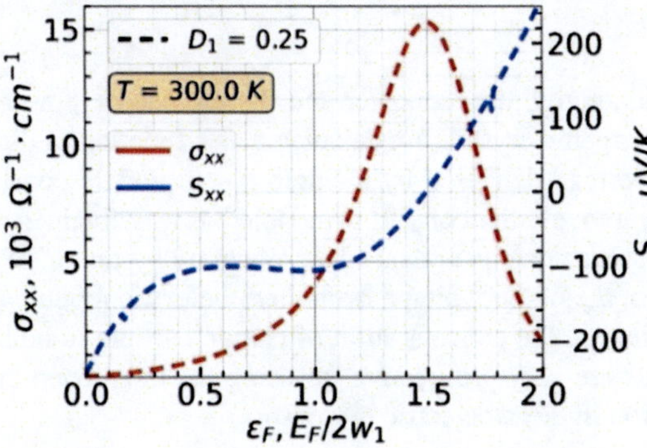

Fig. 2. Electrical conductivity and Seebeck coefficient of TTT(TCNQ)$_2$ crystals as function of dimensionless Fermi energy.

In this case, the optimization procedure consists of additional doping with donors. If the concentration of conducting electrons is increased by 2 times with respect to the stoichiometric one ($\varepsilon_F = 1.04$) one expects $\sigma_{xx} = 4.6 \cdot 10^3\ \Omega^{-1}\text{cm}^{-1}$, $S_{xx} = -105\ \mu\text{V/K}$ and $P_{xx} = 50\ \mu\text{V/K}^2$.

In Fig. 3 the output voltage of a single $p - n$ module, made of a p-type TTT_2I_3 leg and a n − type TTT(TCNQ)$_2$ leg is calculated as a function of dimensionless Fermi energy of electrons in n − leg for three values of dimensionless Fermi energy of holes in p − leg. The parameter describing impurity scattering is considered the same for both crystals. Temperature of the hot and cold sides: $T_{hot} = 320\text{K}$, $T_{cold} = 300$ K.

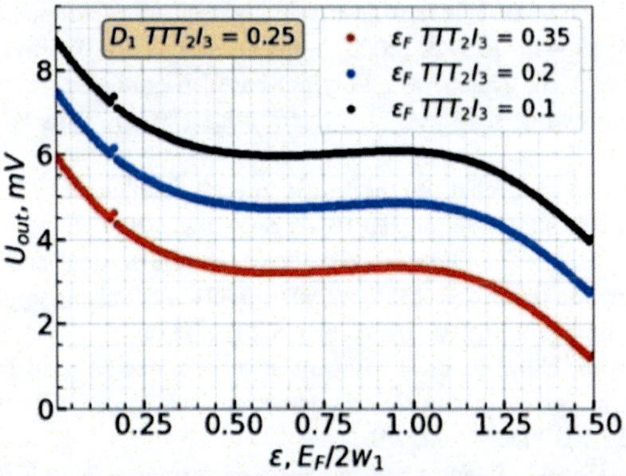

Fig. 3. Thermoelectric voltage generated by a single $p(TTT_2I_3) - n$(TTT(TCNQ)$_2$) module at a temperature difference of 20 K as function of dimensionless Fermi energy of n − leg.

It is observed that for stoichiometric concentration of charge carriers in both p and n legs (i.e. $\varepsilon_F = 0.35$), the TE module can deliver an electrical signal $U_{out} = 3.5\,\text{mV}$. If the optimization is performed, it is predicted $U_{out} = 5\,\text{mV}$.

5 Conclusions

The electrical conductivity, thermopower and thermoelectric power factor of TTT_2I_3 and $TTT(TCNQ)_2$ organic crystals are calculated as a function of dimensionless Fermi energy. The total voltage delivered by a single $p-n$ module constructed from these crystals at a temperature gradient of 20K is modeled. It was obtained that thermoelectric properties of the crystals can be enhanced significantly by tuning of the charge carrier concentration. Also, the output voltage from thermoelectric module varies in the range 3.5–5 mV depending on the concentration of charge carriers in both legs of the $p-n$ module. The results are very prospective for using the mentioned organic crystals for large implementation in thermoelectric biosensors.

Acknowledgement. The authors acknowledge the support of the national project 45/22.10.19 F.

Conflict of Interest. The authors declare that they have no conflict of interest.

References

1. Nohay, J.A.D.: Design and Fabrication of a Portable Solar Powered Thermoelectric Refrigerator for Insulin Storage, report, pp. 150–154 (2020). https://doi.org/10.1109/ICSGRC49013.2020.9232573
2. Csernyus, B., Szabó, Á., Zátonyi, A., et al.: Recent antiepileptic and neuroprotective applications of brain cooling. Seizure (2020). https://doi.org/10.1016/j.seizure.2020.09.018
3. Bhatia, D., Bairagi, S.: Pacemakers charging using body energy. J. Pharm. Bioallied Sci. **2**(1), 51–4 (2010). https://doi.org/10.4103/0975-7406.62713
4. Casian, A.: Violation of the Wiedemann-Franz law in quasi-one-dimensional organic crystals. Phys. Rev. B **81**, 155415–155420 (2000). https://doi.org/10.1103/PhysRevB.81.155415
5. Li, J., Huckleby, A.B., Zhang, M.: Polymer-based thermoelectric materials: a review of power factor improving strategies. J. Materiomics (2021). https://doi.org/10.1016/j.jmat.2021.03.013
6. Sun, Z., Shu, M., et al.: Enhanced thermoelectric performance of PEDOT:PSS. J. Polymer **192**, 122328 (2020). https://doi.org/10.1016/j.polymer.2020.122328
7. Liu, S., Li, H., He, C.: Simultaneous enhancement of electrical conductivity and seebeck coefficient in organic thermoelectric SWINT/PEDOT: PSS nanocomposites. J. Carbon **149**, 25–32 (2019). https://doi.org/10.1016/j.carbon.2019.04.007
8. Jiang, Q., Yang, J., Hing, P., et al.: Recent advances, design guidelines and prospects of flexible organic/inorganic thermoelectric composites. J. Mater. Adv. **5**, 1038–1054 (2020). https://doi.org/10.1039/D0MA00278J
9. Wu, B., Guo, Y., Hou, C., Zhang, Q., et al.: High-performance flexible thermoelectric devices based on all-inorganic hybrid films. Adv. Funct. Mater. **29**, 1900304 (2019). https://doi.org/10.1002/adfm.201900304

10. Karalis, G., Tzounis, L., Lambrou, E.: A carbon fiber thermoelectric generator integrated as a lamina within an 8-ply laminate epoxy composite: efficient thermal energy harvesting by advanced structural materials. Appl. Energy **253**, 113512 (2019). https://doi.org/10.1016/j.apenergy.2019.113512
11. Xu, H., Guo, Y., Wu, B., et al.: Highly integrable thermoelectric fiber. Appl. Mater. Interfaces **12**(29), 33297–33304 (2020). https://doi.org/10.1021/acsami.0c09446
12. Isset, L.C., Perez-Albuerne, E.A.: Low temperature metallic conductivity in bis(tetratiotetracene) triiodide, a new organic metal. Solid State Commun. **21**(5), 433–435 (1977). https://doi.org/10.1016/0038-1098(77)91368-0
13. Buravov, L.I., et al.: Structure and electromagnetic properties of a new high-conductive complex (TTT)+(TCNQ)2-. ZhETF Pis. Red. **20**, 457 (1974)
14. Sanduleac, I., Pflaum, J., Casian, A.: Thermoelectric properties improvement in quasi-one-dimensional organic crystals. J.Appl. Phys. **126**, 175501 (2019). https://doi.org/10.1063/1.5120461
15. Hilti, B., Mayer, C.: Electrical properties of the organic metallic compound bis (Tetrathiotetracene)-Triiodide, (TTT)2I3. Helv. Chim. Acta **61**(40), 501 (1978)

Use of Fractional-Quadratic Approximation Invariant of Nonlinear Characteristic of Magnetoelectric Sensor

A. Penin[1(✉)] and A. Sidorenko[1,2]

[1] Ghitsu Institute of Electronic Engineering and Nanotechnologies, Chisinau, Republic of Moldova

[2] Technical University of Moldova, Chisinau, Republic of Moldova

Abstract. An idea of a sensor of direct magnetic fields using the magnetoelectric effect in a planar ferromagnetic-piezoelectric composite structure is described. It is shown that introduction in the sensor circuit a negative feedback, which contains a current amplifier, a convenient approximation characteristic former, and a compensation coil, results in a widening of the sensor magnetic field region without of linearization of its characteristic. The approximation nonlinear characteristic is close in shape to the original sensor characteristic. Therefore, no deep negative feedback is required and thereby stability is ensured. As this convenient characteristic, a fractional-quadratic relationship is used, for which an invariant is performed between the output voltage value and the magnetic field strength. Such an invariant is a cross ratio (double proportion) for four values or samples of these values. The cross ratio contains the differences of these values and the ratios of these differences. Therefore, additive and multiplicative errors of the output voltage measurements due to the accuracy of the measuring instrument, the noise of the sensor itself and the electronic circuit are mutually reduced.

Keywords: Magnetic field sensor · Nonlinear characteristic · Projective transformation · Invariant · Cross ratio

1 Introduction

Magnetic field sensors using the magnetoelectric *ME* effect in composite planar ferromagnetic-piezoelectric structures have been intensively investigated due to the prospects of their applications in power engineering, transport, navigation systems, geophysics and medicine [1–3]. The output signal vs. magnetic field, as a rule, has a nonlinear shape [4]. In this regard, an urgent task is to develop the methods allowing for widening the operation field range of ME sensors and linearization of their output signal vs. magnetic field characteristic. The inclusion of a compensation circuit in the *ME* sensor measuring scheme makes it possible to expand its operating magnetic field region at least tenfold and to linearize the output voltage vs. measuring field dependence [5].

I. Tiginyanu et al. (Eds.): ICNBME 2021, IFMBE Proceedings 87, pp. 552–559, 2022.
https://doi.org/10.1007/978-3-030-92328-0_71

The fact is that the nonlinear characteristic of a *ME* sensor is a double-valued curve with ascending and descending branches. With sufficiently large values of field strength, the used ascending branch becomes more flat, and then the peak of this characteristic appears. Naturally, measurement errors increase in the flat area of the characteristic. Using only a linear region of this nonlinear characteristic corresponds to a short range of magnetic field strength values.

Introduction of a negative feedback leads to shift of the peak of initial nonlinear characteristic into the region of increased values of magnetic field strength, increase of linear region of ascending branch and decrease of its steepness. This is how the characteristic is linearized. But the more the original nonlinear characteristic differs from the linear dependence, the deeper negative feedback is required. Therefore, measures should be taken to increase the sustainability of feedback.

The present paper demonstrates a possible alternative method of widening the measurement area. For this purpose, a convenient approximation nonlinear characteristic is formed, which is close in shape to the initial or original characteristic of *ME* sensor. Therefore, no deep negative feedback is required and thereby stability is ensured. As this convenient characteristic, a fractional-quadratic relationship is used, for which an invariant is performed between the output voltage value and the magnetic field strength. Such invariant is a cross ratio (double proportion) for four values or samples of these values. The cross ratio contains the differences of these values and the ratios of these differences. Therefore, additive and multiplicative errors of the output voltage measurements are mutually reduced.

So, it is possible to measure and calculate the field strength on the flat area of the characteristic (without linearization) for sufficiently large values of the field strength. Thus, widening of the working area of the ascending branch of the sensor characteristic up to the peak of the initial nonlinear characteristic and stability of operation is achieved.

2 Convenient Approximation Nonlinear Characteristic

We use the sensor parameters [5]. The sensor characteristic is given by

$$U(H) = B \cdot H^b \exp(-C \cdot H^c), \tag{1}$$

where

$$B = 0.7443016,\ b = 0.8702301,$$
$$C = 0.1173849,\ c = 0.9058138.$$

We propose the following expression of the approximation fractional-quadratic characteristic

$$V(H) = 0.429\frac{H}{1 + 0.01H^2}. \tag{2}$$

Maximum values of field strength and voltage correspond to the peak of characteristics

$$H_{\max} = 10,\ V_{\max} = 2.145.$$

The results of calculations of nonlinear characteristics (1), (2) are presented in Fig. 1. The working ascending branch of approximation characteristic is close to the original sensor characteristic.

Let us represent the relationship (2) for negative values of field strength and voltage on the Fig. 2. In this case, two peaks correspond to the maximum voltage values. This results to the single-valued working area between these peaks.

During the test or calibration stage of the sensor, the maximum or base values $H_{\max}$, $V_{\max}$ are determined and the initial or first measurement H^1, V^1 is performed. The next (second) measurement is carried out at the working stage H^2, V^2. Then, the cross ratio [6], page 369, of these four field strength samples has the form

$$m_H^{21} = (-H_{\max}\ H^2\ H^1\ H_{\max}) = \frac{H^2 - (-H_{\max})}{H^2 - (H_{\max})} \div \frac{H^1 - (-H_{\max})}{H^1 - (H_{\max})}. \quad (3)$$

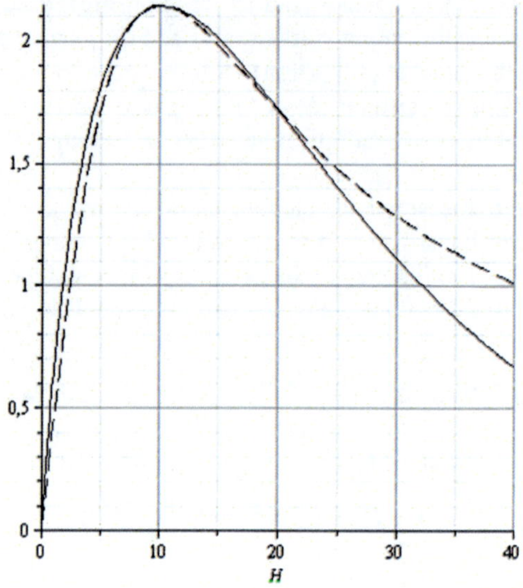

Fig. 1. Sensor characteristic U - solid line, approximation V- dash line.

A similar expression is obtained for the voltage samples

$$m_V^{21} = (-V_{\max}\ V^2\ V^1\ V_{\max}). \quad (4)$$

There is an invariant property

$$m_V^{21} = \left(m_H^{21}\right)^2. \quad (5)$$

The structure of cross ratios shows that additive and multiplicative errors of calculations and measurements are mutually reduced.

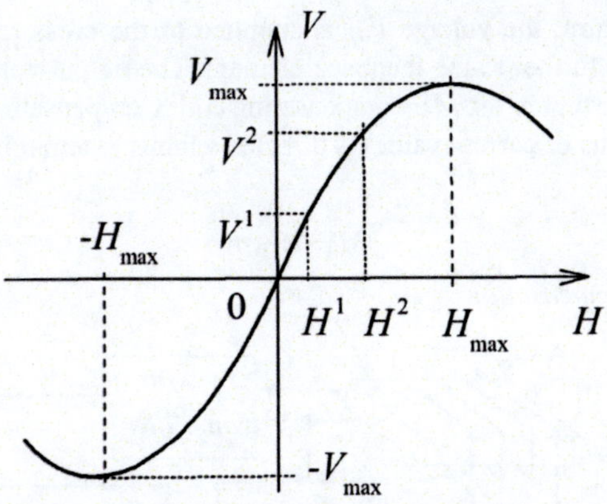

Fig. 2. Single-valued working area of approximation characteristic.

Then, we obtain the calculated value of field strength H^2 through test values and the measured value V^2 taking into account (5)

$$H^2 = H_{\max}\frac{\sqrt{m_V^{21}\frac{H^1+H_{\max}}{H^1-H_{\max}}}-1}{\sqrt{m_V^{21}\frac{H^1+H_{\max}}{H^1-H_{\max}}}+1} = H^2(H^1, m_V^{21}). \tag{6}$$

Additional advantages of the proposed solution are related to group properties of a cross ratio. Let, for example, a third measurement be made. Then, the cross ratio is obtained with respect to the first measurement m_V^{31} or with respect to the second measurement m_V^{32}, wherein the group property is performed through the multiplication operation.

$$m_V^{31} = m_V^{32} \cdot m_V^{21}.$$

Therefore, variants of the calculated field strength expressions H^3 relative to H^1 and H^2 are obtained

$$H^3 = H^3(H^1, m_V^{31}), \quad H^3 = H^3(H^2, m_V^{32}).$$

The processing of the obtained values by known methods (for example, the least squares method) allows improving the measurement accuracy.

3 Device Diagram

Block-diagram of the device [7] is given in Fig. 3. The device operates as follows. Let the *ME* be placed in an external longitudinal constant magnetic field with strength H_{Ex}. The pump generator supplies voltage to the *ME* drive input and to the lock-in amplifier. The *ME* output voltage is also applied to the lock-in amplifier. The obtained voltage *U* is then supplied to the former of the approximation fractional-quadratic nonlinear characteristic. In turn, the voltage V_U is supplied to the cross ratio or field strength H_V calculator and further to the feedback element. The output voltage of the feedback element sets the current in the *ME* compensating coil. Compensating magnetic field H_C (confirms a previous or current value) is directed against external magnetic field H_{Ex}.

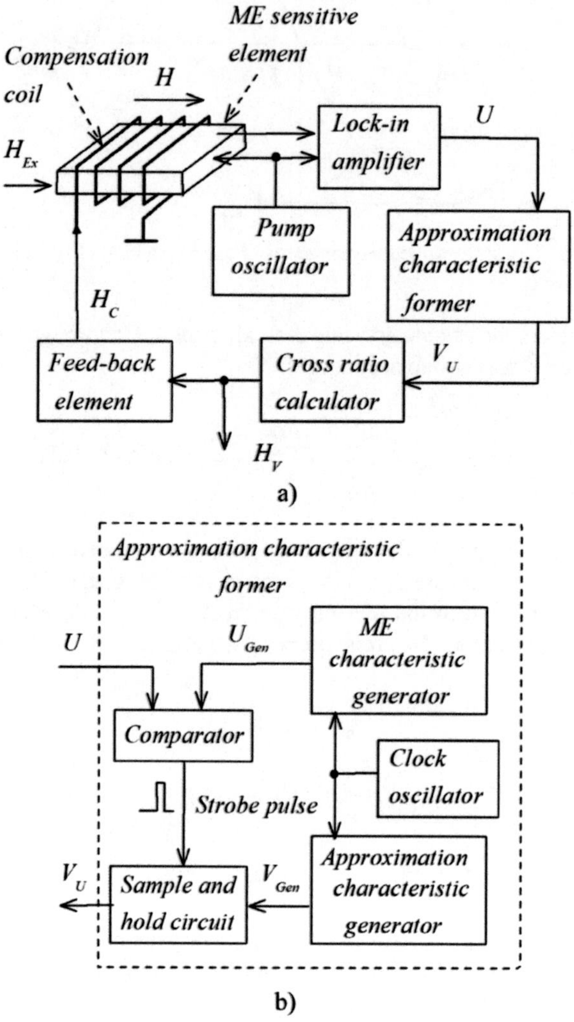

Fig. 3. Block-diagram of the *ME* sensor with compensation coil: a)-general scheme, b)-approximation characteristic former.

Therefore, the above-mentioned *ME* output voltage corresponds to the field strength.

$$H = H_{Ex} - H_C.$$

So, the negative feedback loop closed, which corresponds to the steady state regime at the time interval $0 - t_2$ per Fig. 4.

In turn, the former of the approximation fractional-quadratic nonlinear characteristic operates as follows. The clock oscilator starts the next pulse U_{Gen} of the *ME* initial nonlinear characteristic generator at the time moment t_1 and the next pulse V_{Gen} of the approximation fractional-quadratic characteristic generator at the time interval $t_1 - t_3$ per Fig. 4. The shape of the pulses U_{Gen} corresponds to the ascending branch of the *ME* characteristic, and the shape of the pulses V_{Gen} corresponds to the ascending branch of the approximation fractional-quadratic characteristic. When the voltage values U and U_{Gen} are equal, the comparator generates a short strobe pulse at a time t_2. This strobe pulse opens the sample and hold circuit. Thus, the voltage value V_U of the approximation fractional-quadratic characteristic corresponding to the voltage U and field strength H is stored until to the next search pulse at a time $t_1 + T$. The cross ratio calculator operates according to the expressions (4–6) and can be implemented by a specialized analog microcircuit or by a microprocessor.

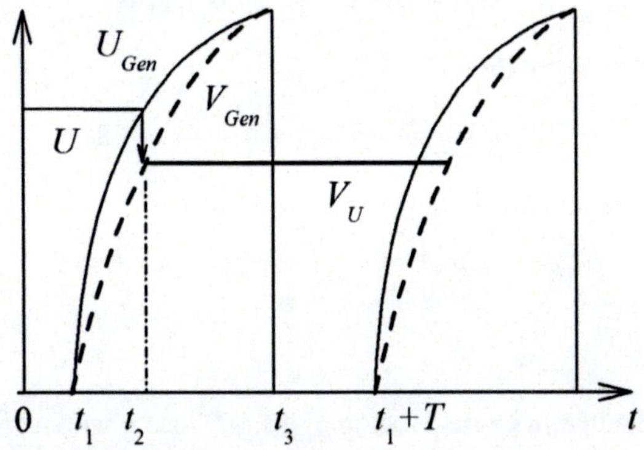

Fig. 4. Voltage diagrams at characteristic points of the *ME* sensor.

4 Example of Calculation

For convenience, we set dimensionless values of field strength and voltages of all incoming elements. Let us assume that in steady state the voltage of H_V will be equal to the strength H. Let the resulting transmission coefficient (of the *ME*, approximation characteristic former, and cross ratio calculator) correspond to $K = 1$. Let the feedback element transmission ratio correspond to the value $\beta = 0.25$. Then, the compensating magnetic field strength $H_C = \beta \cdot H_V$. Therefore, the voltage H_V

$$H_V = \frac{K}{1 + \beta \cdot K} H_{Ex} = \frac{1}{1 + 0.25} H_{Ex}. \tag{7}$$

Let the first measurement correspond to the external field strength $H_{Ex}^1 = 5$. Then, according to (7), $H_V^1 = H^1 = 4$.

The voltage $U^1 = 1.6403$ is calculated according to (1).

Therefore, the pulse U_{Gen} at the time interval $t_3 - t_1$ is generated according to expression (1). In this case, the value H corresponds to time t. Then $V_U^1 = 1.4793$.

Let the second measurement correspond to the external field strength $H_{Ex}^2 = 6$. The results of measurements and calculations are the following:

$$H_V^2 = H^2 = 4.8, U^2 = 1.7815, V_U^2 = 1.6736.$$

We calculate cross ratio (3)

$$m_H^{21} = \frac{4.8 + 10}{4.8 - 10} \div \frac{4 + 10}{4 - 10} = 2.8462 \div 2.3333 = 1.2198.$$

Cross ratio (4)

$$m_V^{21} = \frac{1.6736 + 2.145}{1.6736 - 2.145} \div \frac{1.4793 + 2.145}{1.4793 - 2.145} = 1.4879.$$

Invariant property (5)

$$m_V^{21} = 1.4879 = \left(m_H^{21}\right)^2 = 1.2198^2.$$

Checking of expression (6)

$$H^2 = 10\frac{1.2198 \times 2.3333 - 1}{2.8462 + 1} = 4.8.$$

The external field strength is calculated according to (7)

$$H_{Ex}^2 = 1.25 H_V^2 = 1.25 H^2 = 1.25 \times 4.8 = 6.$$

5 Conclusion

The widening of the operation field region of the *ME* sensor without linearization of its characteristic have been theoretically demonstrated. The approximation characteristic is formed, which is close in shape to the sensor original characteristic.

As this approximation characteristic, a fractional-quadratic relationship is used, for which the invariant is performed between the output voltage value and the magnetic field strength. Such invariant is the cross ratio for four samples of these values.

The use of a cross ratio has made it possible to obtain a formula for calculating the field strength, in which the measurement errors of four resistance samples are mutually reduced.

The results are of interdisciplinary interest in view of similarity of sensor characteristics of different physical nature.

Acknowledgements. The work was carried out within the framework of the Moldova State Program Project «Nanostructuri şi nanomateriale funcţionale pentru industrie şi agricultură» No. 20.80009.5007.11, and financially supported by the European Union Project H2020 "SPINTECH" under the grant agreement No. 810144.

The authors are thankful to prof. Alexander Vakhrushev for useful discussions.

Conflict of Interest. The authors declare that they have no conflict of interest.

References

1. Gao, J., Jiang, Z., Zhang, S., et al.: Review of magnetoelectric sensors. Actuators **10**(6), 109–132 (2021)
2. Zuo, S., Schmalz, J., Özden, M., et al.: Ultrasensitive magnetoelectric sensing system for pico-tesla magnetomyography. IEEE Trans. Biomed. Circuits Syst. **14**(5), 971–984 (2020)
3. Li, M., Dong, C., Zhou, H., et al.: Highly sensitive DC magnetic field sensor based on nonlinear ME effect. IEEE Sensors Lett **1**(6), 1–4 (2017)
4. Srinivasan, G.: Magnetoelectric composites. Annu. Rev. Mater. Res. **40**, 153–178 (2010)
5. Serov, V., Chashin, D., Fetisov, L., et al.: Widening and linearization of dc magnetic field magnetoelectric sensor characteristic using a compensation scheme. IEEE Sens. J. **18**(20), 8256–8260 (2018)
6. Penin, A.: Analysis of Electrical Circuits with Variable Load Regime Parameters. Projective Geometry Method, 3rd edn. Springer International Publishing AG, Switzerland (2020)
7. Penin, A., Sidorenko, A.: Method for measurement of constant magnetic field strength by magnetoelectric sensor with initial nonlinear characteristic. MD Patent Application S20210009 (2021)

Involvement of Contact and Surface Phenomena in Nanolayered Amorphous Te Films for Toxic Gas Detection at Room Temperature

Dumitru Tsiulyanu[1](✉), O. Mocreac[1], and T. Braniste[2]

[1] CIMAN Research Centre, Technical University, Chisinau, Moldova
[2] NCMST Research Centre, Technical University, Chisinau, Moldova

Abstract. A fast responding NO_2 sensitive device operating at room temperature has been realized using the nanolayered amorphous Te (a-Te) grown onto insulating wafer of silicon dioxide (SiO_2) between Pt contact electrodes with larger thickness in a planar arrangement. The structure of the fabricated sensor has been investigated by AFM and SEM but its characterization was realized via studying the current - voltage characteristics, dynamic response, long – term stability and effect of humidity. Explanation of obtained results is given in terms of a model based on simultaneous involvement of contact and surface phenomena for the gas sensing. As the Pt electrode work function (5.43 eV) exceeds the respective value of a-Te (5.03 eV) the ohmic contacts are formed and the current flow is controlled exclusively by bulk resistance of a-Te nanolayer that is known to be controlled by type and concentration of toxic gas of the ambiance. Wherein, as the energetic forbidden gap of a-Te (0.33 eV) is less than the work function difference between contacting materials, at the contacts can arise the degenerate regions of p-type metallic Te, as well as geometric contact gaps originated from microscopically roughness. The gas adsorption inside these contacts gaps leads to increasing the portion of the semiconducting a-Te nanolayer turned into metal of p-type Te and consequently to a fast increasing of the current.

Keywords: Nanolayers · a-Te · Contacts · Gas sorption

1 Introduction

Tellurium thin films can be successfully used for the detection of nitrogen dioxide and hydrogen sulfide at room temperature [1, 2]. The high sensitivity of Te films to nitrogen dioxide attracted special attention because it is a very toxic gas, released by combustion, plants and automobiles and its monitoring in the environment is of great importance. Exemption from stoichiometry problems allowed using of different methods of sensor fabrication including the thermal evaporation, sputtering, chemical precipitation etc. that allowed strongly modifying the film's microstructure. In order to increase the performances in detection of NO_2 at room temperature the new technologies of fabrication were developed allowing the shortening of dimensionalities of Te microcrstallites to nanoscales and further until their total disappearance, accompanied by transition of

I. Tiginyanu et al. (Eds.): ICNBME 2021, IFMBE Proceedings 87, pp. 560–567, 2022.
https://doi.org/10.1007/978-3-030-92328-0_72

films into an amorphous state. For example, the growing of Te nanotubes by direct vapor phase process in an inert atmosphere [3] has allowed enlarging of dynamic range of room temperature operating gas sensitive devices from 1.0 to 400 ppm NO_2, alongside with increasing of the concentration dependent gas sensitivity. Growing of nanocrystalline Te films via rf sputtering in an ultra-high-purity argon onto glass or alumina substrates, [4] or amorphous tellurium one onto pre-nanostructured (porous) dielectric substrates [5] allowed diminishing the response time to approximately 30 and 15 s respectively, maintaining the sensitivity toward NO_2 at ~40% per ppm. The further improvement of the performances of Te based gas sensors can be achieved via utilizing the contact phenomena in the form of Schottky diodes, MOS capacitors or FETs [6]. Recently [7] was find out that the capacitance spectra of glassy tellurium chalcogenide based Pt - $As_2Te_{13}Ge_8S_3$ – Pt structure gives evidence that both surface and contact phenomena being interdependent are controlled by interaction with gases from ambiance. The present work is conducted to show a possibility of enhancing the performance of toxic gas sensing, using the features of Pt/a-Te junction that can comprise a region with p-type metallic conductivity adjacent to contact.

2 Experimental

Pt/a-Te/Pt functional structures, schematically represented on Fig. 1, were realized on thermally oxidized p-type silicon wafers (10 Ω cm) having a size 3 × 8 mm. Two platinum contact lines have been photo lithographically patterned on top of the insulating silicon oxide layer, whereupon the active ultrathin amorphous tellurium film was deposited between them. The last has been grown via thermal evaporation of polycrystalline tellurium (purity 99.999%) in vacuum of 10^{-4} Pa from a tantalum boat without cooling or heating the substrate. Gaseous media with different concentration of NO_2 was obtained by using the calibrated permeation tubes (Vici Metronics, USA), which was introduced into the experimental set-up described elsewhere. Ambient air or dry synthetic air was used as the carrier and reference gas.

The sensing devices were put into a test cell (of 10 ml volume) but the gases were injected with a flow rate of 100 ml/min, parallel to the film surface. Humidity cross-sensitivity to NO_2 was studied via humidification of the carrier gas that was accomplished using different saturated solutions of the salts in water.

I-U characteristics have been carried out in ambient air and its mixture with 1.0 ppm NO_2 at room temperature. The applied voltage varied between −5 V and +5 V in steps, increasing by 20 mV at each step, while the respective values of the current were measured. The delay time between successive measurements was 2 s. Current - transient characteristics have been carried out at a constant (1 V) applied voltage in condition to successive injection of rectangular pulses of 1.0 ppm NO_2 into test cell. Delay time between successive measurements was 40 ms. Data were processed, using a PC and a data acquisition board manufactured by National Instruments Inc. The sensor sensitivity was defined as the relative resistance variation expressed in percent:

$$S = 100\left(R_a - R_g\right)/R_a \quad (1)$$

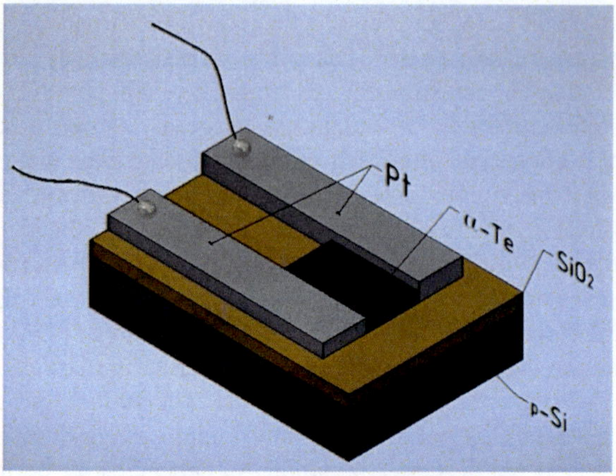

Fig. 1. Schematic of the sensor design

where R_a and R_g are the sensor's resistances in the air and in the presence of NO_2. The response and recovery time were defined as the times taken to reach or to lose the 90% of maximum values of the current respectively.

3 Results

3.1 Morphology and Design Details

The surface morphology of the films was investigated using the scanning electron microscope VEGA TESCAN TS 5130 MM but the design details of the sensor have been studied using an Atomic Force Microscope (AFM) SIS SCAN Control C. Figure 2 shows the AFM scanned profiles of Pt electrodes and Te film, alongside with its SEM image. It can be observed that the thickness of Pt electrodes is around 100 nm but the thickness of Te film is about 30 nm. The SEM image of Te film (Fig. 3c) shows a uniform surface without of morphological structure that confirms its amorphous state.

3.2 Gas Sensing Characterization

a) *Current – voltage characteristics*:

Figure 3 shows the typical I –U characteristic of Pt/a-Te/Pt functional structures in air and in the presence of NO_2 vapor in a semi-logarithmic scale. The influence of NO_2 vapor at bias higher than 0.1 V leads to an increase of the current independently of the direction of the bias voltage.

Alongside at bias less than 0.1 V, the gas induced current increases much stronger, which results in an intersection of I-V curves measured in air and in its mixture with NO_2.

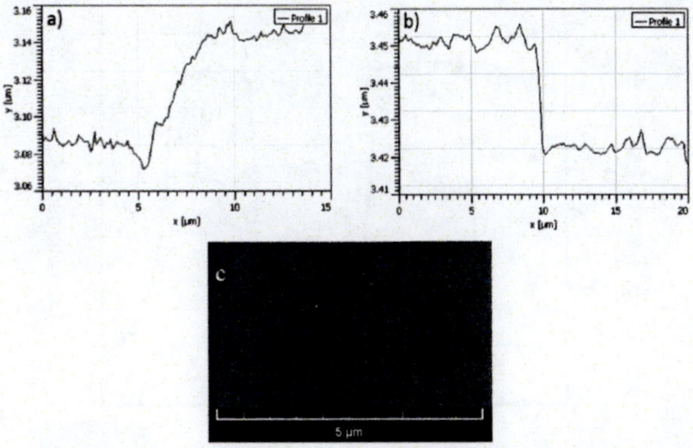

Fig. 2. AFM profiles of Pt electrodes (a); Te film (b) and SEM image of tellurium film (c).

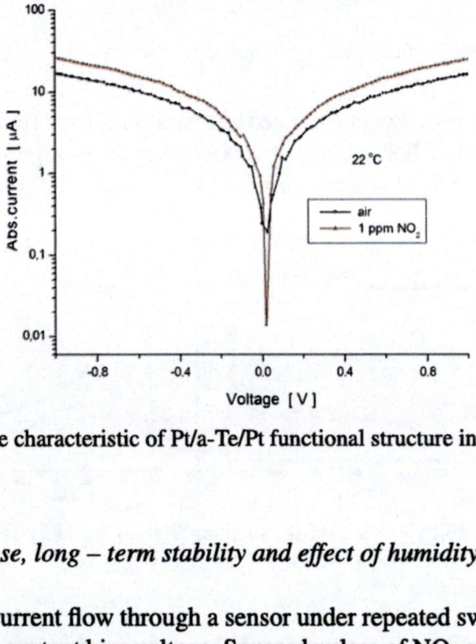

Fig. 3. Current/voltage characteristic of Pt/a-Te/Pt functional structure in air and in the presence of 1.0 ppm of NO_2

b) *Dynamic response, long – term stability and effect of humidity:*

Figure 4 shows the current flow through a sensor under repeated switching on-off of the NO_2 gas mixture at constant bias voltage. Squared pulses of NO_2 vapor with concentration of 1.0 ppm were applied. The dotted line shows the switching schedule. It is seen that the current follows the schedule; the response is fast and consists of around 3–5 s but and the recovery one is longer and is ~60 s. It is significant that there is no baseline drift or noticeable drift of the gas induced current.

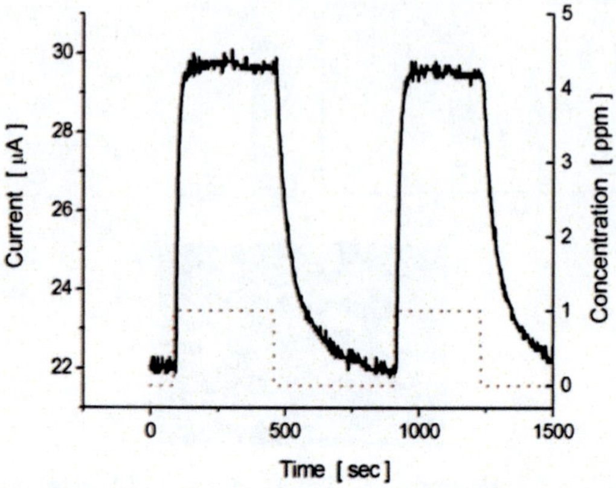

Fig. 4. Transient characteristics by exposure to NO_2 at 22 °C

b) *Long – term stability and effect of humidity*

The long –term stability of the sensor tested by operating it for three months is reported on Fig. 5(a). It is seen that in both air and its mixture with NO_2 vapors there no appreciable changes in its gas sensing characteristics.

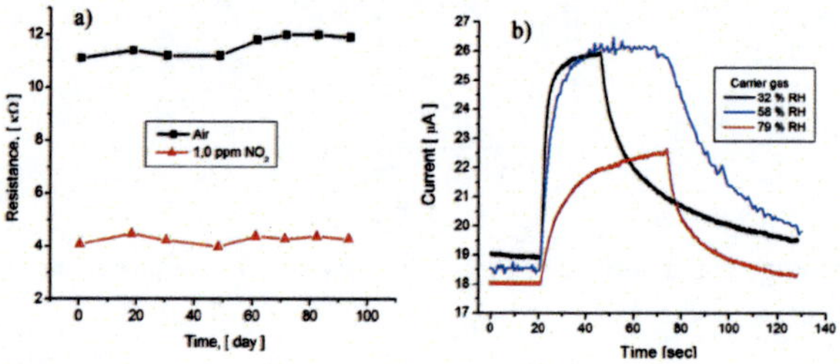

Fig. 5. Long –term stability (a) and effect of humidity (b) by NO_2 sensing at room (22 °C) temperature

The interfering effect of humidity on NO_2 detection with a developed sensor was assessed at room temperature (22 °C). Figure 5(b) illustrates the response of the sensor towards a pulse of 1.0 ppm NO_2 in humid air with 32%, 58% and 79% RH. The effect of NO_2 is to increase the current in the sensor independently on humidity of the carrier gas, but the baseline current, magnitude and velocity of response and recovery times are influenced by water vapor concentration. Increasing the RH diminishes both baseline current and response signal but enlarges the response time, which nevertheless remains in the range of seconds.

4 Model

Figure 6(a) illustrates the possible band diagram of Pt/a-Te junction in normal conditions sketched using the following parameters of contacting materials: Pt and Te work function $\varphi_{Pt} = 5.45$ eV and $\varphi_{Te} = 5.03$ eV respectively [8]; Forbidden gap of a-Te $E_g \approx 0.33$ eV [9]; a-Te electron affinity $\chi \approx 2$ eV [10]; Fermi level position $\eta \approx E_g/2$.

The diagram is sketched according to Schottky - Mott barrier, which assumes that the contact comprises a thin insulating layer (several Angstroms), formed during electrode deposition process [11]. The region between the electrode and semiconducting layer, called the "transition region" serves as a gap where both platinum clusters formed and gas molecules chemisorbed [12]. The diagram given in Fig. 6 shows also that the Pt/Te junction belongs to an extreme one, when the adjacent to contacts regions of *p*-type Te become degenerate, with the Fermi level E_f lying below the valence band edge E_v. Such an unusual situation can occur because the band bending exceeds the Te forbidden gap $\varphi_{Pt} - \varphi_{Te} > E_g$ [13].

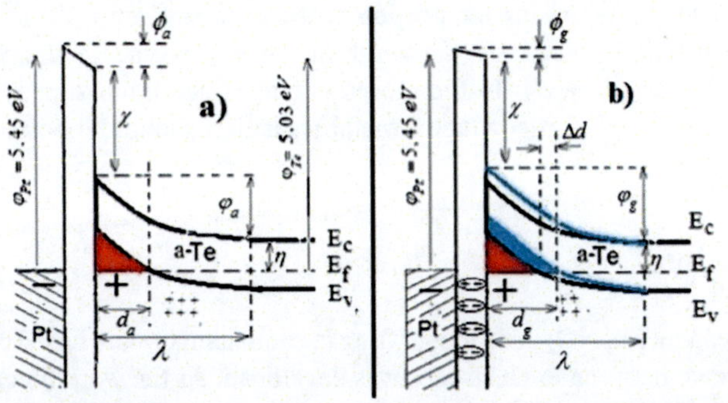

Fig. 6. Possible band diagram of the nonbiased Pt/a-Te junction in air (a) and NO_2 (b) ambiance (see the explanation in the text)

The band diagram shows the relation:

$$\varphi_{Pt} = \phi_a + \chi + \varphi_a + \eta \tag{2}$$

where ϕ_a is the potential drop from the Pt to the a-Te in the interface gap and φ_a is the band bending, both in air.

Let us assume that the examined Pt/a-Te junction is introduced into NO_2 environment and the chemical adsorption of gas molecules occurs both on the surface of the bulk region and contact gaps. Due to peculiarities of NO_2 molecules [14] such simultaneous adsorption leads to different consequences:

1. Absorption on the surface of bulk leads to increase the majority carrier's density in the adjacent surface of the bulk region of Te film, which results in increasing its conductivity, so a parallel shift-up of the linear part of the I-U characteristic should occur.

2. Penetration (adsorption) of polarized NO_2 molecules into contact gaps has to change the potential drop in the interface gap: they will align in the direction controlled by direction of gap field, created as a result of work function difference of contacting materials [12], leading to modulation of both potential drop at the surface and work function of the a-Te. Figure 6 (b) illustrates the effect of NO_2 adsorption into gap of contact transition regions. It is seen that for this case the Eq. (2) can be expressed as:

$$\varphi_{Pt} = \phi_g + \chi + \varphi_g + \eta \tag{3}$$

where ϕ_g and φ_g is the potential drop at the gap and band bending respectively, at exposure to NO_2.

Assuming that the φ_{Pt},χ and a η stay unchanged it can be concluded that the gas molecules modify only the value and curvature of the bands bending due injection of holes from Pt electrodes to restore the electrical neutrality:

$$\Delta\varphi = \varphi_g - \varphi_a = \phi_g - \phi_a = -\Delta\phi \tag{4}$$

The last leads to increasing the portion of the semiconducting Te nanolayer turned into metal of p-type Te, such that the width of metallic p-type Te adjacent to contacts increases as $\Delta d = d_g - d_a$. In the closed – circuit condition supported by voltage application, this partial semiconductor-metal transition should be observed as a sharp increasing of the current.

5 Conclusions

A rapidly – responding NO_2 sensor operating at room temperature based on nanolayered a-Te films grown between Pt electrodes was developed. As the work function difference between contacting materials exceeds the forbidden gap of a-Te at contacts can arise the degenerate (metallic) regions of p-type Te. Gas induced modulation of the width of these regions is assumed to be the reason for the essential diminishing of both response and recovery time.

Acknowledgment. This work is supported by National Agency for Research and Development of Moldova, grant PS 20.80009.5007.21.

Conflict of Interest. The authors declare no conflict of interest.

References

1. Tsiulyanu, D., Marian, S., Miron, V., et al.: High sensitive tellurium based NO2 gas sensor. Sens. Actuators B **73**, 35–39 (2001). https://doi.org/10.1016/s0925-4005(00)00659-6
2. Sen, S., Bhandarkar, V., Muthe, K., et al.: Highly sensitive hydrogen sulphide sensors operable at room temperature. Sens. Actuators B **115**, 270–275 (2006). https://doi.org/10.1016/j.snb.2005.09.013

3. Her, Y., Huang, S.: Growth mechanism of Te nanotubes by a direct vapor phase process and their room temperature CO and NO2 sensing properties. Nanotechnology **24**(9pp), 215603 (2013). https://doi.org/10.1088/0957-4484/24/21/215603
4. Siciliano, T., Di Giulio, M., Tepore, M., et al.: Tellurium sputtered thin films as NO2 gas sensors. Sens. Actuators B **135**, 250–256 (2008). https://doi.org/10.1016/j.snb.2008.08.018
5. Tsiulyanu, D., Mocreac, O., Ciobanu, M., et al.: Peculiarities of ultrathin amorphous and nanostructured te thin films by gas sensing. J. Nanoelectron. Optoelectron. **9**, 282–286 (2014). https://doi.org/10.1166/jno.2014.1585
6. Lundstrom, I.: Solid State Chemical Sensors. Academic press, New York (1985)
7. Ciobanu, M.: Features of contact and surface processes in glassy As2Te13Ge8S3 based structures with Pt electrodes upon interaction with nitrogen dioxide. Mold. J. Phys. Sci. **16**, 234–241 (2017)
8. Michaelson, H.: The work function of the elements and its periodicity. J. Appl. Phys. **48**, 4729–4733 (1977). https://doi.org/10.1063/1.323539
9. Ray, A., Swan, R., Hogarth, C.: Conduction mechanisms in amorphous tellurium films. J. Non-crystall. Solids **168**, 150–156 (1994).https://doi.org/10.1016/0022-3093(94)90131-7
10. Popescu, M., Andrieş, A., Ciumaş, V., et al.: Physics of Chalcogenide Glasses. Stiinţa Publishing house, Chisinau (1996)
11. Yamada, T.: Modeling of carbon nanotube Schottky barrier modulation under oxidizing conditions. Phys. Rev. B **69**, 125408(1)–125408(8) (2004). https://doi.org/10.1103/PhysRevB.69.125408
12. Yamada, T.: Equivalent circuit model for carbon nanotube Schottky barrier: influence of neutral polarized gas molecules. Appl. Phys. Lett. **88**, 083106(1)–083106(3) (2004). https://doi.org/10.1063/1.2177356
13. Walpole, J., Nill, K.: Capacitance-voltage characteristics of metal barriers on p PbTe and p InAs: effects of the inversion layer. J. Appl. Phys. **4**, 5609–5617 (1971). https://doi.org/10.1063/1.1659990
14. Greyson, J.: Carbon, Nitrogen and Sulfur Pollutants and\their Determination in Air and Water. Marcel Dekker Inc., New York (1990)

Biomedical Sensors Based on Micro- and Nanotechnology

B. I. Podlepetsky(✉)

Micro- and Nanoelectronics Department, National Research Nuclear University MEPhI (Moscow Engineering Physics Institute), Moscow, Russia

Abstract. This review paper briefly analyzes the history, development trends, current state of the sensors' developments based on micro-and nanotechnologies for medical and biological research. The classification of such sensors is presented; their design and technological features and applications are considered.

Keywords: Biomedical measurements · Micro-sensors · Micro- and nanotechnology

1 Introduction

The first sensors based on micro-technologies (micro-sensors) for measuring some physical and chemical quantities appeared in the early 1970s [1–7]. Due to the imperfection of technologies and the increased sensitivity of semiconductor elements to the action of external factors, micro-sensors in those years had unsatisfactory characteristics, which limited the possibilities of their wide application. They were mainly used for biomedical measurements, the field of which was characterized by reduced requirements for accuracy, speed and stability of measuring devices. With the development and perfection of micro-technologies, micro-sensors were developed for such unique biomedical measurements that previously could not be carried out in principle. In addition, over the years, the performance characteristics of micro-sensors have improved, which has allowed expanding their application areas. The aim of this work is to analyze the trends and current state of the sensors' developments based on micro-and nanotechnologies for medical and biological research. The analysis was carried out on the basis of generalization of data on the developments of biomedical micro-sensors for 50 years.

2 Characterization of Sensors for Biomedical Research

2.1 Classification and General Characteristics of Sensors

A generalized structural and functional scheme of biomedical research using technical units is demonstrated in Fig. 1. A biological object (BO) can be a human, an animal, their organs, tissues and cells, samples of biological materials (for example, blood, urine, saliva). *Biomedical sensors* (BMS – (1)) are sensors that measure the characteristics of

I. Tiginyanu et al. (Eds.): ICNBME 2021, IFMBE Proceedings 87, pp. 568–576, 2022.
https://doi.org/10.1007/978-3-030-92328-0_73

a BO, and are the main ones in this scheme. Additional sensors (2) can be used to obtain information about environmental parameters (3), control the parameters of generators of artificial factors affecting the BO (7), control the operation of technical units (4) to ensure compliance with the instructions of the BO action (8).

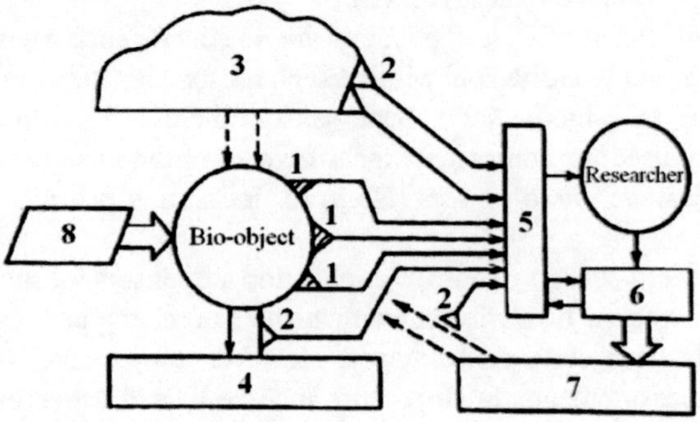

Fig. 1. A generalized structural and functional scheme of biomedical research using technical units and measurement and information system (5)

As a rule, BMS are placed inside or on the surface of the BO (contact measurement methods – *in vivo*). Such sensors yield medical information by measuring directly on the living tissue of a BO.

Contact sensors can be used for non-invasive measurements (without mechanical disturbances of biological tissue) or for invasive measurements when sensors are implanted in the tissue. Note main areas of application of sensors *in vivo*: surface sensors, short - term/medium-term and long-term implants. In any case, the biocompatibility of sensor materials with BO and the safety of BO must be observed, since there are mechanical, thermal, chemical and electromagnetic interactions of sensors with BO. Most contact microelectronic BMS are custom-made products of private use, the need for which is not high.

BMS for non-contact measurement methods, as well as additional sensors are located not on the BO, but in the range of the measured parameters. Such sensors are not subject to special requirements specific to contact BMS, and their selection is usually provided from serial (commercial) sensors.

A feature of many methods of biomedical research is the measurement of the physiological parameters of BO in the conditions of their activity, in particular mobility. In this regard, the requirements for the noise immunity of measurements to motion artifacts and interference of physiological origin are increased. The use of electronic equipment in research poses the task of combating electromagnetic interference. Adequate measurement results are achieved by reducing the mass, dimensions, and power consumption of sensors, developing special designs and methods for attaching sensors, and using implanted and wireless micro-devices. The listed requirements for biomedical measurements and sensors are successfully met in some cases when using BMS based on micro-and nanotechnologies (micro-sensors), which can be divided into four groups.

1. Unique micro-sensors of narrow application are sensors for such measurements, which in principle cannot be carried out without such sensors, but they are unsuitable for other measurements. Such sensors include integrated multi-microsensors (No.1) that are used to measure electric potentials, temperature, magnetic field induction, and ion concentrations *in vivo*, implanted wire-less sensors and neural dust for monitoring BO characteristics (No. 12).
2. Unique micro-sensors of wide application are sensors for such measurements, which in principle cannot be carried out without such sensors, but they can be used for other measurements (non-biomedical). Such sensors include thin-film SQUIDs (No.2) [8], which are used for non-contact measurement of the induction of BO magnetic fields, ion-sensitive micro-sensors (No.8) for measuring pH, pK, pCa, pCl *in vivo* and *in vitro*.
3. Specialized micro-sensors of narrow application are sensors for such measurements, which in principle can be carried out with the help of other types of sensors, although micro-sensors have certain advantages. However, they cannot be used for non-biomedical measurements. Such sensors include hybrid active electrodes (No.3), which are used to measure the electrical potentials and surface temperature of the BO, and hybrid sensors based on optocouplers (No.6) for measuring the power of the luminous flux *in vivo*.
4. Specialized micro-sensors of wide application are such sensors that can be used for both biomedical and other applications. Such sensors include thermal sensors (No. 4), integrated photodetector arrays of IR radiation (No.7), ion-sensitive sensors (No.8), gas concentration sensors (No.9) and enzyme biosensors (No.10), which are used to measure the concentration of organic and inorganic substances in the BO *in vivo* and *in vitro*.

Integrated sensors for measuring pressure, force and acceleration (No. 5) and integrated multiparametric sensors (multisensors No. 11) for simultaneous measurement of various quantities belong to all four groups. To manufacture of biomedical micro-sensors (μBMS) both standard integrated circuit technologies (semiconductor, film and hybrid-film) and special technologies (MEMS, NEMS, MOEMS, LIGA, nano-on-micro, and others) are used.

2.2 Characteristics of the Main Types of MBMS

The general characteristics of MBMS are presented in Table 1. The BMS type numbers correspond to the numbers indicated in parentheses above. The following abbreviations are used: ENG (electroneurogram), ECoG (electrocorticogram), ECG (electrocardiogram), EEG (electro-encephalogram), MCG (magnetocardiogram), PhPG (photoplethysmogram), HR (heart rate), T (temperature).

The cost of microelectronic products depends on the scale of their production, which is determined by its needs. Today, BMS corresponding to numbers 3, 4, 6 and 8 have the greatest need; BMS corresponding to numbers 5 and 10 have an average need; and BMS corresponding to numbers 1, 2, 7, 9 and 11 have a small need.

Table 1. General characteristics of μBMS

BMS type	Measured characteristics	Application areas	Main features
1	ENG, ECoG, T, pH [2, 7, 11–14]	Neurophysiology, experimental medicine	Miniaturization, multi-channel, noise immunity
2	MCG, HR [8]	Cardiological diagnostics	Non-contact measurement
3	ECG, EEG [9]	Diagnosis of the state of the cardio-vascular + central nervous systems	Noise immunity from common-mode interference
4	Body, tissues, blood T; respi-ratory rate [9, 10]	Diagnostics of the state of functional systems of the BO	Miniaturization, high measurement speed
5	Blood pressure, respiratory and pulse parameters [3–5, 16, 17]	Diagnostics and monitoring of the cardiovascular and respiratory systems' parameters	Miniaturization, increased accuracy,and the ability to measure pressure *in vivo*
6	PhPG, HR, blood oxygen saturation, blood pressure, blood glucose level [9]	Diagnosis of the cardio-vascular system state and the psychophysiological state of a person; monitoring the treatment course of diabetes mellitus	Non-invasive measurements of the degree of filling of blood vessels, increased sensitivity to motion artifacts
7	T, thermal fields BO [9, 22]	Assessment of the functional state of BO based on thermal imaging examination	Non-contact measurements, low accuracy
8	Acidity, ionic composition of biological fluids (e.g., blood, urine, saliva) [1, 10, 18–21, 24]	Laboratory analysis of bio-liquids *in vitro*, study of the state of functional systems of BO	Miniaturization, high measurement speed, the possibi-lity of catheter me-asurement *in vivo*
9	Concentrations of gases in the exhaled air [22, 28]	Prevention and diagnosis of stomach diseases, diabetes mellitus and pulmonary tuberculosis	Non-contact measurements, low accuracy
10	Concentrations of glucose, urea, antibiotics, heavy metals, and hydrogen peroxide in bio-liquids [25]	Biochemical analysis, diagnosis of diabetes mellitus, immunological and genetic studies	*In vivo*, *in vitro* and *in situ* measurements, low stability of characteristics

(*continued*)

Table 1. (*continued*)

BMS type	Measured characteristics	Application areas	Main features
11	Simultaneous measurement of various parameters of BO (biopotentials, pressure, T, pH) [9, 23, 26]	Laboratory analysis of biological fluids, diagnostics of diseases and the state of functional systems	*In vivo* and *in vitro* measurements with increased accuracy, manufacturing complexity
12	Wire-less monitoring T and elec-trical potentials BO by implanted BMS [29]	Monitoring nerves and muscles state, to monitor or to stimulate the brain activity	Miniaturization, wireless measuring physiological characteristics or stimulation organs *in vivo*

2.3 Typical MBMS Designs and Technologies

Designs' examples of microprobes for measuring biopotentials and stimulation of neurons (No. 1), pressure μBMS (No. 5), ion-sensitive micro-sensors (No. 8), gas concentration sensors (No. 9), integrated sensors' arrays with multisensors (No. 11), and are shown in Fig. 2, 3, 4, 5, 6, 7 and 8.

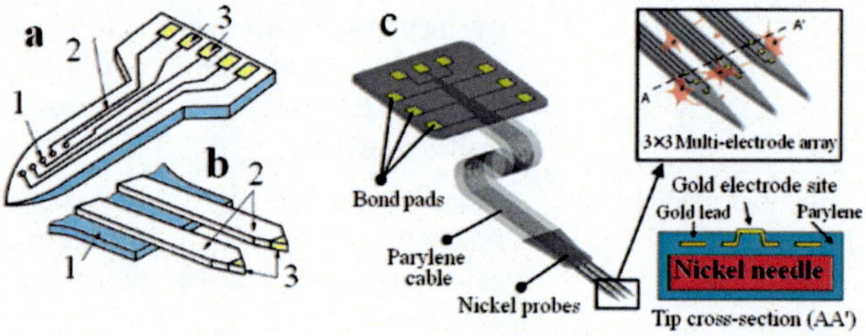

Fig. 2. Multielectrode microprobes' designs: a) Multielectrode silicon probe:1-electrodes, 2-conducting tracks, 3-contact pads; b) Multi-channel electrode with beam terminals: 1- Si substrate, 2- insulated SiO_2 gold petals, 3 - open ends of the electrodes; c) 9-channel multi-microelectrode based on nickel probes with a flexible cable [2, 7, 9]

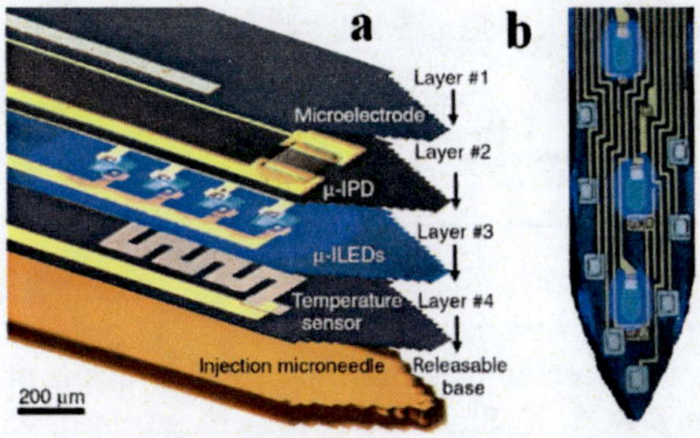

Fig. 3. Fiberless optical stimulation probes using μLEDs: a) GaN μLEDs grown on sapphire wafers and transferred onto a polymer substrate by laserliftoff; b) Monolithic integration of multiple GaN μLEDs on silicon neural probes and capable of a 50 μm pitch [14].

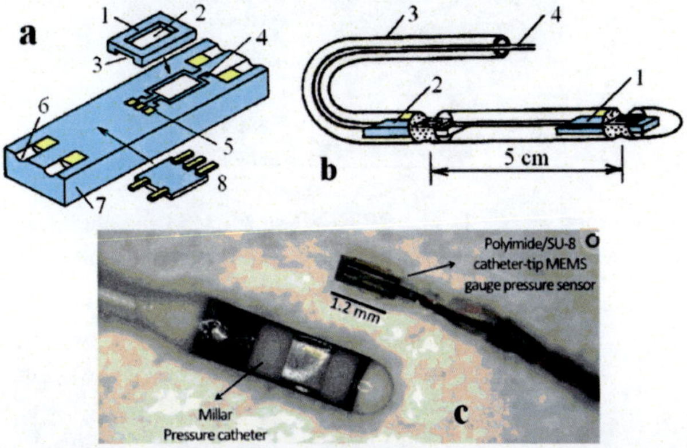

Fig. 4. Integrated pressure μBMS: a) microsensor with integration of 2 chips with substrate: 1-MEMS-technology's Si chip, 2-a thin membrane, 3- a reference pressure cavity, 4-a chip seat mark, 5-contact pads, 6-a recess for connecting wires, 7-a glass substrate, 8-a chip with beam terminals containing secondary conversion devices [9]; b) 2 chip's catheter design: 1- pressure-sensitive chip with strain resistors; 2- secondary conversion's chip; 3 - 0.8 mm diameter catheter, 4-external terminals; c) Catheter designs with Si –chip and Polymide/SU-8 [16]

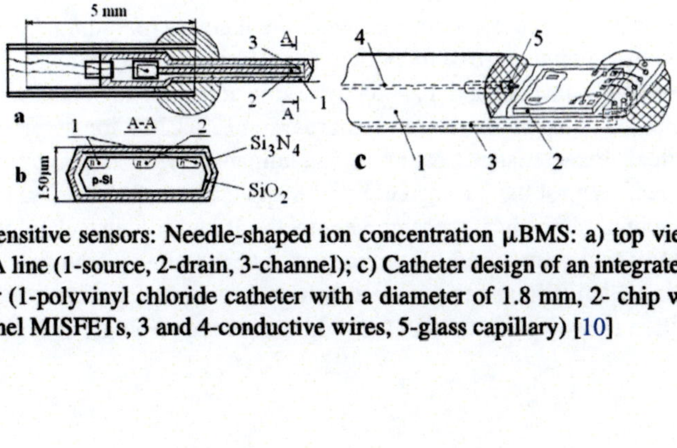

Fig. 5. Ion-sensitive sensors: Needle-shaped ion concentration μBMS: a) top view; b) section along the A-A line (1-source, 2-drain, 3-channel); c) Catheter design of an integrated ion concentration sensor (1-polyvinyl chloride catheter with a diameter of 1.8 mm, 2- chip with 2 ISFETs and 2 n-channel MISFETs, 3 and 4-conductive wires, 5-glass capillary) [10]

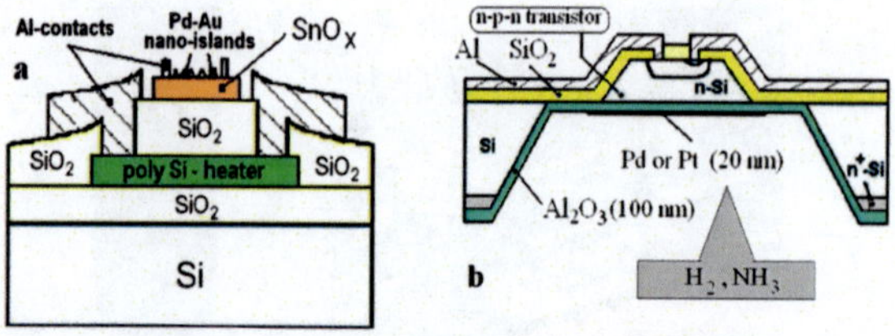

Fig. 6. Semiconductor gas concentration sensors (SGS): a) Structure of ethanol-sensitive sensing element (SE) of integrated SGS; b) Structure of SE of integrated SGS fabricated by MEMS-technology (MEMST) [22]

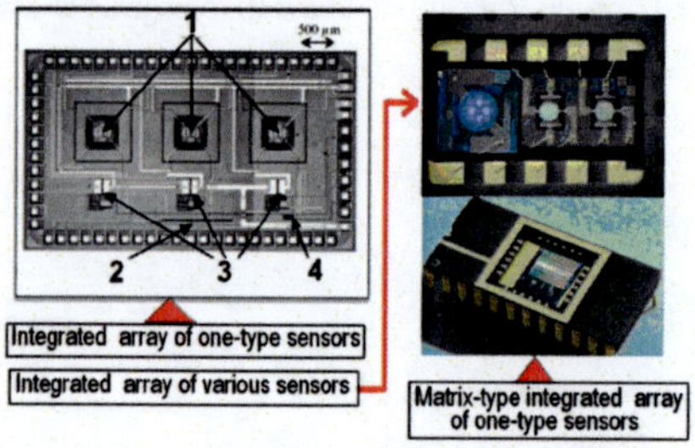

Fig. 7. Integrated sensors' arrays: 1– Gas SE based on MEMST-membrane with micro-heaters; 2 – Substrate temperature sensor; 3 – Temperature controllers; 4 – Counter circuit [22]

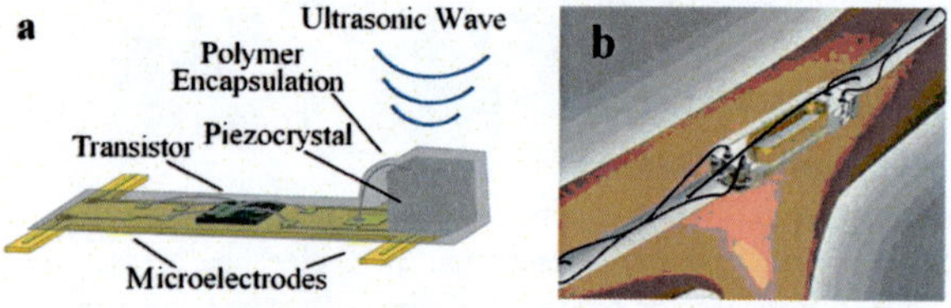

Fig. 8. Implantable μBMS: a) wireless and battery-free sensor [28]; b) Implantable CardioMEMS Champion sensor to monitor a pulmonary artery pressure for treatment of aneurysms [23]

3 Discussion and Conclusions

A characteristic trends of μBMS developments are their *microminiaturization* and *intellectualization* based on the use of micro and nano-technologies [26]. The integration of SEs with actuators, data conversion, and processing units allows to create smart μBMS, μLab-on-chip, small-sized devices and microsystems. μBMS for wireless monitoring of physiological characteristics of animals and humans *in vivo* (implantable chips) have been developed. Against the background of the application of traditional approaches to the development of biomedical measurement methods and silicon technology, the use of nanotechnology (nanostructured films and graphenes) and new materials (organic and bioorganic) in μBMS allowed to improve their performance characteristics and expand their application areas. Today, μBMS are used in neurophysiology, in experimental and practical medicine, in sports and space medicine, in mobile micro-devices for individual self-diagnosis of the state of human functional systems (e.g., in electronic noses) [27, 28]. Implanted wireless and battery-less μBMSs are used for recording biopotentials (brain–computer interface) [29]. BMSs as part of implanted microtransmitters of biotelemetric systems can be used to monitor the physiological characteristics of mobile BO (birds, animals and humans) [30].

Acknowledgment. This work was supported by the MEPhI Academic Excellence Project (contract No. 02.a03.21.0005, 27.08.2013).

Conflict of Interest. The author declares that he has no conflict of interest.

References

1. Bergveld, P.: Development of an ion-sensitive solid-state device for neurophysiological measurements. IEEE Trans. Biomed. Eng. **17**, 70–71 (1970)
2. Wise, K.D., Angel, J.B., Starr, A.: An integrated-circuit approach to extracellular microelectrodes. IEEE Trans. Biomed. Eng. **17**, 238–247 (1970)
3. Franklin, D., Kemper, W., Van Citters, R., Watson, N.: Radio telemetry techniques for measurement of blood pressure and flow in unrestrained animals. UCLA Forum Med. Sci. **10**, 377–382 (1970)
4. Samaun, W.K., Angell, J.: Piezoresistive pressure sensor microcircuits for biomedical devices. IEEE Trans. Biomed. Eng. **20**, 101–109 (1973)
5. Nichols, W., Walker, W.: Experience with the Millar PC-350 catheter-tip pressure transducer. Biomed. Eng. **9**(2), 58–60 (1074)
6. Holmes-Siedle, A.: The space charge dosimeter – general principles a new method of radiation dosimetry. Nucl. Instrm. Methods **121**, 169–179 (1974)
7. Wise, K., James, B., Angell, J.: A low-capacitance multielectrode probe for use in extracellular neurophysiology. IEEE Trans. Biomed. Eng. **22**(3), 212–219 (1975)
8. Indolese, D., Karnatak, P., Kononov, A., et al.: Compact SQUID realized in a double-layer graphene heterostructure. Nano Lett. **20**(10), 7129–7135 (2020)
9. Podlepetskiy, B.: Microelectronic electrodes and sensors for biomedical research. Commun. Equip. Gen. Techn. Ser. **3**, 89–97 (1985). (In Russian)
10. Bergveld, P.: Sensors for biomedical application. Sens. Actuator **10**(2), 165–179 (1986)
11. Chung, T., Wang, J.Q., Wang, J., et al.: Electrode modifications to lower electrode impedance and improve neural signal recording sensitivity. J. Neural Eng. **12**(5), 056018 (2015)
12. Jun, J., et al.: Fully integrated silicon probes for high-density recording of neural activity. Nature **551**, 232 (2017)
13. Fiáth, R., Raducanu, B.C., Musa, S., et al.: A silicon-based neural probe with densely-packed low-impedance titanium nitride microelectrodes for ultrahigh-resolution in vivo recordings. Biosens. Bioelectron. **106**, 86–92 (2018)
14. Seymour, J., Wu, F., Wise, K.D., Yoon, E.: State-of-the-art MEMS and microsystem tools for brain research. Microsyst. Nanoeng. **3**, 16066 (2017)
15. Kim, K., Wu, F., Wise, K.D., Yoon, E.: GaN-on-silicon Micro LEDs for neural interfaces. Semicond. Semimet. **106**, 123–172 (2021)
16. Chau, H.L., Wise, K.D.: An ultraminiature solid state pressure sensor for cardiovascular catheter. IEEE Trans. Electron. Devices **ED-35**(12), 2355–2362 (1988)

17. Hasenkamp, W., Forchelet, D., Pataky, K., et al.: Polyimide/SU-8 catheter-tip MEMS gauge pressure sensor. Biomed. Microdevices **14**(5), 819–828 (2012)
18. Bergveld, P.: Thirty years of ISFETOLOGY: what happened in the past 30 years and what may happen in the next 30 years. Sens. Actuators B Chem. **88**, 1–20 (2003)
19. Sakata, T.: Biologically coupled gate field-effect transistors meet in vitro diagnostics. ACS Omega **4**, 11852–11862 (2019)
20. Cho, S., Cho, W.: High-Sensitivity pH sensor based on coplanar gate AlGaN/GaN metal-oxide-semiconductor high electron mobility transistor. Chemosensors **9**, 42 (2021)
21. Teng, N., Wu, Y., Wang, R., Lin, C.: Sensing characteristic enhancement of CMOS-Based ISFETs with three-dimensional extended-gate architecture. IEEE Sens. J. **21**, 8831–8838 (2021)
22. Shaltaeva, Y., Podlepetsky, B., Pershenkov, V.: Detection of gas traces using semiconductor sensors, ion mobility spectrometry, and mass spectrometry. Eur. J. Mass Spectrom. **23**(4), 217–224 (2017)
23. French, P.: In-Vivo microsystems: a review. Sensors **20**(17), 4953 (2020)
24. Medina-Bailon, C., Kumar, N., Singh Dhar, R., et al.: Comprehensive analytical modelling of an absolute pH sensor. Sensors **21**(15), 5190 (2021)
25. Yan, J., et al.: Electrochemical biosensors for on-chip detection of oxidative stress from immune cells. Biomicrofluidics **5**, 032008 (2011)
26. Pershenkov, V., Podlepetskiy, B., Bocharov, Y., Shagurin, I.: Microelectronics in instrument engineering. Sens. Syst. **1**, 3–22 (2015). (In Russian)
27. Vodička, S., Susič, A., Zelko, E.: Implementation of a savvy mobile ECG sensor for heart rhythm disorder screening at the primary healthcare level: an observational prospective study. Micromachines **12**(1), 55 (2021)
28. Cheng, Z., Warwick, G., Yates, D., Thomas, P.: An electronic nose in the discrimination of breath from smokers and non-smokers: a model for toxin exposure. J. Breath Res. **3**, 036003 (2009)
29. Seo, D., Neely, R.M., Shen, K., et al.: Wireless recording in the peripheral nervous system with ultrasonic neural dust. Neuron **91**(3), 529–539 (2016)
30. Whitford, M., Klimley, A.: An overview of behavioral, physiological, and environmental sensors used in animal biotelemetry and biologging studies. Anim. Biotelemetry **7**, 26 (2019)

Biomaterials for Medical Applications

Mechanical Interactions in Interpenetrating Composites

L. Siebert[1,2(✉)], T. Jeschek[2], B. Zeller-Plumhoff[3], R. Roszak[1], R. Adelung[2], and M. Ziegenhorn[1]

[1] Brandenburg University of Technology Cottbus-Senftenberg, Universitätsplatz 1, 01968 Senftenberg, Germany
lesi@tf.uni-kiel.de
[2] Institute for Materials Science, Kiel University, Kiel, Germany
[3] Institute of Metallic Biomaterials, Helmholtz-Zentrum Hereon GmbH, Geesthacht, Germany

Abstract. Interpenetrating composites are an advanced class of engineering materials where two continuous phases form complex three dimensional interconnected networks. In contrast to traditional composites they can exhibit a variety of advantages such as isotropy and improved coherence between the constituents. In this work the advantages of soft interpenetrating composites made from polymers are explored. A strong interaction between the two phases is found, increasing the compressive modulus of the composite by at least a factor of 2 compared to the unfilled matrix. This increase in stiffness is found in all directions. An additional anisotropy can be introduced by compression of the foam like framework structure prior to infiltration of the soft silicone matrix. Tailoring the stiffness in such a way has promising applications e.g. in artificial cartilage.

Keywords: Interpenetration · Composites · Multiphase · Processing · Soft matter

1 Introduction

There are many different composite materials designed to fulfill advanced tasks in a steadily advancing technological society [1]. The core idea of creating composites is to combine materials in such a way that the favorable properties of each component are transferred to the composite. Exemplary, a common motive is to tailor the mechanical properties of a material, e.g., to be more akin to physiological tissue when using artificial cartilage [2].

One way to do this, is to introduce a stiff material like glass fibers into a soft matrix [3]. These fibers usually have higher stiffness and strength values and thus reinforces the matrix' properties in the direction of their alignment. In such a composite traditionally the modulus of elasticity of the composite material is made up of the Young's moduli of the components, weighted by volume. The composite becomes anisotropic and the modulus of elasticity is limited to that of each component.

In this work, an interpenetrating composite material is investigated that significantly exceeds the elastic moduli of its components. Infiltration of a highly porous, isotropic

I. Tiginyanu et al. (Eds.): ICNBME 2021, IFMBE Proceedings 87, pp. 579–586, 2022.
https://doi.org/10.1007/978-3-030-92328-0_74

sponge-like framework structure with a rubber elastic silicone shows a multiplication of the final elastic modulus as compared to the pristine materials.

Additionally, precompression can introduce a tailor-made anisotropy of the composite material, meaning one axis is stiffer than the others. Numerous applications are conceivable for the resulting composite material, from medical implants to shoes for everyday use.

2 Experimental

Silicone rubber was picked as a matrix material because of its elastic properties. The specific polymer chosen is Ecoflex 00-20. The number refers to the Shore A hardness meaning that this material is a very soft one [4]. As a filler, a piece of Basotect W melamine formaldehyde (MF) foam was used. Its special, open porous framework-like structure allows for infiltration of a liquid phase and the facile formation of an interpenetrating composite. Samples for compression testing of the pure materials were prepared by molding in case of the silicone and by cutting with a sharp blade for the MF foam. The sample geometry was cubes with a side length of 15 mm.

The silicone consists of two components with the mixing ratio of 1:1 by weight. After mixing them together, the mixture was degassed to remove air bubbled and the pre-polymer was placed into a syringe. A custom-designed, airtight mold was 3D printed. This holder contained a cavity with the sponge and a Luer-Lock adapter to fix a syringe to it. Between the syringe and the mold a 3-way valve was inserted to either close or open the connection between syringe and the mold as well as to a vacuum pump. The mold was coated with silicone anti-adhesion fluid to allow for easy demolding.

Composite samples were prepared by placing a piece MF foam into the cavity and closing the mold. A vacuum was pulled on the sponge and afterwards the 3-way valve was turned to enable vacuum infiltration of the MF foam by the liquid silicone pre-polymer. The as-prepared samples were placed in an 80 °C oven for 12 h. After that time full polymerization of the silicone had occurred.

Precompressed samples were fabricated by squeezing a piece of MF foam with overly large dimension into the mold. To achieve a precompression by 25%, the initial sponge was 133% of the height of the mold and compressed down to 75% of its size by restricting the height by the mold. The samples were taken out of the mold and were tested with a custom compression testing apparatus. The compression rate was 0.5 mm/s. During the compression the compressive force was measured and a compression-load diagram was created. The compression was performed dynamically over multiple cycles where the decompression and the compression rate were both kept at 0.5 mm/s.

This measurement was also performed for the pure MF foam and the pure silicone. For the individual cycles the compressive moduli have been determined by linear regression of the initial section below 2% compression. Additionally, the MF foam was investigated via scanning electron microscopy (SEM, FSEM Supra55VP (Zeiss, Germany)) after a sputter coating of gold was applied for conductivity.

3 Results and Discussion

In order to gain an insight into the properties of the composite material, firstly the properties of the individual constituent are determined. The pure MF foam is measured by cyclic compression. The resulting compression-load curves are depicted in (Fig. 1a)).

Two trends can be observed. First, the compressive-load curves show a type of hysteresis, meaning that the compression path is located at higher loads than the decompression. Under normal circumstances such a behavior is attributed to viscoelasticity. On this pretense, the MF foam should exhibit the hysteresis differently strong depending on the compression rate and very low compression rates exhibit no hysteresis at all. The

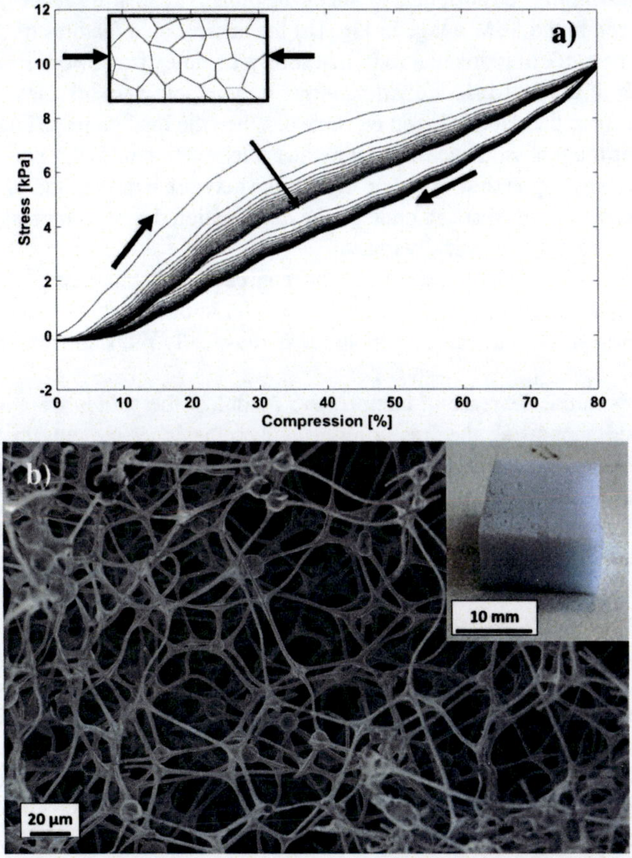

Fig. 1. a) Compression-load curves of a cyclically loaded MF foam. The arrows indicate the direction of the compression. The central arrow indicated the trend upon consecutive cycles. The inset shows a schematic representation of the structure and its macroscopic load application. b) An SEM image of the MF foam microstructure is depicted. It consists of individual rods, with central connection points that always connect four rods. The bonding angle between these can vary leading to a certain degree of the pore size. The inset depicts a macroscopic piece of MF foam.

maximum difference in load between the loading and unloading curve at three compression rates were measured in dependency of the deformation rate and no significant change could be observed, excluding viscoelastic behavior.

Another mechanism to invoke a hysteresis behavior could be the reversible and stiff snapping of individual arms throughout the foam into a second stable position, the snapping back could be delayed until further decompression, leading to a reduction of the load on the decompression curves.

The second phenomenon that can be observed is a settling of the curves into a final shape. The first cycle exhibits much higher compression values than all subsequent cycles. With each cycle, the load and thus the compressive modulus is reduced until the curves settle into a final shape.

This behavior may be attributed to fracturing of individual arms inside the MF foam. As can be seen in the SEM image in Fig. 1b) the microscopic features of the MF foam consist of a network of individual rods, that form a complex framework structure. Upon compression of the network, individual arms in the whole network may experience a higher local force than others based on chance. When the load on these individual arms exceeds their toughness, they will fracture and release the stress, but not contribute to the load resistance upon the next cycle. Due to the network nature of the structure, there are always connections, that can undergo the deformation during compression and thus hold the network macroscopically intact.

This effect can also be seen from the consecutive compressive as depicted in Table 1. The sample was compressed to 80% of its initial height. Especially after the first compression, the compressive modulus is drastically reduced from 27.49 kPa to 23.78 kPa.

After this initial decrease in compressive modulus, the values are not reduced as rapidly, again supporting the fracturing arms hypothesis. Some cuts in the network would naturally result in less resistance against force from outside. The reduction of the compressive modulus in subsequent cycles becomes lower and lower. This means, that even after the first cycle, other places in the network are loaded to the point of fracture. The network may naturally settle into a state where there are never load spikes on individual arms but rather the forces are distributed homogeneously throughout the network, always depending on the maximum compression.

The second constituent of the composites is the very elastic silicone rubber Ecoflex 00-20. The mechanical properties of this silicone was tested like the MF foam.

A compression-load diagram over multiple cycles is depicted in Fig. 2a). In comparison to the pure MF foam, the silicone rubber does not show hysteresis and only a very minor initial compressive modulus change. This may be attributed to contact establishment between the jaws of the compression test setup and the sample upon the first compression. Aside from that, the compressive behavior seems to be purely elastic, showing a straight line with the slope of 26.6 kPa. Usually, silicone rubbers have a certain degree of viscoelasticity to them. This means, that both the pure MF foam and the silicone are in the same regime of mechanical properties, with the silicone being purely elastic and with the MF foam showing also effects that lead to hysteresis.

The purely elastic behavior of the silicone rubber in these tests was verified by performing the compression tests at different compression rates and checking the maximum

Table 1. Compressive modulus of each cycle for a pure MF foam

Cycle number	Compressive Moduli (kPa)
1	27.49
2	23.78
3	22.46
4	22.34
5	21.58
6	21.34
7	21.45
8	21.23
9	20.65
10	20.92
11	19.72
12	19.65
13	19.18

stress difference between load and unload curve. No dependency on the compression rate was found (Fig. 2b)).

One can also see that the stiffness of both materials is in the same regime of around 25 kPa, marking them both as soft materials. For both the silicone rubber and the MF foam no anisotropy was found.

Composites out of the two constituents were fabricated by mixing the two components of the silicone with each other and infiltrating them under vacuum with a custom casting setup. The infiltration of the MF foam is not trivial, since the structure causes turbulences upon infiltration, slowing down the flow of the liquid silicone inside.

A comparison of the compression-load curves of the individual constituents and the composite can be seen in Fig. 3a). The composite exhibits a highly increased stiffness of around 53 kPa, which is an increase by approximately a factor of 2.

This behavior may be related to the structure of the MF foam and the deformation behavior of the individual constituents of the composite. The silicone is incompressible and the individual arms in the MF foam framework are stiff. The arms form a network containing loops which are approximately circular. These circular loops have the lowest circumference per volume ratio. When the volume inside the loops is filled with incompressible silicone and the sample is deformed, the silicone causes a force on the arms which cannot be deformed. Due to the network structure the forces are distributed over the whole composite causing the much higher stiffness. The equal distribution of the framework leads to isotropy of the compressive modulus.

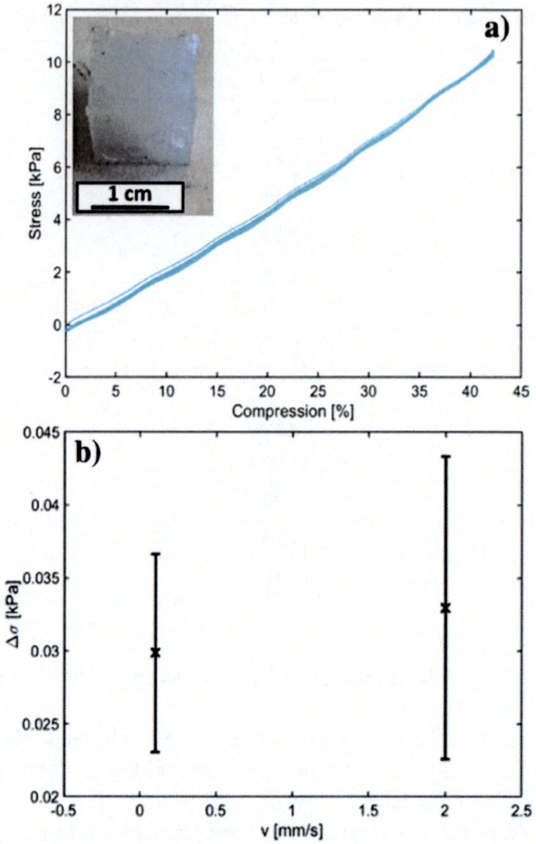

Fig. 2. a) Compression-load curves of a cyclically loaded silicone rubber sample. The consecutive compression curves show no hysteresis-like behavior as it was the case for the MF foam. The inset shows a photograph of a silicone sample. b) The maximum difference of stress between loading and unloading cycles for the silicone rubber for two speeds is shown. The stress difference is negligible and the speed does not show an influence.

Precompression of the pure MF foam by 25% was performed before infiltration of the silicone. The compression-load curve of the precompressed composite reveals a strong anisotropy in the direction of precompression as can be seen in Fig. 3b). The compression in the directions where the sample was not precompressed shows the same compressive modulus as the pure samples at around 25 kPa while the precompressed direction shows the stiffness of the composite.

This effect may be due to the relaxation of the arms in the directions perpendicular to the precompression. Therefore, upon deformation in this direction the silicone can expand without restriction from the MF foam arms, which can just bend out of the way since they do not have to elongate upon deformation. Therefore, this mechanism is unique to the properties of interpenetrating composites and may prove valuable to expand upon in the future.

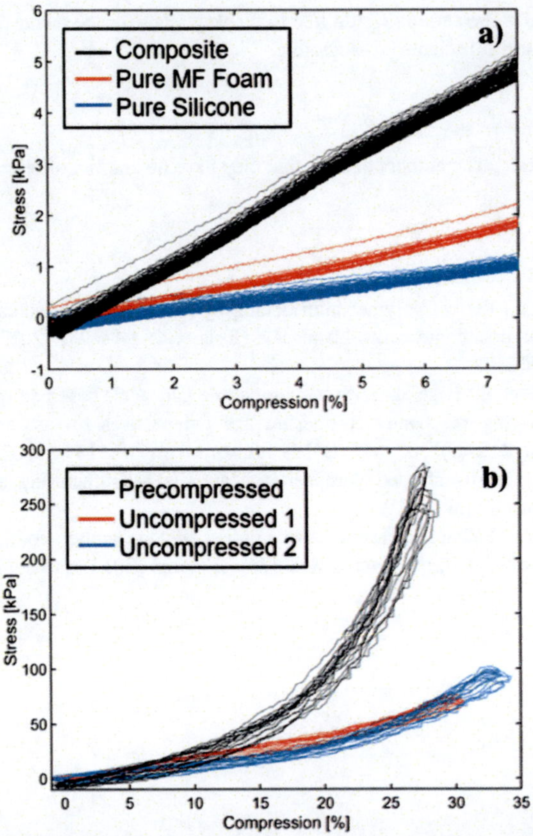

Fig. 3. a) Compression-load curves of the pure materials and a MF foam-silicone composite in comparison. A strongly increased modulus is found for the composite material. b) Compression-load curves of a sample with precompressed prior to infiltration in its three dimensions. Precompression of the MF foam prior to infiltration leads to an increased stiffness in the direction of the precompression.

4 Conclusions

Novel soft interpenetrating composites have been developed and characterized mechanically. The infiltration of soft rubber silicone into framework-like MF foam material led to an increased compressive modulus of a factor of two compared to the pristine material. An anisotropy could be introduced through precompression of the samples with one stiff axis and two soft axes in the same regime as the pure material.

The increased stiffness arising from the unique interaction between the MF foam and the interpenetrating soft rubber silicone and the introduction of an anisotropy *via* precompression of the MF foam may enable the development of novel bioengineering materials with tailored stiffness in the desired axes and flexibility in all other axes. These features may be used in artificial cartilage or in dampening and cushioning applications.

Acknowledgment. The authors would like to thank J. Bahr for the technical assistance and Dr. J. Carstensen for fruitful scientific discussion.

Conflict of Interest. The authors declare that they have no conflict of interest.

References

1. Siebert, L., et al.: Perfect polymer interlocking by spherical particles: capillary force shapes hierarchical composite undercuts. Nanoscale Horiz. **4**(4), 947–952 (2019). https://doi.org/10.1039/C9NH00083F
2. Buyanov, A.L., et al.: High-strength biocompatible hydrogels based on poly(acrylamide) and cellulose: synthesis, mechanical properties and perspectives for use as artificial cartilage. Polym. Sci. Ser. A **55**(5), 302–312 (2013). https://doi.org/10.1134/S0965545X13050027
3. Mallick, P.K.: Fiber-Reinforced Composites: Materials, Manufacturing, and Design, 3rd edn. CRC Press, Boca Raton (2007)
4. Steck, D., et al.: Mechanical responses of Ecoflex silicone rubber: compressible and incompressible behaviors. J. Appl. Polym. Sci. **136**(5), 47025 (2019). https://doi.org/10.1002/app.47025

Imino-Chitosan Hydrogels - Promising Biomaterials for *Candida* Infections' Treatment

Daniela Ailincai[1(✉)], Mihai Mares[2], Andra Cristina Bostanaru[2], and Luminita Marin[1]

[1] "Petru Poni" Institute of Macromolecular Chemistry, Grigore Ghica Voda Alley, Iasi, Romania
ailincai.daniela@icmpp.ro

[2] "Ion Ionescu de la Brad" University of Life Sciences, Iasi, Romania

Abstract. Biocompatible hydrogels were synthesized from chitosan and 2-formylphenylboronic acid (2-FPBA), by the acid condensation reaction of chitosan's amine groups and aldehyde group of 2-FPBA. FTIR and NMR spectroscopy demonstrated that the hydrogelation is a consequence of the formation of reversible imine linkages between the reagents, while wide angle X-ray diffraction proved the highly ordered supramolecular architecture of the obtained hydrogels. The viscoelastic behavior of the hydrogels was evaluated by rheological measurements, performed at human body temperature. The hydrogels were highly elastic, stiff and strong, with a quite high resistance to deformation and a high recovery degree. The morphology investigation by scanning electron microscopy revealed the samples' porosity, forming sponges-like microstructures, with quite uniform pores size distribution. The antifungal activity of the synthesized hydrogels was evaluated on two Candida strains and the obtained results recommend these materials for the treatment of Candida infections.

Keywords: Chitosan · Hydrogels · Antifungal properties · 2-formylphenylboronic acid

1 Introduction

Chitosan based hydrogels are materials intensely used for biomedical application, due to their remarkable properties, given by chitosan's intrinsic properties such as: biocompatibility, biodegradability, antitumoral or hypolipidemic activity [1]. Moreover, due to its moderate hydrophilic character, chitosan facilitates cell adhesion, proliferation and cell differentiation, while its antibacterial activity gives it an added value compared to other natural originating polymers [2]. On the other hand, chitosan presents some disadvantages such as poor mechanical properties, while due to its difficulty of being obtained with a high degree of purity, worsens the mechanical properties of the resulted materials, which usually present the inability to maintain their shape [3]. An approach used for diminishing these drawbacks is the obtaining of chitosan-based derivatives in the form of films or hydrogels. Chitosan-based hydrogels can be prepared by both physical and chemical crosslinking [4]. The physically crosslinked hydrogels have the advantage that they are sensitive to temperature, but their application is limited due to their weak

I. Tiginyanu et al. (Eds.): ICNBME 2021, IFMBE Proceedings 87, pp. 587–594, 2022.
https://doi.org/10.1007/978-3-030-92328-0_75

mechanical properties, as well as their uncontrolled solubilization. For long-term applications, chemically crosslinked hydrogels have proven superior in vivo performance, due to their slower degradation as well as the possibility to control the pores size, by varying the ratio between the polymer and the crosslinker [5].

Taking into consideration that chitosan is a polysaccharide which contains amine functional groups, the main crosslinking pathway is the acid condensation with dialdehydes through reversible imine linkages [7]. For a long time, a special attention was paid to the study of hydrogels based on chitosan cross-linked with glutaraldehyde in view of application in tissue engineering, as drug carriers or for the removal of metals from wastewater. Recently exhaustive studies showed that due to the toxicity of dialdehydes, especially glutaraldehyde, the use of these hydrogels for biomedical purposes is limited and the need to find new non-toxic crosslinking agents is required.

In this context, our goal was to obtain chitosan-based hydrogels, using a new crosslinking agent 2-formylphenylboronic acid, a monoaldehyde known in the literature for its antifungal properties.

2 Materials and Methods

2.1 Materials

2-Formylphenylboronic acid (95%), chitosan (263 kDa, DA: 83%), D-glucosamine hydrochloride, phosphate buffer saline, Yeast Peptone Dextrose Agar (YPD), RPMI-1640, 3-(N-morpholino)propane sulfonic acid (MOPS), 2,3-bis(2-methoxy-4-nitro-5-sulfo-phenyl)-2H-tetrazolium-5-carboxanilide sodium salt (XTT), menadione, calcofluor were purchased from Aldrich and used as received.

2.2 Methods

The hydrogels were synthesized by the acid condensation of chitosan's amine groups and the aldehyde group of 2-formylphenylboronic acid. A model compound (M) has been synthesized by reacting 2-FPBA with D-glucosamine, the structural unit of chitosan (Scheme 1) and used as reference for establishing the hydrogelation mechanism.

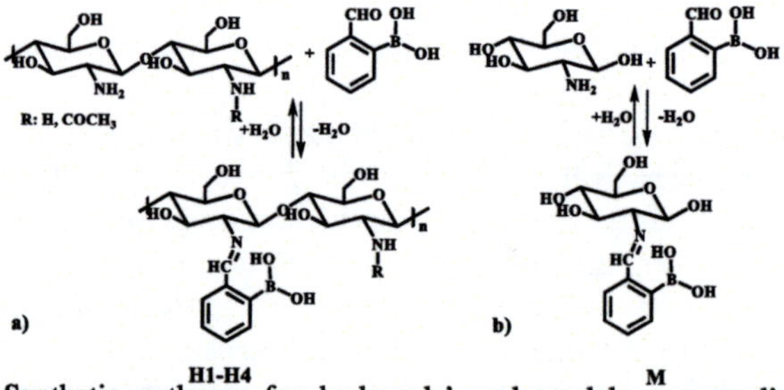

Synthetic pathway for hydrogels' and model compound' synthesis

Two series, each of 6 hydrogels (H1-H3.75), were obtained by varying the molar ratio between the amine groups on chitosan and the aldehyde group of 2-FPBA (Table 1). Because NMR spectroscopy was used to establish the kinetics of the imination reaction, revealing that in one weak, the imine density reached the maximum value, hydrogels which were kept one week before lyophilization were also investigated (H1*-H3.75*). The samples H4 and H4* were synthesized for comparison and characterized by rheological measurements in order to establish the gelation limit related to the aldehyde content.

Table 1. Samples codes

Code	H1	H2	H2.5	H3	H3.5	H3.75	H4
NH_2:CHO (f. p.)	1:1	2: 1	2.5:1	3:1	3.5:1	3.75:1	4:1
Code	H1*	H2*	H2.5*	H3*	H3.5*	H3.75*	H4*
NH_2:CHO (7 days)	1:1	2:1	2.5:1	3:1	3.5:1	3.75:1	4:1

3 Results and Discussions

3.1 FTIR and NMR Spectroscopy

The IR spectra of the obtained hydrogels showed important modifications of their shape compared to chitosan, which indicated the formation of imine bonds and also some supramolecular changes (Fig. 1 a,b). Remarkable was that the absorption maximum which corresponds to the intra-molecular H-bonds from chitosan's spectrum was shifted towards higher wavenumbers in the FTIR spectra of the hydrogels, suggesting the appearance of new intra-molecular H-bonds. Taking into consideration the chemical structure of the formed imino-chitosan derivative, these can be explained by the formation of new intra-molecular H-bonds between the hydrogen from the boric acid and the electron rich nitrogen of the imine units.

The NMR spectra of the model compound and of the hydrogels confirmed the formation of the imine linkages between the reagents, by the appearance of the characteristic chemical shift of the imine proton at 8.7, 8.6 and 8.6, 8.4 ppm respectively (Fig. 1c). Moreover, due to the reaction equilibrium, in the NMR spectra of the hydrogels was observed also the peak corresponding to the unreacted aldehyde, at 9.85 ppm. By comparing the intensities of the peaks corresponding to the imine and aldehyde protons, it could be observed that the reaction equilibrium is shifted to the products over time, a maximum of imine linkages being obtained in 7 days.

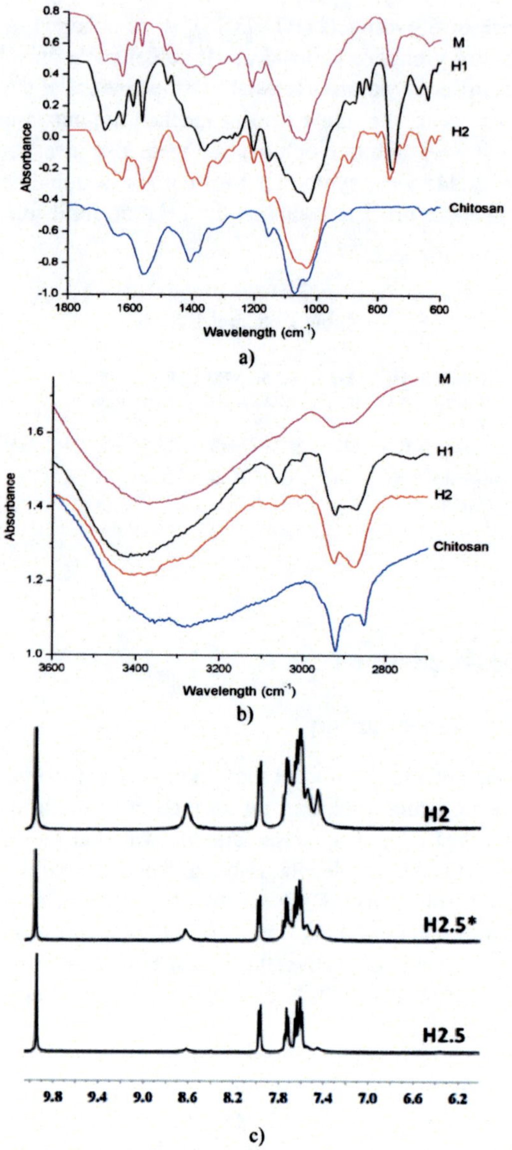

Fig. 1. FTIR (a,b) and NMR (c) spectra for some representative samples

3.2 Wide Angle X-ray Diffraction

The supramolecular architecture of the hydrogels was investigated by Wide Angle X-ray diffraction. Chitosan's diffractogram was recoded and used as reference. As could be observed, it presented a broad band with two maxima, at 12 and 20 two theta degrees (Fig. 2). In the hydrogels' diffractograms three diffraction peaks appeared, suggesting a supramolecular layered architecture, as follows: the reflection in the small angle domain at ~6° corresponds to hydrophilic/hydrophobic segregation, the one at middle angle at 14.2°, corresponds to the inter-chain distances between chitosan's backbones, while the last one at 20° corresponds to the intermolecular distances between two imine units within the same layer.

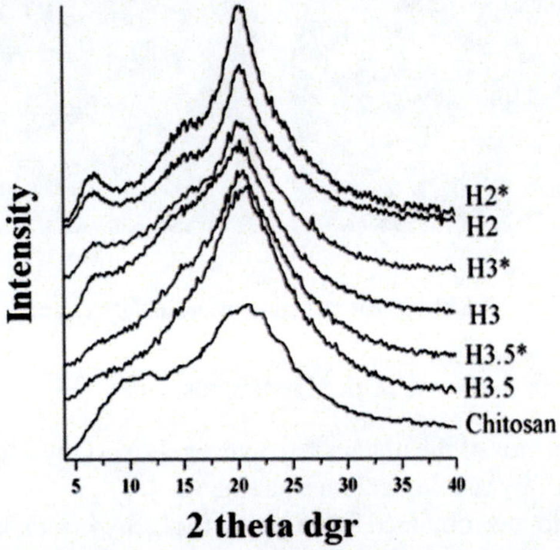

Fig. 2. WXRD diffractograms of chitosan and some representative samples

3.3 Scanning Electron Microscopy

The morphology of the hydrogels was investigated by scanning electron microscopy. All the samples presented a porous morphology, with micrometric pores for which, the size depended on the amount of 2-FPBA used in their synthesis, but also on time. Therefore, the samples containing higher amounts of 2-FPBA presented smaller pores, in comparison with the hydrogels obtained using smaller amount of aldehyde. Regarding the time influence on samples' morphology, a more homogenous microstructure was observed for the samples which were kept 7 days before lyophilization (Fig. 3).

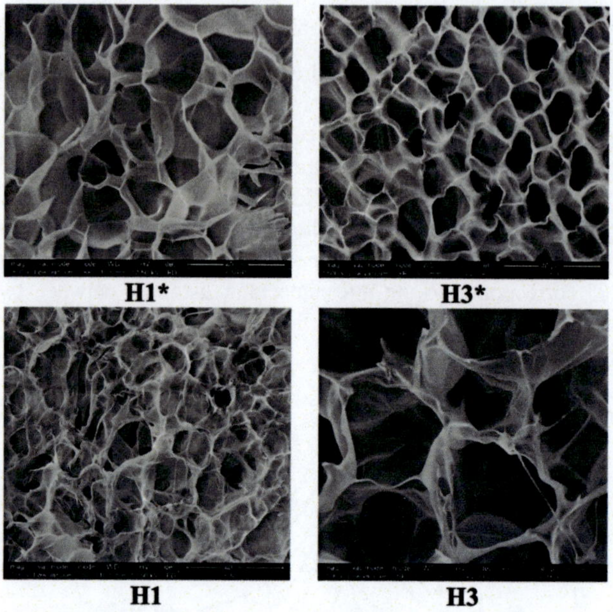

Fig. 3. SEM images of some representative samples

3.4 The Viscoelastic Behavior of the Hydrogels

The viscoelastic behavior of the understudy hydrogels, was investigated at human body temperature of 37 °C, by rheological measurements.

A gel-like behavior was obtained for an amine to aldehyde ratio lower than 3.75, while a liquid-like behavior was evidenced for H4*, in agreement with the visual monitoring of the samples.

A very important aspect which must be highlighted is that by increasing the content of the aldehyde from H3.75 to H2 an increase of rheological parameters G′ and G″ with an order of magnitude happened due to a higher density of imine linkages, in the last case (Fig. 4).

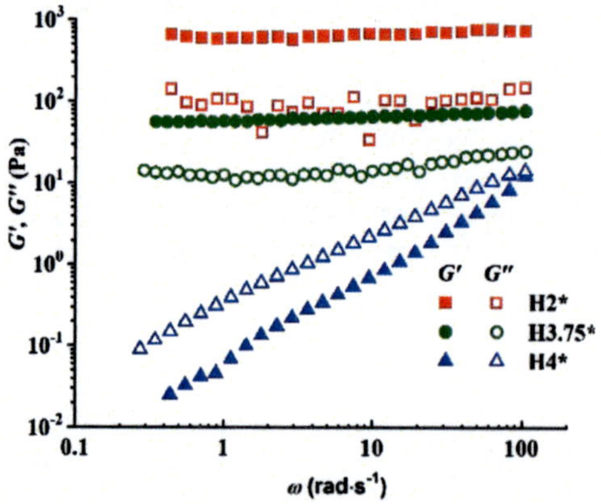

Fig. 4. Rheological parameters of the investigated hydrogels

3.5 The Evaluation of the Antifungal Activity of the Hydrogels

The antifungal activity of the synthesized hydrogels was preliminarily tested on two *Candida* strains - *albicans* and *glabrata* on planktonic yeast and biofilm, mimicking the *in vivo* conditions. The tests were performed also on hydrogels' components- chitosan and 2-FPBA at similar concentration as they have into the hydrogels (Fig. 5).

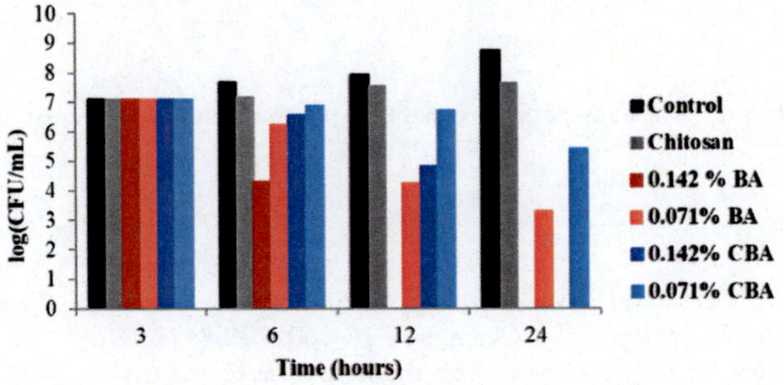

a)

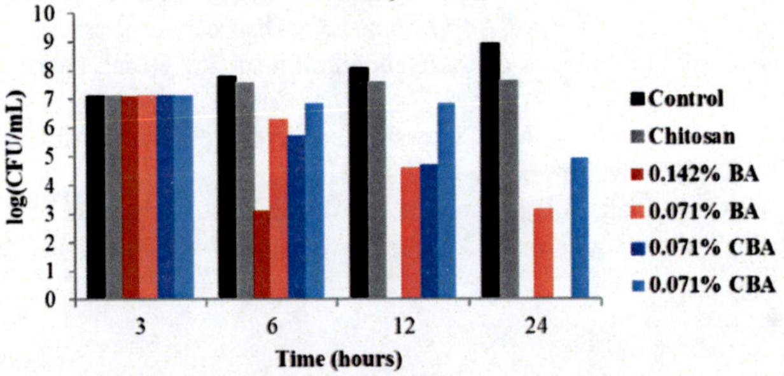

b)

Fig. 5. The antifungal activity of the tested hydrogels and the reference samples on *Candida albicans* (a) and *Candida glabrata* (b)

The tests revealed that the hydrogels presented high antifungal activity (Fig. 5), being able to kill the *Candida* yeast in a controlled manner, due to the presence of the reversible imine linkages in their structure, while the data obtained for chitosan demonstrated its bacteriostatic properties, slowing down the fungi growing but not killing them, according to the literature [1].

4 Conclusions

New chitosan based hydrogels were prepared by the acid condensation reaction with 2-FPBA, a monoaldehyde known to possess strong antifungal activity.

All the synthesized hydrogels proved a layered supramolecular architecture due to hydrophobic/hydrophilic segregation, as revealed by wide angle X-ray diffraction and a highly porous morphology, as was demonstrated by scanning electron microscopy.

Last but not least, the study revealed that the presence of the reversible imine linkages between chitosan and 2-FPBA allowed the gradually and prolonged release of the antifungal aldehyde in the microbiological culture, the obtained hydrogels presenting a very strong antifungal activity on both planktonic yeast and biofilm.

Acknowledgment. This work was supported by a grant of the Ministry of Research, Innovation and Digitization, CNCS/CCCDI – UEFISCDI, project number PD204/2020, within PNCDI III.

Conflict of Interest. The authors declare that they have no conflict of interest.

References

1. Kumar, M.N., Muzzarelli, R.A., Muzzarelli, C., Sashiwa, H., Domb, A.J.: Chitosan chemistry and pharmaceutical perspectives. Chem. Rev. **104**, 6017–6084 (2004)
2. de Sousa Victor, R., et al.: A review on Chitosan's uses as biomaterial: tissue engineering, drug delivery systems and cancer treatment. Materials (Basel, Switzerland) **13**(21), 4995 (2020)
3. Fajardo, A.A., Favaro, S.L., Rubira, A.F., Muniz, E.C.: Dual-network hydrogels based on chemically and physically crosslinked chitosan/chondroitin sulfate. React. Funct. Polym. **73**(12), 1662–1671 (2013)
4. Bhattarai, N., Gunn, J., Zhang, M.: Chitosan-based hydrogels for controlled, localized drug delivery. Adv. Drug Deliv. Rev. **62**(1), 83–99 (2010)
5. Beauchamp, R.O., St Clair, M.B., Fennell, T.R., Clarke, D.O., Morgan, K.T.: A critical review of the toxicology of glutaraldehyde. Crit. Rev. Toxicol. **22**(3–4), 143–174 (1992)

Aqueous Cations and Excess of Translational Vibrations as the Evidences of Charge Transport in Biomaterials

Zarina V. Gagkaeva[1](✉), K. V. Sidoruk[2], B. P. Gorshunov[1], and K. A. Motovilov[1]

[1] Moscow Institute of Physics and Technology, Institutskiy Per., 9, Dolgoprudny, Moscow, Russia
gagkaeva@phystech.edu

[2] State Research Institute of Genetics and Selection of Industrial Microorganisms of the National Research Center "Kurchatov Institute", Moscow, Russia

Abstract. In this study, the terahertz-infrared spectra of extracellular matrix and filaments of S. *oneidensis* bacteria, bovine heart cytochrome c and bovine serum albumin were examined by means of Fourier-transform infrared spectroscopy technique. The absorption lines of water and aqueous cations hydronium H_3O^+ and Zundel $H_5O_2{}^+$ were detected, of the highest intensity in the bacteria extracellular matrix and filaments samples and of the lowest intensity in the albumin samples. We demonstrate that there exists correlation between spectral signatures of aqueous cations and charge transport signs in the investigated materials, which sheds light on the mechanisms of charge transfer.

Keywords: Hydronium · Zundel ion · Charge transport · S. *oneidensis* · Proteins

1 Introduction

The rapid increase in electronical waste production requires an effective replacement of traditional materials utilized in mass consumer electronics by biodegradable alternatives, the so-called "green electronics" [1]. This has provoked an intensification of the studies aimed to find the appropriate substitutes among bacterial extracellular conductive structures, peptide-based supramolecular materials, various melanin/polydopamine materials, etc [2]. However, no matter what transport mechanism is implemented in a given bioorganic material, or which charge carrier type prevails, water will exert, directly or indirectly, significant influence on the charge transport characteristics [3, 4]. Being in contact with polar groups of different condensed materials including those of bioorganic origin, water can exist in different bound states, forming hydration shells and networks *via* hydrogen bonds. When accommodating protons, water molecules unite into various aqueous proton cations H^+- $(H_2O)_n$ such as hydronium (H_3O^+), Zundel ($H_5O_2{}^+$), Eigen ($H_9O_4{}^+$) ions, etc. [5]. These protonic species are now regarded as the major influencers of proton mobility and therefore proton conductivity in various solid state acids [6].

In our previous studies [7, 8], by means of broadband dielectric spectroscopy we showed that S. *oneidensis* extracellular matrix and filaments (EMF) and cytochrome

I. Tiginyanu et al. (Eds.): ICNBME 2021, IFMBE Proceedings 87, pp. 595–601, 2022.
https://doi.org/10.1007/978-3-030-92328-0_76

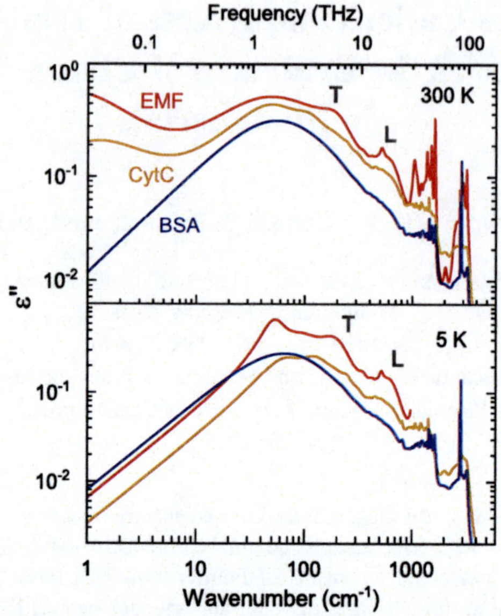

Fig. 1. Terahertz-infrared spectra of dielectric losses of EMF, CytC and BSA with a Debye relaxation of bound water in the low-frequency part of EMF and CytC spectra and with translational (T) and librational (L) vibrations bands in the IR region of the spectra.

c (CytC) obey Drude-like phenomenology of electrical conductivity at temperatures above 250 K. Interestingly, the presence of Drude-like contribution to conductivity $\sigma_1(\nu)$, which is related to the imaginary part of complex dielectric permittivity $\varepsilon^* = \varepsilon' + \mathrm{i}\,\varepsilon''$ as $\sigma_1(\nu) \sim \varepsilon''\,\nu$ (here ν is frequency) always correlated with the emergence at terahertz frequencies of the Debye-type relaxation of bound water and with the dynamics of translational and librational lines of water molecules in the infrared (IR) region. In contrast, in bovine serum albumin (BSA) neither Drude-like response nor Debye relaxation could be observed. However, the exact ways in which water affects conductivity in these materials, as well as possible influence of aqueous cations on charge transport mechanisms were not analyzed. So, within the framework of current study, we took three biological systems with different transport properties: EMF, CytC and BSA proteins, and analyzed their infrared spectral properties in order to investigate the connection between the presence of aqueous protonic species and charge transport mechanisms.

2 Materials and Methods

Experimental Techniques: The vibrational modes of EMF, CytC and BSA samples were studied using Fourier-transform infrared spectrometer Vertex 80V, Bruker, in the frequency range from 100 to 8000 cm^{-1}. For the low temperature measurements down to 5 K a helium flow cryostat was used.

Samples: The *S. oneidensis* MR-1 strain was obtained in Research Institute for Genetics and Selection of Industrial Microorganisms and then its extracellular matrix and filaments were isolated as described in [7]. Bovine serum albumin was purchased from Amresco (code 0332), and cytochrome *c* was purchased from Sigma-Aldrich (code C3131). The samples of EMF, CytC and BSA were lyophilized powders pressed into pellets with a diameter of 1 cm and thickness 1 mm, or prepared as thin films on polyethylene substrates with a thickness of 150 μm films. For hydration control all the samples were kept inside sealed jars above saturated salt solutions until reaching a weight that does not change with time. Thermogravimetry and element analysis were used to characterize the samples.

3 Results and Discussion

3.1 Spectral Signatures of Aqueous Cations and Water

Transmission coefficient spectra of EMF, CytC and BSA samples were processed with a least-square method using Lorentz model for absorption lines:

$$\varepsilon^*(\nu) = \frac{f}{(\nu_0^2 - \nu^2) + i\nu_0\gamma}.$$

Here ε^* is complex dielectric permittivity, f represents the oscillator strength of the resonance, ν_0 and γ are resonance frequency and damping factor, respectively. The fitting results are shown in Fig. 2. While analysis of every single infrared line is beyond the scope of this paper (one may find brief description of these lines parameters in [8]), peaks at frequencies 524 cm^{-1}, 1020 cm^{-1}, 1310 cm^{-1}, 1460 cm^{-1}, 1660 cm^{-1} and 3075 cm^{-1} are of greater interest for current research.

According to [9], the hydronium ion H_3O^+ which is formed as a result of water molecule protonation, has ν_2 vibration-inversion mode, or the so-called umbrella mode, with two bands: $1^+ - 0^-$ transition at frequencies 511–560 cm^{-1} and $1^- - 0^+$ transition at frequencies 954–1075 cm^{-1}. The pronounced peaks observed in the infrared spectra of EMF at 524 cm^{-1} and at 1070 cm^{-1} highly likely correspond to $1^+ - 0^-$ and $1^- - 0^+$ bands of hydronium cation, respectively. Another characteristic group of bands for these cation are δ_{as} (H_3O^+) located at 1597 – 1710 cm^{-1} and associated with H_3O^+ bending, together with $\nu_1(A_1) + \nu_3(E)$ forming a broad intense band at 2800–3200 cm^{-1} referring to the O-H stretching vibrations [10]. These vibration bands also manifested themselves in transmissivity spectra of EMF as both a high-intensity peak with a frequency of 1660 cm^{-1} and a broad band at 3075 cm^{-1}.

The absorption lines at nearly the same frequencies (525 cm^{-1}, 1030 cm^{-1}, 1660 cm^{-1} and 3075 cm^{-1}) but of orders of magnitude lower intensity are also present in the spectra of CytC. In the BSA transmission coefficient spectra their oscillator strengths are even lower, from which one may conclude that the amount of hydronium ions drops dramatically in a row "EMF – CytC – BSA". Shifting of lines of a few cm^{-1} for different biomaterials might indicate different coupling strength of hydronium ions with molecular environment in these materials.

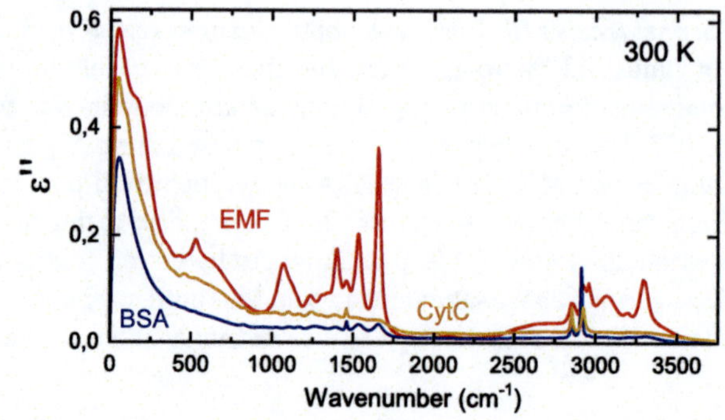

Fig. 2. Infrared spectra of imaginary part of complex dielectric permittivity for three studied biomaterials: EMF, CytC and BSA.

Another aqueous cation, that often participates in charge transport processes in such materials as organic and inorganic proton conductors, is Zundel ion, $H_5O_2^+$, which is formed as a cluster of two water molecules with a proton distributed between them [11]. Its presence in water networks within a material shows up as a group of infrared bands associated with $O - H^+ - O$ group vibrations, including the most intense one corresponding to ν_{as}(OHO) at (1045–1101 cm^{-1}), well-defined one at (1672–1700 cm^{-1}), weaker bands at (1292–1312 cm^{-1}) and (1403–1481 cm^{-1}) due to the out-of- and in-plane deformations γ(OHO) and δ(OHO), respectively, and one more line at (860–995 cm^{-1}) [12]. It is necessary to emphasize that H_3O^+ and $H_5O_2^+$ vibration bands are quite hard to discern due to overlapping with each other and vibrational bands of carboxyl groups. The group of absorption lines meeting all the requirements above can be detected in the IR spectra of each studied material, however, one should treat them with caution. Thus, an intense broad line at 1070 cm^{-1} for EMF and at 1030 cm^{-1} for CytC and BSA might be a result of overlapping of two close lines of H_3O^+ and $H_5O_2^+$; there are also lines at 1310 cm^{-1} and 1460 cm^{-1} and a weak line at 930 cm^{-1} observed for all three materials. The line at 1672–1700 cm^{-1} could also be a part of a complex absorption landscape in that spectral region, together with the δ_{as} (H_3O^+) band of hydronium. As for hydronium spectral features, the conjectural Zundel ion absorption lines are mostly orders of magnitude less intensive in the spectra of CytC and BSA, than in the spectrum of EMF.

Finally, a few words are in order about the broad intense band at 50 cm^{-1}, which can be seen in Fig. 1 for all the three samples. There are several hypotheses about its origin, including one relating this band to long-range protein vibrations [13] and another, according to which this band might be the spectral feature of O-O-O bending mode of polygonal water clusters which often form at the surfaces and in the interstices of different proteins. [14, 15], However, the precise determination of this band nature requires more research.

3.2 Speculations on Charge Transport Mechanisms

The detection of aqueous cations in the samples of EMF, CytC and BSA can be considered as the evidence of the excess protons present within the bulk of the materials. It is known that bound water in protein systems forms the so-called hydration networks [16] and since the protons might not exist in a free state, they travel over these water networks hopping from one water molecule to another and binding with them to form hydronium and Zundel ions [17].

As mentioned above, the free carrier conductivity has been observed in EMF samples hydrated up to 30% mass water content; conductivity values measured for CytC were orders of magnitude lower, and no delocalized charge carriers were detected in the BSA samples despite the equal with CytC level of hydration (11% mass content according to thermogravimetry) [7]. The rise of conductivity values was shown to correlate with the strengthening of bound water contribution to the terahertz absorption of the samples. The same behavior is spotted in the current work for the aqueous cations infrared response.

Although more research is needed to pinpoint the charge carriers type within the studied biological systems, the already existing data demonstrate a strong ability of protons to be dominant in the charge transport processes there. However, the questions still remain, what is the detailed nature of the transport mechanism. Some of the existing models of the behavior of water and protons in condensed media generally omit references to hydronium or the Zundel ion. An example is presented by the "wait-and-switch" model [18], which considers water relaxation as a result of "switching" of states of water molecules due to migration of defects (including ionic defects, as, for example, protons) along hydration networks. Another models consider hydration networks providing proton transfer by diffusion or hopping, as in the cases of vehicle [19] or Grotthuss [20] models, respectively. The last one seems to be more preferable since the activation energy needed for it is lower [21].

Significant differences in the values of conductivity in EMF, CytC, and BSA can be caused by the fact that, due to the different structures of these biomaterials, percolation water networks inside them are organized in different ways. In the case of BSA, water is stronger bound with molecular environment [7] and hydration networks there do not overcome the percolation threshold required for proton transport over relatively long distances by any of the aforementioned mechanisms. Herewith it is not possible to say confidently, whether hydronium or Zundel ion play the major role in this process.

We should also note that the ability of EMF and CytC to conduct charges *via* Drude-like regime correlates with the predominance of water translational vibrations contribution over water librational one to the corresponding dielectric loss spectra of the samples [8]. In case of BSA we see an opposite situation – librations contribution is greater and, correspondingly, no free carrier conductivity is observed [7]. Interestingly, translational vibrations are known to be a key spectral feature of an excess proton transport in water bridge structures. Moreover, the specific spatial water organization responsible for an excess of translational vibrations in water bridges exists only at the interface between bulk water and air [22]. Therefore, there is certain probability that structural peculiarities of water organization at the interfaces within EMF and CytC share similarity with corresponding structures in water bridges yielding specific ability to perform fast proton transfer in electrically constrained water [23–25].

4 Conclusions

Spectral signatures of aqueous cations H_3O^+ and $H_5O_2^+$ observed in the infrared transmissivity spectra of samples of S. *oneidensis* extracellular matrix and filaments (EMF), cytochrome c (CytC) and bovine serum albumin (BSA) indicate the presence of excess protons in all three materials, with a proton concentration orders of magnitude higher in EMF samples than in CytC and in BSA. As per previous studies, this correlates with the transport characteristics of the samples: while free-carrier conductivity observed in EMF is orders of magnitude greater than in CytC, no evidence of charge transport was observed in BSA. Such correlation in delocalized charge carriers and protons behavior allows one to suggest protonic species H_3O^+ and $H_5O_2^+$ to be the main participants of charge transport process within the studied biological materials. Finally, the presence of these cations' infrared absorption lines in the spectrum of biological material might potentially serve as an additional sign of charge transport.

Acknowledgment. The work was supported by the Russian Science Foundation under grant 19-73-10154.

Conflict of Interest. The authors declare that they have no conflict of interest.

References

1. Irimia-Vladu, M.: 'Green' electronics: biodegradable and biocompatible materials and devices for sustainable future. Chem. Soc. Rev. **43**(2), 588–610 (2014). https://doi.org/10.1039/C3CS60235D
2. Mostert, A.B., et al.: Engineering proton conductivity in melanin using metal doping. J. Mater. Chem. B **8**(35), 8050–8060 (2020). https://doi.org/10.1039/D0TB01390K
3. Khodadadi, S., Pawlus, S., Sokolov, A.P.: Influence of hydration on protein dynamics: combining dielectric and neutron scattering spectroscopy data. J. Phys. Chem. B **112**(45), 14273–14280 (2008). https://doi.org/10.1021/jp8059807
4. Amit, M., Appel, S., Cohen, R., Cheng, G., Hamley, I.W., Ashkenasy, N.: Hybrid proton and electron transport in peptide fibrils. Adv. Funct. Mater. **24**(37), 5873–5880 (2014). https://doi.org/10.1002/adfm.201401111
5. Reed, C.A.: Myths about the Proton. The nature of H+ in condensed media. Acc. Chem. Res. **46**(11), 2567–2575 (2013). https://doi.org/10.1021/AR400064Q
6. Kolokolov, D.I., Kazantsev, M.S., Luzgin, M.V., Jobic, H., Stepanov, A.G.: Characterization and dynamics of the different protonic species in hydrated 12-tungstophosphoric acid studied by 2H NMR. J. Phys. Chem. C **118**(51), 30023–30033 (2014). https://doi.org/10.1021/jp510410k
7. Motovilov, K.A., et al.: Observation of dielectric universalities in albumin, cytochrome C and Shewanella oneidensis MR-1 extracellular matrix. Sci. Rep. **7**(1) (2017). https://doi.org/10.1038/s41598-017-15693-y
8. Gagkaeva, Z.V., et al.: Terahertz-infrared spectroscopy of Shewanella oneidensis MR-1 extracellular matrix. J. Biol. Phys. **44**(3), 401–417 (2018). https://doi.org/10.1007/s10867-018-9497-4

9. Liu, D.J., Haese, N.N., Oka, T.: Infrared spectrum of the v2 vibration-inversion band of H 3O+. J. Chem. Phys. **82**(12), 5368–5372 (1984). https://doi.org/10.1063/1.448620
10. Stoyanov, E.S., Kim, K.C., Reed, C.A.: The nature of the H3O+ hydronium ion in benzene and chlorinated hydrocarbon solvents. Conditions of existence and reinterpretation of infrared data. J. Am. Chem. Soc. **128**(6), 1948–1958 (2006). https://doi.org/10.1021/ja0551335
11. Zundel, G.: Hydrate structures, intermolecular interactions and proton conducting mechanism in polyelectrolyte membranes—infrared results. J. Membr. Sci. **11**(3), 249–274 (1982). https://sci-hub.do/10.1016/S0376-7388(00)81250-8. Accessed 28 Aug 2021
12. Stoyanov, E.S., Reed, C.A.: IR spectrum of the H5O2+ cation in the context of proton disolvates L-H+-L. J. Phys. Chem. A **110**(48), 12992–13002 (2006). https://doi.org/10.1021/jp062879w
13. Acbas, G., Niessen, K.A., Snell, E.H., Markelz, A.G.: Optical measurements of long-range protein vibrations. Nat. Commun. **5**(1), 1–7 (2014). https://doi.org/10.1038/ncomms4076
14. Lee, J., Kim, S.-H.: Water poolygons in high-resolution protein crystal structures. PNAS **18**(7), 1370–1376 (2009). https://doi.org/10.1002/pro.162
15. Brudermann, J., Lohbrandt, P., Buck, U., Buch, V.: Surface vibrations of large water clusters by He atom scattering. Phys. Rev. Lett. **80**(13), 2821–2824 (1998)
16. Sasaki, K., Popov, I., Feldman, Y.: Water in the hydrated protein powders: dynamic and structure. J. Chem. Phys. **150**(20), 204504 (2019). https://doi.org/10.1063/1.5096881
17. Garczarek, F., Brown, L.S., Lanyi, J.K., Gerwert, K.: Proton binding within a membrane protein by a protonated water cluster. Proc. Natl. Acad. Sci. U. S. A. **102**(10), 3633–3638 (2005). https://doi.org/10.1073/pnas.0500421102
18. Popov, I., Ben Ishai, P., Khamzin, A., Feldman, Y.: The mechanism of the dielectric relaxation in water. Phys. Chem. Chem. Phys. **18**(20), 13941–13953 (2016). https://doi.org/10.1039/c6cp02195f
19. Kreuer, K.-D., Rabenau, A., Weppner, W.: Vehicle mechanism, a new model for the interpretation of the conductivity of fast proton conductors. Angew. Chemie Int. Ed. Engl. **21**(3), 208–209 (1982). https://doi.org/10.1002/ANIE.198202082
20. de Grotthuss, C.J.T.: Sur la decomposition de l'eau et des corps qu'elle tient en dissolution à l'aide de l'electricit é galvanique. Ann. Chim. **58**, 54–73 (1806)
21. Singh, R.K., Kunimatsu, K., Miyatake, K., Tsuneda, T.: Experimental and theoretical infrared spectroscopic study on hydrated Nafion membrane. Macromolecules **49**(17), 6621–6629 (2016). https://doi.org/10.1021/ACS.MACROMOL.6B00999
22. Teschke, O., Castro, J.R., Soares, D.M.: Translational vibration modes—The spectral signature of excess proton transport in water. Phys. Fluids **30**, 112104 (2018). https://doi.org/10.1063/1.5053483
23. Teschke, O., Castro, J.R., Gomes, W.E., Soares, D.M.: Hydrated excess protons and their local hydrogen bond transport network as measured by translational, librational, and vibrational frequencies. J. Chem. Phys **150**, 234501 (2019). https://doi.org/10.1063/1.5098314
24. Cassone, G.: Nuclear quantum effects largely influence molecular dissociation and proton transfer in liquid water under an electric field. J. Phys. Chem. Lett. **11**(21), 8983–8988 (2020). https://doi.org/10.1021/acs.jpclett.0c02581
25. Fuchs, E.C., Bitschnau, B., Wexler, A.D., Woisetschläger, J., Freund, F.T.: A quasi-elastic neutron scattering study of the dynamics of electrically constrained water. J. Phys. Chem. B **119**(52), 15892–15900 (2015). https://doi.org/10.1021/acs.jpcb.5b10751

Check for updates

GaN Ultrathin Membrane for SERS Detection of Rhodamine B

Vladimir Ciobanu[1(✉)], I. Plesco[1], T. Braniste[1], G. Ceccone[2], P. Colpo[2], and I. Tiginyanu[1]

[1] National Center for Materials Study and Testing, Technical University of Moldova, Stefan cel Mare av. 168, 2004 Chisinau, Moldova
vladimir.ciobanu@cnstm.utm.md

[2] Joint Research Center, European Commission, via E. Fermi 2749, 21027 Ispra, Italy

Abstract. In this paper we demonstrate the fabrication of a SERS detector based on GaN ultrathin membrane. The GaN membranes are elaborated by the so-called Surface Charge Lithography approach. The obtained membranes are functionalized by 20 nm Au nanodots and characterized by different tools in order to demonstrate the material quality and sensitivity enhancement for Rhodamine B detection in the micromolar range.

Keywords: GaN membrane · SERS detector · Rhodamine B sensor · Surface Charge Lithography · Au nanodots

1 Introduction

Optical sensors and biosensors represent an easiest way to detect a very low concentration of specific molecules. There are different types of these kind of optical sensors, but most of them includes very sophisticated and expensive tools. However, the combination of Surface Plasmon Resonance with a Raman Scattering effect proved to be very efficient for the determination of even nM concentration of biogenic amines [1]. Paper based SERS (Surface Enhanced Raman Scattering) sensor [2] which consists of Au nanoparticles deposited on paper substrate proved to be as efficient as solid substrates like quartz, glass or others, but these kinds of sensors are single use ones and not stable in some hazardous solutions compared to GaN substrate which is a very chemically stable material.

GaN is nowadays considered as the second most important semiconductor after Si and it has a wide range of applications like high power electronics [3], high frequency devices [4], biosensors [5], optoelectronics [6], pressure sensors [7], battery energy storage systems [8], memristive systems [9] etc. The nanostructured GaN could enhance some specific properties of the material due to its enhanced surface area as it was previously demonstrated for concrete applications like hydrogen production [10].

Rhodamine B (Rh B) is a widely used dye in the industry like dyeing in textile, leathers, printing and so on. The organic dyes can however cause serious problems for environment or biodiversity. Moreover, the presence of Rh B was reported also in food industry, as it was found in chili powder, sausage, sweets, preserved plums. As Rh B is

I. Tiginyanu et al. (Eds.): ICNBME 2021, IFMBE Proceedings 87, pp. 602–609, 2022.
https://doi.org/10.1007/978-3-030-92328-0_77

considered as potentially carcinogenic [11], it's use in food industry should be strictly controlled by law. Thus, the low concentration and a fast way for the detection of these molecules in food or in the residual water is a very important task.

2 Results and Discussion

2.1 Fabrication of GaN Membranes

For the fabrication of GaN ultrathin membranes, the so-called surface charge lithography approach was used, a technique developed previously at the National Center for Material Study and Testing [12]. A commercial unintentionally doped MOCVD (metal-organic chemical vapor deposition) grown GaN wafer with the concentration of free carriers of $5 \cdot 10^{-17}$ was used. A traditional photolithographic process was done in order to have the desired pattern of the membranes. After the developing process, the membrane is subjected to Ar^{+} plasma treatment. The plasma conditions are very well defined and controlled. The ion dose was maintained at 10^{11} ions/cm^2 in order to have a penetration depth of the ions up to 15 nm in the material according to the previous Monte Carlo simulations [13]. In the last lithographic step, the remained photoresist is dissolved in acetone. The process is followed by a photoelectrochemical etching in a low concentrated KOH solution under UV light illumination. The electrical closed circuit is maintained at 0 V and the current value is continuously monitored by using a Keithley instrument. The fabrication process of the ultrathin GaN membranes is illustrated in the Fig. 1 below.

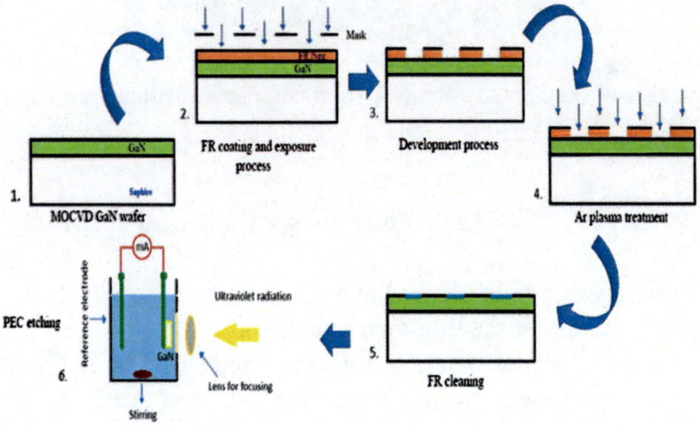

Fig. 1. Fabrication steps of the GaN membranes

2.2 SEM and TEM Analysis of the Membranes

The morphology of the membranes (Fig. 2a, b) was studied by scanning electron microscopy (SEM) Zeiss Gemini Ultra55 Plus working at 10 kV and transmission electron microscopy (TEM) investigations were carried out with FEI Tecnai F30 G2 STwin

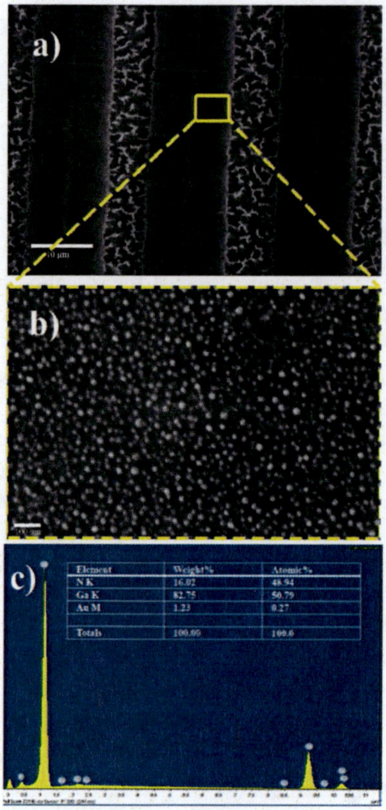

Fig. 2. SEM images of (a) GaN membranes, (b) membrane functionalized with Au dots and (c) EDX analysis

equipment operated at 300 kV (field emission gun, spherical aberration coefficient CS = 1.2 mm).

The EDX (energy dispersive X-ray) analysis shown in Fig. 2c clearly demonstrates the presence of Au on top of the thin membrane.

A thin GaN membrane transparent to e-beam is imaged in Fig. 3a. The high resolution (HR) micrograph Fig. 3b reveals single crystalline texture of membrane. The fast Fourier transformation (FFT) in Fig. 3c was taken from the full region in b and demonstrates wurtzite type $P6_3mc$ GaN structure oriented along [0001] zone axis. At the same time no additional reflections were observed. The FFT matches perfectly to simulated diffraction pattern on Fig. 3d.

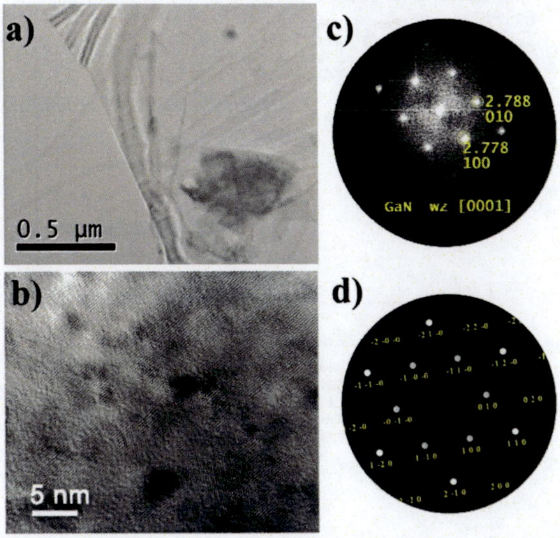

Fig. 3. Crystallographic analysis of GaN nanomembrane: (a) an overview TEM micrograph, (b) HR micrograph, (c) FFT collected from (b), (d) simulated diffraction pattern of wurtzite GaN oriented along [0001] axis.

2.3 XPS Analysis

The surface chemistry was assessed by X-Ray Photoemission Spectroscopy (XPS). XPS analysis was carried out by means of an Axis Ultra-DLD spectrometer equipped with a non-monochromatic Mg Kα source (hν = 1253.6 eV). The take-off angle (ToA) respect to the sample normal was 0° for survey and high-resolution (HR) spectra. Surface charging was compensated using low energy (~4 eV) electrons and adjusted using the charge balance plate on the instrument. Selected samples were also analyzed after ion cleaning. The spectra were calibrated setting hydrocarbon C 1s at 285.0 eV.

Figure 4a shows the survey spectra of the GaN wafer surface. It shows the presence of almost stoichiometric gallium and nitrogen species, carbon due to contamination in air, but also oxygen species which according to the high-resolution analysis of Ga photoemission peaks, a shift to higher binding energy of about 0.5 eV is observed, being attributed to a thin native oxide layer (Fig. 4c).

In case of the GaN membrane after Ar etching for 5 min (Fig. 4b), a higher content of oxygen species was detected. The nitrogen content decreases because the Ar etching causes a preferential etching of nitrogen in GaN thus resulting in formation of metallic gallium.

Even if the oxygen concentration was found higher than in case of GaN wafer, all the Ga photoemission peaks are shifted to lower binding energies suggesting the non-oxidative state. About 7% of Pb and Cu species were additionally found which are attributed to the contamination from silver paste which was used in the photoelectrochemical process for contacting electrically the GaN surface with the electrode.

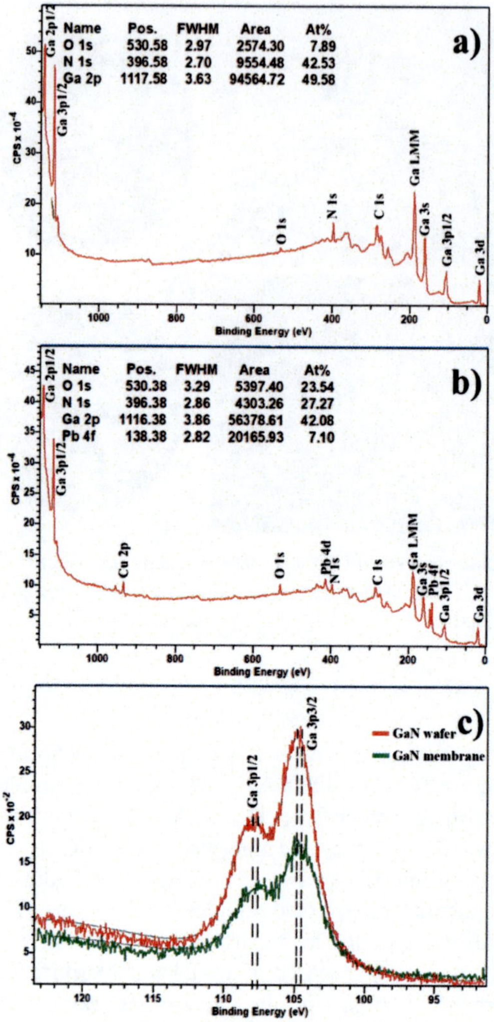

Fig. 4. XPS spectra of: (a) GaN wafer; (b) GaN membrane and (c) a comparison on the position of Ga 3p photoemission peak in case of GaN wafer and GaN membrane

2.4 Raman and SERS Measurements

The Raman instrument used for measurements was an inVia Renishaw equipped with a confocal microscope. The spectra were collected using the 532 nm laser and 100× objective lens with aperture NA = 0.80 for focusing the light on the sample surface. The Raman spectra of a ultrathin GaN membrane is presented in Fig. 5a. The main peak at 568.2 cm^{-1} which is attributed to E2(high) Raman active mode is redshifted with about 0.6 cm^{-1} compared to theoretical study of the GaN vibrational modes. This redshift is attributed to the dislocations and stress induced into the material during the MOCVD growth process on sapphire substrate.

For the functionalization of GaN membrane, a 5 nm Au layer was deposited using a Cressington 108 Sputter Coating System. The formed Au layer is transformed into Au nanodots after a thermal treatment of the sample at 300 °C for 1 h (as shown in Fig. 2b).

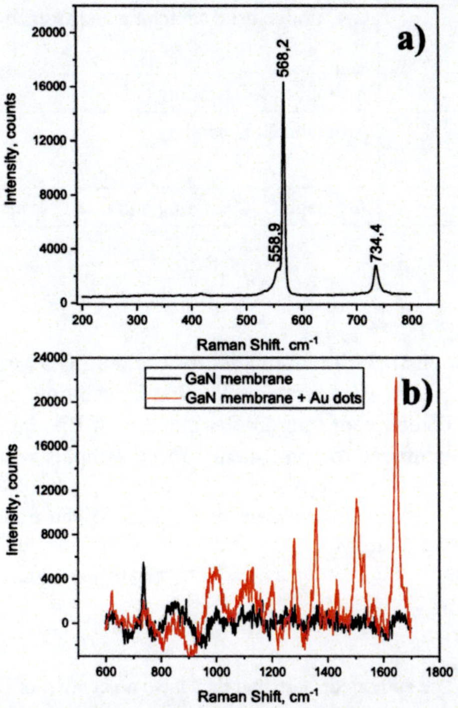

Fig. 5. (a) Raman spectrum of GaN ultrathin membrane and (b) SERS signal for detection of Rhodamine B on a as grown GaN membrane and on a Au dots functionalized membrane

The aqueous solution consisted of 1 μM Rhodamine B dissolved in deionized water (Millipore, MilliQ System). The samples were deepened into the Rh B containing solution for 10 min and then let dry in air at room temperature before Raman measurements. The Raman characterization was performed on 10 different locations on the sample surface and the average result after baseline subtraction is represented in Fig. 5b.

The Raman band located at 619 cm^{-1} is attributed to the xanthene ring puckering mode. The most intense bands in the region from 1100 to 1700 cm^{-1} on the Raman spectrum of Rh B were related to C–C, C–H and C = C [14]. A detailed assignment of the vibrational peaks is given in Table 1. In Fig. 4, characteristic peaks at 1360 cm^{-1}, 1504 cm^{-1}, and 1648 cm^{-1} are relatively obvious on SERS spectra of Rh B aqueous solutions. For the quantitative analysis, the peak located at 1648 cm^{-1} is best suited for the assessment of Rh B and was selected for characterization. As can be seen, a Raman scattering enhancement factor of about 20 in case of sample functionalized with Au nanodots was found. The SERS effect is attributed to the excitation of the Surface Plasmons which causes an enhanced local electromagnetic field around the Au nanodots.

Table 1. Rhodamine B Raman peaks positions

Peak position, cm^{-1}	Assignment
620	Xanthene ring puckering
1200	C–C bridge band stretching and aromatic C–H bending
1281	
1360	Aromatic C–C bending
1504	Aromatic C–H bending
1527	
1648	Aromatic C–C bending and C = C stretching

3 Conclusions

In this work, we demonstrate the fabrication of ultrathin GaN membranes with single crystalline structure which after functionalization by Au dots prove to be efficient as SERS sensor for the detection of micromolar quantity of Rhodamine D molecules. The enhancement factor compared to non-functionalized samples was about 20.

Acknowledgment. The authors acknowledge the financial support from European Commission under the Grant #810652 "NanoMedTwin" and from the Ministry of Education, Culture and Research of the Republic of Moldova under the Grant #20.80009.50007.20.

Conflict of Interest. The authors declare that they have no conflict of interest.

References

1. Buccolieri, A., Bettini, S., Salvatore, L., et al.: Sub-nanomolar detection of biogenic amines by SERS effect induced by hairy Janus silver nanoparticles. Sens. Actuators B Chem. **267**, 265–271 (2018). https://doi.org/10.1016/j.snb.2018.04.028
2. Lin, S., Hasi, W.-L.-J., Lin, X., et al.: Rapid and sensitive SERS method for determination of Rhodamine B in chili powder with paper-based substrates. Anal. Methods **7**(12), 5289–5294 (2015). https://doi.org/10.1039/C5AY00028A
3. Li, H., Li, X., Zhang, Z., et al.: Design consideration of high power GaN inverter. In: 2016 IEEE 4th Workshop on Wide Bandgap Power Devices and Applications (WiPDA), pp. 23–29 (2016). https://doi.org/10.1109/WiPDA.2016.7799904
4. Palacios, T., Mishra, U.K.: 5.06 - GaN-based transistors for high-frequency applications. In: Bhattacharya, P., Fornari, R., Kamimura, H. (eds.) Comprehensive Semiconductor Science and Technology, pp. 242–298. Elsevier, Amsterdam (2011). https://doi.org/10.1016/B978-0-44-453153-7.00021-3
5. Tai, T.-Y., Sinha, A., Sarangadharan, I., et al.: Design and demonstration of tunable amplified sensitivity of AlGaN/GaN high electron mobility transistor (HEMT)-based biosensors in human serum. Anal. Chem. **91**(9), 5953–5960 (2019). https://doi.org/10.1021/acs.analchem.9b00353

6. Dylewicz, R., Patela, S.Z., Paszkiewicz, R.: Applications of GaN-based materials in modern optoelectronics. In: Romaniuk, R.S. (ed.) Photonics Applications in Astronomy, Communications, Industry, and High-Energy Physics Experiments II. SPIE, pp. 328–335 (2004). https://doi.org/10.1117/12.568864
7. Dragoman, M., Ciobanu, V., Shree, S., et al.: Sensing up to 40 atm using pressure-sensitive Aero-GaN. Physica Status Solidi (RRL) Rapid Res. Lett. **13**(6), 1900012 (2019). https://doi.org/10.1002/pssr.201900012
8. Moradpour, M., Ghani, P., Gatto, G.: A GaN-based battery energy storage system for residential application. In: 2019 International Conference on Clean Electrical Power (ICCEP), pp. 427–432 (2019). https://doi.org/10.1109/ICCEP.2019.8890238
9. Dragoman, M., Tiginyanu, I., Dragoman, D., et al.: Learning mechanisms in memristor networks based on GaN nanomembranes. J. Appl. Phys. **124**(15), 152110 (2018). https://doi.org/10.1063/1.5034765
10. Zhang, M., Zhao, S., Zhao, Z., et al.: Piezocatalytic effect induced hydrogen production from water over non-noble metal Ni deposited ultralong GaN nanowires. ACS Appl. Mater. Interfaces **13**(9), 10916–10924 (2021). https://doi.org/10.1021/acsami.0c21976
11. Nestmann, E.R., Douglas, G.R., Matula, T.I., et al.: Mutagenic activity of rhodamine dyes and their impurities as detected by mutation induction in Salmonella and DNA damage in Chinese hamster ovary cells. Can. Res. **39**(11), 4412–4417 (1979)
12. Volciuc, O., Braniste, T., Sergentu, V., et al.: Fabrication of photonic crystal circuits based on GaN ultrathin membranes by maskless lithography. In: Tiginyanu, I.M. (ed.) Nanotechnology VII. SPIE, pp. 8–17 (2015). https://doi.org/10.1117/12.2178525
13. Stevens-Kalceff, M.A., Tiginyanu, I.M., Popa, V., et al.: Cathodoluminescence characterization of suspended GaN nanomembranes. J. Appl. Phys. **114**(4), 43516 (2013). https://doi.org/10.1063/1.4816562
14. Wang, Y., Chen, H., Dong, S., et al.: Surface-enhanced Raman scattering of silver-gold bimetallic nanostructures with hollow interiors. J. Chem. Phys. **125**(4), 44710 (2006). https://doi.org/10.1063/1.2216694

Wettability of Highly Conductive ZnO:Ga:Cl CVT Ceramics with Various Ga Content

G. V. Colibaba[1,2(✉)], N. Costriucova[2], D. Rusnac[1,2], S. Busuioc[3], and E. V. Monaico[3]

[1] Moldova State University, Mateevici 60, Chisinau, Republic of Moldova
gcolibaba@usm.md
[2] Institute of Applied Physics, Chisinau, Republic of Moldova
[3] Technical University of Moldova, Chisinau, Republic of Moldova

Abstract. ZnO:Ga:Cl ceramics were sintered using chemical vapor transport technique. Ga content was varied in a range of 0–10 mol %. The wettability of unpolished and polished surface of ZnO:Ga:Cl ceramics was investigated. The polished and etched surface of ZnO ceramics is in a hydrophilic state. The presence of Ga impurity leads to a strong increase in the water contact angle to 131°. This behavior is attributed to a high concentration of free electrons, which suppress the formation of intrinsic surface defects acting as traps for water molecules. Air pockets on unpolished surfaces of ZnO:Ga:Cl ceramics are an additional factor that increases the water contact angle.

Keywords: ZnO CVT ceramics · Wetting · Hydrophobicity · Electrical conductivity

1 Introduction

Zinc oxide (ZnO) is a well-known material, studied in last two decades for fascinating semiconducting, piezoelectric, optical and emission properties. Nowadays, an increased interest appeared to ZnO nanostructures due to its biomedical applications [1,2]. Controlling the morphology of nanostructures (dimensions, shape, etc.) together with ZnO semiconducting and piezoelectric properties make it promising for biosensors and transducers [3].

It should be mentioned that an important property, namely wettability of water, can expands the applications of solid surfaces such as self-cleaning, corrosion protection, etc. [4].

The c-plane orientated ZnO single-crystalline surface is hydrophilic; the water contact angle (CA) is as low as 31° [5]. The oxygen-related surface defects (−OH radicals adsorbed on oxygen vacancies), acting as traps for water molecules and promoting hydrophilicity of ZnO. The concentration of oxygen vacancies (V_O) can be reduced by storage in the dark [6], by thermal annealing in air or oxygen atmosphere [7, 8], or by deposition of Al atoms acting as donor impurities [9]. Hydrophobicity of the ZnO surface can be significantly increased using various nanowires grown on the surface. The presence of air pockets on the rough surface can increases CA value up to 157° [8, 10, 11].

I. Tiginyanu et al. (Eds.): ICNBME 2021, IFMBE Proceedings 87, pp. 610–616, 2022.
https://doi.org/10.1007/978-3-030-92328-0_78

The chemical vapor transport (CVT) technique is used for an unseeded growth of high-quality ZnO single crystals with controllable growth direction [12]. Recently, this technique was proposed as alternative approach to sintering highly conductive ZnO ceramics. A high pressure of doping gaseous species participated in CVT reactions can contribute to the formation of ceramics, uniformly doped in the gas phase, even at low sintering temperatures (~1050 °C) [13, 14]. This new type of ceramics can be used as magnetron targets for deposition of ZnO thin films with enhanced conductive properties [15]. The wettability of ZnO:Ga:Cl CVT ceramics has not been investigated; the study of this phenomenon is the purpose of the present work.

2 Experiment

$ZnO + Ga_2O_3$ mixed powder was sintered by means of CVT in sealed quartz chambers at 1070 °C during 48 h. After sintering, the furnace with the samples was cooled down at a rate of $\leq$100 °C/h. The technological procedure of sintering in details was described in earlier published works [13, 14]. The dopant concentration (Ga_2O_3) was varied in a range of 0–10 mol %. HCl was used as a transport agent at a loading pressure of 2 atm at the sintering temperature.

For studying electrical properties, the ceramic samples were purified by annealing in a vacuum at 300 °C for 30 min. The resistivity (ρ) and charge carrier concentration ($\boldsymbol{n}$) of the ceramics were calculated from the Hall effect measurements by the van der Pauw method using In droplets fused to the sample surface as the electrical contacts. 2Θ X-ray diffraction (XRD) spectra, recorded using FeK_α radiation (1.936 Å), were used to analyze the composition of the samples. A TESCAN Vega TS 5130 MM scanning electron microscope (SEM) equipped with a secondary and backscattered electron detector was used to study the morphology. For grain size distribution analysis, the obtained ceramics was polished and then etched in HCl aqueous solution to delineate grain boundaries [14]. The mean grain sizes were measured using image analysis software. For wettability experiments, the as-sintered ceramics were polished with a 1 μm diamond paste and then etched in a 0.5% HCl aqueous solution for 30–90 s to remove contaminations and damage defects. Contact angle measurements were carried out with water using the KRUSS DSA25 (drop shape analyzer) equipped with a camera. The sample was illuminated from behind so that the shadow contour of the substrate-drop contact was visible and well highlighted to allow the software to determine the contact angle.

3 Experimental Results

3.1 General Characterization of Ceramics

The obtained ZnO:Ga:Cl ceramics have a high density of 5.3 ± 0.2 g/cm^3, which corresponds to 94.5 ± 3% of the theoretical value [16]. The hardness is 2.0 ± 0.2 GPa, which is close to the hardness of ZnO single crystals [17]. The typical diameter and thickness of samples are 25 and 1.5–2 mm, respectively (see inset in Fig. 1).

XRD spectrum of unintentionally doped ZnO:Cl ceramics is shown in Fig. 1a (ICDD: 01–089-0510). All the peaks are attributed to hexagonal ZnO; there are no significant

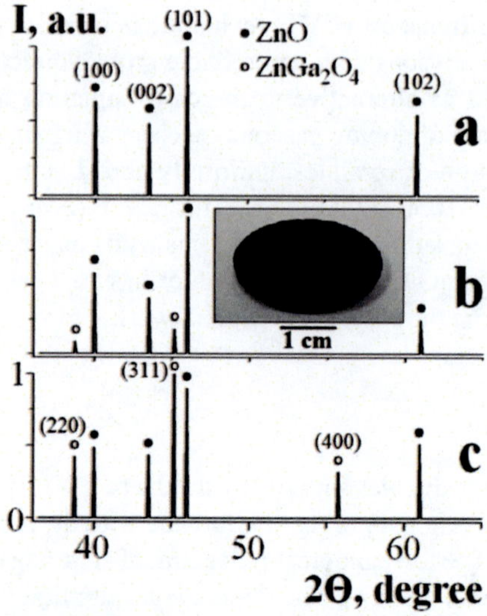

Fig. 1. Normalized XRD spectra of ZnO ceramics with (a) 0, (b) 3, and (c) 10 mol % Ga2O3. Inset illustrates ZnO:Ga:Cl CVT ceramics

reflection lines attributed to Zn, $ZnCl_2$, or spinel phases (for example, $ZnMe_2O_4$). Analysis of the XRD spectra of ZnO:Ga:Cl ceramics with Ga_2O_3 content of 1–2 mol % reveals the same peaks that are observed in the undoped ZnO ceramics. In the case of using 3 mol % Ga_2O_3, X-ray diffraction spectroscopy reveals several peaks of the $ZnGa_2O_4$ spinel phase (PDFcard: 00-038-1240) (Fig. 1b). The XRD spectrum of heavily doped sample (10 mol % Ga_2O_3) reveals several intensive $ZnGa_2O_4$ reflection lines (Fig. 1c). Thus, the solubility limit of Ga_2O_3 in ZnO is estimated at about 3 mol %.

An analysis of the grain size distribution of polished and etched surfaces shown that ZnO:Cl ceramics consists of grains with an average diameter of 7 μm (Fig. 2a). The material with 3 mol % Ga_2O_3 consists of grains having an average size of 19 μm as can be seen in Fig. 2b. A further increase in the dopant concentration (10 mol %) leads to increase in the average grain size to 34 μm (Fig. 2c). The SEM images of the most heavily doped sample (Ga_2O_3 content = 10 mol %), which were recorded with a back scattered electron detector, showed the presence of $ZnGa_2O_4$ inclusions having a mean size of about 5 μm (Fig. 2c, white particles). Unintentionally doped ZnO:Cl ceramics has a relatively smooth unpolished surface (Fig. 2a). ZnO:Ga:Cl samples have many nano- and microvoids on the surface. An increase in the doping level results in rougher unpolished ceramic surface (Fig. 2b, 2c).

The dependence of free electron concentration on Ga_2O_3 content is shown in Fig. 3 (curve 3). The unintentionally doped ZnO:Cl ceramics has n value of 0.6×10^{19} cm^{-3} (resistivity $\rho = 35 \times 10^{-3}$ Ω·cm) [14]. These conductive properties are attributed to Cl impurity present in the ceramics with a concentration of about 10^{19} cm^{-3} [18]. The most conductive ceramics with n value of 5.4×10^{19} cm^{-3} ($\rho = 1.5 \times 10^{-3}$ Ω·cm) can be obtained in the case of loading of 3 mol % Ga_2O_3. Obviously, typical defects in this material are shallow Ga_{Zn} donors [19]. With a further increase in the doping level, when

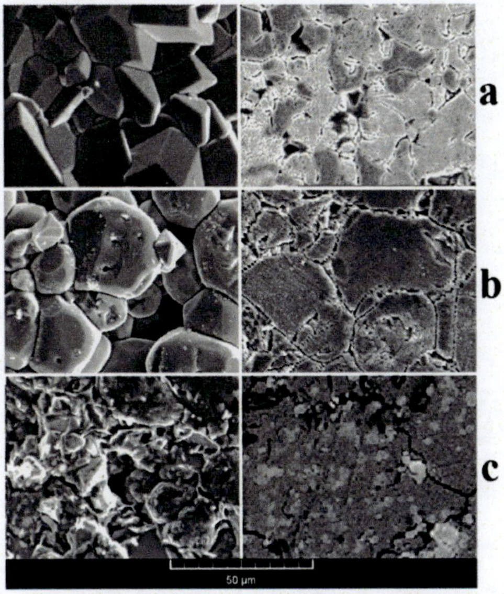

Fig. 2. SEM image of the surface for ZnO:Ga:Cl ceramics with (a) 0, (b) 3, and (c) 10 mol % Ga_2O_3. Unpolished and polished surfaces are presented in left and right sides, respectively

the solubility limit is reached, significant part of the Ga_{Zn} donors should be involved in the Ga_2O_3 or $ZnGa_2O_4$ inclusions, upon slow cooling of the furnace. This phenomenon decreases the concentration of free electrons to 2.8×10^{19} cm^{-3} ($\rho = 19 \times 10^{-3}$ Ω·cm) in the case of 10 mol % Ga_2O_3 (Fig. 3 (curve 3)).

3.2 Wettability of ZnO Ceramics

The external view of water droplets on a polished and etched surface of ZnO ceramics with various Ga_2O_3 content is shown in Fig. 4. The surface of the unintentionally doped ZnO:Cl ceramics showed an hydrophilic behavior: the water contact angle is 76° (Fig. 4a and Fig. 3 (curve 2)). An increase in the Ga content leads to a strong increase in the CA value to 131° (Fig. 4b and Fig. 3 (curve 2)). Therefore, the surface of ZnO:Ga:Cl ceramics with a Ga content corresponding to the solubility limit become a hydrophobic state. With a further increase in the doping level, CA slightly decreases to 117° in the case of 10 mol % Ga_2O_3. CA on the unpolished surfaces is significantly higher for any doping level (Fig. 3 (curve 1)). The difference between CA for unpolished and polished cases is minimal for unintentionally doped ZnO:Cl ceramics and maximum for ceramics heavily doped with Ga.

Two valence electrons of Zn atoms tend to O atoms; thus, the ZnO lattice consists of Zn^{2+} and O^{2-} ions. The reaction of formation of oxygen vacancies, which act as intrinsic donor defects in ZnO, can be written as follows:

$$Zn^{2+} + O^{2-} \leftrightarrow Zn^{2+} + 2n + V_O + 1/2O_2(gas), \quad (1)$$

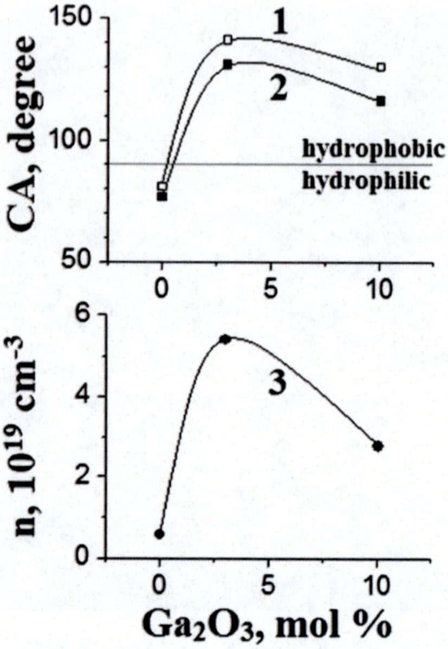

Fig. 3. The influence of Ga_2O_3 content on the water contact angle for (1) unpolished and (2) polished surface of ZnO:Ga:Cl ceramics, and (3) on the free electron concentration

where *n* is the concentration of free electrons in the conduction band, which can recombine to the energy levels of various defects. In accordance with the principal of dynamic equilibrium, the equation for the concentrations and pressures of *substances* participated in the reaction (1) can be written as follows:

$$n^2 \times [V_O] \times \sqrt{P}(O_2) = f(T) \text{ or } [V_O] \propto (n^2 \times \sqrt{P(O_2)})^{-1}, \quad (2)$$

where $[V_O]$ is the concentration of V_O, $P(O_2)$ is partial pressure of oxygen, f(T) is a function of the absolute temperature, which depends on the enthalpy and entropy of the *substances* participated in the reaction (1) [20]. This is typical for semiconductor materials: a high concentration of free electrons promotes the generation of compensating acceptor intrinsic defects (Zn vacancy in ZnO), but suppresses the generation of donor intrinsic defects (oxygen vacancy or Zn interstitial in ZnO) [21].

The dependence of the water contact angle on the doping level correlates with the dependence of the free electron concentration (see Fig. 3 (curve 2 and 3)). It can be assumed that a high concentration of free electron in ZnO:Ga:Cl ceramics suppresses the formation of surface defects (probably, oxygen vacancy), acting as traps for water molecules, causing the switch of ZnO surface behavior from hydrophilic to hydrophobic. Unpolished surfaces of the investigated ceramics should have air pockets, which should be an additional factor increasing CA [8]. The roughness and density of surface voids are minimal for unintentionally doped ZnO:Cl ceramics and are maximum for heavily doped ceramics with 10 mol % Ga_2O_3 (Fig. 2 and Fig. 3 (curve 1 and 2)). The concentration of Cl impurity is relatively low ($\sim 10^{19}$ cm^{-3}); they are not considered in this analysis.

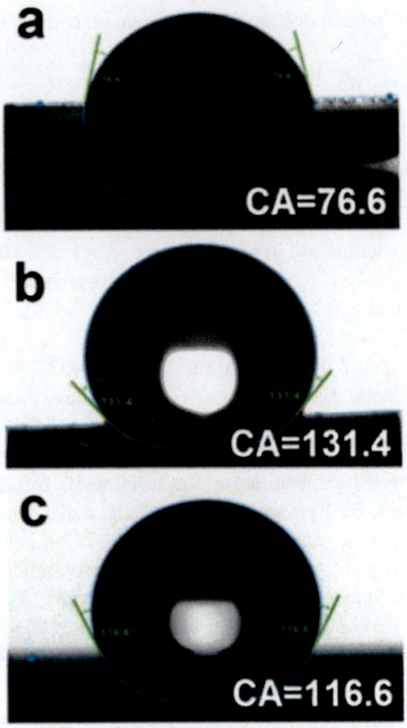

Fig. 4. The external view of water droplets on a polished and etched surface of ZnO ceramics: (a) undoped, (b) 3 mol % and (c) 10 mol % Ga_2O_3

The influence of surface Ga atoms and $ZnGa_2O_4$ inclusions on the wettability of ZnO can be investigated in the further work using ceramics with a higher Ga content.

4 Conclusions

ZnO:Ga:Cl ceramics with various Ga content were sintered using chemical vapor transport technique and HCl as a transport agent. The solubility limit of Ga_2O_3 in ZnO was estimated at about 3 mol %. At this doping level, free electron concentration increases by 9 times. At a higher Ga content, particles of undissolved Ga_2O_3 and $ZnGa_2O_4$ inclusions are revealed. The wettability of unpolished and polished surface of ZnO:Ga:Cl ceramics with various Ga doping level was investigated. The polished and etched surface of ZnO ceramics is in a hydrophilic state; the water contact angle is 76°. The presence of Ga impurity leads to a strong increase in the water contact angle to 131°. Air pockets on unpolished surfaces of ZnO:Ga:Cl ceramics are an additional factor that increases the water contact angle.

Acknowledgment. This work was supported by the Ministry of Education, Culture and Research of Moldova under the project No. 20.80009.5007.16 (Photosensitizers for applications in pharmaceutical medicine and photovoltaics).

Conflict of Interest. The authors declare that they have no conflict of interest.

References

1. Racca, L., Canta, M., Dumontel, B., et al.: Zinc oxide nanostructures in biomedicine. Smart Nanopart. Biomed. Micro Nano Technol. **12**, 171–187 (2018)
2. Marco, C., Barui, S., Cauda, V., Laurenti, M.: Doped zinc oxide nanoparticles: synthesis, characterization and potential use in nanomedicine. Appl. Sci. **10**(15), 5194 (2020)
3. Ghaffaria, M., Moztarzadeha, F., Safavib, M.: A comparative study on the shape-dependent biological activity of nanostructured zinc oxide. Ceram. Int. **45**(1), 1179–1188 (2019)
4. Milionis, A., et al.: Water-based scalable methods for self-cleaning antibacterial ZnO-nanostructured surfaces. Ind. Eng. Chem. Res. **59**(32), 14323–14333 (2020)
5. Pesika, N.S., Hu, Z., Stebe, K.J., Searson, P.C.: Quenching of growth of ZnO nanoparticles by adsorption of octanethiol. J. Phys. Chem. B **106**(28), 6985–6990 (2002)
6. Yadav, K., Mehta, B.R., Bhattacharya, S., Singh, J.P.: A fast and effective approach for reversible wetting-dewetting transitions on ZnO nanowires. Sci. Rep. **6**, 35073 (2016)
7. Velayi, E., Norouzbeigi, R.: Synthesis of hierarchical super-hydrophobic zinc oxide nano-structures for oil/water separation. Ceram. Int. **44**(12), 14202–14208 (2018)
8. Mardosaite, R., Jurkeviciute, A., Rackauskas, S.: Superhydrophobic ZnO nanowires: wettability mechanisms and functional applications. Cryst. Growth Des **21**, 4765–4779 (2021)
9. Chen, C., He, H., Lu, Y., Wu, K., Ye, Z.: Surface passivation effect on the photoluminescence of ZnO nanorods. ACS Appl. Mater. Interfaces **5**(13), 6354–6359 (2013)
10. Chao, C.H., Chi, P.W., Wei, D.H.: Investigations on the crystallographic orientation induced surface morphology evolution of ZnO thin films and their wettability and conductivity. J. Phys. Chem. C **120**(15), 8210–8219 (2016)
11. Li, Q., et al.: Room-temperature nonequilibrium growth of controllable ZnO nanorod arrays. Nanoscale Res. Lett. **6**, 477 (2011)
12. Colibaba, G.V.: Halide-carbon vapor transport of ZnO and its application perspectives for doping with multivalent metals. J. Solid State Chem. **266**, 166 (2018)
13. Colibaba, G.V.: Sintering highly conductive ZnO:HCl ceramics by means of chemical vapor transport reactions. Ceram. Int. **45**, 15843 (2019)
14. Colibaba, G.V., et al.: Low-temperature sintering of highly conductive ZnO:Ga:Cl ceramics by means of chemical vapor transport. J. Eur. Ceram. Soc. **41**, 443–450 (2021)
15. Colibaba, G.V., Rusnac, D., Fedorov, V., et al.: Effect of chlorine on the conductivity of ZnO: Ga thin films. J. Mater. Sci.: Mater. Electron. **32**, 18291–18303 (2021)
16. Huang, H.S., et al.: Highly conductive alumina-added ZnO ceramic target prepared by reduction sintering and its effects on the properties of deposited thin films by direct current magnetron sputtering. Thin Solid Films **518**, 6071 (2010)
17. Jagadish, C., Pearton, S.: Zinc Oxide Bulk, Thin Films and Nanostructures: Processing, Properties, and Applications. Elsevier Science, Amsterdam (2006)
18. Colibaba, G.V., et al.: Effects of impurity band in heavily doped ZnO:HCl. Phys. B: Condens. Matter **553**, 174 (2019)
19. Özgür, Ü., Alivov, Y.I., Liu, C., et al.: A comprehensive review of ZnO materials and devices. J. Appl. Phys. **98**, 041301 (2005)
20. Daniels, F., Alberty, R.A.: Physical Chemistry. Wiley, New York (1961)
21. Aven, M., Prener, J.S.: Physics and Chemistry of II–VI Compounds. North-Holland, Amsterdam (1967)

Antimicrobial Properties of a New Polymeric Material for Medical Purposes Under Conditions of Low-Intensity Current Without External Power Supplies

Roman Chornopyshchuk[1](✉), V. Nagaichuk[1], O. Nazarchuk[2], O. Kukolevska[3], I. Gerashchenko[4], A. Sidorenko[5], and R. Lutkovskyi[1]

[1] Department of General Surgery, National Pirogov Memorial Medical University, Vinnytsya, Ukraine

[2] Department of Microbiology, National Pirogov Memorial Medical University, Vinnytsya, Ukraine

[3] Department of Pharmaceutical Chemistry, National Pirogov Memorial Medical University, Vinnytsya, Ukraine

[4] Department of Biomedical Problems of Surface, Chuiko Institute of Surface Chemistry, NAS of Ukraine, Kyiv, Ukraine

[5] Cryogenics Laboratory, Ghitsu Institute of Electronic Engineering and Nanotechnologies, Chisinau, Moldova

Abstract. Local use of wound dressings still remains an important element of comprehensive wound treatment. Materials capable of dosing the release of drugs, including antimicrobial action, are of the highest priority. The use of composite polymers with antimicrobial properties in low-intensity current without external power supplies can enhance their effectiveness in combating wound infectious agents and improve the results of wound healing in general.

Therefore, the aim of our study was to investigate experimentally the antimicrobial action of a polymeric material based on poly(2-hydroxyethyl methacrylate), modified by creating a porous structure and saturated with antiseptic against strains of gram-negative microorganisms *P. aeruginosa* and *A. baumannii* in low-intensity current without external supplies.

A composite polymer material was synthesized by the method of free radical thermal polymerization of a mixture of liquid monomer 2-hydroxyethyl methacrylate, crosslinking agent triethylene glycol dimethacrylate and polymerization initiator azobisisobutyronitrile. Additionally, distilled water as a pore-forming agent and the antimicrobial agent decamethoxine were added. Suggested composite without an antimicrobial agent, as well as existing materials of synthetic and biological origin that are widely applied for the treatment of wounds were used for comparison. Some of the materials were pre-immersed in 0.02% solution of the decamethoxine. To ensure the action of biogalvanic current, a special device was placed alternately on each of the tested samples. Determination of antimicrobial properties was performed by diffusion method on a dense nutrient medium.

The results of microbiological research allowed to establish the advantages of the suggested material in combating the growth of gram-negative microorganisms

I. Tiginyanu et al. (Eds.): ICNBME 2021, IFMBE Proceedings 87, pp. 617–628, 2022.
https://doi.org/10.1007/978-3-030-92328-0_79

A. baumannii, P. aeruginosa in vitro, as well as the feasibility of its use under conditions of low-intensity current without external power supplies.

Keywords: Wound infection · Poly(2-hydroxyethyl methacrylate) · Wound dressing · Controlled release · Microcurrent

1 Introduction

Historically, the development of medicine has been accompanied by the emergence of new revolutionary technologies that have radically changed existing views and allowed to significantly improve the level of care [1]. The appearance and active introduction of polymeric materials in various fields of medicine can be considered an example of such technologies. The development of wound dressings remains one of the priority areas, which is confirmed by numerous studies and relevant publications, which continue to appear systematically in periodicals [2]. Currently, the medical market is filled with hundreds of different synthetic wound dressings which differ fundamentally only in the chemical structure of the material [3]. Among this variety, poly(2-hydroxyethyl methacrylate) (PHEMA) and the possibility of its combination with other substances is of great interest due to its properties that meet almost all the criteria [4]. Rapid progress in understanding the etiopathogenetic features of the wound process poses new challenges on the developers of wound dressings, the most important of which is to ensure a reliable local antimicrobial effect [5]. Back in the 1970s and 1980s of the last century, experimental, clinical studies were conducted on the effectiveness of combining PHEMA with other polymers and antimicrobial agents in the treatment of burns [6]. The obtained results allowed to establish the undeniable advantages of the studied composite materials in combating microbial pathogens and stimulation of healing processes in general [7]. Not very convenient way to apply the suggested composite by mixing the components directly on the wound surface and a number of other technical features have limited its wide practical application. Moreover, the active emergence of resistant strains of microorganisms, as well as their ability to form biofilms requires the use of highly effective antimicrobial agents of local action. That is why we have developed a material based on PHEMA, modified with a pore-forming agent and saturated with antimicrobial substances, which is characterized by sufficient biocompatibility, low toxicity, no signs of adhesion and a stable profile of drug release into the wound site, providing a reliable pharmacological effect [8]. In addition, the conductivity of the suggested material allows to further use low-intensity current without external power supplies as an effective auxiliary tool to combat pathogens of wound infection [9].

Preliminary experimental *in vitro* studies have shown the ability of the suggested modified polymeric material under the action of low-intensity current without external power supplies to inhibit the growth of museum and clinical strains of *S. aureus* in comparison with existing wound dressings [10]. Today, gram-negative representatives of opportunistic pathogenic microbiota *P. aeruginosa* and *A. baumannii* hold the leading positions in the structure of microbial pathogens of surgical hospitals, and the search for ways to combat them is of paramount importance [11].

Therefore, the aim of our study was to experimentally study the antimicrobial action of the suggested polymeric material modified by creating a porous structure and saturated with antiseptic on strains of gram-negative microorganisms *P. aeruginosa* and *A. baumannii* in low-intensity current without external power supplies.

2 Materials and Methods

Synthesis of the material was carried out by the method of free radical thermal polymerization at a temperature of 70–90 °C in a cylindrical container with a volume of 2.5 ml. For polymerization, a mixture of liquid monomer 2-hydroxyethyl methacrylate (Sigma-Aldrich, Germany), azobisisobutyronitrile initiator (Sigma-Aldrich, Germany), crosslinking agent triethylene glycol dimethacrylate (Sigma-Aldrich, Germany) was used. The fourfold volume of distilled water relative to the volume of monomer (75.4–78.7% by weight of water based on the weight of the sample) as a pore-forming agent and the antimicrobial substance decamethoxine (Pharmchim, Ukraine) were added. The obtained samples, which were easily removed from the container, had an expressed cylindrical shape with a diameter of 6 mm, were transparent, of dense-elastic consistency, cut into disks 2 mm thick.

Samples of the suggested composite without the addition of antimicrobial agent, as well as existing materials that are widely used for the treatment of wounds in the form of discs with a diameter of 6 mm: hydrogel with dimexide and without filler (Ukrtehmed, Ukraine), activated carbon material (ACM) (Dnipro, Ukraine), lyophilized porcine xenograft (Institute of Biomedical Technologies, Ukraine) were also envolved in the study. Before use all the materials except hydrogels were immersed in 0.02% solution of the same antiseptic (Yuria-Pharm, Ukraine).

To ensure the action of bio galvanic current alternately on each of the tested samples a biogalvanization apparatus was placed, which consisted of a copper electrode-donor and an aluminum-magnesium-zinc composite electrode-acceptor with a diameter of 5 mm each, interconnected by a first class conductor.

The study of antimicrobial properties was performed by a classical bacteriological diffusion method on a dense nutrient medium [12]. To that end, Petri dishes filled with a layer of Mueller-Hinton agar medium were inoculated with suspension (0.1 ml) of a daily culture of test strains of microorganisms *A. baumannii* ATCC 19606, *A. baumannii* 56, *P. aeruginosa* ATCC 27853, *P aeruginosa* 71. The microbial load amounted to 10^8 CFU/ml, which was determined densitometrically by the Densi-La-Metr II apparatus (serial number: 266/07). Evaluation of antimicrobial activity was performed 24 h after incubation of the dishes in a thermostat at a temperature of 36–37 °C by measuring the diameter of growth retardation zone of the culture around the test samples. The growth retardation zone less than 10 mm indicated the lack of sensitivity of microorganisms to the test samples.

Statistical processing of the results was performed using the software "Statistica 5.5" (license number: AXXR910A374605FA).

3 Results

Results of the research. The results of microbiological research showed that without the action of biogalvanic currents among all the studied samples only the suggested polymeric material based on PHEMA with the addition of decamethoxine was effective, inhibiting the growth of strains of *A. baumannii* (ATCC 19606) with a diameter of 28.9 $\pm$ 0.3 mm, *A baumannii* 56 – 21.5 $\pm$ 0.5 mm, *P. aeruginosa* (ATCC 27853) – 13.1 $\pm$ 0.2 mm and *P. aeruginosa* 71 – 10.9 $\pm$ 0.3 mm (Fig. 1).

Under the action of low-intensity current without the action of external current supplies, the same material located under the electrode-acceptor did not fundamentally change its antimicrobial properties without fundamental differences in the indicators of growth retardation zones of bacterial cultures: 29.6 $\pm$ 0.4 mm, 21.3 $\pm$ 0.7 mm, 13.7 $\pm$ 0.5 mm, 10.5 $\pm$ 0.5 mm, respectively (Fig. 2).

The exact opposite pattern was observed with a composite material saturated with antimicrobial agent, which was under conditions of biogalvanic current under the electrode-donor. In particular, a significant ($p \leq 0.05$) increase in the size of the growth retardation zone in museum strains of *A. baumannii* ATCC 19606 to 33.7 $\pm$ 0.5 mm (Fig. 3a) and *P. aeruginosa* ATCC 27853 – 17.2 $\pm$ 0.3 mm (Fig. 3c) compared with intact samples was noted. The sensitivity of clinical strains of microorganisms *A. baumannii* 56, *P. aeruginosa* 71 were higher and amounted to 23.1 $\pm$ 0.2 mm (Fig. 3b) and 11.9 $\pm$ 0.4 mm (Fig. 3d), respectively, although did not have significant differences ($p > 0.05$) with previous results.

The appearance of susceptibility of opportunistic pathogens to ACM was also found, which did not exceed 10 mm in diameter to *A. baumannii* strains (Figs. 3a, 3b). Museum and clinical cultures of *P. aeruginosa* were more sensitive, forming growth retardation zones of 10.1 $\pm$ 0.4 mm (Fig. 3c) and 11.3 $\pm$ 0.6 mm (Fig. 3d) in diameter. This phenomenon once again confirms the differences in the mechanisms of influence of low-intensity current on different types of microorganisms with the following consequences, which do not always correspond to the expected results, cause difficulties of interpretation from the standpoint of existing canons of microbiology and require further study. The phenomenon of growth inhibition of all studied strains of microorganisms under the electrode-acceptor located directly on the agar medium without the involvement of any antimicrobial agents, provided that the ACM is under the electrode-donor also remains unclear. Characteristically, around the growth retardation zone, which for *A. baumannii* ATCC 19606 was 10.3 $\pm$ 0.5 mm, *A. baumannii* 56 – 13.6 $\pm$ 0.7 mm, *P. aeruginosa* ATCC 27853 – 19.4 $\pm$ 0.4 mm, *P. aeruginosa* 71 – 24.8 $\pm$ 0.6 mm, a brown edging of different intensity was formed. Such results suggest the ability of low-intensity currents without external power supplies under certain conditions to stimulate the antimicrobial properties of metals, which can not only compete with modern antimicrobials, but also significantly exceed them.

4 Discussion

Wound healing remains a major problem of medicine, far from the final solution [13]. This problem is especially relevant among patients with burns [14]. After all, damage

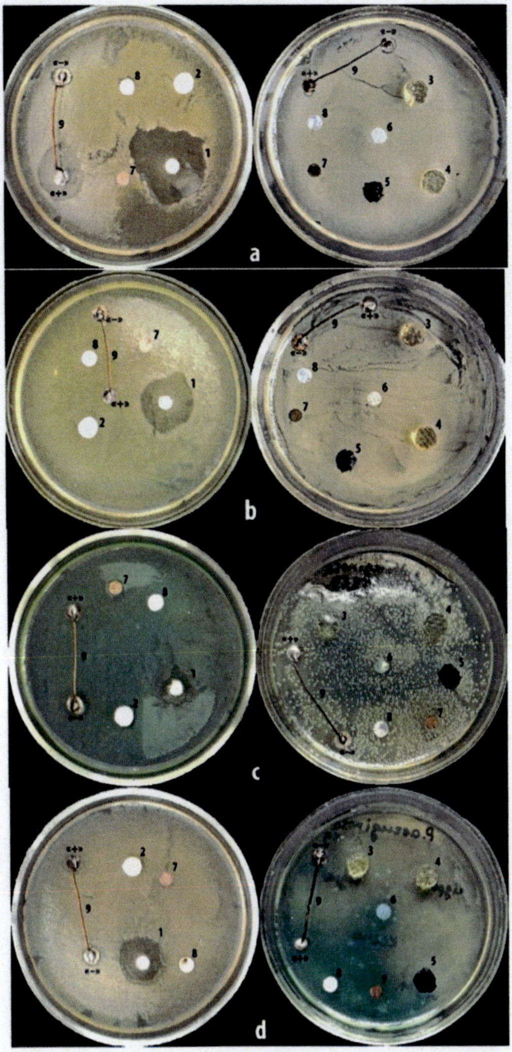

Fig. 1. Results of the study of antimicrobial activity of intact test samples against cultures of *A. baumannii* ATCC 19606 (a), *A. baumannii* 56 (b), *P. aeruginosa* ATCC 27853 (c), *P. aeruginosa* 71 (d). Notes: 1 – modified poly(2-hydroxyethyl methacrylate) with the addition of decamethoxine; 2 – modified poly(2-hydroxyethyl methacrylate) without decamethoxine; 3 – hydrogel material with the addition of dimexide; 4 – hydrogel material; 5 – activated carbon material; 6 – xenograft; 7 – electrode-donor; 8 – electrode-acceptor; 9 – biogalvanic device.

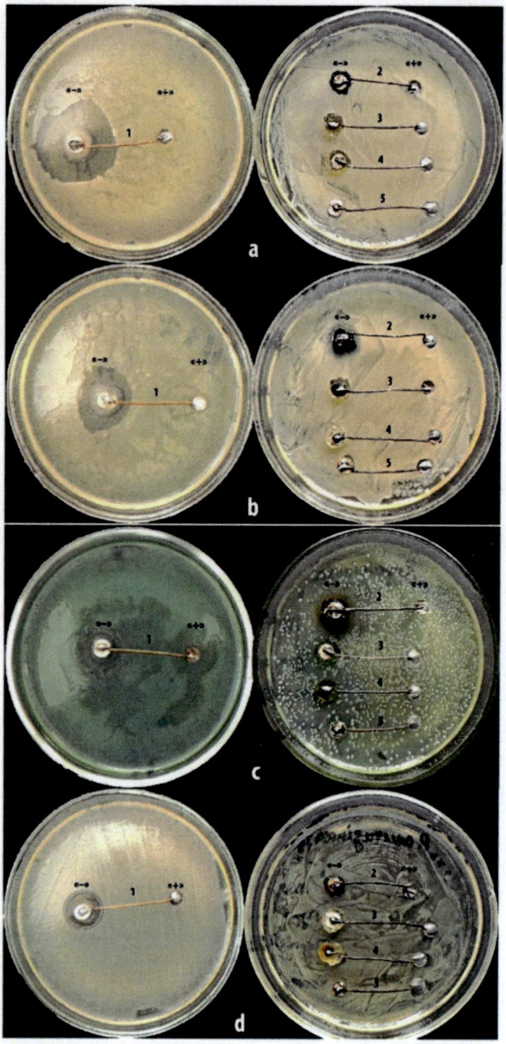

Fig. 2. Results of the study of antimicrobial activity of the test samples located under the electrode-acceptor against cultures of *A. baumannii* ATCC 19606 (a), *A. baumannii* 56 (b), *P. aeruginosa* ATCC 27853 (c), *P. aeruginosa* 71 (d). Notes: 1 – modified poly(2-hydroxyethyl methacrylate) with the addition of decamethoxine; 2 – activated carbon material; 3 – hydrogel material with the addition of dimexide; 4 – hydrogel material; 5 – xenograft.

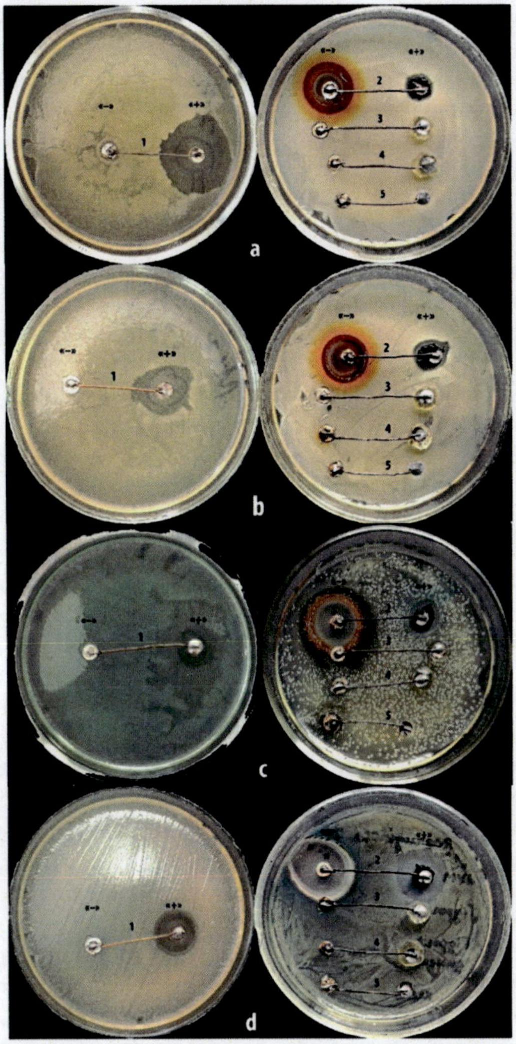

Fig. 3. Results of the study of antimicrobial activity of the test samples located under the electrode-donor against cultures of *A. baumannii* ATCC 19606 (a), *A. baumannii* 56 (b), *P. aeruginosa* ATCC 27853 (c), *P. aeruginosa* 71 (d). Notes: 1 – modified poly(2-hydroxyethyl methacrylate) with the addition of decamethoxine; 2 – activated carbon material; 3 – hydrogel material with the addition of dimexide; 4 – hydrogel material; 5 – xenograft.

to the skin, as one of the main organs of homeostasis, in addition to fluid loss and electrolytes, threatens active penetration of microbial pathogens into the tissues with subsequent generalization of the process [15]. According to the literature, in case of deep thermal injuries with an area of more than 40% of the body surface, 75% of deaths are associated with septic complications caused mainly by infectious factors from the burn wound [16]. Control of infectious agents, especially located directly in the area of injury, remains one of the main etiopathogenetic elements of comprehensive treatment of patients with burns [17]. The appearance of wound dressings with a number of specific physical and chemical properties has significantly improved the treatment of patients with burns, reducing mortality and changing the strategy of care provided for such victims in general [18]. Despite the fact that biological dressings still remain the gold standard for temporary closure of wound defects, materials of synthetic origin are of increasing interest [19]. This became possible due to active development of the chemical industry and fruitful cooperation of physicians with specialists in related fields. The emergence of new dressings capable to provide long-term programmed release of active substances, including those capable to inhibit the growth of microorganisms in the wound, was the result of such a multidisciplinary approach [20]. As part of this concept, we have developed an alternative polymer composite based on PHEMA, modified with a pore-forming agent and saturated with antimicrobial substances. Water was chosen as a pore-forming agent, the introduction of which into the polymerization mixture during synthesis increases the degree of swelling of the material and allows free diffusion from the polymer matrix of the additionally introduced bioactive substances. The developed technology involved the synthesis of several types of samples with different ratio of water-polymer mixture, the study of the kinetics of release of bioactive substances from them and the choice of the optimal proportion at which the maximum desorption of active substances is achieved while maintaining the required structural characteristics of the product [21]. Decamethoxine was chosen as an antimicrobial component, which belongs to Quaternary ammonium compounds and still retains a wide range of antimicrobial and antifungal action when applied topically and has no systemic side effects, as it is almost not absorbed from the wound surface [22]. The results of this study further confirmed the significant antimicrobial properties of the suggested material in comparison with existing wound dressings and traditional dressings, saturated with the above-mentioned antiseptic solution.

Steady dynamic growth of resistance of clinical strains of opportunistic pathogens, their combination with a special form of coexistence forces us to look for other ways to combat them except chemical compounds [23]. That is why there has been recently more information on the use of physical factors: different types of waves, ozonation, negative pressure, electric current, etc. [24]. Potential use of electric current to correct the wound process is of particular interest to scientists [25]. The experience of using this method in medicine has long been known, but its fundamental preclinical study was conducted only in the middle of the twentieth century with numerous experiments *in vitro* and *in vivo*, including models of wound healing of experimental animals of various etiologies [26]. The obtained positive results allowed to implement this technology in the treatment process of patients with wounds, but numerous legal obstacles prevented it from being applied. In particular, independent clinical trials have failed to reach a consensus on the effectiveness of low-intensity current to treat chronic wounds [27]. The emergence of information on the use of low-intensity currents without external power supplies all over the world became an impetus for the revival of this method [28]. The use of only body generated current instead of external supplies have made the profile of its safety unquestionable [29]. Unfortunately, most of the mechanisms of such currents on biological objects are still unspecified, but it is currently known about its ability to affect different stages of the wound process, reducing signs of inflammation, to stimulate vascularization of the lesion in various ways, improving its microcirculation, enhance cell proliferation, promote the processes of granulation growth and epithelization of wounds [30–33]. Based on this, a fundamentally new type of polymer dressings was created, capable of generating microcurrents to stimulate reparative processes in the wound without external power supplies and other supplements [34]. Along with the study of the effect of these dressings on different phases of wound healing, there is no information on their antimicrobial properties except for the theoretical assumption of the ability of the current generated to attract negatively charged microorganisms and reduce their activity [35]. It is doubtful that this will be enough to combat the aggressive wound microflora. Therefore, the combination of this physical method with existing antimicrobial agents appears much more promising, implementing the ability of electric current as directional electrophoresis to create high concentrations of drugs in the wound area, increase their bactericidal and bacteriostatic action, reduce antibiotic resistance of the microflora the cause of which is also considered to be the lack of electron transport chain in microorganisms, which is necessary for the transfer of active substances into the middle of the microbial cell [36]. Stable long-term action of biogalvanic current without external power supplies in the interelectrode space mitigates this condition with the restoration of the sensitivity of microorganisms. Obviously, this is the justification for the effect of increasing the sensitivity of opportunistic pathogens to the test material located under the electrode-donor, which was established by the results of this study. The registered phenomenon of hypersensitivity of gram-negative microflora to the metal electrode without the addition of any antimicrobial substances is quite comparable with the results of previous studies with gram-positive bacteria and adds value to this method indicating the need for further research, especially if we take into account the results of its influence on the strains of *P. aeruginosa*, which currently occupies an absolute leading position among various surgical hospitals and demonstrates persistent resistance to the vast majority of existing antimicrobials [37].

5 Conclusions

On the example of museum and clinical strains of gram-negative microorganisms *A. baumannii, P. aeruginosa* significant antimicrobial properties of the developed composite polymeric material modified by creating a porous structure and saturated with antimicrobial agent decamethoxine were established. The use of this material under conditions of low-intensity current without external power supplies can enhance its antimicrobial properties.

We consider it expedient to continue further study of the influence of the suggested material, including under the conditions of biogalvanization, on the course of the wound process in experimental modeling of skin wounds on animals with possible further introduction of the developed technology into clinical practice of local treatment of patients with wounds of different genesis.

Conflict of Interest. The authors declare that they have no conflict of interest.

References

1. Los', D.M., Shapovalov, V.M., Zotov, S.V.: The use of polymer materials for medical applications. Probl. Health Ecol. **64**(2), 5–13 (2020)
2. Gobi, R., Ravichandiran, P., Babu, R.S., et al.: Biopolymer and synthetic polymer-based nanocomposites in wound dressing applications: a review. Polymers **13**(12), 1962 (2021). https://doi.org/10.3390/polym13121962
3. Sheridan, R.: Closure of the excised burn wound: autografts, semipermanent skin substitutes, and permanent skin substitutes. Clin. Plast. Surg. **36**(4), 643–651 (2009). https://doi.org/10.1016/j.cps.2009.05.010
4. Zare, M., Bigham, A., Zare, M., et al.: pHEMA: an overview for biomedical applications. Int. J. Mol. Sci. **22**(12), 6376 (2021). https://doi.org/10.3390/ijms22126376
5. Shi, C., Wang, C., Liu, H., et al.: Selection of appropriate wound dressing for various wounds. Front. Bioeng. Biotechnol. **8**, 182 (2020). https://doi.org/10.3389/fbioe.2020.00182
6. Nathan, P., Law, E.J., MacMillan, B.G., et al.: A new biomaterial for the control of infection in the burn wound. Trans. Am. Soc. Artif. Intern. **22**(1), 30–39 (1976)
7. Nathan, P., Macmillan, B.G., Holder, I.A.: Effect of a synthetic dressing formed on a burn wound in rats: a comparison of allografts, collagen sheets, and polyhydroxyethylmethacrylate in the control of wound infection. Appl. Microbiol. **28**(3), 465–468 (1974). https://doi.org/10.1128/am.28.3.465-468.1974
8. Kukolevska, O.S., Siora, I.V., Chornopyshchuk, R.M., et al.: Medical and biological research of nanocomposite material «Polydens». Chem. Phys. Technol. Surf. **7**(2), 225–235 (2016). https://doi.org/10.15407/hftp07.02.225
9. Nagaichuk, V., Chornopyshchuk, R., Yunusova, O., Onyshchenko, M.: Use of physical methods as an element of complex treatment of burn wound microbiome. In: Tiginyanu, I., Sontea, V., Railean, S. (eds.) ICNBME 2019. IP, vol. 77, pp. 579–581. Springer, Cham (2020). https://doi.org/10.1007/978-3-030-31866-6_103
10. Nagaichuk, V., Chornopyshchuk, R., Gerashchenko, I., et al.: Experimental substantiation of antimicrobial efficiency of a new composite polymeric material based on Poly(2-Hydroxyethyl Methacrylate) under the action of low-intensity current without external power supplies. In: Materials Science Forum. 2021 International Scientific and Practical Conference «Problems of Emergency Situations», vol. 1038, pp 154–161, Kharkiv, Ukraine (2021).https://doi.org/10.4028/www.scientific.net/MSF.1038.154
11. Pujji, O., Nakarmi, K.K., Shrestha, B., et al.: The bacteriological profile of burn wound infections at a tertiary burns center in Nepal. J. Burn Care Res **40**(6), 838–845 (2019). https://doi.org/10.1093/jbcr/irz096. Official Publication of the American Burn Association
12. Jorgensen, J.H., Ferraro, M.J.: Antimicrobial susceptibility testing: a review of general principles and contemporary practices. Clin. Infect. Dis. **49**(11), 1749–1755 (2009). https://doi.org/10.1086/647952
13. Teot, L., Ohura, N.: Challenges and management in wound care. Plast. Reconstr. Surg. **147**(1S-1), 9S–15S (2021). https://doi.org/10.1097/PRS.0000000000007628

14. Yao, Y., Zhang, A., Yuan, C., et al.: Recent trends on burn wound care: hydrogel dressings and scaffolds. Biomater. Sci. **9**(13), 4523–4540 (2021). https://doi.org/10.1039/d1bm00411e
15. Ladhani, H.A., Yowler, C.J., Claridge, J.A.: Burn wound colonization, infection, and sepsis. Surg. Infect. **22**(1), 44–48 (2021). https://doi.org/10.1089/sur.2020.346
16. Atiyeh, B.S., Gunn, S.W., Hayek, S.N.: State of the art in burn treatment. World J. Surg. **29**, 131–148 (2005)
17. Cancio, L.C.: Topical antimicrobial agents for burn wound care: history and current status. Surg. Infect. **22**(1), 3–11 (2021). https://doi.org/10.1089/sur.2020.368
18. Jiang, N., Rao, F., Xiao, J., et al.: Evaluation of different surgical dressings in reducing postoperative surgical site infection of a closed wound: a network meta-analysis. Int. J. Surg. **82**, 24–29 (2020). https://doi.org/10.1016/j.ijsu.2020.07.066
19. Kumar, S., Gupta, S.K.: Natural polymers, gums and mucilages as excipients in drug delivery. Polimery w medycynie **42**(3–4),191–197 (2012)
20. Hussain, Z., Thu, H.E., Shuid, A.N., et al.: Recent advances in polymer-based wound dressings for the treatment of diabetic foot ulcer: an overview of state-of-the-art. Curr. Drug Targets **19**(5), 527–550 (2018). https://doi.org/10.2174/1389450118666170704132523
21. Kukolevska, O.S., Gerashchenko, I.I., Borysenko, M.V., Pakhlov, E.M., Machovsky, M., Yushchenko, T.I.: Synthesis and examination of nanocomposites based on Poly(2-hydroxyethyl methacrylate) for medicinal use. Nanoscale Res. Lett. **12**(1), 1–9 (2017). https://doi.org/10.1186/s11671-017-1881-7
22. Nazarchuk, O.A., Chereshniuk, I.L., Nazarchuk, H.H.: The research of antimicrobial efficacy of antiseptics decamethoxin, miramistin and their effect on nuclear DNA fragmentation and epithelial cell cycle. Wiad. Lek. **72**(3), 374–380 (2019)
23. Kovalchuk, V.P., Nazarchuk, O.A., Burkot, V.M., et al.: Biofilm forming activity of non-fermenting gram-negative bacteria. Wiad. Lek. **74**(2), 252–256 (2021). https://doi.org/10.36740/WLek202102114
24. Podoinitsyna, M.G., Tsepelev, V.L., Stepanov, A.V.: The use of physical methods in the treatment of the skin burn. Mod. Probl. Sci. Educ. **5**, 184 (2015). https://science-education.ru/ru/article/view?id=22156
25. Levin, M.: Bioelectric mechanisms in regeneration: unique aspects and future perspectives. Semin. Cell Dev. Biol. **20**(5), 543–556 (2009). https://doi.org/10.1016/j.semcdb.2009.04.013
26. Yu, C., Hu, Z.Q., Peng, R.Y.: Effects and mechanisms of a microcurrent dressing on skin wound healing: a review. Mil. Med. Res. **1**, 24 (2014). https://doi.org/10.1186/2054-9369-1-24
27. Barnes, R., Shahin, Y., Gohil, R., et al.: Electrical stimulation vs. standard care for chronic ulcer healing: a systematic review and meta-analysis of randomised controlled trials. Eur. J. Clin. Invest. **44**, 429–440 (2014). https://doi.org/10.1111/eci.12244
28. Isseroff, R.R., Dahle, S.E.: Electrical stimulation therapy and wound healing: where are we now? Adv. Wound Care **1**(6), 238–243 (2012). https://doi.org/10.1089/wound.2011.0351
29. Reid, B., Zhao, M.: The electrical response to injury: molecular mechanisms and wound healing. Adv. Wound Care **3**(2), 184–201 (2014). https://doi.org/10.1089/wound.2013.0442
30. Park, R.J., Son, H., Kim, K., et al.: The effect of microcurrent electrical stimulation on the foot blood circulation and pain of diabetic neuropathy. J. Phys. Ther. Sci. **23**, 515–518 (2011)
31. Kaur, S., Lyte, P., Garay, M., et al.: Galvanic zinc-copper microparticles produce electrical stimulation that reduces the inflammatory and immune responses in skin. Arch. Dermatol. Res. **303**(8), 551–562 (2011). https://doi.org/10.1007/s00403-011-1145-9
32. Tandon, N., Cimetta, E., Villasante, A., et al.: Galvanic microparticles increase migration of human dermal fibroblasts in a wound-healing model via reactive oxygen species pathway. Exp. Cell Res. **320**(1), 79–91 (2014). https://doi.org/10.1016/j.yexcr.2013.09.016
33. Varhus, J.D.: A novel bioelectric device enhances wound healing: an equine case series. J. Equine Vet. Sci. **34**, 421–430 (2014). https://doi.org/10.1016/j.jevs.2013.07.011

34. Harding, A.C., Gil, J., Valdes, J., et al.: Efficacy of a bio-electric dressing in healing deep, partial-thickness wounds using a porcine model. Ostomy/Wound Manag. **58**(9), 50–55 (2012)
35. Blount, A.L., Foster, S., Rapp, D.A., et al.: The use of bioelectric dressings in skin graft harvest sites: a prospective case series. J. Burn Care Res. **33**(3), 354–357 (2012). https://doi.org/10.1097/BCR.0b013e31823356e4. Official publication of the American Burn Association
36. Nahaychuk, V.I.: Influence of bioactivation on staphylococcus hemolytic culture. Hosp. Surg. **4**, 63–66 (2007)
37. Azzopardi, E.A., Azzopardi, E., Camilleri, L., et al.: Gram negative wound infection in hospitalised adult burn patients-systematic review and metanalysis. PLoS One **9**(4), e95042 (2014). https://doi.org/10.1371/journal.pone.0095042

Coordination Compounds of Cu(II), Ni(II) Based on Ethyl 4-Benzoate Thiosemicarbazons Derivatives of Salicyl Aldehyde. Antimicrobial and Antifungal Properties

Anna Rusnac[1(✉)], G. Balan[2], and A. Gulea[1]

[1] Chemistry, State University of Moldova, Chisinau, Republic of Moldova
[2] State University of Medicine and Pharmacy "Nicolae Testemitanu", Chisinau, Republic of Moldova

Abstract. The article publishes in vitro tests of the antibacterial and antifungal properties of new coordination compounds of Cu(II), Ni(II) based on ethyl 4-benzoate thiosemicarbazons derivatives of salicyl aldehyde. Two Gram-positive strains were taken as reference strains: *Staphylococcus aureus* ATCC 25923 and *Bacillus cereus* ATCC 11778; two Gram-negative strains: *E.coli* ATCC 25922 and *Acinetobacter baumannii* BAA-747; three fungal strains: *Candida albicans* ATCC 10231, *Candida krusei* ATCC 6258, *Cryptococcus neoformans* CECT 1043. The best results were recorded at {Ni(HL^1)Cl} complex, a selectivity is observed on Cryptococcus neoformans, with CMI = 0.016, CMF = 0.031 mg/mL and is twice as active as the control substance Nystatin. The purity and structural formula of synthesized compounds was confirmed by thin layer chromatography; IR spectroscopy; ^{1}H, ^{13}C Nuclear Magnetic Resonance Spectroscopy, metal analysis and magnetochemical analysis.

Keywords: Antibacterial · Antifungal activity · Thiosemicarbazons · Coordination compounds

1 Introduction

Thiosemicarbazones are an important class of organic compounds that have attracted significant attention in the pharmaceutical industry due to their rich biological activity such as antibacterial, antifungal, antiviral, antimalarial, antitumor and others [1]. 4-Ethyl aminobenzoate manifests anesthetic properties, used internally on some pathologies as: gastralgia, esophagitis, peptic ulcer of the stomach and duodenum. External: acute inflammation of the middle ear, pain in the external auditory canal, urticaria, itchy skin, perianal fissures, hemorrhoids [2].

The introduction of ethyl 4-benzoate in position 4 of thiosemicarbazones allows the possibility of creating new compounds with antitumor properties, the skeleton of which will consist of two parts: one part - antitumor properties, and other - anesthetic properties. Coordination of thiosemicarbazones to 3d metals, for example copper (II), is beneficial

I. Tiginyanu et al. (Eds.): ICNBME 2021, IFMBE Proceedings 87, pp. 629–636, 2022.
https://doi.org/10.1007/978-3-030-92328-0_80

because it reduces the dose of the minimum inhibition concentration (reduces the dose approximately 5–10 times) compared to uncoordinated thiosemicarbazones [3].

New synthesized molecules have been tested for antibacterial and antifungal properties.

At present, antibiotics are the key elements of modern medicine, being indispensable in the treatment of bacterial diseases. Since the discovery of penicillin and its introduction into medical practice, antibiotics are indispensable not only in the treatment of diseases caused by pathogenic microorganisms, but also in the prevention of infections in patients undergoing surgery, of those with weak immunity or suffering from cancer.

A direct relationship of increasing the level of antibiotic use is antibiotic resistance. Among the most dangerous resistant microorganisms for humans is MRSA -Staphylococcus aureus and Escherichia coli.

So, there is a constant need for new chemotherapeutic compounds with high and selective antibacterial and antifungal activity, which can overcome the resistance of bacteria and fungi.

2 Experimental Section

2.1 Synthesis of Thiosemicarbazones

Two new thiosemicarbazones were synthesized based on ethyl 4-aminobenzoate, according to the scheme below (Fig. 1).

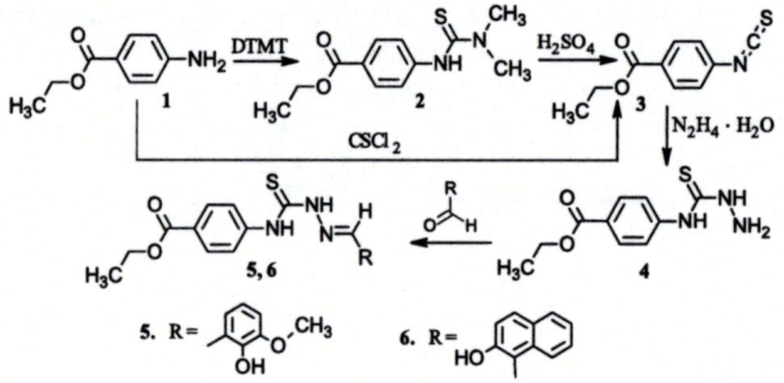

Fig. 1. Synthesis scheme of thiosemicarbazones based on ethyl 4-aminobenzoate.

Ethyl 4-[(hydrazinecarbothioyl)amino]benzoate **(4)** was obtained according to the following steps, ethyl 4-aminobenzoate **(1)** was reacted with tetramethylthiuram disulfide (DTMT) in a 1: 1 molar ratio on heating in dimethylformamide to form ethyl 4-[(dimethylcarbamothioyl)-amino]benzoate **(2)**. In the second stage, compound **2** was subjected to thermal degradation in dioxane with sulfuric acid forming ethyl 4-isothiocyanatobenzoate **(3)**, according to the method [4]. Isothiocyanate **3** can be obtained directly from amine **1** in a single step with $CSCl_2$ (in the presence of $NaHCO_3$) or CS_2 (in the presence of $N(C_2H_5)_3$, ClOCO C_2H_5) reagents [5]. Ethyl 4-[(hydrazinecarbothioyl)amino]-benzoate **(4)** was obtained by dripping the ethanolic solution of isothiocyanate **3** to the ethanolic solution of hydrazine hydrate [6].

Thiosemicarbazones based on ethyl 4-aminobenzoate: ethyl 4-({2-[(2-hydroxy-3-methoxyphenyl)methylidene]-hydrazinecarbothioyl}amino)benzoate **5**, ethyl 4-({2-[(2-hydroxynaphthalen-1-yl)methylidene]hydrazinecarbothio-yl}amino)benzoate **6** were obtained following the condensation reaction between thiosemicarbazide **4** and the corresponding aldehydes (2-hydroxy-3-methoxybenzaldehyde and 2-hydroxynaphthalene-1-carbaldehyde) in ethanol with 3–4 drops of glacial acetic acid. The reaction mixtures were refluxed for 2–3 h, the total consumption of thiosemicarbazide **4** was confirmed by chromatography. The purity and structural formula were confirmed by thin layer chromatography, IR spectroscopy; ^{1}H, ^{13}C Nuclear Magnetic Resonance Spectroscopy. Below are given in the table the characteristics of the thiosemicarbazones obtained (Table 1):

Table 1. Characteristics of thiosemicarbazones based on ethyl 4-aminobenzoate

N	Abbreviation	Reaction time	Yield %	Melting point, °C	R_f (eluent)
5	H_2L^1	2h	84.00	204–205	0,56 (benzene: ethyl acetate, 3:1)
6	H_2L^2	3h	87.00	162–136	0,58 (benzene: ethyl acetate, 3:1)

Based on the synthesized thiosemicarbazones H_2L^1, H_2L^2, 12 coordination compounds were obtained by mixing, the salts of Cu (II) solutions in ethanol and ligand solutions in ethanol ratio of 1: 1 reflux for 0.5 h. Obtained as a green microcrystalline complexes according to the reaction written below (Table 2).

$$H_2L^{1-2} + MX_2 \cdot nH_2O = \left\{M\left(HL^{1-2}\right)X\right\} + nH_2O + HX \quad (1)$$

$$\text{where}: \mathbf{M} = Cu^{2+},\ Ni^{2+}; \mathbf{X} = Cl^-, NO_3^-,\ ClO_4^-; \mathbf{n} = 0 - 6.$$

$$H_2L^{1-2} + M(CH_3COO)_2 \cdot nH_2O = \{M(L^{1-2})H_2O\} + nH_2O + 2CH_3COOH \quad (2)$$

$$\text{where}: \mathbf{M} = Cu^{2+}, Ni^{2+}; \mathbf{n} = 1, 4.$$

Metal analysis showed that the combination ratio to metal and ligand in the complex is 1:1. Magnetochemical research has shown that copper complexes have a monomeric structure because their effective magnetic moments correspond to an uncoupled electron. Nickel complexes are diamagnetic and have a properly planar-quadratic and octahedral structure.

2.2 Materials and Methods

The research took place in the Scientific Research Laboratory "Advanced materials in biopharmaceuticals and technology" of the State University of Moldova. All reagents and solvents used were chemical purity, bought from the company "Aldrich".

Table 2. Characteristics of coordination compounds based on ethyl 4-aminobenzoate thiosemicarbazones

Compound	Molecular formula	Yield %	Metal determ./calc., %	μ_{ef} MB
$Cu(H^1 L)Cl$	$C_{22}H_{38}ClCuN_3O_{10}S$	87.09	10.00/10.38	1.96
$Cu(H^1L)ClO_4$	$C_{20}H_{26}ClCuN_3O_{10}S$	85.06	10.60/10.48	1.75
$Cu(HL^1)NO_3$	$C_{20}H_{31}CuN_4O_{11}S$	82.75	10.61/10.74	1.80
CuL^1H_2O	$C_{20}H_{23}CuN_3O_7S$	79.84	12.39/11.88	1.76
$Ni(HL^1)Cl$	$C_{20}H_{30}ClN_3NiO_8S$	73.21	10.36/10.48	D
NiL^1H_2O	$C_{22}H_{29}N_3NiO_8S$	89.18	10.61/10.98	D
$Cu(HL^2)Cl$	$C_{25}H_{34}ClCuN_3O_7S$	72.92	10.26/10.38	1.96
$Cu(HL^2)ClO_4$	$C_{21}H_{20}ClCuN_3O_8S$	86.14	11.08/11.15	1.87
$Cu(HL^2)NO_3$	$C_{21}H_{26}CuN_4O_{10}S$	74.58	10.83/10.77	1.83
CuL^2H_2O	$C_{23}H_{27}CuN_3O_6S$	83.01	11.83/11.88	1.91
$Ni(HL^2)Cl$	$C_{21}H_{26}ClN_3NiO_7S$	89.62	10.51/10.39	D*
NiL^2H_2O	$C_{23}H_{29}N_3NiO_7S$	90.11	10.67/10.69	D*

* D-diamagnetic.

^1H and ^{13}C Nuclear Magnetic Resonance (NMR) spectra were recorded at room temperature using the Bruker DRX-400 spectrometer. Chemical shifts are shown in ppm relative to SiMe4. DMSO-d_6 was used as solvent. *IR spectra* were recorded on the Bruker Alpha spectrometer, 4000–400 cm^{-1}.

Titrimetric analysis of Cu (II) and Ni (II). 0.05 g of coordinating compound and a mixture of 2–3 drops of sulfuric acid and 5mL of concentrated nitric acid are added to the Kjeldal flask. The contents are heated until the coordinating compound is destroyed. After cooling the residue was added distilled water, after which the excess acid is neutralized with $NaHCO_3$. The solution obtained is quantitatively transferred to a 100 mL volumetric flask. Take an aliquot of 10 mL in a titration flask, bring the ammonia solution to pH = 8–9, add murexid until an intense yellow color is obtained. Titrate with Trilon B until pale purple appears.

The *magnetochemical analysis* was performed according to the *Gouy method* [7].

Antibacterial and antifungal activity The antimicrobial activities of the ligand and complexes were evaluated against four bacterial strains, *Staphylococcus aureus* (ATCC 25923), *Bacillus cereus* (ATCC 11778), *Escherichia coli* (ATCC 25922), *Acinetobacter baumannii* (BAA-747) and three fungal strains, *Candida albicans* (ATCC 10231), *Candida krusei* (ATCC 6258), *Cryptococcus neoformans* CECT 1043. Determination of the MIC (minimum inhibitory concentration, mg/ml) and MBC (minimum bactericidal concentration, mg/ml) and MFC (minimum fungal concentration, mg/ml) was done using the serial dilutions in liquid broth method [8, 9]. The tested substances were dissolved in DMSO in concentration of 10 mg/ml. The next dilutions were made using 2%

of peptonate bullion. Furacillinum and Nystatin was used as the standard antibacterial, antifungal drug.

2.3 Results

Four new synthesized molecules have been tested for antibacterial and antifungal properties (Table 3, 4, 5 and 6).

Table 3. Antibacterial activity of H_2L^1 and coordination compounds of Cu(II), Ni(II) with H_2L^1 on *Staphylococcus aureus* and *Bacillus cereus*

Substance	*Staphylococcus aureus ATCC 25923*		*Bacillus cereus ATCC 11778*	
	MIC mg/mL	MBC mg/mL	MIC mg/mL	MBC mg/mL
H_2L^1	0.5	0.5	0.5	0.5
$\{Ni(HL^1)Cl\}$	0.5	0.5	0.5	0.5
$\{Cu(HL^1)Cl\}$	0.5	0.5	0.125	0.25
$\{Cu(HL^1)ClO_4\}$	0.5	0.5	0.5	0.5
Furacilinum	0.0047	0.0094	0.0047	0.0047

Table 4. Antibacterial activity of H_2L^1 and coordination compounds of Cu(II), Ni(II) with H_2L^1 on *E.coli* and *Acinetobacter baumannii*

Substance	*E.coli ATCC 25922*		*Acinetobacter baumannii BAA-747*	
	MIC mg/mL	MBC mg/mL	MIC mg/mL	MBC mg/mL
H_2L^1	0.5	0.5	0.5	0.5
$\{Ni(HL^1)Cl\}$	0.5	0.5	0.5	0.5
$\{Cu(HL^1)Cl\}$	0.5	0.5	0.5	0.5
$\{Cu(HL^1)ClO_4\}$	0.5	0.5	0.5	0.5
Furacilinum	0.0047	0.0047	0.0047	0.0094

2.4 Discussion

The purity and structural formula of H_2L^1 and H_2L^2 was confirmed by thin layer chromatography; IR spectroscopy; ^1H, ^{13}C Nuclear Magnetic Resonance Spectroscopy.

Table 5. Antifungal activity of H_2L^1 and coordination compounds of Cu(II), Ni(II) with H_2L^1 on *Candida albicans* and *Candida krusei*

Substance	*Candida albicans ATCC 10231*		*Candida krusei ATCC 6258*	
	MIC mg/mL	MFC mg/mL	MIC mg/mL	MFC mg/mL
H_2L^1	0.5	0.5	0.5	0.5
$\{Ni(HL^1)Cl\}$	0.125	0.5	0.25	0.5
$\{Cu(HL^1)Cl\}$	0.5	0.5	0.5	0.5
$\{Cu(HL^1)ClO_4\}$	0.5	0.5	0.5	0.5
Nystatin	0.032	0.064	0.032	0.064

Table 6. Antifungal activity of H_2L^1 and coordination compounds of Cu(II), Ni(II) with H_2L^1 on *Cryptococcus neoforman*

Substance	*Cryptococcus neoformans* CECT 1043	
	CMI, mg/mL	CMF, mg/mL
H_2L^1	0.5	0.5
$\{Ni(HL^1)Cl\}$	0.016	0.031
$\{Cu(HL^1)Cl\}$	0.5	0.5
$\{Cu(HL^1)ClO_4\}$	0.5	0.5
Nistatina	0.032	0.064

H_2L^1, ^{1}H RMN (DMSO-d_6), δ, ppm: 11.98(NH), 10.21(NH), 9.37(OH), 8.54(N = CH), 4.31(CH_2), 3.81(CH_3-O), 1.32(CH_3);

H_2L^2, ^{1}H RMN (DMSO-d_6), δ, ppm: 11.99(NH), 10.30(NH), 9.16(N = CH), 8.52(OH), 4.31(CH_2), 1.32(CH_3);

H_2L^1, ^{13}C RMN (DMSO-d_6), δ, ppm: 175.68(C = S), 163.93(C = O), 148.40(N = C), 61.12(CH_3), 56.37(CH_2), 14.66(CH_3);

H_2L^2, ^{13}C RMN (DMSO-d_6), δ, ppm: 175.93(C = S), 165.88(C = O), 157.34(N = C), 61.05(CH_2), 14.66(CH_3).

The IR spectra of the synthesized complexes suggest that the azomethine functional group ν(C = N, 1597 cm^{-1}) moves at higher wave numbers compared to the uncoordinated ligand ν(C = N, 1580 cm^{-1}). New bands appear ν (O-M, N-M, S-M at 596, 565, 445 cm^{-1}) that are missing in the spectrum of the ligand. The thionic group ν(C = S, 1260, cm^{-1}) moves at lower wave numbers compared to the uncoordinated ligand ν(C = S, 1175, cm^{-1}) suggests the formation of the S-M coordinative bond. The absorption band ν(O-H, 3298 cm^{-1}) disappears from the spectra of the coordinative combinations, which speaks of its deprotonation and the formation of the O-M bond. According to IR data, the ligand coordinates tridentate to the metal, through the sulfur-ionic atom, the azomethine nitrogen atom and the oxygen-phenolic atom (Fig. 2).

In the case of coordination combinations based on copper and nickel acetate salts, $\{CuL^1H_2O\}$, $\{NiL^1H_2O\}$, $\{CuL^2H_2O\}$, $\{NiL^2H_2O\}$ are obtained, which the ligand is double deprotonated, tridentate, confirmed by the appearance of a new vibration band of the imine group $\nu(C = N^2)$, 1600 cm^{-1} and the disappearance of the thionic group $\nu(C = S)$.

In the case of coordination combinations with perchlorate anion $\{Cu(HL^1)H_2O\}ClO_4$, $\{Cu(HL^2)H_2O\}ClO_4$, new vibration bands ν_3(Cl-O from ClO_4, 1170, 1150, cm^{-1}) appear, which confirms the presence of perchlorate ion. In the case of coordination combinations with nitrate anion: $\{Cu(HL^1)\ NO_3\}$, $\{Cu(HL^2)NO_3\}$, new vibration bands ν_1(N-O of NO_3, 1275, cm^{-1}) appear, which confirms the presence of the nitrate anion which monodentate coordinating to the metal (Fig. 2).

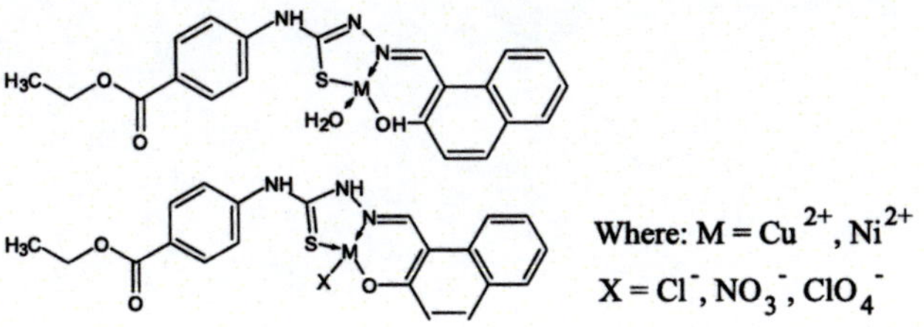

Fig. 2. Coordination modes for H_2L^2.

The results of the antimicrobial activities of the compounds synthesized on *Staphylococcus aureus, Bacillus cereus, Escherichia coli, Acinetobacter baumannii* strains are in the range 0.25–0.5 mg/mL.

The results of the antifungal activities of the compounds synthesized on *Candida albicans* and *Candida krusei* strains are in the range 0.125–0.5 mg/mL. In the $\{Ni(HL^1)Cl\}$ complex, a selectivity is observed on Cryptococcus neoformans, with CMI = 0.016, CMF = 0.031 and is twice as active as the control substance Nystatin (Table 6).

3 Conclusions

Two new ligands and 12 coordination compounds of Cu(II), Ni(II) based on ethyl 4-benzoate thiosemicarbazones derivatives of salicyl aldehyde were synthesized.

The best results of antifungal activity were recorded at $\{Ni(HL^1)Cl\}$ complex. A selectivity is observed on Cryptococcus neoformans, with CMI = 0.016 and CMF = 0.031 mg/mL, and is twice as active as the control substance Nystatin.

Conflict of Interest. The authors have no conflict of interest.

References

1. Shakya, B., et al.: 2-Pyridineformamide N (4)-ring incorporated thiosemicarbazones inhibit MCF-7 cells by inhibiting JNK pathway. Bioorg. Med. Chem. Lett. **29**(13), 1677–1681 (2019). https://doi.org/10.1016/j.bmcl.2019.04.031

2. АКСЕНОВА Э. Н. и др. Фармацевтическая химия (2004). ISBN 5-9231-0438-5
3. Gulea, A.: Noi inhibitori de proliferare a celulelor de cancer. Arad. (1), 102–107 (2013)
4. Barbă, N., et al.: Sinteza 3(4)-izotiocianatobenzoaţilor de etil din esterii acizilor 3(4)-N,N-dimetiltioureidobenzoici. Anale Ştiinţifice USM, pp. 221–223 (2001)
5. Rusnac, R., et al.: Metode de sinteză a unor izotiocianaţi alifatici, aromatici, heterociclici. Integrare prin cercetare şi inovare. Ştiinţe ale naturii şi exacte, pp. 202–206. CEP USM, Chisinau, Republica Moldova (2019). ISBN 978-9975-149-46-4
6. Yogeeswariб, P., et al.: Synthesis and anticonvulsant and neurotoxicity evaluation of N4-phthalimido phenyl (thio) semicarbazides. Eur. J. Pharm. Sci. **20**, 341–346 (2003)
7. Гэрбэлэу Н.В. Физико-химичиские методы исследования. Кишинев. КТУ (1997)
8. Pahonţu, E., et al.: Appl. Organomet. Chem. **32**, 1 (2018)
9. Pahonţu, E., et al.: Molecules **22**, 650 (2017)

Investigation of the Effect of Adding Tantalum on the Microstructure and Mechanical Properties of Biomedical Ti-15Mo Alloy

Hasan Sh. Majdi[1], Amir N. Saud[1,2](✉), Erkan koç[2], and Ameen M. Al Juboori[1]

[1] Biomedical Engineering, Al-Mustaqbal University College, Babylon, Iraq
[2] Biomedical Engineering, Karabuk University, Karabuk, Turkey

Abstract. Titanium alloys have great applications as biomaterials due to their high mechanical strength and density ratio, good corrosion resistance, and biocompatibility. Type β alloys have aroused enormous interest in the development of biomaterials due to their low elastic modulus. This new class of alloys has been formed mainly by adding tantalum, molybdenum, proven not to have biocompatibility. Tantalum is an alloy hardening element, which can increase the mechanical strength of the material. At the same time, molybdenum is a strong β-stabilizer, stabilizing the β phase with 10% quickly. In this work, Ti-15Mo-xTa system alloys were produced by the powder metallurgy method. The result shows the prepared alloy presented the β-phase grain structure, showing more excellent mechanical properties than pure titanium due to hardening in solid solution.

Keywords: Ti-based alloy · Mechanical properties · Microstructure · Biomedical application

1 Introduction

Titanium and alloys are used for various uses, such as biomaterials. The metal has particular characteristics, such as high mechanical strength/density ratios, comparatively low elastic modules, strong resistance to corrosion, and biocompatibility. These properties began to attract interest, particularly after the mid-20th century, when many aerospace and aeronautic titanium alloys were developed and replacing stainless steel and Co-Cr alloys in biomedical applications [1]. Titanium is an allotropic material that transforms from a compact hexagonal configuration (phase α) into a cubic system with a central (phase β) structure, approximately 883 °C. The α process, followed by α + β type alloys, microstructured with coexistence in each phase, was the initial titanium alloy used as biomaterials. The new titanium alloys are currently developing, primarily due to their lower elasticity modulus, to form β-type alloys [2]. The elastic modulus mismatch between biomaterials and surrounding bones is the main reason for failed implantation material fixation [3]. However, the embedded materials must be strong and resilient enough to withstand the physiological pressures and operate for a lifetime without loss or revision surgery. A good mix of strength and stiffness is required to match the bone.

I. Tiginyanu et al. (Eds.): ICNBME 2021, IFMBE Proceedings 87, pp. 637–646, 2022.
https://doi.org/10.1007/978-3-030-92328-0_81

[4]. The Ti-6Al-4V alloy is an aeronautical alloy used in biomedical applications However, long-term application poses challenges. Consequently, some studies show that the ternary alloy's vanadium and aluminum ions can cause cytotoxicity or neurological disorders. Also, over time, this alloy has caused bone resorption and implant loosening. Modern titanium alloys are designed to be non-cytotoxic by adding molybdenum, zirconium, tantalum, and niobium [4, 5]. The remarkable biocompatibility and corrosion resistance of pure Ti and Ta have been tested and recognized [6]. Burke [7] presented pure Ta for sutures, bone screws, and plates in 1940. Pure titanium has been utilized in medicine since the 1950s [7]. Pure Ti and Ta have limited uses because of their weak mechanical properties. Metal alloying improves the mechanical properties of its pure component metals. So Ta can boost strength and lower modulus in Ti alloys. Molybdenum is a refractory metal with a higher melting point than titanium. Thus, lower alloyed element concentration titanium alloys can reduce alloy processing costs. Molybdenum is a promising titanium alloying element. The material outperforms c.p.Ti, Ti, and Ti 6Al-4V in stabilizing the beta phase. Mo is also a non-toxic, non-allergic element that increases the passive range of titanium alloys. The present work aims to develop a low modulus β-type Ti-15Mo-xTa alloy and investigate its properties, such as microstructure and mechanical, after the addition of tantalum to be proposed as an alternative substance aiming at biomedical applications.

2 Experimental

2.1 Materials

Powders of (Titanium, Molybdenum, and Tantalum) were used in this study; these powders were purchased from Lemandou Ltd. co. China. The purity of the used powders mastered using X-ray fluorescent (XRF) and particle size analyzer type (Bettersize) for the starting materials' particle size are shown in Table 1.

Table 1. The purity and the average particle size of the powders used.

Powder	Purity	Average partials size (μm)
Titanium	99.85	55.13
Molybdenum	99.98	19.27
Tantalum	99.99	15.41

2.2 Preparation of Sample

The Ti-15Mo-xTa (x is 0, 10, 15, and 20 wt. %) was produced by the Powder Metallurgy method under an inert atmosphere (Argon). As shown in Table 2, four samples were produced containing all samples of Ti-15Mo-xTa and its composites. Ti, Mo, and Ta powders were mixed using a planetary ball milling machine with zirconia containers and balls. This machine had a 20:1 ball to powder ratio and rotated at 500 rpm. The ball milling was done in an argon-pure atmosphere. An automated powder sample removal machine was employed at 3, 6, 9, and 12 h post milling to characterize powder and evaluate the mechanical alloying process. Grinding generally raises the temperature. The overall heat was lowered by operating the mill in cycles of 90 min milling and 50 min rest. Cold compression at 900 MPa for 1.5 min with a consistent loading rate produced the green compact (0.4 tonnes per second). Six hours at 950 °C, 5 C/min heating, followed by cooling in the vacuum furnace, produced the green compacts.

2.3 Characterization of the Prepared Specimens

The ingots' chemical composition was tested to ensure their purity. ICP OES (induced plasma optical emission spectrometry) was used to analyze the composition. Using an analytical balance and distilled water at room temperature, porosity was determined using Archimedes' principle. The structure was studied using X-ray diffraction and the Rietveld technique for quantitative phase analysis.SEM was used to acquire the microstructure (CARL ZEISS ULTRA PLUS GEMINI FESEM). Initial mechanical properties of the Vickers hardness device were verified using 200 gf for 60 s. Each sample was measured three times in various locations. According to ASTM, the longitudinal and transverse velocities of Elastic Modulus were calculated using an ultrasonic wave instrument (cct-4) (STP 696). Compressive strength was tested using a computer-controlled electronic universal tester (WDW-200KN). ASTM (D695-85) specifies a universal testing machine with a 0.2 mm/min loading rate. All produced specimens were tribologically examined using a Micro-UMT-2 tester's tribometer at room temperature (ball-on-disk). The kinematic form is ASTM G133-05 reciprocating linear motion.

3 Result and Discussions

Table 2 presents the chemical composition of the produced ingots, in percentage by weight, taking the titanium concentration as a balance. The data show that the values were close to the nominal values of the alloys, remaining with a percentage difference of less than 5%. The metallic impurities remained in negligible quantities in the materials, coming from the precursor metals and the melting process.

Table 3 shows the porosity and density values of the prepared alloys. It can indicate the effect of Ta content on the porosity of sintered specimens, and there is a decrease in the porosity values of specimens after sintering. It can be seen that the porosity of sintered specimens decreases as the Ta content increases due to the better inter diffusion caused by these additives. The increase in density with the addition of alloying elements is related to the higher values of density of molybdenum (10.23 g/cm^3) and tantalum (16,6 g/cm^3) compared to titanium (4.51 g/cm^3) [8].

Figure 1 shows a comparison of the diffraction patterns for the alloys. It is observed, predominantly, the presence of peaks of the β phase in alloys with 15% w of Mo. The alloy with 15% w of Mo also exhibited peaks in the β phase. Molybdenum is a strong β-stabilizer, being able to initiate β phase retention at a concentration between 9 and 10% w, ultimately retaining the β phase at concentrations of 15 and 20% w [9]. Tantalum acts slightly to stabilize the β phase when it is in the presence of another β-stabilizing element [10]. From this, it can be seen that both alloy elements are acting as β-stabilizer since there was the complete formation of β phase in the concentration above 15 and 20% w of Mo.

Figure 2 shows micrographs of samples from the Ti-15MoxTa system. According to the literature, martensite with a hexagonal structure presents massive and acicular morphology [11]. The transition from massive hexagonal to acicular martensite occurs with 4% of element alloy [10]. Massive martensite is formed by irregular regions, 50 μm to 100 μm in diameter, with no internal features visible through optical microscopy [11]. Furthermore, it is observed that the morphology of all alloys is characteristic of a predominantly beta phase. However, there are differences in grain size; in agreement with the X-ray diffractograms, it was impossible to observe the acicular formation of α 'martensite due to its small relative quantity.

Table 2. Chemical composition of the alloys used in this work (%p).

Samples	Ti	Mo	Al	Fe	Mn	Ni	Cu	Cr	Si	Ta
Ti-15Mo	Balance	14.87	0.04	0.07	<0,001	0,004	0,002	0,008	<0,01	–
Ti-15Mo-10Ta	Balance	14,84	0,04	0,07	<0,001	0,005	0,003	0,009	<0,01	9,97
Ti-15Mo-15Ta	Balance	14,74	0,02	0,07	<0,001	0,004	0,002	0,004	<0,01	14,25
Ti-15Mo-20Ta	Balance	14,93	0,02	0,07	<0,001	0,006	0,003	0,005	<0,01	19,91

Table 3. Porosity and density for the prepared alloys

Samples	Density (g/cm^3)	Porosity
Ti-15Mo	4.87 ± 0,24	14.78 ± 0,73
Ti-15Mo-10Ta	5,03 ± 0,25	11.32 ± 0,56
Ti-15Mo-15Ta	5,32 ± 0,26	10.06 ± 0,50
Ti-15Mo-20Ta	5,49 ± 0,27	9.02 ± 0,45

After the homogenization heat treatment, the micrographs for the same samples are shown in Fig. 2. Considerable growth of the grains can be observed concerning the condition, which the supply of energy can explain to the sample during the heating, which allowed its growth. It can also be observed that the angles of the grain boundaries are approximately 120°, characteristic of materials in equilibrium.

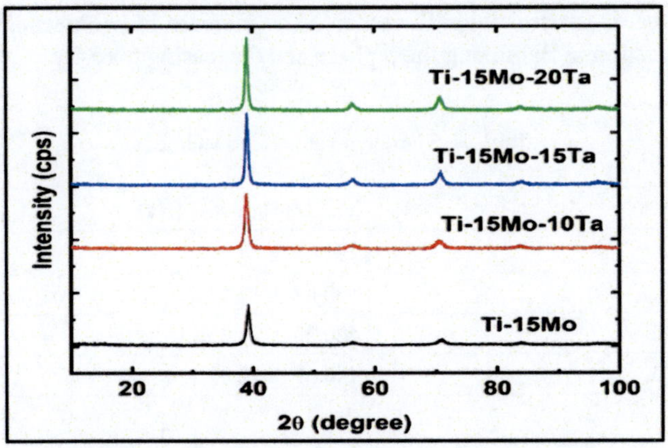

Fig. 1. XRD of the prepared Ti-15MoxTa system.

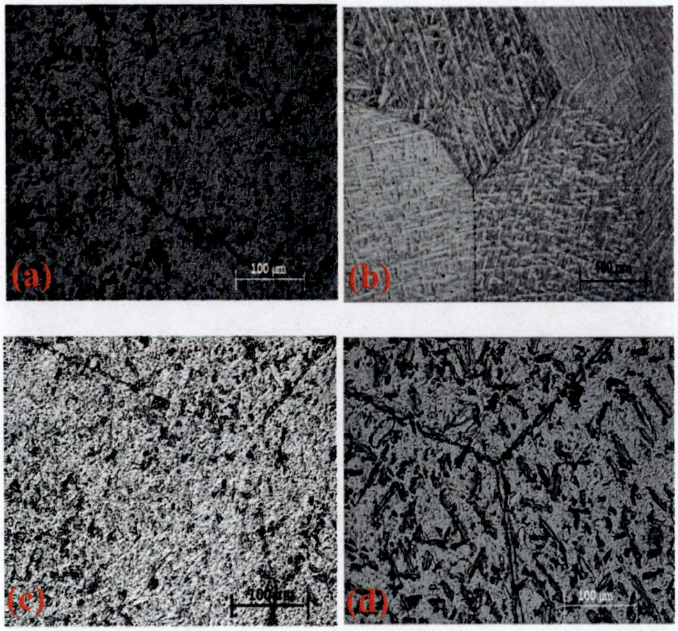

Fig. 2. Micrograph for the samples: (a) Ti-15Mo, (b)Ti-15Mo-10Ta, (c) Ti-15Mo-15Ta, (d) Ti-15Mo-20 Ta with 1000X magnification.

Table 4 shows Vickers hardness values for Ti-15Mo and Ti-15Mo-xTa alloys. Hardness values for Ti-15Mo and Ti-15Mo-X Ta alloys are shown to be dependent on tantalum content. It can be shown that base alloys with Ta addition had substantially greater hardness values than base alloys, base alloys with Mo additive, and base alloys with Mo &

Ta, and that hardness rises with Ta content. The increase in hardness is attributed to tantalum's function in stabilizing the β phase and decreasing porosity.

Table 4. Hardness values for specimens

Sample	Vickers Hardness (Hv)
Ti-15Mo	294.09 ± 14.7
Ti-15Mo-10Ta	350.82 ± 17.54
Ti-15Mo-15 Ta	420.22 ± 21.01
Ti-15Mo-20 Ta	446.67 ± 22.33

The ASTM E9-2000 compression test protocol was used. Table 5 shows the compressive strength values for the produced samples. The compressive strength rises from 313.62 to 402.71 MPa as the tantalum concentration increases to 20%, anticipated since density increases with tantalum content.

Table 5. Compressive strength for prepared samples.

Sample	Compressive strength (MPa)
Ti-15Mo	313.62 ± 15.68
Ti-15Mo-10Ta	350.81 ± 17.54
Ti-15Mo-15 Ta	368.53 ± 18.42
Ti-15Mo-20 Ta	402.71 ± 20.13

Studies of elastic deformation behaviour are essential for biomedical material knowledge [12]. The longitudinal and transverse velocity values in the following equations may be used to estimate material elasticity [13].

$$v = \frac{1 - 2\left(\frac{V_T}{V_L}\right)^2}{2 - 2\left(\frac{V_T}{V_L}\right)^2}. \quad (1)$$

$$E = \frac{V_L^2 \rho (1 + \sigma)(1 - 2\sigma)}{(1 - \sigma)}. \quad (2)$$

Where: ν: Poisson's ratio, VT: Shear (Transverse) velocity (m/sec), VL: Longitudinal velocity (m/sec.), E: Elastic modulus (GPa), ρ: Density (kg/m^3).The elastic properties can be determined, and the results are shown in Table 6.

Table 6. Elastic modulus of the prepared alloys.

Sample	Elastic modulus (GPa)
Ti-15Mo	101.16
Ti-15Mo-10Ta	89.24
Ti-15Mo-15 Ta	80.74
Ti-15Mo-20 Ta	73.14

Increasing the Ta content increases the phase's lattice parameter [7]. The modulus is linked to the crystal structure and the interatomic distances inside the crystal lattice. Thus, increasing the phase's Ta concentration decreases its elastic modulus. Tribological samples were evaluated on a reciprocating wear tester for dry sliding contact performance (in an air at 25 °C temperature). The samples were polished to a mirror finish before testing. Samples having a surface area of 6 mm were exposed to normal loads of 1 N and 3 N and a sliding velocity of 0.01 ms1. The counterface had a 5 mm stroke and a 75 m total sliding distance on the samples. 2-D profilometry is used to look at the samples' surface tracks.

The depth, width, and ultimately cross-sectional area of wear tracks produced on the surfaces of each sample increased with increasing test load (Fig. 3 a–d), as anticipated. The wear rate (mm3 N1 m1) was determined using the per unit test load (N) and per unit sliding distance (m) for a specified volume of wear tracks (cross-sectional area length, mm^3). As shown in Table 7, the wear rates of each sample were consistent across test loads. Table 7 shows that the base alloy Ti-15Mo alloy had the lowest wear resistance among the tested samples. The addition of tantalum reduced the corrosion rate of the prepared alloys. The worn surface topography of the samples may be described as rough and delaminated with signs of plastic deformation along with grooves parallel to the sliding direction (Fig. 4 a–d). In general, worn surfaces showed adhesive wear characteristics associated with material transfer to the counter-face. Material transfer from the produced alloy induced counterface darkening, which intensified with increasing test loads (Fig. 4a–d).

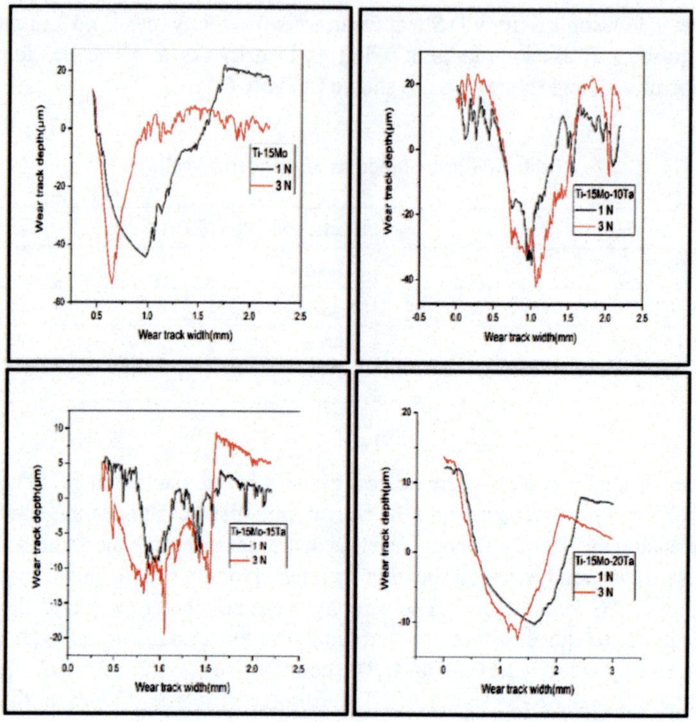

Fig. 3. 2D Wear track profile surface for the prepared alloys.

Table 7. Elastic modulus of the prepared alloys

Test load	Specific wear rate ($mm^3\ N^{-1}\ m^{-1}$)	Wear mechanism
Ti-15Mo		
1	$3.51*10^{-6}$	Abrasive
3	$5.43*10^{-6}$	
Ti-15Mo-10Ta		
1	$3.10*10^{-6}$	Abrasive
3	$5.09*10^{-6}$	
Ti-15Mo-15Ta		
1	$2.71*10^{-6}$	Abrasive
3	$4.54*10^{-6}$	
Ti-15Mo-20Ta		
1	$2.25*10^{-6}$	Abrasive
3	$3.82*10^{-6}$	

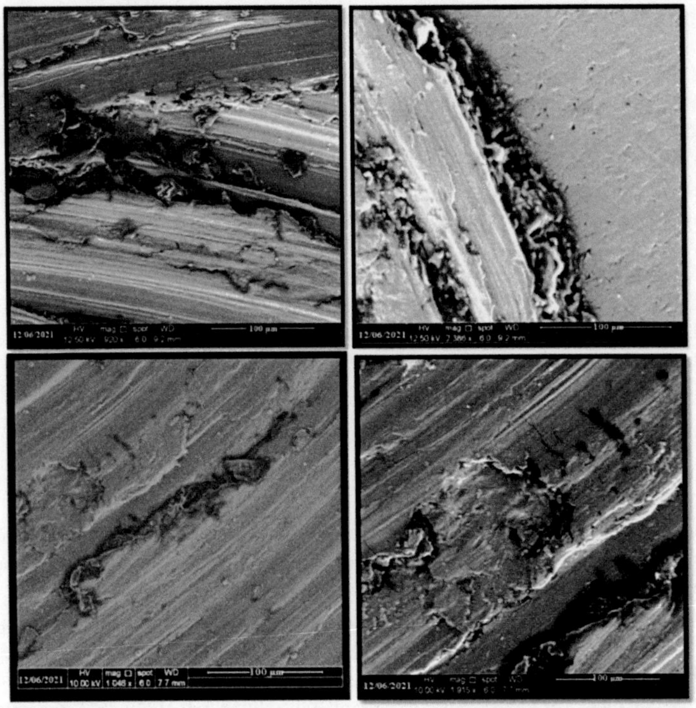

Fig. 4. SEM of wear track formed on the surface of a) Ti-15Mo, b) Ti-15Mo-10Ta, c) Ti-15Mo-15Ta, and d) Ti-15Mo-20Ta.

4 Conclusions

From the results obtained, it is possible to observe a microstructure formed entirely by the β phase. The joint action of tantalum and molybdenum to stabilize the β phase may be interesting to form β titanium alloys with low alloying elements. The hardness of the alloys proved to be higher than commercially pure titanium, which indicates their greater mechanical strength. Mechanical strength is an essential factor for the processing of implants, as it avoids flaws in the material. Based on the results obtained in this work, the Ti-15Mo-20Ta alloy obtained the most significant potential for biomedical application. However, future corrosion tests and biocompatibility can provide better information about the properties of these alloys.

Conflict of Interest. The authors declare that they have no conflict of interest.

References

1. Long, M., Rack, H.J.: Titanium alloys in total joint replacement—a materials science perspective. Biomaterials **19**(18), 1621–1639 (1998)

2. Geetha, M., Singh, A.K., Asokamani, R., Gogia, A.K.: Ti-based biomaterials, the ultimate choice for orthopaedic implants–a review. Prog. Mater Sci. **54**(3), 397–425 (2009)
3. Li, Y., Yang, C., Zhao, H., Qu, S., Li, X., Li, Y.: New developments of Ti-based alloys for biomedical applications. Materials **7**(3), 1709–1800 (2014)
4. Oshida, Y.: Implant application. In: Bioscience and Bioengineering of Titanium Materials, pp. 215–253. Elsevier, New York (2007)
5. Nasab, M.B., Hassan, M.R., Sahari, B.B.: Metallic biomaterials of knee and hip-a review. Trends Biomater. Artif. Organs **24**(1), 69–82 (2010)
6. Matsuno, H., Yokoyama, A., Watari, F., Uo, M., Kawasaki, T.: Biocompatibility and osteogenesis of refractory metal implants, titanium, hafnium, niobium, Tantalum, and rhenium. Biomaterials **22**(11), 1253–1262 (2001)
7. Zhou, Y.L., Niinomi, M., Akahori, T.: Effects of Ta content on Young's modulus and tensile properties of binary Ti–Ta alloys for biomedical applications. Mater. Sci. Eng. A **371**(1–2), 283–290 (2004)
8. Lide, D.R.: Physical constants of organic compounds. CRC Handbook of Chemistry and Physics, 2660 (2005)
9. Ho, W.F., Ju, C.P., Lin, J.C.: Structure and properties of cast binary Ti-Mo alloys. Biomaterials **20**(22), 2115–2122 (1999)
10. Zhao, X., Niinomi, M., Nakai, M.: Relationship between various deformation-induced products and mechanical properties in metastable Ti–30Zr–Mo alloys for biomedical applications. J. Mech. Behav. Biomed. Mater. **4**(8), 2009–2016 (2011)
11. Lütjering, G., Williams, J.C.: Titanium. Springer, Springer (2007). https://doi.org/10.1007/978-3-540-73036-1
12. Nie, L., Zhan, Y., Hu, T., Chen, X., Wang, C.: β-Type Zr–Nb–Ti biomedical materials with high plasticity and low modulus for hard tissue replacements. J. Mech. Behav. Biomed. Mater. **29**, 1–6 (2014)
13. Asmita, F., Wibisono, G., Judawisastra, H., Priambodo, T.A.: Determination of elastic modulus of ceramics using ultrasonic testing. In: AIP Conference Proceedings, vol. 1945, no. 1, p. 020017. AIP Publishing LLC (2018)

Health Informatics, e-Health and Telemedicine

Assessment of Cyber Security Maturity for Critical Domains in Republic of Moldova

Aurelian Buzdugan(✉)

Moldova State University, Chisinau, Moldova

Abstract. In this paper we present the evaluation results based on the model for cyber security maturity assessment upon the cyber security program development for critical infrastructures (CI) in Republic of Moldova [1]. During last year the healthcare domains has become a target of strategic interest amongst cyber criminals. These events can lead not only to security breaches, but also to safety events that affect human lives. The findings confirm the recommendations of previous case studies on evaluation of cyber security requirements for healthcare inclusive nuclear and radiological domain. Therefore, we classify the results based on the dimensions of the recently developed Model for Cyber Security Maturity Assessment in CI [1]. The proposed model focuses on four primary dimensions: policies and administration, education and training, work environment, cyber risk management. It has been developed with the goal to help decision makers analyze the efficiency of information systems used for cyber risk management. The model is multidimensional and can also show the actual state of cyber security maturity in an organization, fact which was also proven by external reviews. The experimental test of this analysis confirm the applicability of the Model for organizations at different stages of cyber security development, as well as from different domains.

Keywords: Critical infrastructure · Cyber assurance · Cyber security · Model

1 Introduction

The digitalization and use of computer systems in various domains has offered tremendous growth and development opportunities for CI domains. Ensuring cyber security for CI is a priority at national [2] as well international level [3, 4]. The most recent report at the United Nations Group of Governmental Experts on advancing responsible state behavior in cyberspace in the context of international security, has also identified the cyber risks towards critical infrastructures, and proposed norms on stopping any attacks against CI infrastructure of other countries, ensuring adequate protection of own CI, as well as cooperating with other states in sharing knowledge and expertise in this sense [5].

The healthcare sector, critical in the past years due to the pandemic, is unfortunately not an exception in this case. Almost one-third of COVID-19-related attacks target public authorities and health care institutions. Only in the pandemic period of 2020, there were multiple ransomware incidents on medical institutions such as in Germany in September

I. Tiginyanu et al. (Eds.): ICNBME 2021, IFMBE Proceedings 87, pp. 649–656, 2022.
https://doi.org/10.1007/978-3-030-92328-0_82

2020 [6] or Finland in October 2020 [7]. While some attacks can be quantified in terms of loss if compared to the ransom that was demanded, or manhours spent in remediating and recovering normal IT operations, it is very hard to predict the actual loss.

Another finding is that many devices in healthcare are outdated from the security perspective [8]. Therefore, devices from this domain are a low hanging fruit for groups looking for targets. This can both be used by commodity malware developers focused on financial fraud, such as via ransomware, or for targeted actors seeking to create harm [8].

Furthermore, healthcare domain can use nuclear or radiological material as part of diagnosis or treatment. This raises the threat severity to a different level, given the harm and impact that an incident involving such materials can have.

The attack vectors vary, from social engineering to exploitation of known or unknown vulnerabilities. Therefore, a holistic view upon all IT systems used in such domains and the vulnerabilities these have, up to policies and standards followed by practitioners in ensuring an adequate level of cyber security.

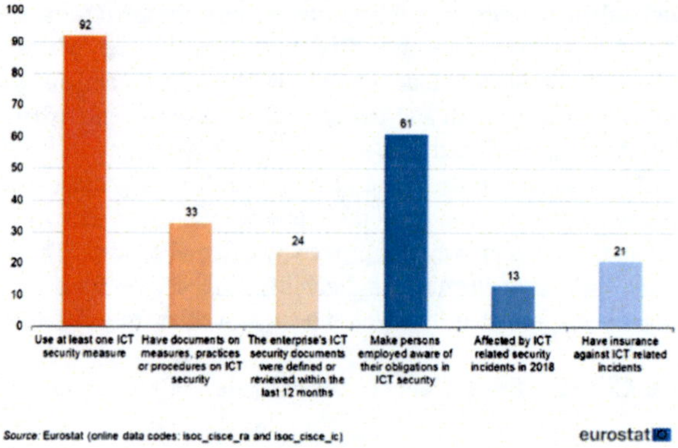

Fig. 1. ICT Security in Enterprises, EU-27, 2019 (% enterprises) [9]

It is worth mentioning that while over 90% of enterprises employed at least one security control, still 13% were affected by a security incident [9]. Figure 1 shows the percentage of security controls, which reveal certain gaps in terms of functions, documents and accuracy of such documents.

In the context of Republic of Moldova, we previously conducted analyses over the cyber security program development for the NR domain, as well as the correlation to international guidance and recommendations in this sense. It is also worth mentioning that cyber-attacks against CI at the international level upon healthcare or nuclear and radiological entities have increased in frequency lately, which require an adequate defense at the national level [10–15]. This urges all states to develop and further improve the capacity to identify and respond to such incidents.

In this paper we review the cyber security assessment in the whole healthcare domain as well as the specific nuclear and radiological (NR) domain using the proposed model [1].

2 Assessment of Cyber Security Maturity in Moldova

One of the first analysis from 2015 has identified the inclusion of cyber security elements in the domestic legislation on safe and security of NR entities [10]. This is strongly correlated to international guidance and recommendations from the IAEA. As the NR domain legislation includes cyber security, this has created necessary premises for a horizontal cooperation with IT specialized bodies. We believe this the best solution in the situation where specialized knowledge in cyber security is still scarce, both in 2015 and at present. We note that vendors are also looking more in depth upon IT security requirements in the CI domain [16, 17], as well as research organizations [18].

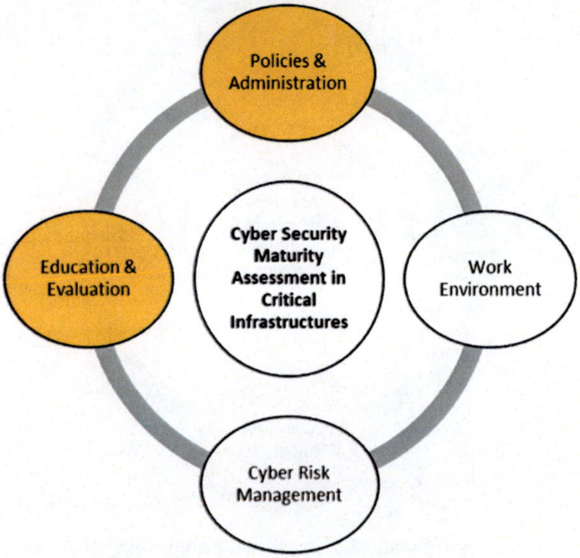

Fig. 2. Cyber security maturity in 2015

This is also dictated via specific technical requirements such as user access, monitoring and incident reported or data confidentiality that are derived from the current NR legislation. This status can be reflected by the proposed Model, when all dimensions have below average ratings, there is a certain awareness and knowledge, however this is highly disrupted across all dimensions [19]. In Fig. 2 we visualize the results of the application of the model against the analysis from 2015 [10].

The dimensions with no colored background represent a very low cyber security maturity, yellow represents low, light green – average, and dark green high to very high. Therefore, a priority in such scenarios is to develop and improve the Policies and Administration section.

A subsequent analysis looks at the interdependence of cyber level at the national level, and the developments of cyber security in the CI domain [12]. The legal frameworks in cyber security has been enhanced in 2015 with the approval of a new National Cyber

Security Program, which covers aspects as functions and responsibilities for information security, training indicators as well as a first attempt to clarify the key terms and build an information security management system [2, 19] Adding these developments to the assessment performed via the Model, we observe a need to develop the Education and Training component (Fig. 2). This is key to ensure that other dimensions, such as cyber risk management and work environment (Fig. 3), will have the necessary prerequisites to be developed. This is directly dictated by the legal framework development, and indirectly by awareness of cyber risks.

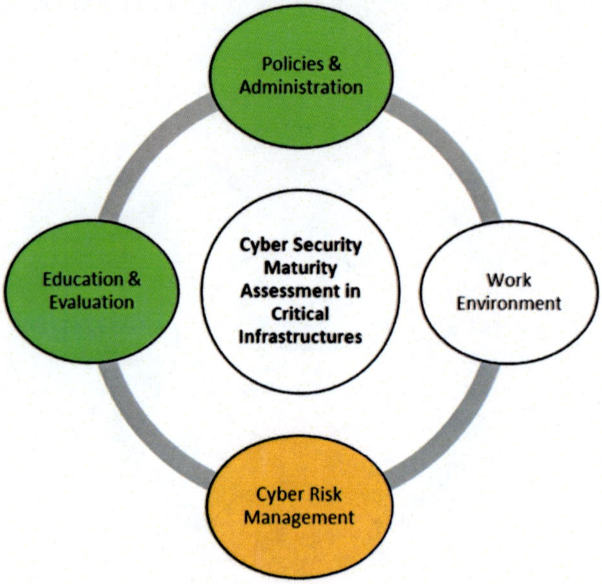

Fig. 3. Development of cyber risk management practices

Another important milestone in the development of the cyber security program in Moldova is the approval on minimum cyber security requirements [20]. This document represents a holistic, but also a technical perspective, upon all areas of cyber security. As it can be deducted from the name of the document, it focuses on technical and functional requirements that promote the cyber security program development. This can be translated to different attributes and dimensions to the proposed Model, covering aspects such as technical tools or configuration, user awareness and training. Therefore, the retroactive assessment of the cyber security maturity via the Model shows a significant improvement compared to 2015 [10]. First of all we note a continuous development in the legislation, which impacts directly to all dimensions of the Model. We believe this scenario and development path can be considered as a standard roadmap.

One last completed milestone in the cyber security program of Moldova is the approval of the Information Security Strategy for 2020–2024 [21]. The strategy identifies the same threats and concerns, however covers comprehensive actions and goals to ensure an adequate cyber security level across all domains of the country. The strategy mentions its scope to cover explicitly organizations from the CI domain, as well as reflects upon the importance that CI institutions have in the process of decision making. We also note the harmonization with EU policies on defining critical infrastructures, which has been mentioned in the strategy. In addition, we support the goals to conduct a cyber security audit for all types of CI, as well as develop plans to secure such type of organizations. We believe the proposed model could also serve for the purpose of auditing/accreditation such type of organizations, given the cyber security maturity and context in Republic of Moldova. The assessment of the cyber security maturity in Moldova via the proposed model is represented in Fig. 4.

In addition, there are national initiatives at development training programs for certain CI domains, such as the nuclear and radiological ones. Recently developed guides and integration of topics in Technical University of Moldova Master curricula, that include inter alia cyber security elements, are also results of the interests to further strengthen the national cyber security program by training experts, raising awareness and the overall security culture [22, 23]. Therefore, in order to ensure the cyber maturity in IC entities, it is necessary the presence and development (in situ or through cooperation) of all dimensions.

The analysis performed for healthcare, in 2016 and respectively 2019, have shown that while technologies exist, these cannot be enforced without national policies and regulations in this sense [14]. We believe this is also strongly correlated to the mindset and overall security culture, as human dimension has a strong influence upon the development of the legal framework. Therefore, we see a correlation in the development phase which is confirmed by the first two dimension of the Model, key for triggering overall technical developments. As solutions to this context, a horizontal cooperation is recommended with domains that are at the same stage, or have went through significant developments in the past years, such as the nuclear and radiological domain. Additionally, this solution could be applied given the use ionizing radiation sources in the healthcare domain, as well as the imminent need to ensure cyber security of healthcare devices in the context of Covid-19 pandemic.

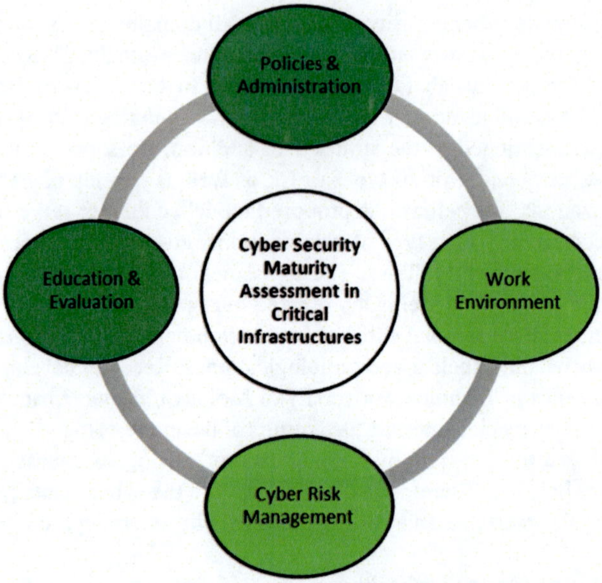

Fig. 4. The four dominant factors that influence cyber security in the economic entity

3 Conclusions

Cyber security program development is a complex and lengthy process that requires a holistic approach. The cyber security legal framework in Republic of Moldova has recorded a positive development cycle, and is being regularly updated in line with international guidance, recommendations or threat landscape. Various methods to evaluate the progress and priorities can be applied, such as via external assessments looking at specific indicators [24] or practical models. The application of the assessment practical methodology proposed in the Model [19], has identified certain findings in relation to the cyber security development roadmap at national levels that directly impacts the CI domain. While the assessment for the case study in Republic of Moldova is represented via above graphs for ease of understanding, we evaluate the current development level as above Average. This means that the current legislative framework can be sufficient to help improve and develop the other dimensions as per the model. However, cyber security is a continuous process that requires adaptations to the current threat landscape, good practices as well as national cyber security program developments. As an example, a recently published implementing guide from the IAEA tackles the integration of computer security for nuclear security. This specific guide covers explicitly the cyber security development as part of the nuclear security at a national level. Therefore, we note that while cyber security can be improved and derived from national regulations, there are still certain CI domains, such as the nuclear one, that requires additional attention. This, on the other hand, is also beneficial for the country, as it can serve as an example for the development of concepts and strategies for other domains, such as healthcare.

Conflict of Interest. The authors declare that they have no conflict of interest.

References

1. Buzdugan, A., Capatana, Gh.: Cyber security maturity model for critical infrastructures. In: 20th International Conference on Informatics in Economy (IE 2021), Bucharest, Romania (2021)
2. Government Decision no. 811 from 29.11.2015 on National Program of Cyber Security of the Republic of Moldova for 2016–2020 years. http://lex.justice.md/index.php?action=view&view=doc&lang=1&id=361818
3. European Commission Joint Research Centre Institute for the Protection and Security of the Citizen (2012). Risk assessment methodologies for Critical Infrastructure Protection. Part I: A state of the art. https://publications.jrc.ec.europa.eu/repository/bitstream/JRC25512/EUR%2020750%20EN.pdf
4. European Commission Joint Research Centre Institute for the Protection and Security of the Citizen (2015). Risk assessment methodologies for critical infrastructure protection. Part II: A new approach. https://publications.jrc.ec.europa.eu/repository/bitstream/JRC96623/lbna27332enn.pdf
5. United Nations - Report of the Group of Governmental Experts on Advancing responsible State behaviour in cyberspace in the context of international security (A/76/135), 14 July 2021
6. The Guardian (2020). Prosecutors open homicide case after cyber-attack on German hospital, viewed 3 August 2021. https://www.theguardian.com/technology/2020/sep/18/prosecutors-open-homicide-case-after-cyber-attack-on-german-hospital
7. The Guardian (2020). Shocking' hack of psychotherapy records in Finland affects thousands, viewed 3 August 2021. https://www.theguardian.com/world/2020/oct/26/tens-of-thousands-psychotherapy-records-hacked-in-finland
8. National Authority for Electronic Certification and Cyber Security (2021). Cyber Security in Health Sector. https://cesk.gov.al/publicAnglisht_html/Publikime/2021/Cybersecurity%20in%20health%20system.pdf
9. Eurostat - ICT security in enterprises (2020). https://ec.europa.eu/eurostat/statistics-explained/index.php?title=ICT_security_in_enterprises
10. Buzdugan, A., Buzdugan, Ar.: Information security development in the Moldovan nuclear and radiological infrastructure. In: IAEA International Conference on Computer Security in a Nuclear World: Expert Discussion and Exchange, Vienna, 1–5 June (2015)
11. Buzdugan, A.: Advances in security requirements for high risk objects. In: 2016 IAEA International Conference on Nuclear Security, Vienna (2016)
12. Buzdugan, A.A., Buzdugan, A.I.: Cyber security in the nuclear and radiological domain: case study of Republic of Moldova. In: Sontea, V., Tiginyanu, I. (eds.) 3rd International Conference on Nanotechnologies and Biomedical Engineering. IP, vol. 55, pp. 541–543. Springer, Singapore (2016). https://doi.org/10.1007/978-981-287-736-9_127
13. Buzdugan, A.: An overview of cybersecurity in the healthcare sector. In: Telecommunications, Electronics and Informatics, 6, Chişinău, 24–27 Mai 2018, pp. 261–262. Tehnica UTM, Republica Moldova (2018)
14. Buzdugan, A.: Integration of cyber security in healthcare equipment. In: Tiginyanu, I., Sontea, V., Railean, S. (eds.) ICNBME 2019. IP, vol. 77, pp. 681–684. Springer, Cham (2020). https://doi.org/10.1007/978-3-030-31866-6_120
15. Buzdugan, A.: Opportunities for improving cyber risk management in Critical Infrastructures. In: National Scientific Conference with International Participation. Institute for Development and Innovation, Moldova State University (2020, in press)

16. Trend Micro (2013). The SCADA That Didn't Cry Wolf. http://www.trendmicro.com/cloud-content/us/pdfs/security-intelligence/white-papers/wp-the-scada-that-didnt-cry-wolf.pdf
17. Siemens (2019). Caught in the Crosshairs: Are Utilities Keeping Up with the Industrial Cyber Threat? https://assets.siemens-energy.com/siemens/assets/api/uuid:38ab2dff-3ee6-48ee-a2a9-29df5e24bfbd/siemens-energy-cybersecurity.pdf
18. MITRE (2020). ATT&CK® for Industrial Control Systems. https://collaborate.mitre.org/attackics/index.php/Main_Page
19. Buzdugan, A.: Model for cyber security maturity assessment in critical infrastructures, Official Catalogue "Cadet INOVA", pp. 154–157, June 2021 (2021). ISSN 2501-3157
20. Government Decision no. 201 from 28.03.2017 on approving minimum requirements obligatory for cyber-security. http://lex.justice.md/index.php?action=view&view=doc&lang=1&id=369772
21. Parliament Decision no. 257 from 2018 on approving Information security strategy of the Republic of Moldova for the 2019–2024 years and on Action Plan for its implementation. http://lex.justice.md/index.php?action=view&view=doc&lang=1&id=378899
22. Buzdugan, A., Buzdugan, A.: Revision of the curriculum on nuclear safety and security in the light of recent international recommendations. In: Tiginyanu, I., Sontea, V., Railean, S. (eds.) ICNBME 2019. IP, vol. 77, pp. 815–819. Springer, Cham (2020). https://doi.org/10.1007/978-3-030-31866-6_145
23. Buzdugan, Ar., Buzdugan, A.: Knowledge and role of technology in university's non-proliferation culture. In: SnT 2021. CTBT Science and Technology Conference, Vienna, Austria, 28 June–02 July 2021
24. Bolun, I., Ciorba, D., Zgureanu, A., Bulai, R., Calin, R., Bodoga, C.: Informatics security assessment in the Republic of Moldova. J. Eng. Sci. **XXVII**(4), 103–119 (2020). https://doi.org/10.5281/zenodo.4288297. ISSN 2587-3474

Features of Telemedicine Technology for Monitoring of Patients with Atopic Dermatitis

K. Kolisnyk[1(✉)], R. Tomashevskyi[1], O. Avrunin[2], V. Kolisnyk[2], A. Trubitcin[2], and V. Klymenko[3]

[1] National Technical University "KhPI", Kharkiv, Ukraine
kolesniknet@ukr.net
[2] Kharkiv National University of Radio Electronics, Kharkiv, Ukraine
[3] Kharkiv National Medical University, Kharkiv, Ukraine

Abstract. The methods of disease remote diagnosis, including the use of diagnostic video information, are becoming more widespread with the development of telemedicine technologies. At the same time, in modern systems of telemedicine video diagnostics, special requirements are imposed on improving the accuracy, reliability, efficiency, and quality of the received video information. The authors consider new approaches in telemedicine diagnostics technologies by digital data visualization using the example of video monitoring of a patient's condition with atopic dermatitis.

Keywords: Telemedicine technologies · Medical diagnostics · Quality of video images

1 Introduction

The COVID-19 pandemic has caused a huge motivation for the telemedicine technologies development and, first of all, teleconsultations for patient examination and making some diagnoses remotely, as well as monitoring the effectiveness of treatment over time [1, 2]. Along with the new capabilities of modern telemedicine technologies, nevertheless, questions arise related to ensuring the identity of conditions and hardware for recording and transferring data, and, accordingly, the repeatability of survey results [2–5]. At the same time, a feature of medical diagnostic data is a high individual anatomical and physical variability of patients and the absence in fact actual standards for the norm conditions and specific pathological conditions at different stages of diseases.

When receiving, processing, and transferring telemedicine video information, a distinctive feature of this process is rather large arrays of digital data, conditioned by their required volumes and quality. Therefore in these processes new methods and technologies are relevant that makes it possible to optimize video data, thereby increasing the efficiency of the software and hardware used [6].

The existing methods for optimizing telemedicine visualization provide for both the creation of new algorithms for processing and transferring video information, and the

I. Tiginyanu et al. (Eds.): ICNBME 2021, IFMBE Proceedings 87, pp. 657–665, 2022.
https://doi.org/10.1007/978-3-030-92328-0_83

optimization of their volume through the use of more informative methods for recording and analyzing this video information. These methods include the methodology for improving the primary data of telemedicine visualization discussed in this paper, as well as the algorithms for their processing proposed by the authors. This paper provides the monitoring example of skin video images to condition control of patients diagnosed with atopic dermatitis. These studies were carried out by the joint efforts of several medical and scientific-education organizations dealing, among other things, with these issues [6, 7].

2 Using of Additional Histogram Features to Improve the Quality of Image Analysis

When diagnosing and monitoring the skin condition of patients with atopic dermatitis at different stages of development and treatment of the disease, it is advisable to use the services of telemedicine. Moreover, in most cases, it is sufficient to take into account only the visual characteristics of the skin condition. So, in [8], a comparison of indicators of immunoglobulins in blood serum and indicators of immunograms, as well as a visual color assessment of the patient's skin, is given. However, when evaluating the color components of skin areas in the HSV system, indicators such as lighting and characteristics of the dermatoscopy model can significantly affect the analyzed data. Therefore, it seems appropriate to consider additional characteristics of the processed images. The paper proposes to introduce such indicators invariant to the shooting conditions as the relative values of histogram features - image contrast and the value of the second (local) maximum of the image intensity histogram [9–11].

An illustration of such characteristics is shown in Figs. 1 and 2, respectively. Thus, Figs. 1a and 1b show photographs of human skin fragments in normal conditions and in the acute stage of atopic dermatitis, respectively (magnification 20 times). Figures 2a and 2b show histograms for the corresponding images of the skin in Fig. 1a and 1b. Determination of contrast was performed as the difference between the maximum and minimum intensity levels on the histogram at a level of 5% of the value of the global extremum [8, 12].

The solving efficiency of monitoring problems the states of objects with random properties, as a rule, depends on the correct choice of the most informative system of parameters (features) that are sensitive to changes in the characteristics of the object. Any control formally implements a testing procedure, the effectiveness of the result of which is determined by reliability - the probability of making the right decision.

This approach is complicated by the fact that with the uncertainty of the properties of the research object, the problem of selecting informative parameters becomes problematic. Especially if the metrological support of information transformations in the structure of the control system is difficult, which often occurs during medical diagnostics.

The choice of the optimal (according to the criterion of maximum reliability) system of information features is a classical problem of statistical synthesis under conditions of a priori uncertainty [8, 12]. The ranking of the features according to the information content is carried out according to the value of the control reliability indicator [8], or the

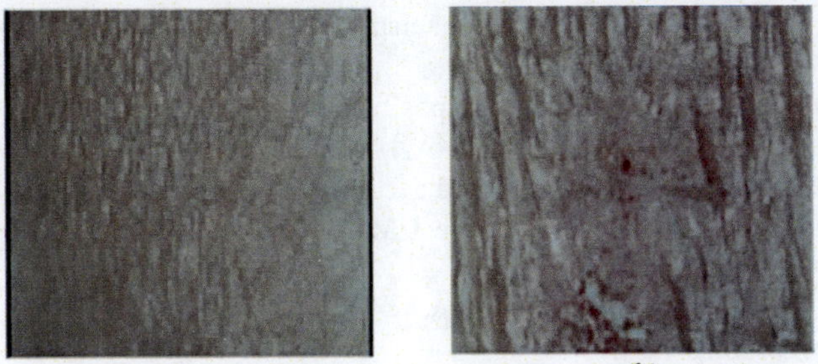

(a) (b)

Fig. 1. Visual images of human skin: a – norm; b – acute stage of atopic dermatitis.

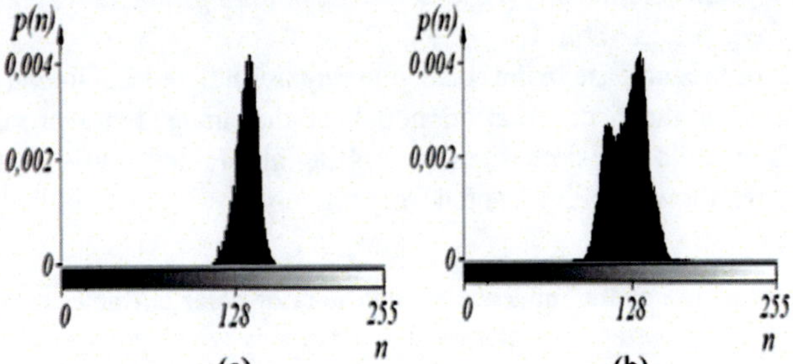

(a) (b)

Fig. 2. Historams of human skin images: a – norm; b – acute stage of atopic dermatitis.

error probability [12]. Quite a lot of studies have been devoted to the diagnosis of atopic dermatitis [13–17] and the processing of medical data [18–20].

Below, the authors assess the possibility of using criteria and models of parametric recognition (discrimination) when comparing the diagnostic capabilities of the image contrast indicators and the value of the second (after the global) of the histogram.

3 Experimental Estimation of Reducing the Error of Diagnostic Decision Making When Adding Additional Diagnostic Parameters

The authors carried out an experimental estimation of the reduction in the error of making a diagnostic decision when adding video parameters.

In this experiment, the five informative parameters X_i $(i = \overline{1, 5})$ were used in the calculations to determine the diagnostic significance of the parameters for the detection of atopic dermatitis at the acute stage. These informative parameters are displayed in ascending order of numbering:

X_1 - color tone of the skin area H (Hue) in the HSV system;

X_2 - color saturation of the skin area S (Saturation) in the HSV system;

X_3 - color intensity of the skin area V (Volume) in the HSV system.

X_4 - contrast of a fragment of a skin image, K;

X_5 - the value of the largest local (in relation to the global) maximum in percent, M %.

For simplicity, in the tables and graphs below, only the number of the informative parameter is indicated.

The above indicators were measured according to the data of digital dermatoscopy using a hand-held digital dermatoscope UM-039 (camera with a 5-megapixel video sensor) at twenty-fold magnification [6].

Patients were divided into two groups - with acute atopic dermatitis and conditional norm (control group).

In the analysis, the authors used the linear discrimination model described by them in [8, 12].

The results of discriminant analysis when estimation the influence of the main visual (color and histogram (in intensity)) characteristics of images of skin areas of patients in these groups are given in Table 1.

The graphs of the increase in the corresponding normalized Euclidean distance when adding parameters to the discrimination model and the change in the probability of error in making a diagnostic decision when controlling atopic dermatitis in the acute stage with the norm are shown in Fig. 3 and 4, respectively.

Table 1. The estimation of the influence of indicators of color characteristics and histogram features of skin areas in normal conditions and in the acute stage of atopic dermatitis

Method type	Conditional norm		Acute stage of atopic dermatitis		Normalized Euclidean distance with accumulation
Parameter	Indicator				
	m	σ_m	m	σ_m	δ_Σ
1. H, °C	18,00	10,52	355,00	7,26	2,17
2. S	50,14	8,43	40,11	6,95	2,47
3. V	98,52	9,30	93,49	8,93	2,54
4. K	47,00	12,28	70,00	18,72	2,84
5. M, %	10,00	6,44	47,00	16,24	2,98
$P_{\Sigma\ error}$, %	0,14				

It follows from these graphs that the greatest contribution to the increase in the normalized Euclidean distance and, accordingly, to the decrease in the probability of error is made by the indicators of changes in the color tone of the skin (reducing the probability of error to 0.27) and then changing the saturation indicator (reducing the probability of error to 0.2). Of the features determined from the image intensity histogram, the contrast index (in fact, the width of the image intensity histogram) plays a significant role, reducing the error probability by 0.04 (to 0.16) at once. The value of the fifth indicator - the largest local (relatively to the global) maximum in percentage terms - reduces the error probability to an additional 0.14.

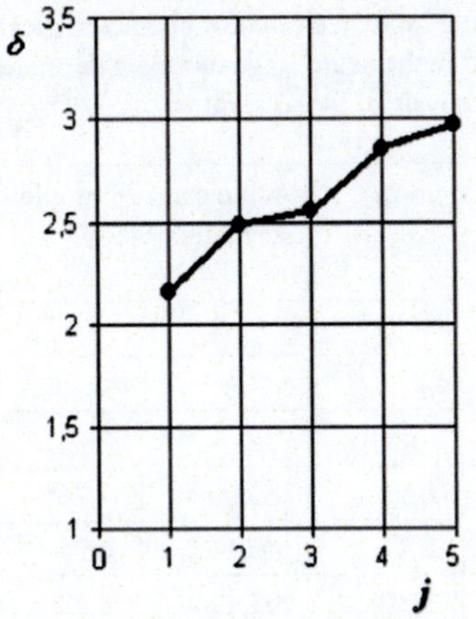

Fig. 3. Graph of the increase in the normalized Euclidean distance when adding parameters (indicators of color characteristics and histogram features) to the discrimination model δ = F (j) in the control of atopic dermatitis in the acute stage with the norm (j = 5 - dimension of the space of informative parameters).

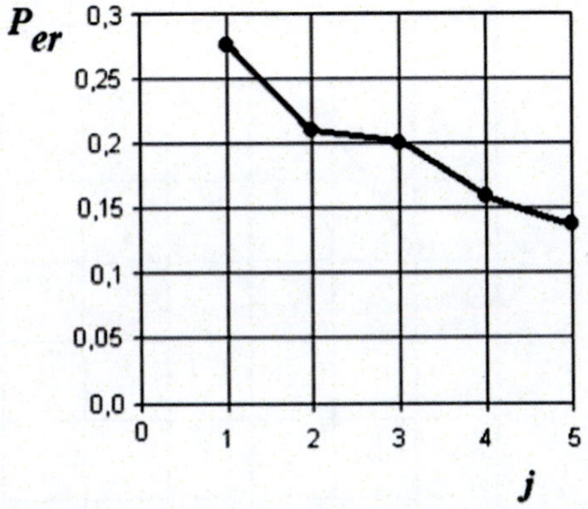

Fig. 4. Graph of the decrease in the error of making a diagnostic decision when adding parameters (indicators of color characteristics and histogram signs) to the discrimination model $P_{er} = f(j)$ when control atopic dermatitis in the acute stage with a norm ($j = 5$ – dimension of the space of informative parameters).

If to consider the indicators influence of visual estimation of dermatoscopic images to make the diagnostic decision in addition to immunological indicators (indicators of serum immunoglobulins and immunogram data (indicator A, Table 2, data taken from [8]) in the discrimination model (the results are shown in Table 2) then from the graphs in Figs. 5 and 6 it can be seen that the total immunological parameters reduce the probability of error to 0.2 [8]; color characteristics reduce the probability of an error in monitoring the state of atopic dermatitis (reducing the error probability to 0.08) and then the histogram features reduce the error probability to 0.05 in total.

Table 2. Estimation of the total influence of color characteristics and histogram features of skin areas in normal conditions and in the acute stage of atopic dermatitis when they added to the total indicator obtained from the analysis of blood serum

Method type	Conditional norm		Acute stage of atopic dermatitis		Normalized Euclidean distance with accumulation
Parameter	Indicator				
	m	σ_m	m	σ_m	δ_Σ
1. $\sum$A					2,60
3. H °C	18,00	10,52	355,00	7,26°	3,38
4. S	50,14	8,43	40,11	6,95	3,59
4. V	98,52	9,30	93,49	8,93	3,63
5. K	47,00	12,28	70,00	18,72	3,84
6. M, %	10,00	6,44	47,00	16,24	3,96
$P_{\Sigma error}$	0,05				

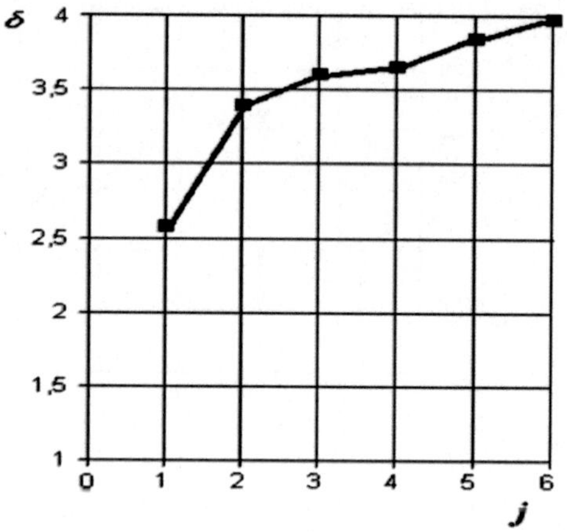

Fig. 5. Graph of the increase in the normalized Euclidean distance when adding parameters to the data of digital dermatoscopy of the skin (color characteristics in the HSV system and indicators of the intensity histogram) in the discrimination model $\delta = F(j)$ in the control of atopic dermatitis in the acute stage with the norm ($j = 6$ - dimension of the space of informative parameters, the first parameter takes into account the integral contribution of immunological data).

Diagnosis of atopic dermatitis only by visual characteristics of the skin (digital dermatoscopy data), color (in the HSV system), and histogram indicators allow monitoring this pathology with an error probability up to 0.14. This allows using the digital dermatoscopy method independently for express diagnostics of atopic dermatitis, without waiting for the data of immunological studies on blood serum.

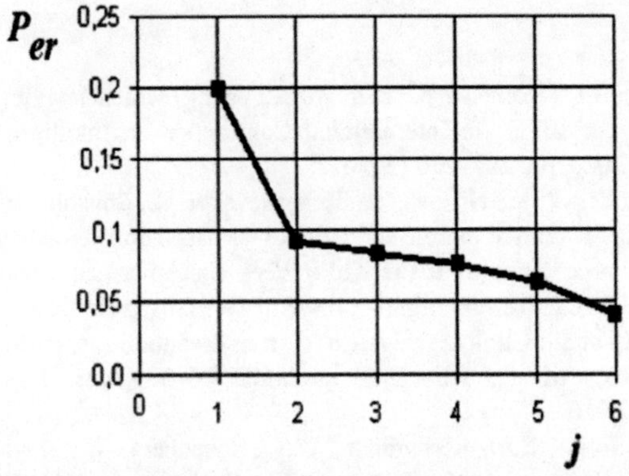

Fig. 6. Diagram of the decrease in the error of making a diagnostic decision when adding parameters to the data of digital dermatoscopy of the skin (color characteristics in the HSV system and indicators of the intensity histogram) in the discrimination model δ = F (j) in the control of atopic dermatitis in the acute stage with the norm (j = 6 - dimension of the space of informative parameters, the first parameter takes into account the integral contribution of immunological data).

Of the features determined from the image intensity histogram, the contrast index (in fact, it is the width of the image intensity histogram) plays a significant role, reducing the error probability by 0.04 (to 0.16) at once. The value of the largest local (relatively to the global) maximum in percent reduces the error probability to 0.14 additionally. It should be borne in mind that, despite the fact that such relative histogram characteristics as contrast and the presence of extrema are quite invariant to lighting conditions, they are significantly affected by the choice of the area of the analyzed skin area, especially the presence and magnitude of local maxima of the histogram.

4 Conclusions

Thus, the use of indicators invariant to the conditions of video shooting for the analysis of digital video information, such as the relative values of histogram features - image contrast and the magnitude of the second (local) maximum of the image intensity histogram, together with the linear discrimination model, made it possible to significantly improve the quality of analysis of transmitted telemedicine information. without a significant increase in its volume.

The authors have shown the effectiveness of this approach on the example of the analysis of telemedicine video information when monitoring the condition of the patient's skin for the diagnosis of atopic dermatitis. However, it is obvious that this method can find application in the analysis of video images of any diagnostic video information, which will significantly improve the quality of telemedicine services.

Conflict of Interest. The authors declare that they have no conflict of interest.

References

1. Sokol, Y., Avrunin, O., Kolisnyk, K., Zamiatin, P.: Using medical imaging in disaster medicine. In: Proceedings of IEEE 4th International Conference on Intelligent Energy and Power Systems, IEPS 2020, pp. 287–290 (2020)
2. Avrunin, O., Kolisnyk, K., Nosova, Y., Tomashevskyi, R., Shushliapina, N.: Improving the methods for visualization of middle ear pathologies based on telemedicine services in remote treatment. In: Proceedings of IEEE KhPI Week on Advanced Technology, KhPI Week, pp. 347–350 (2020). https://doi.org/10.1109/KhPIWeek51551.2020.9250090
3. Palazón-Bru, A., et al.: Clinical relevance of an intervention assessed by a meta-analysis of randomized clinical trials. J. Clin. Epidemiol. **132**, 46–50 (2021). https://doi.org/10.1016/j.jclinepi.2020.12.010
4. Sackett, D.L., Straus, S.E., Richardson, W.S., Rosenberg, W., Haynes, R.B.: Evidence-Based Medicine: How to Practice and Teach EBM. 2nd edn., p. 261. Churchill Livingstone, Edinburgh (2000)
5. Guyatt, G., Rennie, D., Meade, M.O., Cook, D.J.: A Manual for Evidence-Based Clinical Practice. 3rd edn., p. 697. McGraw-Hill Education, New York (2015)
6. Avrunin, O., Kolisnyk, K., Nosova, Y., Tomashevskyi, R., Shushliapina, N.: Reducing the risks of medical diagnosis in an epidemic or pandemic. In: Proceedings of IEEE KhPI Week on Advanced Technology, KhPI Week, pp. 351–356 (2020)
7. Kolisnyk, K., Sokol, Y., Avrunin, O., Nosova, Y.: Improving the quality of telemedicine diagnostic imaging in otolaryngology. In: Proceedings of IEEE International Scientific and Practical Conference "Problems of Infocommunications. Science and Technology" (PIC S&T'2020), Kharkiv, Ukraine, pp. 112–116 (2020). https://doi.org/10.1109/PICST51311.2020.9467974.
8. Kolisnyk, K., Sokol, Y., Avrunin, O., Kolisnyk, V., Klymenko, V.: Discriminant analysis to improve telemedicine diagnostics quality. In: Proceedings of IEEE 19th International Conference on Smart Technologies EUROCON, Ukraine Lviv (2021). In press
9. Avrunin, O., Trubitcin, A., Isaeva, O., Klymenko, V.: Possibilities for assessing the effectiveness of treatment of atopic dermatitis based on analysis of color characteristics of video dermatoscopic images. Eng. Ind. Technol. **2**(12), 127–133 (2020)
10. Avrunin, O., Tymkovych, M., Moskovko, S., Romanyuk, S., Kotyra, A., Smailova, S.: Using a priori data for segmentation anatomical structures of the brain. Przeglad Elektrotech. **93**(5), 102–105 (2017). https://doi.org/10.15199/48.2017.05.20
11. Avrunin, O., Tymkovych, M., Drauil, J.: Automatized technique for three-dimensional reconstruction of cranial implant based on symmetry. In: 2015 Proceedings of ITIB Information Technologies in Innovation Business Conference, pp. 39–42 (2015). https://doi.org/10.1109/ITIB.2015.7355070
12. Avrunin, O., et al.: Possibilities of automated diagnostics of odontogenic sinusitis according to the computer tomography data. Sensors **21**, 1198 (2021). https://doi.org/10.3390/s21041198
13. Hsiao, Y.-P., Wang, H.-C., Chen, S.-H., Tsai, C.-H., Yang, J.-H.: Identified early stage mycosis fungoides from psoriasis and atopic dermatitis using non-invasive color contrast enhancement by LEDs lighting. Opt. Quant. Electron. **47**(7), 1599–1611 (2014). https://doi.org/10.1007/s11082-014-0017-x
14. Yang, G., Seok, J., Kang, H., Cho, Y., Lee, H., Lee, J.: Skin barrier abnormalities and immune dysfunction in atopic dermatitis. Int. J. Mol. Sci. **21**(8) (2020).https://doi.org/10.3390/ijms21082867
15. Tamagawa-Mineoka, R., Katoh, N.: Atopic dermatitis: identification and management of complicating factors. Int. J. Mol. Sci. **21**(8) (2020). https://doi.org/10.3390/ijms21082671

16. Samsoën, M.: Digital "ectoscopy" and daily global dermatology: added values and limits. Nouvelles Dermatol. **24**(2), 60–66 (2005)
17. Menzies, S.W., Zalaudek, I.: Why perform dermoscopy? The evidence for its role in the routine management of pigmented skin lesions. Arch. Dermatol. **142**(9), 1211–1212 (2006). https://doi.org/10.1001/archderm.142.9.1211
18. Menzies, S.W.: Evidence-based dermoscopy. Dermatolog. Clin. **31**(4), 521–524 (2013). https://doi.org/10.1016/j.det.2013.06.002
19. Bodyanskiy, Y.V., Tyshchenko, O.K., Kopaliani, D.S.: Adaptive learning of an evolving cascade neo-fuzzy system in data stream mining tasks. Evol. Syst. **7**(2), 107–116 (2016). https://doi.org/10.1007/s12530-016-9149-5
20. Rashkevych, Y., Peleshko, D., Vynokurova, O., Izonin, I., Lotoshynska, N.: Single-frame image super-resolution based on singular square matrix operator. In: Proceedings of IEEE 1st Ukraine Conference on Electrical and Computer Engineering, UKRCON 2017, pp. 944–948 (2017). https://doi.org/10.1109/UKRCON.2017.8100390

Python Implementation for Brain-Computer Interface Research by Acquiring and Processing the NeuroSky EEG Data for Classifying Multiple Voluntary Eye-Blinks

Oana-Andreea Rușanu(✉)

Product Design, Mechatronics and Environment Department, Transilvania University of Brasov, Brașov, Romania
oana.rusanu@unitbv.ro

Abstract. The Brain-Computer Interface (BCI) is a challenging research field reporting outstanding breakthroughs in biomedical engineering. This paper proposes a new BCI research-related solution by implementing customized Python scripts based on an artificial neural networks model to classify the raw electroencephalographic (EEG) signal detected by the embedded biosensor of NeuroSky portable headset. Achieving this aim is possible by applying features extraction techniques on the raw EEG data to generate the training dataset composed of 3000 recordings corresponding to executing simple, double, or triple voluntary eye-blinks. Detection of their specific EEG patterns resulted in calculating the following seven statistical features: mean, median, standard deviation, route mean square, the sum of values, Kurtosis Coefficient, and skewness. The voluntary eye-blinking proved to be the most precise and easily detected control signal in a BCI application to assist people with neuromotor disabilities. The proposed Python implementation of BCI software is practical, especially for the initial stages of research, by leveraging simple to use, inexpensive, and efficient instruments.

Keywords: BCI · Eye-blink · EEG · Python · ANN

1 Introduction

The Brain-Computer Interface (BCI) research field has considerably evolved during the last decades by proving efficient means of controlling the assistive devices, communication, sleep, and stress monitoring, especially for people with neuromotor disabilities. Their suffering is related to paralysis, spinal cord injuries, cerebral stroke, amyotrophic lateral sclerosis, or locked-in syndrome. These disorders determined the interruption of the neuronal pathway connecting the brain motor areas with the peripheral nerves and muscles. BCI is a promising solution for replacing the natural cerebral path with an artificial route allowing the translation of the thought into commands transmitted to an assistive robotic device. This desire is currently unachievable outside an experimental laboratory due to significantly higher requirements, including the complexity of the real-time neuronal biopotentials processing, the ability of the neuromotor disabled person to

I. Tiginyanu et al. (Eds.): ICNBME 2021, IFMBE Proceedings 87, pp. 666–672, 2022.
https://doi.org/10.1007/978-3-030-92328-0_84

perform BCI related cognitive tasks (motor imagery, concentration, mental calculus) to elicit recognizable patterns and the artificial intelligence techniques necessary to convert the intentions into action.

Therefore, BCI prototypes enabling affordable software and hardware solutions to improve the research results are still welcome and expected to overcome the existent issues by providing maximum accuracy, quick response, and rapid information transfer rate.

According to scientific literature [1, 2], implementing the most straightforward BCI system involves controlling an external device by detecting the voluntary eye-blinking across the electroencephalographic (EEG) signal. Moreover, the easiest way to acquire the raw EEG signal for developing versatile BCI applications is to use a portable headset such as NeuroSky Mindwave Mobile. The majority of the previous scientific articles [3] focused on calling on-the-shelf NeuroSky libraries providing convenient programming methods for achieving simple BCI applications with the expense of technical limitations for accomplishing advanced BCI research instruments. Therefore, several thresholding-based algorithms determined the measurement of the attention and meditation level and the eye-blink strength used as commands in a BCI for controlling: a wheelchair, a mobile robot, a robotic arm, a robotic hand, home appliances, and virtual simulations. Few papers [4] explored the NeuroSky EEG data acquisition and processing by applying different methods based on statistics, wavelet transform, supported vector machines, or artificial neural networks. In addition, the investigation of the mentioned methods targeted a specific BCI application not extendable to a general framework for conducting BCI research by enabling the fundamental phases: EEG data acquisition, processing, and classification.

Thus, it results in the main contribution of this paper by providing simple BCI research automated solutions implemented in Python open-source programming language to enable the NeuroSky EEG data processing, features extraction, training dataset generation, and artificial neural networks (ANN) based classification. Performing these stages resulted in performance assessment of the ANN accuracy for simple, double, and triple voluntary eye-blinks detection. Also, it resulted in an EEG dataset comprising 3000 sequences distributed as follows: 1000 sequences – one eye-blink detected, 1000 sequences – two eye-blinks detected, and 1000 sequences – three eye-blinks detected. Each sequence corresponds to the acquisition of 1024 raw EEG samples and the following seven extracted statistical features: mean, median, standard deviation, the sum of values, skewness, Kurtosis Coefficient, and route mean square (RMS). Unfortunately, the scientific literature reported very few EEG datasets [4] to test the algorithms of voluntary eye-blinking detection. The novelty of this paper resulted from customized Python scripts for enabling the acquisition, analysis, and classification of the raw EEG signal detected by the NeuroSky embedded biosensor. Further, the structure of the paper is: Section II presents an overview of the software and hardware systems used to implement a new BCI research solution, Section III explains the working principle and code instructions underlying the proposed instrument, Section IV discusses the achieved results and Section V summaries and concludes the paper.

2 Hardware and Software

2.1 Hardware System – NeuroSky Mindwave Mobile Headset

The BCI application developed in the current research involved raw EEG signal acquisition from the biosensor of the NeuroSky portable headset. The NeuroSky headset has an embedded sensor located to the prefrontal lobe, on the forehead, at the FP1 position according to the International 10–20 EEG System. Embedding the advanced ThinkGear based, the NeuroSky facilitates the initial stages of BCI research based on EEG data acquisition with a sampling rate of 512. Among the advantages provided by NeuroSky are the simple set-up, the ergonomic design, convenient working principle, and especially, free of charge raw EEG signal acquisition and recording to.csv files for further analysis.

2.2 Software System – Python Implementation Based on Raw EEG Data Acquisition, Processing, and ANN Classification

The current research contributes to the BCI field by implementing customized Python scripts to achieve the following phases (Fig. 1): cerebral biopotentials acquisition, raw EEG graphical displaying, statistical features extraction, EEG dataset generation, and classification of voluntary eye-blinks. There were necessary specific functions included by the following Python libraries: Numpy – to store data in arrays, Matplotlib – to create visualizations, SciPy – to calculate the statistical features, and Winsound – to get a sound.

3 Python-Based Customized Scripts for Enabling Multiple Voluntary Eye-Blinks Classification

3.1 Raw EEG Signal Acquisition and EEG Training Dataset Generation

NeuroPy library [5] was updated to run in Python 3. The rawValue variable stored each value of the raw EEG signal detected by the embedded NeuroSky biosensor. As a drawback noticed to the NeuroPy library, it is not easy to get precision regarding waiting time set to 2 ms between the raw EEG data samples. A healthy subject (girl, 29 years old) participated in the experiment involving 75 sessions of performing one, two, or three voluntary eye-blinks. Each session included 40 sequences representing 40 simple, double or triple voluntary eye-blinks. A recurrent time interval set to 2 s, introduced by a beep sound, determined a sequence of 1024 raw EEG samples.

It resulted in a.csv file comprising 40 sequences with 1024 samples, each organized in 40 rows and 1024 columns representing the raw EEG data. Therefore, the previously mentioned 75 sessions led to obtaining 75.csv files, including the $40 \times 1024 = 40960$ raw samples for each class: one-eye blink, two eye-blinks, and three eye-blinks. Thus, there were $3 \times 25 = 75$.csv files of raw EEG data and 75 graphical representations of the numerical content (40960 EEG raw samples).

It followed the extraction of seven statistical features from the raw EEG data containing 1024 samples and the generation of the EEG dataset necessary to classify the multiple voluntary eye-blinks. The extracted statistical features were: mean, median,

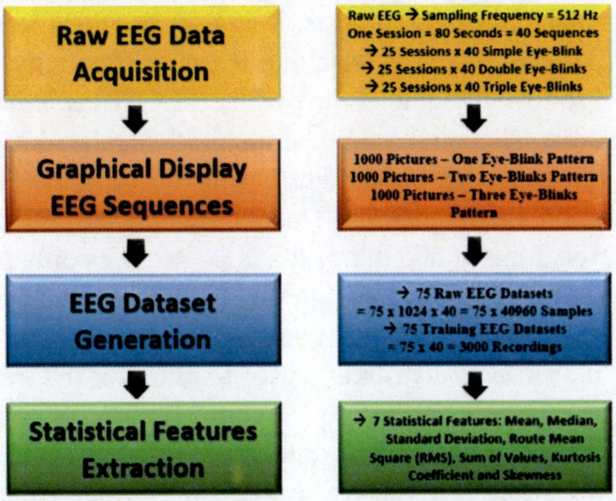

Fig. 1. Python implementation of the main BCI related research stages

standard deviation, route mean square (RMS), the sum of values, skewness, and Kurtosis Coefficient. These seven selected features had to discriminate the executed eye-blinks initially set by the column representing the desired class: one eye-blink detected, two eye-blinks observed, and three eye-blinks recognized.

Thus, a single EEG dataset consisted of 8 columns (7 – features and 1 – class) and 40 rows (40 sequences of 2 s each for recording 1024 EEG samples). 75.csv files represented the 75 EEG datasets to classify the multiple voluntary eye-blinks (Fig. 2): 25 – simple eye-blinks, 25 – double eye-blink, and 25 – triple eye-blink.

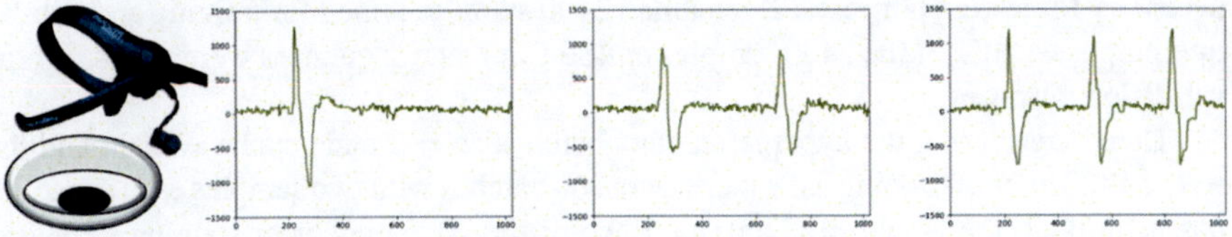

Fig. 2. Graphical displaying a sample from each class: simple (one), double (two) and triple (three) voluntary eye-blinks detected by NeuroSky Sensor

Concatenating all the 75 EEG datasets resulted in a single EEG dataset comprising $75 \times 40 = 3000$ sequences of eye-blinks. Then, the resulted.csv file consisted of 3000 rows and eight columns necessary to train an artificial neural network to classify the multiple voluntary eye-blinks.

3.2 Artificial Neural Networks Based Classification of Multiple Voluntary Eye-Blinks

Another contribution of the current research is implementing a customized Python script to employ the artificial neural network (Fig. 3) for multiple voluntary eye-blinks classification. The following Python libraries included the necessary specific functions: Numpy,

Matplotlib, Pandas – to handle data analysis, Keras – to use the Tensorflow deep learning framework, and Scikit– for enabling Machine learning functions. After importing the essential libraries, it followed the uploading of the EEG dataset, including all the 3000 sequences and the seven calculated statistical features correlated with executing simple, double, and triple voluntary eye-blinks. Further, it followed the EEG dataset splitting into a training subset and a testing subset. Thus, the testing subset constituted 20% of the entire EEG dataset, resulting in 600 out of the 3000 samples only for testing purposes.

An essential stage consisted in creating the artificial neural network composed of four layers. The structure of the first and second hidden layers was the following: number of neurons = 1400, the uniform distribution used to initialize the weights, and activation function = ReLU. The input layer had seven neurons corresponding to the previously mentioned seven statistical features. The output layer had three neurons corresponding to the three classes regarding recognizing simple, double, and triple multiple voluntary eye-blinks.

Compiling the artificial neural network involved setting the following parameters optimizer = SGD (Stochastic Gradient Descent), loss (function) = categorical cross-entropy, and metrics = accuracy. The SGD convex function is used as Optimizer to determine the suitable set of weights by identifying a local minimum of the input function. Thus, it is necessary to set the learning rate to an appropriate value: LR = 0.00001. A momentum = 0.99 was also set to increase the speed of the optimization process.

The following critical stage consisted in fitting the compiled ANN to the generated EEG dataset. The ANN was trained on 1920 samples and validated on 480 samples. A sample is each of the 40 sequences/rows representing the measured values of the seven statistical features from each of the 75 EEG datasets. As mentioned previously, 20% representing 600 samples of the entire EEG dataset containing 3000 samples were necessary for testing purposes. It remained 2400 samples aimed for training and validation purposes. 20% of the 2400 samples or 480 EEG data sequences were necessary for validation purposes.

Then, considering the hyperparameters batch_size = 2 and epochs = 2000 and the rest of 1920 rows of training data, there were 960 batches with two samples each and 2000 passes through the whole EEG dataset. Each of the 2000 epochs or training iterations will involve 960 batches or 960 updates to the ANN-based model. After working through each batch of 2 samples, the internal ANN model parameters are updating. The learning algorithm works 2000 times through the entire EEG dataset.

Finally, the ANN model made predictions on the testing data composed of 600 samples. Also, there resulted in loss and accuracy specific to the testing process necessary to evaluate the performance of the ANN model. Also, the confusion matrix showed detailed results regarding the correctly and incorrectly detected samples from each of the three output classes: simple, double, and triple voluntary eye-blinks.

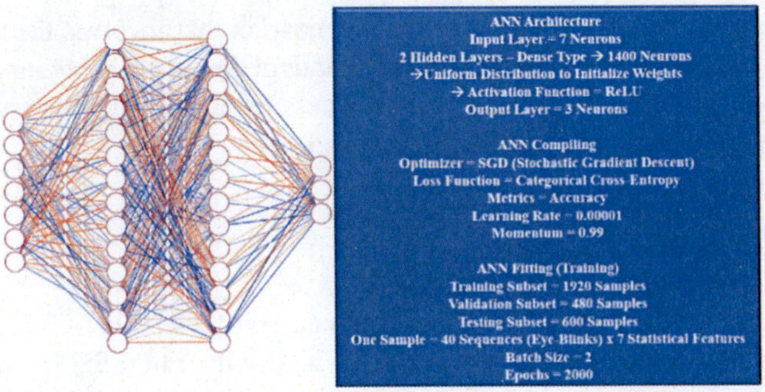

Fig. 3. Python artificial neural network for simple, double and triple voluntary Eye-Blinks classification

4 Results and Discussions

The proposed artificial neural networks-based architecture aimed for multiple voluntary eye-blinks classification reported the results in Table 1.

Table 1. Results for training, validating, and testing the Python ANN

EEG Dataset	Number of Eye-Blinks Sequences	Loss (%)	Accuracy (%)	Correctly Detected Eye-Blinks Sequences
Training	1920	0.56	100.00	1920
Validation	480	33.27	92.92	446
Testing	600	31.21	92.33	554

A video demonstration of the Python-based implementation for voluntary eye-blinks classification is available at the YouTube Unlisted Links: https://youtu.be/tvj8UcGbWa8 and https://youtu.be/MeMMj2JNUfk. Although the obtained results reported high values of accuracy for training, validation, and testing the compiled Python-based ANN, further experiments are necessary to analyze possible particularities implied by a large number of individuals taking into account their age, gender, stress level, degree of disability or the ability to focus on accomplishing the required task – to execute simple, double or triple eye-blinks. The customized ANN-based architecture aims to classify eye-blinks characterized by average amplitude. The ANN may need improvements to differentiate between voluntary eye-blinks of various strengths depending on the effort of the eyeball muscles.

Otherwise, the presented Python automated scripts aim to offer a general-purpose BCI research instrument as long as the conducted experiments involve the following three fundamental stages: raw EEG data acquisition, processing enable by features extraction, and ANN-based classification. Still, the proposed Python implementation aims to enable the ThinkGear embedded chip of the most affordable and portable EEG commercial headset – NeuroSky Mindwave Mobile. In addition, the presented Python software tool proves its usefulness for any application that involves the assessment of the seven statistical features: mean, median, standard deviation, route mean square, the sum of values, Kurtosis Coefficient, and skewness extracted from the raw EEG signal detected by the biosensor placed on the forehead.

5 Conclusions

This paper presented a Python-based implementation of a simple BCI-related research instrument necessary to acquire, process, and classify the raw EEG signal detected by the embedded sensor of the NeuroSky headset. A customized ANN-based architecture classified the multiple voluntary eye-blinks used as control signals in a straightforward brain-computer interface application. It resulted in the generation of a training dataset containing 3000 recordings evenly distributed for detecting simple, double, and triple voluntary eye-blinks. It also involved extracting the following statistical features from the raw EEG signal: mean, median, standard deviation, route mean square, the sum of values, Kurtosis Coefficient, and skewness. The proposed Python application provides simplicity and efficiency to help researchers explore and experiment with the working principle underlying the BCI scientific field.

Future research should update the ANN-based framework to classify voluntary eye-blinks of various strengths: mild, regular, or firm. Also, the Python-based ANN should differentiate between spontaneous, reflexive, and voluntary eye-blinks. Moreover, the BCI instrument requires improvements to detect winks precisely. Extracting additional statistical features will extend the universality of the experiments conducted with the proposed BCI software solution. The ultimate goal is to achieve a real-time running of the Python-based voluntary eye-blinks classification.

Conflict of Interest. The authors declare that they have no conflict of interest.

References

1. Rejer, I., Cieszyński, L.: RVEB—an algorithm for recognizing voluntary eye blinks based on the signal recorded from prefrontal EEG channels. Biomed. Signal Process. Control **59**, 101876 (2020). https://doi.org/10.1016/j.bspc.2020.101876, ISSN 1746-8094
2. Sharma, K., Jain, N., Pal, P.K.: Detection of eye closing/opening from EOG and its application in robotic arm control. Biocybern. Biomed. Eng. **40**(1), 173–186 (2020). https://doi.org/10.1016/j.bbe.2019.10.004, ISSN 0208-5216
3. Prem, S., Wilson, J., Varghese, S.M., Pradeep, M.: BCI integrated wheelchair controlled via eye blinks and brain waves. In: Pawar, P.M., Balasubramaniam, R., Ronge, B.P., Salunkhe, S.B., Vibhute, A.S., Melinamath, B. (eds.) Techno-Societal 2020, pp. 321–331. Springer, Cham (2021). https://doi.org/10.1007/978-3-030-69921-5_32
4. Lo Giudice, M., et al.: 1D Convolutional Neural Network approach to classify voluntary eye blinks in EEG signals for BCI applications. In: 2020 International Joint Conference on Neural Networks (IJCNN), pp. 1–7 (2020). https://doi.org/10.1109/IJCNN48605.2020.9207195
5. Neuropy Library. https://github.com/lihas/NeuroPy.git. Accessed 14 Aug 2021

Low-Cost Telemedicine Platform for Monitoring Patients Suspected of Being Infected with SAR-COV-2

Corneliu Nicolae Druga(✉), I. C. Rosca, I. Serban, and I. Tatulea

Product Design, Mechatronics and Environment Department, Transilvania University of Brasov, Brasov, Romania
druga@unitbv.ro

Abstract. This paper presents the development of a low-cost platform in the field of telemedicine, which can be used to monitor people suspected of being infected with SAR-COV-2. The platform was developed and tested at the Medical Engineering Laboratory of the Transilvania University of Brasov. The first part of the paper is dedicated to the importance of the research topic and the concept of telemedicine. The second part presents the main hardware elements that are used to make this platform. The last part presents the results and conclusions related to the operation of this platform.

Keywords: Arduino · Wi-Fi · Telemedicine · Sensors

1 Introduction

Due to the accelerated degree of evolution of digital technologies, telemedicine has become an area of major importance. The context generated by the SAR-COV-2 (COVID-19) pandemic, especially the obligation to adopt social distancing measures but also the need to respond quickly to medical emergencies of the population, determined the specialists to develop and implement telemedicine systems in medical institutions [1, 2]. Despite the potential of telemedicine, the multiple medical and management benefits, in Romania, its use was still very low. Thus, in the fall of 2020, the general legal framework applicable to telemedicine in Romania was defined, by Emergency Ordinance no. 196/2020 for the amendment and completion of Law no. 95/2006 on health care reform [2]. In view of this regulation, medical and health institutions, both public and private, are allowed to provide medical care services through telemedicine. Also, a positive impact on the development of contemporary telemedicine is imprinted by the development in the field of mobile health [3].

2 Design and Implementation of the Telemedicine Platform

2.1 Description of Hardware Elements

Complete and complex monitoring systems (in the field of Telemedicine) allow continuous observation of the patient's health. This integrated telemedicine device (Fig. 1)

I. Tiginyanu et al. (Eds.): ICNBME 2021, IFMBE Proceedings 87, pp. 673–680, 2022.
https://doi.org/10.1007/978-3-030-92328-0_85

presented in this paper, is a low-cost and effective solution, used for remote and real-time monitoring of vital parameters (temperature, pulse, ECG, respiratory capacity). This platform was built around the Arduino Mega 2560 development board, to which an ECG module, a pulse sensor, a temperature sensor, a differential pressure sensor and a Wi-Fi module were attached. This low-cost telemedicine platform contains the following components:

1. Arduino Mega development board;
2. Arduino Wi-Fi module;
3. IR module MLX90614;
4. XD-58C heart rate sensor;
5. Click ECG module;
6. MPX5010DP differential pressure sensor.

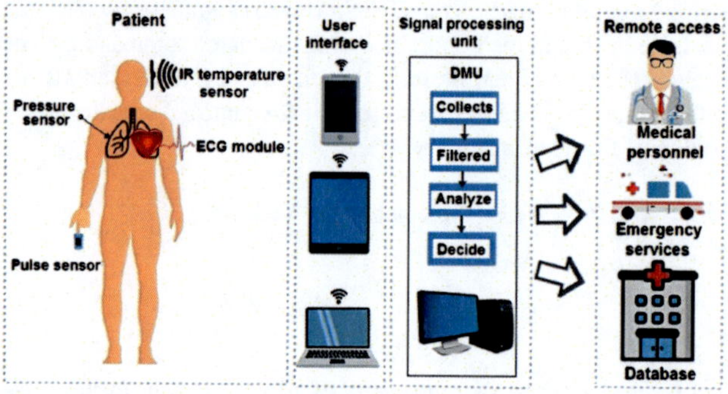

Fig. 1. Concept of low-cost telemedicine platform for monitoring patients suspected of being infected with SAR-COV-2

a) Arduino Mega 2560 development board:

The Arduino Mega 2560 development board is based on the ATmega2560 microcontroller. It has 54 digital input/output pins (of which 15 can be used as PWM outputs), 16 analog inputs, 4 asynchronous UART serial ports, a 16 MHz crystal oscillator, a USB connection, a power plug, a ICSP pin set and a reset button [4, 5].

b) Arduino Wi-Fi module:

The Arduino Wi-Fi module (Fig. 2) allows the Arduino development board to make the wireless connection to the internet network. Programs are written in the Arduino IDE programming environment. They can only run by including the Wi-Fi library. The module also offers the possibility to store data via a micro-SD card. The card reader is integrated on the board and is accessible by including the SD library [4, 5].

The wireless connection is made through the 802.11b and/802.11g communication protocols [6]. For wireless communication, use the HDG204 processor according to the 802.11b/g protocol. The Arduino. board communicates with both the HDG204 processor and the SD card, using the SPI interface on pins 10, 11, 12 (for Arduino Uno) and on pins 50, 51, 52 (for Arduino Mega) [4, 5].

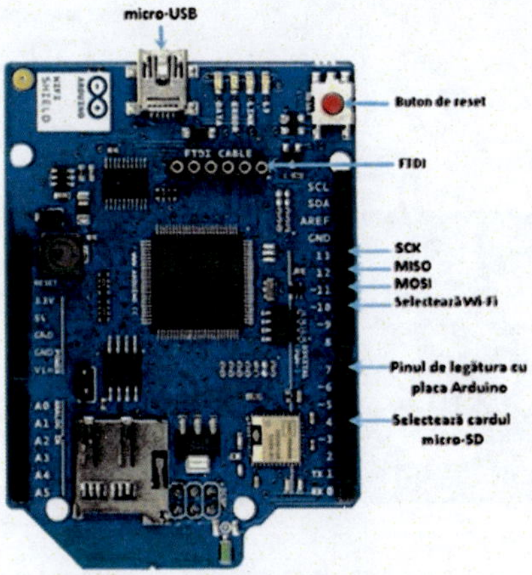

Fig. 2. Arduino Wi-Fi module

c) Description of the MLX90614 IR module:
IR thermometers measure the surface temperature of an object. The device's sensor measures the emitted, reflected and transmitted thermal radiation of an object, and presents this information in the form of a temperature value [7]. The MLX90614 module (Fig. 3) is an IR thermometer, used to perform non-contact temperature measurements. The MLX90614 module is built of 2 chips, developed and manufactured by the company Melexis, as follows:

- MLX81101 chip - IR thermopile detector;
- ASSP MLX90302 chip - used for signal conditioning, specially designed to process the IR sensor output. The main technical specifications of the MLX90614 module are as follows:
- Communication protocol: I2C;
- Measurement accuracy: 0.5 °C;
- Measurement resolution: 0.02 °C;
- Output resolution: 0.14 °C;
- Frequency: 10–100 kHz;
- Supply voltage: 3–5 V [7].

The MLX90614 module is factory calibrated in the following temperature ranges:

- −40–125 °C, for measuring the ambient temperature;
- −70–382.2 °C, for measuring the temperature of an object [7, 9].

The electrical circuit was simulated and tested using Fritzing software and is illustrated in Fig. 3.
d) Description of XD-58C heart rate sensor:
The XD-58C pulse sensor (Fig. 4) is essentially a photoplethysmograph, used for non-invasive heart rate monitoring [10]. Heart rate (or pulse) is the number of beats the heart

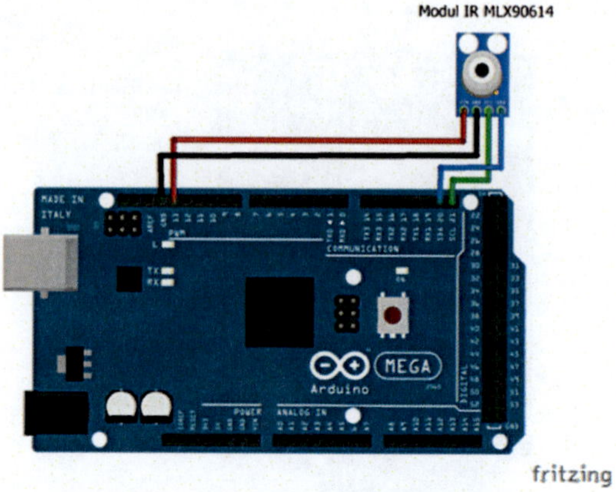

Fig. 3. The MLX90614 module [8]

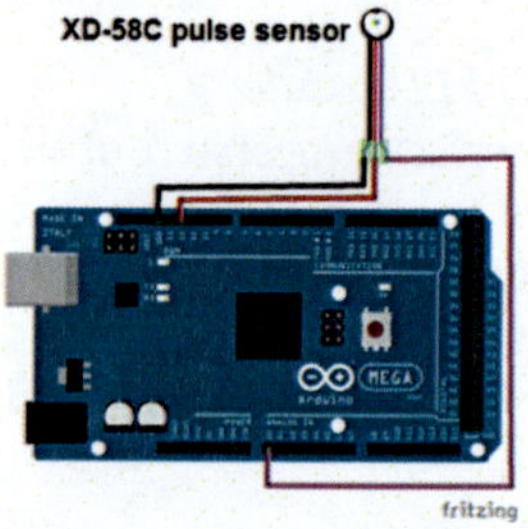

Fig. 4. Electrical circuit for connecting the XD-58C heart rate sensor to the Arduino Mega 2560 platform [10]

makes in a minute to ensure the pumping of blood to other organs.The heart rate signal coming out of a photopletismograph is an analog fluctuation in voltage (V) and has a predictable waveform. The graphical representation of the pulse wave is called the photoplethysogram (PPG).

e) Description of ECG Click module (by Mikroelektronika):

The ECK Click module (Fig. 5), produced by Mikroelektronika, measures the electrical activity of the heart by means of surface adhesive electrodes, like a medical device [11]. The main technical specifications of the Click ECG module are as follows:

- Built-in chips: MCP609 OpAmp, MAX6106;
- PID: MIKROE-2455;
- Supply voltage: 5 V;
- Compatible with Arduino [11].

An ECG is a modified galvanometer that records at high speed the variations of electric current generated by the heart, which it amplifies and transforms into mechanical movements.

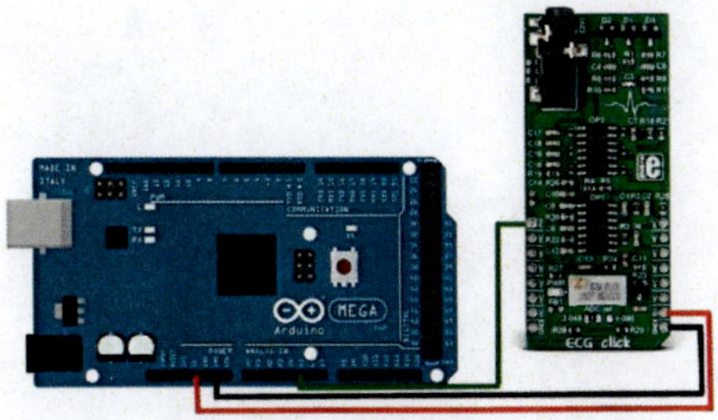

Fig. 5. Electrical circuit for connecting the ECG Click module (by Mikroelektronika) to the Arduino mega 2560 Board

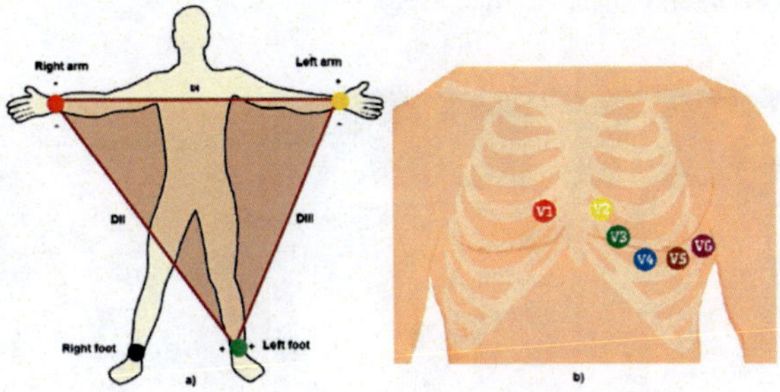

Fig. 6. Placement of ECG electrodes: a) Einthoven's triangle; and b) On the chest [12, 13]

The electric field produced by the heart can be recorded by placing electrodes in different points of the body: on the limbs (Fig. 6a) and on the thorax (Fig. 6b). From these points, the ECG electrodes "look" at the heart from different angles and "translate" the electrical activity of the heart into waves [12]. The Click ECG module has a jumper and a potentiometer to set the output voltage so that it matches the input voltage level of the ADC.

f) Description of the MPX5010DP Differential Pressure Sensor:

The MPX5010DP differential pressure sensor (Fig. 7), produced by NXP Freescale Semiconductors, measures the pressure difference between sources when the pressure is applied differentially (on both sides of the sensor) but also the atmospheric pressure [14, 15].

For the electronic assembly, the connections between the MPX5010DP differential pressure sensor and the Arduino Mega platform were made. The electronic assembly diagram was made using Fritzing software and is illustrated in Fig. 7.

In order to measure respiratory capacity, a mouthpiece was designed in the CATIA V5 software, special for the MPX425DAP differential pressure sensor (Fig. 8). The mouthpiece was obtained using 3D printing technology. For the testing of each sensor, but also of the ECG module, the corresponding libraries were loaded in the Arduino IDE.

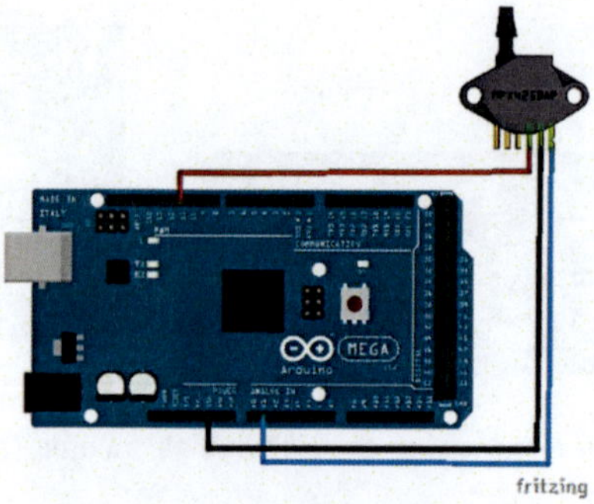

Fig. 7. Electrical circuit for connecting the MPX5010DP differential pressure sensor (by NXP Freescale Semiconductors) to the Arduino Mega

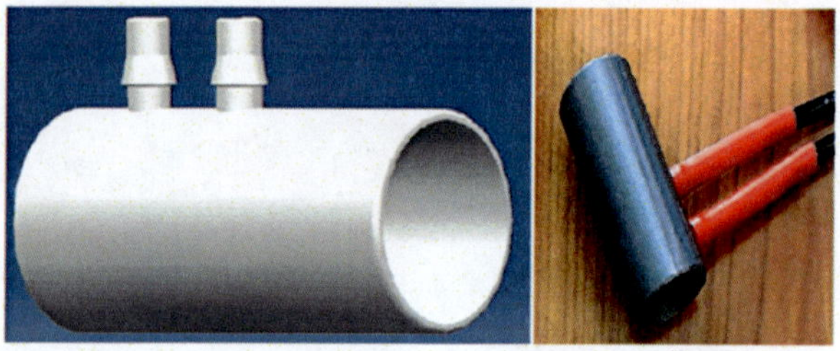

a) b)

Fig. 8. The mouthpiece for MPX5010DP differential pressure sensor: a) CAD model; and b) 3D printed part.

2.2 Realization of the Low-Cost Platform

For the hardware realization of the platform, in the first phase, the Arduino Wi-Fi module was connected to the Arduino Mega board. The pressure, temperature, pulse sensor and Click ECG module were connected to the Arduino Wi-fi module via a test board (Fig. 9).

Fig. 9. Front view with low-cost experimental telemedicine device

In the first phase, the electrical circuits for the temperature sensor, for the ECG module, for the pulse sensor and for the differential pressure sensor were tested separately.

For the testing of each sensor, but also of the ECG module, the corresponding libraries were loaded in the Arduino IDE. Only one code (program) was created from the source codes for each device.

2.3 Testing the Low-Cost Telemedicine Platform

To test the platform, the Arduino Wi-Fi module was connected to the local Wi-Fi network (from the Medical Engineering Laboratory). For this, the following steps have been performed:

- Identification of the wireless network within the Lab. of Medical Engineering - G P 8;
- Identification of the IP address - 192.168.1.103;
- Enter the IP address in a web browser (for example, Google Chrome). The data acquired through the sensors can be viewed by any remote user, by simply entering the IP address in the search bar of a web browser (Fig. 10).

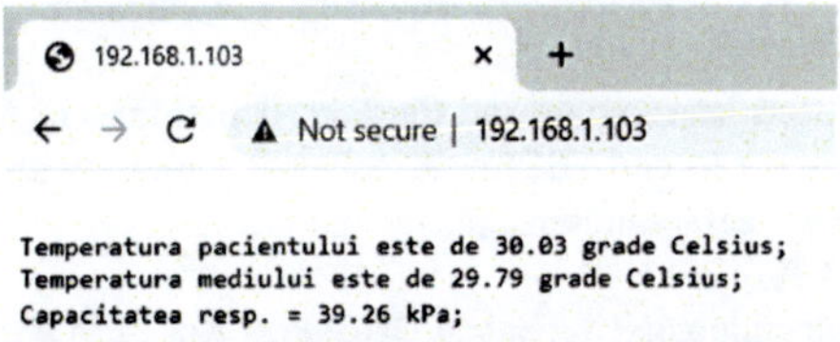

Fig. 10. The data acquired through the sensors

3 Conclusions

Although the constructive solution chosen in this project is not very complex, the integrated telemedicine device has a number of advantages, such as:

- It is a portable device;
- Data can be monitored in real time;
- Access to data can be done from anywhere, through a computer that connects to the Internet.

This platform can be improved by:

- Using the MAX30100 sensor to measure the heart rate but also the blood oxygen concentration at the same time (instead of the XD-58C sensor
- Development of a data storage system purchased by sensors, directly in the cloud.

Conflict of Interest. The authors declare that they have no conflict of interest.

References

1. Galiero, R., Pafundi, P.C., Nevola, R., et al.: The importance of telemedicine during COVID-19 pandemic: a focus on diabetic retinopathy. J. Diabetes Res. **2020**, Article ID 9036847 (2020). https://doi.org/10.1155/2020/9036847
2. Tarlescu, R., Teis, A., Visan, A.: Telemedicina (2020)
3. E-visit: What is telemedicine? Brasov (2021). https://evisit.com/resources/what-is-telemedicine/#1. Accessed June 2021
4. Arduino. Brasov, Romania. https://store.arduino.cc/. Accessed 02 July 2021
5. Robotshop: Arduino Mega 2560 Datasheet, Brasov, Romania. https://www.robotshop.com/media/files/PDF/ArduinoMega2560Datasheet.pdf. Accessed 03 July 2021
6. Zamfira, S.: Protocoale şi interfeţe de Comunicaţie, Comunicaţii fără fir. Universitatea Transilvania din Braşov (2020)
7. Melexis: Data sheet MLX90614 Family, 3901090614 Rev 1.0, 14 September 2006
8. Ardushop: MLX90614 IR Thermometer, Brasov, Romania. :https://ardushop.ro/en/home/896-mlx90614-ir-thermometer.html. Accessed 03 July .2021
9. Ammirato, F., Zayicek, P.: Infrared Thermography Field Application Guide. California, USA (1999)
10. World Famous Electronics, Pulse sensor (2016)
11. Mikroelektronika: ECG Click Moule, Brasov, Romania. https://www.mikroe.com/ecg-click. Accessed 04 July 2021
12. Clinical gate: The Electrocardiogram and Cardiac Arrhythmias in Adults, Iunie (2016)
13. Cable and sensors: 12-Lead ECG Placement Guide with Illustrations, Brasov, Romania. https://www.cablesandsensors.eu/pages/12-lead-ecg-placement-guide-with-illustrations. Accessed 04 July 2021
14. NXP Freescale Semiconductors: Technical Data sheet Rev., 13 October 2012
15. Robofun: MPX5010DP differential pressure sensor. Brasov, Romania. https://www.robofun.ro/senzori/senzor-de-presiune-diferential-mpx5010dp.html. Accessed 04 July 2021

Providing Remote Monitoring of CVD in Specialized Medical and Diagnostic Complexes

Y. Sokol[1], K. Kolisnyk[1](✉), S. Koval[2], and M. Penkova[2]

[1] National Technical University "KhPI", Kharkiv, Ukraine
Kostiantyn.Kolisnyk@khpi.edu.ua
[2] Government Agency "National Institute of Therapy named after L.T. Maloyi", Kharkiv, Ukraine

Abstract. The creation of specialized medical and diagnostic complexes for remote monitoring is currently identified as one of the promising areas of health care structure modernization. However, the efficiency of such complexes directly depends on the telemedicine techniques used, the socio-ethnic characteristics of their location and the general number of patients per unit of time. This paper is analyzed the techniques for remote monitoring of cardiovascular diseases which are necessary for the purpose of their optimal software and hardware support.

Keywords: Cardiovascular diseases · Telemedicine diagnostics · Medical consultative diagnostic complexes

1 Introduction

The relevance of remote monitoring of cardiovascular diseases (CVD) is currently due, first of all, to their widespread prevalence among the population, which is determined by certain changes in the socio-ecological aspect of his life, as well as sharp climatic changes on the planet. In addition, in the context of massive epidemics and pandemics, this also arises the need for distancing in the process of providing medical services and services.

According to the World Health Organization, annually up to 20 million people die from CVD complications in the world, of which about 9.4 million deaths per year are caused by an increase in blood pressure (BP) [1]. The relationship between blood pressure and cardiovascular mortality begins to be significant from a blood pressure level of 115/75 mm Hg - at the age of 40–69 years, an increase in systolic blood pressure by 20 mm Hg and diastolic blood pressure by 10 mm Hg leads to a twofold increase in mortality due to stroke and coronary heart disease [2]. BP in the range 120–139/80–89 mm Hg. is associated with the incidence of cardiovascular disease, ischemic heart disease, strokes and chronic kidney disease, but the greatest relationship is observed at a blood pressure of 130–139/80–89 mm Hg compared with optimal blood pressure.

Currently, the following techniques are used to diagnose CVD [2].

I. Tiginyanu et al. (Eds.): ICNBME 2021, IFMBE Proceedings 87, pp. 681–690, 2022.
https://doi.org/10.1007/978-3-030-92328-0_86

- measurement of blood pressure (BP) by the patient,
- measurement of blood pressure by a doctor,
- daily monitoring of blood pressure,
- stress tests with the study of blood pressure,
- electrocardiography (ECG),
- echocardiography (Echo-KG, ultrasound of the heart),
- examination of the fundus.

The examination of patients with CVD is traditionally based on diagnostics during a visit to a doctor or during a hospital stay. However, a single determination of the diagnostic parameters of CVD provides information only for a single moment in time and does not always reflect the real clinical picture. However, since the end of the last century, new methods of functional diagnostics have appeared that make it possible to assess the cardiovascular system in conditions as close as possible to natural ones - using telemedicine technologies.

Due to the inaccessibility of providing medical services for a number of the population, the creation and implementation of the principles of telemedicine is one of the priority areas of modern medicine. Taking into account the high incidence of cardiovascular pathology, it becomes obvious the need to create remote systems for monitoring the state of the patient's cardiac activity - telemedical specialized medical diagnostic complexes SMDC [2]. The implementation of such monitoring within the framework of telemedicine will make it possible to achieve significant success in the diagnosis and prevention of many diseases, to take a step towards optimizing medical care.

The goal of this paper is to consider the possibility of improving the methods of remote diagnostics of CVDs to ensure comprehensive monitoring of this category of patients, and modern methods of constructing specialized therapeutic and diagnostic complexes (SMDC).

The main tasks and principles of constructing the SMDC for CVD control tasks are considered by the authors in [2–9]. The hardware and software SMDC equipment depends mainly on the regional and ethnic characteristics of the complex location, as well as on the medical methods used for screening and monitoring CVD. Let's consider the basic requirements for these techniques.

2 Features of Remote Diagnostics of CVD

Among the most well-known methods of remote diagnosis of CVD, the most widely used are cardiogram monitoring (Tele-ECG), 24-h blood pressure monitoring (ABPM), as well as some types of laboratory and diagnostic studies. It is these studies that most effectively describe the functioning of the patient's circulatory and cardiovascular systems, the online control of which makes it possible to implement a fairly constructive process of providing therapeutic and diagnostic telemedicine services [10].

2.1 Tele-ECG

The most common way to use telemedicine technologies for patients with cardiovascular disease is tele-ECG.

The main goal of the tele-ECG is to provide quality medical care through remote ECG interpretation, make diagnostic and clinical decisions based on the interpretation results, as well as remote support of the diagnostic and treatment process.

Thus, the tele-ECG system is a hardware and software complex that provides reception, processing, decoding and transmission of conclusions based on electrocardiograms transmitted from remote subscribers (medical institutions or individuals) [11].

Tele-ECG is one of the remote systems for long-term monitoring, is the use of ECG sensors with electronic means of preliminary processing of signals and their transmission via wireless channels to a server in a medical institution. All types of examinations can be carried out on an outpatient basis, without interrupting household chores and work. There is an opportunity to start examining the patient at his workplace. The device does not interfere with work, rest, sports. Thus, operative telemetric and outpatient monitoring of ECG and blood pressure of patients during their stay on vacation, on vacation, at work, at home is possible.

Continuous (Holter) ECG monitoring is recommended when it is necessary to quantify already diagnosed cardiac arrhythmias, heart rate at different times of the day, to identify asymptomatic arrhythmias and cardiac conduction disturbances in patients with symptoms occur daily or almost every day.

In the case of clinical manifestations, more rarely (weekly, monthly or with an even greater frequency), continuous monitoring of the ECG does not make sense. Intermittent ECG recording is much more informative in this case. An important advantage of this technique is the ability to use it to control the pacemaker's work. The value of outpatient ECG tele-monitoring increases significantly if it is carried out in real time and undergoes special analysis in a telemedicine center. In some centers of Western Europe (Germany, Switzerland) there are also telemedicine geographic information systems - Geographic Information system (GIS), which allow not only to detect ECG abnormalities, but also to determine the position of the patient on the ground, sometimes using satellite communications (using GPS -modules), which is very important for the fastest provision of medical care when the patient cannot report himself, for example, being unconscious.

2.2 24-Hour Blood Pressure Monitoring

Another widely used method of remote diagnosis of CVD is 24-h blood pressure monitoring (ABPM), which provides additional information about the blood pressure level outside the doctor's office.

With ABPM, blood pressure is measured during the day at short intervals (15–30 min) during the patient's usual activity during the day and during sleep at night, with further processing of the data obtained on a computer. ABPM allows you to monitor daily fluctuations in blood pressure, identify hypertension, malignant forms of arterial hypertension, leading to strokes and heart attacks [12]. ABPM allows you to carry out the correct hourly selection of drugs and monitor the adequacy of their effects [13].

The physiological interpretation of most of the ABPM indicators is still widely discussed in the literature today [14–16].

The in dictators of ABPM include the following:

- Average values of systolic (SBP), diastolic (DBP), mean and pulse (PBP), as well as average heart rate per day, day and night;
- Hourly average values of blood pressure and heart rate;
- Maximum and minimum values of blood pressure and heart rate for different periods of the day;
- Daily index (the degree of decrease in systolic and diastolic blood pressure at night);
- Indicators: index of time of hypertension (IT), index of measurements, index of area of hypertension (IA);
- Variability of systolic, diastolic, mean and pulse blood pressure and heart rate.

Standard ABPM indicators are calculated automatically by the software and, most often, are provided to the user in the form of a table (Fig. 1).

День 1		Сутки				День				Ночь			
Период	[N]					7:00-0:00				0:00-7:00			
Границы САД	[ммРт.ст.]					140				130			
Границы ДАД	[ммРт.ст.]					90				80			
Границы СрАД	[ммРт.ст.]					130				120			
Граница ЧСС	[уд/мин]					90				90			
Измерения		47				33				14			
		Среднее	Мин	Макс	Станд.	Среднее	Мин	Макс	Станд.	Среднее	Мин	Макс	Станд.
САД	[ммРт.ст.]	122	92	153	16,1	128	92	153	13,7	108	96	136	12,3
ДАД	[ммРт.ст.]	71	47	111	12,5	75	47	96	11,2	63	51	111	15,4
СрАД	[ммРт.ст.]	96	59	137	17,5	100	59	137	17,9	86	72	118	11,6
ЧСС	[уд/мин]	83	64	122	14,5	86	64	122	14,3	76	66	115	12,3
Пульсовое АД	[ммРт.ст.]	51				53				45			
САД Перепад день-ночь						16 %							
ДАД Перепад день-ночь						16 %							
ЧСС Перепад день-ночь						12 %							
САД > границы		7 (15 %)				6 (18 %)				1 (7 %)			
ДАД > границы		2 (4 %)				1 (3 %)				1 (7 %)			
СрАД > границы		2 (4 %)				2 (6 %)				0 (0 %)			
ЧСС > Граница		16 (34 %)				15 (45 %)				1 (7 %)			

Fig. 1. Standard indicators of ABPM (Russia)

The maximum and minimum values of blood pressure and heart rate during the day (Fig. 1), as well as hourly average values of blood pressure and heart rate (Fig. 2) allow a more detailed study of the features of the daily rhythm of blood pressure and, accordingly, select the most optimal antihypertensive therapy.

Evaluation of the features of the circadian rhythm of blood pressure is absolutely necessary to resolve the issue of the need for the appointment and correction of therapy.

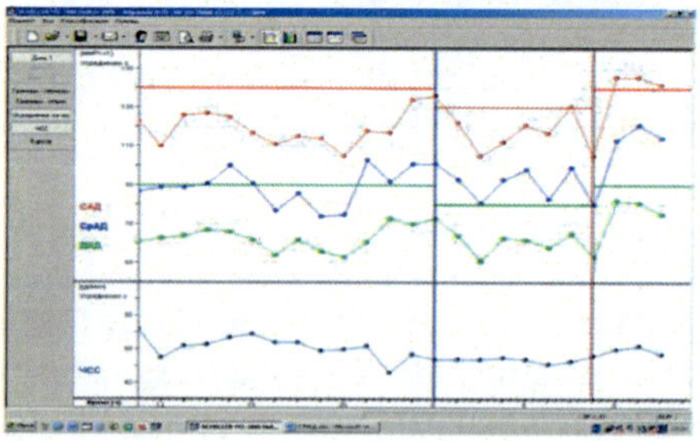

Fig. 2. Hourly average blood pressure and heart rate

Therefore, of undoubted interest is the assessment of the differences between the values of the average daytime and nighttime blood pressure, i.e. the severity of the two-phase rhythm of blood pressure. A healthy person should have a decrease in systolic and diastolic blood pressure at night by 10–20%. The simplest and most widely used method for assessing the circadian rhythm of blood pressure in clinical practice is to calculate the degree of nighttime decrease in blood pressure - the daily index (DI).

The daily index is calculated using the following formula: DI = 100% x (BP day-BP night)/BP day, where BP day is the average BP during wakefulness, BP night is the average BP during sleep.

Depending on the DI value, 4 types of daily blood pressure curves are distinguished (Table 1).

Table 1. Types of daily blood pressure curves

The nature of the night lowering blood pressure	Blood pressure curve type	Daily index, %
Normal	Dipper	10–20
Insufficient	Non-dipper	0–10
Nocturnal hypertension	Night picker	less than 0
Excessive	Over-dipper	Over 20

It is currently considered proven that people with insufficient (less than 10%) decrease in blood pressure at night and nocturnal hypertension have a greater risk of developing cardiovascular complications, i.e. we can talk about the established prognostic value of the initial SI. The informative value of this indicator in relation to assessing the effect of antihypertensive therapy is currently being widely discussed.

To quantify the time during which an elevated blood pressure is recorded, the concept of "pressure loading" has been developed.

Gypertension index time (IT) - the percentage of time during which blood pressure exceeds the critical level (during the day - 140/90 mm Hg, at night - 120/80 mm Hg) for the corresponding time period. This indicator loses its informational content with consistently high blood pressure values. According to various researchers, IV in healthy individuals should not exceed 10–25%. In accordance with the recommendations of the American Hypertensive Society, an IT value of less than 15% is normal, while an excess of IV of more than 30% indicates an undoubtedly increased blood pressure.

Variability of blood pressure (VBP) is understood as natural physiological fluctuations in blood pressure during the day. The daily VBP in healthy individuals increases with age: in the period from 20 to 60 years it is at least 10% of the average level of SBP and DBP. High VBP is stated using the following boundaries: 15 mm Hg for SBP in the daytime and night hours, 14 mm Hg for DBP in the daytime and 12 mm Hg. in nighttime. An increase in VBD against the background of antihypertensive therapy (especially in combination with increased blood pressure values) indicates an unstable effect of the drug.

Sometimes ABPM is performed simultaneously with Holter ECG monitoring. Such a combined study significantly expands the possibilities of diagnosing cardiovascular diseases.

Figure 3 shows an example of 24-h ABPM profile obtained by the authors using the ambulatory blood pressure monitoring system ABPM-04 "MEDITECH" (Hungary). This example shows that a controlled group of patients showed an insufficient decrease in blood pressure at night, which is a risk factor for the development of cardiovascular and cerebrovascular complications (Fig. 3).

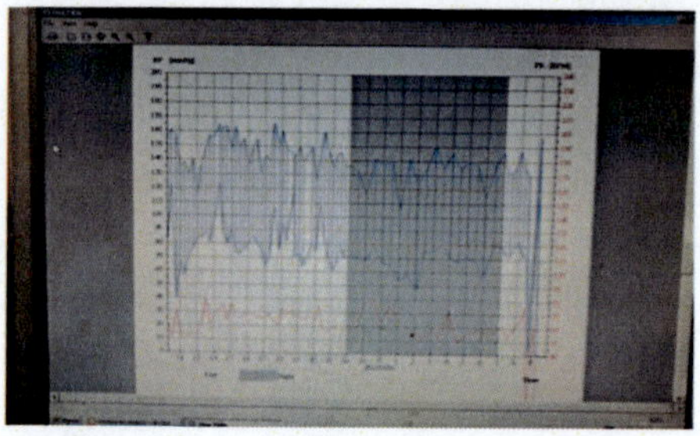

Fig. 3. 24-h blood pressure profile

Figure 3 shows an example of 24-h ABPM profile obtained by the authors using the ambulatory blood pressure monitoring system ABPM-04 "MEDITECH" (Hungary). This example shows that a controlled group of patients showed an insufficient decrease in blood pressure at night, which is a risk factor for the development of cardiovascular and cerebrovascular complications (Fig. 3).

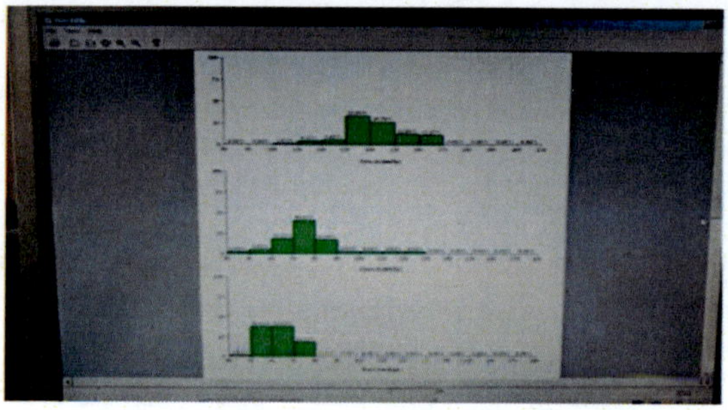

Fig. 4. Average daily SBP, DBP and PBP

It is well known that without nighttime BP control, cardiovascular risk increases exponentially. Figure 4 shows an example of average daily SBP, DBP and PBP. This example shows the distribution of these indicators in a controlled group of patients.

2.3 Laboratory and Diagnostic Tests

The emergence of a mobile interface for health monitoring based on the integration of miniature sensors into mobile communications is one of the actively developed areas in telemedicine. Analysis of data obtained with the help of built-in sensors in mobile systems and image software makes it possible to assess the physiological state of many vital functions without direct contact with the patient.

The developed telemedicine devices allow storing into the device memory not only heart rate, ECG and ABPM, but also sugar level, urine indicators, temperature and other diagnostic parameters, as well as tracking and analyzing them using special applications. Also, the received data can be sent to medical centers to ensure remote monitoring of patients. In medical centers, these data can be analyzed by qualified doctors and recommendations for taking medications and nutrition are issued.

Blood pressure measurement is an oscillometric measurement of blood pressure carried out by means of digital blood pressure monitors equipped with a predominantly wireless data transmission module.

Blood glucose measurement - analysis of the state of carbohydrate metabolism in persons with diabetes mellitus using personal devices - portable digital glucometers. The most modern modifications of such devices are equipped with wired or wireless data transmission modules.

Urine analysis - analysis of the chemical composition of urine using portable digital urine analyzers on test strips. The most modern modifications of such devices are equipped with wired data transmission modules.

Measurement of arterial pulse, pulse oximetry - non-invasive determination of the degree of blood oxygen saturation (by spectrophotometric assessment of the amount of hemoglobin in the blood). It is carried out by means of a digital device - a pulse oximeter equipped with a wired or wireless data transmission module.

Thermometry - measurement of body temperature with a medical maximum thermometer equipped with a wireless data transmission module.

The whole complex of laboratory diagnostic equipment disease [17] allows monitoring the condition of patients with various patterns of diseases.

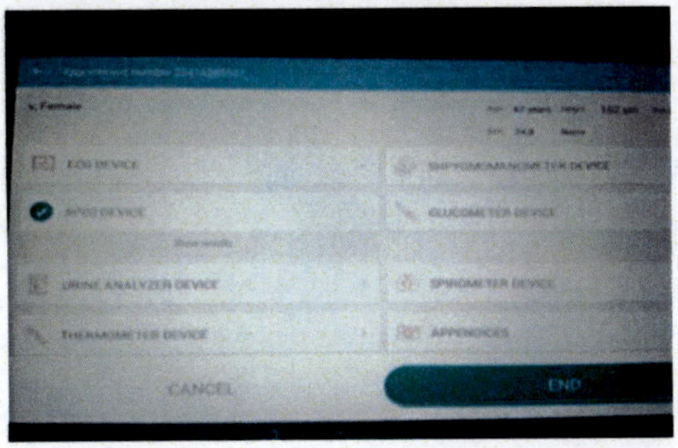

Fig. 5. List of laboratory and diagnostic tests of diagnostic complex IDIS

Figure 5 and 6 shows an example of laboratory diagnostic testing of patients by the authors using the telemedicine mobile diagnostic complex IDIS [18]. Considering these indicators makes it possible to ensure the proper level of medical services for patients with different structures and severity of diseases.

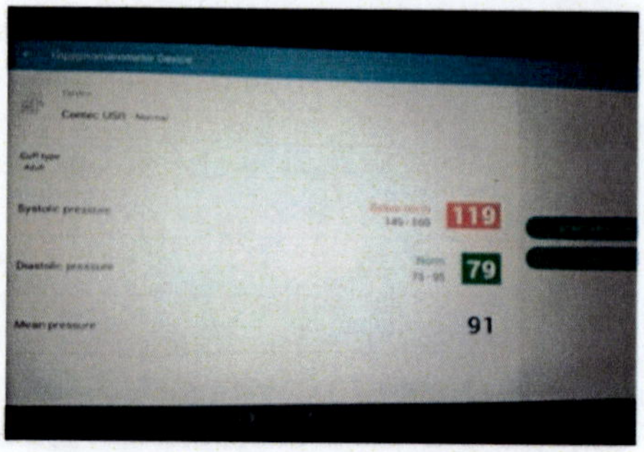

Fig. 6. Example of blood pressure measurement of diagnostic complex IDIS

3 Conclusions

Thus, the comprehensive monitoring of the patient's condition based on the telemedicine means used allows the SLDK to receive all the online information necessary for monitoring the patient's condition. The optimal software and hardware structure and information load on the technical means of tele monitoring are determined depending on the medical techniques used - discussed above, and the total volume of patients per unit of time [19, 20].

Conflict of Interest. The authors declare that they have no conflict of interest.

References

1. Zhou, B., Bentham, J., Di Cesare, M., Bixby, H., Danaei, G., Cowan, M.J., et al.: Worldwide trends in blood pressure from 1975 to 2015: a pooled analysis of 1479 population-based measurement studies with 19·1 million participants. The Lancet. **389**(10064), 37–55 (2017). https://doi.org/10.1016/S0140-6736(16)31919-5
2. Lewington, S., Clarke, R., Qizilbash, N., Peto, R., Collins, R.: Prospective studies collaboration. Age-specific relevance of usual blood pressure to vascular mortality: a meta-analysis of individual data for one million adults in 61 prospective studies. The Lancet **360**(9349), 1903–1913 (2002). https://doi.org/10.1016/S0140-6736(02)11911-8
3. Kolisnyk, K.V., Tomashevskyi, R.S., Sokol, T.V., Koval, S.M., Deineko, D.M.: Creation of regional telemedicine diagnostic and treatment complex. In: Tiginyanu, I., Sontea, V., Railean, S. (eds.) ICNBME 2019. IP, vol. 77, pp. 651–655. Springer, Cham (2020). https://doi.org/10.1007/978-3-030-31866-6_115
4. Gorskaya, L.M.: Functional methods of studying the cardiovascular system (in Russian). http://www.10gkb.by/informatsiya/stati/funktsionalnye-metody-issledovaniya-serdechno-sosudistoj-sistemy
5. Kolesnik, K., Shishkin, M., Koval, S., Lushko, K.: Research of directions of perfection of methods of remote monitoring of patients with arterial hypertension. In: Proceedings I International Scientific and Practical Conference "Information Systems and Technologies in Medicine" ISM-2018, Kharkiv, Ukraine, pp. 160–161 (2018). (in Russian), http://repository.kpi.kharkov.ua/handle/KhPI-Press/38916

6. Vladzimerskiy, A.V.: Telemeditsina (monograf.) . Donetsk: OOO «Digital pr.», p. 477 (2011). (in Russian)
7. Kolesnik, K.V., Shishkin, M.A., Kipenskij, A.V., Sitnikova, O.A.: The use of mobile radio engineering complexes in biometrics and telemonitoring. In: Proceedings Collected scientific works of the 5-th International Forum: Applied Radioelectronics. Status and development prospects. MYFF 2014. Conference: Problems of bioengineering. Science and Technology, Kharkiv, Ukraine, ANPRE, KhNURE, 14–17 October 2014, pp. 166–171 (2014). (in Russian)
8. Sokol, E., Kolisnyk, K., Boyko, V., Zamiatin, P.: Actual problems of information technologies use in the telemedicine services provision in special conditions. In: Proceedings of IEEE International Scientific and Practical Conference: Problems of informations, Science and technology (PIC S&T-2018), Kharkiv, Ukraine, pp. 479–484 (2018). https://doi.org/10.1109/infocommst.2018.8632164
9. Kolisnyk, K., Zamiatin, P.: Perfection of methods for constructing remote monitoring systems for patients in emergency situations. In: Proceedings 2018 IEEE 3rd International Conference on Intelligent Energy and Power Systems (IEPS-2018), Kharkiv, Ukraine, pp. 371–376 (2018). https://doi.org/10.1109/IEPS.2018.8559526
10. Sokol, Y., Shapov, P., Shyshkin, M., Koval, S.M., Tomashevskyi, R.: Automatic diagnosis of atrial fibrillation based on online rr interval analysis. Zeszyty Naukowe Insttytutu Pojazdow/Politechnikawarszawska 2019 2/120, pp. 53–60 (2019). http://bwmeta1.element.xn--baztech-15cbca5d-24214c16-b95965a0f2da44d9-f07zja/
11. Koval, S.M., et al.: Classification of nonstationary cardiac signals based on their spectral and probabilistic properties. In: Information Technology in Medical Diagnostics II – Proceedings International Scientific Internet Conference on Computer Graphics and Image Processing and 48th International Scientific and Practical Conference on Application of Lasers in Medicine and Biology, 2018–2019, pp. 267–274 (2018)
12. Vysotska, O.V., Koval, S.M., Bespalov, Yu.G., Pecherska, A.I., Lytvynova, O.M., Dyvak, A.M.: Mathematical simulation of the structure of pulsed arterial pressure relations with vascular damage factors in patients with arterial hypertension. In: Wójcik , W., Pavlov , S., Kalimoldayev, M. (eds.) Proceedings Information Technology in Medical Diagnostics II 2019, pp. 47–52. Taylor & Francis Group, London (2019). ISBN 978-0-367-17769-0. https://doi.org/10.1201/9780429057618-32
13. Koval, S.M.: Problems of classification and diagnosis of arterial hypertension and stratification of the risk of its complications in the light of the European recommendations of 2018 (commentary to the recommendations). Hypertension **1**(63), 33–41 (2019). (in Ukraine)
14. Penkova, M., Misnichenko, O.: The results of studying the effect of three-component antihypertensive therapy on the nature of the daily profile of blood pressure in patients with hypertension with obesity. In: Proceedings “Achievements and prospects of experimental and clinical endocrinology” (Nineteenth Danilev readings). Proceedings of the scientific-practical conference with international participation, Kharkiv, 27–28 February 2020, pp. 138–139 (2020). (in Ukraine)
15. Penkova, M., Misnichenko, O., Snihurska, I., Koval, S.: Daily blood pressure rhythms in patients with hypertension with insulin resistance under the influence of antihypertensive therapy. Ukr. Therap. J. **3**, 52–56 (2009). (in Ukraine)
16. Koval, S.M., Vysotska, E.V., Strashnenko, A.M., Snihurska, I.O.: Prognosis of essential hypertension progression in patients with abdominal obesity. In: Proceeding of SPIE (USA), vol. 4, pp. 115–120 (2018)
17. Intellectual system of remote diagnostics IDIS (in Russian). https://medstar.ua/ru/resheniya/idis/
18. IDIS. https://medstar.ua/ru/resheniya/idis/

19. Sokol, Y., Lapta, S., Kolisnyk, K., Koval, S., Avrunin, O.: Reducing the risks of medical diagnosis in an epidemic or pandemic. In: Proceedings 2020 IEEE KhPI Week on Advanced Technology (KhPIWeek), Kharkiv, Ukraine, pp. 351–356 (2020)
20. Kolisnyk, K., Sokol, T., Deineko, D., Kutsevlyak, S., Avrunin, O.: Application of modern internet technologies in telemedicine screening of patient conditions. In: Proceedings 2019 IEEE International Scientific-Practical Conference: Problems of Infocommunications Science and Technology, PIC S and T 2019, Kyiv, Ukraine, pp. 459–464 (2019)

Object Locating System by Phone Tracking

I. C. Roşca[1(✉)], C. Drugă[1], I. Şerban[1], and R. D. Necula[2]

[1] Product Design, Mechatronics and Environment Department, Transilvania University, Braşov, Romania
ilcrosca@unitbv.ro
[2] Department of Medical and Surgical Specialties, Transilvania University, Braşov, Romania

Abstract. The paper presents a very simple system for locating and tracking an object/person moving outside a predetermined area, ensuring the transmission of information about the target position through mobile networks. It is highly adaptive to various uses by programming for different areas of action depending on the nature of the target movement - the size of the movement area, the speed of movement, etc. With the help of the device, it is possible to make a location with satisfactory accuracy. Location identification uses common, widespread applications: sending SMS, using Google Maps and Google Earth.

Keywords: Location · GPS · Development board · Arduino

1 Introduction

A location system is actually a navigation system that produces information about the position of a target, fixed or mobile, placed in free space or inside a structure. Concomitant with the dynamics of the development of data capture, transformation and transmission systems, these systems are also developing very fast and very diverse, constituting a very attractive field of research to which many researchers want to contribute.

The location can be approached on two well-defined spaces. In free space (outdoor) the location seems to be based quite solidly by GPS systems and the like but the location inside certain structures - indoor location is found in many areas. We can talk about location and surveillance in the medical field, on the one hand, for protection purposes, for the supervision of physically unsupervised patients and, on the other hand, for the rapid location/detection of instruments or devices in a short time, adapted to time intervention.

The notion also finds its full meaning in the industrial field for the management of devices, tools and other devices or for the efficient location of stored products.

The main purpose of a process of a fixed/mobile target locating/navigating of is to determine its position.

Localization techniques can be separated into two categories depending on how they communicate between nodes:

- centralized that involve the transfer of data to a central node in order to calculate the location of each node;
- decentralized or distributed depend on the ability of each node (sensor) to determine its own location, having limited communication with the nearest nodes.

I. Tiginyanu et al. (Eds.): ICNBME 2021, IFMBE Proceedings 87, pp. 691–698, 2022.
https://doi.org/10.1007/978-3-030-92328-0_87

2 Localisation Methods

The first satellite navigation system was TRANZIT (or Navy Navigation Satellite System - NAVSAT), which became operational in 1965 and operated in the service of the US military fleet until 1996, when it was withdrawn and replaced by GPS. The TRANZIT system initially used 5–6 satellites (later developing to 10) on polar circular orbits with an altitude of 110 km. The determination of the position was based on the Doppler effect.

In principle, the GPS system uses very precise time signals transmitted by itself which is detected by a receiver that synchronizes its own clock with these signals with very good accuracy. Among the received signals is the emission time and the receiver has its own moment of reception. Taking into account the propagation speed of the signal (which is an electromagnetic wave), the distance from the satellite can be determined as a position with known coordinates. This distance represents the radius of a sphere centered on the satellite and on which is located the receiver [1]. If we consider, analogously, two other satellites, which will determine two other spheres, then the receiver can only be placed in a position resulting from the intersection of the three spheres.

This is the principle of satellite navigation that only works theoretically because the determination of the radii of the three spheres is approximate and the positioning would be affected by too large errors. Two other positioning methods are currently used, based on more than three satellites to compensate for errors, namely the absolute method and the differential method.

The absolute positioning method is based on a passive receiver (does not emit any signal) that receives signals from several satellites. In principle, three satellites would be sufficient, but due to errors in measuring time and speed of signal propagation, at least four satellites are required. The more satellites the system has, the better the location accuracy. The obtained position is characterized by three geodetic coordinates: latitude, longitude and altitude.

The differential locating method serves to eliminate unavoidable positioning errors, especially in cases where very high positioning accuracy is required. Differential GPS (DGPS) positioning significantly reduces, up to less than 1 cm, the location errors of a receiver placed at an unknown location by measuring the location errors of a reference receiver from a known location.

To apply this method, it is necessary to have a GPS Reference Station, located in a geometric position established with very good accuracy. The position of the reference station is determined by four satellites - in fact the pseudo-radius errors of each satellite are determined. The user, who also uses a receiver will determine its position according to the same four satellites but with unknown errors. Given the random nature of the factors that generate the errors, several iterative determinations are performed until they become convergent and the errors reach the lowest value.

3 Object Locating System by Phone Tracking

The system presented in this paper locates by GPS positioning, i.e., through the General Positioning System, which bases its operation on geostationary satellite systems placed in known positions above the Earth. Being attached to a person or to an object, it is intended to detect its own position, by specifying its coordinates, and implicitly support position. Further it transmits these coordinates, via SMS in the form of a link to a receiver assigned by programming and opens Google maps indicating the determined position on the map.

The basic components required to perform these functions are schematically represented Fig. 1 and they are:

Arduino GSM shield with M590E module, a SIM card, a GPS module u-blox Neo 6 M, and an Arduino Uno development board.

Arduino GSM shield is a GSM modem that transfers data from a serial port to the GSM network. From the perspective of the mobile communications operator, it is like a mobile phone, and from the perspective of an Arduino it is similar to a modem. It performs operations through a series of AT commands [4]. The library compresses low-level communications between the modem and the SIM card. It relies on the Software Serial library to communicate between the modem and the Arduino. Usually, each separate command is part of a larger series needed to perform a particular function. The library can also receive information and, if necessary, reproduce it. Arduino GSM shield, using the GSM library, makes the Arduino can be used to perform most operations that can be done with a GSM phone: to send SMS and make voice calls, to receive and send SMS, and to connect to the Internet via a GPRS network [2].

The M590E is a compact wireless GPRS module that provides SMS and data services. The M590E module consists of a baseband controller, a flash ROM, an RF section, application interfaces, etc. All sections are coordinated with each other to perform communication functions such as GPRS data and voice.

The M590E module works in the EGSM900/DCS1800 MHz dual-band frequency band, with a sensitivity of less than 107 dBm, the maximum transmitted power is 2 W for class 4 EGSM900 and 1 W for class 1 DCS1800. Its operating protocol is compatible with GSM/GPRS Phase 2/2+. Supports GSM07.07 commands and extended AT commands.

The NEO-6M GPS module is a complete and high-performance GPS receiver with a high power of satellite detection and which allows a wide range of connection options. It has a ceramic antenna. The u-blox 6 positioning motor with 50 channels has a time to the first positioning of less than 1 s. It can perform two million correlations and can perform multiple parallel time/frequency searches, allowing instant detection of satellites. It has a technology that suppresses jamming sources and attenuates multipath effects, which ensures very good performance even in very variable conditions.

The SIM card represents the communication with a communications provider, which also ensures the covered area.

Arduino UNO development board [3] is a microcontroller board, based on the ATmega328P microcontroller (Fig. 1). The board is equipped with sets of digital and analog inputs/outputs that can be interfaced to various expansion boards (shields) or other circuits.

Arduino was created in Italy at the Ivrea Interaction Design Institute and all Arduino boards and associated software are completely open-source.

All components are grouped in a housing made of hard plastic, by 3D printing. That has an aesthetic role and closes the system to make it easy to transport and give it the necessary rigidity during handling.

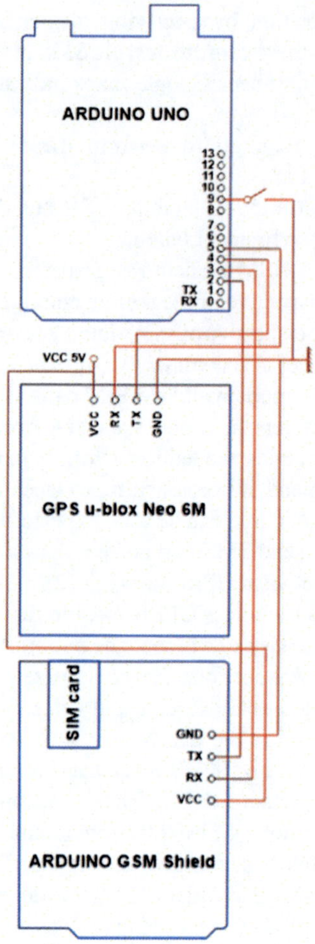

Fig. 1. Schematic representation of the location system

The wiring diagram was developed in EasyEDA software - an integrated tool for making electrical/electronic diagrams, SPICE simulation of circuits and representation of integrated circuits [5].

After mounting the assembly and inserting the SIM card, a series of tests and settings necessary to perform the functions for which the device was built were performed.

Testing the modem and network connection will verify the modem's IMEI identification number. It is unique to each modem and is used to identify valid devices that can connect to a GSM network. As soon as the number has been identified from the modem, Arduino will identify and retain the network to which it is connected as well as the signal strength from the network to the serial port [3]. After connecting to the network with the previous procedure, other functions of the board are tested - connecting to a GSM network and sending an SMS message to an established phone number [3].

To connect to the internet, in addition to the SIM card, some additional information from the provider's cell phone is required. Each mobile phone has an APN (Access Point Name) that connects the cellular network to the internet, via a software routine, which provides the connection to arduino.cc/latest.txt and retains its contents [3].

4 Results

After mounting the location system components and testing their operation and connection to the networks, field tests were performed in known locations.

Acquisition, transmission and displaying the results involves the following steps:

- positioning the device and powering it;
- turning on the device that by programming had already assigned the mobile phone number to which it was to send the SMS message warning of the location;
- receiving messages indicating a link containing the coordinates of the detected location. Following the opening of the respective links, the latitude and longitude coordinates were obtained and the location was placed in Google maps, which also opens automatically by programming.

The tests were performed at Transilvania University of Brasov (Colina street no.1). From the conclusive data obtained, three locations will be mentioned as examples.

First Location

Received message (Fig. 2):

https://www.google.com/maps/?q=45.653144,25.597095.

Fig. 2. Print-screen for the detected first location

In Google Earth, the indicated coordinates on Google maps generated the image of the Fig. 3.

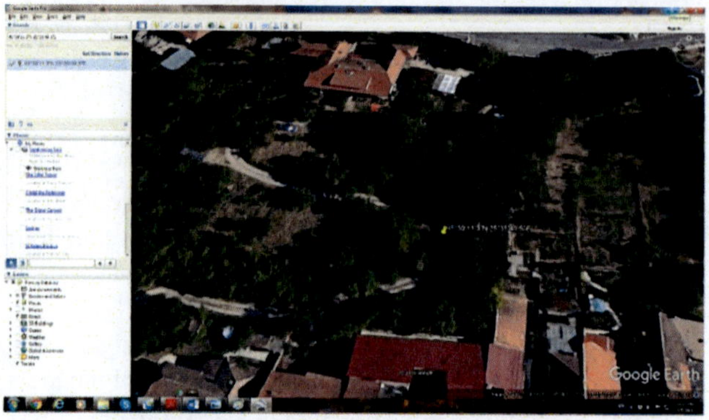

Fig. 3. Google Earth image of the first location

In order to verify the accuracy of the system, the chosen second location was at the same address, in front of the neighbor building, at about 50 m.

Second Location

Received message (Fig. 4):

https://www.google.com/maps/?q=45.654453,25.597843.

The indicated coordinates generate on Google Earth an image as in Fig. 5 was obtained.

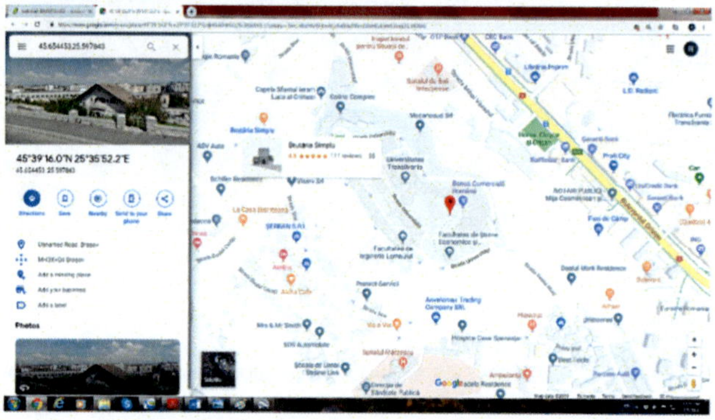

Fig. 4. Print-screen for the detected second location

The third location was chosen at a distance less than 10 m from the second one, and the device indicated a different position, as can be seen in the Figs. 6 and 7.

Fig. 5. Google Earth image of the second location

Third Location
Received message:
https://www.google.com/maps/?q=45.654392,25.597463.

Fig. 6. Print-screen for the detected third location

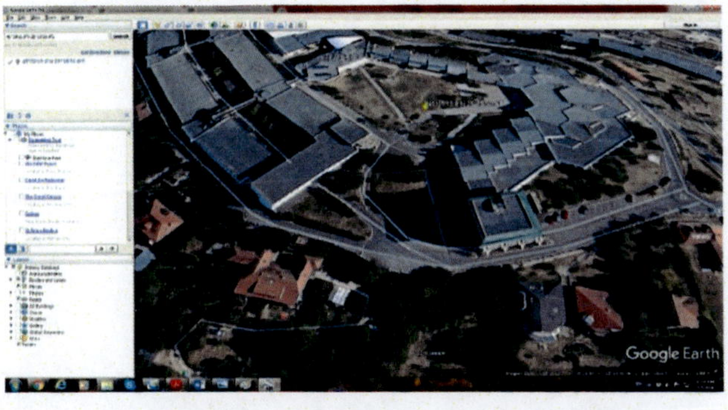

Fig. 7. Google Earth image of the third location

5 Conclusions

The main objective of the study was to create a programmable system for locating and tracking an object/person moving outside a predetermined area, which ensures the transmission of information about the target position through mobile networks. The presented system ensures a high adaptability to various types of uses by programming the system for different areas of action depending on the nature of the target movement - the size of the movement area, the speed of movement, etc. It has relatively small dimensions (118 × 58 × 25 mm) and low weight, less than 100 g and a high autonomy, being powered by an external battery. With the help of the device, it is possible to make a location with satisfactory accuracy as it also results from the locations II and III, in which it can be observed that between the two detected positions there is a small and obvious distance on the positions, including in Google Earth. Location identification uses common, widespread applications: sending SMS, using Google Maps and Google Earth.

Conflict of Interest. The authors declare that they have no conflict of interest.

References

1. Cehan, V.: Principiile sistemelor de radionavigaţie, Iaşi (2008)
2. *** Getting Started with the Arduino GSM Shield. https://www.arduino.cc/en/Guide/ArduinoGSMShield
3. *** Arduino Uno. https://en.wikipedia.org/wiki/Arduino_Uno
4. *** AT GSM. http://ares.utcluj.ro/cm_2018_files/l2.pdf
5. *** SPICE. https://en.wikipedia.org/wiki/SPICE

What Do Family Doctors Think About Patient Safety Culture in the Republic of Moldova?

Galina Buta[1,5](✉), C. Tereanu[2,5], J. Roncali[3,5], S. M. Ghelase[4,5], and M. L. Cara[4,5]

[1] Department of Family Medicine, "N. Testemitanu" University of Medicine and Pharmacy, Chisinau, Republic of Moldova
galina.buta@usmf.md

[2] Department of Hygiene and Prevention, Agenzia di Tutela della Salute, Bergamo, Italy

[3] School of Specialization in Hygiene and Preventive Medicine, University of Brescia, Brescia, Italy

[4] Department of Public Health and Healthcare Management, University of Medicine and Pharmacy, Craiova, Romania

[5] Department of Biomedical Engineering, Tecnical State University, Chisinau, Republic of Moldova

Abstract. The Romanian version of the Medical Office Survey on Patient Safety Culture (MOSOPSC) of the US Agency for Healthcare Research and Quality has been applied for the first time in primary care centers in the Republic of Moldova. The aim was to assess the current status of patient safety culture and then identify strengths of it and areas for improvement.

A cross-sectional study was carried out as part of the IRIS-2 international project on patient safety culture, which also involved Romania and Italia. Data collection in the Republic of Moldova started in February and ended in June 2020. The survey was distributed on paper and via email to a convenience sample of 820 family doctors and 93% of them volunteered to complete and return it. Half of the respondents were from the capital of the country (Chişinău).

The 38 items of the section C-F of the survey were grouped in 10 composites measuring the patient safety culture. Percent positive response was computed for each item and dimension. The percent positive responses (PPRs) per item ranged from 35% to 100% and per composite from 41% to 97%.

The highest developed patient safety culture areas (PPRs > 75%) were: Organizational Learning (97%), Teamwork (95%), Patient Care Tracking/Follow-up (90%), Staff training (87%), Overall Perceptions of Patient Safety and Quality (87%), Office processes and standardization (81%), Communication About Error (81%), Owner/Managing Partner/Leadership Support for Patient Safety (78%). The most critical area was "Work pressure and pace" (41%), while "Communication Openness" (74%) area was between.

In conclusion, although most composites measuring patient safety culture from the point of view of family doctors in the Republic of Moldova showed areas with high development, there is room for further improvement, especially for the "Work pressure and pace" area.

Keywords: Culture · Safety · Family doctors · Republic of Moldova · Baseline assessment

I. Tiginyanu et al. (Eds.): ICNBME 2021, IFMBE Proceedings 87, pp. 699–705, 2022.
https://doi.org/10.1007/978-3-030-92328-0_88

1 Introduction

The Republic of Moldova has 3, 4 million inhabitants (2,9 million are permanent residents in the country). More than half of them are living in rural areas. This is the poorest European country, with a Gross Domestic Product of 4,551$ (2020) [1].

The Government of the Republic of Moldova adopted the first decision on the development of primary health care which introduced the specialization in family medicine in 1997 (no. 1134). The current rules for the organization of primary health care were approved with the Government decision no. 988/2018.

With the implementation of compulsory health insurance, the basic principle in the organization of primary health care in the Republic of Moldova is to guarantee the access of the population to primary health services, according to quantitative and qualitative parameters established by law. In the same time, other principles of compulsory health insurance are implemented in family medicine, such as: free choice of family doctor and medical institution, per capita fee depending on the patient's age for persons registered in the family doctor list, financial autonomy of institutions primary care etc. The population's access to primary health care is universal, regardless of whether or not an individual is a contributor to the compulsory health insurance introduced in 2004.

The compulsory health insurance fund is the main source of funding for the primary care system in the Republic of Moldova. To uniformly develop family medicine throughout the country, a number of incentives have been created. For instance, rural family doctors are better paid than their urban counterparts [2]. Nonetheless, after an initial increasing between 1995 and 2002, the number of family doctors and primary care nurses is now decreasing, especially in the rural areas.

Little is known about quality, clinical risk and patient safety in primary care settings in this country, where the average of the primary care provided is about 3 visits/inhabitant/year [2]. The aim of this research is to improve knowledge about patient safety, based on the family doctors perceptions about patient safety culture in primary health care centers. The safety culture of an organization is "the product of individual and group values, attitudes, perceptions, competencies, and patterns of behavior that determine the commitment to, and the style and proficiency of, an organization's health and safety management" [3]. The objectives of this research are two: to assess the current status of patient safety culture and to identify strengths and areas for patient safety culture improvement, based on the *Medical Office Survey on Patient Safety Culture (MOSOPSC)* composites.

This research carried out in primary care facilities in the Republic of Moldova is part of the IRIS-2 international project, also implemented in Romania and Italy, from May 2019 to May 2021 [4].

2 Material and Methods

2.1 The Medical Office Survey of Patient Safety Culture (MOSOPSC)

The MOSOPSC is a survey designed by the United States of America Agency for Healthcare Research and Quality (AHRQ) specifically for outpatient medical office, to be administered to all providers (physicians, physician assistants, and nurse practitioners who diagnose, treat patients, and prescribe medications) and staff in a medical office (from billing staff and receptionists to nurses and physicians). It asks providers and staff for their opinions about the culture of patient safety and health care quality at their medical offices [5]. Providers who work in more than one medical office are asked to reply thinking about the office where they spend most of their time.

The survey has been translated into Romanian, which is also the official language in the Republic of Moldova, with the AHRQ permission, obtained by the coordinator of the IRIS-2 project. The Romanian version of the MOSOPSC has been tested on a small group of family doctors in both countries, before application, after minimal adjustment with respect to the original US version.

2.2 The Survey Participants

In this cross-sectional study, the MOSOPSC was applied from February to June 2020. Participants were voluntarily recruited, after being shortly informed about the research importance, purposes and contents and about the anonymous mode of presenting the research results. A number of 820 surveys were distributed and 780 providers, who agreed to participate in the study, completed the survey and returned it to the researcher.

During February and June 2020, a number of 480 paper surveys were administered to providers who attended continuous medical education programs (in the classroom) and to other providers in the in primary care centers and n. 473 were completed and returned. From March to May 2020, due to COVID-19 pandemic restrictions, a number of 340 surveys were sent to providers via email and n. 307 were completed and returned.

All development Regions of the country were represented by the participants in the study: Northern, Central and Southern regions as well as Chisinau, the capital of the country. About a third of the respondents represented rural area primary care facilities.

2.3 Data Analysis

Double-registered surveys or with missing response to one or more survey sections, were excluded. Three surveys completed by nurses were also excluded. After the data cleaning process, a number of 776 surveys (completed by doctors only) were included in the final dataset for statistical analysis.

The MOSOPSC has nine sections (from A to I) [5]. The sections C-D-E-F emphasize patient safety and health care quality issues, across 38 items grouped into 10 composite measures (composites). The items are grouped according to the safety culture composites they are intended to measure. The survey also includes questions that ask respondents: about a list of patient safety and quality issues (section A), about problems in exchanging information with other settings (section B); to rate their medical office in five areas of health care quality (patient centered, effective, timely, efficient, and equitable) and to provide an overall rating on patient safety (Section G); to provide personal background information (section H) and to provide free written comments about patient safety (section I). In this paper, we only present sections C-D-E-F and section H results.

Most items in the survey require a response based on a 1 to 5 scale, corresponding to 1-"Strongly disagree", 2-"Disagree", 3-"Neither Agree nor Disagree", 4-"Agree",

5-"Strongly agree" (for the items in section C, E, F) or to 1-"Never", 2-"Rarely", 3-"Sometimes", 4-"Most of the time", 5-"Always" (for the items in section D). Some respondents might not be informed enough to answer all the questions. Therefore the option "Does not Apply/Don't know" was also available in the survey.

Some items are negatively worded, which means that the response "strongly disagree" or "never"/"disagree" or "rarely" indicate a positive patient safety culture.

The frequency of response for each survey item was computed by combining the two lowest response categories (e.g., "Strongly disagree/Disagree" or "Never/Rarely") and the two highest response categories (e.g., "Strongly agree/Agree" or "Most of the Time/Always"). The midpoints of the scales are not reported ("Neither or Sometimes"). Descriptive analyses were carried out, based on percent "positive" response, means, and composites. When calculating this indicators, negatively worded questions have been reverse coded. Besides, "Does not apply/Don't know" and "missing" responses were excluded when computing percentages of response to the survey items.

3 Results

The response rate to the survey was 93%. A number of 776 complete survey were analyzed overall.

Table 1 shows the main actuarial characteristics of the respondents (providers). Two thirds of the respondents were providers in urban area centers (45% in the Chişinău area and 22% in other urban areas across the country) and one third were providers in rural area centers.

Respondents in our study were primary care doctors: 93% specialist doctors in family medicine and 7% specializing doctors or resident doctors (in the last year of the specialization in family medicine). About one third were doctors in chief (29%). Most doctors in chief were also owners/managing partners/or in leadership position with responsibility for making financial decision for his/her medical office.

More than half (60%) had 12-year or more work experience in the current center and 19% of them only had 3-year or less work experience. Half of the respondents (51%) work more than 41 h per week in the current primary care center, followed by 40% of the respondents who work between 33 and 40 h per week in the center.

The percent positive response (PPRs) per item ranged from: 35% (item C6r: "We have too many patients for the number of providers in this office" - negatively worded/ reverse coded) to 100% (item C2: "In this office, there is a good working relationship between staff and provider"). The average PPR for all 38 items was 81%.

As illustrated in Fig. 1, the percent positive response (PPRs) per composite ranged from 41% (composite 2: "Work pressure and peace") to 97% (composite 9: "Organizational Learning". PPRs were higher than 75% for eight composites and lower than 50% for one composite.

For composite 8 "Owner/Managing Partner/Leadership Support for Patient Safety" PPR was computed using a number of 531 respondents only (instead of 776), after excluding who responded "Yes" to the item "Are you an owner, a managing partner, or in a leadership position with responsibility for making financial decisions for your medical office?" and who responded "Doctor in chief" to the item "What is your position in this office?".

Table 1. Actuarial characteristics of the respondents

Characteristic	N	%
Area of the medical office		
Chişinău (capital)	349	45
Rural	253	33
Urban	174	22
Provider's position in the office		
Doctor	499	64
Doctor in chief	227	29
Resident doctor (last year)	50	7
Provider's ownership/managing partnership/leadership position with responsibility for making financial decision		
Yes	214	28
No	562	72
Years worked by the provider in the current medical office		
3 years or less	149	19
Between 4 to 11 years	160	21
12 years or more	467	60
Hours per week worked by the provider in the current medical office		
5 to 32 h per week	71	9
33 to 40 h per week	312	40
41 h per week or more	391	51

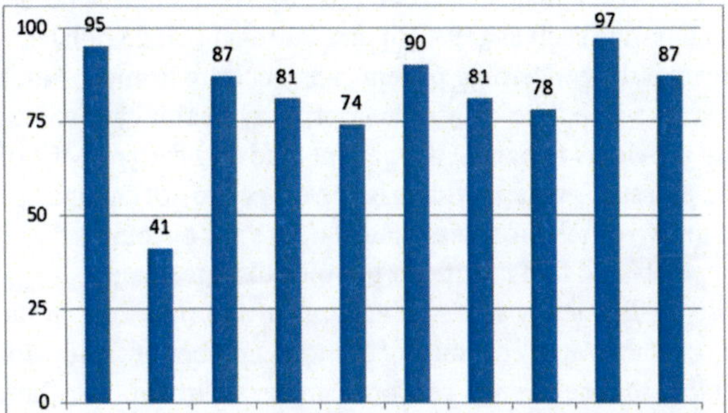

Fig. 1. Distribution of percent positive response for patient safety culture by composite

1. Teamwork (items: C1, C2, C5, C13);
2. Work pressure and pace (items: C3r, C6r, C11, C14r);
3. Staff training (items: C4, C7, C10r);
4. Office processes and standardization (items: C8r, C9, C12r, C15); 5. Communication Openness (items: D1, D2, D4r, D10r);
6. Patient Care Tracking/Follow-up (items: D3, D5, D6, D9); 7. Communication About Error (items: D7r, D8, D11, D12); 8. Owner/Managing Partner/Leadership Support for Patient Safety (items: E1r, E2r, E3, E4r);
9. Organizational Learning (items: F1, F5, F7);
10. Overall Perceptions of Patient Safety and Quality (items: F2, F3r, F4r F6r).
r = negatively worded/reverse coded.

4 Discussion

The Romanian version of the MOSOPSC has been administered for the first time to family doctors in the Republic of Moldova in 2020. This cross-sectional study was based on a convenience non probabilistic sample of 776 voluntary respondents. It had a high response rate (93%) and covered 46% of the family doctors across the country (rural and urban areas). Almost half of the participants were from primary care centers located in the capital of the country (Chisinau).

Most of the 10 patient safety culture composites (based on 38 items) of the MOSOPSC were highly scored by the respondents. These findings indicate a highly developed patient safety culture from the point of view of family doctors for the following areas: Organizational Learning (97%), Teamwork (95%), Patient Care Tracking/Follow-up (90%), Staff training (87%), Overall Perceptions of Patient Safety and Quality (87%), Office processes and standardization (81%), Communication About Error (81%), Owner/Managing Partner/Leadership Support for Patient Safety (78%). Conversely the "Work pressure and pace" area (41%) is critical and needs prompt intervention. The remaining area concerning "Communication Openness" (74%) which has a PPR falling between 50 and 75%, would need further development, as well.

These PPR are higher than those obtained by researchers who applied the same survey in other countries such as USA [6] and Portugal (Madeira Island) [7].

Thus, one main issue concerning patient safety culture from the family doctors point of view is that in the Republic of Moldova primary health care centers there are not enough staff and providers to handle the patient load and therefore the office work pace is hectic. The other issue is that providers in the office are not enough open to staff ideas about how to improve office processes, and staff are not encouraged enough to express alternative viewpoints and find it difficult to voice disagreement.

The strengths of this study are: application for the first time in the country, high response rate, acceptable % of coverage of the total number of family doctors across the country. Conversely, the use of a non-random sample is a limitation to the generalizability of the results.

5 Conclusion

Moldavian, Romanian and Italian researchers collaborated within the IRIS-2 international project to carry out a study to identify areas of patient safety culture needing improvement, based on the perception of family doctors in the Republic of Moldova. Application of the Romanian version of the MOSOPSC pointed out that "Work pressure and pace" area needs urgent intervention. "Communication Openness" is another area which should be considered for improvement.

Acknowledgments. The authors thank all doctors who completed the survey. The authors are grateful to the Latin Association for Analysis of Healthcare Systems (www.alass.org) for their patronage for the IRIS-2 Project.

Conflict of Interest. The authors declare that they have no conflict of interest.

References

1. GDB per capita. http://data.worldbank.org/
2. Zarbailov, N., Gilca, B., Ciocanu, M., et al.: Dezvoltarea medicinii de familie in Republica Moldova - punctul de vedere din exterior. Sănătate Publica, Economie şi Management **1–2**(75–76), 9–13 (2018)
3. Study Group on Human Factors: Organising for safety: Third Report of the ACSNI (Advisory Committee on the Safety of Nuclear Installations). HSE Books, Sudbury (1993)
4. Tereanu, C.: Progetto "IRIDE-2" in SHAM Assicurazioni, ARIS, AIOP, Federsanità, ANCI, 4° Premio per la prevenzione dei rischi 2019, 14° Forum Risk Management in Sanità, 26–29/11/2019, Florence (abstract, p. 114) (2019). http://www.sanita360.it/dev/wp-content/uploads/2019/12/Premio-Sham-2019-Volume-Partecipanti.pdf
5. Sorra, J., Gray, L., Famolaro, T., et al.: AHRQ medical office survey on patient safety culture:user's guide. (Prepared by Westat, under Contract No. HHSA290201300003C). AHRQPublication No. 15(16)-0051-EF (Replaces 08(09)-0059). Agency for Healthcare Research and Quality, Rockville (2016). http://www.ahrq.gov/professionals/quality-patientsafety/patientsafetyculture/medical-office/index.html.
6. Agency for Healthcare Research and Quality. Medical Office Survey on Patient Safety Culture (2016)
7. Ornelas, M.D., Pais, D., Sousa, P.: Patient safety culture in Portuguese primary healthcare. Qual. Prim. Care. **24**(5), 214–218 (2016)

Clinical-Epidemiological Characteristics of Children Hospitalized with COVID - 19 in the Republic of Moldova

Galina Buta[1,2](✉), S. Cojocaru[1], T. Costru[1], R. Puia[1], D. Abdusa-Ganea[1], and A. Ungureanu[1]

[1] Nicolae Testemiţanu State University of Medicine and Farmacy, Chisinau, Republic of Moldova
galina.buta@usmf.md
[2] Technical University of Moldova, Chisinau, Republic of Moldova

Abstract. Despite the spread of SARS-CoV-2 infection, knowledge of the symptomatology, modalities of transmission, severity and prognosis is limited. Based on the complex analysis of the 701 pediatric cases, from 5 public medical-health institutions in the Republic of Moldova, positively diagnosed with SARS - CoV-2 infection by the RT-PCR molecular method, 73% of the children were in close contact with a COVID-19 positive person in the last 14 days until the illness. Most of the hospitalized children develop moderate clinical forms of COVID-19 (84.17% cases). The mean age of children with severe form is lower compared to those with moderate or mild form. The sex of patients does not influence the course of the disease. Children's age influences paternal manifestations: children of preschool age (<7 years) more often show high fever, changes in consciousness, drowsiness, rash, diarrhea and dyspnea, while children > 10 years: headache, pharyngitis, hypo/anosmia, ageusia, myalgia, arthralgia and vertigo. No statistical correlations were found between the form of the disease and the presence of comorbidities. The evolution of pediatric COVID-19 infection is favorable, considering that 97.57% of children do not show clinical manifestations at discharge. Knowledge of various clinical pictures of infection with the SARS-CoV-2 virus would allow the doctor to make an early diagnosis, administer adequate treatment and establish preventive measures.

Keywords: SARS-CoV-2 infection · Pediatric cases · Epidemiology · Comorbidity · Symptoms

1 Introduction

The World Health Organization has declared coronavirus disease 2019 (COVID-19), the disease caused by SARS-CoV-2, a sanitary pandemic [1]. The current rapid worldwide spread of SARS-CoV-2 infection requires continuous improvement of knowledge about the epidemiology and clinical characteristics of COVID-19. After identification of the SARS CoV-2 pathogen, the airway of infection transmission through direct contact with the infected person was also identified [2, 3].

I. Tiginyanu et al. (Eds.): ICNBME 2021, IFMBE Proceedings 87, pp. 706–711, 2022.
https://doi.org/10.1007/978-3-030-92328-0_89

Pediatric and adolescent patients were tested positive on rectal swabs, even after negative nasopharyngeal testing, suggesting viral elimination through the gastrointestinal tract and the possibility of fecal-oral viral transmission. Other routes of transmission of infection are currently being identified and investigated [2, 4]. Pediatric patients acquired COVID-19 infection mainly through the contact with parents or family members [4–8]. Unlike adults, in children there are no risks of the severity of the disease based on sex. However, no data are currently available on the role of comorbidities in the severity of COVID -19 infection in the pediatric population.

In general, pediatric patients with COVID-19 had a good prognosis and recovered within 1–2 weeks after the onset of the disease [9, 10]. The rapid spread of SARS-CoV-2 infection worldwide and the severity of some COVID-19 cases, which mimic that of SARS, justify the global effort to identify effective preventive strategies and optimal medical management of both infections in children and adults, including the implementation of therapeutic objectives. Knowledge of the clinical-epidemiological characteristics of children hospitalized with COVID-19, in particular, primary analysis by age, clinical manifestations, diagnostic and therapeutic management would identify therapeutic strategies and effective preventive measures. The purpose of the study is to evaluate the complex pediatric cases confirmed by SARSCoV-2 infection (COVID-19) being treated in 5 public medical-health institutions in Moldova for the analysis of the impact of the infection on the health status.

2 Materials and Methods

The present study is retrospective, cohort, consisting of a sample of 701 patients with the diagnosis of COVID-19 infection aged 1–18 years treated in 5 Public Medical-Sanitary Institutions in Chisinau municipality, Republic of Moldova between March 1, 2020 and June 30, 2021 and randomly selected. The diagnosis of COVID-19 was made by the detection of SARS CoV-2 RNA by molecular biology tests.

The data included in the study were selected from the patients' medical records, processed and stored in accordance with the requirements of the "Covid-19 electronic patient record Register" software. The severity of COVID-19 infection was assessed according to the criteria of the national clinical protocol PCN-371 "new type coronavirus infection (COVID-19)". COVID-19 mild form: patient with any signs and symptoms (fever, cough, sore throat, malaise, headache, muscle pain) without shortness of breath, dyspnea or abnormal chest images (no pneumonia); COVID-19 moderate form: patient with fever and signs of non-severe pneumonia, no need for additional oxygen; COVID-19 severe form: patient with signs of severe pneumonia.

3 Results

The children were taken to the Municipal Institutions of Public Health (MIPH) included in the study by the team of Emergency Medical Assistance - 200 (28.53%) patients, the medical specialist - 23 (3.28%) patients, the family doctor - 10 (1.43%) patients, at the request of their parents or guardians - 35 (4.99%) patients, the transfer of the COVID triage center or any other MIPH - 424 (60.48%) patients, other modes of hospitalization - 9 (1.29%) the patients.

The epidemiological anamnesis found out that 513 (73.18%) children were in close contact with a COVID-19 positive person in the last 14 days until the illness.

The majority of patients in the study were from urban regions - 487 (69.47%) patients. The mean age of the children was 8.54 years. The 1-year-old group was the largest. The male/female ratio was on average 1.04, the female sex prevailed only in children under 1 year, 9 years and those older than 15 years (Fig. 1).

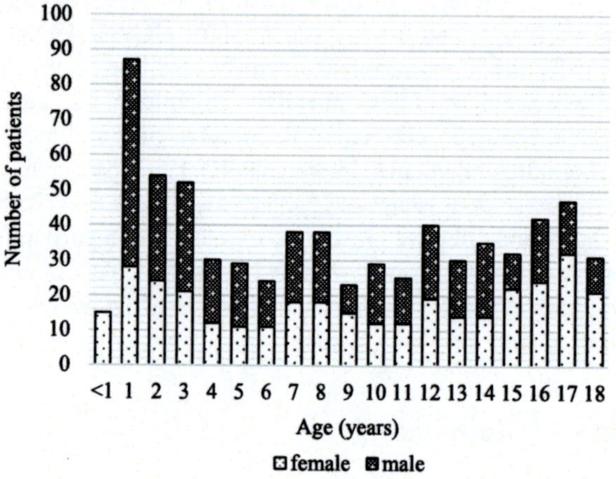

Fig. 1. Age and gender distribution of children

Most children were diagnosed with the moderate form of COVID-19 infection - 590 (84.17%) cases. The mild form of the disease was diagnosed in 52 (7.42%) patients, and the severe form - in 59 (8.42%). No critical cases or deaths were recorded (Fig. 2).

The mean age of children with severe form was lower - 6.97 years, compared to those with moderate or mild form, 8.66 years and 8.92 years respectively. The male / female ratio was greater than 1 for all clinical forms: mild - 1.17 (28/24), moderate - 1.06 (303/287), severe - 1.18 (32/27).

The body temperature above 37 °C was recorded in most cases - 607 (86.59%) patients. Younger children had a higher fever compared to older children. The mean age of patients with temperature above 38 °C, recorded in 249 (35.52%) cases, was 7.49 years, while the mean age of patients with the body temperature 37-38 °C, recorded in 358 (51.07%) cases, was 9.44 years.

In addition to fever, the main complaints of children and/or guardians during hospitalization were fatigue or/and physical asthenia, cough, headache and pharyngitis (Table 1).

Such clinical manifestations as changes in consciousness, drowsiness, rash, diarrhea and dyspnea were more often diagnosed in children of preschool age (<7 years); and headache, pharyngitis, hypo/anosmia, ageusia, myalgia, arthralgia and vertigo – in children aged over 10 years.

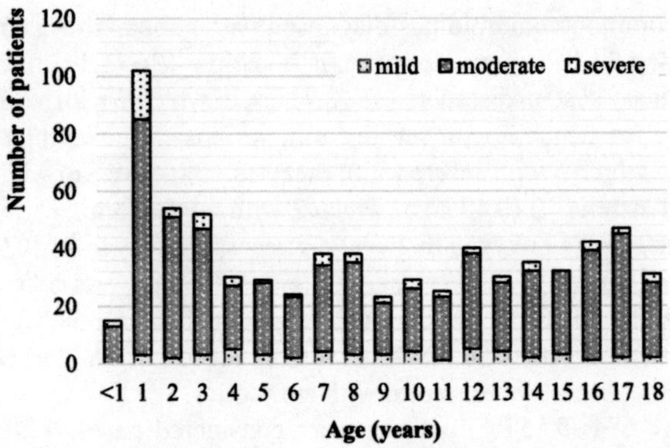

Fig. 2. Distribution of children according to the form of the disease and age

Table 1. The most common clinical manifestations during the course of the disease depending on the age group

Clinical manifestation	No patients	% of patients	Age of patients (years)
Fever > 38 °C	249	35.52	7.49
Fever 37–38 °C	358	51.07	9.44
Fatigue and/or physical asthenia	317	45,22	9.73
Cough	297	42,37	8.78
Headache	172	24,54	12.44
Pharyngitis	146	20.83	11.12
Change in consciousness	76	10.84	3.51
Hypo/anosmia	59	8.42	14.59
Diarrhea	41	5.85	5.27
Myalgia	36	5.14	12.56
Dyspnea	28	3.99	6.46
Drowsiness	23	3.28	3.96
Ageusia	20	2.85	15.15
Rash	13	1.85	4.61
Arthralgia	11	1.57	10.18
Vertigo	10	1.43	12.7

Manifestations often considered pathognomic for COVID such as hypo/anosmia and ageusia were rarely detected in children, in 10.84% and 2.85% cases respectively.

In the imaging investigation of the airways (Rx or/and CT) lung damage was detected in 271 (38.66%) cases with the mean age of patients of 7.39 years, while the mean age of children without imaging changes was 9.26 years.

The most common comorbidities diagnosed were: anemia - 45 (6.42%) cases, of which only 1 patient had severe form; toxic hepatitis - 36 (5.14%), of which 2 with severe form; urinary system disorders - 19 (2.71%), of which 2 with severe form; obesity - 7 (1.0%) cases, of which 3 with severe form; various cardiovascular pathologies - 7 (1.0%) cases, of which 3 with severe form; diabetes - 3 (0.43%) of which 2 with severe form; bronchial asthma - 3 (0.43%) of which 2 with severe form.

The administration of oxygen therapy was necessary only in 21 (2.99%) cases with the mean age of children 12.71 years. Antibiotic therapy was administered in 471 (67.19%) cases: 14 out of 52 children with the mild form (26.92%); 404 out of 590 children with the moderate form (68.47%) and 53 out of 59 (89.83%) children with the severe form of the disease were treated with antibiotics.

At discharge 684 (97.57%) patients were considered cured, without any clinical manifestations. Only 17 (2.42%) children continued to manifest cough or/and fatigue, headache.

Conclusions: The source of infection with SARS-CoV-2 remains unexplained in every 4th case of COVID-19 disease in children, which contributes to the uncontrolled spread of the infection.

Children mainly acquire SARS CoV - 2 infection from their family members, but they seem to experience a less severe form of the disease than adults. The prognosis is favorable and recovery occurs 1–2 weeks after the onset of the disease.

In most cases hospitalized children develop moderate clinical forms of COVID-19. The mean age of children with severe form is lower compared to those with moderate or mild form, the gender of patients does not influence the course of the disease.

Clinical manifestations of COVID-19 in children are nonspecific and do not allow differentiation from other pathologies, the main clinical signs of the disease are fever, fatigue or/and physical asthenia, cough, headache and pharyngitis. Children's age influences paternal manifestations: children of preschool age (<7 years) more often manifest high fever, changes in consciousness, drowsiness, rash, diarrhea and dyspnea, while children aged over 10 years: headache, pharyngitis, hypo/anosmia, ageusia, myalgia, arthralgia and vertigo.

Antibiotic therapy is often given unduly in pediatric COVID-19 infection.

In the case of children, no statistically reliable correlations were found between the form of the disease and the presence of comorbidities. The evolution of pediatric COVID-19 infection is favorable, discharge from MIPHs is done in most cases in the absence of clinical manifestations, children being considered cured by the specialist doctor.

The current rapid worldwide spread of SARS-CoV-2 infection requires continuous improvement of knowledge about the epidemiology and clinical characteristics of COVID-19.

Acknowledgment. The paper was done in the project 20.70086.12/COV – "Assessment of health status and assessment of patients' quality of life including in the electronic Register COVID-19" from Republic of Moldova.

Conflict of Interest. The authors declare that they have no conflict of interest.

References

1. World Health Organization (WHO). WHO Director-General's opening remarks at the media briefing on COVID-19, 11 March 2020 (2020). Accessed 11 Mar 2020. https://www.who.int/dg/speeches/detail/who-directorgeneral-s-opening-remarks-at-the-media-briefingon-covid-19---11-march-2020
2. Xu, Y., Li, X., Zhu. B., et al.: Characteristics of pediatric SARS-CoV-2 infection and potential evidence for persistent fecal viral shedding. Nat Med. (2020). https://doi.org/10.1038/s41591-020-0817-4
3. Infecţia cu coronavirus de tip nou (COVID-19), PCN 371. https://msmps.gov.md/
4. Castagnoli, R., et al.: Severe acute respiratory syndrome coronavirus 2 (SARS-CoV-2) infection in children and adolescents: a systematic review. JAMA Pediatr. **174**(9), 882 (2020). https://doi.org/10.1001/jamapediatrics.2020.1467
5. Chan, J.F., Yuan, S., Kok, K.H., et al.: A familial cluster of pneumonia associated with the 2019 novel coronavirus indicating person-to-person transmission: a study of a family cluster. Lancet **395**(10223), 514–523 (2020). https://doi.org/10.1016/S0140-6736(20)30154-9
6. World Health Organization. Coronavirus disease 2019 (COVID-19): situation report 45 (2020). Accessed 5 Mar 2020. https://www.who.int/docs/default-source/coronaviruse/situation-reports/20200305-sitrep-45-covid-19.pdf?sfvrsn=ed2ba78b_4
7. Chen, Z.M., Fu, J.F., Shu, Q., et al.: Diagnosis and treatment recommendations for pediatric respiratory infection caused by the 2019 novel coronavirus. World J Pediatr. (2020). Accessed 25 Feb 2020, https://doi.org/10.1007/s12519-020-00345-5
8. Wu, Z., McGoogan, J.M.: Characteristics of and important lessons from the coronavirus disease 2019 (covid-19) outbreak in China: summary of a report of 72 314 cases from the Chinese Center for Disease Control and Prevention. JAMA (2020). Accessed 24 Feb 2020, https://doi.org/10.1001/jama.2020.2648
9. Chen, T., Wu, D., Chen, H., et al.: Clinical characteristics of 113 deceased patients with coronavirus disease 2019: retrospective study. BMJ **368**, m1091 (2020). https://doi.org/10.1136/bmj.m1091
10. Rasmussen, S.A., Thompson, L.A.: Coronavirus disease 2019 and children: what pediatric health care clinicians need to know. JAMA Pediatr. (2020). Accessed 3 Apr 202, https://doi.org/10.1001/jamapediatrics.2020.1224

Clinical Engineering, Health Technology Management and Assessment

Biomedical Engineering and Occupational Therapy Approach in Technologies for Enhancement Human Labor and Defense Abilities

Anatolie Jacob Baciu[1](✉), V. V. Fedas[1], I. E. Mereuta[1,2], M. Cecan[3], and L. A. Listopadova[4]

[1] Institution of Physiology and Sanocreatology, MECC, Chisinau, Republic of Moldova
[2] Nicolae Testemitanu State University of Medicine and Pharmacy, Chisinau, Republic of Moldova
[3] Institution of Emergency Medicine, Chisinau, Republic of Moldova
[4] Moldova State University, Doctoral School of Biology and Geonomics, Chisinau, Republic of Moldova

Abstract. A review of published data sources, their synthesis analysis, a description of the developed approach in health-forming technologies, based on the use of natural physical environments of living, daily working and servicing activity, inducing adaptive plastic transformations, optimization of gas-energy exchange in the body, strengthening of sensorimotor integration. Strategically, the development is aimed at enhancement of the working and defense capability, rehabilitation of individuals, i.e. an approach to solving the modern social problem of the growing "disability of society".

Keywords: Working, · Defense · Sensory · Processing · Neuromuscular

1 Introduction

Human enhancement technologies are methods, natural or artificial, that allow not only diagnosing and treatment for disease, rehabilitating a disabled individual, but also improving, increasing the intellectual and physical capabilities of a person. There are three forms of human enhancement: reproductive, physical and mental [1–5]. The World Health Organization (WHO) has developed a Human Capacity-Building Plan [6]. Deterioration of the work and defense capacity of individuals can occur in isolation, primarily due to poor sensory processing and sensorimotor integration. Weakened sensory processing and sensorimotor integration can be combined with or contribute to their development, for example, anxiety, panic disorder, depression, post-traumatic stress syndrome, or even schizophrenia [1, 7]. Learning impairments (mental and physical), attention deficit disorder, developmental disorders, or autism spectrum disorders are also often associated with these conditions [8]. In health-forming technologies, there is one of the methods based on the use of programs for balancing the brain (Brain Balance

I. Tiginyanu et al. (Eds.): ICNBME 2021, IFMBE Proceedings 87, pp. 715–720, 2022.
https://doi.org/10.1007/978-3-030-92328-0_90

Program) [9]. The creation of such programs was inspired by the concept of asymmetric brain activity [10]. The goal is to carry out a literature review and to develop a conceptual description of an approach to health-forming technologies based on natural ways of influencing the sensory and neuromuscular apparatus, providing an increase in work and defense capacity.

2 Materials and Methods

Over a 20-year period, to ensure the fundamentality of our research and the objectivity of the discussion of the results obtained, the following laboratory research methods were used: high-performance liquid chromatography (HPLC) with electrochemical detection for the analysis of monoamines and their metabolites in brain tissue; histochemical determination of the content of nucleic acids in subcellular compartments of neurons and their satellite neurogliocytes in the focal centers of sensory processing and generation of motor control using histochemical analysis; immunohistochemical analysis of the reactivity of the biosynthesis of neurotropic factors; biochemical testing of monoamine oxidase and proteolytic enzymatic activity of cathepsin D in brain centers; complex polysomnographic study of the wakefulness-sleep cycle; neurological, biochemical and functional tests. Experimental modeling of natural living environments on laboratory animals under the action of hypobaric hypoxia and hypercapnia, forced asphyxia. In addition, experimental modeling of sensorimotor exercises was performed on laboratory animals; experimental modeling of desynchronization of the circadian wake-sleep cycle in laboratory animals by total sleep deprivation and selective REM-sleep deprivation (Rapid Eye Movement sleep).

3 Results and Discussion

On the basis of the literature review and synthetic analysis of the material, we make a proposal, which consists in the using of natural physical methods in the technology of human enhancement, based on the harmonious adaptation of an individual to his daily educational, working, servicing activities and, in general, to the living certain environments (habitats). The efforts being made are aimed at developing a strategy for the prevention of degenerative disorders of the sensory and neuromuscular apparatus, based on the disclosure of the fundamental mechanisms of coordination of its activity, depending on the conditions of life cycle, the nature of daily working and servicing activity, energy, plastic metabolism and healthy nutrition. Such development includes a combination of complex methods for assessing and measuring molecular, cellular, tissue indicators of morphological, functional plasticity and degeneration with experimental and occupational modeling of certain types of individual activity.

Assessment of individual's abilities to perform daily vital motor acts and his professional achievements is the initial stage of The Program for Maintaining and Developing Work and Defense Capacity. This assessment is combined with the evaluation of the diet, rest and sleep [8, 11]. This approach allows you to comply with the condition for systematic testing of adaptogenicity, stressfulness or pathogenicity of the degree of applied load. In this case, the concepts of fatigue or extreme fatigue, physical and mental fatigue, psychological fatigue, and occupational fatigue, which do not have a clear division, are important, which makes it difficult to comprehensively study them by methods of physiology and psychophysiology. The state of fatigue formed the basis of the concept that considers the consequences for the body and the sensorimotor integration, in particular, daily activities at the workplace, in sports, in the performance of service duties and responsibilities, associated with a risk to health and life [12].

Based on our results, a selection of a set of various experimental and occupational models was effectuated. These models are suitable for development and rehabilitation of working capacity adequate for the maintenance and perfection of the sensory processing and neuromuscular activities.

Firstly, experimental hypoxic training in a hypobaric chamber as was chosen, followed by an experimentally modeled combination of hypoxemia with hypercapnia by breathing cessation (apnea) during sleep. This experimental combination makes it possible to test and train the system of general and local gas exchange, to start energy and plastic metabolism in functionally active tissues, primarily in the brain.

The next experimental model is based on thermoregulation training, which, in combination with gas exchange training, provides a powerful stimulating effect on angiogenesis, energy and trophic supply of active tissues.

The mechanisms of evolutionary adaptation to specific environment were studied using polysomnographic recordings of the circadian sleep-wake cycle, as well as EEG spectral analysis, which makes it possible to assess the inter-hemispheric and intra-hemispheric brain asymmetry under conditions of forced desynchronosis (sleep deprivation). This forced desynchronosis was induced by communicative sound signals in wild animals. This evolutionary adaptation is characterized by the highest degree of inter-hemispheric asymmetry: unilateral slow-wave (delta) sleep, demonstrates neuroplastic and neuroprotective abilities of the brain, preventing neurodegeneration [13]. Forced desynchronosis of the circadian rhythm is experienced by workers and employees who perform their duties around the clock with a shift schedule of production and service activities. The rehabilitation or maintenance of vital types of labor and educational activities of individuals, groups or communities is implemented, for example, through occupational therapy. Occupational therapy provides assistance to children with disabilities, ensuring their full participation in various social situations, to carry out rehabilitation after injuries and support the elderly who experience deterioration in physical and cognitive ability to work [7]. According to Mary Reilly (1961), occupational therapy can be considered the greatest medical idea of the 20th century [14]. In the period between 1860 and 1910, the international Arts and Crafts Movement significantly influenced the development of occupational therapy [15]. The Arts and Crafts movement promoted and encouraged learning through practical work. Technologies for the development of fine coordination of motor acts based on sensorimotor integration are applicable both in educational activities and in treatment models used in a wide range of client and patient groups. The first step should be the introduction into practice of systematic laboratory screening of the functional status of sensory modulation, sensory integration, neuromuscular and psychosocial deficits in individuals experiencing increased stress in the course of daily work and service. Occupational therapy for individuals with impaired sensory processing and sensorimotor integration is often based on the principles articulated in Ayres Sensory Integration. However, the specific procedures that follow these principles vary greatly depending on the individual history, providing a personalized approach, and are implemented under the auspices of the Sensory Integration Global Network (SIGN). These principles include: 1 – therapeutic intervention based on the skillful use of sensory and motor corrective, aimed at increasing tactile, proprioceptive and neuromuscular capabilities, the ability to plan actions and decision-making, organization of time and space, bilateral integration, acquisition and consolidation of postural ocular skills, as well as adaptability to increased physical activity; 2 – accommodation and adaptation, to wearing protective equipment that change sensory perception, to procedures for increasing sensitivity and improving concentration, self-regulation and self-organization, aimed at increasing the efficiency and productivity of labor and service activities; 3 – a sensory diet program based on building a daily routine associated with the implementation of a plan (menu) of individualized supportive multi-sensory stimulations or deprivations and with the help of equipment (sensory kits, stress balls, a weighted blanket, objects for distraction against the background of musical accompaniment). Such a menu is used throughout the day as needed to help cope with the problems of sensory modulation (proprioceptive, visual and auditory) associated with adverse shifts in emotional status and behavior, with increased anxiety, depression to properly change the model of processing sensory information, minimize the escalation of the crisis, in general, to promote rehabilitation and recovery; 4 – modifications and adaptations to the environment, based on the selection of lighting, the use of equipment that makes white noise, wall painting, furniture and equipment to increase or decrease the sensory stimulation created by the surrounding space. In some cases, sensor rooms, sensor stations, or sensor carts are used to accomplish these tasks; 5 – education (training) of employees, members of their families, disabled patients, as well as managers (administrators) who determine social policy, technologies for prevention, rehabilitation and correction of sensorimotor, mental functions that ensure employee productivity or the defense capability of a serviceman and a reservist [16–21]. However, for the sustainable development of an industrial country, it is important not only the development of human capabilities, but also the harmonious interaction of the human factor with the developing industrial infrastructure, for example, with robotics, which is also applicable in treatment and rehabilitation. For example, in some Toyota factories, new workers are not trained to perform their work or load the necessary materials, but, above all, spend weeks or even months practicing manually manipulations that the robot performs much faster. Such a method of increasing working capacity is used not because Toyota is less automated and robotic, but because its leaders believe that workers grow in their development, intuitively experiencing the transformation of materials into parts, and parts into assembled cars. Thus, employees are trained to appreciate the materials used and to experience more deeply the processes of interaction of materials with various technological devices. Moreover, the very process of learning to assemble a machine by hand develops and reinforces qualities in workers such as imagination, creativity, the ability to find solutions to problems and experimentation [22].

The administrations of most modern companies pay increased attention to the working capacity and qualifications of their workers and employees considering the general

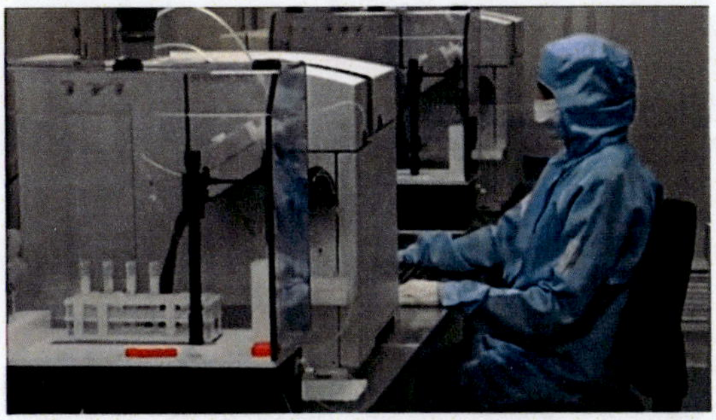

Fig. 1. Equipping a workplace in a neuroengineering manufactory

trend in technology development and the lag in adaptation of human resources to production capacities (Fig. 1). Firm leaders are convinced that the presence of professionals and their ability to work are the key to business success [23]. Therefore, it makes sense to invest in training more and more workers and employees in the most widely used work skills, as well as interacting with automated devices and software. Moreover, well-honed workforce skills help companies create products in a less risky and more predictable manner, with higher speed and productivity, lower costs and less waste [22, 23]. Strategies focused on strengthening the potential of human resources create the foundation for economic innovation and the growth of human capital, which is critical for the conclusion of symbiotic agreements, the organization of perfect production and, as a result, the rise of the national economy [24]. From the educational point of view, training of labor and defense personnel, development of research and development (R&D) programs, cooperation between educational institutions and large companies, departments are of paramount importance to meet industry and government needs.

4 Conclusions

Thus, modern health-forming technologies must make it their primary task to enhance the labor and defense capacity of members of society to prevent its "disability", the development of educational methods for optimal adaptation of workers and employees to their future profession, while showing flexibility.

Modern educational institutions, industries and departments must really pursue an inclusive policy for the socialization of persons with disabilities.

Conflict of Interest. The authors declare that they have no conflict of interest.

References

1. Buchanan, A.: Moral status and human enhancement. Philos. Public Aff. **37**(4), 346–381 (2009)
2. Enhancement Technologies Group European Commission: official site (2020)

3. Antonio, A.M.: Human enhancement: a new issue in philosophical agenda princípios. Princípios **20**(33), 265–304 (2013)
4. Institute for ethics and emerging technologies: official site. Human (2020)
5. Moore, P.: Enhancing Me: The Hope and the Hype of Human. Tech Know Series, 1st edn., p. 276 (2008)
6. World Health Organization: official site. Geneva (2020)
7. About Occupational Therapy. American Occupational Therapy Association: official site. USA (2020)
8. Kielhofner, G.: Conceptual Foundations of Occupational Therapy Practice. F. A. Davis Company, Philadelphia (2009)
9. Brain Balance Programs. Result & Research. Brain Balance Achievement Centers, 36 (2019)
10. Kim, S.M.: Balance deficit and brain connectivity in children with attention-deficit/hyperactivity disorder. Psychiatry Invest. **14**(4), 452–457 (2017)
11. Case-Smith, J.: Occupational Therapy for Children Maryland Heights. Mosby/Elsevier, Amsterdam (2010)
12. Morosan-Danila, L., Bordeianu, O.: Work capacity and fatigue relation in employees activity. Revista Economica **2**, 201–211 (2012)
13. Lyamin, O.I., Baciu, A.Ja., Kosenko, P.O., Listopadova, L.A.: EEG arousal from uni- and bilateral slow-wave sleep in Northern fur seal (*Callorhinus ursinus*). In: Abstracts of III International Conference of Young Scientists "Physiology: from Molecules to Organism", dedicated to the 95th Anniversary of the National Academy of Sciences of Ukraine, Kiev, pp. 39–40 (2013)
14. Canadian Association of Occupational Therapists. Association canadienne des ergothérapeutes, CAOT-ACE: official site. Ottawa (2020)
15. Peloquin, S.M.: Embracing our ethos, reclaiming our heart. Am. J. Occup. Ther. **59**(6), 611–625 (2005)
16. Champagne, T.: Sensory modulation & environment: Essential elements of occupation, 3rd ed. T. Champagne, Sydney (2010). Pearson Assessment
17. Champagne, T.: The influence of posttraumatic stress disorder, depression, and sensory processing patterns on occupational engagement: a case study. Work **38**(1), 67–75 (2011)
18. Champagne, T.: Researching the safety & effectiveness of the weighted blanket with adults during an inpatient mental health hospitalization. Occup. Ther. Ment. Health **31**, 211–233 (2015)
19. May-Benson, T.: Occupational therapy for adults with sensory processing disorder. OT Pract. **14**(10), 15–19 (2009)
20. Mullen, B.: Exploring the safety and effectiveness of the therapeutic use of the weighted blanket with adults. Dickson Occup. Therapy Mental Health **24**, 65–89 (2008)
21. Sensory Integration Global Network. Ayres Sensory Integration: official site
22. Hagel, J.: Skills change, but capabilities endure. Why fostering human capabilities first might be more important than reskilling in the future of work. A report from the Deloitte Center for the Edge. Deloitte Development LLC, 20 (2019)
23. Ebhota, W.: Human capacity building in manufacturing sector: a factor to industrial growth in nigeria. Int. J. Social Manag. Econ. Bus. Eng. **8**(3), 865–870 (2014)
24. Sigam, C.: Human capacity problems in developing countries and local content requirements in the extractive industries. GREAT Insights **1**(5) (2012)

Implementation of a Medical Equipment Inventory at a Regional Healthcare System in Greece

Spilios Zisimopoulos[1](✉), A. Dermitzakis[1], C. Roilos[2], and N. Pallikarakis[1]

[1] Institute of Biomedical Technology, Patras, Greece
[2] 2Nd Regional Health Authority of Piraeus and the Aegean Sea, Athens, Greece

Abstract. Modern medicine is highly dependent on recent developments in new diagnostic and therapeutic methods, techniques and equipment. An inventory of all existing medical equipment is essential to have a clear picture of the technology in use. INBIT undertook a project to create a regional-wide medical equipment inventory of all 85 medical centers belonging to the of the 2nd Regional Health Authority (RHA), most of them at Aegean Sea islands. Since the validity of the final project's results mostly depends on data integrity and uniformity, two points are very critical: the methodology of registration and the use of correct codification and nomenclature. To this scope, parameters such as resources, inventory protocol, codification, and uniformity of the data, along with the adequate software tools to be used were designed and planned. Data to register included the manufacturer, model, serial number, medical device (MD) group, year of put in service, way of procurement, place that it is used/installed etc. A working team of nine biomedical engineers was formed with a specific and clear protocol on the procedure to be followed, with emphasis on the assignment of each device to the correct MD group. Additionally, given that equipment can be found anywhere in the medical centers, the cooperation of the healthcare staff was very crucial. Finally, since the inventory is dynamic, the establishment of update procedures has been designed and prepared and local staff was trained to keep up the inventory alive. The results were delivered to each healthcare unit and to the RHA authorities in an electronic form and uploaded to the web-Praxis medical equipment management system (MEMS), ready for implementation after user training.

Keywords: Medical equipment management · Medical equipment inventory · Biomedical technology

1 Introduction

The availability of reliable data on the installed medical equipment is crucial for making the right decisions regarding the procurement, management and replacement of technology, but also for ensuring that the available equipment is performing as intended by the manufacturer [1, 2]. WHO has published a series of studies to assess the needs of a healthcare system, based on the population covered and the human and material

I. Tiginyanu et al. (Eds.): ICNBME 2021, IFMBE Proceedings 87, pp. 721–728, 2022.
https://doi.org/10.1007/978-3-030-92328-0_91

resources available at National, Regional, or Health Unit level [3, 4]. Applying similar approaches, it is possible to identify the medical technology that should be available at each medical unit, and knowing the existing equipment, estimate the real needs and plan the procurement strategy.

A well-structured medical inventory is the cornerstone for the adoption and use of a medical equipment management system, as it offers a readily available, clear and up-to-date picture of the technology used in a healthcare facility. These systems contribute to the rational management of the equipment, the proper monitoring of the periodic maintenance, as well as the supply and maintenance contracts of the equipment. The adoption of such a management system is expected to reduce the medical equipment downtime (faster fault reporting and access to information), while also recording their cost, type and number of interventions. Finally, it can facilitate investment planning, equipment procurement, development of new departments and the redesign of services.

In the current work the procedure followed in creating a complete medical equipment inventory is presented and the implementation of a web-based MEMS in all the Health Units of the 2nd Regional Health System (RHA) of Piraeus and the Aegean Sea in Greece.

2 Materials and Methods

2.1 Background

The 2nd RHA of Greece consists of 51 Health Centers (HCs) and 34 Autonomous Multipurpose Regional Medical Units, covering the special needs of Greece's sparsely inhabited Aegean Sea islands. For simplicity's sake, both of these unit categories will be referred to as HCs. The registered HCs are scattered across Piraeus, the Western Athens Sector and the Aegean Sea Islands (Fig. 1). Additionally, numerous Local Health Units and Centers exist to serve small communities with no immediate access to a Hospital or a HC. The latter category could not be registered in their entirety by the working team and had to be carried out by the local healthcare professionals, after training.

A working team of nine Biomedical Engineers was formed in September 2020 to carry out the registration and inventorying of the medical equipment. This procedure was performed in three distinct phases: 1) recording of medical equipment data, 2) data entry in a MEMS and 3) user training.

2.2 Medical Equipment Data Recording

The recording of the medical technological equipment was carried out by specialized, highly experienced biomedical engineers, with on-site visits to the HCs and with specific criteria for the kind of medical equipment to be recorded. During the procedure, the medical equipment was marked with indelible labels, and pictures of it, as well as the HCs' areas were taken, using a specially designed and developed smartphone application, adapted to the needs of the project. The working group was scattered simultaneously on different islands and municipalities for faster coverage, according to individual assignments. Individual time plans were created after preliminary contacts with each HC.

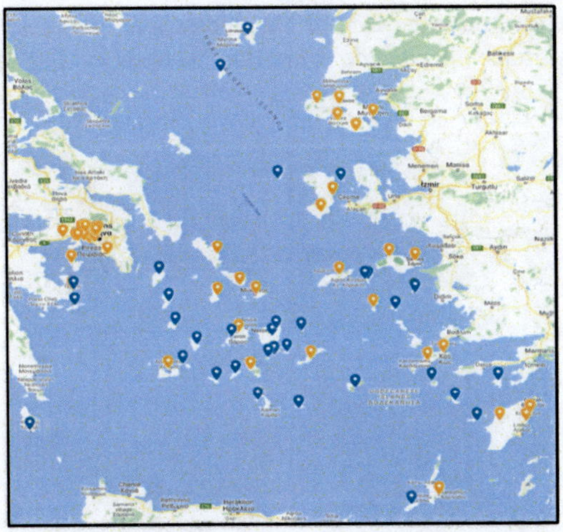

Fig. 1. HCs locations of the 2nd RHA located in the Aegean Sea, Piraeus and West Athens Sector.

The equipment to be recorded was indicated by the HCs' personnel and at least the following pictures were taken: the indelible label with the QR code of the device, the marking label of the manufacturer attached to the machine which usually contains the basic device identification data, and a front view of the device. The indelible label contains a unique combination of a numerical device code, the RHA's and HC's name, the year of the recording and a QR code referring to the full information available on each device (Fig. 2.). Moreover, information like the device status and acquisition method was registered, based on information provided by the local staff. At the same time, the whole process was demonstrated to the staff members of each HC and additional marking labels were delivered, for future recordings.

The onsite visits period spanned from November to December 2020. In total, the medical equipment of 85 HCs and an additional 40 Local Healthcare Units, was registered room-by-room and item-by-item. During this whole process a special effort was made to minimize the inconvenience to the medical and nursing staff, as well as to the patients. The excellent cooperation with the staff proved to be a catalyst in overcoming the additional special constrains imposed by the COVID-19 pandemic.

2.3 Data Entry in a MEMS

Based on the data registered and information acquired during the local visits, data entry of the details of each device was performed in INBIT's developed web-based MEMS, called web-Praxis. The system was previously customized to better fit the purpose of this application. The data entry procedure was performed by the same working group that carried out the registration. The information set corresponding to each device entity in the system comprises of the HC name and code, device code, serial number, model, manufacturer, medical device group, parent system, manufacture year, department of installation, room, ownership department, operational status, supplier, technical support, way of procurement, comments, date of last update and username that performed the last update. The available information is inserted in corresponding text or search fields.

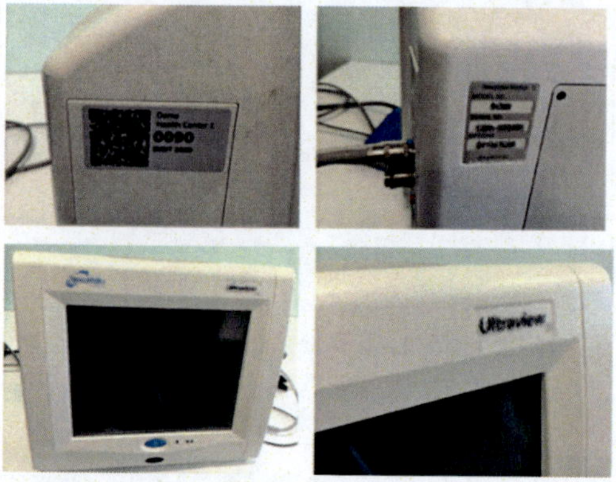

Fig. 2. Example of the set of pictures taken for a medical device, containing the QR code label and the device details.

The medical devices were assigned in MD groups according to the coding system of the Global Medical Device Nomenclature (GMDN) agency [5], after their model identification. This coding is crucial for a complete inventory, as it is the basis for the setting up of the technical specifications of the available equipment in preparation of future supplies. The process was quite complex and took about 8 person months of effort. This was mainly due to the great variety of equipment and manufacture/acquisition time that spanned over 3 decades, in some cases with incomplete data from the manufacturer and the difficulty in matching to a GMDN code, which began to be used only 20 years ago. At the same time, it should be noted that this system presents several drawbacks due to frequent changes in terms and definitions, but also the removal of codes or the creation of new which in some cases are overlapping. For the categorization to be more practical and meaningful to the medical staff, a second, higher level of classification in the form of general medical device groups was created. This was done by assigning different GMDN codes corresponding to devices with similar technical characteristics and use, to the same general group. The general medical device groups of electrocardiographs, defibrillators, sterilizers, ovens, patient monitoring, suction pumps, chemistry analyzers, hematological analyzers, blood gas analyzers, sphygmomanometers, oximeters, incubators, X-ray diagnostic systems, endoscopy tower systems, infusion pumps, invasive and noninvasive ventilators belong to this higher level of classification for easier use of the system from a clinical engineering point of view.

2.4 System Capabilities and User Training

The web-Praxis is a web-based MEMS where the total inventory database is stored. This allows for a centralized database management and facilitates the system updates and backups. Except for the inventory database, various tools and services regarding the everyday work in a Healthcare Unit have been incorporated. In addition, a centralized announcement service allows for immediate informing of possible adverse events and the creation of a vigilance system. A custom made, streamlined workflow regarding device fault reports, repair and maintenance procedures was developed after consultation with the 2nd RHA, in order to cover their specific needs. Each HC will be able to handle the majority of the required steps, without the need for an immediate interaction with the RHA's technical authorities. Moreover, technical documents can now be produced in an automated, electronic form through the web-Praxis system, thus reducing bureaucracy. All in all, the augmented control of daily routine procedures that is granted to the HCs is expected to increase the efficiency and quality of the health services provided. This decentralized medical equipment management model is a novelty and the results of its use are expected to become tangible in a couple of years.

For all that to succeed, a robust training scheme addressing to the web-Praxis users in all HCs has been applied. During the device registration procedure, the staff of the HCs was also informed about the importance of having a correct and always up-to-date inventory of medical equipment, but also of the possibilities that the web-Praxis will provide in its monitoring with constant information on repairs, maintenance, quality control, etc. During the contact, great interest was expressed in the use of the system, especially regarding its fault reporting and inspection - maintenance scheduling aspects.

Following the registration and data entry phases, the training of the users took place. To this scope, a course was developed on an asynchronous education platform. This course was created by the most experienced core of the working group that performed the inventorying, to cover the full needs of a new or advanced user. During the same time, an educational mirror web-Praxis website was developed including a portion of generic, yet realistic data from the total inventory, as well as its full procedure capabilities. Access was given to both tools to the HCs and authorities users for familiarization with the system. Finally, a series of training and technical support webinars were held to ensure the users were ready to use the web-Praxis system under realistic working conditions and address any issues that occurred.

3 Results

This whole procedure led to the creation of an online electronic medical equipment inventory of the 2nd RHA of Piraeus and the Aegean Sea, which is the base for the web-Praxis MEMS to be used. A total number of 4,561 medical devices were registered. In addition, regarding the whole inventory, 1599 different types (models) of medical devices were found, manufactured by 639 different manufacturers and belonging to 284 different MD groups according to the GMDN system.

The recording of the operating condition of the equipment showed that: 75% of the devices are in use, 7% are operational but out of use, 10% are out of use, 3% are stored, 3% are not yet installed/put into operation, and 2% are in withdrawal process.

Regarding the method of acquisition, based on the data that could be collected, it seems that 90% of the medical equipment has been acquired by purchases or through the National Strategic Reference Framework (NSRF), 5% is coming from donations and 3% is under leasing. The remaining 2% was acquired by other methods (e.g. concession, demo etc.). The average age of the 1/3 of the equipment, for which there were data available regarding the year of manufacture and which is mostly new, is 6 years.

Furthermore, more than 1,700 photos and several comments were added to the data recorded. It is noted that these elements are dynamic and are constantly being updated with the use of the web-Praxis MEMS during the daily routine of the Medical Centers.

Additionally, it is worth mentioning that during the first 2 months of use, 48 repair requests have already been made, with immediate response from the 2nd RHA' technical authorities, leading to faster repair times. Every step of the procedure is recorded and the use of hard copies and printed documents has been reduced to a minimum, as the system allows for direct communication.

4 Discussion

Since 2000, many attempts have been made to create national or regional inventories which, albeit successful at the registration level, were not properly maintained and soon went out of date [6, 7]. In the present project, a special effort was made through an extensive training program and the integration of the system in the procedures of the 2nd RHA, to ensure the correct use and continuous updating of the system.

A variety of issues regarding the availability of biomedical equipment in a regional healthcare system with many remote Health Centers were identified and some of the most important are presented below.

The situation in the various HCs, regardless of the number of their medical equipment, is not uniform in many respects. Some of them include the installation and condition of buildings; the medical, nursing, administrative and other staff; the number of medical operations performed, and size of the population served; the organization, safekeeping, etc.; the absence of a clear MD maintenance and inspection plan with contracts.

Although the present project concerns only the registration of medical equipment, for its results to be the basis for the operation of a management system (aiming at its better, safer and more efficient use), the reference and correlation with other factors mentioned above can significantly contribute to immediate improvements. From this point of view, the following remarks can be made:

The registered equipment of the HCs shows that no general model was followed regarding the necessary basic equipment that each primary level Healthcare Unit should have. There is heterogeneity in terms of equipment types, number of devices, their condition, etc. At the same time, in several cases there is a discrepancy between medical staff and available equipment. Many cases were found where radiological equipment, dental systems, ultrasounds, or ophthalmology devices were not used due to lack of staff. In other cases, a lack of equipment was observed in relation to human resources.

In cases where this surplus of equipment is observed, a possible redistribution of equipment between HCs could be examined. It is worth noting that a surplus is frequently observed on remote islands, mainly due to donations not addressing real needs. This equipment is often not used, which in many cases leads to devaluation, highlighting the need for a strategic planning of donation acceptance. Finally, a total of 658 devices were found to be in withdrawal procedure, stored or out of use. It is estimated that 30% of the total above devices will have to be withdrawn. However, one should keep in mind that there is a great heterogeneity within a region with many islands, as they must serve different needs due to different conditions such as population served, with large seasonal variations, difficulty of movement, lack of a hospital etc.

Thus, a properly maintained medical inventory and the use of a MEMS is necessary, as it allows the immediate and real time depiction of the available equipment (Fig. 3).

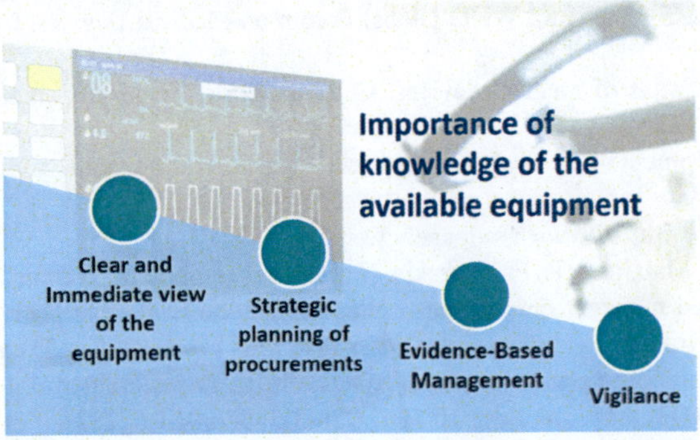

Fig. 3. The importance of knowledge of the available equipment in a Healthcare system. A basis for strategic planning, evidence-based management and vigilance on a healthcare system.

Taking this for granted, an evidence-based management addressing each case's needs can be performed, allowing for a more efficient equipment requirement analysis, distribution, as well as acquisition, maintenance and use costs. Strategic planning of procurements is facilitated, concerning the needs for new purchases or donation acceptances. Finally, a proper adverse event reporting system of can be developed, enabling the immediate detection of faulty equipment, and enacting a vigilance system.

5 Conclusions

The organization deriving from the use of web-Praxis regarding medical equipment management by the HC users, combined with the immediate involvement of centralized technical support, could play an important role in improving many performance indices of the existing medical equipment. Maintenance planning, quality and safety tests, can also provide considerable benefits. At the same time, the central logistics of consumables necessary for the use of some categories of medical equipment with modern methods of supply chain management can yield a positive economic effect, reduce the downtime of the devices and improve health services provided to the patients. The same could be achieved with the withdrawal and replacement of obsolete, unprofitable and in many cases useless equipment. A more comprehensive approach regarding the medical equipment of a RHA would require an analysis per HC, in relation with its staff and the population it serves.

The present project of the 2nd RHA constitutes a pilot application of best practice in Medical Equipment Inventorying in Greece and could be an example for the implementation of such systems throughout the country.

Conflict of Interest. The authors declare that they have no conflict of interest.

References

1. Pallikarakis, N., Bliznakov, Z.: Hospital Technology Management. Clinical Engineering, pp.47–61 (2016). https://doi.org/10.1016/B978-0-12-803767-6.00005-2
2. Pallikarakis, N., Malataras, P., Dermitzakis, A.: Web-Based Medical Equipment Management Systems (MEMS). In: Third WHO Global Forum on Medical Devices, Geneva, Switzerland (2017)
3. WHO, Global atlas of medical devices. Geneva: World Health Organization (2017). ISBN 978–92–4–151231–2
4. WHO, Development of medical device policies. World Health Organization (2011). ISBN 978 92 4 150163 7
5. GMDN agency. https://www.gmdnagency.org
6. Bliznakov, Z., Malataras, P., Pallikarakis, N.: Medical equipment inventorying and installation of a web-based management system – pilot application in the periphery of Crete, Greece. In: Jarm, T., Kramar, P., Zupanic, A. (Eds.) 11th Mediterranean Conference on Medical and Biomedical Engineering and Computing 2007. IFMBE Proceedings, vol. 16. Springer, Berlin, Heidelberg (2007). https://doi.org/10.1007/978-3-540-73044-6_282
7. Malataras, P., Kamarianakis, Z., Dermitzakis, A., Pallikarakis, N.: Nationwide medical device inventory. what to focus on. In: IRTUM – Institutional Repository of the Technical University of Moldova, Tehnica UTM (2016)

Endowment with Medical Devices and Their Conformity Assessment as Key Elements for Improving Access to High Quality Medical Services

Gheorghe Gorceag(✉)

Department of Microelectronics and Biomedical Engineering, Technical University of Moldova, Chisinau, Republic of Moldova
gheorghe.gorceag@mib.utm.md

Abstract. The importance of medical devices in the delivery of high-quality medical services has grown considerably once there was a significant progress of the medical technologies. According to the international experience in the healthcare field, the success of the medical care services is largely due to the joint contribution of physician's proficiency and the appropriate use of medicines and medical devices.

Thus, medical devices are a key component of the medical practice, and subsequently, medical device safety and efficiency are a priority for the health system and the Government overall.

Keywords: Medical devices · Testing devices · Degree of endowment · Performance and safety parameters · Conformity assessment · Periodical assessment · Medical services

1 Introduction

Currently, the technological potential of medical devices in the public medical institutions is physically and morally outdated. This is one of the main reasons that negatively impact the quality of the therapeutic and diagnostic process. The resulting low quality medical services are directly linked to patients' disease evolution, negatively influencing the entire recovery process, starting with the increased probability of faulty diagnosis to the extension of the time for the initiation of the opportune and optimal treatment, as a result increasing the number of people with serious health impairment and potential disability.

An appropriate level of endowment of medical institutions with high-performance medical devices and adequate capacity building of the medical human resources are among the most important factors in the process of securing a proper level of medical services. In turn, this has a direct impact on the beneficiaries' satisfaction degree, functional effectiveness and overall role of the health system.

I. Tiginyanu et al. (Eds.): ICNBME 2021, IFMBE Proceedings 87, pp. 729–734, 2022.
https://doi.org/10.1007/978-3-030-92328-0_92

Based on the international experience in this field, the allocation of the necessary resources and the establishment of structures/units in charge of medical devices management within the medical institutions significantly increase the performance of the medical equipment.

In conclusion, the quality of the provided medical services depends on the available medical technologies, including medical devices, as well as the appropriate human resources that can correctly use and maintain these technologies.

2 COVID-19 Pandemic, a Precondition for Endowment with Medical Devices

COVID-19 infection is a global challenge, being caused by a new SARS-CoV-2 virus, generating serious consequences with severe medical, social and economic impact.

On December 31, 2019, the World Health Organization was informed about cases of pneumonia of unknown etiology in Wuhan, China. The Chinese authorities have established the pathogen as being a new type of Coronavirus, the disease being later named by the World Health Organization as COVID-19.

The aggravation of the epidemiological situation due to COVID-19 outbreak, with the rapid increase in the numbers of infected people and the numbers of lethal outcomes, led the World Health Organization to declare on January 30, 2020 a public health emergency of international concern, and on March 11, 2020, WHO updates COVID-19's status to a global pandemic.

In this context, the Government of the Republic of Moldova, ensured the intersectorial coordination and communication for the implementation of public health measures associated with the emergency and risk of COVID-19 transmission.

The Ministry of Health, Labor and Social Protection performs a continuous needs assessment, launches the mobilization of financial and medicine resources, medical devices, protective equipment and ensures their appropriate distribution to the institutions involved in providing health care assistance [1].

The mechanisms of the financial resources' mobilization involve the allocation of financial and material means for eliminating the consequences of public health emergencies.

Thus, based on World Health Organization scenarios concerning COVID-19 infection, as well as the guidelines for assessing the needs for COVID-19 prevention, the Republic of Moldova performed emergency assessment of the existing medical device endowment and functionality, and established a list of medical devices [2] considered of paramount importance in the fight against COVID-19, as shown in the Table 1.

Subsequently, beginning with 2020, according to the established list of the medical devices used for COVID-19 prevention and control and as a result of joint efforts (donations, acquisitions and projects), it was possible to endow the public medical institutions with additional necessary medical devices, such as: artificial respiration devices/fans; oxygen concentrators; mobile digital radiology devices; infusion/infusion pumps; ultrasonography; electrocardiographs; beds for intensive care; pulse oximeters; patient monitors; laryngoscopes/video laryngoscopes; defibrillators; oxygen generators, RT PCR Analyzer, etc.

Table 1. List of medical devices needed to prevent and combat COVID-19 infection

No	Generic name of the medical device
1	RT PCR Analyzer
2	Analyzer acid-base balance
3	Device for non-invasive assessment of cardiac output
4	Blood gas analyzer
5	Non-infrared thermometer
6	Pulse oximeter
7	Patient monitor
8	Medical gas source (O2 generator, O2 concentrator)
9	Humidifier
10	Laryngoscope/Video Laryngoscope
11	Ventilator & CPAP/BPAP
12	Infusion pump or whit syringe
13	Electrocardiograph
14	Vacuum surgical/medical
15	Defibrillators (automatic, semi-automatic and manual)
16	Endocardial electrocardiostimulator (external pacemaker)
17	Ultrasonograph/Doppler echocardiograph
18	Medical/multifunctional beds
19	ECMO system
20	Magnetic Resonance Imaging/MRI
21	Computed Tomography/CT
22	Angeograph
23	Radiological unit/X-ray
24	Sterilization devices
25	Medical waste processing devices

It should be mentioned that until the pandemic, the ratio of medical devices older than 10 years used in the public medical institutions in the Republic of Moldova constituted about 40%. According to the statistical data, the degree of endowment with medical devices in our country was approximately 1.5 medical devices per hospital bed, compared to the European Union countries, where the average was 8.0 medical devices per bed.

As a result of joint efforts to ensure the medical institutions with the necessary medical equipment, the degree of endowment increased from 1.5 to 3.0 medical devices per bed, and the ratio of outdated medical equipment decreased considerably.

3 Guaranteeing the Quality of Medical Services by Periodical Assessment of Medical Device Conformity

Conformity assessment of the performance parameters, as well as electrical and general safety, functionality, and control of medical devices alarms [3], is another essential component in order to ensure the health system with efficient and safe medical devices.

In this regard, in the Republic of Moldova has been established a system of periodical verification of medical devices [4] which are in use and functioning, a mechanism that aims to increase the safety level of patients and medical device users. These verifications are performed by accredited and recognized conformity assessment bodies, in accordance with the regulatory provisions. The research regarding the necessity for periodical assessment of medical device conformity [5], as well as its benefits have been explained in several previous works such as: Procedures and methods for periodical verification of medical devices that guarantee compliance with performance and electrical safety parameters (2020), Increasing the quality of medical services by periodically checking medical devices (2019), Increasing the quality of medical services by ensuring a high level of performance and security of medical devices (2018), etc.

It is important to mention that the process of periodical verification of the medical devices which are in use and functioning shall be applied after the expiry of the warranty period of the medical device.

At the same time, in accordance with the requirements of the regulations regarding the periodical verification of medical devices, they are to be prepared for the verification procedure, a process that is provided by qualified technical staff, namely the medical bioengineers.

Appropriate devices and tools for testing medical devices are required to perform the above procedures.

Respectively, at the end of 2020 an evaluation on the endowment of the health system with medical bioengineers was performed. As a result, it was found that about 50 medical bioengineers are working in public medical institutions in the Republic of Moldova, however the need is estimated around 250–300 medical bioengineers.

Additionally, some of the largest beneficiaries of medical devices in the COVID-19 context were selected based on the criterion of the medical bioengineer's presence in the public medical institutions, namely: the "Timofei Mosneaga" Republican Clinical Hospital, the Institute of Emergency Medicine, the Institute of Mother and Child, the "Sfanta Treime" Municipal Clinical Hospital and the Balti Clinical Hospital.

A list of tools and devices for testing medical equipment has been developed for these medical institutions, as shown in Table 2.

Based on the presence of the necessary testing devices and tools, the biomedical engineering staff is able to ensure the permanent periodical evaluation of the performance and safety parameters of medical devices, including in the process of preparing them for periodical verification.

Table 2. List of devices for testing medical devices

No	Name of the device for testing
1	Analyzer for ventilators
2	Multiparametric simulator, SpO2 simulator
3	Oxygen analyzer
4	2-channel oscilloscope
5	3D printer
6	Infrared thermometer
7	Test device for defibrillators
8	Infrared soldering station with preheater
9	Test device for infusion pumps
10	Test device for incubators for newborn
11	Electrical safety check device
12	Portable digital multimeter
13	Belt sanding machine
14	Stationary multimeter
15	Power source

4 Conclusions

The increase in the endowment of public medical institutions from 1.5 to 3.0 medical devices per bed, led to the improvement of the population's access to medical services. Additionally, ensuring the process of periodical verification of medical device conformity guarantees the use of safe and efficient medical devices, which in turn, lead to the delivery of high quality medical services.

Conflict of Interest. The author declare that he have no conflict of interest.

References

1. Needs assessment for medical devices. https://www.who.int/publications/i/item/9789241501385
2. Priority medical devices list for the COVID-19 response and associated technical specifications. https://www.who.int/publications/i/item/WHO-2019-nCoV-MedDev-TS-O2T.V2
3. SM EN 62353: Electromedical devices. Recurrent inspection and Inspection after Repair of an Electromedical Device (2016)
4. The Law on Medical Devices no. 102 of June 9, 2017. Official Gazette of the Republic of Moldova, no. 244–251, 389 (2017). https://www.legis.md/cautare/getResults?doc_id=119271&lang=ro

5. Government Decision no. 966 of 14.11.2017 for the approval of the Regulation on the periodical verification of medical devices put into operation and in use. Official Gazette of the Republic of Moldova, no. 399–410, 1076 (2017). http://lex.justice.md/index.php?action=view&view=doc&lang=1&id=372536

The Impact of Positive Blood Alcohol Content on Outcomes of Trauma Patients

E. Corețchi(✉), O. Arnaut, V. Vovc, S. Șandru, S. Cobîletchi, C. Trofimov, V. Mogîldea, R. Baltaga, and I. Grabovschi

Nicolae Testemitanu State University of Medicine and Pharmacy, Chisinau, Moldova
eugeniu.coretchi@usmf.md

Abstract. The association of alcohol use with in hospital trauma deaths remain unclear. This study tries to identify the association of blood alcohol content with outcomes.

Methods: Four electronic databases (MEDLINE, research4life, PubMed, ScienceDirect) were searched from 2015 to 2021 using a search strategy. The search strategy included a combination of medical subject headings (Mesh), and variations of keywords such: trauma, polytrauma, injury, alcohol, ethanol positive, blood alcohol level, and outcome. Some studies concluded that there is a better outcome for BAC positive patients.

Results: A total of 17 studies were published between 2015 and 2021. In total, were included 558577 trauma patients in 17 studies, number of patients ranging from hundreds to tens of thousands. Studies varied in the mechanism of traumatic injury, injury-related to specific body areas, patient ages. Also, studies vary about the level of BAC + (Blood Alcohol Content/concentration level). While other studies suggest an opposite opinion. It seems that data is more informative for TBI (Traumatic Brain Injury) where most studies related to a better outcome for trauma patients with BAC positive.

Conclusion: Some studies present alcohol as a protective factor for trauma patients, while other studies present BAC as a worsening factor for patient outcome. No doubt that is necessary more studies in this direction.

Keywords: Alcohol · Trauma · Outcome

1 Introduction

Alcohol is known to be a psychoactive substance, a Central Nervous system depressant that has highly dependent proprieties. Nowadays, alcoholic beverages are one of the most consumed products daily, some people consume for the recreational purpose at gatherings or parties, while others are dependent on it. Unfortunately, the range of people who consume alcohol is very vast, from teenagers to older adults. More than that, social networks, uploaded pictures with alcoholic beverages, and advertisements have a strong influence to encourage the consumption of alcohol. Over time, serious health problems can occur due to alcohol [1]. In this context, it is very easy to discount the health and social damage caused by alcohol or contributed to by drinking.

I. Tiginyanu et al. (Eds.): ICNBME 2021, IFMBE Proceedings 87, pp. 735–742, 2022.
https://doi.org/10.1007/978-3-030-92328-0_93

Alcohol consumption contributes annually to 3 million deaths and leads to poor health outcomes and disabilities. Overall, the harmful use of alcohol is responsible for 5.1% of the global burden of disease. Harmful use of alcohol is accountable for 7.1% and 2.2% of the global burden of disease for males and females respectively. Alcohol is the leading risk factor for premature mortality and disability among those aged 15 to 49 years, accounting for 10 percent of all deaths in this age group. Low-income countries have higher rates of alcohol-related death and hospitalization. Alcohol has a significant role in the occurrence of more than 230 diseases. Also, causes harm to the well-being and health of people around the consumer. In 2019, the worldwide total consumption was equal to 5,8 L of pure alcohol per person of 15 years and older. Unrecorded consumption accounts for 26% of the worldwide total consumption [2]. In the Republic of Moldova, alcohol consumption was equal to 28,4 L of pure alcohol per person 15 years and older. Distributing by gender is approximative 38 L of pure alcohol for males and 13,4 L of pure alcohol for females [3]. In the Republic of Moldova, dates from 2016 report a prevalence of alcohol use disorders that affects 10.3% of males, 2.2% of females, and 6.0% of both sexes. About alcohol dependence suffering near to 5,5% of males and 1,4% of females [4]. Alcohol is the third leading preventable cause of death in the United States, after tobacco, poor diet, and physical inactivity. National Institute on Alcohol Abuse and Alcoholism (NIAAA) describes the real situation about alcohol-related deaths, where 68.000 men, 27.000 women died because of driving under alcohol influence [5]. In Europe, about 800 adults (5.5% of all deaths = 291000 people) die from alcohol every day. In Moldova, 26.1% of the deaths are caused by alcohol consumption [6].

There is no doubt that alcohol consumption can increase the traumatic frequency, being an important factor that affects the trauma evolution. Also, acute and/or chronic alcohol consumption contributes to the incidence of traumatic brain injury (TBI). However, more than thirty years of research on the impact of alcohol on the neuroanatomical and functional outcomes of TBI have produced ambiguous results [7]. Clinical studies on short-term outcomes such as morbidity, health complications, or mortality during acute care have been associated with a positive admission blood alcohol content. The effects of alcohol consumption increase the risk of injuries during different activities. Also, increases the risk of violence-related injuries, falls, cycling accidents, motor vehicle crashes, head traumas, and assaults. Besides that, increases the risk of cardiovascular disease, mental health issues, and endocrinology diseases [8, 9].

Alcohol intake is linked with an increased incidence of head trauma. Furthermore, the severity of the craniofacial injury is higher in intoxicated patients. Head trauma is commonly caused by assault, falls, or cycling accidents. In cycling accidents, one-third of the injured cyclists were reported to be BAL positive and head injuries were more common in this group [10]. Driving under the influence of alcohol increases the risk of fatal and non-fatal driving accidents. A possible explanation for this is that intoxicated drivers usually take more risks and drive faster. Further, they are also more likely to suffer serious injury or death compared with sober drivers.

The association between alcohol intoxication and injury severity, mortality, and length of stay (LOS) in hospitals is, however, inconsistent in the literature. Indeed, the majority of studies have found no difference in mortality [11], and length of a hospital of stay between BAL positive and BAL negative patients [12]. However, some studies have reported a higher mortality rate, and longer length of hospital stay among BAL positive trauma patients. Based on the findings of some studies, BAL positive patients had higher injury severity scores (ISS) when examined for minor trauma (ISS less than 16), whereas in severely injured patients (ISS greater than or equal to 16) there was no difference in ISS between BAL positive and BAL negative patients [13–17].

2 Methods

This systematic review was conducted following the Preferred Reporting Items for Systematic Reviews and Meta-Analyses (PRISMA) guidelines [18]. Four electronic databases (MEDLINE, research4life, PubMed, ScienceDirect) were searched from 2015 to 2021 using a search strategy. The search strategy included a combination of medical subject headings (Mesh), and variations of keywords such as *trauma, polytrauma, injury, alcohol, ethanol positive, blood alcohol level,* and *outcome*. Chosen studies for analysis met the following inclusion criteria: (a) design –a study of original data involving human participants; (b) population – trauma patients admitted to a hospital or trauma center; and (c) exposure – patients with a positive blood alcohol concentration (BAC) or other evidence of alcohol use on admission. Studies were excluded if they did not report on trauma specifically related to alcohol, or if study patients were presenting for acute intoxication without traumatic injury.

3 Results

A total of 17 studies met all of the inclusion criteria. Table 1, summarizes all the characteristics of each including the study. All 17 studies were published between 2015 and 2021 with six performed in USA [12, 19–23], five in Europe [24–28], four studies in Asia [29–32] and two studies were perform in Australia [33, 34]. With respect to study design, all studies were retrospective in design. Studies were performed both at a single center and also involved major adult trauma referral centers. In total, were included 558577 trauma patients in 17 studies, number of patients ranging from hundreds to tens of thousands. Studies varied in the mechanism of traumatic injury, injury-related to specific body areas, patient ages. Also, studies vary about the level of BAC + (Blood Alcohol Content/concentration/level). There are differences between the time of admission after injury. The most common methods to determine alcohol uses were measurement of BAC levels and evidence from medical records. In 2016, in a cross-sectional study from Taiwan, designed to investigate the effect of alcohol intoxication on the clinical presentation of 929 hospitalized adult trauma patients and to measure the outcome (mortality), Shu-Hui Peng et al. conclude that patients with alcohol intoxication presented with significantly higher short-term mortality [32]. Majid Afshar et al. in a historical cohort of 46222 admissions to the trauma center, found 13372 patients with BAC +. In this study, the authors created three levels of BAC being described as follows: Moderate BAC:1–100 mg/dl, High BAC: 101–230 mg/dl, Very high: > 200 mg/dl. The outcome corresponds with a U-shaped curve which means that moderate BAC-increase mortality and very high BAC-decrease mortality. Alcohol exposure by BAC groups was associated with both injury and in-hospital deaths in an inverse U-shaped distribution [19]. Jennifer S Albrecht et al. in 2018, in a retrospective cohort study on individuals treated for isolated TBI (Trauma Brain Injury), with 282 patients BAC + and 802 patients BAC-, observed that positive blood alcohol was associated with a decreased risk of mortality, among individuals with penetrating injuries [20]. Qiuping Ding et al. in 2018 conclude that positive BAC at the time of admission after TBI reduces mortality rate compared with a negative BAC[29]. Mary K Bryant et al. in 2020, in a retrospective cohort study of injured patients $\geq$ 65 years with BAC testing presenting to a Level 1 trauma center between 2015 and 2018, conclude that any detectable BAC along and increasing thresholds of BAC was not associated with poor in-hospital outcomes of older patients after trauma[21]. In a retrospective cohort study, Nasim Ahmed et al. in 2020, concluded that patients who tested with a BAC above the legal limit sustained a higher injury severity score and higher in-hospital mortality compared to patients who tested negative [22]. In another study, Elizabeth Anne Leonard et al. found that alcohol intoxication in hospitalized major trauma patients is independently associated with decreased injury severity and mortality in patients with TBI. In a retrospective study, Suzanne F. Van Wijck, et al. in 2021 suggest that there was no independent association between alcohol intoxication and higher in-hospital mortality in emergency patients with TBI [23]. In a study with 306 BAC positive patients and 99 BAC negative patients, Rahul Raj et al. in 2015, revealed a significantly reduced risk of six-month mortality [24]. In 2020, Antti Riuttanen et al., in a retrospective study, concluded that one-year mortality was lower among patients with BAC positive [25]. Another study, from Germany, published in 2021, related no difference between BAC positive and BAC negative groups, except patients with TBI which present better outcomes [26]. Il-Jae Wang et al. from Korea, in 2021, after publishing a retrospective study, where BAC positive were considered values of alcohol more than 10.1 mg/dl, related that The patients in the alcohol-positive group had higher mortality rate, higher clotting time, and lower maximum lysis, more fibrinolysis shutdown, and hyperfibrinolysis than those in the alcohol-negative group [30]. Kazuma Sasaki et al., from Japan, noticed that alcohol consumption could be beneficial in severe blunt trauma patients as regards in-hospital mortality[31]. Another study by Majid Afshar et al. in 2016 after data processing of 44502 BAC positive patients with trauma conclude that the patients with a BAC $\geq$ 400 mg/dL also had the lowest proportion of severe injuries and in-hospital death in comparison with the other alcohol groups [12]. In a study, made by James H. Brennan et al. between 2006 and 2011, examining 54882 patients with BAC positive reports that ethanol intoxication was associated with reduced in-hospital mortality [34]. Rahul Raj et al. in a study from Finland with 95941 BAC positive patients with TBI, related lower mortality rates in moderate to severe TBI and BAC positive. Shahin Mohseni et al. concluded that there was a worse long-term functional outcome in BAC negative patients compared to their BAC positive. Positive BAC on admission is associated with a better long-term functional outcome in patients suffering severe traumatic brain injury [28].

4 Discussion

The goal of this systematic review was to examine the outcome of patients with trauma and BAC positive. Some studies concluded that there are better outcome for BAC positive

Table 1. Legend: M- moderate, V.H.- very high, D- decrease, I- increase

Author/Year	Location	Sample size	Type of trauma	
Shu-Hui Peng et al. 2016	Taiwan [32]	929 (BAC +) 10104 (BAC-)	All types	I. mortality
Majid Afshar et al. 2015	USA [19]	13372(BAC +) 21082 (BAC-)	All types	M. BAC-I. mortality V.H-D. mortality
Taa Jennifer S Albrecht et al. 2018	USA [20]	282(BAC +) 802 (BAC-)	Brain injury	D. mortality
Qiuping Ding et al. 2017	China [29]	48537(BAC +) 82487 (BAC-)	Brain injury	D. mortality
Mary K Bryant et al. 2020	USA [21]	111 (BAC +) 772 (BAC-)	All types	No protective effect
Nasim Ahmed et al. 2020	USA [22]	21180 (BAC +) 46003 (BAC-)	All types	I. mortality
E. A. Leonard et al.2016	Australia [33]	60472 (BAC +)	All types	D. mortality most
S. F. Van Wijck, et al. 2021	USA [23]	715 (BAC +) 2150 (BAC-)	Brain injury	D. mortality
Rahul Raj et al. 2015	Finland [24]	306 (BAC +) 99 (BAC-)	Brain injury	D. mortality
A. Riuttanen, et al. 2020	Finland [25]	252 (BAC +) 95 (BAC-)	All types	D. mortality
T. Brockamp et al. 2021	Germany [26]	832 (BAC +) 5271 (BAC-)	Brain injury	No difference /D. mortality for TBI
Il-Jae Wang et al. 2021	Korea [30]	190 (BAC +) 496 (BAC-)	All types	I. mortality
Kazuma Sasaki et al. 2021	Japan [31]	8543 (BAC +) 37818 (BAC-)	All types	D. mortality
Majid Afshar et al. 2016	USA [12]	44502 (BAC +)	All types	D. mortality
J. H. Brennan et al. 2015	Australia [34]	54882 (BAC +)	Brain injury	D. mortality
Rahul Raj et al. 2016	Finland	95941 (BAC +)	Brain injury	D. mortality

patients [20, 23–25, 27–29, 31, 33, 34]. While other studies suggest an opposite opinion [22, 30, 32]. It seems that data is more informative for TBI (Traumatic Brain Injury) where most studies related to a better outcome for trauma patients with BAC positive. Different BAC groups are associated with very different types of injury and BAC needs to be more studied to be considered the only predictive/individual factor for outcome.

Acknowledgment. Our systematic review shows that some studies present alcohol as a protective factor for trauma patients, while other studies present BAC as a worsening factor for patients' outcomes. No doubt that is necessary more studies in this direction. More like this, in the Republic of Moldova, these types of studies are completely absent, although Moldova is among the leaders in alcohol consumption [35], trauma having an important place in the lethality structure [36].

Conflict of Interest. "The authors declare that they have no conflict of interest".

References

1. Homish, G.G., Leonard, K.E.: The social network and Alcohol use. J. Stud. Alcohol Drugs **69**(6), 906 (2008). https://doi.org/10.15288/JSAD.2008.69.906
2. WHO. Hammer, J.H., Parent, M.C.N.C.C., Douglas, M.S.: AHealth, Alcohol Dependence and Harmful Alcohol Use. British Psychological Society (2018). https://www.ncbi.nlm.nih.gov/books/NBK65500/. Accessed 07 Aug 2021
3. WHO, Alcohol, drinkers only per capita (15+ years) consumption in litres of pure alcohol (2016). https://www.who.int/data/gho/data/indicators/indicator-details/GHO/alcohol-drinkers-only-per-capita-(15-)consumption-in-litres-of-pure-alcohol. Accessed 07 Aug 2021
4. WHO, Alcohol Consumption: Levels and Patterns. https://www.who.int/substance_abuse/publications/global_alcohol_report/profiles/mda.pdf?ua=1
5. National Institute on Alcohol and Alcoholism. Alcohol Facts and Statistics | National Institute on Alcohol Abuse and Alcoholism (NIAAA). (2016). https://www.niaaa.nih.gov/publications/brochures-and-fact-sheets/alcohol-facts-and-statistics. Accessed 07 Aug 2021
6. WHO. Hammer, J.H., Parent, M.C., Spiker, D.A.: Global status report on alcohol and health 2018 (2018).
7. Taylor, A.N., Sutton, R.L.: Evidence for beneficial and adverse effects of alcohol in animal models and clinical studies of traumatic brain injury. Brain Neurotrauma Mol. Neuropsychol. Rehabil. Asp. 669–680 (2015). https://doi.org/10.1201/b18126
8. Mayor, S.: Small increase in alcohol duty may cut violence related emergency department visits. BMJ, **354** (2016). https://www.bmj.com/content/354/bmj.i3829
9. Riuttanen, A., Jäntti, S.J., Mattila, V.M.: Alcohol use in severely injured trauma patients. Sci. Rep. **10**(1), 1–8 (2020). https://doi.org/10.1038/S41598-020-74753-Y
10. AiraksinenIS, N.K.: Nurmi-Lüthje JM, Kataja, HPJ, Kröger, PMJ, Lüthje, Cycling injuries and alcohol. Injury **49**(5), 945–952 (2018). https://doi.org/10.1016/J.INJURY.2018.03.002
11. Benson, C., Weinberg, J., Narsule, C.K., Brahmbhatt, T.S.: A comparison of alcohol positive and alcohol negative trauma patients requiring an emergency laparotomy. Am. J. Emerg. Med. **36**(7), 1139–1144 (2018). https://doi.org/10.1016/J.AJEM.2017.11.028
12. Afshar, M., Netzer, G., Salisbury-Afshar, E., Murthi, S., Smith, G.: Injured patients with very high blood alcohol concentrations. Injury **47**(1), 83–88 (2016). https://doi.org/10.1016/J.INJURY.2015.10.063

13. Hadjizacharia, P., O'Keeffe, T., Plurad, D.S., et al.: Alcohol exposure and outcomes in trauma patients. Eur. J. Trauma. Emerg. Surg. **37**, 169–175 (2011). https://doi.org/10.1007/s00068-010-0038-5
14. Draus, J.M., Santos, A.P., Franklin, G.A., Foley, D.S.: Drug and alcohol use among adolescent blunt trauma patients: dying to get high? J. Pediatr. Surg. **43**(1), 208–211 (2008). https://doi.org/10.1016/J.JPEDSURG.2007.09.046
15. Scheyerer, M.J., Dütschler, J., et al.: Effect of elevated serum alcohol level on the outcome of severely injured patients. Emerg. Med. J. **31**(10), 813–817 (2014). https://doi.org/10.1136/EMERMED-2013-202804
16. Xu, B., et al.: Ethanol intoxication alleviates the inflammatory response of remote organs to experimental traumatic brain injury. Int. J. Mol. Sci. **21**(21), 1–15 (2020). https://doi.org/10.3390/IJMS21218181
17. Kowalenko, T., et al.: Alcohol and trauma–in every age group. Am. J. Emerg. Med. **31**(4), 705–709 (2013). https://doi.org/10.1016/J.AJEM.2012.12.032
18. Shamseer, L., et al.: Preferred reporting items for systematic review and meta-analysis protocols (PRISMA-P) 2015: Elaboration and explanation. BMJ, 349. British Medical Journal Publishing Group (2015). https://doi.org/10.1136/bmj.g7647
19. Afshar, M., Netzer, G., Murthi, S., Smith, G.S.: Alcohol exposure, injury, and death in trauma patients. J. Trauma Acute Care Surg. **79**(4), 643 (2015). https://doi.org/10.1097/TA.0000000000000825
20. van Wijck, S.F., Kongkaewpaisan, N., et al.: Association of alcohol with mortality after traumatic brain injury. Am. J. Epidemiol. **187**(2), 233–241 (2018). https://doi.org/10.1093/AJE/KWX254
21. Bryant, M.K., et al.: Does level of blood alcohol content affect clinical outcomes after trauma in older adult patients? Am. Surg. **86**(9), 1106–1112 (2020). https://doi.org/10.1177/0003134820943555
22. Ahmed, N., et al.: Elevated blood alcohol impacts hospital mortality following motorcycle injury: a national Trauma data bank analysis. Injury **51**(1), 91–96 (2020). https://doi.org/10.1016/J.INJURY.2019.10.005
23. van Wijck, S.F., et al.: Association between alcohol intoxication and mortality in severe traumatic brain injury in the emergency department: a retrospective cohort. Eur. J. Emerg. Med. **28**(2), 97 (2021). https://doi.org/10.1097/MEJ.0000000000000754
24. Raj, R., Skrifvars, M.B., Kivisaari, R., Hernesniemi, J., Lappalainen, J., Siironen, J.: Acute alcohol intoxication and long-term outcome in patients with traumatic brain injury. J. Neurotrauma **32**(2), 95 (2015). https://doi.org/10.1089/NEU.2014.3488
25. Riuttanen, A., Jäntti, S.J., Mattila, V.M.: Alcohol use in severely injured trauma patients. Sci. Rep. **10**(1), 17891 (2020). https://doi.org/10.1038/S41598-020-74753-Y
26. Brockamp, T., et al.: Alcohol and trauma: the influence of blood alcohol levels on the severity of injuries and outcome of trauma patients - a retrospective analysis of 6268 patients of the TraumaRegister DGU ®. Scand. J. Trauma. Resusc. Emerg. Med. **29**(1), 101 (2021). https://doi.org/10.1186/S13049-021-00916-Z
27. Raj, R., Mikkonen, E.D., Siironen, J., Hernesniemi, J., Lappalainen, J., Skrifvars, M.B.: Alcohol and mortality after moderate to severe traumatic brain injury: a meta-analysis of observational studies. J. Neurosurg. **124**(6), 1684–1692 (2016). https://doi.org/10.3171/2015.4.JNS141746
28. Mohseni, S., Bellander, B.M., Riddez, L., Talving, P., Thelin, E.P.: Positive blood alcohol level in severe traumatic brain injury is associated with better long-term functional outcome. Brain Inj. **30**(10), 1256–1260 (2016). https://doi.org/10.1080/02699052.2016.1183823
29. Ding, Q., Wang, Z., Shen, M., Su, Z., Shen, L.: Acute alcohol exposure and risk of mortality of patients with traumatic brain injury: a systematic review and meta-analysis (2017). https://doi.org/10.1111/acer.13436

30. Wang, I.-J., et al.: Effect of acute alcohol intoxication on mortality, coagulation, and fibrinolysis in trauma patients. PLoS ONE **16**(3), e0248810 (2021). https://doi.org/10.1371/JOURNAL.PONE.0248810
31. Sasaki, K., Obinata, H., Yokobori, S., Sakamoto, T.: Alcohol does not increase in-hospital mortality due to severe blunt trauma: an analysis of propensity score matching using the Japan Trauma data Bank. Acute Med. Surg. **8**(1), e67 (2021). https://doi.org/10.1002/AMS2.671
32. Peng, S.-H., Hsu, S.-Y., Kuo, P.-J., Rau, C.-S., Cheng, Y.-A., Hsieh, C.-H.: Influence of alcohol use on mortality and expenditure during hospital admission: a cross-sectional study. BMJ Open **6**(11), e013176 (2016). https://doi.org/10.1136/BMJOPEN-2016-013176
33. Leonard, E.A., Buckley, T., Curtis, K.: Impact of alcohol on outcomes in hospitalized major trauma patients: A literature review. J. Trauma Nurs. **23**(2), 103–114 (2016). https://doi.org/10.1097/JTN.0000000000000194
34. Brennan, J.H., Bernard, S., Cameron, P.A., Rosenfeld, J.V., Mitra, B.: Ethanol and isolated traumatic brain injury. J. Clin. Neurosci. **22**(9), 1375–1381 (2015). https://doi.org/10.1016/J.JOCN.2015.02.030
35. "Consumul de Alcool în Cercetări din România, Republica Moldova şi la Nivel Internaţional (2020). https://doi.org/10.1037/0033
36. Oleg, A., Croitoru, D., Grabovschi, I., Sandru, S.: Trauma scoring systems. Mold. Med. J. **63**(1), 64–74 (2020). https://doi.org/10.5281/ZENODO.3685673

Non-invasive Monitoring of Pulse Rate and Desaturation Events with Oximeter in Copd Patients with Cardiovascular Comorbidities

A. Popa(✉), N. Capros, T. Dumitras, O. Corlateanu, M. Dogot, I. Smolenschi, I. Sirbu, and M. Dumitras

Nicolae Testemitanu State University of Medicine and Pharmacy, Chisinau, Republic of Moldova

Abstract. The article reflects a clinical study on non-invasive monitoring of pulse rate and desaturation events with oximeter in chronic obstructive pulmonary disease (COPD) patients with cardiovascular comorbidities, in order to optimise the exacerbations diagnostic, predict the development of the complications and therapeutic approach. The presence of cardiovascular comorbidities in patients hospitalized with chronic obstructive pulmonary disease exacerbations is high and associated with rehospitalizations and deaths. Pulse oximetry is useful for non-invasive monitoring of hypoxemia presence and the severity of the patient's response to supplemental oxygen and other therapies, especially in emergency situations in exacerbated COPD and cardiovascular comorbidities.

Keywords: Chronic obstructive pulmonary disease · Pulse oximetry · Non-invasive monitoring

1 Background

Chronic obstructive pulmonary disease (COPD) is a respiratory condition characterized by persistent airflow limitation. Chronic obstructive pulmonary disease (COPD) is characterized by an abnormal inflammatory response of the lungs to noxious particles or gases. The global prevalence of COPD is approximately 11.7%, and COPD is estimated to become the fourth leading cause of death worldwide by 2030 [1].

COPD is a complex pathology, with a well-defined pulmonary component, but also with multiple extrapulmonary manifestations and important comorbidities, which can emphasize the severity of the disease. Individual exacerbations and comorbidities determine the evolution of the disease, being associated with systemic diseases, heart failure, cardiac arrhythmias and sudden death [2]. It is increasingly recognized that comorbidities are highly prevalent, irrespective of severity of COPD. These comorbidities contribute to the severity of symptoms and progression of the disease [1].

Cardiovascular diseases, including hypertension, coronary artery disease, heart failure, atrial fibrillation, stroke and peripheral arterial disease, are the most common comorbidities in COPD, with the prevalence of cardiovascular diseases in patients with COPD ranges from 13% to 68% [3, 4].

I. Tiginyanu et al. (Eds.): ICNBME 2021, IFMBE Proceedings 87, pp. 743–749, 2022.
https://doi.org/10.1007/978-3-030-92328-0_94

Pulse oximetry is routinely used to non-invasively monitor oxygen saturation levels. A low oxygen level in the blood means low oxygen in the tissues, which can ultimately lead to organ failure. Identification of the oxygen desaturation can support the diagnosis and continuous monitoring of patient pulmonary function to predict deteriorations (prognosis). Specifically, studying the variability of the oxygen saturation signal may provide information on the underlying physiological control systems. Furthermore, it may enhance our understanding of the manifestation and etiology of diseases and identify desaturation for the purpose of health monitoring. The repetitive nocturnal hypoxemia may cause oxidative stress, contributing to the pathogenesis of cardiovascular comorbidities (CVC) [5].

Similarly, patients with advanced chronic obstructive pulmonary disease (COPD) commonly exhibit overnight hypoxemia [6].

In this respect, non-invasive monitoring of pulse rate and desaturation events with oximeter in chronic obstructive pulmonary disease (COPD) patients with cardiovascular comorbidities, in order to optimise the exacerbations diagnostic, predict the development of the complications and therapeutic approach [4].

2 The Aim

Of the study was non-invasive monitoring of pulse rate and desaturation events with oximeter in COPD patients with cardiovascular comorbidities.

3 Methods

The clinical study was conducted on a group of 101 patients with COPD GOLD I-IV, hospitalized for exacerbation in munisipal hospital Holy Trinity. The patients included in the study were aged between 44 and 78 years, of whom 55 were female and 46 were male. According to the GOLD 2016 guide the clinical diagnosis of COPD was based on spirometry, confirming a ratio of postbronchodilator forced expiratory volume in 1 s (FEV_1) and forced vital capacity (FVC) below 0.70. The severity of COPD can be graded according to the degree of airflow limitation into mild ($FEV_1/FVC < 70\%$; $FEV_1 \geq 80\%$ predicted), moderate ($FEV_1/FVC < 70\%$; $50\% \leq FEV_1 < 80\%$ predicted), severe ($FEV_1/FVC < 70\%$; $30\% \leq FEV_1 < 50\%$ predicted) and very severe ($FEV_1/FVC < 70\%$; $FEV_1 < 30\%$ predicted). The current level of patient's symptoms and the risk of exacerbation for classifying COPD patients (GOLD A: low risk, fewer symptoms; GOLD B: low risk, more symptoms; GOLD C: high risk, fewer symptoms; GOLD D: high risk, more symptoms). It was determined the existence of associations between types A, B, C, D of COPD and the presence of CVC.

All patients with acute heart failure (AHF) were shown tomonitor blood pressure, heart rate, body temperature, respiratory rate and ECG, heart rhythm disturbances. Blood pressure monitoring was necessary at the beginning of treatment and was carried out regularly (every 5 min) until effective doses of vasodilators, diuretics and/or inotropic drugs were selected.

Pulse rate (PR) and desaturation events were identifies and recorded with the oximeter data manager - a hand held oximeter, battery powered, that has optional configurable alarms (for SpO2 and heart rate), multiple sensor options (disposable, wrap/bandage, soft "rubber boot", ear clip). SpO2 event (SE) was identified and recorded when SpO2 dropping was over 4% for at least 10 s. Pulse rate event was identified and recorded when PR change was over 6 bpm for at least 8 s. In less severe patients during oxygen therapy, regular pulse oximetry (every hour) was performed.

4 Results and Discussions

According to the diagnostic spirometric criteria of COPD GOLD 2016, in the study group prevailed the patients with GOLD 3 and GOLD 4 severity of the bronchial obstruction: GOLD 3 – 36 (35,64%), GOLD 4 – 31 (30,69%) compared to those with GOLD 2 – 34 (34,66%) (both comparisons $p < 0{,}01$, respectively). By corroborating the anamnestic, clinical and paraclinical data, it was established the presence of risk factors and of the CVC in the studied group, namely smoking, chronic ischemic heart disease, arterial hypertension, type 2 diabetes mellitus and obesity. A large number of patients had high body mass index values, respectively 45 (44,55%) – overweight, 18 (17,82%) – first degree obesity, 10 (9,90%) second degree obesity, and 7 (6,93%) patients had third degree obesity. The smokers in the study were 42 (41,58%) patients, the majority of the male 39 (92,86%), women only 3 (7,14%), the ex-smokers were 15 (14,85%). Type II diabetes mellitus was also frequently encountered in the group of patients studied by us – in 29 (28,71%). The share of cardiovascular diseases in the patients evaluated was 58.42%. The analysis of the presence of CVCs according to types A, B, C, D of COPD shows a predominance of the rate of CVC in subtype D compared to C, B and A (32.26 versus 16.67% versus 8.00% and 6.82%, respectively). Significant statistical correlations between the degree of limitation of air flow in the airways in patients with COPD and the frequency of CVC were not revealed. Patients with the coexistence of COPD and CVC were older (64 ± 5 vs 57 ± 10 years, $p < 0{,}01$) compared to patients without cardiovascular disease. The number of cases detected with the CVC in relation to men: women was 1.1:1. Hypertension was diagnosed in 59 (58,42%) patients, out of which those with stage II predominated (57,69% versus 42,31%, $p < 0{,}01\%$) compared to stage III of COPD.

Ischemic heart disease was present in 46 (45,55%) cases, and in 6 (13,04%) of them in anamnestic was detected old myocardial infarction. Electrocardiographic conduction disorders were found in 42 (41,58%) patients from the study group, of which more frequently – His right bundle block, compared to the left one (71,43% versus 11,91%, $p < 0{,}001$), supraventricular tachycardias – in 6 (13,04%), atrial fibrillation – in 16 (15,84%), without differences by sex. Symptoms and signs of chronic heart failure was revealed in 69 (68,32%). Non-invasive monitoring of pulse rate and desaturation events with oximeter was carried out constantly in 91 (90,09%). The desaturation events were identified in 36 (35,64%), unstable patients receiving oxygen therapy with a concentration of O2 in the inhaled air, pulse rate events in 67 (66,34%). Acute heart failure (AHF) with significant dyspnoea and haemodynamic instability associated with acute coronary syndrome was diagnosed in 5 (4,95%) cases and the initial care was provided in intensive care unit. From them, 3 (2,97%) Patients with moderate acute heart failure presented with the signs and symptoms of hypoperfusion — use of accessory muscles for breathing, respiratory rate > 25/min, heart rate > 130 bpm, SBP < 90 mm Hg showed modest reductions in oxygen saturation (SpO2) < 90% (despite supplemental oxygen), whereas in 2 (1,97%) patients with severe heart failure had severe oxygen desaturation (SpO2 ≤ 60%), even at rest, the immediate resuscitative support and intubation was applied. Understanding the association of the two entities: COPD and CVC can allow improving the prediction of cardiovascular risk in patients with COPD, by identifying people at higher risk of morbidity and CVC mortality [7].

The recent study by Greek researcher Hillas G. (2016) c[learly established that episodes of exacerbation in COPD are a major mortality factor in the case of CVC, especially during and immediately after exacerbation [8].

The ECLIPSE cohort study, the results of which were published by scholar Müllerova H. and her colleagues in the UK (2015), showed that a previous episode of exacerbation of COPD is a risk factor for new exacerbations (RR 2.71; 95% IC:2.24–3.29; p < 0.001). Independently of FEV, the frequency of exacerbations is a separate risk factor for the prognosis, stratifying patients with COPD in the group of those with frequent exacerbations, with poor prognosis (faster decline in lung function, lower quality of life, increased inflammation in the airway, higher mortality) and in the group with rare exacerbations, with lower risk [9].

Often treatment must be administered in parallel with the diagnostic work-up. Although not 'evidence based' in the same way as treatments for chronic HF, the key drugs are oxygen, diuretics, and vasodilators. Opiates and inotropes are used more selectively, and mechanical support of the circulation is required only rarely.

According to Europeans ghuidelines non-invasive ventilation is used commonly in COPD patients with cardiovascular comorbidities, but invasive ventilation is required in only a minority of patients. Systolic blood pressure, heart rhythm and rate, saturation of peripheral oxygen (SpO2) using a pulse oximeter, and urine output should be monitored on a regular and frequent basis until the patient is stabilized. Oxygen may be given to treat hypoxaemia (SpO2, 90%), which is associated with an increased risk of short-term mortality. Oxygen should not be used routinely in non-hypoxaemic patients as it causes vasoconstriction and a reduction in cardiac output [10]. Oxygen therapy is recommended in patients with AHF and 25 breaths/min and SpO2 < 90% or PaO2 < 60 mmHg to correct hypoxaemia. In COPD, hyperoxygenation may increase ventilation perfusion mismatch, suppress ventilation and lead to hypercapnia. During oxygen therapy, acid-base balance and SpO2 should be monitored. Non-invasive positive pressure ventilation, either continuous positive airway pressure and pressure support, improves respiratory failure, increases oxygenation and pH, and decreases the partial pressure of carbon dioxide (pCO2) and work of breathing. Although a large randomized trial had neutral results, meta-analyses suggest it may improve dyspnoea and reduce the need for intubation and mortality, compared with traditional oxygen therapy. Noninvasive positive pressure ventilation should be started as soon as possible in patients with respiratory distress (respiratory rate > 25 breaths/min, SpO2 < 90%) to improve gas exchange and reduce the rate of endotracheal intubation [11]. Non-invasive positive pressure ventilation includes both CPAP and bi-level positive pressure ventilation (PPV). Bi-level PPV also allows inspiratory pressure support that improves minute ventilation and is especially useful in patients with hypercapnia, most typically COPD patients. Congestion affects lung function and increases intrapulmonary shunting, resulting in hypoxaemia [12]. The fraction of inspired oxygen (FiO2) should be increased up to 100%, if necessary, according to oxygen saturation level. Blood pressure should be monitored regularly during non-invasive positive pressure ventilation. The increase in intrathoracic pressure with non-invasive positive pressure ventilation decreases venous return and right and left ventricular preload. It may also decrease cardiac output and BP and should therefore be used with caution in patients with reduced preload reserve and hypotension. The increase in pulmonary vascular resistance and RV afterload may also be detrimental in RV dysfunction. Intubation is recommended for progressive respiratory failure in spite of oxygen administration or non-invasive ventilation [13, 14]. CPAP is a feasible technique in the pre-hospital setting, because it is simpler than pressure support positive end-expiratory pressure (PS-PEEP) and requires minimal training and equipment. On hospital arrival, patients who still show signs of respiratory distress should continue with non-invasive ventilation, preferably PS-PEEP, in case of acidosis and hypercapnia, particularly in those with signs of fatigue [15, 16].

5 Compliance with Ethical Requirements

5.1 Statement of Informed Consent

The study was initiated after signing the informed agreement by all participants (information form and acceptance form), who became familiar with the evaluation methods to be applied, at the time of initiation and during the study the confidentiality and ethical criteria were observed.

5.2 Statement of Human and Animal Rights

The research was conducted during the years 2016–2018 in accordance with the Principles of the Helsinki Declaration - WMA Declaration of Helsinki - Ethical Principles for Medical Research Involving Human Subjects.

The study was approved by the Research Ethics Committee of a State University of medince and Pharmacy "Nicolae Testemițanu", after being examined at the meeting of May 30, 2016, with the issuance of favorable opinion no. 58 of 03.06.2016.

6 Conclusion

The presence of cardiovascular comorbidities in patients hospitalized with chronic obstructive pulmonary disease exacerbations is high and associated with rehospitalizations and deaths. Pulse oximetry is useful for non-invasive monitoring of hypoxemia presence and the severity of the patient's response to supplemental oxygen and other therapies, especially in emergency situations in exacerbated COPD and cardiovascular comorbidities.

Acknowledgment. I would like to express my special thanks of gratitude to our Department director (univ.prof. Sergiu Matcovschi) as well as our principal (Dr. Mihail Lupan) who also helped me in doing a lot of Research.

Conflict of Interest. The authors declare that they have no conflict of interest.

References

1. Vogelmeier, C.F., Criner, G.J., Martinez, F.J., et al.: Global strategy for the diagnosis, management, and prevention of chronic obstructive lung disease 2017 report: GOLD executive summary. Eur. Respir. J. **49**(3), 1700214 (2017)
2. Echave-Sustaeta, J.: Comorbidity in chronic obstructive pulmonary disease. Related to disease severity? Int. J. Chronic Obstructive Pulm. Dis. **9**, 1307 (2014)
3. Houben-Wilke, S., Triest, F.J.J., Franssen, F.M.E., Janssen, D.J.A., Wouters, E.F.M., Vanfleteren, L.E.G.W.: Revealing methodological challenges in chronic obstructive pulmonary disease studies assessing comorbidities: a narrative review. Chronic Obstr Pulm Dis. **6**(2), 166–177 (2019). https://doi.org/10.15326/jcopdf.6.2.2018.0145
4. Levy, J., Álvarez, D., Rosenberg, A.A., et al.: Digital oximetry biomarkers for assessing respiratory function: standards of measurement, physiological interpretation, and clinical use. npj Digit. Med. **4**(1), 1–14 (2021)
5. Lavie, L.: Obstructive sleep apnoea syndrome - an oxidative stress disorder. Sleep Med. Rev. **7**, 35–51 (2003)
6. Budhiraja, R., Siddiqi, T.A., Quan, S.F.: Sleep disorders in chronic obstructive pulmonary disease: etiology, impact, and management. J. Clin. Sleep Med. **11**, 259–270 (2015)
7. Dugac, A.V.: Persistent endothelial dysfunction turns the frequent exacerbator COPD from respiratory disorder into a progressive pulmonary and systemic vascular disease. Med. Hypotheses **84**(2), 155–158 (2015)
8. Hillas, G.: Acute exacerbation of COPD: is it the "stroke of the lungs"? Int. J. Chron. Obstruct. Pulmon. Dis. **11**, 1579–1586 (2016)
9. Müllerova, H.: Hospitalized exacerbations of COPD: risk factors and outcomes in the ECLIPSE cohort. Chest **147**(4), 999–1007 (2015)
10. McDonagh, T.A., Metra, M., Adamo, M., et al.: 2021 ESC Guidelines for the diagnosis and treatment of acute and chronic heart failure developed by the task force for the diagnosis and treatment of acute and chronic heart failure of the European Society of Cardiology (ESC) with the special contribution of the Heart Failure Association (HFA) of the ESC. Eur. Heart J. 1128 (2021). https://doi.org/10.1093/eurheartj/ehab368
11. Park, J.H., Balmain, S., Berry, C., Morton, J.J., McMurray, J.J.: Potentially detrimental cardiovascular effects of oxygen in patients with chronic left ventricular systolic dysfunction. Heart **96**, 533538 (2010)
12. Weng, C.L., et al.: Metaanalysis: noninvasive ventilation in acute cardiogenic pulmonary edema. Ann. Intern. Med. **152**, 590600 (2010)
13. Masip, J., Peacock, W.F., Price, S., Cullen, L.: Acute heart failure study group of the acute cardiovascular care association and the committee on acute heart failure of the heart failure association of the European Society of cardiology. Indications and practical approach to non-invasive ventilation in acute heart failure. Eur. Heart J. **39**, 1725 (2018)
14. Vital, F.M.R., Ladeira, M.T., Atallah, A.N.: Non-invasive positive pressure ventilation (CPAP or bilevel NPPV) for cardiogenic pulmonary oedema. Cochrane Database Syst. Rev. **5**, CD005351 (2013)

15. Mebazaa, A., Yilmaz, M.B., Levy, P., Ponikowski, P.: Recommendations on pre-hospital and early hospital management of acute heart failure: a consensus paper from the HFA of the ESC, the European Society of emergency medicine and the society of academic emergency medicine. Eur. J. Heart Fail **17**, 544 (2015)
16. Ponikowski, P., Voors, A., Anker, S., et al.: 2016 ESC Guidelines for the diagnosis and treatment of acute and chronic heart failure the task force for the diagnosis and treatment of acute and chronic heart failure of the European Society of Cardiology (ESC) developed with the special contribution of the Heart Failure Association (HFA) of the ESC. Eur. Heart J. **37**, 2129–2200 (2016). https://doi.org/10.1093/eurheartj/ehw128

Deep Learning Methods for Tumor Segmentation and Detection in X-Ray Breast Imaging

D. Chatzakis(✉), A. Dermitzakis, and N. Pallikarakis

Biomedical Technology Unit, Department of Medical Physics, School of Medicine, University of Patras, Patras, Greece

Abstract. Recently there have been a series of machine learning methods or deep learning architectures that have been developed and used in the field medical imaging. In this study, we focus on the use of AI in the field of breast imaging and the methods with the highest accuracy results for breast tumor segmentation and classification are presented, achieving robust results in detection. Extensive research which included more than 150 related published papers was performed, containing results published between 2016 to 2020 resulting in a review of 4 selected models all at the forefront of current progress.

Keywords: Deep learning · Tumor segmentation · Tumor classification · Neural networks · Breast cancer

1 Introduction

Breast cancer is the number one cause of death for women below 70 years of age. In order to diagnose it at early stages and thus mitigate death rates, screening tests are advised from ages of 50 onwards. The most common examination is the x-ray mammography which has shown to reduce mortality by 37% in women 50–69 years of age who have been subjected to screening mammography. However, many cases are not that discernible especially in early stages. Now, with the advent of modern technology and neural networks, fast integration in many areas is the golden rush of our today's era aiming to increase tumor detection and classification, especially at an earlier stage. To this date, the main issue is most modern methods of CAD carry a high number of false positives per image [1].

Several methods and architectures have been developed using different annotated databases and a selection of them is presented along with the corresponding metrics and results they have achieved.

2 Methods

In order to create this review, we first searched in the internet using mainly the arxiv.org website for papers using the keywords "Deep learning" and "mammography" resulting

I. Tiginyanu et al. (Eds.): ICNBME 2021, IFMBE Proceedings 87, pp. 750–757, 2022.
https://doi.org/10.1007/978-3-030-92328-0_95

in 59 papers, then with the keywords "deep learning" and "breast cancer" resulting in 153 results and lastly, with the keywords "neural networks" and "breast cancer" resulting in 149 results.

So as to find all the state of the art, all these searches were contained between 2016 and October 2020. Continuing on, we applied a subjective filter where we dismissed several papers due to low achieved accuracy, (accuracy below 80% or Area under Curve (AUC) under 0.85), relevance in the methods employed, namely neural networks and not just machine learning or for achieving worse accuracy relative to another paper with similar architecture. We came upon a number of 7 papers which we then refined to the 4 most important ones based on the accuracy of their results (3.1) innovation, complexity and multitude of methods in their procedure (3.3,3.4) or experimental setup (3.2) where they compare human performance with the performance of standalone A.I. (Fig. 1 below).

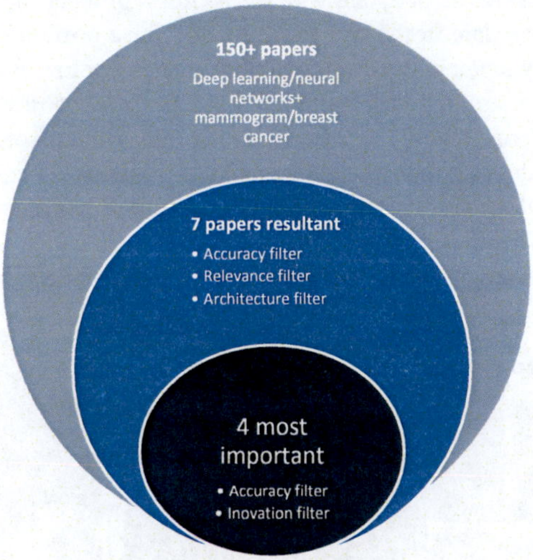

Fig. 1. Methodology of research.

3 Results

3.1 Hidden Markov Model

In [2] a Hidden Markov Tree [3, 4] along with Dual Tree Discrete Wavelet Transform [5] 6 is employed for diagnosis of microcalcifications. The features extracted by this method are then further optimized by genetic algorithms to improve accuracy and then fed to an extreme learning machine [7–9] to perform the final classification. The results of the Area under Curve (AUC) were 0.9856, 0.9941 and 0.9168 on the Nijmegen, MIAS and DDSM datasets of evaluation respectively. The operational scheme of the algorithm is in Fig. 2 below.

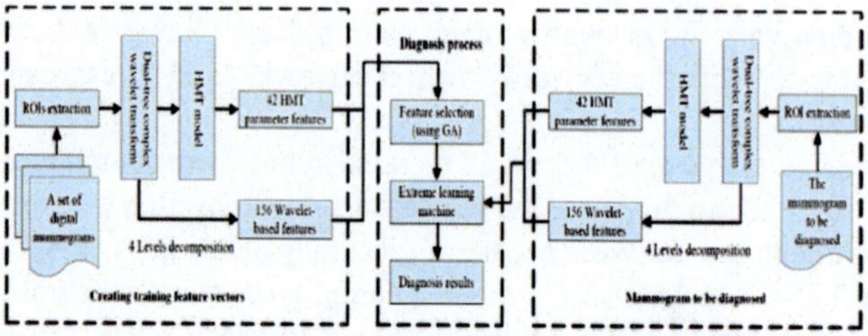

Fig. 2. Process of the model. Image published in [2].

3.2 Resnet-34

In [10], the largest study done to our knowledge thus far, 170230 four-view, full-field, digital mammograms were utilized with the aid of a NN based on the RESNET-34 architecture [11] with stride of 2, and with the addition of batch-instance normalization and deconvolution module in the end for removing false positives. Training was done on two stages, firstly a patch-level training from scratch for low-level features and then at the second stage finetuning was applied to get high level context. The output was an abnormality score (from 0 to 1) as a heatmap. The maximum score of the four views of each mammogram was then used, thresholded, in order to calculate the AUC. In this study AI standalone results were made, along with only human results and human with CAD results. The respective results for the AUC were 0.959, 0.810 and 0.881. The method can be surmised as in Fig. 3 and the network architecture in Fig. 4.

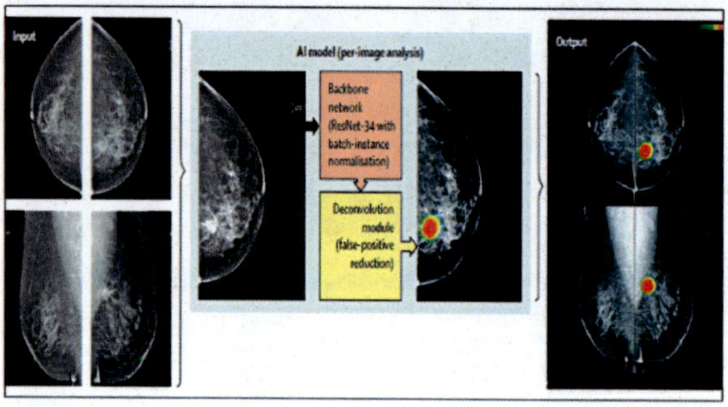

Fig. 3. Model of the process of the model. Image published in [11].

3.3 Deep Belief Network and Gaussian Mixture Model

In [1] another procedure is used as shown in the figure below. The specific research team's mass detection algorithm [12] consists of an ensemble of classifiers whose main aim of each stage is to retain the true positives while at the same time reducing the number of false positives and then enhance the precision of the bounding box detection.

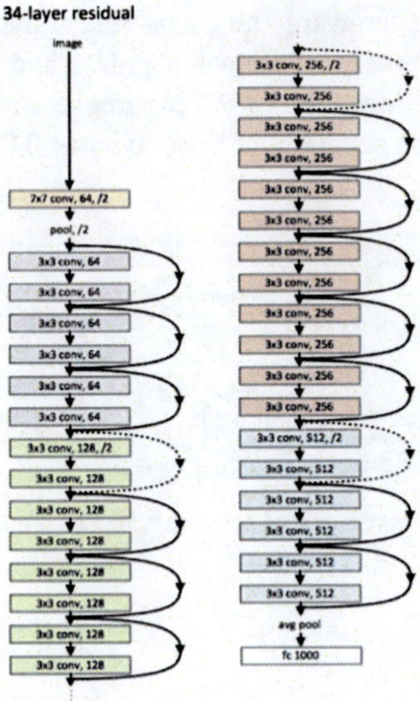

Fig. 4. Resnet 34 architecture. Image published in [12].

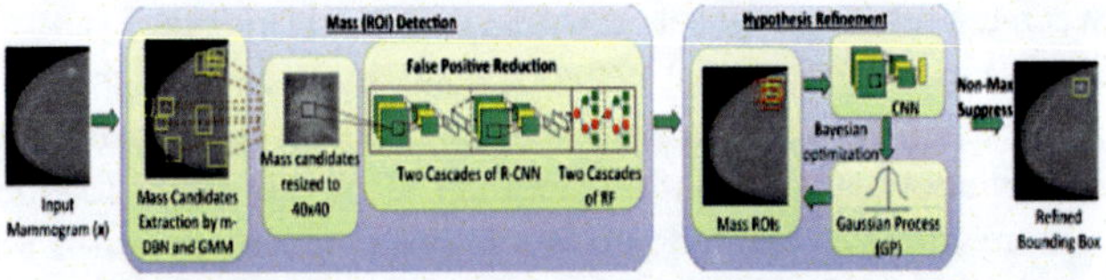

Fig. 5. Process until RoI selection. Image published in [1]

For the ROI extraction a cascade of classifiers is used. The classifier of the first stage consists of a deep belief network (m-DBN) [13] and a Gaussian mixture model (GMM) [14] that selects certain regions to be further processed by an ensemble of deep convolutional neural networks. The regions that are not discarded are then processed by a number of random forest classifiers. Lastly, regions that are maintained by the ensemble of random forest classifiers are used with connected component analysis to produce state-of-the-art results.

The training of the DBN at each resolution level uses a training set of positive patches and negative patches created. The GMM model works exclusively on the finest image resolution with a pixel-wise classification and estimates the likelihood that a pixel grey value represents part of a breast mass, or background. This GMM model will output a posterior probability to produce the final estimated positive and negative labels. False positives among the mass candidates are then reduced by an ensemble of R-CNNs which extracts the features from the last layer of a CNN model and classifies it using a linear SVM. The result is then fed to random forest classifiers (RF) to sort out false positives.

Then the authors propose a refinement of the segmented boxes as presented on below based on Bayesian optimisation [15]. This is then input through a CNN to create refined

Regions of interest (RoIs). Then using the Chan-Vese active contour model [16] we get the final image. The whole process is shown in Figs. 5 and 6.

Results were 90% detection with 1 FPR per image with segmentation accuracy of 85% and with a classification accuracy of Sensitivity of 0.98 and specificity of 0.7.

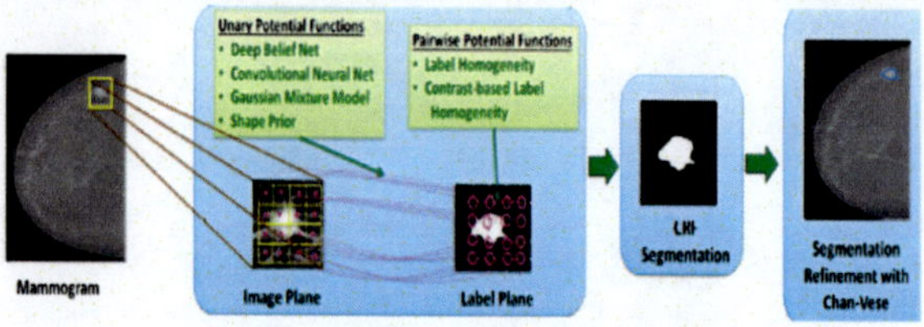

Fig. 6. Process after RoI selection. Image published in [1].

3.4 Faster RCNN

In [16] the OMI-DB and INbreast database is employed [17] to benchmark the performance of the Faster-RCNN. The Faster-RNN proposed here is the result of hybrid approach between the region proposal network (RPN) of [18] (Fig. 7) and the Fast-RNN [19] (Fig. 8). RPN is a convolutional network that simultaneously predicts the bounds of the object and outputs an objectness score at each position. This small network takes as input a spatial window of the input feature map. Each window is mapped to a lower-dimensional feature which is then fed into two fully connected layers—a box-regression layer (reg) and a box-classification layer (cls).

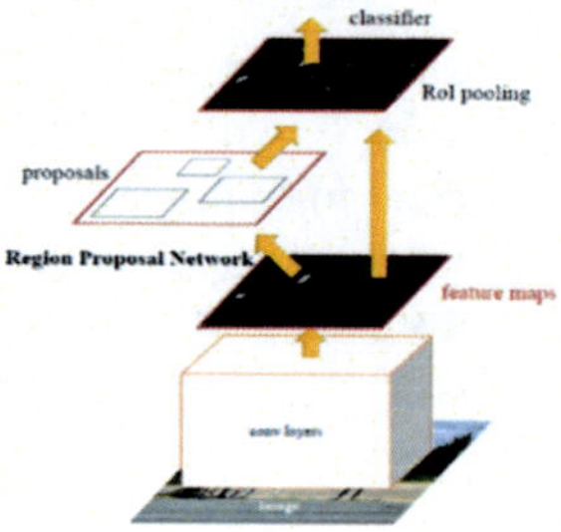

Fig. 7. Region proposal network process. Image published in [19].

Fast RCNN is an RCNN which takes an image as input as well as a number of object proposals. The image is first processed with several convolutional and max pooling layers in order to produce a feature map. Then, a fixed-length feature vector is extracted by a RoI pooling layer. Each vector is then input into a sequence of fully connected layers that continue on to two output layers: one that produces SoftMax probability estimates plus a catch-all "background" class and another layer that outputs 4 real-valued numbers

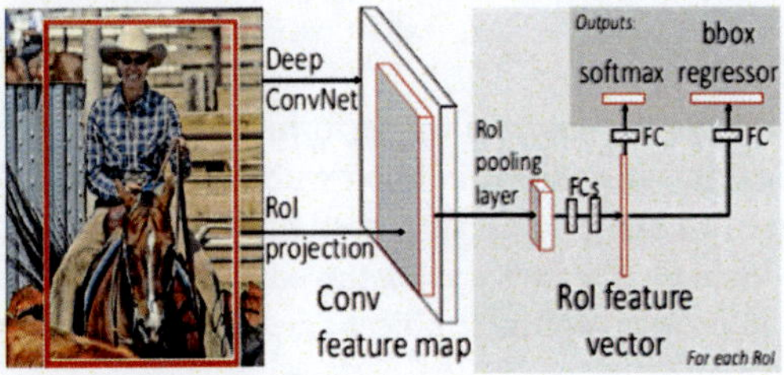

Fig. 8. Fast RCNN architecture. Image source [19].

for each of the N object classes. Each set of values encodes bounding-box positions for one of the classes. The whole procedure of the Faster RCNN is explained below.

In the first step, mammogram images are used as input to the Faster R-CNN. This is put into an InceptionV2 model to produce a feature map. In the next phase, an RPN is built using the extracted features of the CNN and trained to detect masses on the mammograms. A window slides over the feature map and outputs a feature vector connected with two fully convolutional (FC) layers, a box-regression layer (reg) and a box classification layer (cls). Afterwards, the anchors or bounding boxes are built to create region proposals of different shapes and sizes, and are ascribed an objectness score to signify how accurately they can enclose a mass on the image. The best anchors are then passed through to the second stage of the network. Here, the classification and regression issues are tackled to accurately detect any masses, and simultaneously refine the coordinates of the anchors. Finally, the best predictions are obtained by employing suppression on the detected overlapping objects, resulting in the final bounding box with a probability representing how closely it is akin to a mass. Its architecture is represented in Fig. 9.

The results for the OMI-DB are True Positive Rate (TPR) of 0.93 at 0.78 False Positive per Image (FPI) while for the INbreast dataset, obtaining a TPR of 0.99 ± 0.03 at 1.17 FPI for malignant and 0.85 ± 0.08 at 1.0 FPI for benign masses.

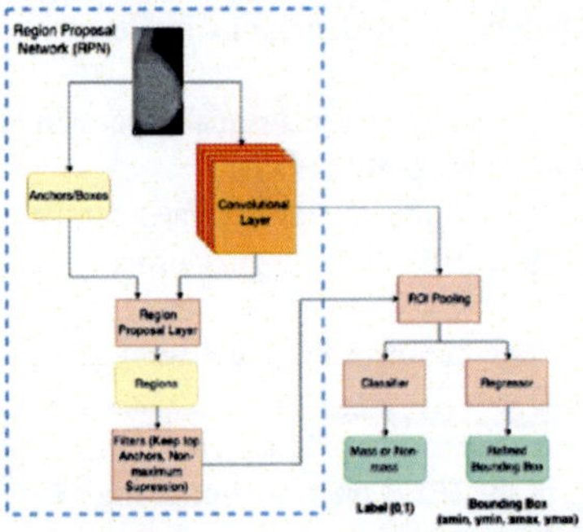

Fig. 9. Faster RCNN architecture. Image published in [17].

4 Conclusion

From our review and from the analysis we performed thus far, we can safely state that the field of biomedical image analysis is filled with creative ideas about deep learning applications that have reached or even surpassed human performance. We can see that in most cases, deep learning methods are employed in an ensemble of various machine learning algorithms in order to optimize object detection and segmentation.

5 Future Work

We are aiming to develop such a new model for breast image segmentation and classification using old and new methods such as employing the NVIDIA CLARA module and utilizing various alterations of the UNet architecture.

Acknowledgements. The research work was supported by the Hellenic Foundation for Research and Innovation (H.F.R.I.) under the "First Call for H.F.R.I. Research Projects to support Faculty members and Researchers and the procurement of high-cost research equipment grant" (Project Number: 4271).

Conflict of Interest. The authors declare that they have no conflict of interest.

References

1. Dhungel, M., et al.: A deep learning approach for the analysis of masses in mammograms with minimal user intervention. Med. Image Anal. (2017)
2. Hu, K., et al.: Microcalcification diagnosis in digital mammography using extreme learning machine based on hidden Markov tree model of dual-tree complex wavelet transform. Experts Syst. Appl. **86**, 135–144 (2017)
3. Choi, H., et al.: Hidden Markov tree modeling of complex wavelet transforms. In: Proceeding of the ICASSP (2000)
4. Crouse, M.S., et al.: Wavelet-based statistical signal processing using hidden Markov models. IEEE Trans. Sig. Proc. **46**(4), 886–902 (1998)
5. Kingsbury, N.G.: The dual-tree complex wavelet transform: a new technique for shift invariance and directional filters. IEEE digital signal processing work- shop (DSP 98) No86 (1998)
6. Selesnick, W.N.G., et al.: The dual-tree complex wavelet transform. IEEE Signal Process. Mag. **22**(6), 123–151 (2005)
7. Huang, G.-B., et al.: Extreme learning machine: a new learning scheme of feedforward neural networks. In: Proceeding of the IEEE Inter- National Joint Conference on Neural Networks, vol. 2, pp. 985–990 (2004)
8. Huang, G.-B., et al.: Extreme learning machine: theory and applications. Neurocomputing **70**(1), 489–501 (2006)
9. Kim, H., et al.: Changes in cancer detection and false-positive recall in mammography using artificial intelligence: a retrospective, multireader study. Lancet Digit. Health **2**, e138–e148 (2020)

10. He, K., et al.: Deep residual learning for image recognition. In: 2016 IEEE Conference on Computer Vision and Pattern Recognition (2016)
11. Dhungel, N., et al.: Automated mass detection in mammograms using cascaded deep learning and random forests. In: Digital Image Computing: Techniques and Applications (DICTA), 2015 International 680 Conference, pp. 1–8 (2015). https://doi.org/10.1109/DICTA.2015.7371234
12. Hinton, G.E., Salakhutdinov, R.R.: Reducing the dimensionality of data with neural networks. Science **313**(5786), 504–507 (2006)
13. Dempster, et al., P.: Maximum likelihood from incomplete data via the em algorithm. J. Royal Stat. Soc. Ser. B (Methodol.), 1–38 (1977)
14. Zhang, Y., et al.: Improving object detection with deep convolutional networks via Bayesian optimization and structured prediction. In: IEEE Conference on Computer Vision and Patter Recognition (CVPR), p. 249 – 258 (2015)
15. Jorstad, A., Fua, P.: Refining mitochondria segmentation in electron microscopy imagery with active surfaces. In: Agapito, L., Bronstein, M.M., Rother, C. (eds.) ECCV 2014. LNCS, vol. 8928, pp. 367–379. Springer, Cham (2015). https://doi.org/10.1007/978-3-319-16220-1_26
16. Agarwal, R., et al.: Deep learning for mass detection in Full Field Digital Mammograms. Elsevier Comput. Biol. Med. **121**, 103774 (2020)
17. Halling-Brown, M.D., et al.: The oncology medical image database (OMI-DB). In: Medical Imaging 2014: PACS and Imaging Informatics: Next Generation and Innovations, vol. 9039, p. 903906. International Society for Optics and Photonics (2014)
18. Szegedy, V., et al.: Rethinking the inception architecture for computer vision. In: Proceedings of the Conference on Computer Vision and Pattern Recognition, pp. 2818–2826. IEEE (2016)
19. Girshick, R.: Fast R-CNN. In: Proceedings of the IEEE International Conference on Computer Vision (2015)

Author Index

I. Tiginyanu et al. (Eds.): ICNBME 2021, IFMBE Proceedings 87, pp. 759–762, 2022.
https://doi.org/10.1007/978-3-030-92328-0